AF683222

Calcium Channel Pharmacology

Calcium Channel Pharmacology

Edited by

Stefan I. McDonough
Marine Biological Laboratory
Woods Hole, Massachusetts

Kluwer Academic / Plenum Publishers
New York, Boston, Dordrecht, London, Moscow

Library of Congress Cataloging-in-Publication Data

ISBN 0-306-47878-1

©2004 Kluwer Academic / Plenum Publishers, New York
233 Spring Street, New York, New York 10013

http://www.wkap.nl/

10 9 8 7 6 5 4 3 2 1

A C.I.P. record for this book is available from the Library of Congress

All rights reserved

No part of this book may be reproduced, stored in a retrieval system, or transmitted in any form or by any means, electronic, mechanical, photocopying, microfilming, recording, or otherwise, without written permission from the Publisher, with the exception of any material supplied specifically for the purpose of being entered and executed on a computer system, for exclusive use by the purchaser of the work.

Permissions for books published in Europe: *permissions@wkap.nl*
Permissions for books published in the United States of America: *permissions@wkap.com*

Printed in the United States of America

PREFACE

Voltage-gated calcium channels are critical regulators of cytoplasmic levels of calcium, the universal signaling ion. As such, calcium channels trigger a wide range of cellular functions, from muscle contraction to neurotransmitter secretion, and are important players in human disease. Prominent in the nervous, cardiovascular, and endocrine systems, members of the calcium channel family are targets for existing antihypertensive and anticonvulsant drugs. In addition, they are emerging targets for drugs to treat an extraordinarily diverse group of disorders, including pain, cerebral ischemia, cardiac arrhythmia, and migraine.

This book reviews the compounds that target individual calcium channel subtypes and the cellular and behavioral functions governed by each different channel. It contains information for basic scientists using calcium channel antagonists as experimental tools, for behavioralists studying animal models of human disease, and for pharmaceutical scientists interested in creating the next generation of calcium channel-targeted drugs.

Several factors make an entire book on calcium channel pharmacology timely. Within the past several years, entire new families of calcium channels have been cloned, and genomic analysis finds ever more splice forms of the same gene with distinct physiology and pharmacology. Calcium channels have been found to be the genetic cause of several forms of human disease, and for almost all calcium channels, the behavioral phenotypes of mice carrying targeted gene disruptions have now been reported. Many more drugs and neurotoxins that target calcium channels are available than is generally known, and the best-known calcium channel blockers are understood in great mechanistic detail, sufficient to serve as a guidepost for the development of future drugs. Calcium channels are increasingly tractable drug targets just as their contributions to a surprising diversity of disorders and of cellular functions have become clear.

Targeting drugs to individual calcium channels, and even to individual channel gating states, offers both a challenge and an opportunity, scientifically and clinically. It is our hope that information presented here will help clarify the issues and the tools.

I am deeply grateful to all my fellow authors and especially to the chapter lead authors: Drs. Steffen Hering, David Triggle, Charles Cohen, Raymond Norton, Eric Ertel, Erika Piedras-Rentería, Tsutomu Tanabe, Yasuo Mori, Gerald Zamponi, and Diane Lipscombe. Your contributions have made the book.

Stefan McDonough

CONTENTS

INTRODUCTION

Voltage-gated calcium channels are integral membrane proteins that open in response to electrical depolarization of the plasma membrane and mediate the entry of calcium ions into the cell. Expressed in every excitable cell, including neurons, myocytes, and pancreatic β cells, calcium channels form the most efficient molecular link between membrane depolarization and intracellular biochemical signaling (Catterall, 2000). At the cellular level, calcium influx mediated by voltage-gated calcium channels governs diverse physiological functions. Muscle contraction, neurotransmitter and hormone release, regulation of membrane excitability (either directly or via calcium-activated potassium and chloride channels), and the regulation of gene expression all depend on calcium channel function. Calcium channels are current or potential targets for drugs treating a wide variety of common disorders, including forms of hypertension, pain, epilepsy, migraine, cerebral ischemia, and cardiac arrhythmia, and calcium channel dysfunction underlies rarer diseases such as hereditary forms of ataxia and night blindness.

The many functions that depend on calcium channels are not governed by the expression of a single channel. Rather, calcium channels come in many biophysically, pharmacologically, and biochemically distinct subtypes. Recent evidence shows that different subtypes of calcium channels support distinct physiological functions, at levels from the single cell to the entire organism. This book deals with calcium channel pharmacology—the study of the physiological roles of different calcium channel subtypes and of the drugs that distinguish among them. The two goals are interrelated: selective drugs are indispensable for the discovery of subtype-specific functions, and the broad finding that calcium channel subtypes govern different functions drives the discovery and further refinement of selective drugs. In addition, characterization of the molecular activity of clinically useful drugs, whether against single calcium channel subtypes, multiple calcium channel subtypes, or a broad spectrum of ion channels, contributes greatly to the understanding and treatment of disease. The purpose of this book is to review the selectivity and biomedical significance of drugs that distinguish among different calcium channel subtypes.

Diversity among calcium channels was first noted in electrophysiological experiments. Multiple components of calcium channel currents within single cells were found to be separable on the basis of biophysical or pharmacological properties (Hagiwara et al., 1975; Carbone and Lux, 1984a, b; Bean, 1985; Nowycky et al., 1985) and were named with a descriptive letter (Nowycky et al., 1985). In parallel with

Table 1. Calcium channel α_1 subunits and corresponding calcium channel currents.

Family	α_1 subunit	Former name	Ca^{2+} current
Ca_v1	$Ca_v1.1$	α_{1S}	L-type
	$Ca_v1.2$	α_{1C}	
	$Ca_v1.3$	α_{1D}	
	$Ca_v1.4$	α_{1F}	
Ca_v2	$Ca_v2.1$	α_{1A}	P (or P/Q)-type
	$Ca_v2.2$	α_{1B}	N-type
	$Ca_v2.3$	α_{1E}	R-type (some)
Ca_v3	$Ca_v3.1$	α_{1G}	T-type
	$Ca_v3.2$	α_{1H}	
	$Ca_v3.3$	α_{1I}	

Calcium channel α_1 subunits are listed using the agreed numeric nomenclature established by sequence homology (Ertel et al., 2000) together with the previous letter nomenclature. See chapter by Lipscombe and Castiglioni for α_1 subunit gene tree and for terminology of splice variants of α_1 subunits. Not all resistant or R-type currents result from $Ca_v2.3$ channels; see chapter by McDonough for detailed description of R-type channels and for explanation of P-type and P/Q-type channels.

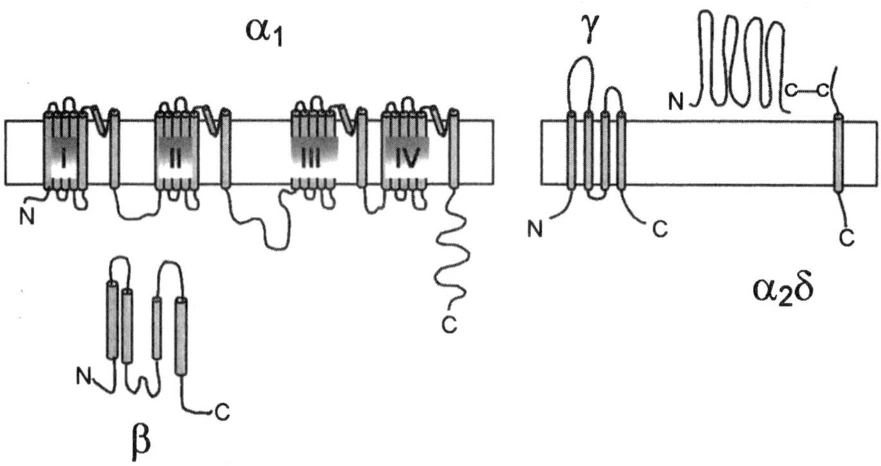

Figure 1. Cartoon hydropathy analysis of calcium channel α_1, β, $\alpha_2\delta$, and γ subunits. The α_1 subunit consists of four homologous membrane-spanning domains (I-IV) each encompassing α-helical structures in domains termed S1 through S6, with a pore-loop region re-entering the membrane between S5 and S6. See Jiang et al., 2003a, b for X-ray structure of a voltage-gated K^+ channel, a tetramer homologous to the four internal domains of the α_1 subunit, that shows the actual location, orientation, structure, and boundaries of the α-helices and how they combine to form the channel voltage sensor. The β subunit is intracellular; the α_2 subunit is extracellular with disulfide link to the transmembrane δ subunit, and the γ subunit is predicted to have multiple membrane-spanning domains.

physiological studies, biochemical and molecular experiments revealed a diversity of calcium channel proteins and genes, starting with the purification (Curtis and Catterall, 1984) and cloning (Tanabe et al., 1987) of the skeletal muscle dihydropyridine receptor. A full calcium channel protein is composed of a pore-forming α_1 subunit, which consists of a fourfold internal repeat of the voltage-gated channel superfamily motif (Hille, 2001), plus additional ancillary β, $\alpha_2\delta$, and γ subunits (Figure 1). The α_2 and δ subunits are derived from the same gene, with the product split into two peptides linked by a disulfide bond (de Jongh et al., 1990). Biochemical and physiological evidence suggests that most calcium channels are composed of α_1, β, $\alpha_2\delta$, and possibly γ subunits in 1:1:1:1 stoichiometry (Takahashi et al., 1987; McEnery et al., 1991; Witcher et al., 1993; Liu et al., 1996; Kang et al., 2001). Involvement and function of the γ subunit is an area of active research (Moss et al., 2002). The multisubunit composition of T-type channel protein is likewise controversial (see Ertel, this volume).

As of this writing, ten α_1 genes have been discovered and are named numerically according to sequence similarity (Ertel et al., 2000). These ten genes may represent the complete suite of α_1 subunits, with the possible exception of a cryptic cardiac calcium channel found after knockout of the $Ca_v1.2\alpha_1$ gene (Seisenberger et al., 2000). Four β subunits (Castellano and Perez-Reyes, 1994), four $\alpha_2\delta$ subunits (Klugbauer et al., 1999; Qin et al., 2002), and eight γ subunits (Klugbauer et al., 2000; Chu et al., 2001; Moss et al., 2002) are currently known. T-type channels excepted, it seems likely that any ancillary subunit can form a channel with any α_1 subunit (Walker and De Waard, 1998), although not all combinations have been tested. In addition to modifying physiological properties of the α_1 subunits, ancillary subunits likely have additional functions in trafficking/ surface expression of calcium channels (Dolphin et al., 1999; Brodbeck et al., 2002) and, surprisingly, of non-calcium channel proteins in at least some cases (Hashimoto et al., 1999; Chen et al., 2000).

Terminology of different calcium channel subtypes is complex. To a first approximation, the main physiological and pharmacological properties of a given channel are determined by the molecular identity of the α_1 subunit, most of which have been linked convincingly with corresponding channel subtypes in native tissue (Table 1). Ca_v3 T-type channels are mostly distinguished by their biophysical properties (see Ertel, this volume), but for Ca_v1 or Ca_v2 channels pharmacological profiles are often used as a guide to the α_1 subunit. Channels sensitive to dihydropyridines are generally called L-type, and those inhibited with high affinity by the peptide conotoxin ω-CTx-GVIA are called N-type. Other subtypes include P-type (or P/Q-type) channels, some forms of which are highly sensitive to the peptide toxin ω-Aga-IVA from spider venom. R-type channels were first referred to as those channels resistant to combined inhibition by blockers of other channels and likely represent a heterogeneous group of channels. For clarity, use of "type" names for channels should be qualified with either molecular characterization of the gene involved or with the pharmacological procedure used to isolate the current.

This book is grouped broadly into two parts; individual chapters are complementary, but can stand individually. The first part covers the specificity and selectivity of drugs that target various classes of calcium channels. Like all ion channels, calcium channels are moving targets. That is, the channel protein itself physically changes shape as depolarization evokes transits among closed, open, and inactivated states, and, more often than not, the potency of a given drug depends dramatically on the precise gating state of the channel. This state-dependent affinity can complicate use of drugs as scientific tools, but the ability to target specific channel gating states often contributes greatly to a drug's therapeutic profile. In the first chapter, Timin, Berjukow, and Hering discuss models for state-dependent drug affinity that underlie all discussion of pharmacology, including open-channel block, the modulated receptor model, and drug-inactivation synergism. In the second chapter, Triggle reviews the pharmacology of Ca_v1 L-type channels, the best-developed pharmacology. Thus far, the success of small molecule blockers for L-type channels has not been repeated for the Ca_v2 family of P-type, N-type, and R-type channels. In the third chapter, Cohen and Kraus review the pharmaceutical interest in finding such molecules to treat pain and migraine and as neuroprotective agents. The most specific blockers for channels of the Ca_v2 family are peptide toxins, mostly derived from the venom of poisonous animals. Peptide toxin inhibition of calcium channels is dealt with in two chapters, focused respectively on channels and on toxins. In the fourth chapter, McDonough reviews toxin selectivity, mechanisms of action, and channel binding sites; in the fifth chapter, Norton, Baell, and Angus review the structure and structure-function relations of peptide toxins, the use of some toxins themselves as drugs to treat pain, and the development of subtype-selective small molecules based on toxin structure. In the sixth chapter, Ertel reviews the physiology, pharmacology, and modulation of T-type calcium channels, the newest family to be cloned (Perez-Reyes et al., 1998).

The second part of the book covers, broadly, the physiological functions of individual calcium channel subtypes. In the seventh chapter, Piedras-Rentería, Mermelstein, and Pitt discuss the coupling of calcium channels to fundamental processes within the cell, especially within neurons: neurotransmitter release, neuronal plasticity, gene expression, the activation of calcium-activated potassium channels, and interactions with calmodulin and CaM kinase II. In the eighth chapter, Saegusa and Tanabe discuss the phenotypes of calcium channel knockout mice. Such mice often have astonishingly specific phenotypes and provide insights into the distinct roles or overlap of calcium channel subtypes at the behavioral level. Knockout mice are also valuable as models of human disorders and disease. In the ninth chapter, Mori, Itsukaichi, Nishida, and Oka review human and animal diseases resulting from mutations in calcium channels, including migraine, ataxia, and forms of epilepsy, with a view towards explaining general mechanisms of how molecular defects cause disease. Although channelopathies generally represent rare forms of human disease (Ashcroft, 2000), there is hope that understanding the genetic forms of disease will lead to treatments for more common forms.

Many existing clinical drugs that exert effects on calcium channels do so indirectly, via G protein-coupled receptors. In the tenth chapter, Beedle and Zamponi review mechanisms of G protein inhibition of high voltage-activated (non-T-type) calcium channels and summarize knowledge about the neurotransmitter and hormone receptors that couple to calcium channels in different cells. Finally, perhaps the newest frontier in calcium channel pharmacology is alternative splicing. In the eleventh chapter, Lipscombe and Castiglioni present the genomic structure of calcium channels, the mechanisms by which potentially hundreds of splice variants are produced, and the contributions of alternative splicing to calcium channel physiology and pharmacology.

It is of great importance to note recent work from the MacKinnon laboratory, the high-resolution X-ray structure of a bacterial voltage-activated K^+ channel from

Aeropyrum pernix (Jiang et al., 2003a; Jiang et al., 2003b; Ruta et al., 2003), work that appeared as this book was in proof. The structure of the K^+ channel, a tetrameric (MacKinnon, 1991) assembly homologous to the fourfold internal repeat of the calcium channel, has major implications for calcium channel topology and pharmacology. By analogy, the predicted transmembrane helices (Figure 1) are not all transmembrane, not all bordered as marked, and not positioned as shown, but rather are buried within the membrane at a variety of angles. Opening of the K^+ channel in response to voltage is accomplished by movement from internal- to external-facing sides of the plasma membrane of a charged helix-turn-helix paddle, made up of the latter part of the previously-named S3 helix plus S4. Paddle movement is transferred to opening of the channel via other helices. For the many drugs and toxins that are voltage-dependent inhibitors of calcium channels, drug access to the channel, the mechanisms of inhibition, and channel mutations affecting drug affinity should be viewed in light of this gating mechanism, and models of channel-drug interactions can now be viewed more physically rather than theoretically.

ACKNOWLEDGMENTS

I thank Dr. Diane Lipscombe for helpful comments.

Stefan I. McDonough

Woods Hole, Massachusetts

REFERENCES

Ashcroft, F. M., 2000, *Ion Channels and Disease*, Academic Press, San Diego, London.

Bean, B. P., 1985, Two kinds of calcium channels in canine atrial cells. Differences in kinetics, selectivity, and pharmacology, *J Gen Physiol* **86**:1-30.

Brodbeck, J., Davies, A., Courtney, J. M., Meir, A., Balaguero, N., Canti, C., Moss, F. J., Page, K. M., Pratt, W. S., Hunt, S. P., Barclay, J., Rees, M., and Dolphin, A. C., 2002, The *ducky* mutation in Cacna2d2 results in altered Purkinje cell morphology and is associated with the expression of a truncated $\alpha 2\delta$-2 protein with abnormal function, *J Biol Chem* **277**:7684-7693.

Carbone, E. and Lux, H. D., 1984a, A low voltage-activated calcium conductance in embryonic chick sensory neurons, *Biophys J* **46**:413-418.

Carbone, E. and Lux, H. D., 1984b, A low voltage-activated, fully inactivating Ca channel in vertebrate sensory neurones, *Nature* **310**:501-502.

Castellano, A. and Perez-Reyes, E., 1994, Molecular diversity of Ca^{2+} channel beta subunits, *Biochem Soc Trans* **22**:483-488.

Catterall, W. A., 2000, Structure and regulation of voltage-gated Ca^{2+} channels, *Annu Rev Cell Dev Biol* **16**:521-555.

Chen, L., Chetkovich, D. M., Petralia, R. S., Sweeney, N. T., Kawasaki, Y., Wenthold, R. J., Bredt, D. S., and Nicoll, R. A., 2000, Stargazin regulates synaptic targeting of AMPA receptors by two distinct mechanisms, *Nature* **408**:936-943.

Chu, P. J., Robertson, H. M., and Best, P. M., 2001, Calcium channel γ subunits provide insights into the evolution of this gene family, *Gene* **280**:37-48.

Curtis, B. M., and Catterall, W. A., 1984, Purification of the calcium antagonist receptor of the voltage-sensitive calcium channel from skeletal muscle transverse tubules. *Biochemistry* 23:2113-2118.

de Jongh, K. S., Warner, C., and Catterall, W. A., 1990, Subunits of purified calcium channels. Alpha 2 and delta are encoded by the same gene, *J Biol Chem* **265**:14738-14741.

Dolphin, A. C., Wyatt, C. N., Richards, J., Beattie, R. E., Craig, P., Lee, J. H., Cribbs, L. L., Volsen, S. G., and Perez-Reyes, E., 1999, The effect of α_2-δ and other accessory subunits on expression and properties of the calcium channel α_{1G}, *J Physiol* **519 Pt 1**:35-45.

Ertel, E. A., Campbell, K. P., Harpold, M. M., Hofmann, F., Mori, Y., Perez-Reyes, E., Schwartz, A., Snutch, T. P., Tanabe, T., Birnbaumer, L., Tsien, R. W., and Catterall, W. A., 2000, Nomenclature of voltage-gated calcium channels, *Neuron* **25**:533-535.

Hagiwara, S., Ozawa, S., and Sand, O., 1975, Voltage clamp analysis of two inward current mechanisms in the egg cell membrane of a starfish, *J Gen Physiol* **65**:617-644.

Hashimoto, K., Fukaya, M., Qiao, X., Sakimura, K., Watanabe, M., and Kano, M., 1999, Impairment of AMPA receptor function in cerebellar granule cells of ataxic mutant mouse stargazer, *J Neurosci* **19**:6027-6036.

Jiang, Y., Lee, A., Chen, J., Ruta, V., Cadene, M., Chait, B. T., and MacKinnon, R., 2003a, X-ray structure of a voltage-dependent K^+ channel, *Nature* **423**:33-41.

Jiang, Y., Ruta, V., Chen, J., Lee, A., and MacKinnon, R., 2003b, The principle of gating charge movement in a voltage-dependent K^+ channel, *Nature* **423**:42-48.

Kang, M. G., Chen, C. C., Felix, R., Letts, V. A., Frankel, W. N., Mori, Y., and Campbell, K. P., 2001, Biochemical and biophysical evidence for $\gamma 2$ subunit association with neuronal voltage-activated Ca^{2+} channels, *J Biol Chem* **276**:32917-32924.

Klugbauer, N., Lacinová, L., Marais, E., Hobom, M., and Hofmann, F., 1999, Molecular diversity of the calcium channel $\alpha_2\delta$ subunit, *J Neurosci* **19**:684-691.

Klugbauer, N., Dai, S., Specht, V., Lacinova, L., Marais, E., Bohn, G., and Hofmann, F., 2000, A family of gamma-like calcium channel subunits, *FEBS Lett* **470**:189-197.

Liu, H., De Waard, M., Scott, V. E., Gurnett, C. A., Lennon, V. A., and Campbell, K. P., 1996, Identification of three subunits of the high affinity ω-conotoxin MVIIC-sensitive Ca^{2+} channel, *J Biol Chem* **271**:13804-13810.

MacKinnon, R., 1991, Determination of the subunit stoichiometry of a voltage-activated potassium channel, *Nature* **350**:232-235.

McEnery, M. W., Snowman, A. M., Sharp, A. H., Adams, M. E., and Snyder, S. H., 1991, Purified ω-conotoxin GVIA receptor of rat brain resembles a dihydropyridine-sensitive L-type calcium channel, *Proc Natl Acad Sci U S A* **88**:11095-11099.

Moss, F. J., Viard, P., Davies, A., Bertaso, F., Page, K. M., Graham, A., Canti, C., Plumpton, M., Plumpton, C., Clare, J. J., and Dolphin, A. C., 2002, The novel product of a five-exon *stargazin*-related gene abolishes $Ca_V2.2$ calcium channel expression, *Embo J* **21**:1514-1523.

Nowycky, M. C., Fox, A. P., and Tsien, R. W., 1985, Three types of neuronal calcium channel with different calcium agonist sensitivity, *Nature* **316**:440-443.

Perez-Reyes, E., Cribbs, L. L., Daud, A., Lacerda, A. E., Barclay, J., Williamson, M. P., Fox, M., Rees, M., and Lee, J. H., 1998, Molecular characterization of a neuronal low-voltage-activated T-type calcium channel, *Nature* **391**:896-900.

Qin, N., Yagel, S., Momplaisir, M. L., Codd, E. E., and D'Andrea, M. R., 2002, Molecular cloning and characterization of the human voltage-gated calcium channel $\alpha_2\delta$-4 subunit, *Mol Pharmacol* **62**:485-496.

Ruta, V., Jiang, Y., Lee, A., Chen, J., and MacKinnon, R., 2003, Functional analysis of an archaebacterial voltage-dependent K^+ channel, *Nature* **422**:180-185.

Seisenberger, C., Specht, V., Welling, A., Platzer, J., Pfeifer, A., Kuhbandner, S., Striessnig, J., Klugbauer, N., Feil, R., and Hofmann, F., 2000, Functional embryonic cardiomyocytes after disruption of the L-type α1C ($Ca_v1.2$) calcium channel gene in the mouse, *J Biol Chem* **275**:39193-39199.

Takahashi, M., Seagar, M. J., Jones, J. F., Reber, B. F. X., and Catterall, W. A., 1987, Subunit structure of dihydropyridine-sensitive calcium channels from skeletal muscle, *Proc Natl Acad Sci U S A* **84**:5478-5482.

Tanabe, T., Takeshima, H., Mikami, A., Flockerzi, V., Takahashi, H., Kangawa, K., Kojima, M., Matsuo, H., Hirose, T., and Numa, S., 1987, Primary structure of the receptor for calcium channel blockers from skeletal muscle, *Nature* **328**:313-318.

Walker, D. and De Waard, M., 1998, Subunit interaction sites in voltage-dependent Ca^{2+} channels: role in channel function, *Trends Neurosci* **21**:148-154.

Witcher, D. R., De Waard, M., Sakamoto, J., Franzini-Armstrong, C., Pragnell, M., Kahl, S. D., and Campbell, K. P., 1993, Subunit identification and reconstitution of the N-type Ca^{2+} channel complex purified from brain, *Science* **261**:486-489.

CONCEPTS OF STATE-DEPENDENT PHARMACOLOGY OF CALCIUM CHANNELS

Evgeni N. Timin[1], Stanislav Berjukow[2], and Steffen Hering[2]

1. INTRODUCTION

For many drugs that target ion channels, drug affinity depends strongly on the precise gating state of the channel. A drug's clinical utility is often due to its state dependence, and state-dependent inhibition is a key consideration for experimental use of a drug. This chapter describes theoretical models of state-dependent inhibition, including use-dependent block, modulated and guarded receptors, and drug-inactivation synergism, and shows how these models can explain the molecular mechanisms of block by several commonly used calcium channel antagonists.

The idea that drugs bind with different strengths to resting closed (R), open (O) and closed inactivated (I) ion channels was originated by Armstrong (1966, 1969) and Strichartz (1973) and soon after was formulated as the theoretical concept of a modulated drug receptor (Hille, 1977; Hondghem and Katzung, 1977). This hypothesis has stimulated numerous biophysical and pharmacological experiments on ion channels leading to a better understanding of the suppression of noceceptive signals by local anesthetics, the antiarrhythmic action of local anesthetics and other sodium channel inhibitors, the mechanism of some antiepileptics, and many other drug effects (Hille, 1992).

In 1980, the modulated receptor hypothesis was applied to explain Ca^{2+} channel inhibition. It provided an elegant explanation for the predominant antihypertensive action of 1,4-dihydropyridines (DHPs) (Bean, 1984; Sanguinetti and Kass, 1984) and the suppression of supraventricular tachyarrhythmias by phenylalkylamines (PAAs) and diltiazem (DIL) (Lee and Tsien, 1983; McDonald et al., 1984).

[1]Vishnevski Institute of Surgery, Bolshaya Serpuchovskaya 27, 113093-Moscow, Russia; and [2]Institut für Biochemische Pharmakologie, Universität Innsbruck, Peter Mayr Straße 1, A-6020 Innsbruck, Austria. Address correspondence to Dr. Hering at Institut für Pharmakologie und Toxikologie, Universität Wien, Pharmaziezentrum, Althanstraße 14, Vienna, Austria.

In spite of the dissimilarity of these pathophysiological events and the different chemical nature of these drugs, many of them share a common feature - the selective inhibition of "abnormal" high frequency spiking without significantly disturbing "normal" action potential generation and propagation. The principal finding that these drugs inhibit channels more efficiently when the channels are "in use" (during an action potential in either the O and/or I states) was designated as "use-dependent" ion channel block (Courtney, 1975).

The mechanism of use-dependent ion channel inhibition is conventionally explained by the "modulated receptor hypothesis" (including the "guarded receptor hypothesis") (Hille, 1977; Starmer and Grant, 1985) or, alternatively, by the "slow inactivation hypothesis" (Khodorov, 1979; Khodorov, 1981). Both concepts refer to the experimental finding that these drugs slow ion channel recovery to rest. The "modulated receptor hypothesis" assumes that in resting channels, the drug binding sites are hidden in the depth of the channel molecule. The receptors are exposed during conformational changes underlying channel activation or inactivation.

A special case of the modulated receptor hypothesis is known as "open channel block" (Starmer and Grant, 1985). In this scenario, the drug interacts with a receptor in the channel pore that is accessible only during channel activation. In this case, slow recovery is interpreted as drug dissociation from the open blocked state.

The "slow inactivation hypothesis" assumes that the drugs accelerate channel transitions into a slow inactivated channel conformation. Drug binding occurs either in a state-dependent manner or irrespective of R, O and I state occupancy (Khodorov, 1979; Khodorov, 1981). In this case, use-dependent block is explained by slow recovery of the channels from the slow inactivated state.

2. USE-DEPENDENT CHANNEL BLOCK

The typical features of use-dependent inhibition of a Ca^{2+} channel during a pulse train are illustrated in Figure 1A. Under control conditions, the $Ca_v2.1$ channels (see Ertel et al., 2000) recover completely between the individual pulses of a train, whereas in the presence of mibefradil a cumulative current inhibition is observed (see panel B). Fast recovery in the absence (broken line) and slow recovery in the presence of mibefradil (filled circles) is illustrated in Figure 1C.

The "modulated receptor hypothesis" and the "slow inactivation hypothesis" predict a very similar chain of events during the train (Figure 2): before the first pulse is applied, all channels dwell in the resting state. The first depolarizing pulse activates the majority of channels (fraction illustrated as white circles) and the drug molecules either inhibit a small fraction of ion channels in the O or I states ("modulated receptor hypothesis") or, alternatively, the channels enter the slow inactivated state. These channels (grey circles) are not available for activation. Before application of the next pulse the blocked channels (or, alternatively, the channels that have entered the slow inactivated state) slowly recover at the resting potential. However, some of them remain unavailable (black circles) due to the slow recovery at rest. The subsequent test pulse will elicit a smaller inward current, and during the subsequent pulse a new portion of channels is added to the "slowly recovering" non-available state. This process continues until a balance between the inhibited and recovered channel fractions is achieved—that is, until accumulation in the non-available states equals recovery.

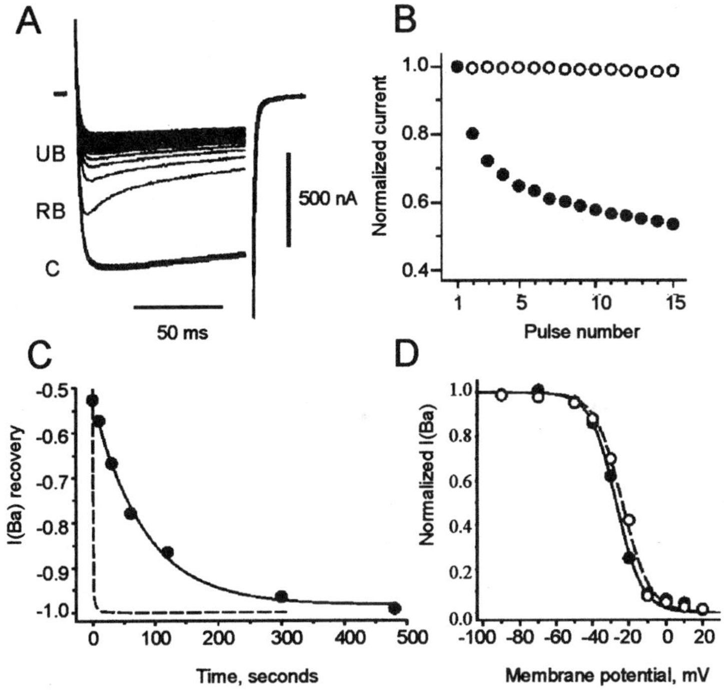

Figure 1. Resting state and use-dependent block of $Ca_V2.1$ Ca^{2+} channels by mibefradil. (A) Oocytes expressing $Ca_v2.1(\beta_{1a})$ channels were incubated for 3 minutes in 100 µM mibefradil. 15 depolarizing test pulses (100 ms, 0.1 Hz) were applied from -80 mV to 20 mV before (C) and after drug application in 40 mM barium solution. Peak barium current (I_{Ba}) inhibition during the first test pulse was defined as resting state block (RB). Use-dependent (UB) block was subsequently measured as cumulative current inhibition during the train. **(B)** Normalized peak I_{Ba} during the pulse train (A) in the absence (open circles) and presence (black circles) of mibefradil. **(C)** Slow recovery from use-dependent block by mibefradil. I_{Ba} was elicited by 100 ms test pulse trains (see A). The time course of peak I_{Ba} recovery from use-dependent block by mibefradil (50 µM) at -80 mV is illustrated by the filled circles. Broken line represents I_{Ba} recovery after a 3 second depolarizing pulse in the absence of drug. The curves represents monoexponential fits to the experimental points by the function: $I_{Ba,recovery} = 1-\exp(-t/\tau_{recovery})$ yielding recovery time constants of 81.4 seconds in drug (●) and 0.29 ± 0.03 seconds in control (broken line). **(D)** Inactivation curve (O-control, ●- 50 µM mibefradil) of $Ca_v2.1$ estimated from normalized peak I_{Ba} during a 200 ms test pulse to 20 mV following a 30 second conditioning prepulse to the indicated voltages (inter-pulse interval 3 ms). Curves are drawn according to the equation $I/I_{max} = \{1+\exp[(V-V_{0.5})/k]\}^{-1}$, yielding in control $V_{0.5} = -23.6$ mV and k = 7.4 mV (O) and $V_{0.5} = -26.9$ mV and k =7.0 mV in drug (●). Data reproduced with permission from Aczel et al., 1998.

2.1. Quantitative Analysis of Use-Dependent Channel Block

If B_N is the fraction of non-available channels at the Nth pulse (accumulated in previous pulses), then the fraction of available channels is $(1 - B_N)$. We assume that in each pulse a certain fraction β of activated channels enters a non-available state. Thus $\beta\cdot(1 - B_N)$ of channels are added to the non-available fraction. At the end of Nth pulse the fraction of non-available channels is $B_N + \beta\cdot(1 - B_N)$.

Between the pulses the fraction ρ has recovered. Consequently, the fraction of non-available channels at the application of the $(N + 1)^{th}$ pulse is $B_{N+1} = (1 - \rho)\cdot[B_N + \beta\cdot(1 - B_N)]$. Finally we obtain a mono-exponential time course of current inhibition during a train:

$$B_N = B_\infty (1 - \sigma^{N-1})$$

with a steady-state level of inhibition:

$$B_\infty = \frac{(1-\rho)\cdot\beta}{1-(1-\rho)\cdot(1-\beta)}$$

and a coefficient of the rate of current inhibition during a train:

$$\sigma = (1 - \rho)\cdot(1 - \beta).$$

The fractions of the channels that are inhibited during a pulse (β) and the fraction of channels that recover between two pulses (ρ) can be calculated from B_∞ and σ. Both parameters can be estimated experimentally:

$$\beta = \frac{B_\infty \cdot (1-\sigma)}{\sigma + B_\infty \cdot (1-\sigma)}$$

$$\rho = (1 - B_\infty) \cdot (1 - \sigma)$$

From the above model the number of pulses required to approach the steady state level of current inhibition, defined arbitrarily as two successive current amplitudes that differ by less then 5%, can be calculated:

$$N_{SS} = 1 - \frac{\ln[1 + (1-\sigma)/0.05]}{\ln(\sigma)}$$

To illustrate a practical application of these deliberations we have simulated two examples of use-dependent current inhibition (Figure 2B). In one case the peak current inhibition reaches its steady state (B_∞) within a few pulses (white circles), and in the other case channel block develops at a much slower rate (black circles). As illustrated in Table 1, the different kinetics reflect different association and dissociation kinetics of the drug.

The faster kinetics reflect not only faster current inhibition but also faster recovery of channels between the pulses. So far we have outlined the macroscopic picture of ion channel inhibition without considering a particular molecular mechanism of the drug-channel interaction. This type of analysis gives a first estimate of the channel fraction entering a non-available state during a pulse (or action potential) and the recovery between pulses (or action potentials). The two parameters β and ρ are sufficient for a first evaluation of the "therapeutic potency" of a given drug. They can be estimated from the kinetics of use-dependent current inhibition.

2.2. Accounting for the Molecular Mechanism of Channel Block

In order to interpret use-dependent inhibition in terms of rate constants of the drug-channel interaction, we have to make some assumptions about mechanism of drug binding. Such an approach was first formulated by Starmer and Grant (1985) and later developed by Timin and Hering (1992) for "open channel block."

The total population of ion channels is divided into 3 parts: blocked (B), protected from interaction with drug (U_p), and accessible to the drug molecules (U_a). With those notations the balance equations are:

$$U_a + U_p + B = 1$$

$$\frac{dB}{dt} = k_1[D] \cdot U_a - k_{-1} \cdot B$$

where [D] is the drug concentration, and k_1 and k_{-1} are, respectively, rate constants of drug binding and of dissociation of the drug-channel complex. In the frame of our model the dissociation of drug-channel complex occurs irrespective of the channel state (Timin and Hering, 1992). We did not account for details of drug dissociation because experimental evidence for individual steps of dissociation, such as drug trapping in Ca^{2+} channels, is insufficient for a quantitative description.

The fraction of channels accessible for drug binding U_a is a fraction of open channels. Substituting $U_a = g(t) \cdot (1-B)$, where g(t) is the fraction of open channels, we obtain:

$$\frac{dB}{dt} = k_1[D] \cdot g(t) \cdot (1 - B) - k_{-1} \cdot B$$

The solution of this equation for repetitive stimulation with period T gives a mono-exponential decrease of the peak currents during a train:

$$B_N = B_\infty [1 - \sigma^N],$$

where B_N is the fraction of blocked channels after the Nth pulse, B_∞ is the steady state level of blockade, and σ the steepness of the current amplitude inhibition during the train. Fractions of blocked channels increase from pulse to pulse from 0 to B_∞; that is, the relative current amplitudes decrease from 1 to 1 - B_∞. Both the steady-state level B_∞ and the steepness σ depend on the rates of binding and dissociation. Apparent rate constants can be deduced from B_∞ and σ according to:

$$k_1[D] \approx \frac{\ln(1 - B_\infty + B_\infty / \sigma)}{\int g(t)dt} \qquad k_{-1} \approx -\frac{\ln[\sigma \cdot (1 - B_\infty) + B_\infty]}{T}$$

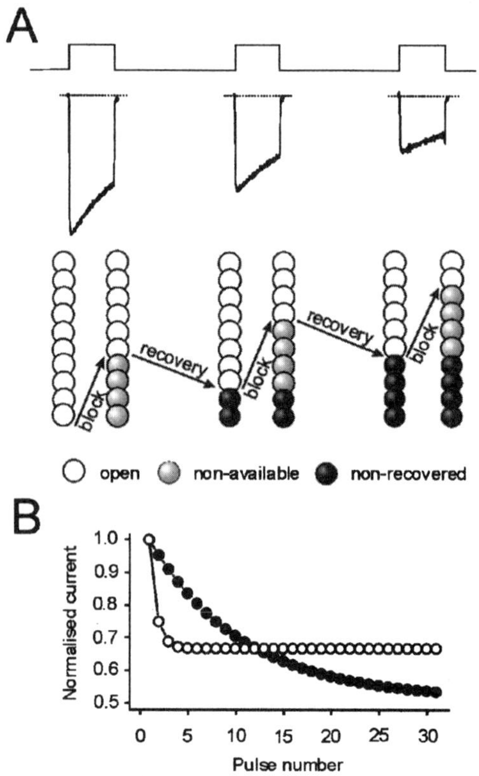

Figure 2. Use-dependent ion channel inhibition during a pulse train. (A) The upper panel illustrates the depolarizing test pulses (0.1 Hz, 100ms) and the corresponding barium currents through $Ca_v1.2$ in the presence of 10 µM µM (-)-devapamil. The lower panel exemplifies the fractions of activated Ca^{2+} channels entering a non-available channel state (gray circles). The black circles illustrate the channel population that did not recover after the previous pulses (accumulated channel inhibition). **(B)** Simulation of different kinetics (peak current inhibition) of use-dependent channel block. The different rates of channel block during a train for a given rate of channel block during a pulse and a given rate of recovery between pulses are indicated in **Table 1** below. The parameters of gradual current inhibition during a train are calculated as a function of the rate of channel inhibition during a single pulse and recovery between pulses. The rate of channel inhibition was taken as fraction of activated channels entering the non-available state during each test pulse. The rate of recovery was taken as fraction of non-available channels released between two subsequent pulses.

Table 1

	β inhibition in one pulse	ρ recovery between two pulses	σ resulting rate of current inhibition in train	B_∞ steady state inhibition	N_{SS} number of pulse close to steady state
●	0.01	0.01	0.98	0.5	18-20
O	0.5	0.5	0.25	0.33	3

For long pulses $\int g(t) \cdot dt$ can be reduced to $\tau_{current\ decay}$. For 'short' ($t_{pulse} << \tau_{current\ decay}$) pulses, $\int g(t) \cdot dt$ can be reduced to t_{pulse}. Furthermore, for short pulses the expression for steady state inhibition (Timin and Hering, 1992) may be approximated by:

$$B_{\infty} = \frac{1}{1 + \frac{k_{-1}}{k_1[D]} \cdot \frac{T}{t_{pulse}}}$$

This expression demonstrates that the steady-state level of current inhibition during a train may serve as an index of drug affinity. The relationship between the pulse length and the period of stimulation, however, has to be taken into account.

3. STATE DEPENDENT DRUG ACTION

It is widely accepted that calcium channels dwell in at least three conformational states: R (resting, nonconducting, but available for excitation), O (open, conducting, or excited), and I (inactivated, nonconducting, and not available for excitation). First order transitions between states depend on the membrane potential. During a depolarization channels move from R via O to I with rate constants α_m and β_h, and during a hyperpolarization (i.e., following return to the resting potential) back to the R state with rate constants α_h and β_m. These notations were chosen to keep some consistency with the Hodgkin-Huxley formalism.

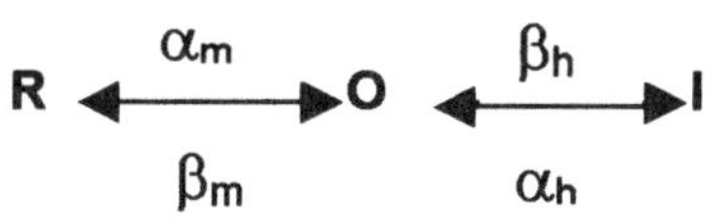

Before discussing more complex gating mechanisms, let us analyse whether the simple scheme above will enable us to distinguish some characteristic features of Ca^{2+} antagonist interaction with the different states (R, O, I).

3.1. R-State Dependent Inhibition

The mechanism of resting state-dependent block is not sufficiently understood. This interaction is feasible, but it would imply that transitions to open or inactivated states initiate drug dissociation from resting channels. At least for Ca^{2+} channels there is no evidence for a 'use-dependent release' from block. A further difficulty with the interpretation of R-state dependent block is illustrated in Figure 1A. Peak current inhibition by mibefradil during the first test pulse of the train (the initial block) may be interpreted as selective inhibition of resting channels (RB). However, if the drug significantly accelerates the current decay, the reduced peak current amplitude could partially result from changes in channel gating.

3.2. "Fingerprints" of O and I State Dependent Channel Inhibition

The mechanism of open channel block is illustrated by the kinetic scheme in Figure 3A. Drug binding occurs only in the O-state and interrupts the current flow (Figure 3B). The

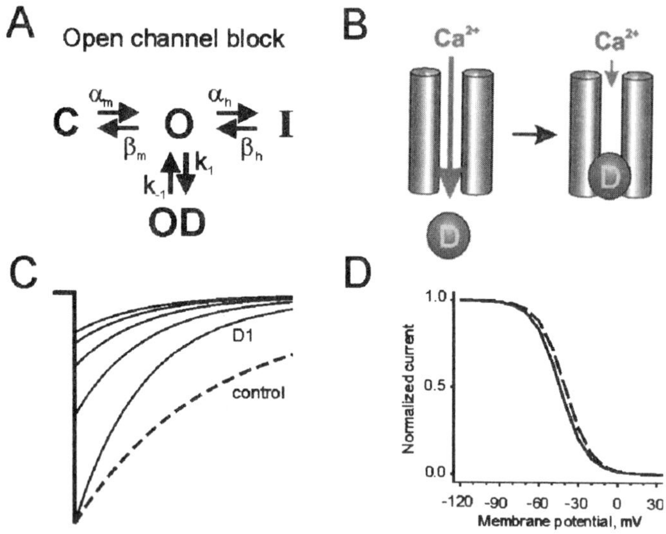

Figure 3. The "open channel block" model. **(A)** Kinetic scheme of Ca^{2+} channel state transitions and their selective inhibition in the open state (see Appendix). A cartoon illustrating the underlying molecular mechanism is shown in **(B)**. **(C)** Simulated changes in current kinetics characteristic for open channel inhibition and use-dependent current inhibition. The current traces in Figures 3C and 4C (see facing page) were simulated with identical rate constants of channel activation and inactivation and identical rates constants of drug association and dissociation. Channel activation (assumed to be very fast) is not resolved on this time scale. The absolute values of the rate constants of inactivation are not important (and time scale is therefore not shown), but the relationship $\alpha_h = 20 \cdot \beta_h$ is essential. The rate constants of drug binding is $k_1 = \alpha_h$ and the rate of drug dissociation $k_{-1} = k_1/60$. Note the marked acceleration of the current decay in the open channel block model compared to the negligible changes in inactivation channel block. Trace D1 illustrates the kinetics of the first current trace in drug compared to control. **(D)** Changes in the availability curves corresponding to the open channel block model.

model predicts cumulative suppression of the current during a train (Figure 3C). The channels recover with a two-exponential time course: the fast phase describes recovery from inactivation (I) and the slow phase reflects the dissociation of the drug-channel complexes (OD; not shown). In the presence of drug in concentrations close to saturation, the fast phase is expected to be almost undetectable. The resulting changes in channel availability are simulated in Figure 3D.

The I-state inhibition model (or "slow inactivation model", Figure 4A,B) produces a very similar picture of "use-dependent" current inhibition (Figure 4C). The recovery process occurs with a two-exponential time course, with one exponential describing the recovery from fast inactivation and the other recovery from the slow inactivated state (not shown). The model predicts a pronounced shift of the availability of the channels in the hyperpolarizing direction.

These simulations are helpful to distinguish the basic features of the two mechanisms. First, the "open channel block" model predicts a more pronounced acceleration of current decay than the "slow inactivation model" (Figure 3C and 4C). Theoretically, drug-binding to I state and successive channel block should also accelerate the current decay. A comparison of the simulated currents through identical "theoretical

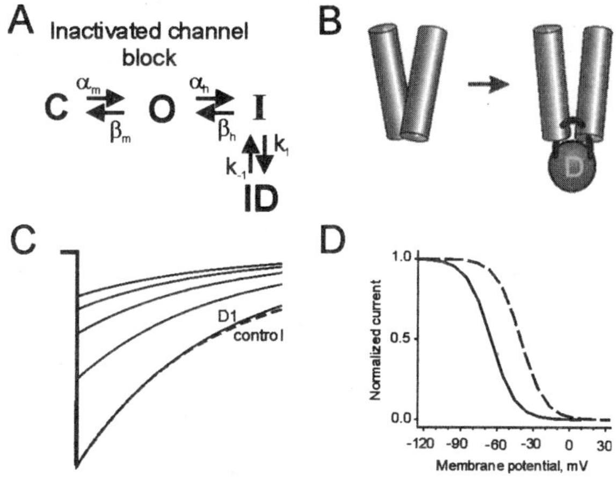

Figure 4. The "inactivated channel block" model. (A) Kinetic scheme of Ca^{2+} channel state transitions and their selective inhibition in the inactivated state (see Appendix) (cartoon shown in **B**). **(C)** Simulated current kinetics characteristic for inhibition of inactivated channels and use-dependent current inhibition (similar rate constant for as in Figure 3C for open channel block). Trace D1 illustrates the kinetics of the first current trace in drug compared to control. **(D)** Changes in the availability curves corresponding to the inactivated channel block model. Only the inactivated channel block results in a pronounced shift of the availability curve to more hyperpolarized potentials (compare with Figure 3D).

channels" under comparable conditions (i.e., with identical rate constants) reveals, however, only an insignificant effect (Figure 4C) compared to the prominent acceleration in the O state block scenario (Figure 3C). This model appears, therefore, appropriate to explain the key features of $Ca_v2.1$ inhibition by mibefradil (Aczel et al., 1998) (Figure 1). Second, both models predict different drug effects on the voltage dependence of the channel availability. I state dependent block causes a remarkable shift of the curve in the hyperpolarizing direction (Figure 4D). The shift on the voltage axis is related to the applied drug concentration:

$$\delta V = \psi \cdot \ln(1 + k_1 \cdot [D]/k_{-1});$$

where ψ is the steepness of the availability curve in the middle point.

Open channel block also affects channel availability. However, these changes are less prominent and cannot be reduced to a simple scaling and shift of curve (Figure 3D). This criterion, the shift of the availability curve on the voltage axis, is frequently applied to distinguish between the two mechanisms: drug binding to either open or inactivated channel states.

It is important to note that this model implies that drug binding does not affect channel gating. In other words, the drug inhibits the channels exclusively by "freezing" them in a non-conducting state. In alternative models such as drug-inactivation synergism (see below), when drug binding changes channel gating, neither acceleration of the current decay nor a shift of the availability curve can be used as criteria for predominant interaction with a certain channel state. Moreover, in many cases drug

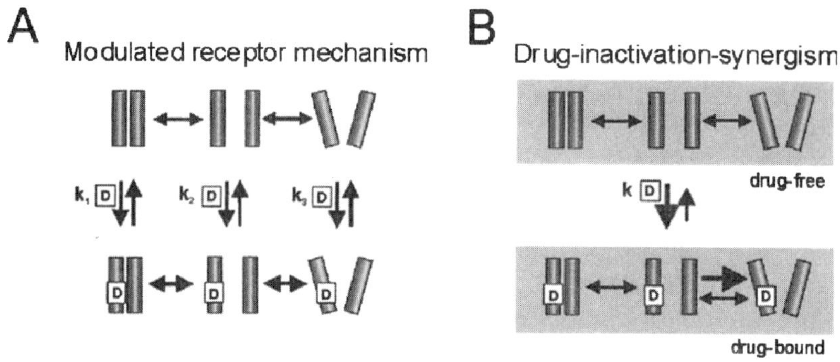

Figure 5. Receptor concepts. (A) Modulated receptor model. The modulated receptor hypothesis assumes that a drug binds with individual rate constants to resting (R) open (O) and inactivated (I) channels. Ca^{2+} antagonists are believed to bind with high affinity to the inactivated state (Bean, 1984; Sanguinetti and Kass, 1984). There is no consensus about predominant binding of the different classes of Ca^{2+} antagonists to a certain channel state: DHP Ca^{2+} antagonists are believed to bind with high affinity to the inactivated channel conformation (Bean, 1984; Sanguinetti and Kass, 1984; but see Berjukow et al., 2000). PAAs have been proposed to bind with high affinity to open (Timin and Hering, 1992) or inactivated channels (Nawrath and Wegener, 1997; Hockerman et al., 2000). Mibefradil is believed to bind either with high affinity to inactivated channels (Bezprozvanny and Tsien, 1995; Jimenez et al., 2000) or, alternatively, to block predominantly open Ca^{2+} channels (Aczel et al., 1998).

In HVA Ca_v channels containing an "accelerating" β-subunit, transitions to the inactivated state (I) increase the amount of inactivated channels during a pulse. A modulated receptor model would explain the stronger use-dependent inhibition of fast inactivating channel constructs by favored drug binding and inhibition of inactivated channels. Conversely, "decelerating" β_{2a}-subunits apparently reduce the amount of inactivated channels, which would reduce drug affinity (Hering, 2002).

(B) Drug-inactivation synergism model. In a drug-inactivation synergism model, the drugs are allowed to bind irrespective of the channel state (vertical arrow). Drug binding promotes inactivation and delays recovery. In this scenario, enhanced use-dependent channel block results from synergism between drug-induced and intrinsic (e.g. "accelerating" β-subunit-induced) inactivation (right panel). Reduced apparent drug sensitivity of Ca_v composed of a decelerating β_{2a} subunit would reduce net transitions towards inactivation and thereby hinder drug-induced inactivation. A simulation of the different drug affinities in Ca_v composed of different β-subunits is illustrated in Figure 6.

interaction changes the slope of the availability curve. These effects cannot be reduced to "scaling and shift" as predicted by the "I state inhibition model" (Figure 4D). Significant changes in the slope of the availability curves in the presence of DHPs indicate that these drugs modulate the process of channel inactivation (Berjukow et al., 2000; Lacinova et al., 2000).

3.3. New Tools for Molecular Studies on Ca^{2+} Channel Inhibition

Recent pharmacological studies on mutant and recombinant Ca_v channels of different subunit composition provided new important insights into the molecular events underlying Ca^{2+} channel block. The properties of high voltage activated (HVA) Ca_v channels (Catterall, 2000; Ertel et al., 2000) are strongly regulated by their subunit composition. In particular the inactivation of HVA is strongly modulated by the four β subunits (β_1, β_2, β_3, β_4). This is exemplified in Figure 6 displaying currents through a $Ca_v1.2$ incorporating either a β_{1a} subunit, which accelerates inactivation, or a β_{2a} subunit, which decelerates inactivation. Co-assembly of HVA α_1-subunits with different β-

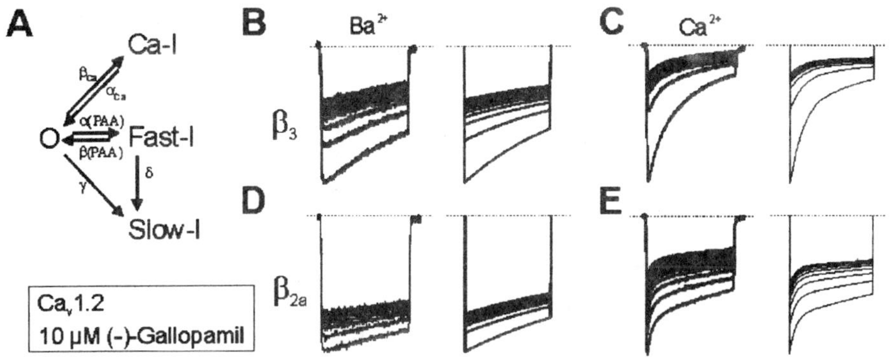

Figure 6. Simulation of use dependent inhibition of Ca$_v$1.2 by (-)-Gallopamil. (A) The kinetic model of channel inactivation incorporates Ca^{2+} dependent (Ca-I), fast voltage-dependent (Fast-I) and slow inactivation (Slow-I). **(B)** Simulation of use dependent inhibition of Ca$_v$1.2(β_3) and Ca$_v$1.2(β_{2a}) channels by 10 μM (-)gallopamil with Ba^{2+} or Ca^{2+} as a charge carrier. Experimental traces on the left are shown for comparison. Only transitions between O and Fast-I states (α and β) are assumed to be modified by drug. The effect of the β_{2a} subunit is simulated as a destabilization of the Fast-I state by decreasing α and increasing β rates approximately two-fold. The effect of (-)gallopamil is simulated by increasing α by six-fold and decreasing β by two-fold, and 50% of channels are assumed to be modified by drug (see http://circres.ahajournals.org/cgi/content/full/hh2001.098983/ DC1 for the full set of parameters). Difference in apparent PAA sensitivity of α_{1C}/β_3 and α_{1C}/β_{2a} channels does not require the assumption that the β subunit affects drug-channel interaction and may be reproduced solely by interplay of the α and β rates. Data are reproduced with permission from Sokolov et al. (2001).

subunits changes the fraction of open and inactivated channels, apparently key conformational states for interaction with Ca^{2+} antagonists. Thus Ca$_v$ channels composed of different β subunits have emerged as interesting tools to clarify the molecular mechanism of Ca^{2+} channel inhibition (Hering, 2002). Furthermore, point mutants with significantly changed inactivation properties have served as interesting instruments to analyze state-dependent ion channel block (Hering et al., 2003, *in press*).

3.4. Drug-Inactivation Synergism

There is increasing evidence from studies on recombinant Ca^{2+} channels that Ca^{2+} antagonists modulate the mechanism of inactivation (Berjukow et al., 2000; Sokolov et al., 2001). PAAs, mibefradil, and DIL appear to accelerate the time course of channel inactivation in a manner similar to point mutations or "accelerating β subunits." To explain these data a "drug-inactivation synergism" model was proposed (Figure 5) (Hering, 2002). In the frame of this concept, drug binding is not necessarily dependent on a certain channel state. The model accounts for two channel fractions: one with kinetics of drug-free channels, the other one displaying the kinetics of drug-bound (modified) channels (Figure 5B). The simulated currents, therefore, decay with a bi-exponential time course. The model also describes the bi-exponential recovery from inactivation, reflecting fast recovery of drug-free channels and slow recovery of drug-modified channels (Sokolov et al., 2001). If drug-modified channels recover

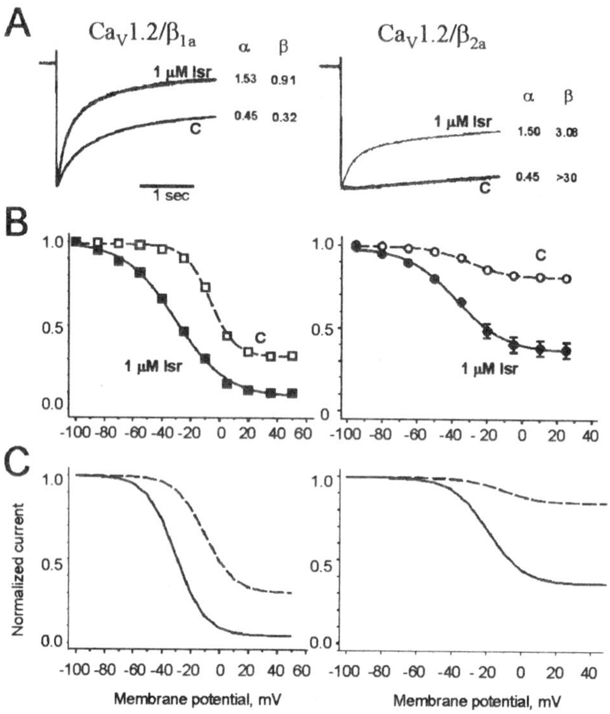

Figure 7. Simulation of the (+)-isradipine effect on the voltage dependent gating kinetics of $Ca_v1.2$. (A) Representative scaled I_{Ba} through $Ca_v(\beta_{1a})$ (left panel) and $Ca_v(\beta_{2a})$ (right panel) constructs during 3 second depolarization from -80 to 20 mV in control and in the presence of 1 μM (+)-isradipine. The theoretical curves were calculated using a model described by Sokolov et al. (2001). During a membrane depolarization channels move from the open state to the fast and slow inactivated states:

(scheme 7A)

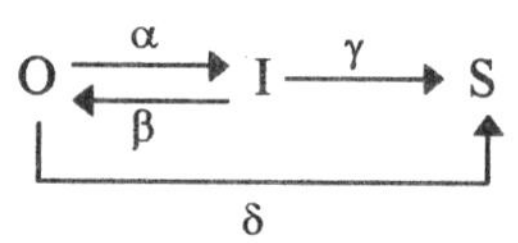

DHP effects on the current kinetics were simulated by a 3-fold increase of the forward (α) and 3-fold decrease of the backward (β) rate constants:

	α	β	γ	δ
		$Ca_v(\beta_{1a})$ (left panel)		
Control	0.45	0.32	0.03	0.03
DHP	1.53	0.1	0.03	0.03
		$Ca_v(\beta_{2a})$ (right panel)		
Control	0.45	10	0.03	0.03
DHP	1.5	3.1	0.03	0.03

(B) Voltage dependence of $Ca_v(\beta_{1a})$ (left panel) and $Ca_v(\beta_{2a})$ (right panel) inactivation during a 3 second depolarizing test pulses. Experimental points were fitted to a Boltzmann function:

$$H_{\infty} = H_{SS} + \frac{1 - H_{SS}}{1 + \exp(\frac{V - V_H}{k})}$$

(C) Simulation of the voltage dependence of $Ca_v(\beta_{1a})$ (left panel) and $Ca_v(\beta_{2a})$ (right panel) by means of the drug-inactivation synergism concept. Isradipine was assumed to accelerate the O→I transition (rate constant α on scheme 7A) and decelerate I→O transition (rate constant β on scheme 7A). The ratio α/β was assumed to increase 9 times in the presence of the DHP compared to control at all membrane voltages. The Boltzmann equation fit in control was recalculated for DHP action:

$$H_{\infty,DHP} = \frac{1 + H_{ss}\exp(\frac{V - V_H}{k})}{1 + (H_{SS} + \chi - \chi \cdot H_{ss}) \cdot \exp(\frac{V - V_H}{k})}$$

where $\chi = 9$ is the coefficient of changes of the α/β relationship. The parameters of the Boltzmann curve in control were:

	H_{SS}	V_H, mV	k, mV
$Ca_v(\beta_{1a})$	0.31	-10	7
$Ca_v(\beta_{2a})$	0.85	-30	7

Data in (A) are reproduced with permission from Berjukow et al., 2000. Other data are from Berjukow et al. (unpublished).

sufficiently slowly, "use-dependent" channel block is predicted (Figure 6). The voltage dependence of the modified availability resembles the shift produced by the I state model. Drug-induced changes in channel availability, however, reflect changes in channel gating but not enhanced drug binding to the inactivated state (Figures 4D and 7C). Thus, some of the basic findings of "state-dependent" pharmacology can be reduced to different extents of acceleration of channel inactivation and a deceleration of channel recovery at rest.

4. MOLECULAR MECHANISMS OF DRUG ACTION

4.1. PAA, DIL, and Mibefradil Action

The crucial determinants of the putative receptor sites for PAAs, DIL and DHPs were localized in pore forming transmembrane segments IIIS5, IIIS6 and IVS6 (Hering et al., 1996; Degtiar et al., 1997; Kraus et al., 1998; Striessnig et al., 1998; Motoike et al., 1999; Hockerman et al., 2000). Subsequent studies revealed, however, that many mutations within the putative drug-receptor region are accompanied by changes in inactivation kinetics (Hering et al., 1997; Hering, 2002). Furthermore, in the majority of cases mutations affecting channel inactivation simultaneously affected the apparent affinity for PAA, DIL, and mibefradil, even when their location was far outside the putative receptor site. These studies suggest that state dependent inhibition of Ca^{2+} channels is closely linked to the process of channel inactivation (Hering et al., 1997; Hering, 2002). Some of these recent findings on recombinant channels, however, are difficult to explain in terms

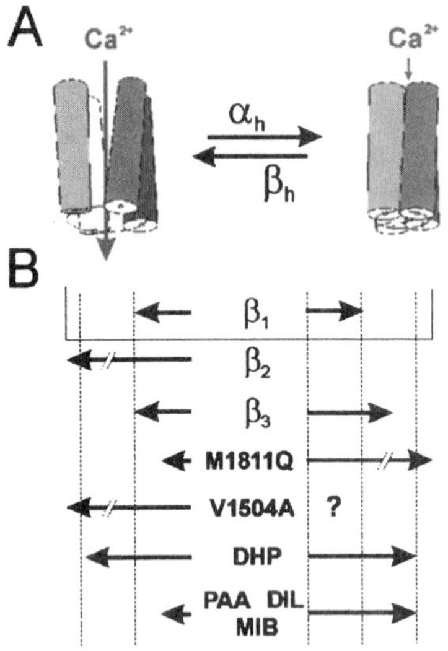

Figure 8. Modulation of channel distribution between available and non-available channel states by β-subunits, point mutations and drugs. The concept of drug-inactivation synergism is based on the finding that "accelerating" β-subunits and point mutations in the α_1-subunit accelerate channel inactivation like DHPs, PAAs, DIL and mibefradil (MIB) (Aczel et al., 1998; Berjukow et al., 1999; Sokolov et al., 1999; Berjukow et al., 2000; Jimenez et al., 2000; Sokolov et al., 2001). Structural changes reducing apparent drug sensitivity are, in contrast, often associated with a deceleration of the time course of inactivation. The arrows in the lower panel illustrate the modulatory effects of different β subunits (Sokolov et al., 2001), mutations M1811Q (in $Ca_v2.1$) (Berjukow et al., 2001), V1504A (in $Ca_v1.2$) (Berjukow et al., 1999), and Ca^{2+} antagonists on the transition rates between the open and inactivated channel state.

of state-dependent pharmacology. Mibefradil, for example, displays the hallmarks of an open channel blocker: the drug-induced acceleration of the current decay is accompanied by only insignificant changes in channel availability (compare Figure 1 with Figure 3C,D). Co-expression of the mibefradil-sensitive calcium channel α_1 subunit with the β_{2a} subunit prolongs the time that the channels spend in the open state. Consequently, these channels are expected to be more efficiently inhibited by 'open channel blockers'. Jimenez et al., however, have clearly demonstrated that co-expression of the $\alpha_1 2.1$ subunit with a "decelerating" β_{2a}-subunit decreased, rather than enhanced, mibefradil sensitivity. This finding is in contradiction to the open channel block model. A similar observation was made for $Ca_v1.2$ inhibition by PAAs (Sokolov et al., 1999; Sokolov et al., 2001) (Figure 6).

4.2. DHP Action

Co-expression of α_{1C} or a DHP-sensitive α_{1A}-mutant ($\alpha_{1A\text{-}DHP}$) with the inactivation-accelerating β_{1a} subunit or the inactivation-decelerating β_{2a} subunit produced significant changes in the steady state inactivation at a holding potential of -80 mV. Enhanced steady state inactivation in the channels composed of β_{1a}, however, did not significantly affect channel block. This finding is in contradiction to high affinity binding of DHPs to the inactivated state (Berjukow et al., 2000). Furthermore, it is well established that DHPs induce a pronounced tonic channel block while simultaneously shifting the availability curve to more hyperpolarized potentials. Tonic block by DHPs can neither be enhanced nor weakened by repetitive stimulation, and also can not be removed by strong membrane hyperpolarization. Apparently, tonic Ca^{2+} channel inhibition by DHPs occurs irrespective of the predominant sojourn of the channels in a certain state.

The shift of the availability curve is conventionally interpreted as high affinity DHP-binding and block of inactivated channels (Bean, 1984). However, the DHP-induced acceleration of the current decay (most prominent with barium as charge carrier, Figure 7A) is characteristic of open channel inhibition (Figure 3C). Thus in terms of "state dependent pharmacology" we have to account for three different mechanisms simultaneously: state-independent (tonic), state-dependent open-channel inhibition, and state-dependent inactivated-channel inhibition.

A drug-inactivation synergism model offers an alternative explanation for some of the most intriguing findings of DHP action. The accelerated current decay (Figure 7A) may reflect a DHP-induced acceleration of the process of Ca^{2+} channel inactivation (Berjukow et al., 2000; Berjukow and Hering, 2001). The slight deceleration of the recovery from inactivation is in correspondence with the deceleration of the backward transition from the fast inactivated (I_{fast}) to the open channel state (Berjukow et al., 2000; Berjukow and Hering, 2001). Due to the faster recovery of drug-bound channels, use-dependent block of Ca^{2+} channels by DHPs is less pronounced compared to PAAs, DIL and mibefradil. A further interesting result of the simulation illustrated in Figure 7 was that an acceleration of the forward and a deceleration of the backward transitions reproduced the changes in steady state availability (Figure 7 C, D).

5. CONCLUSIONS

A modulated receptor mechanism presumes that drug-bound channels travel between different states, resulting in multiple transitions between drug-bound and drug-free channel states. Cycles on such a kinetic scheme have to obey the principle of thermodynamic reversibility. Consequently, changes in the transition rates within the channel inevitably result in different drug affinities for R-, O-, and I-states. This very attractive idea, however, leads to complicated quantitative description. Even a simplified scheme, as illustrated in Figure 5A, is described by five differential equations predicting five exponential time courses that are hardly distinguishable. The "O state," "I state" (Figures 3 and 4) and the "drug-inactivation synergism" models appear to be more suitable for a quantitative analysis of the kinetics of channel inhibition (Figures 5B, 6, 7). There is also growing evidence from studies on recombinant Ca_v channels that drug binding of some Ca^{2+} channel inhibitors might occur irrespective of the channel state (Berjukow et al., 2000; Berjukow and Hering, 2001; Sokolov et al., 2001; Hering, 2002).

Results from several groups have convincingly demonstrated that β subunit interaction with an intracellular loop between domains I - II has an important role in Ca^{2+} channel block (Sokolov et al., 1999; Jimenez et al., 2000). Thus, in HVA Ca^{2+} channels, allosteric sites remote from the pore have major effects on the apparent sensitivity for Ca^{2+} antagonists. A "drug-inactivation synergism" model explains these findings by drug-induced inactivation without the necessity of assuming enhanced drug-binding to a privileged channel state or a direct modulation of drug affinity by the β subunits (Hering, 2002). This model reproduces the changes in current kinetics characteristic for open channel block, inactivated channel block, and also the corresponding changes in Ca^{2+} channel availability (Figures 6 and 7).

ACKNOWLEDGMENTS

This work was supported by FWF grants P12828, P15527, and P15914.

APPENDIX

1. The kinetic scheme of the O-state-dependent block (see Figure 3) is described by the set of equations (rate constants are shown on the scheme) which reflects the mass balance:

$$\frac{d[R]}{dt} = -\alpha_m \cdot [R] + \beta_m \cdot [O]$$

$$\frac{d[I]}{dt} = -\alpha_h \cdot [I] + \beta_h \cdot [O]$$

$$\frac{d[OD]}{dt} = k_1 \cdot [D] \cdot [O] - k_{-1} \cdot [OD]$$

$$[R] + [O] + [I] + [OD] = 1$$

The solution under assumption that activation is faster than inactivation ($\alpha_m + \beta_m >> \beta_h + \alpha_h$) and inactivation in turn is faster than binding and dissociation of drug ($\beta_h + \alpha_h >> k_1[D] > k_{-1}$) gives for the fraction of channels available for excitation ($[H] = [R] + [O]$):

$$H(t) = H_\infty + A_{fast} \cdot \exp(-\frac{t}{\tau_{fast}}) + A_{slow} \cdot \exp(-\frac{t}{\tau_{slow}})$$

Where:

$$\tau_{fast} = \frac{1}{\alpha_h + \beta_h \cdot \dfrac{\alpha_m}{\alpha_m + \beta_m}}$$

$$\tau_{slow} = \frac{1}{k_{-1} + \dfrac{k_1[D]}{1 + \dfrac{\beta_m}{\alpha_m} + \dfrac{\beta_h}{\alpha_h}}}$$

$$H_\infty = \frac{1}{1 + \dfrac{\alpha_m}{\alpha_m + \beta_m}(\dfrac{k_1[D]}{k_{-1}} + \dfrac{\beta_h}{\alpha_h})}$$

2. The set of equations describing the mass balance for the I-state-dependent block (see Figure 4) is:

$$\frac{d[R]}{dt} = -\alpha_m \cdot [R] + \beta_m \cdot [O]$$

$$\frac{d[O]}{dt} = \alpha_m \cdot [R] - \beta_m [O] - \beta_h \cdot [O] + \alpha_h \cdot [I]$$

$$\frac{d[ID]}{dt} = k_1 \cdot [D] \cdot [I] - k_{-1} \cdot [ID]$$

$$[R] + [O] + [I] + [ID] = 1$$

The solution under the assumption that activation is faster than inactivation ($\alpha_m + \beta_m >> \beta_h + \alpha_h$) and that inactivation in turn is faster than binding and dissociation of drug ($\beta_h + \alpha_h >> k_1[D] > k_{-1}$) gives for the fraction of channels available for excitation:

$$H(t) = H_\infty + A_{fast} \cdot \exp(-\frac{t}{\tau_{fast}}) + A_{slow} \cdot \exp(-\frac{t}{\tau_{slow}})$$

Where:

$$\tau_{fast} = \frac{1}{\alpha_h + \beta_h \cdot \frac{\alpha_m}{\alpha_m + \beta_m}}$$

$$\tau_{slow} = \frac{1}{k_{-1} + \frac{k_1[D]}{1 + \frac{\beta_m}{\alpha_m} + \frac{\beta_h}{\alpha_h}}}$$

$$H_\infty = \frac{1}{1 + (1 + \frac{k_1[D]}{k_{-1}}) \cdot \frac{\beta_h}{\alpha_h \cdot (1 + \frac{\beta_m}{\alpha_m})}}$$

REFERENCES

Aczel, S., Kurka, B., and Hering, S., 1998, Mechanism of voltage- and use-dependent block of class A Ca^{2+} channels by mibefradil, *Br J Pharmacol* **125**:447-454.

Armstrong, C. M., 1966, Time course of TEA^+-induced anomalous rectification in squid giant axons, *J Gen Physiol* **50**:491-503.

Armstrong, C. M., 1969, Inactivation of the potassium conductance and related phenomena caused by quaternary ammonium ion injection in squid axons, *J Gen Physiol* **54**:553-575.

Bean, B. P., 1984, Nitrendipine block of cardiac calcium channels: high-affinity binding to the inactivated state, *Proc Natl Acad Sci U S A* **81**:6388-6392.

Berjukow, S., Gapp, F., Aczel, S., Sinnegger, M. J., Mitterdorfer, J., Glossmann, H., and Hering, S., 1999, Sequence differences between α1C and α1S Ca^{2+} channel subunits reveal structural determinants of a guarded and modulated benzothiazepine receptor, *J Biol Chem* **274**:6154-6160.

Berjukow, S., Marksteiner, R., Gapp, F., Sinnegger, M. J., and Hering, S., 2000, Molecular mechanism of calcium channel block by isradipine. Role of a drug-induced inactivated channel conformation, *J Biol Chem* **275**:22114-22120.

Berjukow, S. and Hering, S., 2001, Voltage-dependent acceleration of $Ca_V1.2$ channel current decay by (+)- and (-)-isradipine, *Br J Pharmacol* **133**:959-966.

Berjukow, S., Marksteiner, R., Sokolov, S., Weiss, R. G., Margreiter, E., and Hering, S., 2001, Amino acids in segment IVS6 and beta-subunit interaction support distinct conformational changes during $Ca_V2.1$ inactivation, *J Biol Chem* **276**:17076-17082.Bezprozvanny, I. and Tsien, R. W., 1995, Voltage-dependent blockade of diverse types of voltage-gated Ca^{2+} channels expressed in Xenopus oocytes by the Ca^{2+} channel antagonist mibefradil (Ro 40-5967), *Mol Pharmacol* **48**:540-549.

Catterall, W. A., 2000, Structure and regulation of voltage-gated Ca^{2+} channels, *Annu Rev Cell Dev Biol* **16**:521-555.

Courtney, K. R., 1975, Mechanism of frequency-dependent inhibition of sodium currents in frog myelinated nerve by the lidocaine derivative GEA, *J Pharmacol Exp Ther* **195**:225-236.

Degtiar, V. E., Aczel, S., Doring, F., Timin, E. N., Berjukow, S., Kimball, D., Mitterdorfer, J., and Hering, S., 1997, Calcium channel block by (-)devapamil is affected by the sequence environment and composition of the phenylalkylamine receptor site, *Biophys J* **73**:157-167.

Ertel, E. A., Campbell, K. P., Harpold, M. M., Hofmann, F., Mori, Y., Perez-Reyes, E., Schwartz, A., Snutch, T. P., Tanabe, T., Birnbaumer, L., Tsien, R. W., and Catterall, W. A., 2000, Nomenclature of voltage-gated calcium channels, *Neuron* **25**:533-535.

Hering, S., Sokolov, S., Berjukow, S., Marksteiner, R., and Margreieter, E., 2003, *in press*.

Hering, S., Aczel, S., Grabner, M., Doring, F., Berjukow, S., Mitterdorfer, J., Sinnegger, M. J., Striessnig, J., Degtiar, V. E., Wang, Z., and Glossmann, H., 1996, Transfer of high sensitivity for benzothiazepines from L-type to class A (BI) calcium channels, *J Biol Chem* **271**:24471-24475.

Hering, S., Aczel, S., Kraus, R. L., Berjukow, S., Striessnig, J., and Timin, E. N., 1997, Molecular mechanism of use-dependent calcium channel block by phenylalkylamines: role of inactivation, *Proc Natl Acad Sci U S A* **94**:13323-13328.

Hering, S., 2002, beta-Subunits: fine tuning of Ca^{2+} channel block, *Trends Pharmacol Sci* **23**:509-513.

Hille, B., 1977, Local anesthetics: hydrophilic and hydrophobic pathways for the drug-receptor reaction, *J Gen Physiol* **69**:497-515.

Hille, B., 1992, Ionic Channels of Excitable Membranes, in ed., Sinauer Associates, Sunderland, MA.

Hockerman, G. H., Dilmac, N., Scheuer, T., and Catterall, W. A., 2000, Molecular determinants of diltiazem block in domains IIIS6 and IVS6 of L-type Ca^{2+} channels, *Mol Pharmacol* **58**:1264-1270.

Hondghem, L. M. and Katzung, B. G., 1977, A unifying molecular model for the interaction of antiarrhythmic drugs with cardiac sodium channels: application to quinidine and lidocaine, *Proc West Pharmacol Soc* **20**:253-256.

Jimenez, C., Bourinet, E., Leuranguer, V., Richard, S., Snutch, T. P., and Nargeot, J., 2000, Determinants of voltage-dependent inactivation affect Mibefradil block of calcium channels, *Neuropharmacology* **39**:1-10.

Khodorov, B., 1979, Some aspects of the pharmacology of sodium channels in nerve membrane. Process of inactivation, *Biochem Pharmacol* **28**:1451-1459.

Khodorov, B. I., 1981, Sodium inactivation and drug-induced immobilization of the gating charge in nerve membrane, *Prog Biophys Mol Biol* **37**:49-89.

Kraus, R. L., Hering, S., Grabner, M., Ostler, D., and Striessnig, J., 1998, Molecular mechanism of diltiazem interaction with L-type Ca^{2+} channels, *J Biol Chem* **273**:27205-27212.

Lacinova, L., Klugbauer, N., and Hofmann, F., 2000, State- and isoform-dependent interaction of isradipine with the α1C L-type calcium channel, *Pflugers Arch* **440**:50-60.

Lee, K. S. and Tsien, R. W., 1983, Mechanism of calcium channel blockade by verapamil, D600, diltiazem and nitrendipine in single dialysed heart cells, *Nature* **302**:790-794.

McDonald, T. F., Pelzer, D., and Trautwein, W., 1984, Cat ventricular muscle treated with D600: characteristics of calcium channel block and unblock, *J Physiol* **352**:217-241.

Motoike, H. K., Bodi, I., Nakayama, H., Schwartz, A., and Varadi, G., 1999, A region in IVS5 of the human cardiac L-type calcium channel is required for the use-dependent block by phenylalkylamines and benzothiazepines, *J Biol Chem* **274**:9409-9420.

Nawrath, H. and Wegener, J. W., 1997, Kinetics and state-dependent effects of verapamil on cardiac L-type calcium channels, *Naunyn Schmiedebergs Arch Pharmacol* **355**:79-86.

Sanguinetti, M. C. and Kass, R. S., 1984, Voltage-dependent block of calcium channel current in the calf cardiac Purkinje fiber by dihydropyridine calcium channel antagonists, *Circ Res* **55**:336-348.

Sokolov, S., Weiss, R. G., Kurka, B., Gapp, F., and Hering, S., 1999, Inactivation determinant in the I-II loop of the Ca^{2+} channel α1-subunit and β-subunit interaction affect sensitivity for the phenylalkylamine (-) gallopamil, *J Physiol* **519 Pt 2**:315-322.

Sokolov, S., Timin, E., and Hering, S., 2001, On the role of Ca^{2+}- and voltage-dependent inactivation in $Ca_V1.2$ sensitivity for the phenylalkylamine (-)gallopamil, *Circ Res* **89**:700-708.

Starmer, C. F. and Grant, A. O., 1985, Phasic ion channel blockade. A kinetic model and parameter estimation procedure, *Mol Pharmacol* **28**:348-356.

Strichartz, G. R., 1973, The inhibition of sodium currents in myelinated nerve by quaternary derivatives of lidocaine, *J Gen Physiol* **62**:37-57.

Striessnig, J., Grabner, M., Mitterdorfer, J., Hering, S., Sinnegger, M. J., and Glossmann, H., 1998, Structural basis of drug binding to L Ca^{2+} channels, *Trends Pharmacol Sci* **19**:108-115.

Timin, E. N. and Hering, S., 1992, A method for estimation of drug affinity constants to the open conformational state of calcium channels, *Biophys J* **63**:808-814.

PHARMACOLOGY OF Ca_V1 (L-TYPE) CHANNELS

David J. Triggle[1]

1. INTRODUCTION

Calcium is a critical cation both physiologically and pathologically. Its control is of fundamental significance to the life and death of the cell. Calcium is a key physiological requirement for both sperm motility and the fertilization response, and calcium is a pathological contributor to cell death processes. Thus, in a very real sense we are conceived in a moment of calcium-mediated enthusiasm, we die from an excess of calcium and we are laid to rest under a tombstone of calcium carbonate. Voltage-gated calcium channels are important contributors to both the life and death of the cell.

1.1. Calcium: The Control of a Charismatic Cation

Calcium is the most abundant terrestrial metal ion and one of the most abundant elements in the body. Calcium serves both physiological and pathological roles and its concentrations, movements and storage are highly regulated. Under tightly controlled conditions calcium is a critical cellular messenger mediating multiple stimulus-response coupling events (Carafoli, 2002; Carafoli and Klee, 1999; Putney, 2000): when uncontrolled, calcium is a pathological signal mediating cell death and destruction (Verkhratsky et al., 2002). The control of cellular calcium is thus a substantial preoccupation of the cell, as observed early by Ringer, who established the role of calcium in cardiac contractility, and by Heilbrunn, who described the relationship between calcium, membrane integrity and cellular permeability (Campbell, 1983; Heilbrunn, 1943; Ringer, 1883; Triggle, 1972).

The physiological role of calcium depends essentially upon the following set of cellular conditions:

1. The concentration of free ionized calcium within the cell is typically maintained at approximately 10^{-8} M during the resting state;

[1] 463 Hochstetter, State University of New York at Buffalo, Buffalo, NY 14260.

2. This concentration is maintained in the presence of large inwardly directed concentration and electrochemical gradients and large intracellular stores of calcium;
3. During cell stimulation, the intracellular calcium concentration rises to levels between 10^{-6} and 10^{-3} M in a spatially and temporally heterogeneous manner;
4. The elevated calcium concentration interacts with calcium binding proteins, including the ubiquitous calmodulin, to complete the linkage between stimulus and cellular response.

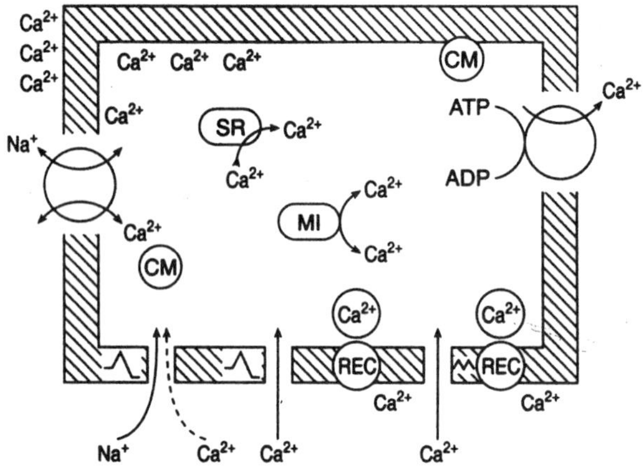

Figure 1. Cellular calcium regulation. Depicted are plasma membrane and intracellular events including (plasma membrane) voltage-gated calcium channels, receptor-operated calcium channels, Ca^{2+} - ATPase, Na^{+}-Ca^{2+}-exchanger, and a non-specific cation channel.

These conditions are satisfied in physiological situations because the cell maintains a set of processes that control calcium entry, efflux, mobilization from intracellular stores, and sequestration (Figure 1). It is this role of calcium as an information-bearing second-messenger *together* with the control processes that make possible the existence of calcium controlling drugs. Cellular calcium regulation is in turn linked to whole body calcium regulation whereby plasma calcium levels are maintained at approximately 2.5 millimolar and stable depots of calcium are maintained as bone. A triumvirate of hormones – vitamin D, parathyroid hormone and calcitonin – that serve to regulate calcium absorption, calcium excretion and bone deposition and resorption (remodeling), achieves this control. Some lines of evidence suggest that some cardiovascular disorders, including hypertension, may be linked to serum calcium deficiency with elevated urinary calcium excretion and compensating changes in the three calcium regulatory hormones. These changes are in accord with postulates that hypertension, elevated vascular smooth muscle reactivity and sensitivity to calcium antagonists are associated with elevated intracellular calcium levels and a "leakiness" of the plasma membrane to calcium (Bucher et al., 1996; Hatton et al., 1995; Luft and McCarron, 1991; Resnick, 1994).

1.2. The Calcium Channel Antagonists

In principle, each of the sites depicted in Figure 1 should have a class of drugs that acts specifically to control calcium movements and storage. In practice, this is true *therapeutically* for one site only – the voltage-gated calcium channel - where a heterogeneous group of agents known as the calcium antagonists, including the first-generation verapamil, diltiazem and nifedipine, act (Figures 2 and 3). Verapamil is a member of the phenylalkylamine class of molecules, well represented in biogenic amine-type structures, that contains linked phenethylamine structures; nifedipine contains the 1,4-dihydropyridine nucleus, a pharmacophore capable of being directed towards a variety of structures, and diltiazem is a benzothiazepinone originally synthesized as part of as search for centrally active molecules of the benzodiazepine type. These drugs are employed almost exclusively for their actions in the cardiovascular system, where they serve as first-line anti-hypertensive, anti-anginal and (class selectively) Class IV anti-arrhythmic agents (Tables 1 and 2).

Table 1. Cardiovascular profile of calcium antagonists.

	Nifedipine	Diltiazem	Verapamil
Coronary vessels			
Tone	--	--	--
Flow	+++	++	++
Peripheral vasodilation	+++	+	++
Heart rate	++	-	-
Contractility	0,+	0,-	0,-
A-V node conduction	0	-	-
A-V node ERP	0	-	-

+ = increase, - = decrease, 0 = no effect

Table 2. Therapeutic uses of calcium channel antagonists.

	Antagonist		
Use	Verapamil	Nifedipine	Diltiazem
Angina:			
Exertional	+++[a]	+++	+++
Prinzmetal's	+++	++	+++
Variant	+++	+	+++
Arrhythmias:			
paroxysmal supraventricular tachyarrhythmias	+++	-	++
Atrial fibrillation and flutter	++	-	++
Hypertension	++	+++	++
Hypertrophic cardiomyopathy	+	-	-
Raynaud's phenomenon	++	++	++
Cerebral vasospasm (post-hemorrhage)	-	+[b]	-

[a]Number of plus signs indicates extent of use: +++ indicates very common, - indicates not used.
[b]Nimodipine: a nifedipine analog.

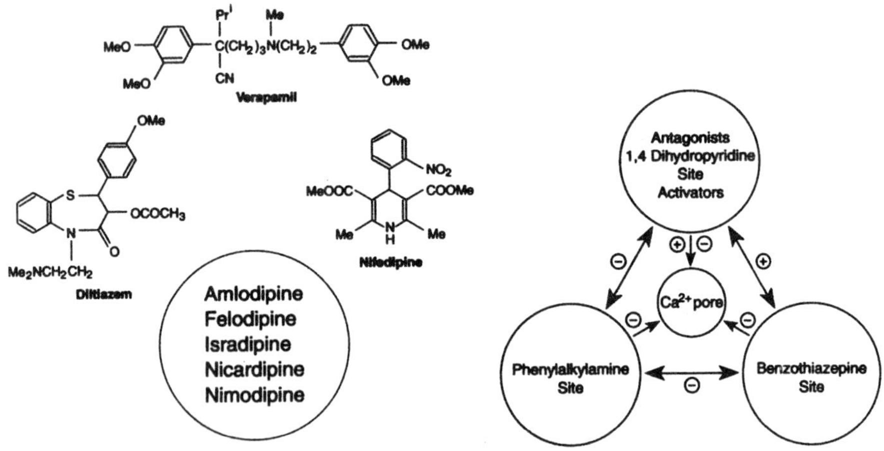

Figure 2. Drug structures. *Left,* structural formulae of first-generation clinically used calcium channel antagonists – verapamil, nifedipine and diltiazem – and (inset) second-generation 1,4-dihydropyridines. *Right,* arrangement of receptor sites at the L-type calcium channel, depicting allosteric linkages among the receptor sites for different drug classes and their relation to the channel pore.

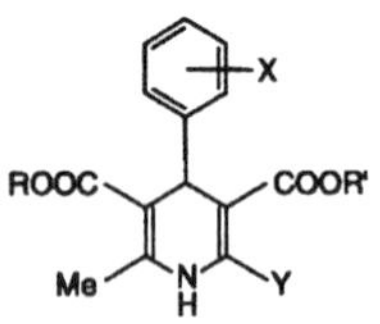

Amlodipine	X=2Cl; R=Me; R'=Et; Y=$CH_2OCH_2CH_2NH_2$
Benidipine	X=$3NO_2$; R=Me; R'= (piperidin-3-yl, N-CH_2PH); Y=Me
Cilnidipine	X=$3NO_2$; R=$MeOCH_2CH_2$; R'=CH_2CH=CH•C_6H_5(E); Y=Me
Felodipine	X=2,3Cl_2; R=Me; R'=Et; Y=Me
Isradipine	X= (benzoxadiazole); R=Me; R'=$CHMe_2$; Y=Me
Lacidipine	X=CH=CHCOOBu^t; R=Et; R'=Et; Y=Me
Lercanidipine	X=$3NO_2$; R=Me; R'=$CMe_2CH_2NMeCH_2CH_2CHPh_2$; Y=Me
Manidipine	X=$3NO_2$; R=Me; R'=CH_2CH_2N(piperazine)N$CHPh_2$; Y=Me
Nicardipine	X=$3NO_2$; R=Me; R'=CH_2NMeCH_2Ph; Y=Me
Nifedipine	X=$2NO_2$; R=Me; R'=Me; Y=Me
Nilvadipine	X=$3NO_2$; R=Me; R'=$CHMe_2$; Y=Me
Nimodipine	X=$3NO_2$; R=$CHMe_2$; R'=CH_2CH_2OMe; Y=Me
Nisoldipine	X=$2NO_2$; R=Me; R'=CH_2CHMe_2; Y=Me
Nitrendipine	X=$3NO_2$; R=Me; R'=Et; Y=Me

Figure 3. Structural formulae of 1,4-dihydropyridine calcium channel antagonists.

The calcium antagonist class of drugs is one of several principal classes of drugs used worldwide in the control of hypertension – diuretics, β-blockers, ACE inhibitors, AII receptor antagonists, α-blockers, α-agonists, aldosterone agonists, vasodilators, and calcium antagonists. The choice amongst these agents is significantly dependent upon cardiovascular risk profile, but the calcium antagonists are important for the treatment of hypertension in the presence of complicating disorders, including diabetes, asthma and peripheral vascular complications (Chalmers, 1999; Sheps, 1997; Singh et al., 1999). The recognition of these agents and their development owes much to the work of Albrecht Fleckenstein (Fleckenstein, 1983a,b; Triggle, 1990), who observed that verapamil mimicked the effects of calcium withdrawal on cardiac contractility and blocked E-C coupling, effects that could be reversed by the addition of calcium. Subsequent work showed that these properties were shared by other agents including the more potent, and structurally unrelated, nifedipine.

The worldwide sales of these agents approximate $6 billion (US): the late Senator from Illinois, Everett Dirksen, would have doubtless commented, *" A billion dollars here, and a billion dollars there, and pretty soon you have real money."* This market, the largest of any drug class that interacts directly with an ion channel, is thus made possible by the specific ability of these agents to interact selectively with one class of calcium channel to restrict access of calcium to intracellular sites that control cardiac rate and cardiac and vascular smooth muscle contraction. The widespread distribution of voltage-gated calcium channels in the cardiovascular system, non-vascular smooth muscle, secretory, and neuronal systems indicates that it is appropriate to consider a multiplicity of actions of appropriately selective calcium antagonists for disorders from asthma through migraine to urinary incontinence and vertigo (see Section 6 on future directions).

Given the ubiquity of calcium as a messenger and the widespread distribution of voltage-gated calcium channels in virtually all cell and organ types, the critical issue of selectivity of drug action arises. The available calcium antagonists owe their pharmacological and therapeutic profile to their ability to interact with considerable selectivity at *one* of the calcium mobilization processes depicted – a member of the voltage-gated calcium channel family, the L-type, or Ca_v1, channel subclass. This selectivity of action is a major determinant of the observed pharmacological and therapeutic selectivity of action observed for the agents depicted in Figures 2 and 3. It is worthy of observation that thus far it is *only* the Ca_v1 subclass of calcium channel that has yielded small molecule therapeutically active agents: the other channel classes are well defined by potent and selective toxins that are a prominent part of Nature's armamentarium, but despite considerable effort have not yielded to the more useful small molecule agents. In any event, the chemical structures underlying verapamil, nifedipine and diltiazem were discovered by serendipity from drug discovery programs directed at targets quite distinct from calcium channels (see Section 6).

1.3. The Actions of Calcium Channel Antagonists

The therapeutic agents referred to as "calcium channel antagonists," "calcium antagonists," and "calcium channel blockers" are a chemically, pharmacologically and therapeutically heterogeneous group of agents that have as their primary target a specific class of voltage-gated calcium channel – the L-type channel of the Ca_v1 family – itself a subgroup of a family of voltage-gated calcium channels, Ca_v1 – Ca_v3, which is in turn a member of a "super-family" of voltage-gated ion channels. The calcium channel antagonists, including the first-generation verapamil, nifedipine and diltiazem (Figure 2), are qualitatively and quantitatively distinct entities with substantially distinct individual selectivity of therapeutic action (Tables 1, 2) and are not simply interchangeable. This distinctiveness extends to second- and third-generation agents (Figure 3) and arises from several mechanisms, including the modes of calcium mobilization in the cell or tissue in question, differentiation of receptor site, the class and subclass of voltage-gated calcium channel used, state-dependent interactions, pharmacokinetic properties, and channel regulation during disease (Epstein, 2002; Triggle, 2002).

Consistent with the chemical heterogeneity of the agents depicted in Figure 2 it was demonstrated early that the three classes of chemical agents – 1,4-dihydropyridine (nifedipine), phenylalkylamine (verapamil), and benzothiazepinone (diltiazem) – interact at discrete loci at the calcium channel (Figure 2; Adachi-Akahane and Nagao, 2000; Catterall, 2000; Catterall and Striessnig, 1992; Glossmann et al., 1987; Janis and Triggle, 1983; Janis et al., 1987; Striessnig et al., 1998; Triggle, 1999, 2002; Voradi et al., 1995), sites subsequently defined by molecular biology studies to be close to the permeation and gating machinery of the channel (Catterall, 2000; Hockerman et al., 1997a,b,c). The binding sites depicted in Figure 2 are discrete but are linked allosterically one to the other and to the machinery of the channel. Occupancy of any one receptor not only alters the gating properties of the channel, but also influences positively (increases interaction and response) or negatively (decreases interaction and response) the interaction of a second drug with its receptor site. Thus, nifedipine potentiates (positive allosteric effect) the binding of diltiazem and, in reciprocally linked manner, diltiazem potentiates the binding of nifedipine. These allosterically linked events translate to response whereby the cardiovascular effects of combined diltiazem and nifedipine are not simply additive (Balwierczak et al., 1986; Yousif and Triggle, 1985).

These voltage-gated calcium channels, and other ion channels, are conveniently regarded as pharmacological receptors with the following general properties:

1. Specific binding sites for antagonist and activator ligands with defined structure-activity relationships, including stereoselectivity;
2. Coupling of the binding sites to the permeation and gating machinery of the channel;
3. Regulation by homologous and heterologous influences;
4. Alteration of expression and function in disease states.

These expectations have been realized in full for the several classes of voltage-gated calcium channels, which can thus be regarded as a homologous group of receptors defined by localization, function and specific pharmacological and electrophysiological criteria. These receptors are carried on the major pore-forming α_1 subunit with which are associated the disulfide-linked $\alpha_2\delta$-subunit and the intracellular β-subunit. The skeletal muscle L-type channel carries an additional γ-subunit. Although the α_1-subunit is the dominant partner in this heteromeric channel complex, its electrophysiological, expression and pharmacological properties are significantly influenced by the auxiliary subunits, notably the β-subunit, whose co-expression increases 1,4-dihydropyridine binding and affinity (Catterall, 2000; Hofmann et al., 1999; Mitterdorfer et al., 1998; Striessnig, 1999; Wei et al., 1997). Hence, the activity of a calcium channel activator or antagonist is determined not only by the subtype of α_1-subunit or its alternatively spliced form, but also by the other subunits expressed. This provides an additional complexity to the interpretation of drug action at these receptors.

The principal focus of this chapter will be on the pharmacological properties of the L-type or Ca_v1 class of voltage-gated channels. A number of existing reviews are available (Adachi-Akahane and Nagao, 2000; Catterall et al., 2002; Hockerman et al., 1997b; Catterall, 2000; Striessnig et al., 1998; Triggle, 1999, 2002).

Table 3. The Ca_V1 family.

	$Ca_V1.1$	$Ca_V1.2$	$Ca_V1.3$	$Ca_V1.4$
Other names	α_{1S} skeletal	α_{1C} cardiac/ smooth	α_{1D} neuroendocrine	α_{1F} ---
Localization	skeletal muscle T-tubules	cardiac muscle smooth muscle endocrine cells neurons	endocrine cells sensory cells neurons heart hair cells	retina
Function	E-C coupling	E-C coupling cardiac AP hormone release synaptic plasticity	NT release hormone release synaptic plasticity	NT release in retinal cells
Therapeutic significance	?	target of available drugs	?	?

Table 4. Vascular selectivity of calcium channel antagonists in human coronary and cardiac tissue (data from Godfraind et al., 1992).

Antagonist	Selectivity ratio
Diltiazem	~ 1
Verapamil	~ 1
Nifedipine	~ 10
Amlodipine	~ 10
Felodipine, isradipine, nitrendipine, nicardipine, lacidipine	~ 100
Nisoldipine	~ 1000

2. THE Ca_v1 CLASS OF CHANNEL

The Ca_v1 channel is a major subclass of the voltage-gated calcium channel family and is composed of four principal members – $Ca_v1.1$-1.4 - (Table 3) with a number of splice variants. This channel class is widely distributed both peripherally and centrally, but $Ca_v1.1$ and $Ca_v1.4$ are restricted to skeletal muscle and retina respectively, whereas $Ca_v1.2$ is widely distributed, particularly in cardiac and smooth muscle where it serves to mediate the therapeutic effects of the clinically available calcium antagonists (Epstein, 2002).

The Ca_v1 subclass has been best characterized pharmacologically by the phenylalkylamine, benzothiazepine and 1,4-dihydropyridine class of structures, but there is a multiplicity of other structures, both synthetic and naturally occurring, that have demonstrable actions at this channel (Janis, Silver and Triggle, 1987; Rampe and Triggle, 1990, 1993; Spedding et al., 1995; Triggle, 1999; Triggle et al., 1991).

Although these agents exhibit a common biochemical and pharmacological endpoint to mediate their therapeutic effects – inhibition of the entry of calcium into cardiac and vascular smooth muscle – they do exhibit significant qualitative and quantitative differences in activity (Tables 1, 2). Nifedipine and second- and third-generation 1,4-dihydropyridines are more potent vasodilators than verapamil or diltiazem and lack the cardiodepressant properties of the latter. Additionally, significant differences exist between members of the 1,4-dihydropyridine series in their relative activities in cardiac and smooth muscle (Table 4). The greater vascular selectivity of the more recently introduced 1,4-dihydropyridines is one of the factors that improve their therapeutic profile and impact contraindication considerations (Epstein, 2002, 2002a; Koshy and Bakris, 2000; Mahon and McKenna, 1998; Montoyler and Follath, 2001; Neal, 2000; Opie and Schall, 2002; Robertson, 2000; Triggle, 2002).

These differences in tissue selectivity of the calcium antagonists are important with respect to definition of their sites and mechanisms of action at the Ca_v1 channel class and to their pharmacological and therapeutic activities and uses. In principle, the differences can arise from a number of factors alone or in combination:

1. Mode of calcium mobilization;
2. Class and subclass of calcium channel involved;
3. Location of drug binding site;
4. Activator/antagonist properties of the drug;
5. State-dependent interactions – voltage- and frequency-dependence;
6. Pathological state of the tissue and channel regulation – "channelopathies."

3. PHARMACOLOGY OF THE Ca_V1 CHANNEL CLASS

3.1. Major Structural Classes of Drugs

The three principal categories of calcium channel drugs are represented by the 1,4-dihydropyridines (including both activators and antagonists), the phenylalkylamines, and the benzothiazepinone structures (Figure 2). For therapeutic applications the 1,4-dihydropyridine class is represented by a number of agents (Figure 3), but the phenylalkylamine and benzothiazepinone classes are limited to one member each, verapamil and diltiazem respectively. However, there are many other structures, both synthetic and naturally occurring, that have actions at the L-type calcium channel although they have not achieved therapeutic significance.

3.2. Other Classes of Drugs

Despite the focus on the 1,4-dihydropyridine structure, there is no lack of other structures that have defined and, in some cases, potent and selective actions at L-type channels (Janis et al., 1987; King et al., 1988, 1989; Rampe and Triggle, 1993; Spedding et al., 1995; Triggle, 1999; Triggle et al., 1991). Representative structures are shown in Figures 4-7. The benzothiazine HOE 166 is one of a series of agents including semotiadil

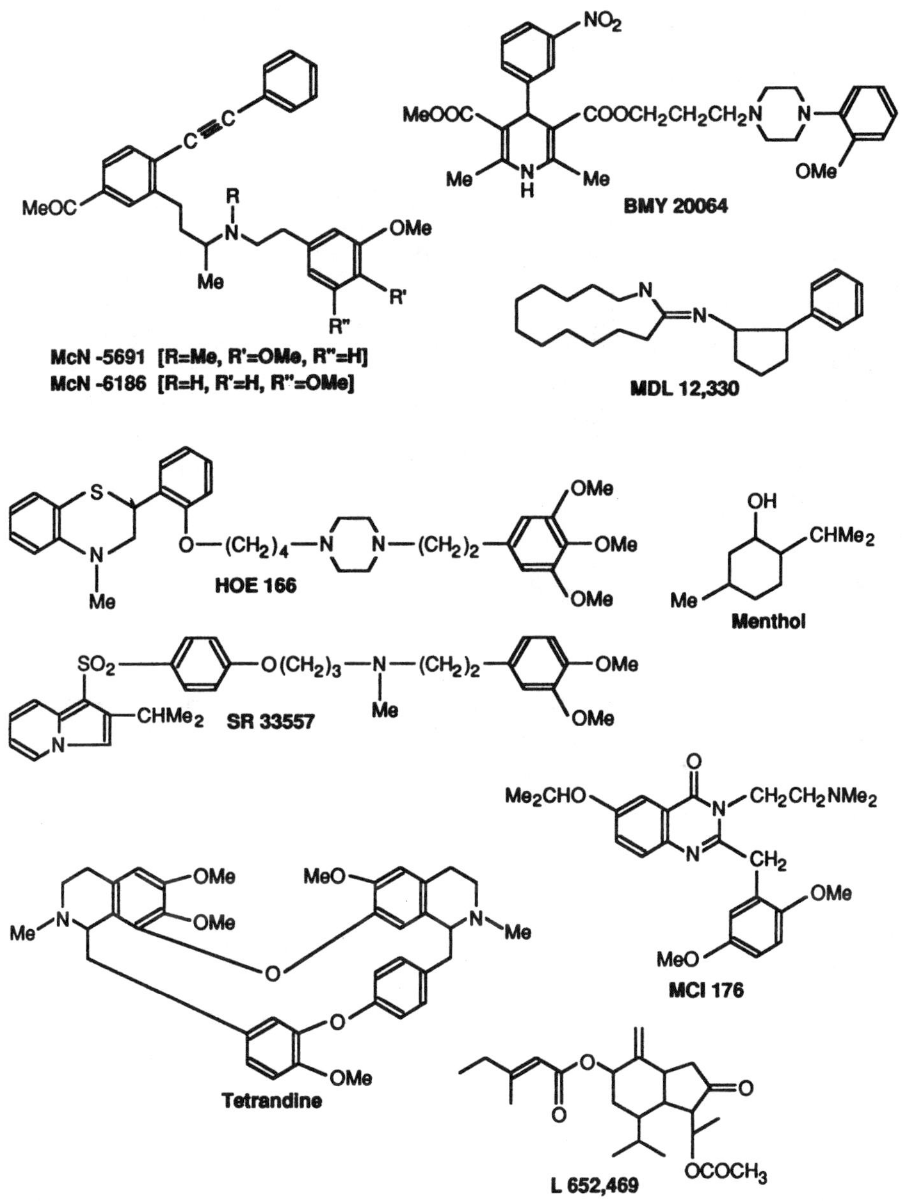

Figure 4. L-type channel inhibitors. Structures of a representative collection of molecules that serve as blockers of the L-type calcium channel.

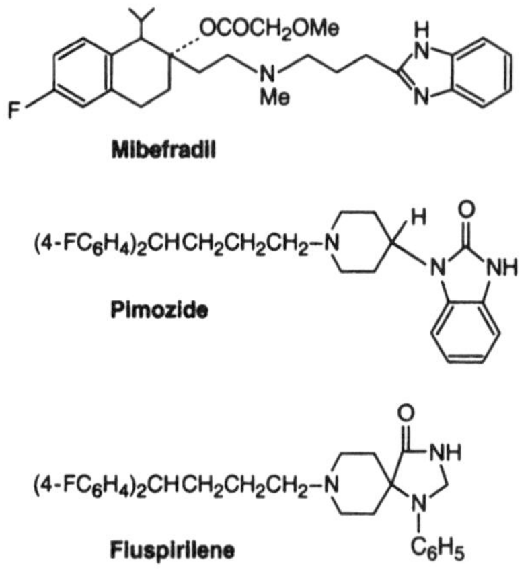

Figure 5. Non-selective L-type channel inhibitors. Molecules with non-selective and potent antagonist activity at the L-type calcium channel – mibefradil, pimozide and fluspiriline.

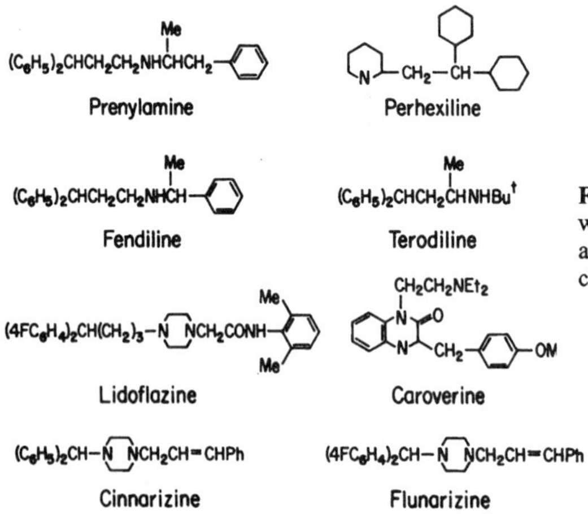

Figure 6. Weak L-type inhibitors. Molecules with non-selective and relatively weak antagonist activity at the L-type calcium channel.

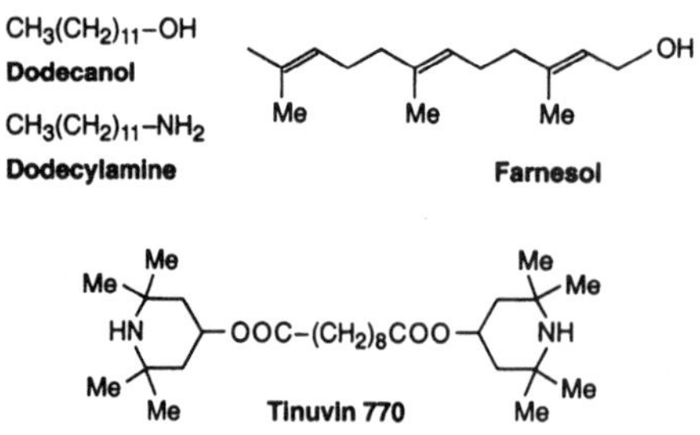

Figure 7. Aliphatic inhibitors. Aliphatic molecules with antagonist activity at L-type calcium channels.

(SD 3211) that appear to define a fourth specific binding site at the L-type channel that is negatively coupled to the phenylalkylamine, 1,4-dihydropyridine, and benzothiazepinone sites (Fujita et al., 1990; Nakayama, 1996; Nakayama et al., 1994; Striessnig et al., 1988). BMY 20064 (Stanton et al., 1988) is representative of a significant number of molecules where the 1,4-dihydropyridine nucleus – a "privileged structure" – is capable of interacting or directing interaction to receptor and channel sites other than the L-type calcium channel (Rampe and Triggle, 1993; Triggle, 2002a; see Section 3.9). Other structures with potent, albeit non-selective, actions at L-type channels (Figures 5 and 6) are the diphenylbutylpiperidine class that includes pimozide and fluspirilene (Enyeart et al., 1992; Galizzi et al., 1986; Kenny et al., 1990; King et al., 1989), the diphenylalkylamines cinnarizine and flunarizine (Chaisewikul et al., 2001; Fleckenstein, 1983; Hunter, 1997; Janis et al., 1987; Singh et al., 1985; Spagnoli and Tognoni, 1983; Todd and Benfield, 1989), the ethynylbenzenealkanamines McN 6186 and McN 5691 (Carson et al., 1988; Flaim et al., 1988), the lactamimide MDL 12330 (Rampe et al., 1987), the indolizinesulfone SR 335557 (Nokin et al., 1990; Schmid et al., 1989), MCI 176 (Ishibashi et al., 1987), and the benzimidazole mibefradil (Bezprozvanny and Tsien, 1995; Clozel et al., 1997). Related to the diphenylalkylamines is the antifungal agent clotrimazole (1-[(2-chlorophenyl)diphenylmethyl]imidazole), which is antiproliferative against a variety of cell types (Benzaquern et al., 1995; Fearon et al., 2000; Thomas et al., 1999), and has neuroprotective effects in cultured neurons (Isaev et al., 2002).

A series of simple aliphatic molecules have also been described as L-type channel blockers and to be active in the micromolar range (Figure 7). These include farnesol (and a number of its derivatives), a 15-C isoprenoid derived from farnesyl pyrophosphate, an intermediate in the mevalonic acid pathway, and simple aliphatic alcohols and amines including dodecanol and dodecylamine (Beedle and Zamponi, 2000; Roullet et al., 1996, 1997, 1999). Similarly, *bis*(2,2,6,6-tetramethyl-4-piperidyl)sebacate (Tinuvin 770), a component of plastic materials, is also effective with nanomolar affinity as an L-type blocker (Glossmann et al., 1993). Menthol, a principal constituent of peppermint oil, is active at both L- and T-type channels and this property may underlie its gastrointestinal smooth muscle relaxant properties (Hawthorn et al., 1988).

The precise nature of the binding interactions of this diverse class of molecules with the Ca_v1 class of channels has not for the most part been established. It is probable, however, that they share binding sites in common with those already established, since there is likely a "hot spot" for drug binding in a specific region of the channel (see Section 3.7).

3.3. Channel Activators

Although a diverse collection of molecules function as antagonists at the L-type channel, comparatively few activator structures exist. Activators exhibit effects that are essentially the opposite of those exhibited by antagonists at this class of channel. Activators are quite generally hypertensive, vasoconstrictive, with positive inotropic and chronotropic effects on the heart. Additionally, activators exhibit a variety of neuronal and neurosecretory effects since they serve to release neurotransmitters and hormones both peripherally and centrally. In these latter effects, activators behave differently from antagonists that are generally inactive in neuronal and neuroendocrine cells. An explanation of this effect lies in the preferential interaction of antagonists with inactivated channels, a state favored by more or less prolonged depolarization not typically seen under physiological conditions in neuronal and neurosecretory cells cell types (see Section 4.4). The majority of

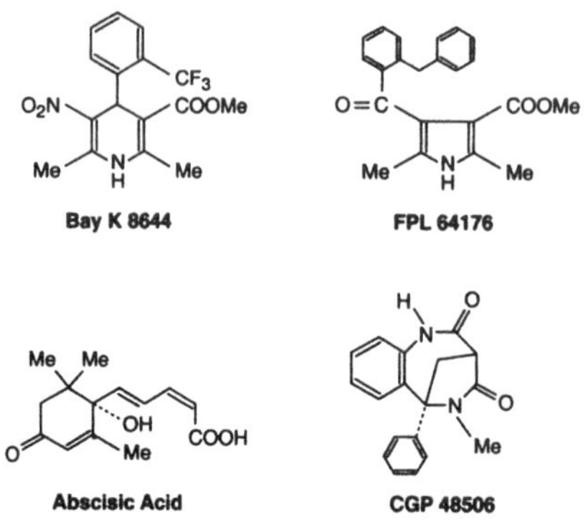

Figure 8. L-type agonists. Structures of molecules with activator activity at L-type calcium channels.

activators belong to the 1,4-dihydropyridine class, including Bay K 8644 (Figure 8; Bechem et al., 1997; Goldmann and Stoltefuss, 1991; Hof et al., 1985; Patmore et al., 1990; Priego et al., 1993; Schramm et al., 1983; Shan and Knaus, 1999; Takenaka et al., 1988; Zheng et al., 1992). Of the non-1,4-dihydropyridine structures the benzoylpyrroles typified by FPL 64176 (Figure 8) are of particular interest (McKechnie et al., 1989). These activators interact at a site partially distinct from the 1,4-dihydropyridine site and linked in allosteric manner (Rampe and Lacerda, 1991; Rampe et al., 1993; Zheng et al., 1991). Consistent with this there are differences in calcium channel activation: Bay K 8644 and FPL 64176 both increase current amplitude, shift activation to more negative potentials and prolong tail currents; however, the benzoylpyrrole significantly slows current activation likely due to an increase in channel mean open time (Fan et al., 2000; Kunze and Rampe, 1992). The benzodiazocine derivative CGP 48506 (Figure 8) is an activator at L-type channels, but is significantly less potent, 10^{-4} to 10^{-5} M, and with effects distinct from those of 1,4-dihydropyridine activators, since it does not shift activation to more negative potentials (Herold et al., 1995; Herzig, 1996). The phytohormone abscisic acid (Figure 8) has prominent calcium channel activating properties in plant cells (White, 2000) and this may extend to mammalian systems (Huddart et al., 1986).

3.4. Peptide Toxins

The L-type calcium channel was the first (and best) characterized pharmacologically and has been characterized largely by small molecule synthetic ligands, in contrast to the neuronal N-type, P/Q-type (and some R-type) channels that are still characterized almost exclusively by peptide toxins. Some toxins, however, do exhibit potent and specific effects on L-type channels. Calciseptine is a 60 amino acid peptide with four disulfide bridges (Figure 9) isolated from venom of the black mamba snake that blocks L-type current and cardiac and smooth muscle contractions and interacts competitively at the 1,4-dihydropyridine site (De Weille et al., 1992; Yasuda et al., 1993). Its homolog FS2 acts similarly (Watanabe et al., 1995). However, it serves as a calcium channel activator at the L-type channels (α_{1S}) in skeletal muscle (Garcia et al., 2001). The putative calcium channel binding site in these peptides has been identified as a four residue sequence, 43-46, flanked by two signature proline residues. This sequence expressed as an 8-residue peptide possesses calcium channel blocking properties, albeit with lower potency than the parent toxins (Kini et al., 1998). In contrast, the green mamba toxin calcicludine, also a 60 amino acid peptide (Figure 9) blocks a variety of calcium channel types with affinity in the 10-100 nM range, with blockade of the L-type current in cerebellar granule neurons at 0.2 nM (Schweitz et al., 1994). However, tested on heterologously expressed high voltage-activated channels, the toxin was a partial blocker

(1) RICYHKASLPRATKTCVENT
CYKMFIRTQREYISERGCGC
PTAMWPYQTGCCKGDRCNK (60)

Calciseptine

A-(P)TAMW(P)-A

L-Calchin

(1) WQPPWYCKEPVRIGSCKKQF 20
CGSFLFPLCKKATWKFYFSS
GGNANRFQTIGECRKKCLGK

Calcicludine

Figure 9. Peptide toxins. Primary amino acid sequences of several toxins active at L-type calcium channels.

only of α_{1C}-mediated current and actually up-regulated current density in α_{1B} (N-type) channels (Stotz et al., 2000).

Specific conotoxins and agatoxins have been employed as specific blockers of N-type and P-type channels respectively (Olivera et al., 1994). At least one conotoxin, however, ω-conotoxin-TxVII, is specific for L-type channels, blocking currents in the 10^{-7} – 10^{-5} M range, a potency lower that that typically seen for ω-conotoxins. ω-Conotoxin-TxVII has the characteristic disulfide-bridged structure of this class, but uncharacteristically carries a net negative charge and is very hydrophobic (Fainzilber et al., 1996). ω-Agatoxin IIIA, a 76-residue member of the agatoxin family, blocks both L- and N-type currents with nanomolar potency in a voltage-independent manner (Cohen et al., 1992; Ertel et al., 1994a,b) and also blocks P-type channels (Mintz, 1994). A limited structure-activity study of related toxins indicates that congeners ω-AgaIIIB, IIIC and IIID retain their N-type channel potency but are significantly less active at L-type channels (Ertel et al., 1994b). KP4 is a 105 residue polypeptide encoded by the UMV4 virus that infects the fungus *Ustilago maydis*. It blocks L-type channels with some specificity in a voltage-independent manner with a potency similar to that of calciseptine (Gage et al., 2002). It is likely that all of these toxins serve to block channel function in a non-voltage-dependent manner by interacting with residues at the channel mouth (see McDonough, this volume).

3.5. Structure-Activity Relationships

The structure-activity relationships of the 1,4-dihydropyridine class have been particularly well investigated and are summarized broadly in Figure 10 (Bossert and Vater, 1989; Coburn et al., 1988; Goldmann and Stoltefuss, 1991; Hemmateenejad et al., 2002; Mager et al., 1992; Schliefer and Tot, 2002; Triggle, 1989, 1999; Triggle et al., 1989). Structure-activity relationships of drugs active at ion channels present particular difficulties of interpretation since there are quantitative and qualitative differences according to the state – open, closed – of the channel for which the relationship is elucidated (Kwon and Triggle, 1991; Triggle, 1994). Radioligand binding studies typically employ cellular membrane preparations that are dead, depolarized, and dephosphorylated, and likely represent an approximation to the inactivated state of the channel favored by persistent depolarization. In contrast, whole cell binding studies can mimic more closely interactions with resting and inactivated channel states according to the membrane potential imposed (Wei et al., 1989; Zheng et al., 1992). Tight control of membrane potential can be achieved in electrophysiological studies, but typically such studies do not lend themselves to the high throughput necessary for study of structure-

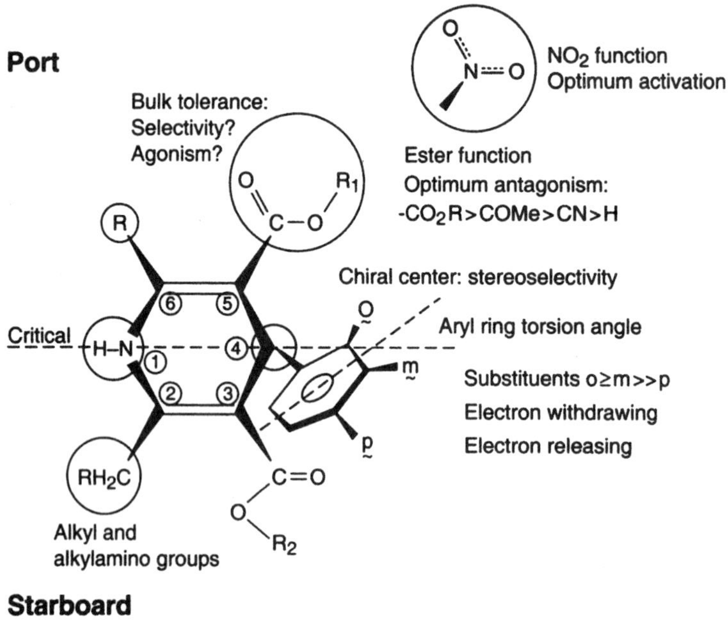

Figure 10. Molecular basis of activity. Structure-activity relationships for 1,4-dihydropyridine antagonist and activator activities at the L-type calcium channel.

activity relationships. Pharmacological preparations, typically isolated tissues, also lend themselves to structure-activity relationships particularly when K^+-depolarized; however, the use of receptor-directed agonists as stimulants is fraught with danger since receptor activation may mobilize calcium through pathways other than the voltage-gated calcium channel.

A number of features are worthy of note in the structure-activity discussion and can be discussed most conveniently in terms of the components of the structure of Figure 10. Ester substituents at C3 and C5 confer optimum antagonist activity, although considerable variation is possible. The ester groups may be dissimilar, thus generating a chiral center at C4 with attendant stereoselectivity of interaction. Quite generally, the greater the difference in the size of the ester groups the greater the resultant stereoselectivity of interaction (Janis et al., 1987; Triggle et al., 1989). Thus nitrendipine (Figure 3) has a modest stereoselectivity index of between 5-10 in both radioligand binding and tissue studies: in contrast, amlodipine has a stereoselectivity index (-) > (+) of approximately 1000, and the four stereoisomers of 2-(2-pyrrolidinylmethylthio) methyl-3-carboethoxy-5-carbomethoxy-6-methyl-4-(3-nitrophenyl) 1,4-dihydropyridine have stereoselectivity ratios of approximately 1000-fold (Arrowsmith et al., 1986; Bolger et al., 1983; Janis et al., 1987; Kwon et al., 1990; Triggle et al., 1989). Different ester functions appear to be associated with different degrees of vascular selectivity, with different degrees of state-dependent interaction, and/or with preferential interaction with L-type calcium channel subtypes (see Sections 3.6. and 4.4). Nimodipine (Figure 3) exhibits selectivity for the cerebral vasculature, hence its use in the treatment of the sequelae of subarachnoid hemorrhage. The third-generation agent lercanidipine (Figure

3) has a bulky hydrophobic ester substituent that confers a prolonged duration of action and vascular selectivity (Herbette et al., 1997). A series of molecules with a positively charged membrane-impermeant ester function and a variable-length carbon chain (Figure 11) has been used as a molecular ruler to determine the location of the 1,4-dihydropyridine binding site within the membrane-spanning α_1 subunit (Baindur et al., 1993; Bangalore et al., 1994; Peri et al., 2000).

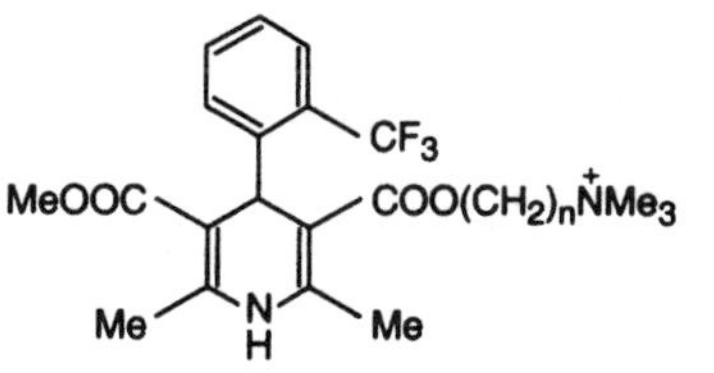

Figure 11. Molecules used to measure the DHP binding site. A series of charged 1,4-dihydropyridines used as molecular probes of the L-type calcium channel 1,4-dihydropyridine receptor, in which n is an integer reflecting the variable length of the carbon chain.

The aromatic ring at C4 and its substituents are of particular importance. Optimum activator or antagonist activity is associated with a phenyl ring at position C4. Electron withdrawing *ortho-* and *meta-*substituents in the phenyl ring are associated with increased or maintained activity, whereas *para*-substituents are invariably associated with a large decrease in activity. The third-generation agent lacidipine (Figure 3) contains a bulky hydrophobic substituent in the ortho position that confers a particularly prolonged duration of action (Herbette et al., 1994). Replacement of the 4-phenyl ring by alkyl or cycloalkyl substituents reduces activity, as does substitution with heterocyclic nuclei such as pyridyl, furyl, thienyl, etc (*inter alia*, Iqbal et al., 1998).

The 1,4-dihydropyridine structure is remarkable for encompassing both potent antagonist and activator properties with relatively small structural change. Activator activity is associated with the presence of electron withdrawing groups, notably nitro, at the C3 position. Weaker activator properties are associated with lactone substituents. Of particular interest is the remarkable chirality of interaction, whereby the S-enantiomers are activators and the R-enantiomers are antagonists (Hof et al., 1985; Schramm et al., 1983; Zheng et al., 1992; Figure 12). Such *qualitative* differences in stereoselectivity of interaction are consistent with the enantiomers interacting with different conformations of the channel protein associated with the open (activator) and inactivated (antagonist) states (Triggle, 2002a).

The C-2 methyl group of nifedipine can be replaced with a variety of substituents. Basic substituents at C2 generate molecules of the amlodipine class (Figure 2) with slow onset and offset of action and a very high stereoselectivity of interaction, with eudismic

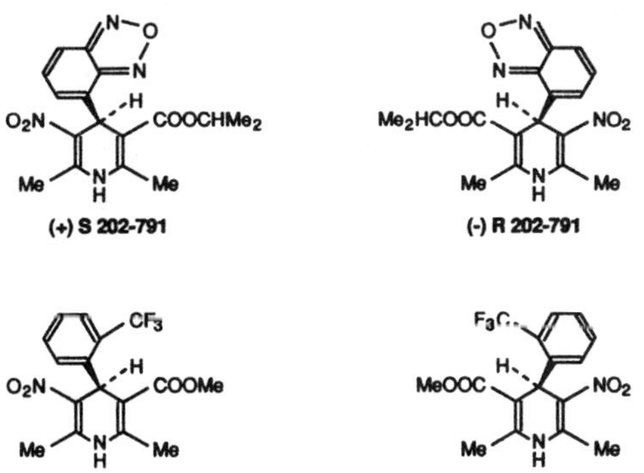

Figure 12. L-type activator and antagonist structures. Stereoselective 1,4-dihydropyridine activator- antagonist activities at the L-type calcium channel.

ratios of one thousand and higher (Arrowsmith et al. 1986; Kwon et al., 1990; Triggle, 2002a).

The NH moiety of the 1,4-dihydropyridine ring is essential. Replacement by Me or higher substitution abolishes activity. This suggests that this residue makes a critical H-bonding interaction at the receptor site (Langs et al., 1990, 1991; Zhorov et al., 2001). The 1,4-dihydropyridine ring (or its isosteric replacement, including 1,4-dihydropyrimidines (Figure 13), Atwal et al., 1990a,b, 1991; Cho et al., 1989; Rovnyak et al., 1992, 1995) is essential. Dihydropyrimidines are approximately equipotent to corresponding members of the 1,4-dihydropyridine series: they have a solid state conformation very similar to that of the 1,4-dihydropyridines (Atwal et al., 1990a). Oxidation or reduction to the pyridine or piperidine structures respectively completely abolishes activity.

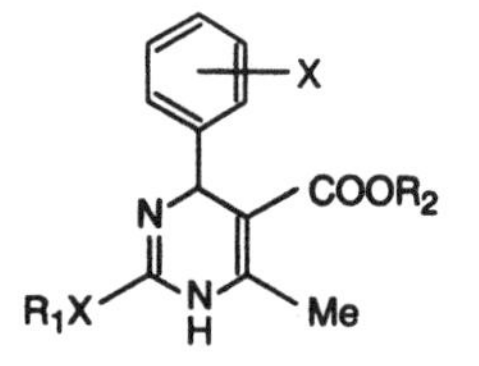

X=S,O; R_2=alkyl; R_3=NO_2,CF_3,F etc.

Figure 13. Dihydropyrimidines. The dihydropyrimidine nucleus: an isostere of the 1,4-dihydropyridine structure.

There have also been extensive studies on the conformational requirements for 1,4-dihydropyridine action (discussed in Fossheim et al., 1981, 1988; Goldmann and Stoltefuss, 1991; Langs and Triggle, 1985; Langs et al., 1990, 1991; Rovnyak et al., 1995; Triggle et al., 1980, 1989). These studies indicate that the 1,4-dihydropyridine nucleus has a boat conformation with the substituted phenyl ring in the pseudoaxial orientation. This defines "*port*" and "*starboard*" space (Figure 10) where substituent interactions defining activator and antagonist properties occur. In principle, substituents on the phenyl ring may occupy antiperiplanar (*ap*) and synperiplanar (*sp*) orientations (Figure 14). The study of rigid analogs indicates that the *sp* orientation is associated with optimum activity (Goldmann and Stoltefuss, 1991; Janis et al., 1987; Rovnyak et al., 1988; Seidel et al., 1985; Triggle et al., 1989). Similarly, the 3,5 ester substituents (or their equivalent) can be oriented in the *cis*- or *trans*-conformations (Figure 14): here also the study of rigid analogs suggests that the *cis*-orientation is preferred. A correlation exists between the degree of planarity of the 1,4-dihydropyridine ring and biological activity. Increasing planarity is associated with increased activity over a wide range of activity, suggesting that this ring is associated with a stacking interaction with residues at the receptor site (Figure 15; Langs et al., 1990, 1991; Triggle et al., 1989).

Interpretations of these observations in terms of binding orientations at the 1,4-dihydropyridine receptor site of the L-type channel are still, of necessity, speculative. It has been proposed, however, that the orientation of Figure 10 is necessary, with an essential occupancy of the port side space by an ester, lactone or nitro group determining the activator/antagonist nature of the interactions (Langs et al., 1990,1991; Langs et al., 1989; Holtje, 1992; Holtje and Marrer, 1987). Alternatively, it has been argued that the essential nature of the port side (left hand) space for occupancy by a *cis*-oriented carbonyl of an ester group will generate two potential bound orientations where the direction of the pseudoaxial phenyl ring will determine activation or antagonism (Rovnyak et al., 1995: Figure 16).

Although nifedipine and the other clinically available 1,4-dihydropyridines are all vascular selective, albeit to significantly different degrees (section 4), the structure-activity relationship of the 1,4-dihydropyridines does accommodate cardiac selective agents. Thus, derivatives of amlodipine actually show both neutral and cardiac selective

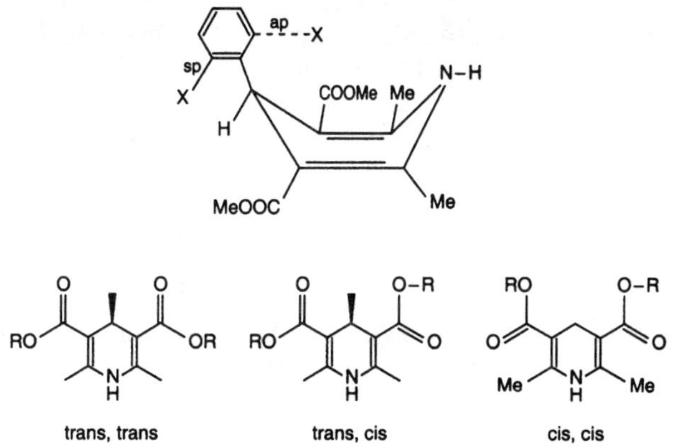

Figure 14. DHP isomers. Conformational arrangements of ester and aromatic substituents in 1,4-dihydropyridines (see text for further details).

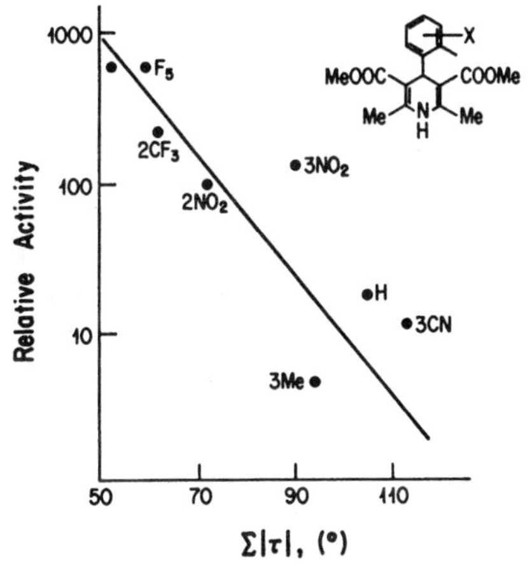

Figure 15. Ring planarity and drug potency. Relationship between 1,4-dihydropyridine ring planarity (sum of torsion angles) and calcium channel blocking activity for a series of substituted 1,4-dihydropyridines. Data from Fossheim *et al.,* 1981, 1988; Triggle et al., 1980, 1989.

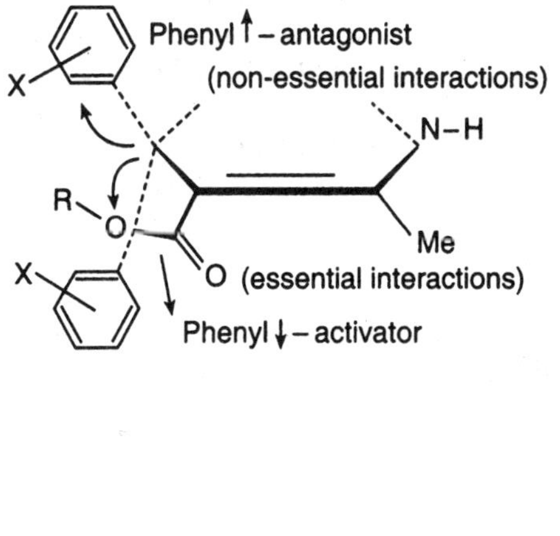

Figure 16. Model of activator-antagonist activity. Representation of 1,4-dihydropyridine interaction at receptor site with the orientation of the 4-phenyl group serving as a "molecular switch" for activator-antagonist activity. Concept from Rovnyak et al., 1995.

properties (Alker et al., 1990; Table 5). The fluorenone 1,4-dihydropyridine fluodipine (Figure 17) is also a selective cardiodepressant agent (Budriesi et al., 1998). A series of 1,4-dihydropyridines bearing a 4-(2-pyridyl) group in place of the 4-phenyl residue revealed both activator and antagonist properties, being cardiostimulatory and vasorelexant in their profile, a potential benefit in the treatment of congestive heart failure (Table 6; Iqbal et al., 1998). The differing vascular:cardiac selectivity of 1,4-dihydropyridines likely arises from a number of factors including differential interaction with L-type channel subtypes and state-dependent interactions (see Section 4).

Table 5. Vascular:cardiac selectivity of 1,4-dihydropyridines (Alker et al., 1990).

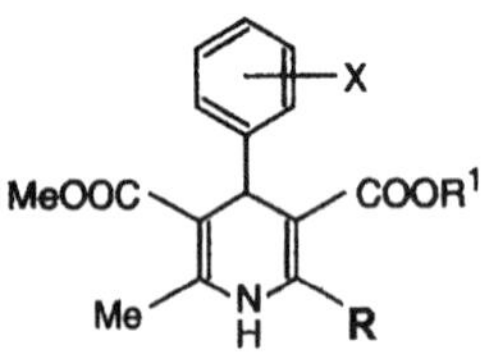

	X	R	R^1	Vascular: cardiac
Felodipine	2,3-Cl_2	Me	Et	~ 100:1
Nifedipine	2-NO_2	Me	Me	~ 20:1
Amlodipine	2-Cl	$CH_2OCH_2CH_2NH_2$	Et	~ 10:1
	2-Cl	$CH_2OCH_2CH_2NHCH_2$	Et	~ 1:1
	2-Cl	$H_2NCH_2CH_2NHCO$	Et	~ 1:10

Table 6. Activator and antagonist properties of 4-(2-pyridyl)-substituted 1,4-dihydropyridines (Iqbal et al., 1998).

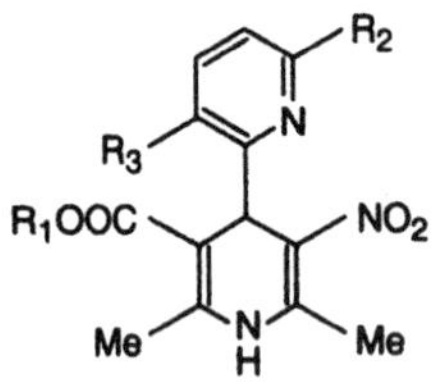

R_1	R_2	R_3	EC_{50}, M Antagonist, smooth muscle	EC_{50}, M Activator, heart
Pr^i	Me	H	2.5×10^{-5}	1.8×10^{-6}
Et	Me	H	1.7×10^{-5}	1.2×10^{-5}
Pr^i	H	Me	2.8×10^{-6}	9.9×10^{-6}
Pr^i	Cl	H	1.7×10^{-5}	5.9×10^{-6}

3.6. The Pharmacology of L-Type Channel Subtypes

The L-type channel has four major subtypes ($Ca_v1.1 - 1.4$) and is generally recognized as high voltage-activated and 1,4-dihydropyridine-sensitive; although quantitative differences in these parameters exist, they have been generally assumed to be small. Nonetheless, the issue of comparative subtype pharmacology is of considerable interest to the pharmacologic and therapeutic profile of the calcium antagonists. Although detailed structure-activity relationships are not available, sufficient evidence exists to indicate that significant and important differences exist among the subtypes that reflect both affinity and voltage-dependent changes (Abernethy and Soldatov, 2002; Lacinova et al., 2000; Triggle, 2002).

3.6.1. Unique pharmacology of $Ca_v1.3$ (a_{1D})

The $Ca_v1.3$ channel is widely expressed in neurons and neuroendocrine cells and appears to show a lower sensitivity to 1,4-dihydropyridines than $Ca_v1.2$ channels (Koschak et al.,

2001; Platzer et al., 2000; Zidanic and Fuchs, 1995). More detailed investigations indicate that the $Ca_v1.3$ channel is distinctive in several ways: it activates at negative membrane potentials some 25 mV more hyperpolarized than $Ca_v1.2$ channels, and it is some 20-fold less sensitive to nimodipine and nitrendipine (Xu and Lipscombe, 2001). This same channel also plays an important role in sinoatrial node dysfunction (Lipscombe, 2002; Platzer et al., 2000; Zhang et al., 2002). Null mutant $Ca_v1.3^{-/-}$ mice have sinus bradycardia, and these channels are postulated to be important in the generation of the spontaneous action potential in SA nodal cells and in the control of nodal firing rate. This same mutation produces congenitally deaf mice because of the absence of these channels in hair cells (Platzer et al., 2000).

Table 7. Differential interaction of calcium channel antagonists with L-type calcium channel splice forms.

	K_D −100 mV (nM)		K_D −50 mV (nM)		K_I (binding) (nM)	
	$\alpha_{1C\text{-}a}$	$\alpha_{1C\text{-}b}$	$\alpha_{1C\text{-}a}$	$\alpha_{1C\text{-}b}$	$\alpha_{1C\text{-}a}$	$\alpha_{1C\text{-}b}$
Nifedipine	47	10	6.9	3.7	2.9	1.9
Nisoldipine	2.1	0.56	0.33	0.15	0.16	0.08
(+)PN200 110	15	2.2	0.82	0.46	0.41	0.19
SDZ 207-180*	91	100	20.0	32.0	6.2	11.0

*Charged 1,4-dihydropyridine; data from Morel et al., 1998.

3.6.2. Cardiac and Smooth Muscle Splice Forms of $Ca_v1.2$

At least 50 exons code for the $Ca_v1.2$ subunit, and the expression of this subunit is regulated by alternative splicing. There exist alternate cardiac and smooth muscle splice forms of the $Ca_v1.2$ channel, $Ca_v1.2a$ and $Ca_v1.2b$ respectively. Although the splice site (in the IS6 segment) is remote from the dihydropyridine binding site, the cardiac and smooth muscle splice forms show differential sensitivity to nisoldipine (Welling et al., 1993, 1997). The smooth muscle splice form was more sensitive to low concentrations of nisoldipine, showed greater block at negative membrane potentials, and developed block faster at depolarized membrane potentials. In contrast, the cardiac splice was less sensitive to nisoldipine, and recovery from inactivated state block was observed only with this subunit. These data are consistent with earlier observations of greater resting state block in smooth muscle channels relative to cardiac channels (Bean et al., 1986; Worley et al., 1987; Yatani et al., 1987). A comparison of the activities of several 1,4-dihydropyridines at native $Ca_v1.2a$ and $Ca_v1.2b$ subunits is provided in Table 7, demonstrating greater differences in activity at polarized membrane potentials (Morel et al., 1998). A series of seven splice variants of the $Ca_v1.2$ channel revealed 2-10-fold differences in isradipine affinity (Zuhlke et al., 1998). Because of the prominent state-dependent interactions of 1,4-dihydropyridines (Triggle, 1989; Wei et al., 1989), affinity differences for 1,4-dihydropyridines may arise from binding site differences or from differences in the transitions between resting, open, and inactivated states. However, Lacinova et al. (2000) found three- to five-fold lower IC_{50} values for isradipine in the $Ca_v1.2b$ channel than for the $Ca_v1.2a$ channel, with no differences in the transitions of the channels to the inactivated state in the presence or absence of drug or in the recovery from inactivation under control conditions. Hu and Marban (1998) came to a similar conclusion using nisoldipine as the 1,4-dihydropyridine.

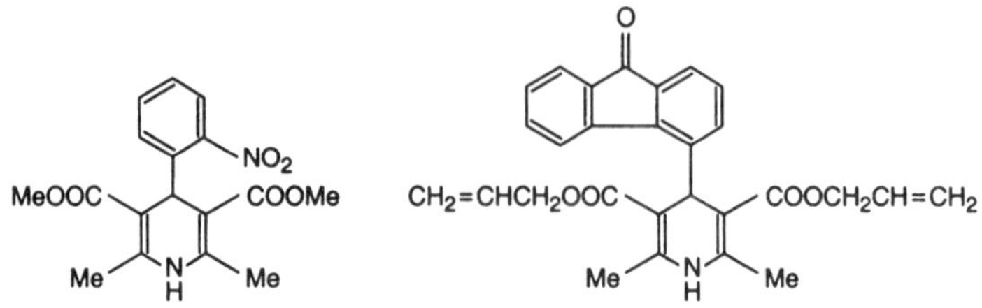

Nifedipine Fluodipine

Figure 17. DHP structure. Structures of nifedipine (vascular-selective) and fluodipine (cardioselective) 1,4-dihydropyridines.

3.7. Localization and Definition of Receptor Sites

Although there is now a considerable precision of information concerning the localization of the receptor sites on the L-type calcium channel and even of specific amino acids whose presence is essential to antagonist and activator binding, the structural basis of stereoselectivity of interaction and of the voltage-dependence of blockade of a specific receptor state awaits determination of a three-dimensional structure of the channel. This is not available, although the recent determination of potassium channel structures indicates the progress that is being made in this area (Doyle et al., 1998; Yellen, 2002; Zhou et al., 2001).

3.7.1. Distinct Receptor Sites for Different Drugs

Early studies employing radioligand binding and pharmacological techniques indicated that the receptor sites for the 1,4-dihydropyridines, phenylalkylamines, and benzothiazepines were distinct, allosterically linked one to the other, and likely localized close to the channel pore and its calcium permeation machinery (*inter alia*, Adachi-Akahane and Nagao, 2000; Bolger et al., 1983; Catterall, 2000; Hockerman et al., 1997a,b,c; Hofmann et al., 1999; Luchowski et al., 1984; Mitterdorfer et al., 1995; Striessnig, 1999; Striessnig et al., 1998; Triggle, 1999). Additionally, studies with permanently charged and membrane impermeant derivatives of nifedipine, verapamil and diltiazem indicated that the receptor sites for 1,4-dihydropyridines and benzothiazepines were located on the external approaches of the channel, whereas verapamil bound to an intracellular location (Affholter and Coronado, 1980; Bangalore et al., 1994; Hering et al., 1993; Hescheler et al., 1982; Kass et al., 1991; Kurokawa et al., 1997; Seydl et al., 1993).

More specific localization of these receptor sites on the α_1 subunit of the L-type channel has been achieved by labeling techniques that localized sites to the IIIS5, IVS5 and IVS6 regions of the subunit (*inter alia*, Catterall and Striessnig, 1992; Glossman et al., 1987; Nakayama et al., 1991; Regulla et al., 1991; Striessnig et al., 1990, 1991, 1998; Watanabe et al., 1993). Confirmation of this localization has been provided by several studies of "gain-of-function" mutants in which sequences of 1,4-dihydropyridine-sensitive L-type channels have been transferred into non-sensitive channels to render them 1,4-dihydropyridine-sensitive (Grabner et al., 1996; Schuster et al., 1996; Striessnig et al., 1998; Tang et al., 1993). Similar approaches for the phenylalkylamine and benzothiazepine receptors have been reported (Hering et al., 1996; Doring et al., 1996). Collectively, these data show that small stretches only of sequence, less than 10% of the total, principally localized around IIIS6, IVS5 and IVS6, are sufficient to generate activator- and antagonist-sensitive receptors.

3.7.2. Individual Residues Critical for 1,4-Dihydropyridine Function

Individual amino acids critical for receptor recognition and function in the $Ca_v1.2$ (α_{1C}) subunit have been identified in a series of studies by site-directed mutagenesis, including alanine scanning (Bodi et al., 1997; He et al., 1997; Hockerman et al., 1997a,b,c; Ito et al., 1997; Peterson et al., 1997; Sinnegger et al., 1997; Striessnig et al., 1998). The replacement of eight residues (using α_{1E} numbering) in the $Ca_v2.3$ (α_{1E}) subunit by Thr^{1295}, Gln^{1299} (IIIS5), Ile^{1406}, Ile^{1409}, Met^{1414} (IIIS6), Tyr^{1706}, Met^{1707} and Ile^{1714} (IVS6) yielded a chimera with high affinity dihydropyridine block and stimulation of calcium currents (Ito et al., 1997). Similarly, substitution of nine amino acid residues in the $Ca_v2.1$ subunit (α_{1A}) defined the 1,4-dihydropyridine receptor by the residues (using α_{1C} numbering) Thr^{1066}, Gln^{1070} (IIIS5), Ile^{1180}, Ile^{1183}, Met^{1188} (IIIS6), Tyr^{1490}, Met^{1491}, Ile^{1497} and Ile^{1498} (IVS6) as indicated in Figure 18 (Hockerman et al., 1997a,c; Sinnegger et al., 1997). These "reconstructed" channels possess the principal characteristics of 1,4-dihydropyridine activator and antagonist properties, stereoselectivity, and voltage-dependence of action. These channels generate approximately 90% of the ligand binding energy of the native $Ca_v1.1$ or $Ca_v1.2$ channels. The location of these critical residues on IIIS5, IIIS6, and IVS6 is consistent with a domain model in which the binding site for 1,4-dihydropyridines is located at the interface of domains III and IV (Catterall and Striessnig, 1992; Lipkind and Fozzard, 2003).

These studies are important for the potential construction of a 1,4-dihydropyridine receptor, but they do not necessarily prove that these residues interact directly in 1,4-dihydropyridine binding, and they ignore the role of other residues that are conserved between dihydropyridine -sensitive and -insensitive channels (Hockerman et al., 1997a,b,c). For example, Asn^{1374} is a conserved residue and is essential for 1,4-dihydropyridine action (Peterson et al. 1997). Tyr^{1048} (in $Ca_v1.1$ IIIS6) and Tyr^{1174} (in $Ca_v1.2$ IIIS6), conserved between 1,4-dihydropyridine-sensitive and -insensitive channels, are essential for drug action. Replacement by alanine reduces 1,4- dihydropyridine

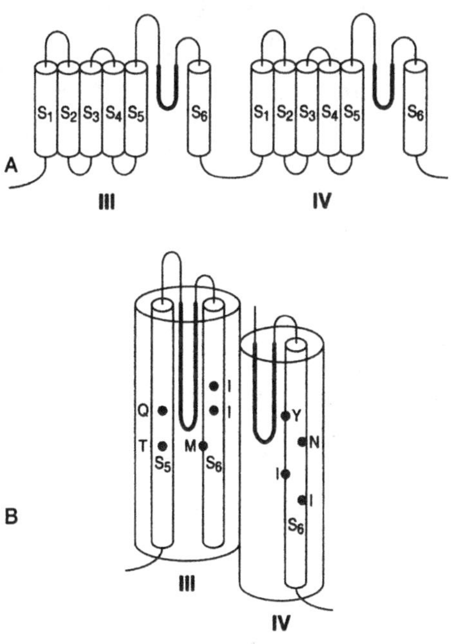

Figure 18. Channel residues critical for DHP action. Arrangement of amino acid residues of IIIS5, IIIS6 and IVS6 of the L-type calcium channel involved in 1,4-dihydropyridine binding. *A*, thick traces indicate location of the S5-S6 loops of the third and fourth pseudosubunits of the channel. *B*, residues within the S5-S6 loops whose mutation affects DHP action (see text for further details).

sensitivity by between 50 and 1000-fold (Peterson et al., 1996; Ito et al., 1997). Conserved serine residue Ser^{1115} in the IIIS5-S6 linker is also important for 1,4-dihydropyridine action (Yamaguchi et al., 2000). Similarly, the ascidian channel TuCa1 has been shown to have all nine residues determined as critical for 1,4-dihydropyridine interaction, yet is only weakly sensitive to 1,4-dihydropyridine antagonists and insensitive to activators (Okamura, 1999). The CyCaa1 channel from jellyfish carries all but two of the residues deemed critical for 1,4-dihydropyridine action, yet is also insensitive (Jeziorski et al., 1998). Additionally, although available data indicate that 1,4-dihydropyridine activators and antagonists share the same binding site, it is likely that there are differential contributions from different residues. Thus, Thr^{1066} and Gln^{1070} ($Ca_v1.2$ numbering) are both required for 1,4-dihydropyridine interaction, but activator interaction is critically dependent upon the glutamine residue (Mitterdorfer et al., 1996).

3.7.3. Individual Residues Critical for Phenylalkylamine and Benzothiazepine Function
Similar studies have served to localize the receptor sites for phenylalkylamines and benzothiazepines in regions partially common to those used by the 1,4-dihydropyridines (Hockerman et al., 1997b; Striessnig et al., 1998). Phenylalkylamine interactions are dependent upon interactions within domains IIIS6 and IVS6 and in the pore regions of domains III and IV. Involvement with binding sites in the pore regions, specifically glutamates 1118 and 1419 of the $Ca_v1.2$ subunit, is consistent with the presumed mechanism of interaction of phenylalkylamines at the calcium channel (and local anesthetics at the sodium channel; *inter alia*, Ragsdale et al., 1994) of binding in the pore region according to the "modulated receptor" model (Hille, 1977; Hondeghem and Katzung, 1985; Triggle, 1989; reviewed by Timin et al., this volume). In IIIS6 four residues were critical to phenylalkylamine binding – Tyr^{1152}, Ile^{1153}, Phe^{1164} and Val^{1165} – of which only Ile^{1153} is unique to the $Ca_v1.2$ channel. Similarly, the conserved glutamate residues in the P-loops in domains III and IV associated with calcium binding and translocation (Tang et al., 1993; Mitterdorfer et al., 1995) are essential for phenyalkylamine binding (Hockerman et al., 1997a,b). This may represent direct interaction of the cationic heads of phenylalkylamines and benzothiazepines with the negatively charged carboxylate residues. Four residues in IVS6 of $Ca_v1.2$ – Tyr^{1463}, Met^{1464}, Ala^{1467} and Ile^{1470} - located on adjacent turns of a putative α-helix and not found in non-L-type channels contribute significantly to phenylalkylamine interactions and may be part of the binding site (Hering et al., 1996; Hockerman et al., 1995; Schuster et al., 1996).

Benzothiazepine interactions with the calcium channel have much in common with phenylalkylamine interactions, including frequency-dependent blockade, although the two drug classes access their binding sites through different pathways. They share significant elements of a binding site in domains IIIS6 and IVS6 (Hockerman et al., 1997a,b,c; Striessnig et al., 1998). In IVS6, the Tyr, Ala and Ile residues are shared by both benzothiazepines and phenylalkylamines (Hering et al., 1996). In domain IIIS6, conserved Phe and Val residues are important in controlling both diltiazem sensitivity and inactivation (Kraus et al., 1998): these residues are also important to phenylalkylamine interactions. The benzothiazepine and phenyalkylamine interaction sites share some residues in IVS6 – Tyr^{1463}, Met^{1464} and Ile^{1470} – in common with dihydropyridines, consistent with much biochemical and pharmacological evidence for a set of allosteric and overlapping interactions between the three putative receptor sites.

3.7.4. Structural Basis of State-Dependent Binding

The definition of the location of the receptor site(s) for calcium channel antagonists and activators and of the specific amino acid residues involved in the construction of the receptor site is but one step in the construction of a complete structure-activity relationship for these molecules. Ultimately, it will be necessary to factor these contributions into those that contribute to molecular recognition (binding affinity) and those that contribute to activation and blockade of the channel. Since it is believed that the 1,4-dihydropyridine activators and antagonists interact with and stabilize open and open/inactivated states respectively, a significant understanding of channel structure and dynamics will be necessary. This is not yet available. Success with the potassium channel, however, is a likely prelude to success with the calcium channel. A number of mutational studies, for example, are defining locations and residues that contribute to the state-dependent interactions of the 1,4-dihydropyridine, phenylalkylamine and benzothiazepinone classes of antagonists (Timin et al., this volume; Berjukow et al., 1999; Bodi et al., 1997, 2002; Cai et al., 1997; Hering et al., 1997; Johnson et al., 1996; Kraus et al., 1998; Motoike et al., 1999). These studies, together with existing structure-activity studies and extrapolations from the known structure of a potassium channel, make possible construction of homology models of the calcium channel and its drug binding sites (*inter alia,* Holtje, 1992, Holtje and Marrer, 1987; Langs et al, 1990, 1991; Schleifer, 1999; Zhorov et al., 2001).

3.8. Other Sites of Drug Action at the Channel

3.8.1. The $\alpha_2\delta$ Subunit

The α_1 subunit is the major structural and functional component of the voltage-gated calcium channel, carrying the permeation and gating machinery as well as the binding sites for activators and antagonists. However, since the channel is a multimeric complex, other sites of drug interaction likely exist, although these have not been well investigated. Gabapentin (Figure 19; Neurontin) is an anticonvulsant agent employed clinically and a potential treatment for neurogenic pain (Field et al., 2000; Jones, 2002; McLean, 1995; Rosner et al., 1996; Taylor, 1993). The mechanism of action of gabapentin is not clear since, despite its structure similar to the neurotransmitter GABA, it does not appear to interact directly with the GABA system (Taylor et al., 1998). However, it does bind with high affinity and with a structure-activity relationship (including stereoselectivity) to the calcium channel $\alpha_2\delta$ subunit (Bryans et al., 1998; Gee et al. 1996; Marais et al., 2002; Wang et al., 1999). Gabapentin

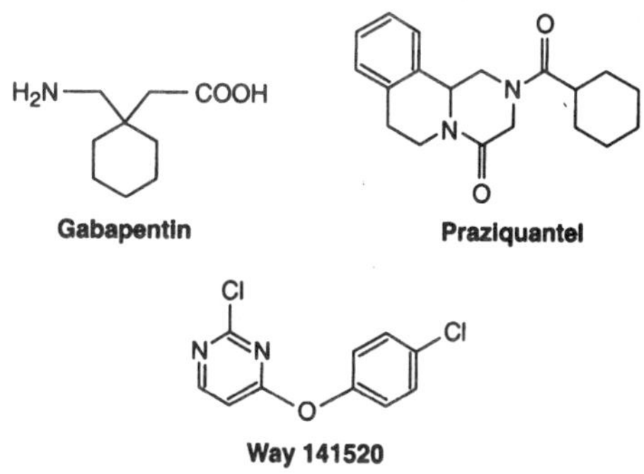

Figure 19. Molecules that interact at non-α_1 subunits of the calcium channel.

blocks high voltage activated calcium currents in neurons (Alden and Garcia, 2001; Stefani et al., 2001), and since the $\alpha_2\delta$ subunit is upregulated in models of pain, this molecular action may underlie its clinical use (Luo et al., 2001; Sutton and Snutch, 2002).

3.8.2. The β Subunit

This interaction between the α_1- and β- subunits of the heteromeric calcium channel complex is reversible, characterized by an 18 residue sequence in the intracellular loop separating domains I and II of the α_1 subunit, and by a 30 residue sequence in the β-subunit. This protein-protein interface is a potential site for drug interaction that should increase or inhibit channel function by promoting the association or dissociation of the β-subunit (Bichet et al., 2000; De Waard et al., 1994; Pragnell et al., 1994). Peptides 21 and 68 residues in length containing the α_1 interaction sequence exhibit affinity for the β-subunit and promote dissociation of the $\alpha_1\beta$ complex (Bichet et al., 2000). This interaction site may be the locus of action of the antischistosomal drug praziquantel (Figure 19) that is efficacious against flatworms of the genus *Schistosoma* by promoting calcium influx and muscle paralysis (Jim and Triggle, 1979; Redman et al., 1996). In schistosomes, the interaction of the β-subunit appears to be inhibitory to channel function. Praziquantel reverses this inhibitory interaction (Kohn et al., 2001). The small molecule inhibitor WAY141520 (Figure 19) was identified in a yeast two-hybrid screen as blocking at micromolar concentrations the α_{1B}-β_3 neuronal interaction and serving as a neuronal calcium antagonist (Young et al., 1998).

3.9. Other Sites of Action of Calcium Antagonists

The underlying basis for the pharmacological and therapeutic activities of the calcium channel antagonists derives from their selective interactions at the L-type channel. However, the several structural classes do interact with other receptors and sites with varying degrees of affinity: of the available classes of agents employed clinically, the 1,4-dihydropyridine structure affords the greatest potency and the greatest selectivity of action. Thus the phenylalkylamine and benzothiazepine classes interact with a variety of receptors including neurotransmitter receptors, typically with affinities 10 – 1000 times lower than those for interaction with the L-type channel (Janis, Silver and Triggle, 1987; Triggle, 1990).

3.9.1. Interactions with Non-Ion Channel Targets

At high concentrations, the calcium antagonists exhibit antioxidant and antiatherogenic properties, but with different structure-activity relationships from those exhibited at calcium channels (Henry, 1991; Rojstaczer and Triggle, 1994, 1996; Triggle, 1988). In principle, these anti-atherogenic effects may arise at a variety of sites and by a variety of mechanisms. Although well demonstrated in animal studies their clinical significance is not yet clear (Mancini, 2002). Amongst the possible mechanisms are anti-oxidant effects, inhibition of smooth muscle growth and movement, increased NO supply (Section 3.10), decreased cholesterol ester deposition, and improved vascular dynamics. In clinical conditions, the INTACT and the MHINT trials (with nifedipine and nicardipine respectively) revealed no difference in the progression or regression of stenoses over the treatment period; however, a reduction in the number of new minimal lesions was found. In the PREVENT trial with amlodipine, however, a significant reduction in the progression of atherosclerosis in carotid arteries was observed (Mancini, 2000).

Table 8. 1,4-Dihydropyridine block of T-type calcium channels

PREPARATION	DIHYDROPYRIDINE[a]									
	1	2	3	4	5	6	7	8	9	Reference
	IC$_{50}$ or K$_I$, x 10^{-6} M									
Rat hypothalamic neuron				3.5	5.0	7.0				e
Rat amygdaloid neuron				2.5						f
Rat hippocampal CA1				1.6						g
Rat thalamic nucleus					2.6					h
Rat Meynert neurons				>3.0						i
Mouse dorsalroot ganglion				2.0				>2.0	>2.0	j
Neuroblastoma-glioma				2.5	39	0.24	9.8			k
Adrenal glomerulosa								13.4 0.16[b]		l
Rat portal vein								0.05		m
Rat aorta				0.6						n
Canine atria								~10		o
GP atria myocytes				0.013					0.02[c]	p
GP atrial myocytes						0.018[d]				q
GP ventricle myocytes		~1.0								r
Human heart Ca$_V$3.2	5.7									s
Human thyroid tumor α_1H							10			t
Mouse spermatocyte					~10					u
Mouse spermatocyte					0.4					v
GH3 cells			0.7							w

a: 1 = amlodipine; 2 = arandipine; 3 = felodipine; 4 = nicardipine; 5 = nifedipine; 6 = niguldipine; 7 = nimodipine; 8 = nitrendipine; 9 = PN 200 110.
b: calculated affinities according to "modulated receptor" model
c: non-stereoselective effects of the (-) and (+) enantiomers
d: stereoselective effects of niguldipine (+) > (-).

References: e, Akaike et al., 1989; f, Kaneda and Akaike, 1989; g, Takahashi and Akaike, 1991; h, Tarasenko et al., 1997; i, Rhee et al., 1999; j, Richard et al., 1991; k, Stengel et al., 1998; l, Cohen et al., 1988; m, Loriand et al., 1989; n, Kuga et al., 1990; o, Bean, 1985; p, Cohen et al., 1992; q, Romanin et al., 1992; r, Masumiya et al., 2000; s, Perchenet et al., 2000; t, Williams et al., 1999; u, Santi et al., 1996; v, Cohen et al., 1988.

Amlodipine is a more lipophilic agent with an enhanced tendency to partition into plasma membranes, and this property may contribute to the measured antiatherogenic potency (Section 4.2).

3.9.2. Interactions with Non-L-Type Calcium Channels

In addition to these more or less non-specific actions of the calcium antagonists there is evidence for more specific interactions, particularly for the 1,4-dihydropyridine class of structure (Rampe and Triggle, 1993; Triggle, 2003). The term "privileged structure" was introduced for the benzodiazepine nucleus in 1988 by Evans and co-workers who postulated that, *"judicious modification of such structures could be a viable alternative in the search for new receptor agonists and antagonists"* (Evans et al., 1988). Such privileged structures are templates that interact with common protein domains or structures and that when appropriately "decorated" can interact with selectivity at a variety of receptors (Wiley and Rich, 1993; Patchett and Nargund, 2000). The 4-phenyl-1,4-dihydropyridine structure is one such entity that interacts with a variety of structures, including calcium channels (Triggle, 2003).

3.9.2a. Interactions of 1,4-Dihydropyridines with T-type Channels. 1,4-DHP interactions have been reported at a number of T-type channels (Table 8), and although the interactions reported are quite generally of low affinity, several important limitations to the data exist. First, the 1,4-dihydropyridines studied are those whose structures have been optimized for interaction at L-type channels, and second, the interactions of 1,4-dihydropyridines are state-dependent. The data of Table 8 do reveal, however, evidence for specificity of interaction. Thus, niguldipine is some 2000 times more potent at T-type channels than is nifedipine, but the two are approximately equi-effective at L-type channels. Furthermore, niguldipine exhibits stereoselectivity of interaction at the T-type channel, consistent with specific interactions. The influence of state-dependent interactions is seen clearly with felodipine interactions with L- and T-type channels (Cohen et al., 1992). Felodipine is actually more potent against inactivated T-type channels (K_D = 13 nM) than against resting L-type channels (K_D ~ 1000 nM). Efonidipine interacts at both L- and T-type channels with relatively low (10-fold) selectivity for the L-type channel: this may underlie its significant negative chronotropic activity (Masumiya et al., 1998). A new series of 1,4-dihydropyridines containing a 2,6-disubstituted phenyl residue has significant T:L selectivity, approximately 40-fold (Figure 20; Kumar et al., 2002).

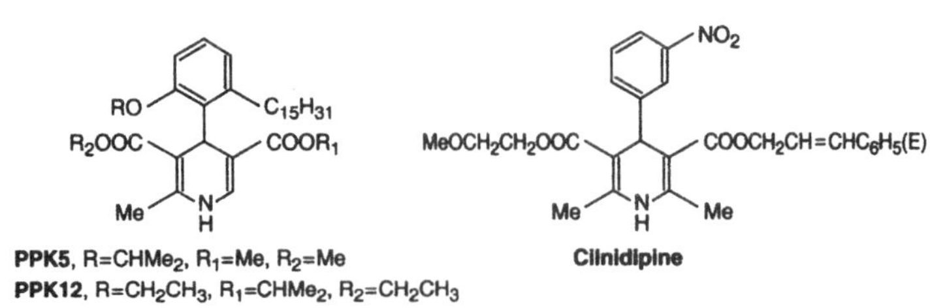

Figure 20. DHPs that block non-L-type channels. 1,4-Dihydropyridine structures with activities at T-type calcium channels (PPK 5 and PPK 12) and N-type channels (cilnidipine).

3.9.2b. Interactions of 1,4-Dihydropyridines with N- and P/Q-type Channels. 1,4-DHPs also interact at N-type channels, generally with no selectivity over L-type channels (Furukawa et al., 1999; Triggle, 2002b). However, cilnidipine (Figure 20) has actions on both L- and N-type channels and, although more selective for L-type channels, was able to block N-type channels in neurons at concentrations only some 10-fold higher than those needed to block L-type channels (Murai et al., 2000; Uneyama et al., 1999). This N-type channel blocking activity may underlie the absence of reflex sympathetic stimulation during blood pressure reduction (Uneyama et al., 1999) and the antinociceptive effects of this agent (Murakami et al., 2000). Similarly, the ability of amlodipine and some other 1,4-dihydropyridines, including cilnidipine, benidipine, barnidipine, and nicardipine, to block N- and P/Q-type channels may define some aspects of their pharmacology, notably their vasodilating capacity without accompanying reflex increase in heart rate, and the ability to reduce catecholamine output (Furukawa et al., 1999). Such dual N- and L-type blockade may be of benefit during cardiovascular therapy (Zhou et al., 2002).

3.10. Clinical Action Due to Interaction at Non-Channel Receptors

Despite the overall powerful selectivity of action of the clinically available 1,4-dihydropyridines for the L-type channel, some evidence suggests that a component of their vascular selective and vasodilating properties may arise from an ability to stimulate nitric oxide release from vascular endothelial cells (Dhein et al., 1999). Such an action would add to the vasodilation produced by direct blockade of the calcium channels in vascular smooth muscle cells. Endothelial nitric oxide synthase (eNOS) converts substrate L-arginine to nitric oxide, which diffuses to the vascular smooth muscle cells and promotes vascular relaxation (Furchgott and Zawadzki, 1980; Lamas et al., 1992; Thomas et al., 2001). The nitric oxide releasing actions of the 1,4-dihydropyridines appear to be a class action, to be exerted at therapeutic concentrations and to involve the stimulation of calcium influx into endothelial cells through a non-L-type process (Dhein et al., 1995; Zhang and Hintze, 1998; Crespi et al., 2002).

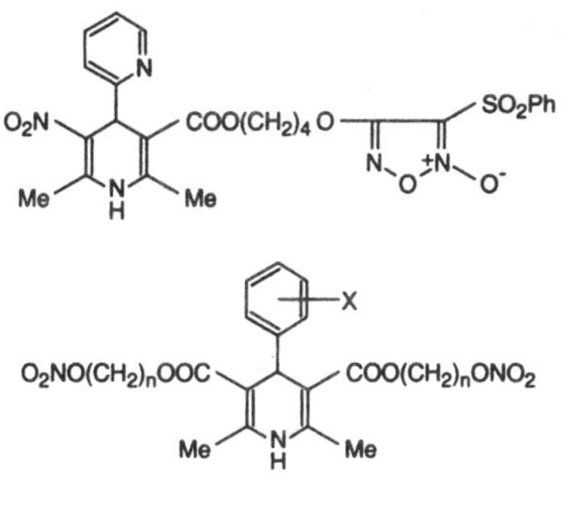

Figure 21. 1,4-Dihydropyridines with NO-releasing activities.

1,4-Dihydropyridines may have at least two other effects on the nitric oxide-mediated vasorelaxant system. Pranidipine (0PC-13340, methyl-3-phenyl-2(E)-propenyl 1,4-dihydro-2,6-dimethyl-4(3-nitrophenyl)-3,5-pyridinedicarboxylate) enhances nitric oxide-mediated vasorelaxation by a mechanism apparently distinct from NO release (Mori et al., 2001). Additionally, a series of 1,4-dihydropyridines exist with "dual" pharmacological properties blockade of L-type calcium channels and direct release of nitric oxide (Figure 21; Cena et al., 2002; Ogawa et al., 1993; Vo et al., 2002).

4. SELECTIVITY OF ACTION OF CALCIUM ANTAGONISTS

The several classes of voltage-gated calcium channels, including the L-type channel, enjoy a widespread distribution, yet the existing calcium antagonists of the phenylalkylamine, benzothiazepine and 1,4-dihydropyridine structural classes exhibit their therapeutic effects primarily in the cardiovascular system. Major actions are not seen in the peripheral or central nervous system, although L-type channels are present as determined by both electrophysiological and biochemical measurements and these appear to be qualitatively similar to the functional sites found in cardiac and smooth muscle preparations. Thus, the high affinity 1,4-dihydropyridine binding sites in smooth muscle correlate on a 1:1 basis with pharmacological activity, but the same or similar sites found in neuronal preparations do not correlate with measurable pharmacological activities (Janis et al., 1987; Janis and Triggle, 1984a,b; Triggle, 1989). Additionally, these antagonists exhibit both qualitative and quantitative differences in their actions in the cardiovascular system: nifedipine (and other 1,4-dihydropyridines) is quantitatively more potent than verapamil or diltiazem as a vasodilator, whereas the latter agents exhibit significant cardiac depressant properties that are essentially absent in the 1,4-dihydropyridine class (Tables 1, 2, 4, 9). These differences underlie their cardiovascular applications and the role of verapamil as a class IV antiarrhythmic. Additionally, drugs within a single structural class have significant differences in cardiovascular profile: in the 1,4-dihydropyridine series major differences exist in vascular:cardiac selectivity in both human and animal tissues (Tables 1, 2, 5, 9). These issues also link to the contraindications and safety profiles of the three different structural classes of antagonist.

In principle, these differences in tissue selectivity may have a number of origins either alone or in combination. Some of these have already been discussed:

1. Mode of calcium mobilization – intracellular and extracellular sources;
2. Pharmacokinetic factors;
3. Class and subclass of calcium channel activated – more than one channel may be activated by a depolarizing stimulus;
4. State-dependent interactions – drug interactions are modified in potential-dependent manner;
5. Pathologic state of tissue that affects channel expression and function – "channelopathies."

Table 9. Vascular selectivity of 1,4-dihydropyridines in isolated rabbit tissues.

1,4-Dihydropyridine	Selectivity Ratio**
Lercandipine	730
Lacidipine	193
Amlodipine	95
Felodipine	6
Nitrendipine	3

Data from Angelico et al., 1999; ** indicates ratio of IC_{50} values for inhibition of contraction of rabbit aorta and rabbit ventricle.

4.1. Mode of Calcium Mobilization

Cellular calcium mobilization in response to stimuli is derived in principle from a variety of sources, alone or in cooperation (Figure 1). L-type calcium antagonists have fairly good selectivity for L-type channels and hence affect predominantly calcium influx through L-type channels. However, some cross-over interactions to other channel classes may occur (Section 3.9.2).

Tissues of the cardiovascular system can support stimulus-response coupling through multiple pathways, thereby generating differential sensitivity to calcium antagonists. In the kidney the renal afferent and efferent arteries that control glomerular flow are both contracted by angiotensin II. However, only the afferent response is sensitive to nifedipine and other calcium antagonists since the efferent arteriole mobilizes calcium from intracellular sources (Carmines and Navar, 1989; Loutzenhiser and Loutzenhiser, 2000). Calcium antagonists may differ, however, in their ability to modulate blood flow through the renal afferent and efferent arterioles. In preclinical studies, lercanidipine, a third-generation antagonist, dilates both efferent and afferent arterioles (Messerli, 2002; Sabbatini et al., 2000). An action on both pre- and post-capillary vasodilatation may underlie the reduced clinical incidence of pedal edema, the most common side effect of calcium antagonists, seen with lercanidipine relative to other 1,4-dihydropyridines (Fogari et al., 2000; Messerli, 2002).

4.2. Pharmacokinetic Factors

The extent and the rate at which a drug is absorbed, distributed and metabolized can profoundly alter response. Differences in these factors occur between different drugs and between different formulations of the same drug. The calcium antagonists can be classified into three distinct structural classes according to their overall pharmacokinetics (Table 10; Leenen, 2002; Leenen et al., 1996). Thus the hemodynamic changes of nifedipine given as a rapid infusion include a rapid fall in blood pressure and a large reflex increase in heart rate associated with a rapid increase to a peak plasma level. Slow infusion of the same preparation of nifedipine, however, results in a less rapid and smaller decrease in blood pressure and virtually no change in heart rate (Kleinbloesem et al., 1987). In contrast, amlodipine has a very long duration of action and a long plasma half-life that is intrinsic to its chemical structure. The third-generation 1,4-dihydropyridines lacidipine and lercanidipine are extremely hydrophobic and partition extensively into tissues (Angelico et al., 1999; Epstein, 1999; Hall et al., 1991; McLellan and Jarvis, 2000). The contrasting selectivity of the 1,4-dihydropyridines nifedipine and nimodipine with the latter having cerebroparenchymal vascular selectivity likely arises from the higher transfer rate of nimodipine across the blood-brain barrier and its larger volume of distribution in the brain (Van den Kerchoff and Drewes, 1989).

Drug partitioning into biological membranes is a non-homogeneous process. Both specific and non-specific interactions of the drug and membrane components influence the extent of partitioning (Herbette et al., 1989, 1991). Thus, significantly different partition coefficients for 1,4-dihydropyridines are measured in octanol and in membrane preparations (Table 11; Herbette et al., 1994). The extensive partitioning into cell membranes of the most hydrophobic agents, lacidipine and lercanidipine, likely underlie observations that their duration of action is significantly longer than their plasma half-life of elimination (Herbette et al., 1993, 1994, 1997). Of particular interest is the influence of membrane composition on partition coefficients—the cholesterol:phospholipid molar ratio has a significant influence (Table 12; Herbette et al., 1987). It is thus possible, from these model studies, that different membrane compositions regulated by both physiological and pathological factors influence the partitioning behavior of 1,4-dihydropyridines and so contribute to observed differential tissue selectivity.

4.3. Class and Subclass of Calcium Channel Activated

Although detailed comparative structure-activity relationships are not yet available, it is clear that the availability of three different α_1 subunit splice forms – $Ca_V1.2a$-c in the cardiovascular system – with differential pharmacologic sensitivity plays an important role in control of the cardiac:vascular sensitivity differences that are apparent both pharmacologically and therapeutically (Section 3.6).

4.4. State-Dependent Interactions

A critical component of the definition of the selectivity of action of calcium antagonists is the state-dependent interaction of drug and its channel receptor. These interactions arise because drugs may exhibit differential affinity for, or access to, their receptor sites in the resting, open or inactivated states (or families of states) of the channel – the "modulated receptor hypothesis" (Hille, 1977; Hondeghem and Katzung, 1985; Triggle, 1989). Transitions between these states are determined by changes in membrane or chemical potential and the kinetics of the channel response. Drugs that interact preferentially with, for example, the open or inactivated state will exhibit an increased apparent affinity under stimulus conditions that increase the availability of the open or inactivated states of the channel. Additionally, the structure of the drug and its overall physicochemical properties – hydrophilic/hydrophobic and charged/uncharged – may control the access route of the drug to its receptor site. Hydrophilic and charged drugs may access the binding site through a polar and hydrophilic pathway, including the channel vestibule and pore, while nonpolar and hydrophobic species access their binding sites through membrane pathways (for further details see Timin et al., this volume).

Both verapamil and diltiazem exhibit frequency-dependent interactions, whereas nifedipine and other 1,4-dihydropyridines exhibit voltage-dependent interactions (*inter*

Table 10. Pharmacokinetic classification of calcium antagonists.

Class	Properties
Rapid acting (nifedipine capsules, original formulation)	Very rapid absorption; plasma levels rise and fall rapidly; hemodynamic changes are swift and abrupt
Slow- or controlled-release formulations (nifedipine GITS, felodipine ER)	Slower absorption with correspondingly slow rise and fall in plasma levels and slow hemodynamic changes
Slow-onset, long-acting, slow-offset intrinsic properties (amlodipine, lacidipine, and lercanidipine)	Hemodynamic effects develop very slowly (days), slow elimination; very smooth and gradual changes in hemodynamic parameters

Table 11. 1,4-Dihydropyridine partition coefficients in biological membranes and octanol/ buffer.

Drug	Octanol/Buffer	Biological Membrane
Amlodipine	30	19000
Nisoldipine	40	13000
Nimodipine	290	6300
Nifedipine	--	3000

Data from Herbette et al., 1994.

Table 12. Membrane partition coefficients of 1,4-dihydropyridines in bovine cardiac phosphatidylcholine membranes as a function of cholesterol:phospholipid ratio.

1,4-Dihydropyridine	Cholesterol:Phospholipid Molar Ratio	$K_p$mem
Lacidipine	0:1	136,000
Lacidipine	0.3:1	103,000
Lacidipine	0.45:1	83,000
Lacidipine	0.6:1	62,000
Lercanidipine	0:1	315,000
Lercanidipine	0.3:1	230,000
Amlodipine	0:1	21,300
Amlodipine	0.3:1	19,000

Data from Herbette et al., 1994.

Table 13. State-dependent interactions of 1,4-dihydropyridines.

Tissue	Ligand	K_D, Resting (M)	K_D, Inactivated (M)	Reference
Cardiac Purkinje	Nisoldipine	1.3×10^{-6}	1.0×10^{-9}	a
Ventricle	Nitrendipine	7.3×10^{-7}	2.5×10^{-9}	b
Dorsal root ganglion	Nimodipine	4.0×10^{-8}	1.3×10^{-9}	c
GH3	(+)RBayK-8644	$>1.0 \times 10^{-8}$	4.0×10^{-9}	d
Skeletal muscle	(+)Isradipine	1.3×10^{-8}	1.5×10^{-9}	e
Smooth muscle (A7r5)	S12968	4.0×10^{-7}	5.0×10^{-8}	f
Saphenous vein	Nitrendipine	1.0×10^{-8}	2.5×10^{-10}	g
Mesenteric artery	Nisoldipine	1.2×10^{-8}	7.0×10^{-11}	h
Mesenteric artery	Nitrendipine	2.2×10^{-7}	4.6×10^{-10}	i
Uterus	(+)Isradipine	2.3×10^{-9}	1.0×10^{-10}	j

References: a, Sanguinetti and Kass, 1984; b, Bean, 1984; c, McCarthy, 1989; d, McCarthy and Cohen, 1986; e, Cognard et al., 1986; f, Randle et al., 1990; g, Yatani et al., 1987; h, Nelson and Worley, 1989; i, Bean et al., 1986; j, Honore et al., 1989.

Table 14. 1,4-Dihydropyridine binding in polarized and depolarized cells.

Tissue	[^{3}H] Ligand	K_D (M) (polarized)	K_D (M) (depolarized)	Reference
Rat cardiomyocytes	(+)Isradipine	3.5×10^{-9}	6.3×10^{-11}	a
Rat cardiomyocytes	(+)Isradipine	5.2×10^{-8}	5.5×10^{-11}	b
Rat aorta	(+)Isradipine	2.5×10^{-10}	6.8×10^{-11}	c
Bovine coronary	Nifedipine	2.0×10^{-9}	1.4×10^{-10}	d
Rat mesenteric	(+)Isradipine	2.5×10^{-11}	4.1×10^{-10}	e,f
	Nifedipine	3.2×10^{-9}	1.2×10^{-9}	

References: a, Wei et al., 1989; b, Ferrante et al., 1989; c, Morel and Godfraind, 1991; d, Sumimoto et al., 1988; e, Morel and Godfraind, 1988; f, Morel and Godfraind, 1987.

alia, Bayer et al., 1975; Bean, 1984; Hondeghem and Katzung, 1985; McDonald et al., 1984; Sanguinetti and Kass, 1984; Talajic et al., 1989; Triggle, 1989). These differences define in significant part the antiarrhythmic properties of verapamil and diltiazem and the overall vascular selectivity of the clinically available 1,4-dihydropyridines, which reflect selective interactions with open and inactivated states of the channel respectively. Verapamil and diltiazem are charged species at physiological pH, whereas nifedipine and other 1,4-dihydropyridines are nonpolar species that likely access their receptor site through membrane-delineated pathways (Chester et al., 1987).

Evidence for the state-dependent interactions of the calcium antagonists is available from electrophysiological, radioligand binding, and pharmacological studies (for reviews see *inter alia* Hondeghem and Katzung, 1985; Triggle, 1989, 2002). The electrophysiological data of Table 13, though obtained under different recording conditions and not directly comparable, do indicate quite generally the higher affinity of 1,4-dihydropyridines for the inactivated state relative to the resting state of the channel. These differences can be substantial, up to 1000-fold. The high affinity 1,4-dihydropyridine (and other calcium antagonists) binding observed in the cellular membrane fractions customarily employed in radioligand binding reflects drug interaction under depolarized (and dephosphorylated) conditions that approximate the inactivated state of the channel. The excellent 1:1 correlations observed between 1,4-dihydropyridine binding affinities and pharmacologic activities for inhibition of K^+ depolarization-induced smooth muscle responses provide further evidence of the preferential interaction of these agents with an inactivated state of the channel favored by depolarization (Bolger et al., 1983; Janis et al., 1987; Yousif et al., 1987; Sun and Triggle, 1995; Triggle, 1989). Radioligand binding studies in intact cells under polarized and depolarized conditions also provide evidence for these voltage-dependent interactions of 1,4-dihydropyridines (Table 14). The excellent linear correlation observed for a series of clinically available 1,4-dihydropyridines – amlodipine, nifedipine, nisoldipine, felodipine, nimodipine, nicardipine, and nitrendipine – between vascular selectivity and voltage-dependent binding provides further evidence that the prominent state-dependent interactions of the 1,4-dihydropyridines are a significant contribution to their vascular selectivity (Sun and Triggle, 1995).

The structure of the 1,4-dihydropyridine controls both its activator and antagonist potency at the calcium channel and the extent to which it exhibits voltage-dependent binding. This is very clearly seen in the data of Figure 22. For three series of 1,4-

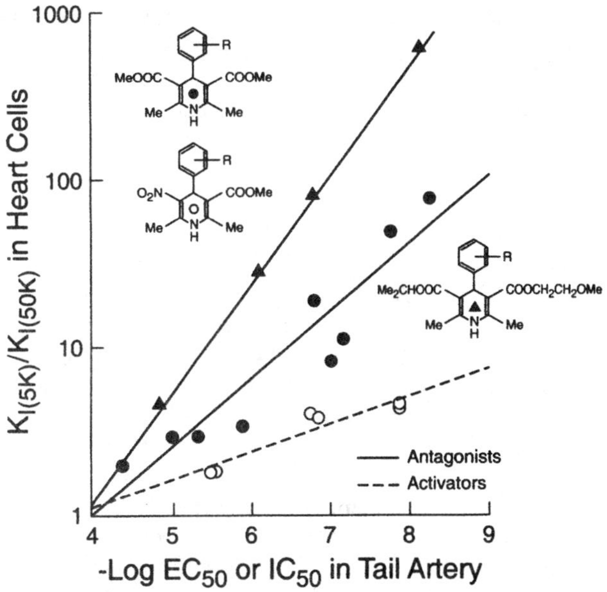

Figure 22. Binding correlates with potency. Correlation between voltage-dependent binding of 1,4-dihydropyridines (ratio of binding affinities in intact cardiac cells at 5 mM and 50 mM K^+) and activator (EC_{50}) and antagonist (IC_{50}) activities in vascular smooth muscle for two series of dihydropyridine antagonists (nifedipine and nimodipine series) and one series of activators (Bay K 8644).

dihydropyridines there exists a linear relationship between the extent of voltage-dependent binding and potency as an activator or antagonist in vascular smooth muscle (Wei et al., 1986: Zheng et al., 1992). There is little voltage-dependent interaction with the activator series, and this is consistent with observations that the activator/antagonist profile of these 1,4-dihydropyridines is voltage-dependent, switching from activator to antagonist with increasing cellular depolarization (Kass, 1987).

4.5. Channel Pathology

Voltage-gated calcium channels, and other families of ion channels, are regulated by homologous and heterologous factors, including pathology (reviewed in Catterall, 2000; Ferrante and Triggle, 1990; Goplalakrishnan and Triggle, 1990; Triggle, 2002). Little evidence exists, however, for "withdrawal" or "rebound" phenomena subsequent to calcium antagonist chronic administration (Raftery, 1984; Triggle, 2002). Additionally, an increasing number of genetic defects are known to alter calcium channel expression and function (Ashcroft, 2001; Muth et al., 2002; see Mori et al., this volume, and Saegusa and Tanabe, this volume). Such changes can impact tissue sensitivity to calcium channel activators and antagonists.

5. CONTRAINDICATIONS OF CALCIUM CHANNEL ANTAGONISTS

Some of the factors that control the selectivity of action of the calcium antagonists are of direct relevance to their contraindication profile. A detailed discussion is outside the scope of this chapter, but is available elsewhere (Epstein, 2002, 2002a,b; Kizer and Kimmerl, 2002; Opie and Schall, 2002; Opie et al., 2000; Packer, 1989; Sleight, 1998). Two principal issues arise. It is likely that the original rapid-acting formulation of nifedipine has deleterious effects on the cardiovascular system: the rapid acting agent may be proischemic, and the reflex sympathetic stimulation induced may produce excessive activation of the neurohormonal axis, situations that are detrimental to patients with compromised cardiac function (Leenen, 2002). These properties likely underlie, in part, three reports that the clinical use of calcium channel antagonists for the treatment of hypertension was associated with elevated mortality (Furberg et al., 1995; Pahor et al., 1995; Psaty et al., 1995). These reports were, however, based on observational studies and suffer from all of the limitations of that approach (Epstein, 2002a; Kizer and Kimel, 2002). More recently, *meta*-analysis of randomized clinical trials of the safety of sustained and extended release nifedipine did not reveal safety concerns (Stason et al., 1997). More importantly, several large randomized trials, including the Systolic Hypertension in Europe (Syst-Eur) trial, have demonstrated the safety of calcium antagonists as initial hypertensive therapy (Staessen et al., 1997, 2002). From the Framingham Heart Study an assessment of the mortality risk associated with calcium channel antagonists did not support risk associated with their use (Abascal et al., 1998). The second- and third-generation 1,4-dihydropyridines are improved agents, with their slow onset and offset of action and their enhanced vascular selectivity.

Nonetheless, the calcium channel antagonists available need to be used with reference to the profiles and side-effects of other classes of cardiovascular drugs. Thus, it is likely that ACE inhibitors exhibit a better profile than calcium channel antagonists in diabetic subgroups and in congestive heart failure and that beta blockers should be first-line therapy for patients with acute coronary syndromes and stable angina. However, cardiovascular disease presents complex scenarios and it is possible that multiple drug therapy is appropriate to maintain the tight control of blood pressure that is desirable, particularly in at-risk populations (Epstein, 2002b).

6. PERSPECTIVE AND FUTURE DIRECTIONS

The available calcium channel antagonists are clearly a highly successful and major therapeutic group of drugs directed primarily at cardiovascular diseases, particularly at hypertension. There are important lessons to be learned for the drug discovery process from these agents. First, these drugs were discovered before substantial knowledge existed of the voltage-gated calcium channel. Knowledge of channel classification or structure contributed nothing to this drug discovery process. It is clear, however, that these drugs have proved to be invaluable molecular tools with which to dissect channel localization, structure and function. In turn, this knowledge contributes to a better understanding of drug action, which translates in turn to more rational therapeutic applications. Second, the drugs were discovered in biological assay systems with a direct relationship between the measured response – vascular smooth muscle dilatation and tension – and the desired therapeutic goal of anti-anginal and anti-hypertensive activity. Third, the cardiovascular system is dominated by the activity of the Ca_v1 class of channel; hence, differential functional contributions by other classes of channel are relatively unimportant, and fine-tuning of drug action occurs within one channel class. Finally, the prototypical molecules all show state-dependent interactions that contributed significantly to an immediately observable desirable pharmacological profile. Had the molecules been state-independent in their actions they likely would have been non-selective and toxin-like in their properties, and further research may well have been discontinued.

It is likely that further development of existing compounds will, together with our expanding knowledge of Ca_v1 subtype pharmacology, yield therapeutically superior agents to those currently on the market. In particular, increased vascular selectivity could lead to agents directed at specific vascular beds including the cerebral, pulmonary and renal vasculature. Such selectivity of action applied to other systems may have relevance to the design of new human contraceptive agents. It has long been known that the fertilization response is a calcium-dependent process (Whitaker and Steinhardt, 1982) and more specifically that voltage-gated calcium channels are an important component of this response (Darszon et al., 1999). Evidence has been provided for roles for T-type and L-type channels (Benoff, 1998; Goodwin et al., 1997, 2000; Kanwar et al., 1993; Son et al., 2000; Perez-Reyes, 2003) as well as sperm-specific channels (CatSper1 and 2) that resemble voltage-gated calcium channels in their structure (Ren et al., 2001; Quill et al., 2001). Greater definition of the roles of these channels and the design of specific antagonists might represent a valuable new tool for both human and animal fertility control.

There are lessons to be learned from the discovery of L-type calcium channel drugs here that are immediately applicable to the discovery of drugs acting at other channel classes – N-type, P/Q-type, and R-type – where there are a large number of potential neuronal disorders from affective and cognitive disorders, epilepsy and pain, tinnitus and vertigo that should be susceptible to appropriately selective agents. However, the challenge is large. Multiple channel types are expressed in most neurons and with differential localization. Moreoever, there is not necessarily correlation in channel type and distribution between animal and human neurons. The molecular probes available have been essentially complex toxins that, while potent and selective, are often largely state-independent in their function and hence offer less subtlety of pharmacological action. Finally, the observed linkages between selective neuronal channel blockade and therapeutic endpoint are frequently still obscure. Much, however, can now be learned from transgenic studies that provide through over-expression or knockout new insights into calcium channel function (Muth et al., 2001; see Mori et al., this volume, and Saegusa and Tanabe, this volume).

Calcium channels do not operate in isolation – they are activated and modulated by a variety of signals. Hence, the signaling and topographical relationships between the channel and signal generator are also important to the definition of drug action (Zamponi and Snutch, 2003). It is likely, in fact, that specific associations exist between calcium channels and receptors and other macromolecular entities (Catterall, 2000). In the heart, activation of β-adrenoceptors and adenylate cyclase via the G protein system, with downstream potentiation of $Ca_v1.2$ channel activity, leads to increased cardiac rate and contractility. These component proteins – β_2-adrenoceptor, adenylyl cyclase, cAMP-dependent protein kinase, phosphatase PPA2 and the $Ca_v1.2c$ subunit – are all linked as one signaling unit (Davare et al., 2001). The exocytotic process by which calcium entry through calcium channels stimulates the release of neurotransmitters and hormones involves the SNARE protein complex that includes syntaxin, SNAP-25 and VAMP/synaptobrevin. The vesicle protein synaptotagmin serves as a calcium sensor and interacts with the calcium channel, as does syntaxin through a synaptic protein interaction (synprint) site on the intracellular II-III loop of $Ca_v2.1$ and $Ca_v2.2$ channels (Bruns and Jahn, 2002; Catterall, 2000; Sheng et al., 1998; see Piedras Rentería et al., this volume). It is likely that channels of the Ca_v1 class are also similarly regulated. The synprint site represents a site, albeit intracellular, for potential selective modulation of neurotransmitter and hormone release (see also Section 3.8).

A further consideration, and speculative, is that of the selectivity of action of drug action at calcium channels. The search for drugs typically focuses on potency and selectivity for a specific selected target, it being argued that selectivity and potency will be important, albeit not exclusively so, in reducing unwanted side effects of the drug. Two examples illustrate this point. The antihypertensive Posicor (mibefradil) is a calcium channel antagonist with actions at both L- and T-type channels, and its overall cardiovascular profile has been associated with both actions (Clozel et al., 1997; Marsh and Antman, 2000; Triggle, 1998; van der Vring, 1998). Although the drug was withdrawn from clinical use because of adverse drug reactions through cytochrome P4503A4, it has been suggested that the combination of L-and T-channel blockade does offer therapeutic advantage over selective L-type blockers in the management of stable angina pectoris (Thadani, 2002; Lee et al., 2002). A more complex example is that of ischemic brain damage, where calcium overload, a key player in the resultant cell death and destruction, is derived from number of discrete pathways and stores, including voltage-gated calcium channels, sodium channels, and excitatory amino acid-gated channels (Scriabine, 1997; Yu et al., 2001; Koroshetz and Moskowitz, 1996; Small et al., 1997; O'Neill et al., 1997). If it is important to block simultaneously all or several of these pathways, then either a "cocktail" of drugs or a drug(s) with multiple pharmacological activities must be administered. Such agents exist (*inter alia,* O'Neill et al., 1997; Brown et al., 1995), but they are not drugs readily sought in screening searches that emphasize selectivity and potency. This is an interesting challenge for stroke therapy, a pathology where close to some 200,000 die per year and many more are partially or totally incapacitated and for which drug therapy other than thrombolytic therapy has scarcely progressed over the past fifty years. A similar challenge may exist for antiarrhythmic therapy. Many disappointments in therapy have led some to conclude that, "..antiarrhythmic drug development strategy has slowly begun to evolve from a search for selective ion channel blockers to agents that exhibit multiple pharmacological actions in cardiac tissue" (Pugsley, 2002).

ACKNOWLEDGMENTS

My interest in calcium began in the late nineteen sixties when I wanted to know what made the smooth muscle systems that we used as pharmacological assays for our work on adrenergic and cholinergic receptor ligands contract. It was not a large step to seeking agents that might interfere with calcium mobilization. As so often happens in science, it was a moment whose time had come. But none of this work would have been possible were it not for the collaboration with as good a group of undergraduate and graduate students and post-doctoral fellows as one could hope for. They are too numerous to mention individually – but they know who they are and they all have made a significant impact in the calcium field.

REFERENCES

Abernethy, D. R., and N. M. Soldatov, Structure-functional diversity of human L-type Ca^{2+} channel: perspectives for new pharmacological targets,. J. Pharmacol. Exp. Therap. 300, 724-728 (2002)

Abascal, V. M., M. G. Larson, J. C. Evans, A. T. Blohm, K. Poli and D. Levy, Calcium antagonists and mortality risk in men and women with hypertension in the Framingham Heart Study, Ann. Int. Med. 158, 1882-1886 (1998).

Adachi-Akahane, S., and T. Nagao, Ca^{2+} channel antagonists and agonists, in, Pharmacology of Ionic Channel Function, Activators and Inhibitors, edited by M. Endo, Y. Kurachi and M. Mishira (Springer-Verlag, Berlin and Heidelberg, 2000), pp. 119-154.

Affolter, H., and R. Coronado, Sidedness of reconstituted calcium channels from muscle transverse tubules as determined by D600 and D890 blockade, Biophys. J. 49, 767-771 (1980).

Akaike, N., P. G. Kostyuk and Y. V. Osipchuk, Dihydropyridine-sensitive low threshold calcium channels in isolated rat hypothalamic neurons, J. Physiol. 412, 181-195 (1989).

Alden, K. J., and J. Garcia, Differential effect of gabapentin on neuronal and muscle currents. J. Pharmacol. Exp. Therap. 297, 727-735 (2001).

Alker, D., S. F. Campbell, P. E. Cross, R. A. Burges, A. J. Carter and D. G. Gardiner, Long-acting dihydropyridine calcium antagonists. 4. Synthesis and structure-activity relationships for a series of basic and non-basic derivatives of 2-[(2-aminoethoxy)methyl]-1,4-dihydropyridine calcium antagonists. J. Med. Chem. 33, 585-591 (1990).

Angelico, P., L. Guarneri, A. Leonardi and R. Testa, Vascular-selective effect of lercanidipine and other 1,4-dihydropyridines in isolated rabbit tissues, J. Pharm. Pharmacol., 51, 709-714 (1999).

Arnoult, C., M. Villazz and H. M. Forman, Pharmacological properties of the T-type Ca^{2+} current of mouse spermatogenic cells, Mol. Pharmacol. 53, 1104-1111 (1998).

Arrowsmith, J. E., S. F. Campbell, P. E. Cross, J. K. Stubbs, R. A. Burges, D. G. Gardiner and K. J. Blackburn, Long-acting dihydropyridine calcium antagonists. I. 2-Alkoxymethyl derivatives incorporating basic substituents, J. Med. Chem., 29, 1696-1702 (1986).

Ashcroft, F. M., Ion Channels and Disease (Academic Press, London, 2001).

Atwal, K. S., G. C. Rovnyak, J. Schwartz, S. Moreland, A. Hedberg, J. Gougolas, M. Malley and D. Floyd, Dihydropyrimidine calcium channel blockers: 2-heterosubstituted-4-aryl-1,4-dihydro-6-methyl-5-pyrimidinecarboxylic acid esters as potent mimics of dihydropyridines, J. Med. Chem. 33, 1510-1515 (1990a).

Atwal, K. S., G. C. Rovnyak, S. D. Kimball, D. M. Floyd, S. Moreland, N. N. Swanson, J. Z. Gougotas, J. Schwartz, K. M. Smillies and M. F. Malley, Dihydropyrimidine calcium channel blockers. 2. 3-Substituted 4-aryl-1.4-dihydro-6-methyl-5-pyrimidinecarboxylic acid esters as potent mimics of dihydropyridines, J. Med. Chem., 33, 2629-2635 (1990b).

Atwal, K. S., B. N. Swanson, S. E. Unger, D. M. Floyd, S. Morleand, A. Hedberg and B. C. O'Reilly, Dihydropyrimidine calcium channel blockers. 3. 3-Carbamoyl-4-aryl-1,2,3,4-tetrahydro-6-methyl-5-pyrimidinecarboxylic acid esters as orally effective antihypertensive drugs. J. Med. Chem. 34, 806-811 (1991).

Baindur, N., A. Rutledge and D. J. Triggle, A homologous series of permanently charged 1,4-dihydropyridine novel probes designed to localize drug binding sites on ion channels, J. Med. Chem. 36, 3743-3745 (1993).

Balwierczak, J., I.L. Grupp, G. Grupp and A. Schwartz, Effects of bepridil and diltiazem on [^{3}H]nitrendipine binding to canine cardiac sarcolemma. Potentiation of pharmacological effects of nitrendipine, J. Pharmacol. Exp. Therap., 237, 40-48 (1986).

Bangalore, R., N. Baindur, A. Rutledge, D. J. Triggle and R. S. Kass, L-Type calcium channels: asymmetrical intramembrane binding domain revealed by variable length, permanently charged 1,4-dihydropyridines, Mol. Pharmacol. 46, 660-666 (1994).

Bayer, R., R. Hennecke s, R. Kaufmann and R. Mannhold, Inotropic and electrophysiological actions of verapamil and D600 in mammalian myocardium, Naunyn-Schmied. Arch. Pharmacol. 290, 49-68 (1975).

Beedle, A. M., and G.W. Zamponi, Block of voltage-dependent calcium channels by aliphatic monoamines, Biophys. J. 79, 260-270 (2000).

Bean, B. P., Nitrendipine block of cardiac calcium channels. High-affinity binding to the inactivated state, Proc. Nat. Acad. Sci. USA 81, 6388-6392 (1984).

Bean, B. P., Two kinds of calcium channels in canine atrial cells. Differences in kinetics, selectivity and pharmacology, J. Gen. Physiol. 86, 1-30 (1985).

Bean, B. P., M. Sturek, A. Puga and K. M. Hermsmeyer, Calcium channels in muscle cells isolated from rat mesenteric arteries: modulation by dihydropyridine drugs, Circ. Res. 59, 229-235 (1986).

Bechem, M., S. Goldmann, R. Gross, S. Hallerman, S. Hebisch, J. Hutter, H.-P. Rounding, M. Schramm, J. Stoltefuss and A. Straub, A new type of calcium channel modulation by a novel class of 1,4-dihydropyridines, Life Sci. 68, 107-118 (1997).

Benoff, S., Voltage-dependent calcium channels in mammalian spermatozoa, Frontiers in Bioscience 3, 1220-1240 (1998).

Benzaquan, L. R., C. Brugnara, H. R. Byerssss, S. Gattani-Delli and J.A. Halperin, Clotrimazole inhibits cell proliferation *in vitro* and *in vivo*, Nature Med. 1, 534-540 (1995).

Berjukow, S., F. Gapp, S. Aczel, M.J. Sinneger, J. Mitterdorfer, H. Glossmann and S. Hering, Sequence differences between α1C and α1S Ca^{2+} channel subunits reveal structural determinants of a guarded and modulated benzothiazepinone receptor, J. Biol. Chem. 274, 6154-6160 (1999).

Bezprozvanny, I., and R.W. Tsien, Voltage-dependent blockade of diverse types of voltage-gated Ca^{2+} channels expressed in Xenopus oocytes by the Ca^{2+} channel antagonist mibefradil (Ro 40-5967), Mol. Pharmacol. 48, 540-549 (1995).

Bichet, D., C. Lecomte, J.-M. Sabatier, R. Felix and M. De Waard, Reversibility of the Ca^{2+} channel α-β subunit interaction, Biochem. Biophys. Res. Commun. 277, 729-735 (2000).

Bodi, I., H. Yamaguchi, M. Hara, M. He, A. Schwartz and G. Varadi, Molecular studies on the voltage-dependence of dihydropyridine action on L-type Ca^{2+} channels. J. Biol. Chem. 272, 24952-24960 (1997).

Bodi, I., S. E. Koch, H. Yamaguchi, G. P. Szigeti, A. Schwartz and G. Varadi, The role of region IVS5 of the human cardiac calcium channel in establishing inactivated channel conformation, J. Biol.Chem. 277, 20651-20659 (2002).

Bolger, G. T., P. Gengo, R. Klockowski, E. Luchowski, H. Siegel, R. A. Janis, A. M. Triggle and D. J. Triggle, Characterization of the binding of the Ca^{2+} channel antagonist [^{3}H]nitrendipine to guinea pig ileal smooth muscle, J. Pharmacol. Exp. Therap. 225, 291-310 (1983).

Bossert, F., and W. Vater, 1,4-Dihydropyridines – a basis for developing new drugs, Med. Res. Revs. 9, 291-324 (1989).

Brown, C. M., W. R. Rush and H. A. Colquhoun, Lifarizine: a blocker of inactivated voltage-dependent sodium channels and calcium channels with cerebral neuroprotective action, CNS Drug Reviews 1, 149-167 (1992).

Bryan, J. S., N. Davies, N. S. Gee, V. V. K. Dissonayake, G. S. Ratcliffe, D.C. Hornwell, C. O. Keen, A. I. Morrell, R. J. Oles, J.C. O'Toole, G. M. Perkins, L. Singh, N. Suman-Chauhan and J. O'Neill, Identification of novel ligands for the gabapentin binding site on the $\alpha_2\delta$ subunit of a calcium channel and their evaluation as anticonvulsant agents, J. Med. Chem. 41, 1838-1845 (1998).

Bruns, D., and R. Jahn, Molecular determinants of exocytosis, Pflug. Arch. Eur. J. Physiol. 443, 333-338, 2002.

Bucher, H. C., R. J. Cook, G. H.Guyatt, J.D. Lang, D. J. Cook, R. Hatala and D. L. Hunt, Effects of dietary calcium supplementation on blood pressure. A meta-analysis of randomized clinical trials, JAMA 275, 1016-1022 (1996).

Budriesi, R., G. Aicardi, G. Campana, S. Sapapinato,A. Zaza, A. Bisa, A. Rampa, P. Valenti and A. Chiarini, Selective cardiodepressant activity of fluodipine, a fluorenone-1,4-dihydropyridine derivative, Eur. J. Pharmacol. 359, 161-170, (1998).

Cai, D., J. G. Mulle and D.T. Yue, Inhibition of recombinant Ca^{2+} channels by benzothiazepines and phenylalkylamines: class-specific pharmacology and underlying molecular determinants, Mol. Pharmacol. 51, 872-881 (1997).

Campbell, A. K., Intracellular Calcium (Wiley, New York, 1983).

Carafoli, E., and C. Klee, editors, Calcium as A Cellular Regulator (Oxford University Press, New York, 1999).

Carafoli, E., Calcium signaling: a tale for all seasons, Proc. Nat. Acad. Sci. USA 99, 1115-1122 (2002).

Carmines, P. K., and L. G. Navar, Disparate effects of Ca channel blockade on afferent and efferent arteriolar responses to angiotensin II, Amer. J. Physiol. 256, F1015-F1020 (1989).

Carson, J. R., H. R. Almond, M.D. Brannen, R. J. Carmosin, S. F. Flaim, A. Gill, M. M. Gleason, S. L. Keely, D. W. Ludovici, P. M. Pitis, M. C. Rebarchak and F. J. Villeri, 2-Ethynylbenzenealkanamines. A new class of calcium entry blocker, J. Med. Chem. 31, 630-636 (1988).

Catterall, W. A., Structure and regulation of voltage-gated Ca^{2+} channels, Ann. Rev. Cell Dev. Biol. 16, 521-555 (2000).

Catterall, W. A., K. G. Chandy and G. A. Gutman, IUPHAR Compendium of Voltage-gated Ion Channels (IUPHAR Media, University of Leeds, United Kingdom, 2002).

Catterall, W. A., and J. Striessnig, Receptor sites for Ca^{2+} channel antagonists, Trends Pharmacol. Sci. 13: 256-262, 1992.

Cena, C., S. Visentin, A. Di Stilo, D. Boschi, R. Fruttero and A. Gasio, Studies on agents with mixed NO-dependent and calcium channel antagonistic vasodilating activators, Pharm. Res. 18, 157-165 (2002).

Chaisewikul, R., N. Baillie and A. G. Marson, Calcium antagonists as add-on therapy for drug-resistant epilepsy, Cochrane Database of systematic reviews, 4, CD002750, (2001).

Chalmers, J., 1999 World Health Organization and International Society of Hypertension guidelines for the management of hypertension, J. Hypertension 17, 151-183 (1999).

Chester, D. W., L. G. Herbette, R. P. Mason, A. F. Joslyn, D. J. Triggle and D. E. Koppel, Diffusion of dihydropyridine calcium channel antagonists in cardiac sarcolemmal lipid multilayers, Biophys. J. 52, 1021-1030 (1987).

Cho, H., M. Ueda, K. Shima, A. Mizuno, M. Hayashimatsu, Y. Ohnaka, Y. Takeuchi, M. Hamaguchi, K. Aisaka, T. Hidaka, M. Kawai, M. Takeda, T. Ishihara, K. Funahashi, F. Satoh, M. Morita and T. Noguchi, Dihydropyrimidines: novel calcium antagonists with potent and long-lasting vasodilative and antihypertensive activity, J. Med.Chem. 32, 2399-2406 (1989).

Clozel, J. P., E. A. Ertel and S. I. Ertel, Discovery and main pharmacological properties of mibefradil (Ro 40-5967), the first selective T-type calcium channel blocker, J. Hypertension 15, S17-S25 (1997).

Coburn, R. A., M. Wierzba, M. J. Suto, A. J. Solo, A. M. Triggle and D. J. Triggle, 1,4-Dihydropyridine antagonist activities at the calcium channel: a quantitative structure-activity approach, J. Med. Chem. 31, 2103-2107 (1988).

Cohen, C. J., E. A. Ertel, M. M. Smith, V. J. Venema, M. E. Adams and M. D. Liebowitz, High affinity block of myocardial L-type calcium channels by the spider toxin ω-agatoxin IIIA: advantages over 1,4-dihydropyridines, Mol. Pharmacol. 42, 947-951 (1992).

Cohen, C. J., S. Spires and D. van Skiver, Block of T-type Ca channels in guinea pig atrial cells by antiarrhythmic agents and Ca^{2+} channel antagonists, J. Gen. Physiol. 100, 703-728 (1992).

Cohen, C. J., R. T. McCarthy,P. Q. Barrett and H. Rasmussen, Ca channels in adrenal glomerulosar cells: K^+ and angiotensin II increase T-type Ca channel current, Proc. Nat. Acad. Sci. USA 85, 2412-2416 (1988).

Cognard, C., G. Romey, J.-P. Galizzi, M. Fosset and M. Lazdunski, Dihydropyridine-sensitive Ca^{2+} channels in mammalian muscle cells in culture: electrophysiologic properties and interaction with Ca^{2+} channel activators (Bay K 8644) and antagonist (PN 200 110), Proc. Nat. Acad. Sci. USA 83, 1518-1522 (1986).

Crespi, F., E. Vecchiato, C. Lazzarini , M. Andreoli and G. Gaviraghi, Evidence that lacidipine at nonsustained antihypertensive doses activate nitrogen monoxide system in the endothelium of salt-loaded Dahl-S rats, J. Cardiovas. Pharmacol. 39, 471-477 (2002).

Darszon, A., P. Labarca, T. Nishigaki and F. Espinosa, Ion channels in sperm physiology, Physiol. Revs. 79, 481-510 (1999).

Davare, A., V. Avdonin, D. D. Hall, E. M. Peden, A. Burette, R. J. Weinberg, M C. Horne, T. Hishi and J.W. Hell, A β_2 adrenergic receptor signaling complex assembled with the Ca^{2+} channel $Ca_V1.2$, Science, 293, 98-01, 2001.

DeWaard, M., M. Pragnell, and K. P. Campbell, Ca^{2+} channel regulation by a conserved β-subunit domain, Neuron, 13, 495-503 (1994).

de Weille, J. R., H. Schweitz, P. Moes, A. Tartor and M. Lazdunski, Calciseptine, a peptide isolated from black mamba venom is a specific blocker of the L-type calcium channel, Proc. Nat. Acad. Sci. USA 88, 2437-2440 (1991).

Dhein, S., A. Salameh, R. Berkels and W. Klaus, Dual mode of action of dihydropyridine calcium antagonists. A role for nitric oxide. Drugs 58, 397-404. (1999).

Dhein, S., Y. Zhao, S. Simsek, A. Salameh and W. Klaus, Actions of dihydropyridines in isolated mesenteric vascular beds, J. Cardiovas. Pharmacol. 26, 784-791 (1995).

Doyle, D. A., J. M. Cabral, R. A. Pfuetzner, A. Kuo, J. M. Gulbis, S. L. Cohen, B. T. Chait and R. M. McKinnon, The structure of the potassium channel: molecular basis of K^+ conduction and selectivity, Science, 280, 69-77 (1998).

During, F., V. E. Degtiar, M. Grabner, J. Striessnig, S. Herig and H. Glossmann, Transfer of L-type calcium channel IVS6 segment increases phenylalkylamine sensitivity of α_{1A}, J. Biol. Chem. 271, 11745-11749 (1996).

Enyeart, J. J., B. A. Biagi and B. Mlinar, Preferential block of T-type calcium channels by neuroleptics in neural crest-derived rat and human cell lines, Mol. Pharmacol. 42, 364-372 (1992).

Epstein, M., Role of a third generation calcium antagonist in the management of hypertension, Drugs 57(Suppl. 1), 1-10 (1999).

Epstein, M., editor, Calcium Antagonists in Clinical Medicine, 3rd Ed. (Hanley and Belfus, Philadelphia, 2002).

Epstein, M., Safety of calcium antagonists as antihypertensive agents: an update., In, Calcium Antagonists in Clinical Medicine, ed. M. Epstein, pp. 807-832, Hanley and Belfus, Philadelphia,PA (2002a).

Epstein, M., Recent landmark clinical trials: how do they modify the therapeutic paradigm? Amer .J. Hypertension, 15, 82S-84S (2002b).

Ertel, E. A., M. M. Smith, M. D. Leibowitz and C. J. Cohen, Isolation of myocardial L-type channel gating currents with the spider toxin ω-AgaIIIA, J. Gen. Physiol. 103, 731-753 (1994a).

Ertel, E. A., V. A. Warren, M. E. Adams, P. K. Griffin, C. J. Cohen and M. M. Smith, Type III ω-agatoxins: a family of probes for similar binding sites on L- and N-type calcium channels, Biochemistry 33, 5098-5108 (1994b).

Evans, B. E., K. E. Rittle, M. G. Bock, R. M. DiPardo, R. M. Friedinger, W. L. Whitter, G. F. Lindell, D. F. Veber, P. S. Anderson, R. S. L. Chang, V. J. Lotti, D. J. Cerino, T. B. Chen, P.J. Kling, K. A. Kunkel, J. P. Springer, J. Hirschfield, Methods for drug discovery: development of potent, selective orally effective cholecystokinin antagonists, J. Med. Chem. 31, 2235-2246 (1988).

Fainzilber, M., J. C. Lodder, R. C. Van den Schors, K. W. Li, Z. Yu, A. L. Burlingame, W. P. M. Geraerts and K. S. Kits, A novel hydrophobic ω-conotoxin blocks molluscan dihydropyridine-sensitive calcium channels, Biochemistry 35, 8748-8752 (1996).

Fan, J.-S., Y. Yuan and P. Palade, Kinetic effects of FPL 64176 on L-type Ca^{2+} channels in cardiac myocytes, Naunyn-Schmied. Arch. Pharmacol. 361, 465-476 (2000).

Fearon, I. M., S. G. Ball and C. Peers, Clotrimazole inhibits the recombinant cardiac L-type Ca^{2+} channel α_{1C} subunit, Brit. J. Pharmacol. 129, 547-555 (2000).

Ferrante, J., and D. J. Triggle, Drug- and disease-induced regulation of voltage-dependent Ca^{2+} channels, Pharmacol. Rev. 42, 29-44 (1990).

Ferrante, J., E. M. Luchowski, A. Rutledge and D. J. Triggle, Binding of a 1,4-dihydropyridine calcium channel activator, (-)S Bay K 8644, to cardiac preparations, Biochem. Biophys. Res. Commun. 158, 149-154 (1989).

Field, M. J., J. Hughes and L. Singh, Further evidence for the role of the $\alpha_2\delta$ subunit of voltage-dependent calcium channels in models of neuropathic pain, Brit. J. Pharmacol. 131, 282-286 (2000).

Flaim, S. F., M. T. Stranieri, M. T. Gill, J. R. Carson and M. D. Brannen, Effects of the novel calcium blocking agent, McN-5691, on cardiocirculatory dynamics and cardiac output distribution in conscious spontaneously hypertensive rats, J. Cardiovas. Pharmacol. 11, 489-500 (1988).

Fleckenstein, A., Calcium Antagonists in Heart and Smooth Muscle (Wiley and Sons, New York, NY, 1983a).

Fleckenstein, A., History of calcium antagonists, Circ. Res. 52(Suppl. I), 3-26, (1983b).

Fogari, R., G. D. Malamani, A. Zoppi, P. Preti, A. Vanasia, E. Rogari and A. Mugellini, Comparative effect of lercanidipine and nifedipine gastrointestinal therapeutic system on ankle volume and subcutaneous interstitial pressure in hypertensive patients: a double-blind, randomized parallel-group study, Curr.Ther. Res. 61, 850-862 (2000).

Fossheim, R., A. Joslyn, A. J. Solo, E. Luchowski, A. Rutledge and D. J. Triggle, Crystal structures and pharmacologic activities of 1,4-dihydropyridine calcium channel antagonists of the isobutyl methyl-2,6-dimethyl-4-(substitutedphenyl)-1,4-dihydropyridine-3,5-dicarboxylate (nisoldipine) series, J. Med. Chem, 31, 300-305 (1988).

Fossheim, R., K. Svarteng, A. Mostad, C. Romming, E. Shefter and D. J. Triggle, Crystal structures of and pharmacological activities of calcium channel antagonists, J. Med. Chem. 25, 126-131, (1981).

Fujita, M., S. Ito, A. Ota, N. Katu, K. Yamamoto, Y. Kawashima, H. Yamauchi and J. Iwao, Synthesis and Ca^{2+} antagonistic activity of 2-[[2-(aminoalkyl)oxy]-5-methoxyphenyl]-3,4-dihydro-4-methyl-3-oxo-2H-1,4-benzothiazines, J. Med. Chem. 33, 1898-1905 (1990).

Furberg, C. D., B. M. Psaty and J.V. Meyer, Nifedipine: dose-related increase in mortality in patients with coronary heart disease, Circulation 92, 1326-1331 (1995).

Furchgott, R. F., and J.V. Zawadaki, The obligatory role of endothelial cells in the relaxation of arterial smooth muscle by acetylcholine, Nature 288, 373-379 (1980).

Furukawa, T., T. Yamakawa, T. Midera, T. Sagawa, Y. Mori and T.Nukada, Selectivities of dihydropyridine derivatives in blocking Ca^{2+} channel subtypes expressed in Xenopus oocytes, J. Pharmacol. Exp. Therap. 291, 464-473 (1999).

Gage, M. J., S. G. Rane, G. H. Hockerman and T. J. Smith, The virally encoded fungal toxin KP4 specifically blocks L-type voltage-gated calcium channels, Mol. Pharmacol. 61, 936-944 (2002).

Garcia, M. C., Z. Hernandez-Gallegos, J. Escamilla and J. A. Sandez, Calciseptine, a Ca^{2+} channel blocker has agonist actions on L-type Ca^{2+} currents of frog and mammalian skeletal muscle, J. Mem. Biol. 184, 121-129 (2001).

Galizzi, J.-P., M. Fosset, G. Romey, P. Laduron and M. Lazdunski, Neuroleptics of the diphenylbutylpiperidine series are potent calcium channel inhibitors, Proc. Nat. Acad. Sci. USA 83, 7513-7517 (1986).

Gee, N. S., J. P. Brown, V. U. K. Dissanayakee, J. Offord, R. Thurlow and G. N. Woodruff, The novel anticonvulsant drug, gabapentin (Neurontin) binds to the $\alpha_2\delta$ subunit of a calcium channel, J. Biol. Chem. 271, 5768-5776 (1996).

Glossmann, H., D. R. Ferry, F. Luebbecke, R. Mewes and F. Hofmann, Calcium channels: indirect identification with radioligand binding studies, Trends Pharmacol. Sci 3, 431-437 (1987).

Glossmann, H., S. Hering, A. Savchenko, W. Berger, K. Friedrich, M.L. Garcia, M. A. Goetz, J. M. Liesch, D. L. Zink and G, J. Kaczorowski, A light stabilizer (Tinuvin 770) that elutes from polypropylene plastic tubes is a potent L-type Ca^{2+} channel blocker, Proc. Nat. Acad. Sci. USA 90, 9253-9257 (1993).

Godfraind, T., S. Salamone, C. Dessy, B. Verhelst, R. Dion and J. C. Schoevaerts, Selectivity scale of calcium antagonists in the human cardiovascular system based on *in vitro* studies, J. Cardiovas. Pharmacol. 20 (Suppl. 5), S34-S41 (1992).

Goldmann, S., and J. Stoltefuss, 1,4-Dihydropyridines: effects of chirality and conformation on the calcium antagonist and calcium agonist activities. Angew. Chemie Internat. Ed. 30, 1559-1578 (1991).

Goodwin, L. O., N. B. Leeds, I. Hurley, F. S. Mandel, R. G. Pergolizzi and S. Benoff, Isolation and characterization of the primary structure of testis-specific L-type calcium channel: implications for contraception, Mol. Human Reprod. 3, 255-268 (1997).

Goodwin, L. O., D. S. Karabinus, R. G. Pergolizzi and S. Benoff, L-Type voltage-dependent calcium channel α1C subunit mRNA is present in ejaculated human spermatazoa, Mol. Human Reprod. 6, 127-136 (2000).

Gopalakrishnan, M., and D. J. Triggle, The regulation of receptors, ion channels and G proteins in congestive heart failure, Cardiovas. Drug Revs. 8, 255-302 (1990).

Grabner, M., Z. Wang, S. Hering, J. Striessnig and H. Glossmann, Transfer of 1,4-dihydropyridine sensitivity for L-type to class A (B1) calcium channels, Neuron 16, 2017-218 (1996).

Hall, S. T., S. M. Harding and G. L. Evans, Clinical pharmacology of lacidipine, J. Cardiovas. Pharmacol. 17(suppl. 4), S9-S13 (1991).

Hatton, D. C., Q. Yue and D. A. McCarron, Mechanism of calcium's effects on blood pressure, Seminar Nephrol. 15, 593-602 (1995).

Hawthorn, M., J. Ferrante, E. Luchowski, A. Rutledge, X.-Y. Wei and D. J. Triggle, The actions of peppermint oil and menthol on calcium channel-dependent processes in intestinal, neuronal and cardiac preparations, J. Alim. Pharmacol. 2, 101-118 (1988).

He, M., I. Body, G. Mikali and A. Schwartz, Motif IIIS5 of L-type calcium channels is involved in the dihydropyridine binding site, J. Biol. Chem., 272, 3629-3633 (1997).

Hemmateenejad, B., R. Miri, M. Akhond, and M. Shamsipur, Quantitative structure-activity relationship study of recently synthesized 1,4-dihydropyridine calcium channel antagonists. Application of the Hansch analysis method. Arch. Pharm. 10, 472-480 (2002).

Henry, P. D., Antiperoxidative actions of calcium antagonists and atherogenesis. J. Cardiovas. Pharmacol. 18(Suppl.1), S6-S10 (1991).

Herbette, L. G., R. P. Mason, G. Gaviraghi, T. N. Tulenko and R. P.Mason, The molecular basis for lacidipine's unique pharmacokinetics: optimal hydrophobicity results in membrane interactions that may facilitate the treatment of hypertension, J. Cardiovas. Pharmacol. 23(Suppl.5), S16-S25 (1994).

Herbette, L. G., G. Gaviraghi, . T. N. Tulenko and R.P. Mason, Molecular interactions between lacidipine and biological membranes. J. Hypertension 11, S13-S19 (1993)..

Herbette, L. G., Y. M. H. Vanterve and D. G. Rhodes, Interaction of 1,4-dihydropyridine calcium channel antagonists with biological membranes: lipid bilayer partitioning could occur before drug binding to receptors, J. Mol. Cell. Cardiol. 21, 187-201 (1989).

Herbette, L. G., M. Vechiarelli and A. Leonardi, Lercandipine: short plasma half-life, long duration of action: a molecular model to rationalize its pharmacological properties, J. Cardiovas. Pharmacol. 29 (Suppl.1), S19-S24,(1997).

Hering, S., S. Aczel, M. Grabner, F. Doring, S. Berjukow, J. Mitterdorfer, M. J. Sinnegger, J .Striessnig, V. E. Degtiar, Z. Wang and H. Glossmann, Transfer of high sensitivity for benzothiazepines for L-type class A (B1) calcium channels, J. Biol. Chem. 271, 24471-24475 (1996).

Hering, S., S. Aczel, R.L. Kraus, S. Berjukow, J. Striessnig and E. N. Timin, Molecular mechanisms of use-dependent calcium channel block by phenylalkylamines: role of inactivation, Proc. Nat. Acad. Sci.USA 94, 13323-13328 (1997).

Hering, S., A. Savchenko, C. Strubing, M. Lakitsch and J. Striessnig, Extracellualr localization of the benzothiazepine binding domain of L-type Ca^{2+} channels, Mol. Pharmacol. 43, 820-826 (1993).

Herold, P., J. W. Herzig, T.Leubert, P. Zbinder, W. Fuhrer, S. Stutz, K. Schenkar, M. Meier and G. Rihs, 5-Methyl-6-phenyl-1,3,5,6-tetrahydro-3,6-methano-1,5-benzodiazocino-2,4-dione (BA #1879); representative of a novel class of purely calcium sensitizing agents, J. Med. Chem. 38, 2946- (1995).

Herzog, S., Ca^{2+} channel activation by CGP 48506, a new positive inotropic benzodiazocine agent, Eur. J. Pharmacol. 295, 113-117 (1996).
Heilbrunn, L. V., An Outline of General Physiology, Editions 1,2 and 3 (W. B. Sunders, Philadelphia ,PA, 1937-1952).
Hescheler, J. D., D. Pelzer, G. Trube and W. Trautwein, Does the organic calcium channel blocker D600 act from inside or outside on the cardiac cell membrane?, Pflug.Arch. 393, 287-291 (1982).
Hille, B., Local anesthetics: hydrophilic and hydrophobic pathways for the drug-receptor reaction, J. Gen. Physiol. 69, 497-515 (1977).
Hockerman, G. H., B. D. Johnson, T. Scheuer and W. A. Catterall, Molecular determinants of high affinity block of L-type calcium channels, J. Biol. Chem. 270, 22119-22122 (1995).
Hockerman, G. H., B.D. Johnson, M. R. Abbott, T. Scheuer and W. A. Catterall, Molecular determinants of high affinity phenylalklylamine block of L-type calcium channels in transmembrane segment IIIS6 in the pore region of the $\alpha 1$ subunit, J. Biol. Chem. 272, 18759-18765 (1997a).
Hockerman, G. H., B. Z. Peterson, B. D. Johnson, and W. A. Catterall, Molecular determinants of drug binding and action on L-type calcium channels, Ann. Rev. Pharmacol. Toxicol., 37, 361-396, (1997b).
Hockerman, G. H., B. Z. Peterson, E. Sharp, T. N. Tanada, T. Scheuer and W. A. Catterall, Construction of a high-affinity receptor site for dihydropyridine agonists and antagonists by single amino acid substitutions in a non-L-type Ca^{2+} channel, Proc. Nat. Acad. Sci USA 94, 14906-14911 (1997c).
Hof, R. P., U.T. Ruegg, A. Hof and A. Vogel, Stereoselectivity of the Ca^{2+} channel: opposite action of the enantiomers of a 1,4-dihydropyridine, J. Cardiovas. Pharmacol. 7, 689-693 (1985).
Hofmann, F., L. Lacinova and N. Klugbauer, Voltage-dependent calcium channels: from structure to function, Rev. Physiol. Biochem. Pharmacol. 139, 33-87 (1999).
Holtje, H.-D., and S. A. Marrer, Molecular graphics study on structure-activity relationships of calcium antagonistic and agonistic 1,4-dihydropyridines, J. Comp. Aid. Mol. Des., 1,20-30 (1987).
Holtje, H.-D., Molecular modeling studies on 1,4-dihydropyridines acting at the Ca channel. Quant. Struct. Act. Relat. 11, 224-232 (1992).
Hondeghem, L. M., and B. G. Katzung, Anti-arrhythmic agents: the modulated receptor mechanism of action of sodium and calcium channel blocking drugs, Ann. Rev. Pharmacol. Toxicol., 24, 387-423 (1985).
Honore, E., T. Amedee, C. Martin, C. Daquet, C. Mironneau amd J. Mironneau, Calcium channel current and its sensitivity to (+)isradipine in cultured pregnant rat myometrial cells, Pflug.Arch. 414, 477-483 (1989).
Hu, H., and E. Marban, Isofluran-specific inhibition of L-type calcium channels by dihydropyridines is independent of isoform-specific gating properties, Mol. Pharmacol. 53, 902-907 (1998).
Huddart, H., R.J. Smith, P. D. Langston, A. M. Hetherington and T. A. Mansfield, Is abscisic acid a universally active calcium agonist? New Phytol. 104, 161-173 (1986).
Hunter, A. J., Calcium antagonists: their role in neuroprotection, Int. Rev. Neurobiol. 40, 95-108 (1997).
Hwang, S. B., M. N. Chang, M. L. Garcia, O. Q. Han, L. Huang, V. F. King, G. J. Kaczorowski and R. J. Winquist, L-652,469 – a dual receptor antagonist of platelet activating factor and dihydropyridines from *Tussilago farfara L.*, Eur. J.Pharmacol. 141, 269-281 (1987).
Iqbal, N., M. R. Akula, D. Vu, W. C. Matowe, C.-A. McEwen, M. W. Wolowyck and E. E. Knaus, Synthesis, rotamer orientation, and calcium channel modulation activities of alkyl- and 2-phenethyl-1,4-dihydro-2,6-dimethyl-3-nitro-4(3- or 6-substituted-2-pyridyl)-5-pyridine carboxylates, J. Med. Chem. 41, 1827-1837 (1998).
Isaev, N. K., E. V. Stelmashook, U. Dirnagl, N. A. Andreeva, L. Manuhova, V. S. Vorobjev, I. N. Sharanova, V. B. Skrebitsky, I. V. Victorov, J. Katchanov, M. Weig and D. B. Zorov, Neuroprotective effects of the antifungal drug clotrimazole, Neroscience 113, 47-53 (2002).
Ishibashi, A., and D. Horii, Effect of MCI-176, a new calcium antagonist on the calcium-induced contractions of isolated porcine coronary arteries, Jap. J. Pharmacol. 43, 234-236 (1987).
Ito, H., N. Klugbauer and F. Hofmann, Transfer of the high affinity dihydropyridine sensitivity from L-type to non-L-type calcium channel, Mol. Pharmacol. 52, 735-740 (1997).
Janis, R. A., and D. J. Triggle, New developments in Ca^{2+} antagonists, J. Med. Chem. 26, 775-785 (1983).
Janis, R. A., and D. J.Triggle, 1,4-Dihydropyridine Ca^{2+} channel antagonists and agonists. A comparison of binding characteristics with pharmacology, Drug Dev. Res. 4, 257-285 (1984a).
Janis, R. A., and D. J. Triggle, Calcium channel antagonists: new perspectives from the radioligand binding assay, in, Modern Methods in Pharmacology, edited by N. Back and S. Spector, Volume 2, Alan R. Liss, New York, pp. 1-28 (1984b).
Janis, R. A., P. Silver and D. J. Triggle, Drug action and cellular calcium regulation, Adv. Drug Res., 16, 309-591 (1987).
Jeziorski, M. C., R. M. Greenberg, K. S. Clark and A. V. Anderson, Cloning and functional expression of a voltage-gated calcium channel alpha1 subunit from jellyfish, J. Biol. Chem., 273, 22792-22799 (1998).

Jim, K., and D. J. Triggle, The action of praziquantel and 1-methyladenine in guinea pig ileal longitudinal muscle, Can. J. Physiol. Pharmacol. 57, 1460-1462, (1979).

Johnson, B., G. H. Hockerman, T. Scheuer and W. A. Catterall, Distinct mutations in transmembrane segment IVS6 on block of L-type calcium channels by structurally similar phenylalkylamines, Mol. Pharmacol. 50, 1388-1400 (1996).

Kaneda, M., and N. Akaike, Low-threshold Ca current in isolated amygdaloid neurons in the rat, Brain Res. 497, 187-190 (1989).

Kanwar, U., R.J. K.Anand and S. N. Sanyal, The effect of nifedipine, a calcium channel blocker, on human spermatazoal functions, Contraception 48, 453-470 (1993).

Kass, R. S., Voltage-dependent modulation of cardiac calcium channel current of optical isomers of Bay K 8644. Implications for channel gating, Circ. Res. 61(Suppl. I), 1-5 (1987).

Kass, R. S., J. P. Arena and S. Chin, Block of L-type calcium channels by charged dihydropyridines, J. Gen. Physiol. 98, 63-75 (1991).

Kawashima, T., T. Hayakawa, Y. Miwa, H. Oohka, T. Suzuki, K. Fujimotyo, T. Ogina and Z. Chen, Structure and hypotensive relationships of tetrandrine derivatives in stroke-prone spontaneously hypertensive rats, Gen. Pharmacol. 21, 343-347 (1990).

Kenny, B. A., S. Fraser, A. T. Kilpatrick and M. Spedding, Selective antagonism of calcium channel activators by fluspirilene, Brit. J. Pharmacol. 100, 211-216 (1990).

King, V. F., M.L. Garcia, D. Himmel, J. P. Reuben, Y. K. Lem, J. X. Pan, G. Q. Han and G. J. Kaczorowski, Interaction of tetrandrine with slowly inactivating calcium channels. Characterization of calcium channel modulation by an alkaloid of Chinese medicinal herb origin, J. Biol. Chem. 263, 2238-2244 (1988).

King, V. F., M. L. Garcia, J. L. Shevell, R. S. Slaughter and G. J. Kaczorowski, Substituted diphenylbutylpiperidines bind to a unique high affinity site on the L-type calcium channel, J. Biol. Chem. 264, 5633-5644, (1989).

Kini, R. M., R. A. Caldwell, Q. Y. Wu, C. M. Baumgartan, J. J. Feher and H. J. Evans, Flanking proline residues identify the L-type Ca^{2+} channel binding site of calciseptine on FS2, Biochemistry 37, 9058-9063 (1998)

Kizer, J. R., and S. E. Kimmerl, Epidemiologic review of the calcium channel blocker drugs: an up-to-date perspective on the proposed hazards, Arch. Int. Med. 161, 1145-1158 (2001).

Kleinbloesem, B. H., P. van Brummelen, M. Danhof, H. Tabor, J. Urquart and D. D. Briemer, Rate of increase in the plasma concentration of nifedipine as a major determinant of its hemodynamic effects in humans, Clin. Pharmacol. Ther. 41, 26-30 (1987).

Kohn, A. B., P. A. V. Anderson, J. M. Roberts-Misterly and R. M. Greenberg, Schistosome calcium channel β–subunits. Unusual modulatory effects and the potential role in the action of the antischistosomal drug praziquantel, J. Biol. Chem. 276, 36873-36876 (2001).

Koroshetz, W. J., and M. A. Moskowitz, emerging treatments for stroke in humans, Trends Pharmacol. Sciences, 17, 227-233, 1996.

Koschak, A., D. Reimer, I. Huber, M. Grabner, H. Glossmann, J. Engel and J. Striessnig, α_{1D} ($Ca_V1.3$) Subunits can form L-type Ca^{2+} channel activity at negative voltages, J. Biol. Chem. 276, 22100-22106 (2001).

Koshy, S., and G. L. Bakris, Therapeutic approaches to achieve desired blood pressure goals: focus on calcium channel blockers, Cardiovas. Drugs Therap. 14, 295-301 (2000).

Kraus, R. L., S. Hering, M. Grabner, D. Ostler and J. Striessnig, Molecular mechanism of diltiazem interaction with L-type Ca^{2+} channels, J. Biol. Chem., 273, 27205-27212 (1998).

Kuga, T., J. I. Sadoshima, H. Tomoike, N. Akaike and M. Nakamura, Actions of Ca^{2+} antagonists on two types of Ca^{2+} channels in rat aorta smooth muscle cells in primary culture, Circ. Res. 67, 469-480, (1990).

Kumar, P. H., S. C. Stotz, R. Paramasiuappa, A. M. Reedle and G. W. Zamponi, Synthesis and evaluation of a new class of nifedipine analogs with T-type calcium channel binding activity, Mol. Pharmacol. 61, 649-658 (2002).

Kunze, D., and D. Rampe, Characterization of the effects of a new Ca^{2+} channel activator FPL 64176 on GH3 cells, Mol. Pharmacol. 42, 666-670 (1992).

Kurokawa, J., S. Adachi-Akahane and T. Nagao, 1,5-Benzothiazepine binding domain is located on the extracellular side of the cardiac L-type Ca^{2+} channel, Mol. Pharmacol. 51, 262-268 (1997).

Kwon, Y.-W., and D. J. Triggle, Chiral aspects of drug action at ion channels, Chirality 3, 393-404 (1991).

Kwon, Y.-W., Q. Zhong, X.-Y. Wei and D. J. Triggle, The interaction of 1,4-dihydropyridines bearing a 2-(2-aminoethylthio)methyl substituents at voltage-dependent Ca^{2+} channels of smooth muscle, cardiac muscle and neuronal tissues. Naunyn-Schmied. Arch. Pharmacol. 341, 128-136 (1990).

Lacinova, L., N. Klugbauer and F. Hofmann, State-dependent and isoform-dependent interaction of Isradipine with the α_{1C} L-type calcium channel, Pflug. Arch. 440-, 50-60 (2000).

Lamas, S., P. A. Marsdenn, G. K. Li, P. Tempst and T. Michel, Endothelial nitric oxide synthase, Proc. Nat. Acad. Sci. USA 89, 6348-6352 (1992).

Langs, D., and D. J. Triggle, Conformational features of calcium channel agonist and antagonist analysis of nifedipine, Mol. Pharmacol. 27, 544-548 (1985).

Langs, D. A., P. D. Strong and D. J. Triggle, Receptor model for the molecular basis of tissue selectivity of 1,4-dihdyropyridine calcium channel drugs, J. Comp. Aid. Mol. Design 4, 215-230 (1990).

Langs, D. A., Y.-W. Kwan, P. D. strong and D. J. Triggle, Molecular level model for the agonist/antagonist selectivity of the 1,4-dihydropyridine calcium channel receptor, J. Comp. Aid. Mol. Design, 5, 95-106 (1991).

Lee, D. S., S. Goodman, D. M.Dean, J. Lenis, P. Ma, P. B. Gervais, A. Langer and PRIDE Investigators, Randomized comparison of T-type versus L-type calcium channel blockade on exercise duration in stable angina: Results of the Posicor Reduction of Ischemia During Exercise (PRIDE) trial, Amer. Heart J. 144, 60-67, 2002.

Leenen, F. H., Clinical relevance of 24-hour blood pressure control by 1,4-dihydropyridines, Amer. J. Hypertension, 9, 97S-104S (1996).

Leenen, F. H., Sympathoinhibitory and sympathoexcitatory effects of dihydropyridine calcium antagonists, in, Calcium Antagonists in Clinical Medicine, Edited M. Epstein, 3rd Ed. pp. 771-794, Hanley and Belfus, Philadelphia,PA (2002).

Lipkind, G. M., and H. A. Fozzard, Molecular modeling of interactions of dihydropyridines and phenylalkylamines with the inner pore of the L-type Ca^{2+} channel. Mol Pharmacol 63, 499-511 (2003).

Lipscombe, D., L-Type calcium channels. Highs and new lows, Circ. Res. 90, 933-935 (2002).

Loriand, G., C. Mirronneau, J. Mirronneau and P. Pacaud, Two types of calcium currents in single smooth muscle cells from rat portal vein, J. Physiol. 412, 333-349 (1989).

Loutzenhiser, K., and R. Loutzenhiser, Angiotensin II-induced Ca^{2+} influx in renal afferent and efferent arterioles: differing roles of voltage-gated and store-operated Ca^{2+} entry. Circ. Res. 87, 551-557 (2000).

Luchowski, E. M., F. Yousif, D. J. Triggle, S. C. Mauser, J. G. Sarmiento and R. A. Janis, The effect of metal cations and calmodulin antagonists on [^{3}H]nitrendipine binding in smooth and cardiac muscle, J. Pharmacol. Exp. Therap. 230, 607-619 (1984).

Luft, F. C., and D. A. McCarron, Heterogeneity of hypertension: The diverse role of electrolyte intake, Ann. Rev. Med. 42, 347-355 (1991).

Luo, Z. D., S. R. Chaplan, E. S. Higuera, L. S. Sorkin, K. A. Stauderman, M. E. Williams and T. L. Yaksh, Upregulation of dorsal root ganglion $\alpha_2\delta$ calcium channel subunit and its correlation with allodynia in spinal nerve-injured rats, J. Neurosci., 21, 1868-1875 (2001).

Mager, P., R. A. Coburn, A. J. Solo, D. J. Triggle and H. Rotke, QSAR, diagnostic statistics and molecular modeling of 1,4-dihydropyridine calcium channel antagonists: a difficult road ahead, Drug Dev. Res. 8, 273-289 (1992).

Mahon, N., and W. K. McKenna, Calcium channel blockers in cardiac failure, Prog. Cardiovas. Drugs 41, 191-206, 1998.

Mancini, G. B. J., Antiatherosclerotic effects of calcium channel blockers, Prog. Cardiovas. Diseases, 45, 1-20, 2002.

Marais, E., N. Klugbauer and F. Hofmann, Calcium channel $\alpha_2\delta$ subunits – structure and gabapentin binding, Mol. Pharmacol. 59, 1243-1248 (2002).

Marsh. J. D., and E. M. Antman, T-Type calcium channel blockade in the management of chronic ischemic heart disease, Cardiovas. Drugs and Therap. 14, 459-461, 2000.

Mason, R. P., D. G. Rhodes and L. G. Herbette, Reevaluating equilibrium and kinetic binding parameters for lipophilic drugs based on a structural model for drug interaction with biological membranes, J. Med. Chem., 34, 869-877 (1991).

Masumiya, H., S. Shijuki, H. Tanka and K. Shigneobu, Inhibition of myocardial L- and T-type Ca^{2+} currents by efonidipine: possible mechanism for its chronotropic effect, Eur. J. Pharmacol. 349, 351-357 (1998).

Masumiya, H., Y. Tanaka, H. Tanaka and K. Shigenobu, Inhibition of T-type and L-type Ca^{2+} currents by aranidipine, a novel dihydropyridine Ca^{2+} antagonist, Pharmacology, 61, 57-61 (2000).

McCarthy, R. T., Nimodipine block of L-type calcium channels in dorsal root ganglion cells, In, Nimodipine and Central Nervous System Function: New Vistas, edited by J. Traber and W. Gispen, Stuttgart, Schattauer 1989 ,pp. 35-51 (1989).

McCarthy, R. T., and C. J. Cohen, The enantiomers of Bay K 8644 have differential effects on Ca^{2+} channel gating in rat anterior pituitary cells, Biophy. J., 49, 432A (1986).

McDonald, T. F., D. Peltzer and W. Trautwein, Cat ventricle muscle treated with D600: characteristics of calcium channel block and unblock, J. Physiol. 352, 217-244 (1984).

McKechnie, K., P.G. Killinback, I. Naga, S. E. O'Connor, G. W. Smith, D.G. Wattam, E. Wells, Y. M. Whitehead and G. E. Williams, Calcium channel activator properties in a novel non-dihydropyridine FPL 64176, Brit. J. Pharmacol. 98, 673P (1989).

McLean, M. J., Gabapentin, Epilepsia 36, S73-S86 (1995).

McLellan, K. J., and B. Jarvis, Lercanidipine: a review of its use in hypertension, Drugs 60, 1123-1140 (2000).

Messerli, F. H., Calcium antagonists in hypertension: from hemodynamics to outcomes, Amer. J. Hypertension 15, 94S-97S, 2002.

Mintz, I. M., Block of Ca channels in rat central neurons by the spider toxin w-Aga-IIIA, J Neurosci 14, 2844-2853 (1994).

Mitterdorfer, J., M. J. Sinnegger, M. Grabner, J. Striessnig and H. Glossmann, Coordination of Ca^{2+} by the pore region glutamates is essential for high affinity dihydropyridine binding to the cardiac Ca^{2+} channel α_1 subunit, Biochemistry 34, 9350-9355 (1995).

Mitterdorfer, J., Z. Wang,, M. J. Sinnegger, S. Hering , J. Striessnig and H. Glossmann, Two amino acid residues in the IIIS5 segment of L-type calcium channels differentially contribute to 1,4-dihydropyridine sensitivity, J. Biol. Chem., 271, 30330-30335 (1996).

Mitterdorfer, J., M. Grabner, R. L. Knaus, S. Hering, H. Prinz, H. Glossmann and J. Striessnig, Molecular basis of drug interaction with L-type Ca^{2+} channels, J. Bioeng. Biomem., 30, 319-334 (1998).

Montoyler, J., and M. Follath, Calcium channel blockers in treatment of hypertension, Prog. Cardiovas. Dis. 44, 207-216 (2001).

Morel, N., and T. Godfraind, Prolonged depolarization increases the pharmacological effects of dihydropyridines and their binding affinity for calcium channels of vascular smooth muscle, Eur. J.Pharmacol. 243, 711-715 (1987).

Morel, N., and T. Godfraind, Selective modulation by membrane potential of the interaction of some calcium entry blockers with calcium channels in mesenteric artery, Brit. J. Pharmacol. 95, 252-258 (1988).

Morel, N., and T. Godfraind, Characterization in rat aorta of the binding sites responsible for the blockade of noradrenaline-evoked calcium entry by nisoldipine, Brit. J. Pharmacol. 102, 4657-477 (1991).

Morel, N., V. Buryi, O. Feren, J.-P. Gomez, M.-O. Christen and T. Godfraind, The action of calcium channel blockers on recombinant L-type calcium channel α_1 subunits, Brit. J. Pharmacol. 125, 1005-1012 (1998).

Mori, T., H. Takase, K. Toide, T. Hirano, T. Kambe, N. Nakayama and A. Schwartz, Pranidipine, a 1,4-dihydropyridine calcium channel blocker that enhances nitric oxide-induced vascular relaxation, Cardiovas. Drug Revs., 19, 1-8 (2001).

Motoike, H. K., I. Bodi, H. Nakayam, A. Schwartz and G. Varadi, A region in IVS5 of the human cardiac L-type calcium channel is required for the use-dependent block by phenylalkylamines and benzothiazepines, J. Biol.Chem. 274, 9409-9420 (1999).

Murai, Y., H. Uneyama, H. Ishibashi, K. Takahama and N. Akaike, Preferential inhibition of L- and N-type calcium channels in the rat hippocampal neurons by cilnidipine, Brain Res. 854, 6-10 (2000).

Murakami, N., O. Nakagawasai, S. Fujii, M. Hosono, S. Hozumi, A. Esashi, R. Taniguchi, T. Okamura, T. Suzuki, H.Sasano, T. Yanagisawa, K. Tasano and K. Kisara, Antinociceptive effect of cilnidipine, a novel N-type calcium channel antagonist, Brain Res. 868, 123-127 (2000).

Muth, J. N., G. Varadi and A. Schwartz, Use of transgenic mice to study voltage-dependent Ca^{2+} channels, Trends Pharmacol. Sci. 22, 526-532 (2001).

Nakayama, H., M. Taki, J. Striessnig, H. Glossmann, W. A. Catterall and Y. Kanaoka, Identification of 1,4-dihydropyridine binding regions within the α_1 subunit of skeletal muscle Ca^{2+} channels by photoaffinity labeling with diazipine, Proc. Nat. Acad. Sci. USA 88, 9203-9201 (1991).

Nakayama, K., Y. Nozawa and K. Fukuta, Allosteric interaction of semotiadil fumarate, a novel benzothiazine, with 1,4-dihydropyridines, phenylalkylamines, and 1,5-benzothiazepines at the Ca^{2+} channel antagonist binding sites in canine skeletal muscle membranes, J. Cardiovas. Pharmacol. 23, 731-740 (1994).

Nakayama, K., Pharmacological profile and therapeutic perspective of semotiadil fumarate (SD-3211): a novel benzothiazine Ca^{2+} antagonist, Cardiovas. Drug Revs., 14, 97-133 (1996).

Neal, B., (on behalf of the blood pressure lowering treatment trialists collaboration), Effect of ACE inhibitors, calcium antagonists and other blood pressure lowering drugs: results of prospectively designed overview of randomized trials, Lancet 356, 1955-1964 (2000).

Nelson, M. T., and J. F. Worley, Dihydropyridine inhibition of single calcium channels and contraction in rabbit mesenteric artery depends on voltage, J. Physiol. 412, 65-91 (1989).

Nokin, P., M. Clinet, P. Polster, P. Beaufort, L. Meysmans, J. Gougat and P. Chatelain, SR 33557, a novel calcium antagonist: interaction with $[^3H]$(+/-)nitrendipine and $[^3H]$-(-)desmethoxyverapamil binding sites in cerebral membranes, Naunyn-Schmied. Arch. Pharmacol. 339, 31-36 (1989).

Ogawa, T., A. Nakazatu, K. Tschchida and K. Hatayama, Synthesis and antihypertensive activities of new 1,4-dihydropyridine derivatives containing nitrooxyalklylester moieties at the 3- and 5-positions, Chem. Pharm. Bull., 41, 1049-1054 (1993).

Okamura, Y., Functional expression of a protochordate L-type Ca^{2+} channel, Biophys. J. 76, A340 (1999).

Olivera, B. M., G. P. Miljanich, J. Ramachandran and M. E. Adams, Calcium channel diversity and neurotransmitter release: the *omega*-conotoxins and the *omega*-agatoxins, Ann. Rev. Biochem. 63, 823-867 (1994).

O'Neill, M. J., C. P. Bath, C.P. Dell, C. A. Hickes, J. Gilmore, S. J. Ambler, M. A. Ward and D. Bleakman, Effects of Ca^{2+} and Na^{+} channel inhibitors *in vitro* and in global cerebral ischemia *in vivo*, Eur. J. Pharmacol. 332, 121-131 (1997).

Opie, L. H., and R. Scholl, Evidence-based evaluation of calcium channel blockers for hypertension, J. Amer. Coll. Cardiol. 39, 315-322 (2002).

Opie, L. H., S. Yusuf and W. Kubler, Current status of safety and efficacy of calcium channel blockers in cardiovascular diseases. A critical analysis based on 100 studies. Prog. Cardiovas. Dis. 43, 171-196 (2000).

Packer, M., Pathophysiological mechanisms underlying the adverse effects of calcium channel-blocking drugs in patients with chronic heart failure, Circulation 80(Suppl. IV), 59-67 (1989).

Pahor, M., J. M. Guralnik, M. C. Corti, D. J. Foley, P. Carbonin and R.J. Haulik, Long-term survival and use of anti-hypertensive medications in older persons, J. Am. Geriatric Soc. 43, 1191-1197 (1998).

Patchett, A. A., and R. P. Nargund, Privileged structures: an update, Ann. Rep. Med. Chem. 35, 289-298 (2000).

Patmore, L., G. P. Duncan, B. Clarke, A. J. Anderson, R. Greenhouse and J. G. Pfuster, RS 30026: a potent and effective calcium channel agonist, Brit. J. Pharmacol. 99, 687-694 (1990).

Perchenet, L., A. Bernardeau and E. A. Ertel, Pharmacological properties of $Ca_V3.2$, a low voltage-activated Ca^{2+} channel cloned from human heart, Naunyn-Schmied. Arch. Pharmacol. 362, 596-599 (2000).

Peri, R., S. Padmanabhan, A. Rutledge, S. Singh and D. J. Triggle, Permanently charged chiral 1,4-dihydropyridines, molecular probes of L-type calcium channels: synthesis and pharmacological characterization of methyl ω-trimethylalklyammonium –1,4-dihydro-2,6-dimethyl-4-(3-nitrophenyl)-3,5-pyridinedicarboxylate iodide, calcium channel antagonists, J. Med. Chem. 43, 2906-2914.

Peterson, B. Z., T. N. Tanada and W. A. Cattterall, Molecular determinants of high affinity 1,4-dihydropyridine binding in L-type calcium channels, J. Biol. Chem. 271, 5293-5296 (1996).

Peterson, B. Z., B. D. Johnson, G.H. Hockerman, M.Acheson and W. A. Catterall, Analysis of the dihydropyridine receptor of L-type calcium channels by alanine -scanning mutagenesis, J. Biol. Chem. 272, 18752-18758 (1997).

Platzer, J., J. Engel, A. Schrott-Fischer, K.Stephen, S. Bova, H. Chen, H. Zheng and J. Striessnig, Congenital deafness and sinoatrial node dysfunction in mice lacking class D L-type Ca^{2+} channels, Cell 102, 89-97 (2000).

Pragnell, M., M. De Waard, Y. Mori, T. Tanabe, T. P. Snutch and K. P. Campbell, Calcium channel β-subunit binds to a conserved motif in the I-II cytoplasmic loop in the α_1-subunit, Nature 368, 67-70 (1994).

Priego, J., M.A. Gonzales-Morales, F. J. Cillero, M. Villaroya, C. Sunkel, M. F. de Casa-Juana, M. G. Lopez, C. R. Artalejo, R. de Pascual and A. G. Garcia, PCA 50941, a novel Ca^{2+} channel agonist, Eur. J. Pharmacol. 243, 25-34 (1993).

Psaty, B. M., S. R. Heckbert, T. D. Koepsell, D. S. Siscovick, T. E. Raghunathan, N. S. Weiss, F. R. Rosendaal, R. N. Lemaitre, N. L. Smith and P. W. Wahl, The risk of myocardial infarction associated with antihypertensive drug therapies, JAMA 274, 620-625 (1995).

Pugsley, M. K., Antiarrhythmic drug development: historical review and future perspective, Drug Dev. Res. 55, 3-17 (2002).

Putney, J. W., editor, Calcium Signaling (CRC Press, Boca Raton, Florida, 2000).

Qar, J., J. Barhanin, G. Romey, R. Henning, U. Lerch, R. Okonomopulos, H. Urbach and M. Lazdunski, A novel high affinity class of Ca^{2+} channel blockers, Mol. Pharmacol. 33, 363-369 (1988).

Quill, T. A., D. Ren, D. E. Clapham and D. L. Garbers, A voltage-gated ion channel expressed specifically in spermatozoa, Proc. Nat. Acad. Sci. USA 98, 12527-12531 (2001).

Raftery, E. B., Cardiovascular drug withdrawal syndromes: a potential problem with calcium antagonists? Drugs 28, 371-374 (1984).

Ragsdale, D. S., J. C. McPhee, T. Scheuer and W. A. Catterall, Molecular determinants of state-dependent block of Na^{+} channels by local anesthetics, Science 265, 1724-1728 (1994).

Rampe, D., and A.E. Lacerda, A new site for the activation of cardiac calcium channels defined by the non-dihydropyridine FPL 64176, J. Pharmacol. Exp. Therap. 259, 982-987 (1991).

Rampe, D., and D. J. Triggle, New ligands for L-type channels, Trends Pharmacol. Sci, 11, 112-115 (1990).

Rampe, D., and D. J. Triggle, New synthetic ligands for L-type voltage-gated calcium channels, Prog. Drug Res. 40, 191-238 (1993)

Rampe, D., B. Anderson, V. Raoier-Pryor, T. Li and R. C. Dage, Comparison of the *in vitro* and *in vivo* cardiovascular effects of two structurally distinct Ca^{2+} channel activators, Bay K 8644 and FPL 64176, J. Pharmacol. Exp. Therap. 265, 1125-1130 (1993).

Rampe, D., T. Poder, Z.-Y. Zhao and W. P. Schilling, Calcium channel agonist and antagonist binding in a highly enriched sarcolemmal preparation obtained from canine ventricle, J. Cardiovas. Pharmacol. 13, 547-566 (1989).

Rampe, D., A. Skattebol, D. J. Triggle and A. M. Brown, Effects of McN-6186 on voltage-dependent Ca^{++} channels in heart and pituitary cells, J. Pharmacol. Exp.Therap. 248, 164-169 (1989).

Rampe, D., D. J. Triggle and A. M. Brown, Electrophysiologic and biochemical studies on the putative Ca^{++} channel blocker MDL 12,330A in an endocrine cell J. Pharmacol. Exp.Therap. 243, 402-407 (1987).

Randle, J. C. R., A. Lambet, N. Najel, Ca^{2+} channel inhibition by a new dihydropyridine derivative, S11568 and its enantiomers S12967 and S12968, Eur. J.Pharmacol. 60, 523-525 (1990).

Redman, C. A., A. Robertson, P. G. Fallon, J. Modha, J.R. Kusel, M. J. Doenhoff and R. J. Martin, Praziquantel: an urgent and exciting challenge, Parasitol. Today 12, 14-220 (1996).

Regulla, S., T. Schneider, W. Nastainczyk, H. E. Meyer and F. Hofmann, Identification of the site of interaction of the dihydropyridine channel blockers nitrendipine and azidopine with the calcium channel α_1 subunit, EMBO J 10, 45-49 (1991).

Resnick, L. M., Calciotropic hormones in salt-sensitive essential hypertension: 1,25-dihydroxyvitamin D and parathyroid hypertensive factor, J. Hypertension 12(Suppl 1): S3-S((1994).

Ren, D., B. Navarro, G. Perez, A. C. Jackson, S. Hsu, Q. Shi, J. L. Tilly and D.E. Clapham, A sperm ion channel required for sperm motility and male fertility, Nature 413, 603-609 (2001).

Rhee, J.-S., H. Ishibashi and N. Akaike, Calcium channels in the GABAergic presynaptic nerve terminals projecting to Meynert neurons in the rat, J. Neurochem. 72, 800-807 (199).

Richard, S., S. Diochot, J. Nargeot, M. Baldy-Moulinier and J. Valmier, Inhibition of T-type calcium currents by dihydropyridines in mouse embryonic dorsal root ganglion neurons, Neurosci. Lett. 132, 229-234 (1991).

Ringer, S., A further contribution regarding the influence of the different constituents of the blood on the contractions of the heart, J. Physiol. 4, 29-42 (1893).

Robertson, J. I. S., Which antihypertensive drug classes have been shown to be beneficial? What are these benefits? A critique of antihypertensive treatment trials. Cardiovas. Drugs Therap. 14, 357-366 (2000).

Rojstaczer, N., and D. J. Triggle, Ca^{2+} antagonists as antioxidants, Cardiovas. Drug Revs. 12, 70-84 (1994).

Rojstaczer, N., and D. J. Triggle, Structure-function relationship of calcium antagonists. Effect on oxidative modification of low density lipoprotein, Biochem. Pharmacol. 51, 141-150 (1996).

Romanin, C., K. Seydl, H. Glossmann and H. Schindler, The dihydropyridine niguldipine inhibits T- type Ca^{2+} currents in atrial myocytes, Pflug. Arch 420, 410-412 (1992).

Rose, M. A., and P. C. Kam, Gabapentin: pharmacology and its use in pain management, Anesthesia 57, 451-462 (2002).

Rosner, H., L. Rubin and A. Kestenbaum, Gabapentin adjunctive therapy in neuropathic pain states, Clin. J. Pain 12, 56-58 (1996).

Roulett, J.-B., H. Xue, J. Chapman, P. McDougal, C. M. Roulett and D. A. McCarron, Farnesyl analogs inhibit vasoconstriction in animal and human arteries, J. Clin. Invest. 97, 2384-2390 (1996).

Roulett, J.-B., U.C. Luft, H. Xue, J. Chapman, R. Bychkov, C. M. Roulett, F.C. Luft, H. Haller and D. A. McCarron, Farnesol inhibits L-type Ca^{2+}channels in vascular smooth muscle cells, J. Biol. Chem. 272, 32240-32446 (1997).

Roulett, J.-B., R.L. Spaetgens, T. Burlingame, Z.-P. Feng and G. W. Zamponi, Modulation of neuronal voltage-gated calcium channels by farnesol, J. Biol. Chem. 274, 25439-25466 (1999).

Rovnyak, G. C., N. Anderson, J. Gougotas, A. Holberg, S. Kimball, M. Malley, S. Moreland, M. Porubean and A. Pudzianowski, Studies directed towards ascertaining the active conformation of 1,4-dihydropyridine calcium entry blockers, J. Med. Chem. 31, 936-944 (1988).

Rovnyak, G. C., K. S. Atwal, A. Hedberg, S. D. Kimball, S. Moreland, J. Z. Gougotas, B. C. O'Reilly, J. Schwartz and M. F. Malley, Dihydropyrimidine calcium channel blockers. 4. Basic 3-substituted –4-aryl-1,4-dihydropyrimidine-5-carboxylic acid esters. Potent antihypertensive agents, J.Med.Chem. 35, 3254-3263 (1992).

Rovnyak, G. C., S. D. Kimball, B.Beyer, G. Cucinotta, J.D. DiMario, J. Gougotas, A. Headberg, M. Madley, J..P. McCarthy, R. Zhang and S. Moreland, Calcium entry blockers and activators: conformational and structural determinants of dihydropyrimidine calcium channel modulators, J. Med. Chem. 38, 119-129 (1995).

Sabbatini, M., A. Leonardi, R. Testa, L. Vitaioli and F. Amenta, Effect of calcium antagonists on glomerular arterioles in spontaneously hypertensive rats, Hypertension, 35, 775-779 (200).

Sanguinetti, M. C., and R. S. Kass, Voltage-dependent block of calcium channel current in calf cardiac Purkinje fibers by dihydropyridine calcium antagonists, Circ. Res. 55, 336-348 (1984).

Santi, B. M., A. Darszon and A. Hernandez-Cruz, A dihydropyridine-sensitive T-type Ca^{2+} current is the main Ca^{2+} current carrier in mouse primary spermatocytes, Amer. J. Physiol. 271, C1583-C1593 (1996).

Schleifer, K.-J., Steroseletive characterization of the 1,4-dihydropyridine binding site at L-type calcium channels in the resting state and the opened/inactivated state, J. Med. Chem. 42, 2204-2211, 1999.

Schleifer, K.-J., and E. Tot, CoMFA, CoMSIA, and GRID/GOLPE studies on calcium entry blocking 1,4-dihydropyridines. Quant. Struct. Act. Relat. 21, 239-248 (2002).

Schmid, A., G. Romey, J. Barbarin and M. Lazdunski, SR 33557, an indolizinesulfone blocker of Ca2+ channels: identification of receptor sites and analysis of is mode of action, Mol. Pharmacol. 35, 766-773 (1989)

Schramm, M., G. Thomas and G. Frankowiak, Novel dihydropyridines with positive inotropic action through activation of Ca^{2+} channels, Nature 303, 535-537 (1983).

Schuster, A., L. Lacinova, N. Klugbauer, H. Ito, L. Birnbaumer and F. Hofmann, The IVS6 segment of the L-type calcium channel is critical for the action of dihydropyridines and phenylalkylamines, EMBO J 15, 2365-2370 (1996).

Schweitz, H., C. Heurteaux, P. Bois, D. Moinier, G. Romey and M. Lazdunski, Calcicludine, a venom peptide of the Kunitz-type protease inhibitor family is a potent blocker of high-threshold Ca^{2+} channels with a high affinity for L-type channels in cerebellar granule neurons, Proc. Nat. Acad. Sci. USA 91, 878-882, (1994).

Scriabine, A., Calcium channel antagonists as neuroprotective agents, In, Neuroprotection in CNS diseases, editors P. R. Bar and M. F. Beal, pp. 27-51. Dekker, New York, 1997.

Seidel, W., H. Meyer, L. Born, S. Kazda and W. Dompert, in QSAR Strategies Des. Bioact. Compd. Proc. Eur. Symp. Quant. Struct.-Act. Relat. 5th. (VCH Weinheim, 1985) pp. 366-369.

Seydl, K., D. Kimball, H. Schindler and C. Romanin, The benzazepine/benzothiazepine binding domain of the cardiac L-type Ca^{2+} channel is accessible only from the extracellular side, Pflug. Arch. 424, 552-554, (1993).

Shan, R., and E. E. Knaus, The design of (-)-(S)-2-nitroxyethyl-1,4-dihydro-2,6-dimethyl-3-nitro-4-(2-trifluoromethylphenyl)pyridine-5-carboxylate: a cardioselective positve inotropic derivative of Bay K 8644, Bioorg,. Med.Chem.Lett 9, 2613-2614 (1999).

Sheng, Z.-H., R. E. Westenbroek and W. A. Catterall, Physical link and functional coupling of presynaptic calcium channels and the synaptic docking/fusion machinery, J. Bioenergetics and Biomembranes, 30, 335-345, 1998.

Sleight, P., Calcium antagonists during and after myocardial infarction, Amer. Heart J. 135, 406-413 (1998).

Sheps, S. G., The Sixth Report of the Joint National Committee on Prevention, Detection, Evaluation and Treatment of high blood pressure, Arch. Int. Med. 157, 2413-2446 (1997).

Singh, B. N., S. Baky and K. Nademanee, Second-generation calcium antagonists: search for greater selectivity and versatility, Amer. J. Cardiol. 55, 214B-221B, 1985.

Singh, V., J. Christiana and W. H. Frishman, How to use calcium antagonists in hypertension. Putting the JNC-VI guidelines into practice, Drugs 58, 579-587 (1999).

Sinnegger, M. J., Z. Wang, M. Grabner, S. Hering, J. Striessnig, H. Glossmann and J. Mitterdorfer, Nine L-type amino acid residues confer full 1, 4-dihydropyridine sensitivity to a neuronal calcium channel α_{1A} subunit, J. Biol. Chem. 272, 27686-27693 (1997).

Small, D. L., R. Monette, A. M. Buchan and P. Morley, Identification of calcium channels involved in neuronal injury in rat hippocampal slices subjected to oxygen and glucose deprivation, Brain Res. 753, 209-218, 1997.

Soltadov, N. M., A. Bouron and H. Reuter, Different voltage-dependent inhibition by dihydropyridines of human Ca^{2+} channel splice variants, J. Biol. Chem. 270, 10540-10543 (1995).

Son, W.-Y., J.-H. Lee, J.-H. Lee and C.-T. Han, Acrosome reaction of human spermatozoa is mainly mediated by α1H-type calcium channels, Mol. Human Reprod. 6, 893-897 (2000).

Spagnoli, A., and G. Tognoni, 'Cerebroactive' drugs. Clinical pharmacology and therapeutic role in cerebrovascular disorders, Drugs, 26, 44-69 (1983).

Spedding, M., B. Kenny and P. Chatelain, New drug binding sites in Ca^{2+} channels, Trends Pharmacol. Sci. 16, 139-142 (1995).

Stason, W. B., C. H. Schmid, D. Niedzwieecki, G.W. Whiting, J.F. Caubet,D. Luo, S.D. Ross and T.C. Chalmers, Safety of nifedipine in patients with hypertension: a meta-analysis, Hypertension 30, 7-14 (1997).

Staessen, J. A., R. Fagard, L. Thijs and the Systolic Hypertension in Europe (Syst-Eur) Trial Investigators, Randomized double-blind comparison of placebo and active treatment for older patients with isolated systolic hypertension, Lancet 350, 757-764 (1997).

Staessen, J. A., J-G. Wang and L. Thijs, Calcium-channel blockade and cardiovascular prognosis: recent evidence from clinical outcome trails, Amer. J. Hypertension, 15, 85S-93S (2002).

Stanton, H. C., L. B. Rosenberger, R. C. Hanson, J. S. Fleming and G. Poindexter, Pharmacology of BMY 20064, a potent Ca^{2+} entry blocker and selective alpha-adrenoceptor antagonist, J. Cardiovas. Pharmacol. 11, 387-395 (1988).

Stefani, A., F. Spadoni, P. Giacomini, F. Lavaroni and G. Bernardi, The effects of gabapentin on different ligand- and voltage-gated currents in isolated cortical neurons, Epilepsy Res. 43, 239-248 (2001).

Stengel, W., M. Jainz and K. Andreas, Different potencies of dihydropyridine derivatives in blocking T-type but not L-type Ca^{2+} channels in neuroblastoma-glioma cells, Eur. J. Pharmacol. 342, 339-345 (1998).

Stotz, S. C., R. L. Spaetgens and G.W. Zamponi, Block of voltage-dependent calcium channel by the green Mamba toxin calcicludine, J. Mem. Biol. 174, 157-165 (2000).

Striessnig, J., E. Meusburger, M. Graber, H. G. Knaus, H. Glossmann, J. Kaiser, B. Scholkens, R. Becker, W. Linzon and R. Hering, Evidence for a distinct Ca^{2+} antagonist receptor for the novel benzothiazine compound HOE 166, Naunyn-Schmied. Arch. Pharmacol. 337, 331-340 (1988).

Striessnig, J., H. Glossman and W. A. Catterall, Identification of a phenylalkylamine binding region with the α_1 subunit of skeletal muscle Ca^{2+} channels, Proc. Nat. Acad. Sci. USA 87, 9108-9112 (1990).

Striessnig, J., B. J.Murphy and W. A. Catterall, Dihydropyridine receptor for L-type Ca^{2+} channels: identification of binding domains for [^{3}H](+)PN 200 110 and [^{3}H]azidopine with the α_1 subunit, Proc. Nat. Acad. Sci. USA 88, 10769-10773 (1991).

Striessnig, J., M.Grabner, J. Mitterdorfer, S. Hering, M. J. Sinnegger and H.Glossmann, Structural basis of drug binding to L Ca^{2+} channels, Trends Pharmacol. Sci 19, 108-115 (1998).

Striessnig, J., Pharmacology, structure and function of cardiac L-type calcium channels, Cell. Physiol. Biochem. 9, 242-269 (1999).

Sumimoto, K., M. Hiraka and H. Kuriyama, Characterization of [^{3}H]nifedipine binding to intact vascular smooth muscle cells, Amer. J. Physiol. 254, C45-C52 (1988).

Sun, J. P., and D J. Triggle, Calcium channel antagonists: cardiovascular selectivity of action. J. Pharmacol. Exp. Therap. 274, 419-428 (1995).

Sutton, K. G., and T. P. Snutch, Gabapentin: a novel analgesic targeting voltage-gated calcium channels. Drug Dev. Res. 54, 167-172 (2002).

Takahashi, K., and N. Akaike, Calcium antagonist effects on low-threshold (T-type) calcium current in rat isolated hippocampal CA1 pyramidal neurons, J. Pharmacol. Exp. Therap. 256, 169-175 (1991).

Takenaka, T., O. Inagaki, M. Terai, M. Asano and K. Kubo, New 1,4-dihydropyridines with vasoconstrictor action through activation of Ca^{++} channels, J. Pharmacol. Exp. Therap. 244, 699-708 (1988).

Talajic, M., M. Nayebpour, W. Jing and S. Nattel, Frequency-dependent effects of diltiazem at the atrioventricular node during experimental atrial fibrillation, Circulation 80, 380-389 (1989).

Tang, S., G. Mikala, A. Bachinski, A. Yatani, G. Varadi and A. Schwartz, Molecular localization of ion selectivity sites within the pore of a human L-type cardiac calcium channel, J. Biol. Chem. 13026-13029 (1993).

Tang, S., A. Yatani, A. Bahinski, Y. Mori, and A. Schwartz, Molecular localization of regions in the L-type calcium channel critical for dihydropyridine action, Neuron 11, 1013-1021 (1993).

Tarasenko, A. N., P. G. Kostyuk, A. V. Eremin and D. S. Isaev, Two types of low-voltage-activated Ca^{2+} channels in neurons of rat laterodorsal thalamic nucleus, J. Physiol. 499, 77-86 (1997).

Taylor, C. P., The role of Gabapentin, In, New Trends in Epilepsy Management, Ed. D. Chadwick (Royal Society of Medicines Services, London, 1993), pp. 13-40.

Taylor, C. P., N. S. Gee, T. Z. Su, J. D. Kocsis, D. F. Welby, J. P. Brown, D. J. Dooley, P. Boden and L. Singh, A summary of mechanistic hypotheses of gabapentin pharmacology, Epilepsy Res. 29, 233-249 (1998).

Thadani, U., Selective L-type, T-type and nonspecific calcium-channel blockers for stable angina pectoris, Amer. Heart J. 144, 8-10, 2002.

Thomas, G. D., W. Zhang and R. Victor, Nitric oxide deficiency as a cause of clinical hypertension: promising new drug targets for refractory hypertension, JAMA 285, 2055-2057 (2001).

Thomas, G. P., M. Karmazyn, A. C. Zygmunt, C. Antzelevitch and N. Narayan, The antifungal antibiotic clotrimazole potently inhibits L-type calcium current in guinea-pig ventricular myocytes, Brit. J. Pharmacol. 126, 1531-1533 (1991).

Todd, P. A., and P. Benfield, Flunarizine: a reappraisal of its pharmacological properties and therapeutic use in neurological disorders, Drugs 38, 481-499 (1989).

Triggle, A. M., E. Shefter and D. J. Triggle, Crystal structures of calcium antagonists: 2,6-dimethyl-3,5-dicarbomethoxy-4-(2-nitro-, 3-cyano-, 4-(dimethylamino)- and 2,3,4,5,6,-pentafluorphenyl)-1,4-dihydropyridine, J. Med. Chem. 23, 1442-1445 (1980).

Triggle, D. J., Effects of calcium on excitable membranes and neurotransmitter action, Prog. Surf. Mem. Sci. 5, 267-331 (1972).

Triggle, D. J., Calcium antagonists in atherosclerosis, Cardiovas. Drug Revs. 6, 320-335, (1988).

Triggle, D. J., Structure-function correlations of 1,4-dihydropyridine calcium channel antagonists and activators, in Molecular and Cellular Mechanisms of Antiarrhythmic Agents edited by L. M. Hondeghem (Futura Pub. Mt. Kiscoe, NY, 1989), pp. 269-291.

Triggle, D. J., Calcium antagonists. History and perspective, Stroke 21 (Suppl. IV), 49-58 (1990).

Triggle, D. J., Calcium antagonists, in Cardiovascular Pharmacology, edited by M. Antonaccio (Raven Press, NY 1990), pp. 107-160.

Triggle, D. J., On the other hand: the stereoselectivity of drug action at ion channels, Chirality 6, 58-62 (1994).

Triggle, D. J., The physiological and pharmacological significance of cardiovascular T-type, voltage-gated calcium channels, Amer. J. Hypertension 11, 80S-87S, 1998.

Triggle, D. J., The pharmacology of ion channels with particular reference to voltage-gated calcium channels, Eur. J. Pharmacol. 375, 311-325 (1999).

Triggle, D. J., Mechanisms of action of calcium antagonists, in Calcium Antagonists in Clinical Medicine, edited by M. Epstein, Third edition (Hanley and Belfus, Philadelphia, PA, 2002) pp.1-32.

Triggle, D. J., Chirality in drug design and development, in, Chirality in Natural and Applied Science, eds. W. J. Lough and I. W. Wainer, pp. 109-138. Blackwell, London (2002a).

Triggle, D. J., The 1,4-dihydropyridine nucleus: a pharmacophoric template, Mini-Rev. Med. Chem 3, 137-147 (2003).

Triggle, D. J., M. Hawthorn, M. Gopalakrishnan, A. Minarini, S. Avery, A. Rutledge, R. Bangalore, W. Zheng, Synthetic organic ligands active at voltage-gated calcium channels, Ann. New York Acad. Sci. 635, 123-138 (1991).

Triggle, D. J., D. A. Langs and R. A. Janis, Ca^{2+} channel ligands: structure-function relationship of the 1,4-dihydropyridines, Med.Res. Revs. 9, 123-180 (1989).

Uneyama, H., H. Uchida, T. Kanda, R. Yoshimoto and N. Aikake, Selectivity of dihydropyridines for cardiac L-type and sympathetic N-type Ca^{2+} channels, Eur. J. Pharmacol. 373, 93-100 (1999).

Vo, D., J.-T. Nguyen, C.-A. McEwen, R. Shan and E. E. Knaus, Synthesis, calcium channel agonist-antagonist modulatory effects and nitric oxide release studies of [3-(benzenesulfonyl)furoxan-4-yloxy]alkyl-1,4-dihydro-2,6-diemthyl-5-nitro-4-(2-trifluoromethylphenyl)benzofurazan-4-yl-1, 2-,3-, or 4-pyriidyl)-3-pyridine carboxylates Drug Dev. Res. 56, 1-16 (2002).

Van den Kerchoff, W., and L. R. Drewes, Transfer of nimodipine and another calcium antagonist across the regional blood-brain barrier and their regional distribution *in vivo*, in, The Diagnosis and Treatment of Senile Dementia, edited by M. Bergerer and L. Reichberg (Springer-Verlag, Berlin and Heidelberg, 1989), pp. 308-321.

van Der Vring, J. A., T. J. Cleophas, E.E. vander Wall, and M.G. Niemeyer (on behalf of the T-channel calcium channel blocker study group), T-Channel-selective calcium channel blockade: a review of published data and therapeutic profile, Current Ther. Res. 59, 754-761, 1998.

Verkhratsky, A., P. W. Landfield and O. Thibault, Ca^{2+} and neuronal pathology, Eur. J. Pharmacol. 447, 115-296 (2002).

Voradi, G., Y. Mori, G. Mikala and A. Schwartz, Molecular determinants of Ca^{2+} channel function and drug action, Trends Pharmacol. Sci. 16, 43-49 (1995).

Wang, M., J. Offord, D. L. Oxender and T.-Z. Su, Structural requirements of the calcium channel subunit $\alpha_2\delta$ for gabapentin binding, Biochem. J. 342, 313-320 (1999).

Watanabe, T., H. Kalasz, H. Yabana, A. Kumiyasu, P.L. Vaghy, K. Naito, H. Nakayama and A. Schwartz, Azidobutyrylclentiazem, a new photoactivatable benzothiazine binding sites in the α_1 subunit of the skeletal muscle calcium channel, FEBS Lett, 334, 261-264 (1993).

Watanabe, T. X., Y. Itahara, H. Kuroda, Y. N.Chen, T. Kimura and S. Sakakibara, Smooth muscle relaxing and hypotensive activities of synthetic calciseptine and the homologous snake venom peptide FS2, Jap. J. Pharmacol. 68, 305-313 (1995).

Wei, X.-Y., E. M. Luchowski, A. Rutledge, C. M. Su, and D. J. Triggle, Pharmacologic and radioligand binding analysis of the actions of 1,4-dihydropyridine activator-antagonist pairs in smooth muscle, J. Pharmacol. Exp. Therap., 239, 144-153 (1986).

Wei, X.-Y., A. Rutlege and D. J. Triggle, Voltage-dependent binding of 1,4-dihydropyridine Ca^{2+} channel antagonists and activators in cultured neonatal rat ventricular myocytes, Mol. Pharmacol. 35, 541-552 (1989).

Wei, X.-Y., S. Pan, H.Kim. T. Schneider, E.Perez-Reyes and L. Birnbaumer, Molecular determinants of cardiac Ca^{2+} channel pharmacology. Subunit requirements for the high affinity and allosteric regulation of dihydropyridine binding, J. Biol. Chem. 270, 27106-27111 (1997).

Welling, A., Y. W. Kwan,E. Bosse, V. Flockerzi, F. Hofmann and R. S. Kass, Subunit-dependent modulation of recombinant L-type calcium channels. Molecular basis for dihydropyridine tissue selectivity, Circ. Res. 73, 974-980 (1993).

Welling, A., A. Ludwig, S. Zimmer, N. Klugbauer, V. Flockerzi and F. Hofmann, Alternatively spliced IS6 segments of the alpha1C gene determine the tissue-specific dihydropyridine sensitivity of the cardiac and vascular smooth muscle L-type Ca^{2+} channels, Circ. Res. 81, 526-532 (1997).

Whitaker, M. J., and R. A. Steinhardt, Ionic regulation of egg activation, Q. Revs. Biophys. 15, 593-666 (1982).

Wiley, R. A., and D. H. Rich, Peptidomimetics derived from natural products, Med. Res. Revs, 13, 327-384 (1993).

White, P. J., Calcium channels in higher plants. Biochim. Biophys. Acta 1465, 171-189 (2000).

Williams, M. E., M.S. Washburn, M. Hans, A. Urrutia, P.F.Brust, P. Prodanovich, M.M. Harpold and I Stauderman, Structure and functional characterizarion of a novel human low v oltage-activated calcium channel, J. Neurochem. 72, 791-799 (1999).

Worley, J. R., J. W. Datmer and M.T. nelson, Single nisoldipine-sensitive calcium channels in smooth muscle cells isolated from rabbit mesenteric artery, Proc. Nat. Acad. Sci. USA 83, 5746-5750 (1987).

Xu, W., and D. Lipscombe, Neuronal $Ca_V1.3a$, L-type channels activate at relatively hyperpolarized membrane potentials and are incompletely inhibited by dihydropyridines, J. Neurosci. 21, 5944-5951 (2001).

Yamaguchi, S., Y. Okamura, T. Nagao and S. Adachi-Akahane, Serine residues in the IIS5-S6 linker of the L-type Ca^{++} channel α_{1C} subunit is the critical determinant of the action of dihydropyridine Ca^{2+} channel agonist, J. Biol. Chem. 275, 4150-4151 (2000).

Yasuda, O., S. Morimoto, Y. Chen, B. Jiang, T. Kimura, S. Satickibera, E. Koh, K. Fukuo, S. Kitano and T. Ogihara, Calciseptine binding to a 1,4-dihydropyridine recognition site of the L-type calcium channel of rat synaptosomal membranes, Biochem. Biophys. Res. Commun. 194, 587-594 (1993)

Yatani, A., C. L. Seidel, J. Allen and A. M. Brown, Whole cell and single channel calcium currents of isolated smooth muscle cells from saphenous vein, Circ. Res. 60, 523-533 (1987).

Yellen, G., The voltage-gated potassium channels and their relatives, Nature 419, 35-42 (2002).

Young, K., S. Lin, E. Lee, M. Modi, S. Hellings, M. Husbands, B. Ozenberger and R. Franco, Identification of a calcium channel modulator using a high throughput yeast two-hybrid screen, Nature Biotech.16, 946-950 (1998).

Yousif, F. B., and D. J. Triggle, Functional interactions between organic calcium channel antagonists in smooth muscle, Can. J. Physiol. Pharmacol. 63, 193-195 (1985).

Yousif, F. B., and D. J. Triggle, Inhibitory actions of a series of Ca^{2+} channel antagonists against agonist- and K^+ depolarization-induced responses in smooth muscle: an assessment of selectivity of action, C an. J. Physiol. Pharmacol. 64, 273-283 (1987).

Yu, S. P., L. M. Canzoniero and D. W. Choi, Ion homeostasis and apoptosis, Current Opinion in Cell Biology 13, 405-411 (2001).

Zamponi, G. W., and T. P. Snutch, Modulating modulation: crosstalk between regulatory pathways of presynaptic calcium channels. Mol. Intervent. 2, 476-478 (2002).

Zeng, F., D. H. Saha and R. I. Ogilvie, Kinetic disposition and hemodynamic effects of nitrendipine in anesthetized dogs, J. Cardiovas. Pharmacol. 7, 1034-1039 (1985).

Zhang, X., and T. H. Hinke, Amlodipine releases nitric oxide from canine coronary microvessels: an unexpected mechanism of action of a calcium channel-blocking agent, Circulation 97, 576-580 (1998).

Zhang, Z., Y. Xu, H. Song, J. Rodriguez, D. Tuleja, Y. Nanking, H.-S. Shin and N. Chiamvimonvat, Functional roles of $Ca_V1.3$ (α1D) calcium channel in sinoatrial nodes. Insight gained using gene-targeted null mutant mice, Circ.Res. 90, 981-987 (2002).

Zheng, W., D. Rampe and D. J. Triggle, Pharmacological, radioligand binding and electrophysiological characteristics of FPL 64176, a novel nondihydropyridine Ca^{2+} channel activator, in cardiac and vascular preparations, Mol. Pharmacol. 40, 734-741 (1991).

Zheng, W., J. Stoltefuss, S. Goldmann and D. J. Triggle, Pharmacological and radioligand binding studies of 1,4-dihydropyridines in rat cardiac and vascular preparations: stereoselectivity and voltage-dependence of antagonist and activator interactions, Mol. Pharmacol. 41, 535-541 (1992).

Zhorov, B. S., E. V. Folkmann and V.S. Ananthanarayanana, Homology model of dihydropyridine receptor: implications for L-type Ca^{2+} channel modulation by agonists and antagonists, Arch. Biochem. Biophys. 393, 22-41, 2001.

Zhou, M., J. H. Morals-Cabral, S. Mann and R. MacKinnon, Potassium channel receptor site for the inactivation gate and quaternary amine inhibitors, Nature 411, 657-661 (2001).

Zhou, X., H. Ono, Y. Ono and E. Frohlich, N- and L-Type calcium channel antagonist improves glomerular dynamics, reverses severe nephrosclerosis, and inhibits apoptosis and proliferation in an LNAME/SHR model, J. Hypertension 20, 993-1000 (2002).

Zidanic, M., and P. A. Fuchs, Kinetic analysis of barium currents in chick cochlear cells, Biophys J. 68, 1323-1336 (1995).

Zuhlke, R. D., A. Bouron, N.M. Soldatov and H. Reuter, Ca^{2+} channel sensitivity towards the blocker Isradipine is affected by alternate splicing of the human α_{1C} gene, FEBS Lett 427, 220-224 (1998).

THE THERAPEUTIC UTILITY OF TARGETING Ca_V2 CHANNELS

Charles J. Cohen[1] and Richard L. Kraus[1]

1. INTRODUCTION

The Ca_v2 family of voltage-gated calcium channels consists of 3 main subtypes ($Ca_v2.1$, $Ca_v2.2$ and $Ca_v2.3$), defined by their α_1 subunit ($\alpha_12.1$, $\alpha_12.2$ and $\alpha_12.3$, respectively; also known as α_{1A}, α_{1B} and α_{1E}). The use of selective peptide toxins has allowed the association of $Ca_v2.1$, 2.2 and 2.3 with P- or Q-, N-, and R-type calcium currents, respectively (Catterall, 2000; Ertel et al., 2000; Jarvis and Zamponi, 2001). These currents are found almost exclusively in the CNS, PNS and neuroendocrine cells and constitute the predominant forms of presynaptic voltage-gated calcium current. Presynaptic action potentials cause channel opening, and neurotransmitter release is steeply dependent upon the subsequent calcium entry (Meir et al., 1999; Sabatini and Regehr, 1999). Presynaptic calcium entry is modulated by many types of G-protein coupled receptors (GPCRs), and studies with agonists for these receptors indicate that modulation of Ca_v2 channels is a widespread and highly efficacious means of regulating neurotransmission. In this review, we describe the prospects for targeting Ca_v2 channels with novel therapeutic agents.

The subunit compositions of $Ca_v2.1$ and $Ca_v2.2$ have been determined by biochemical isolation of the complex. The $\alpha_12.1$ and $\alpha_12.2$ subunits contain high affinity binding sites for the peptide toxins ω-conotoxin-MVIIC and ω-conotoxin-GVIA, respectively. Isolation of the receptors for these toxins indicates that each channel contains an α_1, β, $\alpha_2\delta$ and γ subunit (McEnery et al., 1991; Witcher et al., 1993; Liu et al., 1996; Kang et al., 2001). Similar biochemical isolation of $Ca_v2.3$ has not been possible because of the lack of a selective high affinity ligand. Most therapeutic drug discovery has focused on the α_1 subunit, as this subunit forms the pore and contains the voltage-sensing gates and most known drug binding sites (Hockerman et al., 1997; Striessnig, 1999). Among the Ca_v2 channels, most therapeutic drug discovery has

[1] Department of Membrane Biophysics and Biochemistry, Merck Research Laboratories-San Diego, 3535 General Atomics Court, San Diego, CA 92121

Calcium Channel Pharmacology, Edited by McDonough
Kluwer Academic/Plenum Publishers, 2004

focused on identifying blockers of $Ca_v2.2$ (N-type calcium channels), which contain the $\alpha_1 2.2$ subunit. The primary goal is to identify $Ca_v2.2$ antagonists suitable for treating various forms of pain. The $\alpha_2\delta$ subunit of Ca_v2 channels has also been targeted for drugs to treat neuropathic pain because this subunit has a high affinity binding site for gabapentin, a drug that is effective against neuropathic pain and is also an anticonvulsant (Taylor et al., 1998; Sutton and Snutch, 2001). In addition, there is interest in modulating $Ca_v2.1$ channels for treating migraine because some mutations of these channels are associated with familial hemiplegic migraine (FHM) (Ophoff et al., 2001). We will summarize the rationale for targeting $Ca_v2.1$ and $Ca_v2.2$ for the indications mentioned above as well as for neuroprotection.

The pharmacological effect of blocking Ca_v2 channels has often been inferred from studies with selective peptide toxins, especially ω-agatoxin-IVA for studies of $Ca_v2.1$ and ω-conotoxin-GVIA and ω-conotoxin-MVIIA for $Ca_v2.2$. However, these peptides are often treated as perfectly selective channel blockers, when they are not, and we will discuss the limitations of studies with peptide toxins. Possible indications for drugs targeting $Ca_v2.3$ can be inferred from studies with knockout mice, but there are no complementary studies of channel pharmacology with a selective blocker. Consequently, the possible utility of modulators of $Ca_v2.3$ is poorly defined and the review of $Ca_v2.3$ is brief.

2. $Ca_v2.2$ CHANNEL ANTAGONISTS

2.1. Treatment of Neuropathic Pain

Multiple types of voltage-gated calcium channels are found in nociceptive pathways, but the isoform most clearly established as important for nociception is $Ca_v2.2$. The $\alpha_1 2.2$ subunit of these channels is found in the ventral and dorsal horns of the spinal cord and in dorsal root ganglia neurons, with the greatest levels in the outer layers of the dorsal horn (Westenbroek et al., 1998; Cizkova et al., 2002). All primary sensory afferents provide input to neurons in the dorsal horn, and calcium influx through $Ca_v2.2$ channels triggers the release of neurotransmitters from presynaptic nerve terminals in the spinal cord. Thus, block of $Ca_v2.2$ channels is expected to be broadly efficacious because these channels are in a common pathway downstream from the wide variety of receptors that mediate pain responses (Julius and Basbaum, 2001). Indeed, intrathecal injection of a selective peptide antagonist (ω-conotoxin-MVIIA, also known as SNX-111 and ziconotide) is broadly effective against both neuropathic and inflammatory pain in laboratory animals and man (Bowersox et al., 1996; Jain, 2000; Vanegas and Schaible, 2000; see Norton et al., this volume). Ziconotide is currently in phase III clinical trials for treatment of severe cancer pain by intrathecal infusion. AM336, a similar conopeptide also known as ω-conotoxin-CVID, is entering phase II clinical trials and reportedly has a better therapeutic index (Scott et al., 2002; Smith et al., 2002).

Additional evidence for efficacy of $Ca_v2.2$ antagonists in neuropathic pain is provided by studies with $\alpha_1 2.2$ and β_3 knockout mice. $\alpha_1 2.2$ knockout mice are viable and have normal CNS and motor function, but are resistant to development of neuropathic pain in a spinal nerve ligation (SNL or Chung) model (Hatakeyama et al., 2001; Kim et al., 2001; Saegusa et al., 2001; see Saegusa and Tanabe, this volume). These mice also have blunted responses in pain caused by injection of formalin into the paw and in visceral pain tests. The phenotype of $\alpha_1 2.2$ knockout mice differs dramatically from that of $\alpha_1 2.1$ knockouts: although both types of subunits are found in presynaptic nerve terminals in the CNS and PNS, the $\alpha_1 2.1$ knockouts display severe ataxia and live only 3-4 weeks (Jun et al., 1999), while the $\alpha_1 2.2$ knockouts are viable and relatively unimpaired. The main deficit in $\alpha_1 2.2$ knockout mice is impaired sympathetic innervation of the cardiovascular system. The association of ataxia with block of $Ca_v2.1$ but not $Ca_v2.2$ is expected to extend to humans because neuromuscular transmission is blocked by ω-agatoxin-IVA, but not by ω-conotoxin-GVIA (Protti et al., 1996).

Knock out of β_3 subunits also diminishes $Ca_v2.2$ in dorsal root ganglia and confers resistance to the development of pain without gross behavioral abnormalities (Namkung et al., 1998; Murakami et al., 2002). A majority of $Ca_v2.2$ channels in the spinal cord and dorsal root ganglia neurons contain β_3 subunits (Scott et al., 1996). β subunits have multiple functions, one of which is as a necessary partner for surface expression of the α_1 subunit (Bichet et al., 2000). Thus, knockout of the β_3 subunit diminishes $Ca_v2.2$ involved in nociception and produces a phenotype similar to knockout of $\alpha_1 2.2$ subunits.

Calcium entry via $Ca_v2.2$ channels is functionally upregulated in pain states and, simultaneously, endogenous mechanisms for limiting presynaptic calcium entry via $Ca_v2.2$ are diminished. This dual modulation occurs in virtually all types of pain and is an important reason why $Ca_v2.2$ antagonists are broadly efficacious. The upregulation of $Ca_v2.2$ in pain states occurs as a consequence of the modulation of the channels by many different types of inhibitory G-protein coupled receptors (Dolphin, 1995; Jones and Elmslie, 1997; Catterall, 2000; Jarvis and Zamponi, 2001). Most commonly, activation of inhibitory GPCRs releases βγ subunits of G-proteins ($G_{\beta\gamma}$) that bind directly to the channel and inhibit opening (Herlitze et al., 1996; Ikeda, 1996; De Waard et al., 1997). This mechanism is most pronounced for $\alpha_1 2.2$, but is also observed with $\alpha_1 2.1$ and to a lesser extent with $\alpha_1 2.3$ (Bourinet et al., 1996; Meza and Adams, 1998; Colecraft et al., 2000). This inhibition is manifest as block of presynaptic calcium entry and is an important mechanism for regulating neurotransmission. Examples of GPCRs that inhibit presynaptic calcium entry by close coupling to Ca_v2 channels are opioid, $GABA_B$, adenosine, cannabinoid, α_2 adrenergic, muscarinic, metabotropic glutamate and somatostatin (see Beedle and Zamponi, this volume). The binding of $G_{\beta\gamma}$ subunits to $\alpha_1 2.2$ is inhibited by phosphorylation of the channel by protein kinase C (PKC) and by increases in electrical spiking activity (Swartz, 1993; Shapiro et al., 1996; Brody et al., 1997; Hamid et al., 1999; Herlitze et al., 2001); both of these conditions are associated with a variety of pain states. Among the mediators of inflammatory pain that cause activation of PKC are substance P, bradykinin, ATP and glutamate (acting via group I metabotropic glutamate receptors). In addition to diminishing inhibition via $G_{\beta\gamma}$, phosphorylation of the $\alpha_1 2.2$ subunit by PKC also increases basal calcium current. Functional upregulation of $Ca_v2.2$ can be observed in the spinal nerve ligation (or Chung) model for neuropathic pain; this nerve injury increases the sensitivity of dorsal horn neurons to inhibition by ω-conotoxin-GVIA (Matthews and Dickenson, 2001). It is possible to block the functional upregulation via PKC with selective antagonists for each type of GPCR, but the antagonists will only be effective if one type of GPCR agonist is dominant in mediating pain. The clinical experience with NK1 antagonists indicates that this strategy can fail, presumably because inflammation most often involves a mixture of many agents (Hill, 2000; Julius and Basbaum, 2001). $Ca_v2.2$ antagonists that act directly on the channel act downstream of inflammatory agents and can override the stimulation of $\alpha_1 2.2$ by PKC. In particular, ziconotide is effective against neuropathic pain that is resistant to morphine (Wang et al., 2000), suggesting that $Ca_v2.2$ antagonists are a viable alternative to narcotics as anti-nociceptive agents.

In addition to functional upregulation of $Ca_v2.2$ in pain states, the $\alpha_1 2.2$ subunit is also physically upregulated four- to seven-fold in parallel with the development of mechanical or thermal allodynia in the chronic constriction injury model (Cizkova et al., 2002).

The key issue in targeting $Ca_v2.2$ channels for treating neuropathic pain is expected to be attaining an adequate therapeutic index. There are clinical reports of severe side effects associated with intrathecal infusion of ziconotide, such as nystygmus, profound ataxia, and orthostatic hypotension (Penn and Paice, 2000). Similar side effects are observed following intrathecal infusion in rats (Bowersox et al., 1992; Malmberg et al., 2003), although it is possible to achieve antinociceptive activity without motor impairment with a therapeutic index of 2 (Malmberg and Yaksh, 1995; Scott et al., 2002). Ziconotide also produces hypotension and impairs baroreceptors in rabbits (Wright et al., 2000). Likewise, $\alpha_1 2.2$ knockout mice lack normal baroreflexes, although their motor activity and CNS function appear normal. These knockout mice also have blunted inotropic responses to sympathetic innervation of myocardium, but there is no cardiac hypertrophy, a condition associated with impaired sympathetic innvervation (Ino et al., 2001). Impaired motor activity due to block of $Ca_v2.2$ could be inferred from the earliest studies with ω-conotoxin-GVIA, in which the toxin produced "shaking" in mice following intra-cerebral injection (Olivera et al., 1999).

It is commonly assumed that all of the adverse effects associated with ziconotide are due to block of $Ca_v2.2$ channels in the sympathetic nervous system, the basal ganglia, brainstem, and motor neurons. However, the worst side effects occur when ziconotide is infused at concentrations far in excess of the IC_{50} for block of $Ca_v2.2$. In voltage clamp studies, the IC_{50} is 1-40 nM (Sanger et al., 2000; Kaneko et al., 2002) and, at concentrations >1 μM, ziconotide blocks other types of ion channels, particularly other types of voltage-gated calcium channels (Mermelstein and Surmeier, 1997; Lewis et al., 2000), neuronal nicotinic receptors (Fernandez et al., 1995; Herrero et al., 1999), and perhaps P2X receptors (Lalo et al., 2001). The conotoxin AM336 (also known as ω-conotoxin-CVID) is about 100-fold more selective for $\alpha_1 2.2$ over $\alpha_1 2.1$ (Lewis et al., 2000) and has a therapeutic index of 10 vs. 2 for ziconotide (Scott et al., 2002). Thus, some of the reported side effects are not mechanism based. Impairment of baroreceptor reflexes, which is observed both with ziconotide and in $\alpha_1 2.2$ knockout mice, is almost certainly a mechanism based side effect. The therapeutic index of AM336 is incompletely defined, and it is particularly important to quantify possible effects on baroreceptor reflexes. If the therapeutic index of 10 for AM336 withstands closer scrutiny, then antagonists of $Ca_v2.2$ can indeed have an adequate therapeutic index, and the challenge is to find small molecules with high selectivity for this channel.

Another significant problem in targeting $Ca_v2.2$ for therapeutic drug discovery is likely to be selectivity over $Ca_v2.1$, the most similar isoform. The $\alpha_1 2.1$ subunit is ≈ 64% identical to $\alpha_1 2.2$ and ≈ 73% identical within the 24 transmembrane segments (Williams et al., 1992b). Moreover, the residues lining the pore in all S5 and S6 segments are nearly completely identical, and these are thought to contain the likely binding sites for therapeutically useful drugs that block voltage-gated calcium channels or sodium channels (Hockerman et al., 1997; Striessnig et al., 1998). Block of $Ca_v2.1$ is likely to result in ataxia (see section on $Ca_v2.1$ below).

There are two main strategies for obtaining $Ca_v2.2$ antagonists with a better therapeutic index than ziconotide: 1) selectively block $Ca_v2.2$ during patterns of electrical activity associated with pathological conditions; and 2) identify sub-types of $Ca_v2.2$ and selectively block those in the target tissue, such as the spinal cord.

The first strategy consists of identifying $Ca_v2.2$ antagonists that selectively block calcium channels in partially depolarized cells or during rapid electrical spiking activity. This is the paradigm used for the therapeutically useful calcium antagonists, such as nifedipine and diltiazem, and for anti-arrhythmic and analgesic agents such as lidocaine (Hille, 1977; Bean, 1984; Ertel and Cohen, 1994). Thus, both tetrodotoxin and lidocaine block neuronal sodium channels, but only lidocaine is therapeutically useful. Ziconotide, like tetrodotoxin, plugs the channel, and block is independent of electrical activity in nerve (McDonough et al., 2002). The therapeutically useful channel antagonists interact with the voltage sensors of α_1 subunits and block is enhanced by channel opening (so called "use-dependent" block). It seems likely that a similar paradigm will work for drugs targeting $\alpha_1 2.2$. Painful stimuli clearly produce the required rapid electrical spiking activity and partial depolarization. Indeed, some dorsal horn neurons exhibit a progressive increase in spiking activity in response to stimulation (known as "wind-up"). The increase in firing rate often occurs in response to partial depolarization (< 20 mV) caused by the afferent input (Sivilotti et al., 1993; Vikman et al., 2001). It is likely that wind up contributes to some instances of central sensitization, in which dorsal horn neurons are hyperexcitable.

Although voltage-dependent block of $Ca_v2.2$ is a promising approach to attaining an adequate therapeutic index, it will also be necessary to achieve an appropriate rate of development of block during depolarizations. For example, ω-conotoxin-GVIA has been reported to produce voltage-dependent block of $Ca_v2.2$ expressed in *Xenopus* oocytes (Stocker et al., 1997). However, at concentrations that produce selective block of inactivated channels, the rate of development and recovery from block is so slow that fractional inhibition changes little during bursts of spiking activity.

The second strategy for obtaining a high therapeutic index requires inhibition of a subset of $Ca_v2.2$ channels. $Ca_v2.2$ channels vary in the splice variant of the $\alpha_1 2.2$ subunit and in the composition of auxiliary subunits and associated proteins, such as $G_{\beta\gamma}$, regulators of G-protein signaling (RGS) proteins, calmodulin, synaptotagmin and syntaxin (Herlitze et al., 1996; Stea et al., 1999; Catterall, 2000; Feng et al., 2001; Lu et al., 2001; Kaneko et al., 2002; Lipscombe et al., 2002). All of these variables can result in pharmacological distinctions, especially if one is using a drug that produces time- and voltage-dependent block. Variants of $Ca_v2.2$ vary widely in the time- and voltage-dependence of activation and inactivation and undergo switching between modes that differ in rates of gating transitions (Delcour et al., 1993; Rittenhouse and Hess, 1994). Drugs that bind more potently to channels in the open or inactivated state will preferentially block variants that open or inactivate at more negative voltages.

Splice variants of the $\alpha_1 2.2$ subunit have been known since the original cloning of the gene (Dubel et al., 1992; Williams et al., 1992a), and differences between the splice variant found in brain and spinal cord are known (Lin et al., 1997). Almost all electrophysiological studies have characterized the channels in neuronal soma, but recent studies indicate that the channels in dendrites and presynaptic active zones have distinctly different physiological and pharmacological properties, apparently because of alternative splicing. Compared to $Ca_v2.2$ channels in the soma of superior cervical ganglia neurons, those in dendrites activate at more negative voltages, are more sensitive to modulation by inhibitory GPCRs, and are less sensitive to block by ω-conotoxin-GVIA (Delmas et al., 2000). Splice variants that contain a long C-terminal segment are targeted to synapses (Maximov and Bezprozvanny, 2002) and are more sensitive to modulation by GPCRs (Simen et al., 2001). Activation at more negative voltages, coupled with greater sensitivity to GPCR modulation, is caused by a single substitution in transmembrane segment IS6 (Zhong et al., 2001). Splice variants that lack a segment of the linker between domains II-III are less sensitive to block by ziconotide or ω-conotoxin-GVIA (Kaneko et al., 2002). The relative potency of ziconotide, ω-conotoxin-GVIA and ω-conotoxin-MVIIC varies among different presynaptic nerve preparations (Boot, 1994; Sanger et al., 2000), suggesting the existence of additional splice variants of $\alpha_1 2.2$. Thus, pharmacologically distinct subtypes of $Ca_v2.2$ channels have been identified that could be selectively targeted. At present, this heterogeneity is more of a problem than an opportunity. A detailed understanding of the localization of subunit combinations and of splice variations is not available, so it is not clear which subtype should be targeted. The channels most often studied *in vitro* are those in the soma, while presynaptic channels may be pharmacologically dissimilar.

A number of pharmaceutical companies have claimed small molecule $Ca_v2.2$ antagonists for treatment of pain (see WO9910438 and WO0236567 by Astra Zeneca, WO0222588 by Ajinomoto, WO0149670 by Neuromed, US6,423,689 by Warner Lambert/ Pfizer/ Neurex/Elan and EP1099692 by Eisai; see also Snutch et al., 2001b). However, none of these compounds have been characterized in detail. It is not clear whether any of these compounds is highly selective for $Ca_v2.2$ over $Ca_v2.1$, $Ca_v2.3$ and other receptors, or whether they completely suppress pain responses without effect on baroreceptor reflexes and motor coordination. Detailed information is available for piperidines such as fluspirilene, haloperidol and penfluridol (Zamponi et al., 1996; Zamponi, 1999). The potency of block by these compounds is extraordinarily sensitive to cytoplasmic factors and is less potent in cells that are not dialyzed for whole cell patch voltage clamp studies. No similar finding has been made for the therapeutically useful calcium antagonists that target the $\alpha_1 1.2$ subunit, as most workers implicitly assume that cell dialysis does not alter the potency of channel block. Many of the compounds patented by Neuromed are structurally related piperidines, and Neuromed claims efficacy with a high therapeutic index (Snutch et al., 2001a).

2.2. Treatment of Stroke

Ziconotide (see Norton et al., this volume) is highly effective as a neuroprotective agent in rat models of global or focal ischemia (Valentino et al., 1993; Buchan et al., 1994; Yenari et al., 1996; Colbourne et al., 1999). In all of these studies, ziconotide was administered intravenously 6 or 24 hours after the ischemic insult, and efficacy was determined 5-7 days later. These results are noteworthy because stroke is usually treated several hours after the ischemic event. Ziconotide is one of the few agents effective when administered after a long delay. Since the site of action is presumably the central nervous system, and brain penetration of this peptide is very low (about 0.03%; Valentino et al., 1993), extremely high concentrations are administered (typically 5 mg/kg dissolved in 1 ml of saline, equivalent to about 600 μM peptide). Ziconotide is far more potent when tested *in vivo* by injection directly into the lateral ventricle (Valentino et al., 1993) or *in vitro* against hypoxia-induced neurodegeneration (Pringle et al., 1996). Thus, it is reasonable to conclude that the site of action following intravenous administration is $\alpha_1 2.2$ in the CNS. Nonetheless, the very high concentrations of ziconotide used intravenously could affect off-site targets; it will be necessary to obtain highly brain-penetrant compounds to clearly validate $Ca_v2.2$ antagonists as neuroprotective agents.

Although ziconotide reduces ischemic injury when evaluated 7 days after the insult, it is not effective after 28 days (Colbourne et al., 1999). This is an intriguing result that could have great importance for stroke research, but a difficult claim to support with histological measurements because parts of the infarct may be lost after 28 days. The limited duration of efficacy of ziconotide indicates that neurodegeneration develops slowly after an ischemic insult. It is not clear whether slowly developing neurodegeneration is a continuation of cell necrosis or due to a second process, such as apoptosis or neuroinflammation. Repeated dosing with ziconotide should be effective in the first case, but not in the latter. Brain penetrant $Ca_v2.2$ antagonists should be useful tools in defining the etiology of neuronal damage.

2.3. Treatment of Cardiovascular Disease

$Ca_v2.2$ channels are found in the periphery and mediate catecholamine release from sympathetic neurons and adrenal chromaffin cells. Some forms of hypertension result from elevated sympathetic tone, and $Ca_v2.2$ antagonists could be particularly effective in treating this disorder. Although complete block of $Ca_v2.2$ can cause hypotension or impair baroreceptor reflexes, partial inhibition by $Ca_v2.2$ antagonists might reduce hypertension with minimal reflex tachycardia (Uneyama et al., 1999). Cilnidipine is a 1,4-dihydropyridine anti-hypertensive agent that reportedly blocks both $Ca_v1.2$ (an L-type calcium channel containing the $\alpha_1 1.2$ subunit) and $Ca_v2.2$ (see Triggle, this volume). Cilnidipine causes less reflex tachycardia than nifedipine or amlodipine and is more effective at blocking stress-induced hypertension (Uneyama et al., 1999), properties attributed to block of $Ca_v2.2$ channels. Likewise, mibefradil, an anti-hypertensive agent structurally derived from verapamil, lowers blood pressure without reflex tachycardia and inhibits sympathetic outflow in human atria (Gothert and Molderings, 1997). ω-conotoxin-GVIA also inhibits sympathetic outflow, but not nifedipine, leading to the suggestion that mibefradil also blocks $Ca_v2.2$ at therapeutic concentrations (Gothert and Molderings, 1997). Although the absence of reflex tachycardia with mibefradil treatment is usually attributed to block of Ca_v3 (T-type) channels, some reports show that mibefradil blocks $Ca_v2.1$ and $Ca_v2.3$ with the same potency as Ca_v3 (IC_{50} of 20-60 nM in partially depolarized cells; Jimenez et al., 2000); it is reasonable to expect that $Ca_v2.2$ will be blocked with similar potency. Subsequent studies with mibefradil have both supported (Van der Lee et al., 2000) and disputed (Xi and Angus, 2001) mibefradil block of noradrenaline release. The studies with cilnidipine and mibefradil are inconclusive with respect to anti-hypertensive activity of $Ca_v2.2$ antagonists. Both agents are potent inhibitors of $Ca_v1.2$, but block is highly voltage-dependent and, in well polarized cells, appears to be no more potent than block of Ca_v2 or Ca_v3 channels. The steep voltage dependence of calcium channel block makes it difficult to assess the relative potencies for block of $Ca_v1.2$ versus $Ca_v2.2$. Although both cilnidipine and mibefradil have a pharmacological profile different from nifedipine, it is possible that this derives from different kinetics of channel block rather than a different site of action. Note that mibefradil does not impair baroreceptor reflexes, yet plasma levels and *in vivo* activity are consistent with block of $Ca_v2.2$.

3. $Ca_v2.1$ CHANNEL MODULATORS

Voltage-gated $Ca_v2.1$ calcium channels are widely distributed in synapses of the central and peripheral nervous system (Hillman et al., 1991; Mintz et al., 1992; Uchitel et al., 1992; Stea et al., 1994). Two types of calcium currents mediated by the $\alpha_1 2.1$ subunit were originally described in cerebellar Purkinje cells and granule cells, namely P-type and Q-type. These two calcium currents primarily differ in sensitivity to the peptide toxin ω-agatoxin-IVA (see McDonough, this volume) and in the kinetics of inactivation. It is now recognized that alternative splicing of the $\alpha_1 2.1$ subunit gene results in channel isoforms displaying distinct kinetic and pharmacological properties resembling P- and Q-type channels (Bourinet et al., 1999; Hans et al., 1999b).

Like $Ca_v2.2$ channels, $Ca_v2.1$ channels mediate depolarization-induced calcium influx in presynaptic nerve terminals tightly coupled to neurotransmitter release (Dunlap et al., 1995). Opening of both channel types can be inhibited in a voltage-dependent manner by different types of activated GPCRs, through the binding of $G\beta\gamma$ subunits to distinct sites on the α_1 subunit. Furthermore, calcium/calmodulin, the neuronal calcium binding protein CaBP1, and SNARE proteins are involved in regulating $Ca_v2.1$ channels (Lee et al., 1999; Catterall, 2000; Lee et al., 2002a; see Piedras-Rentería et al., this volume).

The physiological function of $Ca_v2.1$ channels has largely been defined by the use of the selective inhibitor ω-agatoxin-IVA, by characterization of $\alpha_1 2.1$ knockout mice, and by $Ca_v2.1$-channelopathies. It has been shown that $Ca_v2.1$ channels play only a minor role in pain. $\alpha_1 2.1$ channel ablation in the central and peripheral nervous system in mouse models has neuropathological consequences and, even more importantly, $\alpha_1 2.1$ mutations inducing channel dysfunction lead to severe neurological disorders in humans.

3.1. $Ca_v2.1$ Channels and Pain

The localization of $Ca_v2.1$ channels in the superficial laminae of the dorsal horn of the spinal cord suggests involvement of these channels in the perception and maintenance of certain forms of pain (Westenbroek et al., 1998; Vanegas and Schaible, 2000). Knockout mice were generated to assess the role of $Ca_v2.1$ channels in neurotransmission, nociception and motor function. Complete elimination of P-type and Q-type calcium currents alters synaptic transmission, resulting in severe ataxia. Animals die at the age of four weeks (Jun et al., 1999; Fletcher et al., 2001). Unfortunately, due to their short lifespan, $\alpha_1 2.1$ deficient mice have not been tested in pain models.

The involvement of $Ca_v2.1$ channels in nociception in animal models of inflammatory and neuropathic pain has also been studied by blocking channels with intrathecal ω-agatoxin-IVA. Although there is evidence for a role of $Ca_v2.1$ channels in the generation and maintenance of inflammatory pain, ω-agatoxin-IVA has limited efficacy. Blocking $Ca_v2.1$ channels with the peptide has no effect on neuropathic pain (Vanegas and Schaible, 2000).

3.2. $Ca_v2.1$ Channelopathies

The human CACNA1A gene encoding the pore forming $\alpha_1 2.1$ subunit is located on chromosome 19p13 (Diriong et al., 1995). At least three different human neurological diseases are associated with mutations in CACNA1A: episodic ataxia type 2 (EA-2), spinocerebellar ataxia type 6 (SCA-6), and familial hemiplegic migraine (FHM). By briefly describing the phenotype of these diseases we illustrate how changes in $Ca_v2.1$ channel function lead to neurological disorders in humans (see also Mori et al., this volume). Hypothetical options for the treatment of migraine will be discussed.

Patients affected by episodic ataxia type 2 (EA-2) experience recurring attacks of cerebellar ataxia, headaches and cerebellar atrophy. Several $\alpha_1 2.1$ mutations have been identified in EA-2 patients (Van Den Maagdenberg et al., 2002). All mutants are believed to form nonfunctional channels, thereby causing the EA-2 clinical phenotype. Consistent with this hypothesis, $\alpha_1 2.1$ mutants expressed in mammalian cells show complete loss of channel function (Guida et al., 2001; Jouvenceau et al., 2001; Wappl et al., 2002).

Patients affected by spinocerebellar ataxia type 6 (SCA-6) also develop progressive cerebellar ataxia. An insertion in the CACNA1A gene generates a variety of $\alpha_1 2.1$ isoforms with extended cytoplasmic C-termini containing stretches of polyglutamine (Zhuchenko et al., 1997). In contrast to EA-2, these $\alpha_1 2.1$ mutants form functional ion channels. Several laboratories report that mutant channels have abnormal biophysical properties when heterologously expressed in mammalian cells (Matsuyama et al., 1999; Restituito et al., 2000; Toru et al., 2000). It is particularly interesting that pathogenic polyglutamine expansions increase current density by enhancing the expression of $Ca_v2.1$ channels on the cell surface. A resulting elevation of calcium influx as well as an accumulation of mutant $\alpha_1 2.1$ protein in the cytosol probably contributes to the neurodegenerative pathogenesis of SCA-6 (Ishikawa et al., 1999; Piedras-Renteria et al., 2001).

Familial Hemiplegic Migraine (FHM) is a rare autosomal dominant subtype of migraine with aura, associated with ictal hemiparesis and, in some families, with progressive cerebellar atrophy. The symptoms of headache and the aura experienced by FHM patients are almost identical to more common forms of migraine. Recent genetic studies suggest the involvement of the FHM locus also in the generation of normal migraine with and without aura (May et al., 1995; Terwindt et al., 2001). Consequently, FHM has been considered a useful model to unravel the pathophysiology of migraine. It should be mentioned that migraine is likely to be a polygenetic multifactorial disorder. Besides CACNA1A, other novel susceptibility genes have been mapped to regions on chromosomes 1 and 6 (Ducros et al., 1997; Carlsson et al., 2002; Fusco et al., 2003), and there are a variety of other molecular components that may also be involved in the pathophysiology of migraine (Goadsby et al., 2002). In this chapter we exclusively review $Ca_v2.1$ channels as a potential drugable target.

About 50% of the families suffering from FHM carry a point mutation in one allele of the gene encoding the $\alpha_1 2.1$ subunit. At least 13 FHM mutations have been identified that result in amino acid substitutions in the voltage sensor or pore-forming region of the $\alpha_1 2.1$ protein (Ophoff et al., 2001). The effects of seven FHM mutations on channel function have been analyzed. Initial expression studies, using *Xenopus laevis* oocytes or non-neuronal mammalian cells, indicated that FHM mutations increase channel open probability and have variable effects on channel inactivation, conductance and density of functional $Ca_v2.1$ channels in the membrane (Kraus et al., 1998; Hans et al., 1999a; Kraus et al., 2000). Surprisingly, the FHM mutations do not produce a consistent change in calcium influx; depending on the mutation, the substitution can give rise to either increased or decreased calcium influx into nerve terminals. This observed inconsistency might be due to the recombinant expression systems that have been used. It seems likely that one cannot extrapolate properties in oocytes or non-neuronal cells to an *in vivo* situation because native $Ca_v2.1$ channels interact with modulatory proteins, such as calmodulin, synaptic vesicle proteins, and G-protein coupled receptors. For example, the mutation R192Q, located in the voltage sensor of the channel, reduces G-protein mediated inhibition of $Ca_v2.1$ currents (Melliti et al., 2003). Furthermore, a mutation in the $Ca_v2.1$ pore region of *tottering (tg)* mice that results in absence epilepsy (Fletcher et al., 1996), not only reduces the $Ca_v2.1$ current density in Purkinje cells (Wakamori et al., 1998) but also potentiates low voltage activated calcium currents in thalamic neurons (Zhang et al., 2002) involved in epileptogenesis. Consequently, to study the role of $Ca_v2.1$ channel mutants in migraine pathology, mutants should be analyzed in a more *in vivo*-like environment.

In order to achieve a more native environment, Tottene et al. (2002) expressed FHM mutants in cerebellar granule cells from $\alpha_1 2.1$ knockout mice. This study uncovered two functional effects on channel properties. All tested $\alpha_1 2.1$ mutants activate at more negative potentials than the wild type $\alpha_1 2.1$ subunit and lead to decreased maximal $Ca_v2.1$ current density in neurons. The authors present an interesting hypothesis: a shift of $Ca_v2.1$ channel activation to more negative potentials could cause neuronal hyperexcitability and thereby favor FHM attacks, even though $Ca_v2.1$ current density in the neuron is decreased (Tottene et al., 2002). Very similar effects on $Ca_v2.1$ channel properties have been found by Cao and coworkers (2002) for a set of FHM mutants expressed in cultured hippocampal cells of $Ca_v2.1$ knockout mice (Cao et al., 2002). Clearly, more relevant results will come from transgenic animals expressing the respective FHM mutation.

For the time being, it is reasonable to assume that these recent studies in transfected neurons reflect the *in vivo* situation in neurons of a migraine patient. What kind of drug does one need to compensate the postulated activation of $Ca_v2.1$ channels at weak depolarizations, underlying neuronal hyperexcitability?

Any effort to target $Ca_v2.1$ channels with an antagonist is very likely to fail because of undesirable effects, such as ataxia. However, it is theoretically possible to develop useful therapeutic agents that only slightly modulate $Ca_v2.1$ channel activation. Targeting the $\alpha_1 2.1$ subunit directly with a channel gating modifier might be one option. In fact, existing $Ca_v2.1$ channel inhibitors that do not bind to the pore of the channel, such as the peptide ω-agatoxin-IVA, cause a dramatic shift of channel activation to more positive voltages (McDonough, this volume). Small molecules binding to the same extracellular site on the channel could have similar but more moderate effects on $Ca_v2.1$ channel gating. Alternatively, targeting any other protein that modulates $Ca_v2.1$ channel activation, such as GPCRs could be a viable approach. Like ω-agatoxin-IVA, G-proteins shift $Ca_v2.1$ channel activation to more positive potentials (Herlitze et al., 1996). A GPCR agonist could indirectly inhibit channel opening preferentially during weak depolarizations to counteract neuronal hyperexcitability.

Interestingly, block of voltage gated calcium channels (Richter et al., 2002) and decreased calcium influx through mutated $Ca_v2.1$ channels in *tottering* mice (Ayata et al.,

2000) affect cortical spreading depression, a critical event in the initiation of migraine (Bolay et al., 2002). Elevating the threshold for cortical spreading depression by dampening calcium channel activity would be an elegant way to prevent the onset of migraine.

Clearly, the safety range for $Ca_v2.1$ drugs is narrow. More research needs to be done to validate $Ca_v2.1$ channels as targets for the treatment of common forms of migraine and to justify the development of a rather sophisticated $Ca_v2.1$ channel gating modifier.

4. $Ca_v2.3$ CHANNEL MODULATORS

The function of $Ca_v2.3$ channels is much less well understood compared to $Ca_v2.1$ and $Ca_v2.2$, primarily because there are no highly selective blockers of this channel type. The pore-forming $\alpha_1 2.3$ (α_{1E}) subunit is found throughout the brain and in kidney, spleen and pancreatic islet cells (Soong et al., 1993; Williams et al., 1994). Most of the neuronal channels in the spinal cord are on the soma and dendrites, rather than at nerve terminals (Westenbroek et al., 1998). A component of high-voltage activated calcium current has been reported in CNS neurons that is insensitive to block by 1,4-dihydropyridines (such as nimodipine), ω-conotoxin-GVIA and ω-conotoxin-MVIIC; this drug-resistant component of calcium current is commonly known as R-type (Randall and Tsien, 1995). The time- and voltage-dependence and ionic permeability of this component of calcium current are similar to those produced by heterologous expression of $\alpha_1 2.3$. Antisense to $\alpha_1 2.3$ largely eliminates R-type current (Piedras Renteria and Tsien, 1998; Tottene et al., 2000), and the R-type current is greatly reduced in $Ca_v2.3$ knockout mice (Wilson et al., 2000; Lee et al., 2002b). Unfortunately, it is rarely feasible to evaluate the physiological function of $Ca_v2.3$ by first blocking all other types of Ca_v1 and Ca_v2 channels.

The spider toxin SNX-482 has been touted as a selective blocker of $Ca_v2.3$ (Newcomb et al., 1998). However, there appear to be splice variants of $Ca_v2.3$ that differ in sensitivity to this peptide, and the toxin blocks $Ca_v1.2$ with about the same potency as the less sensitive component of $Ca_v2.3$ (Tottene et al., 2000; Bourinet et al., 2001). It is possible to demonstrate a role of $Ca_v2.3$ in presynaptic calcium entry and neurotransmission by carefully using low concentrations of SNX-482 (Gasparini et al., 2001), but the maximal effect cannot be clearly defined. Thus, the rationale for targeting $Ca_v2.3$ channels for therapeutic drug discovery rests almost entirely on studies using knockout mice.

Studies with mice lacking $Ca_v2.3$ leave one confused as to whether an agonist or antagonist is desired. On the one hand, loss of function reduces pain responses and eliminates cocaine-induced hyperactivity (Saegusa et al., 2000; Han et al., 2002), suggesting that an antagonist is desired. However, the knockout mice develop abnormally large infarcts and elevated increases in cell calcium in response to focal ischemia induced by permanent occlusion of the middle cerebral artery (Toriyama et al., 2002) and have elevated fear responses and anxiety (Lee et al., 2002b). The latter results suggest that an agonist of channel activity would be more desirable. At this time, there is no clear path forward in targeting $Ca_v2.3$ for therapeutic agents.

5. GABAPENTIN

Gabapentin (Neurontin®, 1-(aminomethyl)cyclohexaneacetic acid) has been used clinically for many years as add-on therapy for the treatment of epilepsy. In recent years, it has emerged as a leading treatment for neuropathic pain, with efficacy in clinical trials for the treatment of post-herpetic neuralgia, diabetic neuropathy, trigeminal neuralgia, migraine, and fibromyalgia (Bryans and Wustrow, 1999; Mellegers et al., 2001; Sutton and Snutch, 2001). Pregabalin (S-(+)-3-isobutyl-GABA), an analog of gabapentin, is in phase III clinical trials for treatment of neuropathic pain, epilepsy, and general anxiety disorder. Both gabapentin and pregabalin are active in multiple animal models for neuropathic pain (Bryans and Wustrow, 1999). Although gabapentin has been in clinical use for many years, the mechanism of action is incompletely understood and has been the subject of controversy.

Gabapentin was designed as a metabolically stable GABA mimetic, but most studies find no effect on GABA receptors (Taylor et al., 1998; Lanneau et al., 2001). The $\alpha_2\delta$ subunit of voltage-gated calcium channels has been identified as a high affinity binding site for gabapentin in the CNS (Gee et al., 1996). There are 4 isoforms of $\alpha_2\delta$, called $\alpha_2\delta$-1, -2, -3 and –4, and gabapentin binds with high affinity to $\alpha_2\delta$-1 and $\alpha_2\delta$-2 (Marais et al., 2001; Qin et al., 2002). These subunits are components of Ca_v1 and Ca_v2 calcium channels and enhance heterologous expression levels of the α_1 subunit (Williams et al., 1992b; Felix et al., 1997; Felix, 1999). Loss of function of $\alpha_2\delta$-2 in mice dramatically reduces calcium channel expression (Brodbeck et al., 2002). Expression of the $\alpha_2\delta$-1 subunit is increased as much as 18-fold in dorsal root ganglia neurons in some models for neuropathic pain (Luo et al., 2001; Luo et al., 2002). Thus, it seems plausible that gabapentin could inhibit neurotransmission in the spinal cord by interfering with the function of $\alpha_2\delta$ subunits, thereby inhibiting presynaptic calcium currents. The type of calcium channel block that could most readily account for the anti-nociceptive activity is block of $Ca_v2.2$, as described above. Likewise, inhibition by gabapentin of presynaptic calcium channels in the CNS could account for the anti-convulsant activity and efficacy in migraine, and block of $Ca_v2.1$ might be more relevant (Fink et al., 2000; Dooley et al., 2002). Several laboratories have tested this hypothesis, but until recently, there were no reports of calcium channel block at concentrations comparable to the K_d for binding to the $\alpha_2\delta$ subunits ($K_d \approx 50$ nM), and most studies have failed to detect any direct calcium channel block by gabapentin (Taylor et al., 1998). Several recent publications report the expected high affinity calcium channel block in cultured dorsal root ganglia neurons; maximal inhibition of $Ca_v2.2$ is $\approx 40\%$, and the potency of inhibition is consistent with the binding affinity to $\alpha_2\delta$. (Martin et al., 2002; Sutton et al., 2002). Other laboratories have reported conflicting results, however, even when reporting calcium channel modulation by gabapentin. Two laboratories report no acute block of calcium currents in dorsal root ganglia neurons (Rock et al., 1993; Alden and Garcia, 2001), but one of these groups found voltage-dependent block at high concentrations after dosing for 60 minutes (Alden and Garcia, 2001). In another report, gabapentin did not cause acute inhibition of heterologously expressed $Ca_v2.1$ channels, but chronic treatment enhanced current by slowing inactivation (Kang et al., 2002). It is possible that calcium channel block by gabapentin requires synaptic proteins absent in the heterologous expression system or that gabapentin block is modified by channel phosphorylation (Martin et al., 2002). Unfortunately, it is unclear how binding of drug to the non-pore forming $\alpha_2\delta$ subunit would inhibit calcium currents through the channel, with or without auxiliary subunits. The hypothesis that gabapentin effects are mediated via $\alpha_2\delta$ does not yet account for the tissue specificity and excellent safety profile of this drug. The $\alpha_2\delta$ subunit is a component of both muscle and nerve calcium channels, yet gabapentin has no cardiovascular side effects and apparently no effect on contraction of smooth or striated muscle. Furthermore, changes in calcium currents due to chronic treatment do not account for the acute efficacy of drug in animal pain models.

It is very important to clarify the mechanism of action of gabapentin for the purpose of developing drugs that target Ca_v2 channels. The major concern regarding neuronal calcium antagonists is whether efficacy can be achieved without significant side effects. If gabapentin does function by inhibiting Ca_v2 channels, then a better understanding of its mechanism of action could define a strategy for obtaining safe and effective Ca_v2 antagonists.

ACKNOWLEDGMENTS

We thank Drs. Stephen Hess and William Martin for helpful and thoughtful comments on the manuscript.

REFERENCES

Alden, K. J. and Garcia, J., 2001, Differential effect of gabapentin on neuronal and muscle calcium currents, *J Pharmacol Exp Ther* **297**:727-735.

Ayata, C., Shimizu-Sasamata, M., Lo, E. H., Noebels, J. L. and Moskowitz, M. A., 2000, Impaired neurotransmitter release and elevated threshold for cortical spreading depression in mice with mutations in the α_{1A} subunit of P/Q type calcium channels, *Neuroscience* **95**:639-645.

Bean, B. P., 1984, Nitrendipine block of cardiac calcium channels: High affinity binding to the inactivated state, *Proc Natl Acad Sci USA* **81**:6388-6392.

Bichet, D., Cornet, V., Geib, S., Carlier, E., Volsen, S., Hoshi, T., Mori, Y. and De Waard, M., 2000, The I-II loop of the Ca^{2+} channel α_1 subunit contains an endoplasmic reticulum retention signal antagonized by the β subunit, *Neuron* **25**:177-190.

Bolay, H., Reuter, U., Dunn, A. K., Huang, Z., Boas, D. A. and Moskowitz, M. A., 2002, Intrinsic brain activity triggers trigeminal meningeal afferents in a migraine model, *Nat Med* **8**:136-142.

Boot, J. R., 1994, Differential effects of ω-conotoxin GVIA and MVIIC on nerve stimulation induced contractions of guinea pig ileum and rat vas deferens, *Eur J Pharmacol* **258**:155-158.

Bourinet, E., Soong, T. W., Stea, A. and Snutch, T. P., 1996, Determinants of the G protein-dependent opioid modulation of neuronal calcium channels, *Proc Natl Acad Sci U S A* **93**:1486-1491.

Bourinet, E., Soong, T. W., Sutton, K., Slaymaker, S., Mathews, E., Monteil, A., Zamponi, G. W., Nargeot, J. and Snutch, T. P., 1999, Splicing of α_{1A} subunit gene generates phenotypic variants of P- and Q-type calcium channels, *Nat Neurosci* **2**:407-415.

Bourinet, E., Stotz, S. C., Spaetgens, R. L., Dayanithi, G., Lemos, J., Nargeot, J. and Zamponi, G. W., 2001, Interaction of SNX482 with domains III and IV inhibits activation gating of α_{1E} ($Ca_V2.3$) calcium channels, *Biophys J* **81**:79-88.

Bowersox, S. S., Singh, T., Nadasdi, L., Zukowska Grojec, Z., Valentino, K. and Hoffman, B. B., 1992, Cardiovascular effects of ω-conopeptides in conscious rats: mechanisms of action, *J Cardiovasc Pharmacol* **20**:756-764.

Bowersox, S. S., Gadbois, T., Singh, T., Pettus, M., Wang, Y. X. and Luther, R. R., 1996, Selective N-type neuronal voltage-sensitive calcium channel blocker, SNX-111, produces spinal antinociception in rat models of acute, persistent and neuropathic pain, *J Pharmacol Exp Ther* **279**:1243-1249.

Brodbeck, J., Davies, A., Courtney, J. M., Meir, A., Balaguero, N., Canti, C., Moss, F. J., Page, K. M., Pratt, W. S., Hunt, S. P., Barclay, J., Rees, M. and Dolphin, A. C., 2002, The ducky mutation in *Cacna2d2* results in altered Purkinje cell morphology and is associated with the expression of a truncated $\alpha2\delta$-2 protein with abnormal function, *J Biol Chem* **277**:7684-7693.

Brody, D. L., Patil, P. G., Mulle, J. G., Snutch, T. P. and Yue, D. T., 1997, Bursts of action potential waveforms relieve G-protein inhibition of recombinant P/Q-type Ca2+ channels in HEK 293 cells, *J Physiol* **499**:637-644.

Bryans, J. and Wustrow, D., 1999, 3-substituted GABA analogs with central nervous system activity: a review, *Med Res Rev* **19**:149-177.

Buchan, A. M., Gertler, S. Z., Li, H., Xue, D., Huang, Z. G., Chaundy, K. E., Barnes, K. and Lesiuk, H. J., 1994, A selective N-type Ca^{2+}-channel blocker prevents CA1 injury 24 h following severe forebrain ischemia and reduces infarction following focal ischemia, *J Cereb Blood Flow Metab* **14**:903-910.

Cao, Y., Piedras-Renteria, E. S., Smith, G. B., Teitge, S. A., Harata, N. and Tsien, R. W., 2002, Properties of P/Q-type calcium channel mutants associated with familial hemiplegic migraine (FHM) as expressed in cultured hipocampal neurons, *Soc Neurosci Abstr* **32**:Abstr. 251.213.

Carlsson, A., Forsgren, L., Nylander, P. O., Hellman, U., Forsman-Semb, K., Holmgren, G., Holmberg, D. and Holmberg, M., 2002, Identification of a susceptibility locus for migraine with and without aura on 6p12.2-p21.1, *Neurology* **59**:1804-1807.

Catterall, W. A., 2000, Structure and regulation of voltage-gated Ca^{2+} channels, *Annu Rev Cell Dev Biol* **16**:521-555.

Cizkova, D., Marsala, J., Lukacova, N., Marsala, M., Jergova, S., Orendacova, J. and Yaksh, T. L., 2002, Localization of N-type Ca^{2+} channels in the rat spinal cord following chronic constrictive nerve injury, *Exp Brain Res* **147**:456-463.

Colbourne, F., Li, H., Buchan, A. M. and Clemens, J. A., 1999, Continuing postischemic neuronal death in CA1: influence of ischemia duration and cytoprotective doses of NBQX and SNX-111 in rats, *Stroke* **30**:662-668.

Colecraft, H. M., Patil, P. G. and Yue, D. T., 2000, Differential occurrence of reluctant openings in G-protein-inhibited N- and P/Q-type calcium channels, *J Gen Physiol* **115**:175-192.

De Waard, M., Liu, H., Walker, D., Scott, V. E., Gurnett, C. A. and Campbell, K. P., 1997, Direct binding of G-protein $\beta\gamma$ complex to voltage-dependent calcium channels, *Nature* **385**:446-450.

Delcour, A. H., Lipscombe, D. and Tsien, R. W., 1993, Multiple modes of N-type calcium channel activity distinguished by differences in gating kinetics, *J Neurosci* **13**:181-194.

Delmas, P., Abogadie, F. C., Buckley, N. J. and Brown, D. A., 2000, Calcium channel gating and modulation by transmitters depend on cellular compartmentalization, *Nat Neurosci* **3**:670-678.

Diriong, S., Lory, P., Williams, M. E., Ellis, S. B., Harpold, M. M. and Taviaux, S., 1995, Chromosomal localization of the human genes for α_{1A}, α_{1B}, and α_{1E} voltage-dependent Ca^{2+} channel subunits, *Genomics* **30**:605-609.

Dolphin, A. C., 1995, The G.L. Brown Prize Lecture. Voltage-dependent calcium channels and their modulation by neurotransmitters and G proteins, *Exp Physiol* **80**:1-36.

Dooley, D. J., Donovan, C. M., Meder, W. P. and Whetzel, S. Z., 2002, Preferential action of gabapentin and pregabalin at P/Q-type voltage-sensitive calcium channels: inhibition of K^+-evoked [^{3}H]-norepinephrine release from rat neocortical slices, *Synapse* **45**:171-190.

Dubel, S. J., Starr, T. V., Hell, J., Ahlijanian, M. K., Enyeart, J. J., Catterall, W. A. and Snutch, T. P., 1992, Molecular cloning of the α_1 subunit of an ω-conotoxin-sensitive calcium channel, *Proc Natl Acad Sci U S A* **89**:5058-5062.

Ducros, A., Joutel, A., Vahedi, K., Cecillon, M., Ferreira, A., Bernard, E., Verier, A., Echenne, B., Lopez de Munain, A., Bousser, M. G. and Tournier-Lasserve, E., 1997, Mapping of a second locus for familial hemiplegic migraine to 1q21-q23 and evidence of further heterogeneity, *Ann Neurol* **42**:885-890.

Dunlap, K., Luebke, J. I. and Turner, T. J., 1995, Exocytotic Ca^{2+} channels in mammalian central neurons, *Trends Neurosci* **18**:89-98.

Ertel, E. A. and Cohen, C. J., 1994, Voltage-dependent interactions: The influence and significance of membrane potential on drug-receptor interactions, *Drug Develop Res* **33**:203-213.

Ertel, E. A., Campbell, K. P., Harpold, M. M., Hofmann, F., Mori, Y., Perez Reyes, E., Schwartz, A., Snutch, T. P., Tanabe, T., Birnbaumer, L., Tsien, R. W. and Catterall, W. A., 2000, Nomenclature of voltage-gated calcium channels, *Neuron* **25**:533-535.

Felix, R., Gurnett, C. A., De Waard, M. and Campbell, K. P., 1997, Dissection of functional domains of the voltage-dependent Ca^{2+} channel $\alpha_2\delta$ subunit, *J Neurosci* **17**:6884-6891.

Felix, R., 1999, Voltage-dependent Ca^{2+} channel $\alpha_2\delta$ auxiliary subunit: Structure, function and regulation, *Receptors Channels* **6** (5):351-362.

Feng, Z. P., Hamid, J., Doering, C., Bosey, G. M., Snutch, T. P. and Zamponi, G. W., 2001, Residue Gly^{1326} of the N-type calcium channel α_{1B} subunit controls reversibility of ω-conotoxin GVIA and MVIIA block, *J Biol Chem* **276**:15728-15735.

Fernandez, J. M., Granja, R., Izaguirre, V., Gaonzalez-Garcia, C. and Cena, V., 1995, ω-conotoxin GVIA blocks nicotine-induced catecholamine secretion by blocking the nicotinic receptor-activated inward currents in bovine chromaffin cells, *Neurosci Lett* **191**:59-62.

Fink, K., Meder, W., Dooley, D. J. and Gothert, M., 2000, Inhibition of neuronal Ca^{2+} influx by gabapentin and subsequent reduction of neurotransmitter release from rat neocortical slices, *Br J Pharmacol* **130**:900-906.

Fletcher, C. F., Lutz, C. M., O'Sullivan, T. N., Shaughnessy, J. D., Jr, H., R., Frankel, W. N., Copeland, N. G. and Jenkins, N. A., 1996, Absence epilepsy in tottering mutant mice is associated with calcium channel defects, *Cell* **87**:607-617.

Fletcher, C. F., Tottene, A., Lennon, V. A., Wilson, S. M., Dubel, S. J., Paylor, R., Hosford, D. A., Tessarollo, L., McEnery, M. W., Pietrobon, D., Copeland, N. G. and Jenkins, N. A., 2001, Dystonia and cerebellar atrophy in Cacna1a null mice lacking P/Q calcium channel activity, *FASEB J* **15**:1288-1290.

Fusco, M. D., Marconi, R., Silvestri, L., Atorino, L., Rampoldi, L., Morgante, L., Ballabio, A., Aridon, P. and Casari, G., 2003, Haploinsufficiency of ATP1A2 encoding the Na^+/K^+ pump α2 subunit associated with familial hemiplegic migraine type 2, *Nat Genet* **33**:192-196.

Gasparini, S., Kasyanov, A. M., Pietrobon, D., Voronin, L. L. and Cherubini, E., 2001, Presynaptic R-type calcium channels contribute to fast excitatory synaptic transmission in the rat hippocampus, *J Neurosci* **21**:8715-8721.

Gee, N. S., Brown, J. P., Dissanayake, V. U. K., Offord, J., Thurlow, R. and Woodruff, G. N., 1996, The novel anticonvulsant drug, gabapentin (Neurontin), binds to the $\alpha_2\delta$ subunit of a calcium channel, *J Biol Chem* **271**:5768-5776.

Goadsby, P. J., Lipton, R. B. and Ferrari, M. D., 2002, Migraine--current understanding and treatment, *N Engl J Med* **346**:257-270.

Gothert, M. and Molderings, G. J., 1997, Mibefradil and ω-conotoxin-GVIA-induced inhibition of noradrenaline release from the sympathetic nerves of the human heart, *Naunyn Schmiedebergs Arch Pharmacol* **356**:860-863.

Guida, S., Trettel, F., Pagnutti, S., Mantuano, E., Tottene, A., Veneziano, L., Fellin, T., Spadaro, M., Stauderman, K., Williams. M, Volsen, S., Ophoff, R., Frants, R., Jodice, C., Frontali, M. and Pietrobon, D., 2001, Complete loss of P/Q calcium channel activity caused by a CACNA1A missense mutation carried by patients with episodic ataxia type 2, *Am J Hum Genet* **68**:759-764.

Hamid, J., Nelson, D., Spaetgens, R., Dubel, S. J., Snutch, T. P. and Zamponi, G. W., 1999, Identification of an Integration Center for Cross-talk between Protein Kinase C and G Protein Modulation of N-type Calcium Channels, *J Biol Chem* **274**:6195-6202.

Han, W., Saegusa, H., Zong, S. and Tanabe, T., 2002, Altered cocaine effects in mice lacking $Ca_v2.3$ (α1E) calcium channel, *Biochem Biophys Res Comm* **299**:299-304.

Hans, M., Luvisetto, S., Williams, M. E., Spagnolo, M., Urrutia, A., Tottene, A., Brust, P. F., Johnson, E. C., Harpold, M. M., Stauderman, K. A. and Pietrobon, D., 1999a, Functional consequences of mutations in the human α_{1A} calcium channel subunit linked to familial hemiplegic migraine, *J Neurosci* **19**:1610-1619.

Hans, M., Urrutia, A., Deal, C., Brust, P. F., Stauderman, K., Ellis, S. B., Harpold, M. M., Johnson, E. C. and Williams, M. E., 1999b, Structural elements in domain IV that influence biophysical and pharmacological properties of human α_{1A} -containing high-voltage-activated calcium channels, *Biophys J* **76**:1384-1400.

Hatakeyama, S., Wakamori, M., Ino, M., Miyamoto, N., Takahashi, E., Yoshinaga, T., Sawada, K., Imoto, K., Tanaka, I., Yoshizawa, T., Nishizawa, Y., Mori, Y., Niidome, T. and Shoji, S., 2001, Differential nociceptive responses in mice lacking the α_{1B} subunit of N-type Ca^{2+} channels, *Neuroreport* **12**:2423-2427.

Herlitze, S., Garcia, D. E., Mackie, K., Hille, B., Scheuer, T. and Catterall, W. A., 1996, Modulation of Ca^{2+} channels by G-protein βγ subunits, *Nature* **380**:258-262.

Herlitze, S., Zhong, H., Scheuer, T. and Catterall, W. A., 2001, Allosteric modulation of Ca^{2+} channels by G proteins, voltage-dependent facilitation, protein kinase C, and $Ca_v\beta$ subunits, *Proc Natl Acad Sci U S A* **98**:4699-4704.

Herrero, C. J., Garcia-Palomero, E., Pintado, A. J., Garcia, A. G. and Montiel, C., 1999, Differential blockade of rat α3β4 and α7 neuronal nicotinic receptors by ω-conotoxin MVIIC, ω-conotoxin GVIA and diltiazem, *Br J Pharmacol* **127**:1375-1387.

Hill, R., 2000, NK1 (substance P) receptor antagonists - why are they not analgesic in humans?, *Trends Pharmacol Sci* **21**:244-246.

Hille, B., 1977, Local anesthetics: hydrophilic and hydrophobic pathways for the drug-receptor reaction, *J Gen Physiol* **69**:497-515.

Hillman, D., Chen, S., Aung, T. T., Cherksey, B., Sugimori, M. and Llinas, R. R., 1991, Localization of P-type calcium channels in the central nervous system, *Proc Natl Acad Sci U S A* **88**:7076-7080.

Hockerman, G. H., Peterson, B. Z., Johnson, B. D. and Catterall, W. A., 1997, Molecular determinants of drug binding and action on L-type calcium channels, *Annu Rev Pharmacol Toxicol* **37**:361-396.

Ikeda, S. R., 1996, Voltage-dependent modulation of N-type clacium channels by G-protein $\beta\gamma$ subunits, *Nature* **380**:258-262.

Ino, M., Yoshinaga, T., Wakamori, M., Miyamoto, N., Takahashi, E., Sonoda, J., Kagaya, T., Oki, T., Nagasu, T., Nishizawa, Y., Tanaka, I., Imoto, K., Aizawa, S., Koch, S., Schwartz, A., Niidome, T., Sawada, K. and Mori, Y., 2001, Functional disorders of the sympathetic nervous system in mice lacking the α_{1B} subunit Ca_v 2.2 of N-type calcium channels, *Proc Natl Acad Sci USA* **98**:5323-5328.

Ishikawa, K., Fujigasaki, H., Saegusa, H., Ohwada, K., Fujita, T., Iwamoto, H., Komatsuzaki.Y., Toru, S., Toriyama, H., Watanabe, M., Ohkoshi, N., Shoji, S., Kanazawa, I., Tanabe, T. and Mizusawa, H., 1999, Abundant expression and cytoplasmic aggregations of α_{1A} voltage-dependent calcium channel protein associated with neurodegeneration in spinocerebellar ataxia type 6, *Hum Mol Genet* **8**:1185-1193.

Jain, K. K., 2000, An evaluation of intrathecal ziconotide for the treatment of chronic pain, *Exp Opin Invest Drugs* **9**:2403-2410.

Jarvis, S. E. and Zamponi, G. W., 2001, Interactions between presynaptic Ca^{2+} channels, cytoplasmic messengers and proteins of the synaptic vesicle release complex, *Trends Pharmacol Sci* **22**:519-525.

Jimenez, C., Bourinet, E., Leuranguer, V., Richard, S., Snutch, T. P. and Nargeot, J., 2000, Determinants of voltage-dependent inactivation affect Mibefradil block of calcium channels, *Neuropharmacology* **39**:1-10.

Jones, S. W. and Elmslie, K. S., 1997, Transmitter modulation of neuronal calcium channels, *J Membr Biol* **155**:1-10.

Jouvenceau, A., Eunson, L. H., Spauschus, A., Ramesh, V., Zuberi, S. M., Kullmann, D. M. and Hanna, M. G., 2001, Human epilepsy associated with dysfunction of the brain P/Q-type calcium channel, *Lancet* **358**:801-807.

Julius, D. and Basbaum, A. I., 2001, Molecular mechanisms of nociception, *Nature* **413** (6852):203-210.

Jun, K., Piedras Renteria, E. S., Smith, S. M., Wheeler, D. B., Lee, S. B., Lee, T. G., Chin, H., Adams, M. E., Scheller, R. H., Tsien, R. W. and Shin, H. S., 1999, Ablation of P/Q-type Ca^{2+} channel currents, altered synaptic transmission, and progressive ataxia in mice lacking the α_{1A}-subunit, *Proc Natl Acad Sci U S A* **96**:15245-15250.

Kaneko, S., Cooper, C. B., Nishioka, N., Yamasaki, H., Suzuki, A., Jarvis, S. E., Akaike, A., Satoh, M. and Zamponi, G. W., 2002, Identification and characterization of novel human $Ca_v2.2$ (α_{1B}) calcium channel variants lacking the synaptic protein interaction site, *J Neurosci* **22**:82-92.

Kang, M.-G., Chen, C.-C., Felix, R., Letts, V. A., Frankel, W. N., Mori, Y. and Campbell, K. P., 2001, Biochemical and Biophysical Evidence for γ_2 Subunit Association with Neuronal Voltage-activated Ca^{2+} Channels, *J Biol Chem* **276**:32917-32924.

Kang, M.-G., Felix, R. and Campbell, K. P., 2002, Long-term regulation of voltage-gated Ca^{2+} channels by gabapentin, *FEBS Lett* **528**:177-182.

Kim, C., Jun, K., Lee, T., Kim, S. S., McEnery, M. W., Chin, H., Kim, H. L., Park, J. M., Kim, D. K., Jung, S. J., Kim, J. and Shin, H. S., 2001, Altered nociceptive response in mice deficient in the α_{1B} subunit of the voltage-dependent calcium channel, *Mol Cell Neurosci* **18**:235-245.

Kraus, R. L., Sinnegger, M. J., Glossmann, H., Hering, S. and J., S., 1998, Familial hemiplegic migraine mutations change α_{1A} Ca^{2+} channel kinetics, *J Biol Chem* **273**:5586-5590.

Kraus, R. L., Sinnegger, M. J., Koschak, A., Glossmann, H., Stenirri, S., Carrera, P. and Striessnig, J., 2000, Three new familial hemiplegic migraine mutants affect P/Q-type Ca^{2+}channel kinetics, *J Biol Chem* **275**:9239-9243.

Lalo, U. V., Pankratov, Y. V., Arndts, D. and Krishtal, O. A., 2001, [Omega]-conotoxin GVIA potently inhibits the currents mediated by P2X receptors in rat DRG neurons, *Brain Res Bull* **54**:507-512.

Lanneau, C., Green, A., Hirst, W. D., Wise, A., Brown, J. T., Donnier, E., Charles, K. J., Wood, M., Davies, C. H. and Pangalos, M. N., 2001, Gabapentin is not a GABAB receptor agonist, *Neuropharmacology* **41**:965-975.

Lee, A., Wong, S. T., Gallagher, D., Li, B., Storm, D. R., Scheuer, T. and Catterall, W. A., 1999, Ca^{2+}/calmodulin binds to and modulates P/Q-type calcium channels, *Nature* **399**:155-159.

Lee, A., Westenbroek, R. E., Haeseleer, F., Palczewski, K., Scheuer, T. and Catterall, W. A., 2002a, Differential modulation of $Ca_v2.1$ channels by calmodulin and Ca^{2+}-binding protein 1, *Nat Neurosci* **5**:210-217.

Lee, S. W., Choi, S., Lee, T., Kim, H.-L., Chin, H. and Shin, H.-S., 2002b, Molecular basis of R-type calcium channels in central amygdala neurons of the mouse, *Proc Natl Acad Sci USA* **99**:3276-3281.

Lewis, R. J., Nielsen, K. J., Craik, D. J., Loughnan, M. L., Adams, D. A., Sharpe, I. A., Luchian, T., Adams, D. J., Bond, T., Thomas, L., Jones, A., Matheson, J.-L., Drinkwater, R., Andrews, P. R. and Alewood, P. F., 2000, Novel ω-Conotoxins from *Conus catus* Discriminate among Neuronal Calcium Channel Subtypes, *J Biol Chem* **275**:35335-35344.

Lin, Z., Haus, S., Edgerton, J. and Lipscombe, D., 1997, Identification of functionally distinct isoforms of the N-type Ca^{2+} channel in rat sympathetic ganglia and brain, *Neuron* **18**:153-166.

Lipscombe, D., Pan, J. Q. and Gray, A., 2002, Functional Diversity in Neuronal Voltage-Gated Calcium Channels by Alternative Splicing of $Cav\alpha_1$, *Mol Neurobiol* **26**:21-44.

Liu, H., De Waard, M., Scott, V. E., Gurnett, C. A., Lennon, V. A. and Campbell, K. P., 1996, Identification of three subunits of the high affinity ω-conotoxin MVIIC-sensitive Ca^{2+} channel, *J Biol Chem* **271**:13804-13810.

Lu, Q., AtKisson, M. S., Jarvis, S. E., Feng, Z. P., Zamponi, G. W. and Dunlap, K., 2001, Syntaxin 1A supports voltage-dependent inhibition of α_{1B} Ca^{2+} channels by $G\beta\gamma$ in chick sensory neurons, *J Neurosci* **21**:2949-2957.

Luo, Z. D., Chaplan, S. R., Higuera, E. S., Sorkin, L. S., Stauderman, K. A., Williams, M. E. and Yaksh, T. L., 2001, Upregulation of dorsal root ganglion $\alpha_2\delta$ calcium channel subunit and its correlation with allodynia in spinal nerve-injured rats, *J Neurosci* **21**:1868-1875.

Luo, Z. D., Calcutt, N. A., Higuera, E. S., Valder, C. R., Song, Y. H., Svensson, C. I. and Myers, R. R., 2002, Injury type-specific calcium channel $\alpha_2\delta$-1 subunit up-regulation in rat neuropathic pain models correlates with antiallodynic effects of gabapentin, *J Pharmacol Exp Ther* **303**:1199-1205.

Malmberg, A. B. and Yaksh, T. L., 1995, Effect of continuous intrathecal infusion of ω-conopeptides, N-type calcium-channel blockers, on behavior and antinociception in the formalin and hot-plate tests in rats, *Pain* **60**:83-90.

Malmberg, A. B., Gilbert, H., McCabe, R. T. and Basbaum, A. I., 2003, Powerful antinociceptive effects of the cone snail venom-derived subtype-selective NMDA receptor antagonists conantokins G and T, *Pain* **101**:109-116.

Marais, E., Klugbauer, N. and Hofmann, F., 2001, Calcium channel $\alpha_2\delta$ subunits - Structure and gabapentin binding, *Mol Pharmacol* **59** (5):1243-1248.

Martin, D. J., McClelland, D., Herd, M. B., Sutton, K. G., Lee, K., Pinnock, R. D. and Scott, R. H., 2002, Gabapentin-mediated inhibition of voltage-activated Ca^{2+} channel currents in cultured sensory neurones is dependent on culture conditions and channel subunit expression, *Neuropharmacology* **42**:353-366.

Matsuyama, Z., Wakamori, M., Mori, Y., Kawakami, H., Nakamura, S. and Imoto, K., 1999, Direct alteration of the P/Q-type Ca^{2+} channel property by polyglutamine expansion in spinocerebellar ataxia 6, *J Neurosci* **19**:RC14.

Matthews, E. A. and Dickenson, A. H., 2001, Effects of spinally delivered N- and P-type voltage-dependent calcium channel antagonists on dorsal horn neuronal responses in a rat model of neuropathy, *Pain* **92**:235-246.

Maximov, A. and Bezprozvanny, I., 2002, Synaptic Targeting of N-Type Calcium Channels in Hippocampal Neurons, *J Neurosci* **22**:6939-6952.

May, A., Ophoff, R. A., Terwindt, G. M., Urban, C., van Eijk, R., Haan, J., Diener, H. C., Lindhout, D., Frants, R. R., Sandkuijl, L. A. and Ferrari, M. D., 1995, Familial hemiplegic migraine locus on 19p13 is involved in the common forms of migraine with and without aura, *Hum Genet* **96**:604-608.

McDonough, S. I., Boland, L. M., Mintz, I. M. and Bean, B. P., 2002, Interactions among Toxins That Inhibit N-type and P-type Calcium Channels, *J Gen Physiol* **119**:313-328.

McEnery, M. W., Snowman, A. M., Sharp, A. H., Adams, M. E. and Snyder, S. H., 1991, Purified ω-conotoxin GVIA receptor of rat brain resembles a dihydropyridine-sensitive L-type calcium channel, *Proc Natl Acad Sci U S A* **88**:11095-11099.

Meir, A., Ginsburg, S., Butkevich, A., Kachalsky, S. G., Kaiserman, I., Ahdut, R., Demirgoren, S. and Rahamimoff, R., 1999, Ion Channels in Presynaptic Nerve Terminals and Control of Transmitter Release, *Physiol Rev* **79**:1019-1088.

Mellegers, M. A., Furlan, A. D. and Mailis, A., 2001, Gabapentin for neuropathic pain: systematic review of controlled and uncontrolled literature, *Clin J Pain* **17**:284-295.

Melliti, K., Grabner, M. and Seabrook, G., 2003, The familial hemiplegic migraine mutation R192Q reduces G-protein-mediated inhibition of P/Q-type ($Ca_V2.1$) calcium channels expressed in human embryonic kidney cells, *J Physiol* **546**:337-347.

Mermelstein, P. G. and Surmeier, D. J., 1997, A calcium channel reversibly blocked by ω-conotoxin GVIA lacking the class D α_1 subunit, *Neuroreport* **8**:485-489.

Meza, U. and Adams, B., 1998, G-Protein-dependent facilitation of neuronal α_{1A}, α_{1B}, and α_{1E} Ca channels, *J Neurosci* **18**:5240-5252.

Mintz, I. M., Adams, M. E. and Bean, B. P., 1992, P-type calcium channels in rat central and peripheral neurons, *Neuron* **9**:85-95.

Murakami, M., B.Fleischmann, Felipe, C. D., M.Freichel, C.Trost, A.Ludwing, U.Wissenbach, H.Schwegler, F.Hofmann, J.Hescheler, V.Flockerzi and A.Cavalie, 2002, Pain perception in mice lacking the β_3 Subunit of voltage-active calcium channels, *J Biol Chem* **277**:40342-40351.

Namkung, Y., Smith, S. M., Lee, S. B., Skrypnyk, N. V., Kim, H. L., Chin, H., Scheller, R. H., Tsien, R. W. and Shin, H. S., 1998, Targeted disruption of the Ca^{2+} channel β_3 subunit reduces N- and L-type Ca^{2+} channel activity and alters the voltage-dependent activation of P/Q-type Ca^{2+} channels in neurons, *Proc Natl Acad Sci U S A* **95**:12010-12015.

Newcomb, R., Szoke, B., Palma, A., Wang, G., Chen, X. h., Hopkins, W., Cong, R., Miller, J., Urge, L., Tarczy Hornoch, K., Loo, J. A., Dooley, D. J., Nadasdi, L., Tsien, R. W., Lemos, J. and Miljanich, G., 1998, Selective peptide antagonist of the class E calcium channel from the venom of the tarantula *Hysterocrates gigas*, *Biochem* **37**:15353-15362.

Olivera, B. M., Cruz, L. J. and Yoshikami, D., 1999, Effects of *Conus* peptides on the behavior of mice, *Curr Opin Neurobiol* **9**:772-777.

Ophoff, R. A., van den Maagdenberg, A. M., Roon, K. I., Ferrari, M. D. and Frants, R. R., 2001, The impact of pharmacogenetics for migraine, *Eur J Pharmacol* **413**:1-10.

Penn, R. D. and Paice, J. A., 2000, Adverse effects associated with the intrathecal administration of ziconotide, *Pain* **85**:291-296.

Piedras Renteria, E. S. and Tsien, R. W., 1998, Antisense oligonucleotides against α_{1E} reduce R-type calcium currents in cerebellar granule cells, *Proc Natl Acad Sci U S A* **95**:7760-7765.

Piedras-Renteria, E. S., Watase, K., Harata, N., Zhuchenko, O., Zoghbi, H. Y., Lee, C. C. and Tsien, R. W., 2001, Increased expression of α_{1A} Ca^{2+} channel currents arising from expanded trinucleotide repeats in spinocerebellar ataxia type 6, *J Neurosci* **21**:9185-9193.

Pringle, A. K., Benham, C. D., Sim, L., Kennedy, J., Iannotti, F. and Sundstrom, L. E., 1996, Selective N-type calcium channel antagonist ω conotoxin MVIIA is neuroprotective against hypoxic neurodegeneration in organotypic hippocampal-slice cultures, *Stroke* **27**:2124-2130.

Protti, D., Reisin, R., Mackinley, T. and Uchitel, O., 1996, Calcium channel blockers and transmitter release at the normal human neuromuscular junction, *Neurology* **46**:1391-1396.

Qin, N., Yagel, S., Momplaisir, M. L., Codd, E. E. and MR, D. A., 2002, Molecular cloning and characterization of the human voltage-gated calcium channel $\alpha_2\delta$-4 subunit, *Mol Pharmacol* 62:485-496.

Randall, A. and Tsien, R. W., 1995, Pharmacological dissection of multiple types of Ca^{2+} channel currents in rat cerebellar granule neurons, *J Neurosci* **15**:2995-3012.

Restituito, S., Thompson, R. M., Eliet, J., Raike, R. S., Riedl, M., Charnet, P. and Gomez, C. M., 2000, The polyglutamine expansion in spinocerebellar ataxia type 6 causes a β subunit-specific enhanced activation of P/Q-type calcium channels in Xenopus oocytes, *J Neurosci* **20**:6394-6403.

Richter, F., Ebersberger, A. and Schaible, H. G., 2002, Blockade of voltage-gated calcium channels in rat inhibits repetitive cortical spreading depression, *Neurosci Lett* **334**:123-126.

Rittenhouse, A. R. and Hess, P., 1994, Microscopic heterogeneity in unitary N-type calcium currents in rat sympathetic neurons, *J Physiol* **474**:87-99.

Rock, D. M., Kelly, K. M. and Macdonald, R. L., 1993, Gabapentin actions on ligand- and voltage-gated responses in cultured rodent neurons, *Epilepsy Res* **16**:89-98.

Sabatini, B. L. and Regehr, W. G., 1999, Timing of synaptic transmission, *Annu Rev Physiol* **61**:521-542.

Saegusa, H., Kurihara, T., Zong, S., Minowa, O., Kazuno, A., Han, W., Matsuda, Y., Yamanaka, H., Osanai, M., Noda, T. and Tanabe, T., 2000, Altered pain responses in mice lacking α_{1E} subunit of the voltage-dependent Ca^{2+} channel, *Proc Natl Acad Sci USA* **97**:6132-6137.

Saegusa, H., Kurihara, T., Zong, S., Kazuno, A., Matsuda, Y., Nonaka, T., Han, W., Toriyama, H. and Tanabe, T., 2001, Suppression of inflammatory and neuropathic pain symptoms in mice lacking the N-type Ca^{2+} channel, *EMBO J* **20**:2349-2356.

Sanger, G. J., Ellis, E. S., Harries, M. H., Tilford, N. S., Wardle, K. A. and Benham, C. D., 2000, Rank-order inhibition by ω-conotoxins in human and animal autonomic nerve preparations, *Eur J Pharmacol* **388**:89-95.

Scott, D. A., Wright, C. E. and Angus, J. A., 2002, Actions of intrathecal ω-conotoxins CVID, GVIA, MVIIA, and morphine in acute and neuropathic pain in the rat, *Eur J Pharmacol* **451**:279-286.

Scott, V. E., De Waard, M., Liu, H., Gurnett, C. A., Venzke, D. P., Lennon, V. A. and Campbell, K. P., 1996, β subunit heterogeneity in N-type Ca^{2+} channels, *J Biol Chem* **271**:3207-3212.

Shapiro, M. S., Zhou, J. and Hille, B., 1996, Selective disruption by protein kinases of G-protein-mediated Ca^{2+} channel modulation, *J Neurophysiol* **76**:311-320.

Simen, A. A., Lee, C. C., Simen, B. B., Bindokas, V. P. and Miller, R. J., 2001, The C terminus of the Ca channel α_{1B} subunit mediates selective inhibition by G-protein-coupled receptors, *J Neurosci* **21**:7587-7597.

Sivilotti, L. G., Thompson, S. W. and Woolf, C. J., 1993, Rate of rise of the cumulative depolarization evoked by repetitive stimulation of small-caliber afferents is a predictor of action potential windup in rat spinal neurons in vitro, *J Neurophysiol* **69**:1621-1631.

Smith, M. T., Cabot, P. J., Ross, F. B., Robertson, A. D. and Lewis, R. J., 2002, The novel N-type calcium channel blocker, AM336, produces potent dose-dependent antinociception after intrathecal dosing in rats and inhibits substance P release in rat spinal cord slices, *Pain* **96**:119-127.

Snutch, T. P., Feng, Z. P., Doering, C., Cayabyab, F., Janke, D., Parker, D. B., Belardetti, F., Morimoto, B., Vanderah, T., Zamponi, G. W. and Porreca, F., 2001a, A new class of N-type calcium channel blocker efficacious in animal models of chronic pain, *Soc Neurosci Abstr* **31**:Abstr. 465.461.

Snutch, T. P., Sutton, K. G. and Zamponi, G. W., 2001b, Voltage-dependent calcium channels--beyond dihydropyridine antagonists, *Curr Opin Pharmacol* **1**:11-16.

Soong, T. W., Stea, A., Hodson, C. D., Dubel, S. J., Vincent, S. R. and Snutch, T. P., 1993, Structure and functional expression of a member of the low voltage-activated calcium channel family, *Science* **260**:1133-1136.

Stea, A., Tomlinson, W. J., Soong, T. W., Bourinet, E., Dubel, S. J., Vincent, S. R. and Snutch, T. P., 1994, Localization and functional properties of a rat brain α_{1A} calcium channel reflect similarities to neuronal Q- and P-type channels, *Proc Natl Acad Sci U S A* **91**:10576-10580.

Stea, A., Dubel, S. J. and Snutch, T. P., 1999, α_{1B} N-type calcium channel isoforms with distinct biophysical properties, *Ann N Y Acad Sci* **868**:118-130.

Stocker, J. W., Nadasdi, L., Aldrich, R. W. and Tsien, R. W., 1997, Preferential interaction of ω-conotoxins with inactivated N-type Ca^{2+} channels, *J Neurosci* **17**:3002-3013.

Striessnig, J., Grabner, M., Mitterdorfer, J., Hering, S., Sinnegger, M. J. and Glossmann, H., 1998, Structural basis of drug binding to L Ca^{2+} channels, *Trends Pharmacol Sci* **19**:108-115.

Striessnig, J., 1999, Pharmacology, structure and function of cardiac L-type Ca^{2+} channels, *Cell Physiol Biochem* **9**:242-269.

Sutton, K. G. and Snutch, T. P., 2001, Gabapentin: A novel analgesic targeting voltage-gated calcium channels, *Drug Develop Res* **54** (3):167-172.

Sutton, K. G., Martin, D. J., Pinnock, R. D., Lee, K. and Scott, R. H., 2002, Gabapentin inhibits high-threshold calcium channel currents in cultured rat dorsal root ganglion neurones, *Br J Pharmacol* **135**:257-265.

Swartz, K. J., 1993, Modulation of Ca^{2+} channels by protein kinase C in rat central and peripheral neurons: disruption of G protein-mediated inhibition, *Neuron* **11**:305-320.

Taylor, C. P., Gee, N. S., Su, T. Z., Kocsis, J. D., Welty, D. F., Brown, J. P., Dooley, D. J., Boden, P. and Singh, L., 1998, A summary of mechanistic hypotheses of gabapentin pharmacology, *Epilepsy Res* **29** (3):233-249.

Terwindt, G. M., Ophoff, R. A., van Eijk, R., Vergouwe, M. N., Haan, J., Frants, R. R., Sandkuijl, L. A. and Ferrari, M. D., 2001, Involvement of the *CACNA1A* gene containing region on 19p13 in migraine with and without aura, *Neurology* **56**:1028-1032.

Toriyama, H., Wang, L., Saegusa, H., Zong, S., Osanai, M., Murakoshi, T., Noda, T., Ohno, K. and Tanabe, T., 2002, Role of $Ca_v2.3$ (α_{1E}) Ca^{2+} channel in ischemic neuronal injury, *Neuroreport* **13**:261-265.

Toru, S., Murakoshi, T., Ishikawa, K., Saegusa, H., Fujigasaki, H., Uchihara, T., Nagayama, S., Osanai, M., Mizusawa, H. and Tanabe, T., 2000, Spinocerebellar ataxia type 6 mutation alters P-type calcium channel function, *J Biol Chem* **275**:10893-10898.

Tottene, A., Volsen, S. and Pietrobon, D., 2000, α_{1E} subunits form the pore of three cerebellar R-type calcium channels with different pharmacological and permeation properties, *J Neurosci* **20**:171-178.

Tottene, A., Fellin, T., Pagnutti, S., Luvisetto, S., Striessnig, J., Fletcher, C. and Pietrobon, D., 2002, Familial hemiplegic migraine mutations increase Ca^{2+} influx through single human $Ca_V2.1$ channels and decrease maximal $Ca_V2.1$ current density in neurons, *Proc Natl Acad Sci U S A* **99**:13284-13289.

Uchitel, O. D., Protti, D. A., Sanchez, V., Cherksey, B. D., Sugimori, M. and Llinas, R., 1992, P-type voltage-dependent calcium channel mediates presynaptic calcium influx and transmitter release in mammalian synapses, *Proc Natl Acad Sci U S A* **89**:3330-3333.

Uneyama, H., Takahara, A., Wakamori, M., Mori, Y. and Yoshimoto, R., 1999, Pharmacology of N-type Ca^{2+} channels distributed in cardiovascular system (Review), *Int J Mol Med* **3**:455-466.

Valentino, K., Newcomb, R., Gadbois, T., Singh, T., Bowersox, S., Bitner, S., Justice, A., Yamashiro, D., Hoffman, B. B., Ciaranello, R., Miljanich, G. and Ramachandran, J., 1993, A selective N-type calcium channel antagonist protects against neuronal loss after global cerebral ischemia, *Proc Natl Acad Sci USA* **90**:7894-7897.

Van Den Maagdenberg, A. M., Kors, E. E., Brunt, E. R., Van Paesschen, W., Pascual, J., Ravine, D., Keeling, S., Vanmolkot, K. R., Vermeulen, F. L., Terwindt, G. M., Haan, J., Frants, R. R. and Ferrari, M. D., 2002, Episodic ataxia type 2 Three novel truncating mutations and one novel missense mutation in the CACNA1A gene, *J Neurol* **249**:1515-1519.

Van der Lee, R., Pfaffendorf, M., De Mey, J. G. and van Zwieten, P. A., 2000, Inhibitory effect of mibefradil on contractions induced by sympathetic neurotransmitter release in the rat tail artery, *Naunyn Schmiedebergs Arch Pharmacol* **361**:74-79.

Vanegas, H. and Schaible, H., 2000, Effects of antagonists to high-threshold calcium channels upon spinal mechanisms of pain, hyperalgesia and allodynia, *Pain* **85**:9-18.

Vikman, K. S., Kristensson, K. and Hill, R. H., 2001, Sensitization of dorsal horn neurons in a two-compartment cell culture model: wind-up and long-term potentiation-like responses, *J Neurosci* **21**:Rc169.

Wakamori, M., Yamazaki, K., Matsunodaira, H., Teramoto, T., Tanaka, I., Niidome, T., Sawada, K., Nishizawa, Y., Sekiguchi, N., Mori, E., Mori, Y. and Imoto, K., 1998, Single tottering mutations responsible for the neuropathic phenotype of the P-type calcium channel, *J Biol Chem* **273**:34857-34867.

Wang, Y. X., Gao, D., Pettus, M., Phillips, C. and Bowersox, S. S., 2000, Interactions of intrathecally administered ziconotide, a selective blocker of neuronal N-type voltage-sensitive calcium channels, with morphine on nociception in rats, *Pain* **84**:271-281.

Wappl, E., Koschak, A., Poteser, M., Sinnegger, M. J., Walter, D., Eberhart, A., Groschner, K., Glossmann, H., Kraus, R. L., Grabner, M. and Striessnig, J., 2002, Functional consequences of P/Q-type Ca^{2+} channel $Ca_v2.1$ missense mutations associated with episodic ataxia type 2 and progressive ataxia, *J Biol Chem* **277**:6960-6966.

Westenbroek, R. E., Hoskins, L. and Catterall, W. A., 1998, Localization of Ca^{2+} channel subtypes on rat spinal motor neurons, interneurons, and nerve terminals, *J Neurosci* **18**:6319-6330.

Williams, M. E., Brust, P. F., Feldman, D. H., Patthi, S., Simerson, S., Maroufi, S., McCue, A. F., Velicelebi, G., Ellis, S. B. and Harpold, M. M., 1992a, Structure and functional expression of an ω-conotoxin-sensitive human N-type calcium channel, *Science* **257**:389-395.

Williams, M. E., Feldman, D. H., McCue, A. F., Brenner, R., Velicelebi, G., Ellis, S. B. and Harpold, M. M., 1992b, Structure and functional expression of α_1, α_2, and β subunits of a novel human neuronal calcium channel subtype, *Neuron* **8**:71-84.

Williams, M. E., Marubio, L. M., Deal, C. R., Hans, M., Brust, P. F., Philipson, L. H., Miller, R. J., Johnson, E. C., Harpold, M. M. and Ellis, S. B., 1994, Structure and functional characterization of neuronal α_{1E} calcium channel subtypes, *J Biol Chem* **269**:22347-22357.

Wilson, S. M., Toth, P. T., Oh, S. B., Gillard, S. E., Volsen, S., Ren, D., Philipson, L. H., Lee, E. C., Fletcher, C. F., Tessarollo, L., Copeland, N. G., Jenkins, N. A. and Miller, R. J., 2000, The status of voltage-dependent calcium channels in α_{1E} knockout mice, *J Neurosci* **20**:8566-8571.

Witcher, D. R., De Waard, M., Sakamoto, J., Franzini Armstrong, C., Pragnell, M., Kahl, S. D. and Campbell, K. P., 1993, Subunit identification and reconstitution of the N-type Ca^{2+} channel complex purified from brain, *Science* **261**:486-489.

Wright, C. E., Robertson, A. D., Whorlow, S. L. and Angus, J. A., 2000, Cardiovascular and autonomic effects of ω-conotoxins MVIIA and CVID in conscious rabbits and isolated tissue assays, *Br J Pharmacol* **131**:1325-1336.

Xi, Q. and Angus, J. A., 2001, Evidence against an action of mibefradil at N-type voltage-operated calcium channels, *Naunyn Schmiedebergs Arch Pharmacol* **364**:430-436.

Yenari, M. A., Palmer, J. T., Sun, G. H., de Crespigny, A., Mosely, M. E. and Steinberg, G. K., 1996, Time-course and treatment response with SNX-111, an N-type calcium channel blocker, in a rodent model of focal cerebral ischemia using diffusion-weighted MRI, *Brain Res* **739**:36-45.

Zamponi, G. W., Soong, T. W., Bourinet, E. and Snutch, T. P., 1996, β subunit coexpression and the α1 subunit domain I-II linker affect piperidine block of neuronal calcium channels, *J Neurosci* **16**:2430-2443.

Zamponi, G. W., 1999, Cytoplasmic determinants of piperidine blocking affinity for N-type calcium channels, *J Membr Biol* **167**:183-192.

Zhang, Y., Mori, M., Burgess, D. L. and Noebels, J. L., 2002, Mutations in high-voltage-activated calcium channel genes stimulate low-voltage-activated currents in mouse thalamic relay neurons, *J Neurosci* **22**:6362-6371.

Zhong, H., Li, B., Scheuer, T. and Catterall, W. A., 2001, Control of gating mode by a single amino acid residue in transmembrane segment IS3 of the N-type Ca^{2+} channel, *Proc Natl Acad Sci U S A* **98**:4705-4709.

Zhuchenko, O., Bailey, J., Bonnen, P., Ashizawa, T., Stockton, D. W., Amos, C., Dobyns, W. B., Subramony, S. H., Zoghbi, H. Y. and Lee, C. C., 1997, Autosomal dominant cerebellar ataxia (SCA6) associated with small polyglutamine expansions in the α_{1A}-voltage-dependent calcium channel, *Nat Genet* **15:**62-69.

PEPTIDE TOXIN INHIBITION OF VOLTAGE GATED CALCIUM CHANNELS

Selectivity and Mechanisms

Stefan I. McDonough[1]

1. INTRODUCTION

Peptide toxins from animal venom are some of the most potent and selective antagonists currently available for voltage-gated calcium channels, especially of the $Ca_v2.1\alpha_1$, $2.2\alpha_1$, and $2.3\alpha_1$ families (P-type, N-type, and R-type channels). Such toxins are commonly used to investigate the involvement of different channel subtypes in calcium-mediated signaling and may form the basis for drugs targeted against calcium channel-related disorders including severe pain, stroke, and migraine. This chapter discusses the specificity of known toxins for different subtypes and splice forms of vertebrate and invertebrate calcium channels, mechanisms by which the best-characterized toxins inhibit calcium influx, toxin binding sites on the channel, and experimental considerations for the use of peptide toxins in physiology experiments.

Voltage-gated calcium channels are expressed in the plasma membrane of every excitable cell. They come in many genetically, physiologically, and pharmacologically separate subtypes, often with distinct patterns of expression (for review see Catterall, 1998, 2000; Hille, 2001). A full channel protein is made up of α_1 and β subunits, a disulfide-linked $\alpha_2\delta$ subunit, and in at least some cases a γ subunit. Thus far, ten different calcium channel α_1 genes are known: $Ca_v1.1$-1.4, $Ca_v2.1$-2.3, and $Ca_v3.1$-3.3 (Ertel et al., 2000), each of which is likely subject to extensive alternative splicing (Lipscombe and Castiglioni, this volume). In studies of native tissue, where precise genotype and subunit combination is uncertain, the corresponding channels are referred to with a "type" terminology. Although channel subtypes were originally assigned on the basis of activation voltages, inactivation kinetics, or pharmacology of native channels, the sequence of the corresponding α_1 subunit is now the accepted way to classify channels (Ertel et al., 2000). Pharmacological criteria, however, are a reasonable, though certainly

[1] BioCurrents Research Center, Marine Biological Laboratory, 7 MBL Street, Woods Hole, MA 02543.

not infallible, way to classify native channels according to the α_1 subunit. Channels of the Ca_v1 family are L-type, and channels of the Ca_v3 family (which may not incorporate additional ancillary subunits) are T-type. Within the Ca_v2 family, $Ca_v2.1$ channels are P-type, $Ca_v2.2$ channels are N-type, and at least some $Ca_v2.3$ channels are R-type. Calcium channels of the Ca_v2 gene family mediate fast neurotransmission and regulate membrane excitability, but are generally not coupled to gene expression (Piedras-Rentería et al., this volume). They are commonly modulated by G protein-coupled receptors (Dolphin, 1998).

The most potent, selective inhibitors of calcium channels of the Ca_v2 gene family to date are polypeptide toxins isolated from animal venom. The many peptide components of a given venom likely represent a combinatorial genetic strategy, with presynaptic calcium channel antagonists acting in concert with toxins affecting other pre- or post-synaptic channels to generate motor paralysis in the prey (Olivera, 1997; Conticello et al., 2001). [For review of cone snail biology and toxins, see Olivera et al., 1990, Olivera et al., 1994, Olivera, 1997, Espiritu et al., 2001; for review of spider toxins, see Hamilton and Perez, 1987, Olivera et al., 1994, Rash and Hodgson, 2002; for review of scorpion toxins, see Miller, 1995, Garcia et al., 1999, Possani et al., 2000, Gurevitz et al., 2001, Srinivasan et al., 2002.] Toxins with selectivity for $Ca_v2.1$, $Ca_v2.2$, and, increasingly, $Ca_v2.3$ channels have been invaluable for uncovering physiological functions of these channels and may contribute to the development of drugs for Ca_v2-related disorders (Cohen and Kraus, this volume). Such disorders include intractable pain (Jain et al., 2000; Saegusa et al., 2001), migraine (Ophoff et al., 1996; Knight et al., 2002), some forms of ataxia (Jun et al., 1999; Matsuyama et al., 1999; Piedras-Renteria et al., 2001; Guida et al., 2001), ischemic brain damage (Miljanich and Ramachandran, 1995; Bowersox and Luther, 1998), and Lambert-Eaton myasthenic syndrome/ small-cell lung carcinoma (Lennon et al., 1995; Takamori et al., 2000). Peptide toxins offer an advantage over small-molecule calcium channel ligands in that they can be used to map the physical dimensions or movements of the channel via dual mutagenesis studies. In addition to their selectivity for channels of a given α_1 subunit, some peptide toxins are known to distinguish among α_1 subunit splice variants. This makes them a starting point for investigating splice form-specific calcium channel functions.

For each of these uses, it is important to consider the precise mechanism by which the toxin inhibits calcium current. A toxin's selectivity and potency usually depend on the voltage used to open the channel and on the precise ionic conditions. When using toxins to map channel physical dimensions, an approach used with success for the pore vestibule of voltage-gated sodium and potassium channels, it is obviously important to know where and how the toxin acts on the channel. Moreover, knowledge of the mechanism by which a toxin acts allows different toxins to be grouped functionally as well as structurally, likely to be a key consideration in the development of small molecules that mimic toxin selectivity and potency. The purpose of this chapter is to review the specificity, mechanisms, and binding sites of toxins on different calcium channels, and to serve as a guide for use of the toxins in physiological assays.

2. TOXIN SELECTIVITY AMONG CALCIUM CHANNEL SUBTYPES

Table 1 lists commonly used peptide toxins and the calcium channel subtypes that they are known to inhibit. Sequences and alignments, structure, and residues critical for

Table 1. Selectivity of peptide toxins for calcium channel subtypes.

Toxin	Species	α_1 target(s)	IC_{50}	Likely mechanism	Selected references
ω-CTx-GVIA (GVIB) (GVIC)	*C. geographus*	2.2	10 - 100 pM (binding) ~ 100-700 pM (current)	pore block	a,b,c,d,e
ω-CTx-GVIIA (GVIIB)	*C. geographus*		~ 10 nM	pore block	b, f
ω-CTx-MVIIA (MVIIB)	*C. magus*	2.2	10 - 100 pM	pore block	b, g, h
ω-CTx-MVIIC	*C. magus*	2.1 2.2	1 - 5 nM ~ 3 nM	pore block	i, j
ω-CTx-MVIID	*C. magus*	2.1 2.2	~ 1 nM ~ 500 nM	pore block	k, l
ω-CTx-SVIA	*C. striatus*	2.2		pore block	m
ω-CTx-SVIB	*C. striatus*	2.1 2.2	1.7 nM 13 nM	pore block	n,o
ω-CTx-TxVII	*C. textile*	mollusk	5 μM		p,q
ω-CTx-CVIA	*C. catus*	2.1 2.2	~ 2 μM ~ 2 nM	pore block	r
ω-CTx-CVIB	*C. catus*	2.1 2.2	~ 20 nM ~ 20 nM	pore block	r
ω-CTx-CVIC	*C. catus*	2.1 2.2	~ 100 nM ~ 50 nM	pore block	r
ω-CTx-CVID	*C. catus*	2.1 2.2	~ no block ~ 500 pM	pore block	r
ω-CTx-TVIA	*C. tulipa*	2.2	250 pM		h
MrVIA	*C. marmoreus*	mollusk	1- 10 μM (agonist and antagonist)		s
MrVIB	*C. marmoreus*	mollusk	1 – 10 μM (antagonist)		s
ω-CTx-PnVIA	*C. pennaceus*	mollusk	~ 5 μM		t
ω-CTx-PnVIB	*C. pennaceus*	mollusk	~ 5 μM		t
ω-Aga-IA (IB, IC)	*A. aperta*	insect 1 2.2 (?)	~ 7- 50 nM ~ 10 nM ~ 10 nM		u, v, w, x

Toxin	Species	α_1 target(s)	IC_{50}	Likely mechanism	Selected references
ω-Aga-IIA	*A. aperta*	insect 2.2	~ 20 – 45 nM < 100 nM		u, y
ω-Aga-IIB	*A. aperta*	1 3	~ 40 nM ~ 40 nM		hhh
ω-Aga-IIIA	*A. aperta*	locust 1.2 2.1 2.2 2.3	~ 10 nM 10 - 350 pM 5 pM 10 - 170 pM 10 pM	pore block (1.2: full; 2.1, 2.2: partial)	z, aa, bb, cc, dd
ω-Aga-IIIB	*A. aperta*	locust 1.2 2.2	~ 200 nM 47 nM 220 pM		z
ω-Aga-IIIC	*A. aperta*				z
ω-Aga-IIID	*A. aperta*	1.2 2.2	130 nM 330 pM		z
ω-Aga-IVA	*A. aperta*	2.1 2.2 insect	1 nM (some) > 1 μM 10 – 20 nM	gating modifier	ee, ff, gg, rr
ω-Aga-IVB	*A. aperta*	2.1	3 nM (some)	gating modifier	hh
kurtoxin	*P. transvaalicus*	3 1 2.1 2.2	15 – 50 nM 72 nM 14 nM 450 nM	gating modifier	ii, jj
ω-PTx-IIA	*P. nigriventer*	1 2.1 2.2 2.3	28 nM ~ 50 pM ~ 50 pM 67 nM	pore block	kk, ll
SNX-325	*S. florentina*	1.2 2.1 2.2 2.3	~ 2 μM < 2 μM < 20 nM > 2 μM		mm
SNX-482	*H. gigas*	1.2 2.2 2.3	~ 1 μM > 500 nM ~ 30 nM	gating modifier	nn, oo, pp
HwTx-I	*S. huwena*	2.2			qq
ω-ACTx-Hv2a	*H. versuta*	insect	~ 1 nM		rr
calcicludine	*D. angusticeps*	1.2 2	88 nM (see text)		ss, tt
calciseptine/ FS2	*D. polylepis polylepis*	1	10 – 1000 nM	DHP-like	uu, vv

Toxin	Species	α_1 target(s)	IC_{50}	Likely mechanism	Selected references
curtatoxins	*H. curta*	insect			ww
PMP-C, PMP-D2	*L. migratoria*	2.2	~ 1 μM		xx
KP4	*U. maydis* (virally encoded)	1.2	1.2 μM		yy
Ptu-I	*P. turpis*	1.2 2.1 2.2	> 300 nM > 300 nM 300 nM		zz
grammotoxin	*G. spatulata*	2.1 2.2	20 - 180 nM 270 nM	gating modifier	aaa, bbb
DW13.3	*F. hibernalis*	1.2 2.1 2.2 2.3	27 nM 4 nM 2 – 22 nM 96 nM	 partial partial 	ccc
ω-PnTx3-3	*P. nigriventer*	1 2.1 2.2 2.3	~ 80 nM ~ 20 nM ~ 80 nM ~ 30 nM	 partial 	ddd
PLTx-II	*P. tristes*	insect	< 1 nM		eee, fff
Helothermine	*H. horridum horridum*	1 2.1 2.2	250 nM		ggg
Glycerotoxin	*G. convoluta*	2.2		gating modifier	iii
Agelenin	*A. opulenta*	lobster			jjj

Listed for each toxin are the species from which the toxin was isolated, the calcium channel subtypes (α1 subunit) blocked, approximate half-maximal inhibitory concentration, likely mechanism of inhibition, and key references. Blank space indicates not determined or not known. Key: "some" indicates only some forms of the given α1 subunit are inhibited with the given affinity; "partial" indicates that at maximal effect, the toxin does not inhibit currents fully; "binding" indicates IC_{50} or K_D determined from binding assays, "current" indicates IC_{50} or K_D determined from inhibition of current. Channel subtypes are referred to by $Ca_V\alpha_1$ subunit, regardless of whether measurements were made from native or cloned channels, where 2.1 is P-type (or P/Q-type), 2.2 is N-type, 2.3 is R-type, 1.2 is α_{1C} L-type, 1 is multiple or undetermined L-type, and 3 is T-type of undetermined isoform. See text for further details on insect, mollusk, and locust channels. A range of inhibitory concentrations indicates differences among research groups or the values determined with different assays, including inhibition of radiolabeled calcium flux, binding assays, competition assays, and electrophysiological inhibition of currents. Likely mechanism is assigned on the basis of mechanistic studies or of effects that overlap those of toxins with known mechanism.

Selected references: a Olivera et al., 1984; b Olivera et al., 1985; c Wagner et al., 1988; d Boland et al., 1994; e Cruz and Olivera, 1986; f Abe and Saisu, 1987; g Olivera et al., 1987; h Wang et al., 1998; i Hillyard et al., 1992; j McDonough et al., 1996; k Monje et al., 1993; l Gandia et al., 1997; m Ramilo et al., 1992; n Ramachandran et al., 1993; o Woppmann et al., 1994; p Fainzilber et al., 1996; q Sasaki et al., 1999; r Lewis et al., 2000; s Fainzilber et al., 1995; t Kits et al., 1996; u Adams et al., 1990; v Bindokas and Adams, 1989; w Scott et al., 1990; x Venema and Adams, 1990; y Bindokas et al., 1991; z Mintz et al., 1991; aa Ertel et al., 1994; bb Mintz, 1994; cc Yan and Adams, 2000; dd McDonough et al., 2002; ee Mintz et al., 1992b; ff McDonough et al., 1997b; gg Sidach and Mintz, 2000; hh Adams et al., 1994; ii Chuang et al., 1998; jj Sidach and Mintz, 2002; kk Dos Santos et al., 2002; ll Cassola et al., 1998; mm Newcomb et al., 1995; nn Newcomb et al., 1998; oo Bourinet et al., 2001; pp Tottene et al., 2000; qq Peng et al., 2001; rr Wang et al., 2001; ss Schweitz et al., 1994; tt Stotz et al., 2000; uu de Weille et al., 1991; vv Teramoto et al., 1996; ww Stapleton et al., 1990; xx Scott et al., 1997; yy Gage et al., 2002; zz Corzo et al., 2001; aaa Lampe et al., 1993; bbb McDonough et al., 1997a; ccc Sutton et al., 1998; ddd Leao et al., 2000; eee Leung et al., 1989; fff Branton et al., 1993; ggg Nobile et al., 1996; hhh Ertel et al., 1993; iii Meunier et al., 2002; jjj Hagiwara et al., 1990.

function are presented in Norton et al. (this volume). Toxin selectivities and potencies listed must be taken with several caveats in mind. First, the IC_{50} values shown are for mammalian channels, with most experiments performed on rat tissue or on expressed rat or rabbit clones. Many cases are known in which small differences in channel primary sequence, even changes in a single residue, alter toxin potency or kinetics dramatically. The potency of many toxins is also affected by the particular genetic splice variant, subunit combination, or post-translational modification of the channel under study. Thus is it unwise to conclude in general that pharmacological phenotype is a reliable guide to channel genotype (Miller, 1995), and in particular that the specificity of a given toxin for an α_1 subunit is necessarily conserved across species boundaries. Second, toxin potency depends on the precise assay used to measure it. For example, toxin affinity measured from binding assays is often higher than toxin affinity measured from the inhibition of calcium fluxes or currents. Such differences can be due to predominant channel occupancy of inactivated states in the membrane fractions used for binding experiments (e.g., Bean, 1984), or to the use of supersaturating toxin concentrations and nonequilibrium conditions in electrophysiological experiments due to the slow on rate of many toxins. Third, as detailed below, toxin affinity and kinetics are sensitive to experimental parameters such as voltage and the concentration of divalent ions.

Given these cautions, used at appropriate concentrations, toxins are effective tools for selective inhibition of α_1 subunits, particularly of the Ca_v2 gene family. The following section summarizes the tools for selective inhibition of a given $Ca_v\alpha_1$ channel.

2.1. P-type and P/Q-type ($Ca_v2.1$) Channels: ω-CTx-MVIIC and ω-Aga-IVA

In binding studies, ω-CTx-MVIIC binds with subnanomolar affinity to $Ca_v2.1$ channels (Figure 1; Hillyard et al., 1992; Nielsen et al., 1999; Lewis et al., 2000). Although ω-CTx-MVIIC at tens of nanomolar also inhibits current through N-type channels, the toxin unbinding from N-type channels is so fast (Figure 2) that for practical purposes, ω-CTx-MVIIC is specific for $Ca_v2.1$ channels in a binding experiment. As a probable pore blocker (Section 4 below), ω-CTx-MVIIC can be expected to be effective against most forms of $Ca_v2.1$, since the selectivity filter and pore vestibule of the channel are unlikely sites of alternative splicing (Lipscombe et al., 2002). For electrophysiological and other short-term studies, however, the potent inhibition of N-type channels by ω-CTx-MVIIC and the extremely slow on-rate of P-type inhibition can make ω-CTx-MVIIC use problematic. Conotoxins from *C. catus* are potent inhibitors of P-type (as well as N-type) channels (Figure 1; Lewis et al., 2000), but have not been characterized mechanistically. ω-Aga-IVA is the other toxin widely used to inhibit $Ca_v2.1$ currents, and when used at up to several hundred nanomolar has the best selectivity for P-type channels among available toxins (Mintz et al., 1992a; Bargas et al., 1994; Brown et al., 1994; Foehring and Scroggs, 1994; Fisher and Bourque, 1995; Desmadryl et al., 1997; Churchill and Macvicar, 1998; Connor and Christie, 1998; Sidach and Mintz, 2000). At micromolar concentrations, however, ω-Aga-IVA also inhibits some inward current through ω-CTx-GVIA-sensitive N-type channels (but not through L- or T-type calcium channels, or through voltage-gated sodium or potassium channels; Sidach and Mintz, 2000; Delmas et al., 2000). The mechanism of inhibition means that channels bound to ω-Aga-IVA can open with sufficient depolarization, and that continued depolarizations cause rapid dissociation of the toxin from the channel (Section 3.2 below). Moreover, ω-Aga-IVA only inhibits certain splice forms of $Ca_v2.1$ channels with nanomolar affinity, and cell type-specific post-translational modifications likely contribute to the high affinity binding site (Section 4.4 below).

2.2. N-type ($Ca_v2.2$) Channels: ω-CTx-GVIA and ω-CTx-MVIIA

The gold standard for selective and potent inhibition of native and cloned $Ca_v2.2\alpha_1$ N-type channels is the original conotoxin, ω-conotoxin-GVIA (Figure 1; Olivera et al., 1984; Kerr and Yoshikami, 1984; Feldman et al., 1987; Wagner et al., 1988; Williams et al., 1992a; Fujita et al., 1993; Witcher et al., 1993b). Toxin binds to the channel and inhibits currents with 1:1 stoichiometry (Cruz and Olivera, 1986; Wagner et al., 1988; Hillyard et al., 1992; Boland et al., 1994; Lu and Dunlap, 1999; Lewis et al., 2000; Kaneko et al., 2002). The block is fast, complete, and irreversible within many hours under physiological ionic conditions (Cruz and Olivera, 1986; Boland et al., 1994). In most neurons examined, L-type channels are completely insensitive to even high concentrations of ω-CTx-GVIA (Aosaki and Kasai, 1989; Jones and Marks, 1989; Plummer et al., 1989; Regan et al., 1991; Cox and Dunlap, 1992; Kasai and Neher, 1992; Mintz et al., 1992a; Wu and Saggau, 1994). The onset of ω-CTx-GVIA inhibition of chick N-type channels is ~ tenfold slower than rat channels, due to the expression of differently spliced forms of $Ca_v2.2$ (Lu and Dunlap, 1999). In cultured rat SCG neurons, dendritic N-type channels have somewhat lower ω-CTx-GVIA sensitivity than somatic N-type channels (40 nM vs. 2 nM), one of several indications of N-type variants possibly formed by alternative splicing (Delmas et al., 2000).

ω-CTx-MVIIA is also used for specific inhibition of N-type channels. It has specificity comparable to ω-CTx-GVIA in mammalian tissue, and it unbinds faster than ω-CTx-GVIA (Olivera et al., 1987), but is still essentially irreversible on the timescale of an electrophysiological experiment (e.g., Feng et al., 2001a). ω-CTx-MVIIC also blocks N-type channels potently (Woppmann et al., 1994), and this block is quickly reversible, but this toxin also blocks P-type channels potently, slowly, and irreversibly (Figure 2; McDonough et al., 1996). Binding studies show that the *C. catus* toxin ω-CTx-CVID has the highest selectivity for N-type over P-type channels (Figure 1; Lewis et al., 2000) and may target a pharmacologically distinct subset or splice variant (Adams et al., 2003).

It should be noted that even for ω-CTx-GVIA, there are reports of less than total selectivity for $Ca_v2.2$ channels. In early embryonic chick neurons, L-type antagonists (nitrendipine) and agonists (FPL-64176) clearly target the same channel population as ω-CTx-GVIA, although this mixed pharmacological phenotype disappears with development (White et al., 1997). Early reports of the effects of ω-CTx-GVIA (again on avian neurons) suggested some overlap between dihydropyridine-sensitive and ω-CTx-

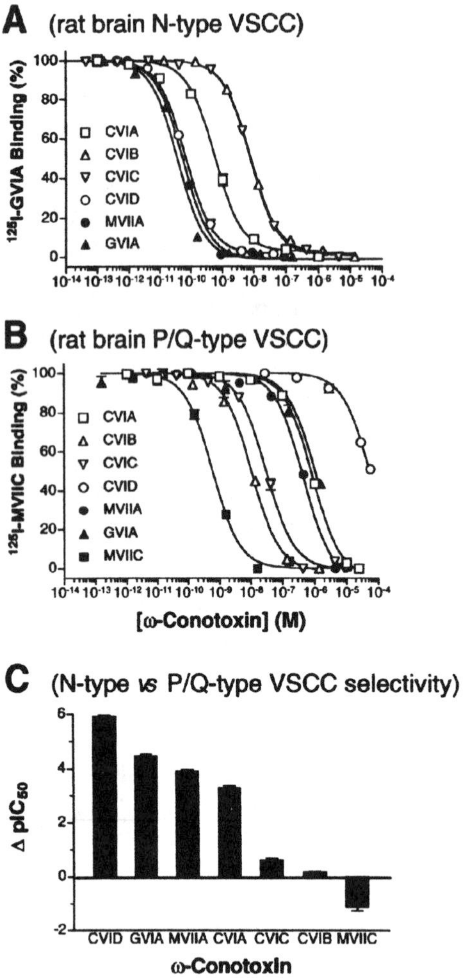

Figure 1. Potency and selectivity of ω-conotoxins for (A) N-type and (B) P-type channels, determined by displacement of radiolabeled ω-CTx-GVIA and ω-CTx-MVIIC binding respectively. (C) summary of toxin selectivity for N-type and P-type channels. Reprinted with permission from Lewis et al., 2000.

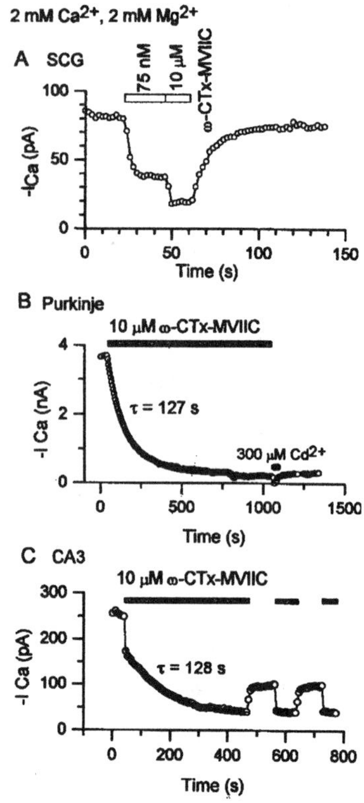

Figure 2. ω-CTx-MVIIC inhibits N-type channels with fast kinetics and P-type channels with slow kinetics. Currents recorded from (A) neuron isolated from superior cervical ganglion (mostly N-type channels); (B) cerebellar Purkinje neuron (mostly P-type channels); and (C) hippocampal pyramidal neuron (mixed channel population). Reprinted with permission from McDonough et al., 1996; copyright 1996 by the Society for Neuroscience.

GVIA-sensitive currents (Kasai et al., 1987; McCleskey et al., 1987). In rat neurons, however, ω-CTx-GVIA as high as 3 μM has no effect on pure L-type tail currents isolated with agonists (Regan et al., 1991), suggesting that ω-CTx-GVIA has highest selectivity for mammalian $Ca_v2.2$ channels (Olivera et al., 1994). Only very weak effects (~ 10% inhibition) have been reported for ω-CTx-GVIA at several micromolar tested on expressed cloned $Ca_v1.3$ (Williams et al., 1992b) and $Ca_v2.3$ channels (Ellinor et al.,

1993; Williams et al., 1994). ω-CTx-GVIA also binds with low affinity to nicotinic acetylcholine receptors (Horne et al., 1991), but this is unlikely to be generally true of ω-conotoxins, since ω-CTx-SVIA and ω-CTx-SVIB do not (Ramilo et al., 1992).

2.3. R-type ($Ca_v2.3$) Channels: SNX-482

"R-type" channels were originally named for their resistance to combined inhibition by ω-Aga-IVA, ω-CTx-MVIIC, dihydropyridines, and ω-CTx-GVIA. Such channels were presumed to be made up of $Ca_v2.3$ (α_{1E}) channels, since $Ca_v2.3$ channels from several species expressed in oocytes were similarly resistant to these inhibitors (Ellinor et al., 1993; Soong et al., 1993; Zhang et al., 1993; Randall and Tsien, 1997). The name "R-type" has stuck to describe $Ca_v2.3$ (α_{1E}) channels, even though the peptide toxin SNX-482 from tarantula venom is now available to inhibit at least some forms of $Ca_v2.3$ channels with some selectivity over other calcium channel subtypes (Newcomb et al., 1998).

In addition to $Ca_v2.3$ channels in heterologous expression, SNX-482 also inhibits current in several types of neuron, with IC_{50} in the range of tens of nanomolar (Newcomb et al., 1998; Wang et al., 1999; Tottene et al., 2000; Wilson et al., 2000). This current almost certainly is carried by $Ca_v2.3$ channels. In recordings from cerebellar granule neurons, however, Pietrobon and colleagues (Tottene et al., 2000) found that SNX-482 inhibited two separate components of resistant current with different potency and had no effect on a third component. Antisense experiments showed that all three components were made up of $Ca_v2.3$ R-type channels. The different forms may represent different splice variants of $Ca_v2.3$: SNX-482 inhibits inward current by modifying channel gating (Bourinet et al., 2001; see Section 3.2 below), suggesting binding to an extracellular S3-S4 linker of the channel, which is a common site of alternative splicing. Thus it seems likely that SNX-482 is a good inhibitor of some, but not all, splice forms of $Ca_v2.3$ channels. Saturating (200 nM) concentrations of SNX-482 also alter the gating of expressed $Ca_v1.2$ L-type channels, although by much less than the gating of $Ca_v2.3$ channels, suggesting a different binding site (Bourinet et al., 2001).

Not all resistant calcium channel current in native cells is likely to be made up of $Ca_v2.3$ R-type channels. $Ca_v1.3$ (α_{1D}) L-type channels and certain splice forms of $Ca_v2.1$ channels are also resistant to ω-conotoxins, ω-Aga-IVA, and dihydropyridines (Bourinet et al., 1999; Xu and Lipscombe, 2001; Koschak et al., 2001). Thus a distinction needs drawn between $Ca_v2.3$ channels (often still referred to as R-type) and "resistant" channels. SNX-482 is a good inhibitor of some, but not all, forms of $Ca_v2.3$ channels. Likewise some, but not all, channels "resistant" to combined inhibition by dihydropyridines, ω-CTx-GVIA, ω-Aga-IVA, and ω-CTx-MVIIC are $Ca_v2.3$.

2.4. L-type ($Ca_v1.1$-1.4) Channels

Peptide toxins that target mammalian L-type channels include calciseptine, its homolog FS2, calcicludine, and ω-Aga-IIIA (see detailed descriptions below). Relatively few peptide toxins are known that target L-type channels, either because L-type channels generally do not trigger neurotransmission and so present an unfavorable target for paralytic venoms, or because the great success of other L-type inhibitors means that venoms are not widely screened against L-type channels. The most potent and selective compounds are small organic molecules such as the dihydropyridines, discussed in detail by Triggle (this volume).

Although the four L-type clones are grouped together by sequence homology, pharmacology of the different clones is far from identical. Dihydropyridine compounds bind well to $Ca_v1.1$ (α_{1S}) skeletal muscle protein (Tanabe et al., 1987; Rios and Brum, 1988) and inhibit all current through $Ca_v1.2$ (α_{1C}) cardiac-type channels (Triggle, this volume). $Ca_v1.3$ (α_{1D}) channels, however, are inhibited only incompletely by high concentrations of nifedipine and nitrendipine (Xu and Lipscombe, 2001). The only compound described thus far to inhibit $Ca_v1.3$ currents completely is isradipine (Koschak et al., 2001; Scholze et al., 2001), and only at holding potentials sufficiently positive to produce significant fractional inactivation. $Ca_v1.4$ (α_{1F}) channels are expressed most prominently in the retina (Bech-Hansen et al., 1998; Strom et al., 1998), especially in the synaptic terminals of rod photoreceptors (Morgans, 2001). Recordings of these cells suggest that $Ca_v1.4$ channels are inhibited completely by some dihydropyridines, but it is not yet clear whether all dihydropyridines are equally effective and whether the standard L-type agonists are also effective. Calcium channels from cone inner segments isolated from lizard retina were blocked 70% by 3 μM nitrendipine and enhanced 140% by 3 μM BayK-8644 (Maricq and Korenbrot, 1988). Similarly, calcium channels from salamander inner segments were blocked by 10 μM nitrendipine (Barnes and Hille, 1989), and noninactivating (likely $Ca_v1.4$) calcium channels recorded from rat rod bipolar cells were blocked 92% by 1-10 μM nifedipine or isradipine (Protti and Llano, 1998).

2.5. T-type ($Ca_v3.1$-3.3) Channels

Thus far, T-type channels appear to be remarkably insensitive to peptide toxins, even to those toxins that are relatively nonspecific blockers of other calcium channel subtypes. T-type channels are unaffected by even high concentrations of ω-Aga-IIIA (Cohen et al., 1992; Mintz, 1994), ω-Aga-IIIB and ω-Aga-IIID (Ertel et al., 1994), ω-Aga-IVA (Mintz et al., 1992a; Sidach and Mintz, 2000), grammotoxin (Avery and Johnston, 1996), calciseptine (de Weille et al., 1991), calcicludine (Schweitz et al., 1994), HwTx-I (Peng et al., 2001), SNX-482 (Newcomb et al., 1998; Morita et al., 2002), ω-PnTx3-3 (Leao et al., 2000), DW13.3 (Sutton et al., 1998), and PMP-D2 (Scott et al., 1997). No ω-conotoxin is known to affect T-type channels (Olivera et al., 1994; Avery and Johnston, 1996; Beck et al., 1997; Perchenet et al., 2000; Hirooka et al., 2002; Morita et al., 2002), suggesting an entirely different pore architecture for Ca_v3 channels compared to Ca_v1 and Ca_v2. Small molecules are the best inhibitors currently available (Ertel, this volume).

Two toxins are known to inhibit T-type channels: the spider toxin ω-Aga-IIB (Ertel et al., 1993) and the scorpion toxin kurtoxin, homologous to α-scorpion toxins. Kurtoxin inhibits inward current through native and cloned T-type channels, but it also slows sodium channel inactivation and affects strongly the gating of non-T-type calcium channels recorded from neurons (Chuang et al., 1998; Sidach and Mintz, 2002; section 3.2.2 below). There is no peptide toxin with reliable specificity for T-type channels, and currently no way to distinguish among T-type channels made up of $Ca_v3.1$, $Ca_v3.2$, and $Ca_v3.3$ subunits.

2.6. Toxins that Inhibit Invertebrate Calcium Channels

Most interest in calcium channel-blocking polypeptide toxins comes from the medical perspective, for which the effects of toxins on mammalian channels are obviously the most interesting. High-affinity blockers of invertebrate calcium channels, however, may be useful for the development of novel insecticides (Hall et al., 1994; Wang et al., 1999) and for use in neurobiological model systems, such as locust (Laurent, 2002) or molluscan systems including *Aplysia* and *Lymnaea* (Smit et al., 2001).

2.6.1. Insect Calcium Channels

Unsurprisingly, given their prey, spider venoms are a rich source of compounds that target insect ion channels. In addition to peptide toxins, spider venom contains many other active components, including non-peptide acylpolyamines (Jackson and Usherwood, 1988; Olivera et al., 1994). Toxins that block insect calcium channels are comparatively uncharacterized, to the extent that toxins isolated from arachnid venoms (e.g., grammotoxin, SNX-482, DW13.3, SNX-325, HwTx-I) that are potential insect channel inhibitors have only been assayed on mammalian or avian preparations. Thus there are almost certainly more toxins currently available than those described below, and many more from uninvestigated venoms.

Biochemically unrelated peptides from spider venom, *Hololena curta* toxin (HoTX) and toxin from *Plectreurys tristes* (PLTX) were originally isolated by their ability to block neurotransmission in *Drosophila* (Branton et al., 1987; Bowers et al., 1987). When its sequence was determined, PLTX (PLTX-II) was found to have an amidated and acylated C-terminal threonine residue (Branton et al., 1993). Block of neurotransmission by PLTX-II and HoTx was potent and irreversible, with full block reached with ~ 1 nM toxin. Nifedipine, diltiazem, and verapamil do not block calcium current from these *Drosophila* neurons (Byerly and Leung, 1988), suggesting that HoTX and PLTX-II act against non-L-type channels, although the correspondence between insect and mammalian Ca_v1 channels is uncertain at best and the pharmacology may not be conserved. Likewise, the targets of HoTX and PLTX-II are non-N-type: both toxins are ineffective against neuromuscular transmission in the frog, which is controlled by ω-CTx-GVIA-sensitive channels (Kerr and Yoshikami, 1984; Sano et al., 1987; Zengel et al., 1993). Conversely, ω-CTx-GVIA does not block neuromuscular transmission in *Drosophila* (McCleskey et al., 1987). One possibility is that HoTX and PLTX-II target an insect form of the $Ca_v2.1$ channel, the channel that triggers postnatal mammalian neuromuscular transmission (Ousley and Froehner, 1994; Bowersox et al., 1995; Protti et al., 1996; Takamori et al., 2000). An additional class of toxins, curtatoxins, were also purified from *H. curta* venom and found to cause flaccid paralysis in cricket, suggesting neuromuscular blockade (Stapleton et al., 1990). It would be very interesting to test these toxins against mammalian $Ca_v2.1$ P-type channels.

The ω-agatoxins, isolated by Adams and colleagues, represent a family of peptide toxins from American funnel-web spider venom, some of which are active against insect ion channels (Olivera et al., 1994). ω-Aga-IA is the most abundant ω-agatoxin in the venom of *Agelenopsis aperta*, at 0.6% of total venom protein (Adams et al., 1990). At a concentration of tens of nM, it blocks neuromuscular transmission (housefly and frog) and the calcium component of action potentials in locust neurosecretory neurons (Bindokas and Adams, 1989). ω-Aga-IA was also found to inhibit calcium channels recorded from rat dorsal root ganglion neurons and cultured cerebellar neurons, consistent with activity against mammalian L-, N-, and perhaps P- and R-type channels, plus possibly slight activity against T-type channels (Scott et al., 1990). Interestingly, ω-Aga-IA and its congeners ω-Aga-IB and ω-Aga-IC have no effect on ω-CTx-GVIA binding to chick synaptosomal membranes (Adams et al., 1990; Bindokas et al., 1991). As ω-Aga-IA inhibits ω-CTx-GVIA-sensitive channels in mammalian neurons (Scott et al., 1990), this suggests that either ω-Aga-IA is selective for mammalian over chick $Ca_v2.2$ channels, or that it binds to a site different from the ω-conotoxin binding site. ω-Aga-IA does not inhibit calcium influx into rat brain synaptosomes (Venema et al., 1992), suggesting that it is ineffective against mammalian $Ca_v2.1$ channels.

Type II ω-agatoxins (ω-Aga-IIA and ω-Aga-IIB), which are not close structural homologs to ω-Aga-IA, are active against both insect and vertebrate calcium channels. Each blocks neurotransmission in housefly and calcium potentials in locust neurons (Bindokas et al., 1991), but unlike ω-Aga-IA, ω-Aga-IIA does prevent ω-CTx-GVIA binding to chick membranes (Adams et al., 1990). Thus ω-Aga-IIA is active against insect channels and against some vertebrate N-type channels, although toxin activity has not been tested directly on voltage-clamped calcium currents. ω-Aga-IIA seems likely to inhibit mammalian $Ca_v2.1$ channels as well, since it inhibits calcium influx into rat brain synaptosomes (Venema et al., 1992), but inhibition of current has not been tested directly.

ω-Aga-IIIA and ω-Aga-IVA were isolated from *A. aperta* venom by their ability, respectively, to prevent ω-CTx-GVIA binding to membranes (Venema et al., 1992) and to prevent calcium influx into rat brain synaptosomes (Mintz et al., 1992b). They and congeners ω-Aga-IIIB and ω-Aga-IIID are inactive against presynaptic calcium channels of the fly neuromuscular junction (Bindokas et al., 1991; Ertel et al., 1994), but are notably potent and specific on mammalian ion channels. They are effective against other invertebrate channels: ω-Aga-IIIA and ω-Aga-IIIB inhibit locust channels potently (Ertel et al., 1994), and ω-Aga-IVA inhibits calcium channels from bee brain neurons (Wang et al., 2001) and cockroach neurons (Benquet et al., 1999) with IC_{50} 10 nM to 20 nM.

The newest toxins blocking insect calcium channels are ω-atracotoxins, a set of five homologous toxins isolated from several species of Australian funnel-web spiders (Wang et al., 2000, 2001). ω-ACTx-Hv2a, from *Hadronyche versuta*, inhibits all inward current through calcium channels from bee brain neurons, with EC_{50} of ~ 1 nM (Wang et al., 2001). It is structurally homologous to ω-Aga-IVA despite fairly low sequence identity (Norton et al., this volume). Remarkably, at concentrations up to 1 μM, ω-ACTx-Hv2a was ineffective against calcium channels from rat trigeminal ganglion neurons, suggesting almost complete specificity for insect channels over mammalian channels (Figure 3). Sequences of the four related ω–atracotoxins, ω-Hi2a and ω-Hi2b from *Hadronyche infensa* and ω-As2a and ω-As2b from *Atrax* sp. Illawarra, were inferred from cDNA libraries. Their activity against calcium channels has not yet been assayed.

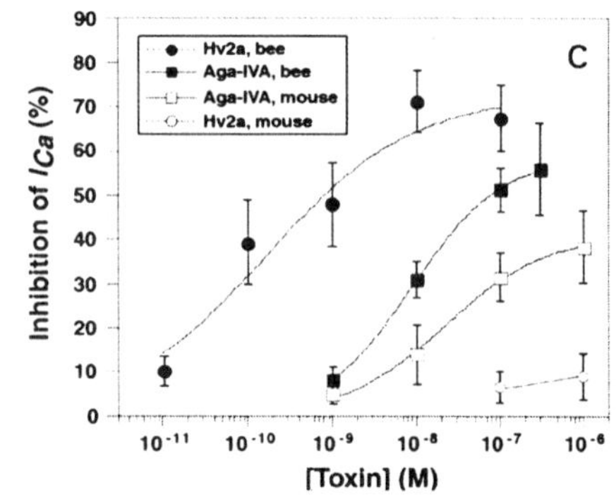

Figure 3. Inhibition of calcium currents from mammalian and insect neurons by ω-Aga-IVA and ω-atracotoxin-Hv2a. Reprinted with permission from Wang et al., 2001.

Like mammalian neurons, individual insect neurons clearly can contain multiple types of pharmacologically distinct calcium channels, although the few studies of toxins and the large number of experimental preparations mean there is no clear terminology for pharmacological classification of channels. In embryonic *Drosophila* neurons, for example, PLTX blocks essentially all calcium channel current, whereas HoTX selectively spares a more rapidly inactivating component of the total current (Leung et al., 1989). And in high calcium concentrations, type I and type II ω-agatoxins each produce partial inhibition of synaptic transmission in houseflies, but together produce complete inhibition (Bindokas et al., 1991).

2.6.2. Molluscan Calcium Channels

As some *Conus* species are molluscivorous, it is not surprising that their venoms contain peptides targeting mollusk ion channels. Several peptide toxins have been assayed for inhibition of calcium channels from caudodorsal cells of the freshwater snail *Lymnaea stagnalis*. As with insect calcium channels, results indicate that peptide toxins (ω-conotoxins in this case) are capable of distinguishing among multiple channel subtypes coexpressed in the same cells. Channels in these neurons are thus far divided into two components: a transient component that inactivates within ~ 50 ms and is relatively insensitive to dihydropyridines, and a sustained component that persists during a depolarization of several hundred ms and is inhibited by low concentrations of dihydropyridines (Dreijer and Kits, 1995; Kits et al., 1996). The dihydropyridine sensitivity of the sustained component suggests some correspondence to the mammalian Ca_v1 L-type gene family.

Conotoxins from several species have been found with action against these channels. ω-conotoxins PnVIA and PnVIB, purified from *C. pennaceus* venom, have divergent sequences but similar effects: each inhibits the transient current fully (IC_{50} 1 to 5 μM) and reversibly, with minimal effect on the dihydropyridine-sensitive persistent current (Kits et al., 1996). This persistent current was blocked with some selectivity by ω-CTx-TxVII from *C. textile* (IC_{50} ~ 5 μM; Fainzilber et al., 1996; Sasaki et al., 1999). This toxin is ineffective against nifedipine-sensitive calcium channels in the rat-derived PC-12 cell line, suggesting selectivity for molluscan over at least some forms of mammalian L-type channels (Sasaki et al., 1999). Toxins MrVIA and MrVIB were isolated from *C. marmoreus* (Fainzilber et al., 1995; McIntosh et al., 1995). They share a cysteine framework with ω-conotoxins, but also inhibit sodium channels and so are termed μO-conotoxins by McIntosh and colleagues for this functional similarity. MrVIB inhibited the persistent (nifedipine-sensitive) component of calcium current, with complete inhibition at 10 μM (Fainzilber et al., 1995). Effects of MrVIA were complex: the transient component was selectively enhanced at low concentrations, but at higher toxin concentrations the transient and sustained calcium current were both inhibited. The enhancement implies an effect on channel gating, and the dual effect implies multiple toxin binding sites on the channel, but these toxins in general have not been investigated mechanistically, or in detail against mammalian channels. Both are potent blockers of molluscan voltage-gated sodium channels (Fainzilber et al., 1995; Terlau et al., 1996).

3. MECHANISMS OF INHIBITION

How exactly do these different toxins inhibit the flux of calcium ions through the calcium channel? To a first approximation, calcium channel-blocking toxins can be grouped into

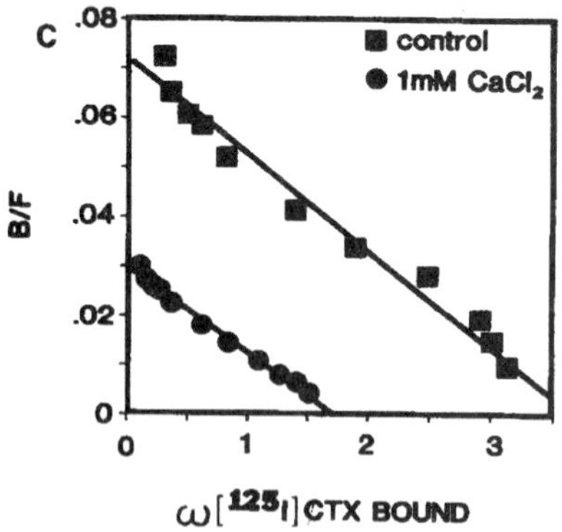

Figure 4. Calcium inhibits ω-CTx-GVIA binding. Scatchard analysis of ω-CTx-GVIA binding to rat brain membranes shows non-competitive inhibition by calcium ions. Reprinted with permission from Wagner et al., 1988; copyright 1988 by the Society for Neuroscience.

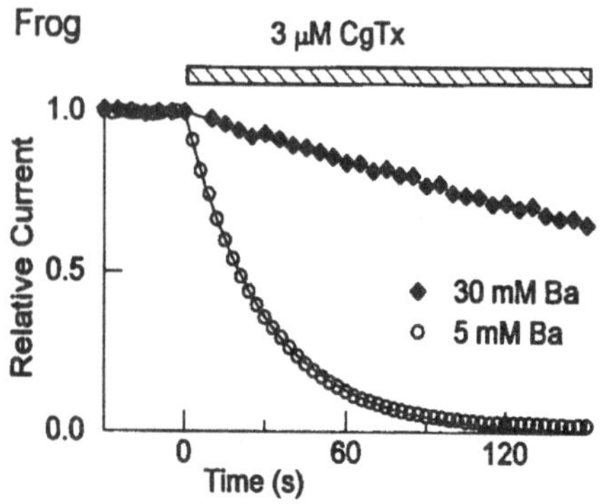

Figure 5. High barium concentrations slow block by ω-CTx-GVIA. 30 mM barium slows the inhibition of N-type channels recorded from a frog neuron. Reprinted with permission from Boland et al., 1994; copyright 1994 by the Society for Neuroscience.

two classes by mechanism—those that prevent ion permeation, either directly or allosterically, and those that interfere with the gating machinery of the channel to prevent it from opening in response to a physiological stimulus. Precedents from non-calcium channels are known for both mechanisms. The distinction between gating and permeation is not absolute: several toxins likely affect both mechanisms, and many toxins, including some putative pore-blockers, have different affinity for closed, open, and inactivated channel gating states (see Timin et al., this volume, for a general discussion of state-dependent drug affinity).

3.1. Inhibition of Permeation: the Pore Vestibule

As shown by structural experiments on several classes of ion channel, a portion of the channel is exposed to the extracellular solution and forms an access path to the actual selectivity filter of the ion channel located within the transmembrane region. The inside and outside of this pore vestibule form obvious targets for the action of paralytic venoms, and many species have evolved toxins that disrupt channel function by binding at or within the channel pore to disrupt the conduction of nerve impulses. Such pore-blocking toxins include μ-conotoxins that target voltage-gated sodium channels (Cruz et al., 1985; Catterall, 1992; Cestele and Catterall, 2000) and toxins from scorpion venom that target potassium channels (Miller, 1995). α-bungarotoxin also targets the pore vestibule of nicotinic acetylcholine receptors (nAChRs), in this case by binding to the outside of the pore vestibule to occlude directly acetylcholine binding (Moise et al., 2002).

Structural measurements show that the diameter of the pore vestibule is more or less consistent among ion channels for which measurements are available. The pore vestibule is most striking for nAChRs: a large extracellular portion of the channel sequence forms a cylindrical pore vestibule, with walls measuring ~ 65 Å long and ~ 20 Å internal diameter (Toyoshima and Unwin, 1988; Mitra et al., 1989; Unwin, 1998; Brejc et al., 2002). Calcium channels are more likely to be related in structure to voltage-gated sodium and potassium channels, which are unlikely to have such large walls. The pore vestibule for potassium channels is shaped like a shallow cone, with the apex formed by the orthogonal intersection of two grooves on the channel due to its fourfold subunit symmetry (MacKinnon, 1991). This shallower vestibule measures ~ 25 to 35 Å wide and 4 to 12 Å deep, based on contact with residues of pore-blocking scorpion toxins (Goldstein et al., 1994; Hidalgo and MacKinnon, 1995; Aiyar et al., 1995; Ranganathan et al., 1996) and confirmed by crystal structure of a bacterial potassium channel (Doyle et al., 1998; MacKinnon et al., 1998). Similarly, the sodium channel pore vestibule apparently measures ~ 25 Å across, based on multiple channel contacts with the pore-blocking μ-conotoxins of this size (Lancelin et al., 1991; Wakamatsu et al., 1992; Hill et al., 1996; Chang et al., 1998; Li et al., 2000, 2001). Surprisingly, lower-resolution (19 Å) structural studies of the entire sodium channel from *Electrophorus electricus* eel showed that the sodium channel pore vestibule does not open directly to the extracellular solution. Rather, an aqueous pathway splits into four branches around a part of the channel that forms a cap atop the central vestibule (Sato et al., 2001; Catterall, 2001). It is not known if calcium channels share this structure.

3.1.1. ω-CTx-GVIA

ω-CTx-GVIA is one of the best-studied toxins mechanistically, and it shares a binding site with many other peptide toxins (Section 4.2 below), suggesting a shared mechanism. Several biophysical features of ω-CTx-GVIA inhibition of $Ca_v2.2$ N-type channels suggest that it blocks ion permeation, either allosterically or by direct binding within the pore vestibule. Binding affinity (Figure 4; Abe et al., 1986; Wagner et al., 1988; Feigenbaum et al., 1988) and the kinetics and potency of inhibition of calcium channel currents (Figure 5; Carbone and Lux, 1988; Witcher et al., 1993a; Boland et al., 1994; Liang and Elmslie, 2002) are strongly affected by the concentration of divalent calcium or barium ions in the extracellular solution, suggesting an interaction between the toxin and ion permeation. The effects of divalent ions are likely to be on the channel, rather than on toxin structure, since the ω-CTx-GVIA shape is constrained by a rigid backbone structure of three pairs of disulfide-linked cysteine residues (Norton et al., this volume), unlike the smaller conantokin-G conotoxin that alters its conformation upon calcium binding (Rigby et al., 1997a,b). The interaction between toxin and divalent ions is unlikely to be at the selectivity filter of the channel, since calcium inhibits ω-CTx-GVIA binding noncompetitively (Figure 4; Wagner et al., 1988). Site-directed mutagenesis studies suggest that the site of interaction is outside the selectivity filter, in a part of the sequence predicted to contribute to the pore vestibule (Feng et al., 2001a, Section 4.1 below). Consistent with this, divalent ion occupancy within the selectivity filter of the pore itself had no effect on the stability of ω-CTx-GVIA binding (Boland et al., 1994).

ω-CTx-GVIA inhibits outward, monovalent ion-carried currents at voltages up to +170 mV (Boland et al., 1994; McDonough et al., 2002). Again, this is consistent with inhibition of ion permeation, unless the toxin inhibits depolarization-induced channel gating more strongly than any gating modifier toxin known. This inhibition of outward currents at even strong depolarizations argues against a competition between ω-CgTx-GVIA and permeant ions within the pore, in contrast to scorpion toxin inhibition of potassium channels, which is relieved by outward potassium currents (MacKinnon and Miller, 1988; Park and Miller, 1992; Ranganathan et al., 1996). ω-CTx-GVIA binds more strongly to the inactivated state of channels expressed in oocytes (Stocker et al., 1997) and does not seem to require channel opening to bind: at high concentrations, it inhibits currents nearly fully within a second, before delivery of the first test depolarization to open channels and assay the extent of inhibition (Boland et al., 1994). Although the precise affinity for open and closed channels has not been tested, this suggests that ω-CTx-GVIA reaches its binding site when the channel selectivity filter is closed—again suggesting binding at a medium depth within the pore vestibule, rather than within the selectivity filter of the channel. This is not necessarily a common feature of all toxins that bind within the pore vestibule: for example, two peptides structurally homologous to ω-CTx-GVIA, PMP-C and PMP-D2, are use-dependent blockers of N-type channels that require channel opening (Scott et al., 1997). ω-CTx-GVIA binding within the pore, yet not within the selectivity filter, would be reminiscent of μ-conotoxins, which bind within the sodium channel pore but not as deeply as tetrodotoxin or saxitoxin (Chang et al., 1998; Li et al., 2000). Based on biophysical data, however, it is still uncertain whether ω-CTx-GVIA binds within the pore vestibule and acts as a physical plug, or whether it binds to the outside of the pore vestibule and constricts the permeation pathway via an allosteric mechanism. Mutational analysis of the channel (Section 4.1 below) suggests the former.

3.1.2. ω-Aga-IIIA

The spider toxin ω-Aga-IIIA (Venema et al., 1992) is structurally unrelated to the conotoxins, yet likely shares a mechanism of binding within the pore vestibule (or of allosteric coupling to constriction of the pore). Several features of this large (76 residue) peptide are particularly noteworthy. ω-Aga-IIIA inhibits completely current through L-type channels, unusual among peptide toxins, and its affinity is extremely high: $IC_{50} \sim 1$ nM for inhibition of currents (Mintz et al., 1991; Cohen et al., 1992; Ertel et al., 1994) and ~ 10 pM for binding to membrane fractions (Yan and Adams, 2000). It has no effect on L-type gating currents (Ertel et al., 1994), consistent with an effect entirely on permeation. The inhibition of inward N-type and P-type currents, although equally potent, is only partial (~ 70%). The remaining currents open at slightly more depolarized potentials, possibly due to surface charge-screening effects of the highly charged toxin (Mintz et al., 1991; Mintz, 1994). Non-stationary fluctuation analysis of the remaining currents shows that the partial inhibition is due to a reduction in single channel conductance, rather than to complete block of a sub-population of channels (McDonough et al., 2002). In addition to L, N, and P-type channels, ω-Aga-IIIA also binds to and inhibits R-type channels with subnanomolar affinity (Randall and Tsien, 1997; Yan and Adams 2000).

3.2. Alteration of Channel Voltage Dependence

For a given toxin, knowing only that it inhibits inward calcium current, there is no reason to assume that the toxin blocks the channel pore rather than altering channel gating so that channels do not open in response to the depolarization. Initial observations of interactions between toxins and channel gating were made with scorpion toxins on sodium (Catterall, 1977a,b; Mozhayeva et al., 1980) and potassium (Carbone et al., 1987) channels. Much subsequent work has found toxins from many different venoms that alter voltage-dependent channel gating—both to inhibit and to promote channel opening (Swartz and MacKinnon, 1997; Cestele and Catterall, 2000). Among toxins acting against calcium channels, the effects on gating of ω-Aga-IVA, the closely related ω-Aga-IVB (Adams et al., 1993; also called ω-Aga-TK; Teramoto et al., 1993), ω-grammotoxin-

SIA, SNX-482, and kurtoxin have been studied in detail. These seem to stabilize closed states of the channel and produce a depolarizing shift in activation voltage. Glycerotoxin, from worm, is the only peptide known to produce a hyperpolarizing shift in activation voltage (Meunier et al., 2002). In general, movement of the channel charge sensor and the coupling of this movement to channel opening and inactivation are complex and interrelated processes, each involving multiple steps and conformational rearrangements of the channel protein. As a toxin may interact with any step in the gating process, it is likely that mechanistic descriptions of toxin effects on channel gating will always require some oversimplification.

3.2.1. Depolarizing Shifts in Activation Voltage: ω-Aga-IVA, Grammotoxin, and SNX-482

ω-Aga-IVA inhibits all inward current through P-type channels (Mintz et al., 1992b). Strong depolarizations, however, evoke outward Cs^+- or K^+-carried currents, followed by an inward Ba^{2+} or Ca^{2+} tail current (Figure 6). Since a moderate depolarization delivered immediately after a strong depolarization still evokes no inward current (not shown), the interpretation is that sufficiently large depolarizations open the channel while the toxin remains bound (McDonough et al., 1997b). Tail current activation curves in control and in toxin show that in toxin, channels open as a smooth Boltzmann function of voltage, but only in response to larger depolarizations than control and, importantly, with less sensitivity to voltage. The decreased sensitivity to voltage is quantified as an increase in the slope factor k in the Boltzmann equation describing channel activation as a function of voltage: fractional activation = 1 / {1 + exp[- $(V-V_{1/2})/k$]} where V is voltage, $V_{1/2}$ is half-maximal activation voltage, and k is the slope factor. (Fitting channel activation with a single Boltzmann function is an approximation.) In addition to its alteration of channel voltage dependence, ω-Aga-IVA slows channel opening in response to sufficiently strong voltages and speeds channel closing (deactivation). All these observations are consistent with a qualitative picture in which toxin stabilizes closed states of the channel compared to open states.

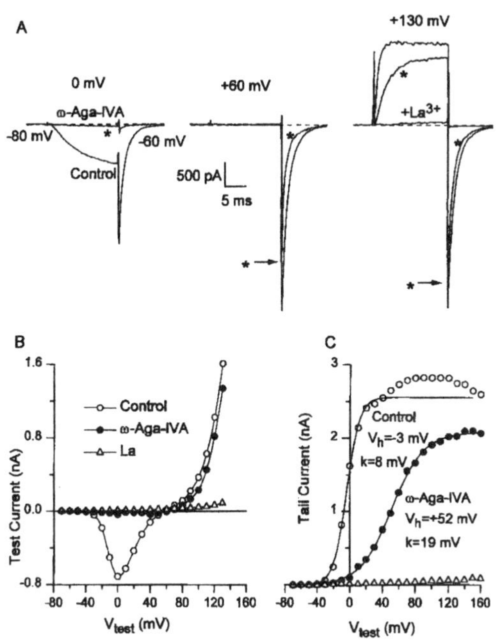

Figure 6. ω–Aga-IVA inhibits inward current by altering channel gating. Currents recorded from a single cerebellar Purkinje neuron in control (hollow circles), ω-Aga-IVA (asterisk and filled circles), and lanthanum (triangles). (A) currents at different test voltages; (B) current as a function of test voltage; (C) tail current as a function of test voltage. Reprinted with permission from McDonough et al., 1997b, copyright 1997 by the Biophysical Society.

Similar protocols show that grammotoxin, SNX-482, and kurtoxin (T-type channels only; see Section 3.2.2) each inhibit channels by a mechanism comparable to that of ω-Aga-IVA. Each of these four toxins increases the half-maximal activation voltage by ~ +70 mV, increases the slope factor by 10 to 15 mV (i.e., makes the channel less sensitive to voltage), and slows the speed of channel opening (McDonough et al., 1997a,b; Chuang et al., 1998; Bourinet et al., 2001; Sidach and Mintz, 2002). The dramatic slowing of channel opening by some toxins, most notably by grammotoxin, means that during a current-voltage curve defined with short test pulses as in Figure 6, steady-state channel activation is not reached, and the shift in activation voltage measured with such a protocol is overstated. The true shift in channel activation by ω-Aga-IVA and grammotoxin is ~ +40 mV, but the channel is still less sensitive to voltage (Boltzmann slope factor increases by ~ 10 mV; McDonough et al., 1997a,b). For ω-Aga-IVA, grammotoxin, and SNX-482, following a pulse strong enough to open toxin-bound channels, instantaneous current-voltage curves are unchanged compared to control, implying no toxin effect on ion permeation.

For each of these toxins, the ~ +70 mV shift in gating voltage measured with short test pulses means that channels that normally start to open at ~ -20 mV instead open at ~ +50 mV, close to the ionic reversal potential. The net effect is that these toxins inhibit inward divalent ion-carried currents, whereas outward monovalent ion-carried currents are not blocked (Figure 6). This is immediately reminiscent of scorpion toxins that block the potassium channel pore and compete with outwardly flowing potassium ions (MacKinnon and Miller, 1988), suggesting that the calcium channel gating-modifier toxins may also bind within the pore. Such is not the case. In frog sympathetic neurons, when divalent cations were removed from the external solution, the ionic reversal potential was –20 mV, and yet grammotoxin strongly inhibited outward Cs^+-carried currents as well as the inward Na^+-carried currents due to the change in channel voltage dependence (McDonough et al., 1997a).

Since strong depolarizations will indeed open channels bound to gating modifier toxins, it is a question whether these toxins inhibit inward current completely in the physiological situation, in which action potentials have peak overshoots of ~ +30 mV. At steady-state, a depolarization of +30 mV is indeed sufficient to give some degree of channel opening (Figure 6). Several experiments argue that these gating modifier toxins do indeed inhibit all inward current in response to an action potential. ω-Aga-IVA blocks synaptic transmission completely at several synapses (Regehr and Mintz, 1994; Mintz et al., 1995), although the power law of calcium entry and transmitter release, especially steep at P-type channels, makes this less than definitive. However, inward calcium-carried currents evoked by voltage clamping with an action potential waveform were indeed inhibited fully by ω-Aga-IVA (McDonough et al., 1997b). Since the peak voltage reached by the action potential was strong enough to open toxin-bound channels at steady-state, it seems likely that slowing of channel opening by toxin contributes to the prevention of calcium entry in the physiological situation.

Toxin stabilization of channel closed states means that, qualitatively, the toxin-channel complex is energetically more stable with the channel closed than open. A corollary to this mechanism is that toxin binds more tightly to closed than open channels. Depending on the exact difference in free energy between three cases—toxin bound to closed channel, toxin bound to open channel, and toxin not bound to channel—this

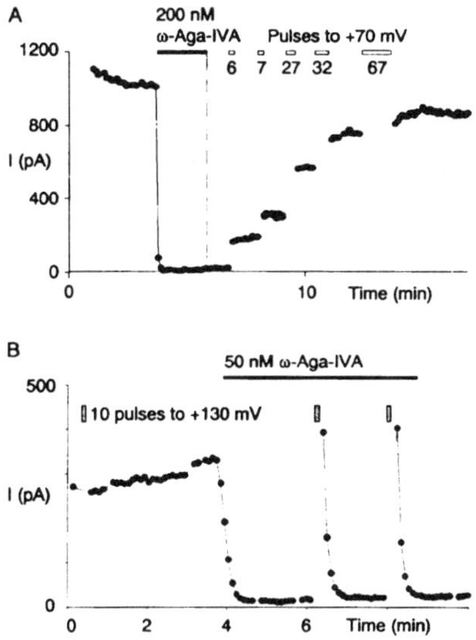

Figure 7. Knockoff of ω-Aga-IVA. Strong depolarizations induce fast reversal ("knockoff") of ω-Aga-IVA inhibition of inward P-type currents. (A) strong depolarizations spped reversal of toxin inhibition. (B) depolarizations reverse inhibition in the continual presence of toxin. Reprinted from Mintz et al., 1992a, with permission from Elsevier.

suggests that given a depolarization large enough to open the channel, the affinity of the toxin for the channel might decrease and toxin might unbind rapidly. Depolarization-induced toxin unbinding, sometimes termed "knockoff," has indeed been observed (Figure 7) with SNX-482 (Bourinet et al., 2001), grammotoxin (McDonough et al., 1997a), and ω-Aga-IVA (Mintz et al., 1992a; McDonough et al., 1997b; Bourinet et al., 1999). Knockoff of ω-Aga-IVA and grammotoxin is extraordinarily slow with temperature cooled to ~ 12 °C, which enables the separation of knockoff from the opening of toxin-bound channels. Knockoff does not require the flow of outward current (McDonough et al., 1997a,b), again consistent with a toxin effect limited to channel gating. For grammotoxin, the voltage dependence of knockoff matches closely the voltage dependence of opening of toxin-bound channels, supporting the interpretation that toxin has low affinity for the major channel open state. For ω-Aga-IVA, stronger depolarizations are required for knockoff than to open toxin-bound channels, possibly reflecting low toxin affinity for a second open state of P-type channels reached only via strong or long depolarization (Usowicz et al., 1992; McDonough et al., 1997b; McFarlane, 1997). Thus depending on the precise protocol, long depolarizations, such as the seconds-long application of high potassium solutions often used to study induction of gene expression in cultured neurons, could open toxin-bound channels or unbind the toxin completely. For such experiments it is very important to ensure that the precise protocol used is consistent with the voltages at which the toxin is an effective inhibitor of inward current.

What does an alteration in channel gating correspond to physically? One strong possibility is that binding of toxin hinders movement of the channel charge sensor, such that it moves only with input of additional energy. This seems likely with toxins such as ω-Aga-IVA that bind to an S3-S4 linker in the channel sequence (Section 4.4.1 below), directly adjacent to the S4 segment that physically moves during gating (Mannuzzu et al., 1996; Glauner et al., 1999; Cha et al., 1999; Bezanilla, 2000; Bezanilla, 2002). The alteration in gating is more likely to be produced by steric hindrance of the charge sensor, rather than by direct electrostatic repulsion between positively charged residues on the toxin and the channel, since the same gating shift of P-type channels is seen with ω-Aga-IVA (net charge +7) and ω-Aga-IVB (no net charge; Adams et al., 1993). Inhibition of the movement of the channel charge sensor should be visible as a slowing or reduction in gating currents, which are transient currents produced by the physical movement of charged residues within the transmembrane electric field during channel activation and deactivation (Bezanilla, 2002). Several toxins that modify the gating of sodium and potassium channels are known to alter gating current. On voltage-gated sodium channels, site 3 (e.g., anthopleurin-A toxin; see Cestele and Catterall, 2000) and site 4 (Css IV scorpion toxin) toxins bind to sites that incorporate S3-S4 linkers and alter gating currents, suggesting direct toxin effects on the channel charge sensors (Sheets and Hanck, 1995; Rogers et al., 1996; Cestele et al., 1998; Sheets et al., 1999; Cestele et al., 2001). Similarly, the tarantula toxin hanatoxin, closely related to grammotoxin, alters the gating of voltage dependent potassium channels, binds to the extracellular S3-S4 linker, and inhibits the movement of potassium channel gating charge (Swartz and MacKinnon, 1997a,b; Li-Smerin and Swartz, 2000, 2001; Swartz, 2001). Effects of calcium channel toxins on gating currents have not yet been reported.

Although there is clear precedent for toxins that alter voltage dependent gating to affect the movement of channel charge sensors, this is not necessarily the case. An increase in the non-electrical work required to open the channel would shift the voltage dependence of gating without affecting charge movement (Hille, 2001). Similarly, channel opening could be slowed by alterations in a voltage-independent, final, closed to open channel transition (e.g., Zagotta et al., 1989; Zagotta and Aldrich, 1990), rather than by slowing the movement of channel charge sensors. A more reliable predictor of whether a toxin affects gating current is a decrease in channel sensitivity to voltage (i.e., increase in Boltzmann slope factor). Kurtoxin, ω-Aga-IVA, grammotoxin, and SNX-482 all produce such a decrease, suggesting an effect on gating currents. The alteration could be quite subtle, however, if toxin affects the charge movement of just one of the four pseudo-subunits, as seems likely for ω-Aga-IVA (Section 4.4.1). Profound effects on the voltage dependence of channel opening, with minimal effects on gating current, could also be produced by toxin alteration of late, cooperative steps in the gating pathway that depend only weakly on voltage, as described for the *shaker* potassium channel (Smith-Maxwell et al., 1998a,b; Ledwell and Aldrich, 1999).

3.2.2. Kurtoxin: Subtype-Specific Alteration of Gating

Kurtoxin, a 63-residue toxin from scorpion venom, was first isolated and characterized on the basis of its ability to inhibit potently (IC_{50} = 15 nM) inward currents through $Ca_v3.1$ T-type channels expressed in oocytes (Chuang et al., 1998). It was inferred to act by shifting the voltage dependence of channel opening to more depolarized potentials, analogous to ω-Aga-IVA, SNX-482, and grammotoxin (Figure 8). Kurtoxin also inhibited currents through $Ca_v3.2$ channels, but at 350 nM was completely ineffective on high voltage-activated channels expressed in oocytes. The effects of kurtoxin were strikingly different, however, in a detailed study of toxin effects on calcium channels recorded from acutely dissociated neurons (Sidach and Mintz, 2002). Although its effects on native T-type channels were analogous to its effects on cloned channels, kurtoxin also had potent and complex effects on the gating of pharmacologically identified L-type, N-type, P-type, and R-type or resistant channels from native neurons (Figure 9). The effects of kurtoxin on the different high voltage-activated subtypes differed dramatically, although in each case the effects were consistent with 1:1 binding stoichiometry. BayK-8644-enhanced L-type currents were inhibited partially (~ 80%) with IC_{50} ~ 70 nM, N-type currents were inhibited biphasically with IC_{50} ~ 450 nM, and P-type currents were slowed dramatically and at large depolarizations actually increased

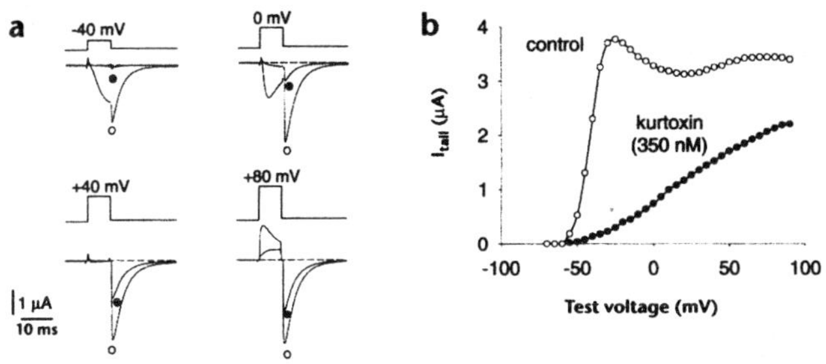

Figure 8. Kurtoxin on cloned T-type channels. Kurtoxin inhibits inward current through T-type channels by altering the voltage dependence of gating. (A) currents through $Ca_V3.1$ channels expressed in oocytes before (hollow circles) and after (filled circles) exposure to kurtoxin; (B) tail currents as a function of voltage. Reprinted with permission from Chuang et al., 1998.

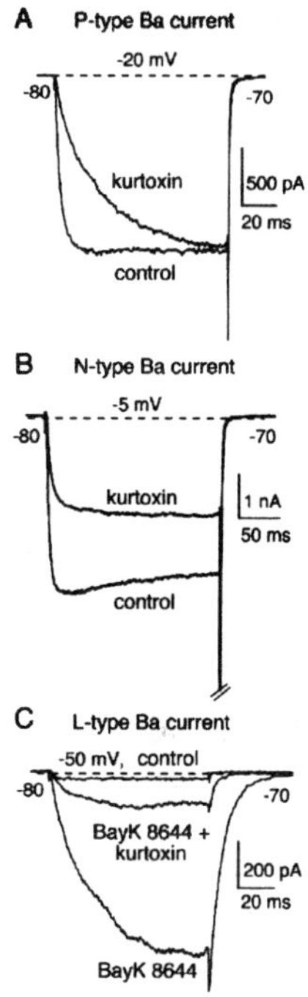

Figure 9. Kurtoxin on calcium channels from neurons. Kurtoxin slows or inhibits inward current through high voltage-activated (A) P-type, (B) N-type, and (C) L-type channels recorded from native neurons. Reproduced with permission from Sidach and Mintz, 2002. Copyright 2002, Society for Neuroscience.

by toxin. The increase may be due to an alteration in the relative prevalence of multiple open states that exist for Purkinje neuron P-type and expressed $Ca_v2.1$ channels (Usowicz et al., 1992; Bourinet et al., 1999; McFarlane, 1997; McDonough et al., 1997b). It is worth noting that kurtoxin inhibition of L-type currents enhanced by BayK-8644, presumably a mixture of $Ca_v1.2$ and $Ca_v1.3$ channels, was much stronger at more negative potentials. This suggests that kurtoxin might inhibit the low voltage-activated $Ca_v1.3$ L-type channel much more potently than the high voltage-activated $Ca_v1.2$ L-type channel.

It is not yet clear how to reconcile the total lack of kurtoxin effect on cloned high voltage-activated channels expressed in oocytes (Chuang et al., 1998) with its dramatic effects on the gating of high voltage-activated channels recorded from native neurons (Sidach and Mintz, 2002). Probably the potency of kurtoxin, like ω-Aga-IVA, depends on cell type-specific post-translational modifications or on alternative splicing. If the gating alterations of kurtoxin can be replicated on cloned channels, mutational analysis will be quite informative for understanding several different forms of gating modification.

3.2.3. Independent Binding of Gating Modifier Toxins and ω–Conotoxins

Given that these gating modifier toxins have a mechanism of action inconsistent with pore block, it is reassuring that to a first approximation they operate independently of putative pore-blocking conotoxins. For example, grammotoxin does not displace ω-CTx-GVIA binding to chick synaptosomes (Lampe et al., 1993), and ω-Aga-IVA does not compete with ω-CTx-MVIIC binding to rat brain membranes (Hillyard et al., 1992). (It should be noted, however, that at micromolar concentrations, ω-CTx-MVIIC does displace ω-Aga-IVA binding to rat brain membranes, suggesting a low-affinity interaction or conotoxin-induced changes in channel gating states; Adams et al., 1993.) The independence of binding of gating modifier toxins and pore-blocking toxins is supported by measurements of current through the channels: ω-CTx-MVIIC and ω-CTx-GVIA each inhibited all current through channels pre-bound to ω-Aga-IVA or to grammotoxin (McDonough et al., 2002). Likewise, partial inhibition of inward currents by the putative pore-blocking spider toxins DW13.3 (Sutton et al., 1998) and ω-Aga-IIIA (Mintz, 1994) did not prevent inhibition of the remaining currents by ω-Aga-IVA.

3.2.4. Dihydropyridine-like Inhibition by Calciseptine

Calciseptine is a 60-residue peptide isolated from the venom of the black mamba, *Dendroaspis polylepis polylepis* (de Weille et al., 1991). It and the closely related peptide FS2 (Yasuda et al., 1994) are specific for L-type channels, with no effects on N-type currents from chick neurons (de Weille et al., 1991), ω-CTx-GVIA binding to membranes (Yasuda et al., 1993, 1994), or T-type currents from neuroblastoma (de Weille et al., 1991). Calciseptine has no effects on neostriatal synaptic transmission affected by ω-CTx-GVIA, ω-Aga-IVA, and ω-CTx-MVIIC (Bargas et al., 1998), suggesting that it has no effects on $Ca_v2.1$ or $Ca_v2.2$ channels. Its effects on different isoforms of L-type channels vary considerably, with the strongest effects generally on cardiovascular L-type channels: IC_{50} of 15 nM for inhibition of atrial contraction, 430 nM for inhibition of smooth muscle L-type currents, and 1 μM for L-type currents from chick neurons (de Weille et al., 1991). These values are in general agreement with the $IC_{50} \sim$ 10 - 75 nM determined for inhibition of smooth muscle channels (Yasuda et al., 1994; Watanabe et al., 1995; Teramoto et al., 1996) and $K_D \sim 290$ nM for binding to rat brain membranes (Yasuda et al., 1994). In skeletal or striated muscle, by contrast, some investigators find species-dependent inhibition (Rogers and Brown, 2001), and others find that calciseptine actually increases inward currents ~ 1.5-fold and increases total charge movement by 60% (Garcia et al., 2001). It seems likely that in addition to affecting skeletal muscle $Ca_v1.1$ and smooth muscle $Ca_v1.2$ channels, calciseptine has some inhibitory effects on $Ca_v1.3$ low voltage-activated L-type channels, since it reduces the sustained, low voltage-activated component of calcium current in hippocampal pyramidal neurons (Avery and Johnston, 1996).

Surprisingly for a large peptide toxin that only acts from the extracellular side (Teramoto et al., 1996), calciseptine seems to share a mechanism with dihydropyridines, since it inhibits nitrendipine (and diltiazem) binding to membranes (Yasuda et al., 1993, 1994). Moreover, in a single-channel study of guinea pig smooth muscle, Teramoto et al. (1996) found that the IC_{50} decreased from ~ 60 nM to ~ 3 nM with more depolarized holding potentials. Single channel conductance was unaffected by toxin, whereas channel open times were somewhat shortened. This suggests toxin action on channel gating and that, like dihydropyridines, calciseptine has higher affinity for inactivated than resting channels. The agonist action of calciseptine on mammalian myotubes and skeletal muscle, however, is not precisely analogous to that of the dihydropyridine agonist (-)-BayK-8644: calciseptine did not slow tail currents (Garcia et al., 2001), and (-)-BayK-8644 does not enhance charge movement of $Ca_v1.1$ skeletal muscle channels (Lamb and Walsh, 1987), as did calciseptine.

3.3. Calcicludine: Effects on Gating and Permeation?

Calcicludine, a 60 residue toxin isolated from green mamba snake venom, is, like ω-Aga-IIIA, reported to inhibit calcium channels potently, incompletely, and irreversibly, although reports of its selectivity and potency differ. Its homology is less to other calcium channel-blocking toxins than to protease inhibitors and to the K^+ channel-blocking dendrotoxins, although it has no effect on voltage dependent Na^+ or K^+ channels (Schweitz et al., 1994). Tested on calcium channels recorded from several types of cells (Schweitz et al., 1994), calcicludine inhibited completely inward currents through L-type, N-type, and P-type, but not T-type channels, with IC_{50} ranging from 5 nM to 80 nM. On expressed cloned calcium channels, however, calcicludine gave potent (IC_{50} 88 nM), irreversible, and only partial inhibition of $Ca_v1.2$ channels expressed in tsA-201 (HEK) cells, and had almost no inhibitory effect on currents through expressed $Ca_v2.1$, $Ca_v2.2$, and $Ca_v2.3$ channels (Stotz et al., 1999). Instead, calcicludine caused a small (-3 mV to -5 mV) but consistent shift in the voltage dependence of activation. Effects of 10 nM calcicludine on the survival of immature neurons were mimicked by nifedipine, but not by N- and P-type blockers, supporting a specific L-type action (Pardo and Honegger, 1999).

4. TOXIN BINDING SITES ON CALCIUM CHANNELS

In conjunction with mechanistic experiments, assays of toxin affinity for expressed mutant channels help to define a toxin binding site. Especially for toxins binding in the pore vestibule of the channel, it seems likely that many toxin:channel contacts ("microsites") are made within the larger framework of an overall binding site (the "macrosite," Olivera et al., 1991; Olivera et al., 1994). The location of the primary determinants of channel permeation and gating in the α_1 subunit (Hille, 2001) make the α_1 subunit perhaps the most logical area for macrosites that bind inhibitory toxins. Although α_1, $\alpha_2\delta$, and γ subunits each present binding epitopes accessible to soluble toxins, it is harder to see how toxin binding to an ancillary subunit could generate specificity for α_1 subunits, since each ancillary subunit is likely capable of forming functional channels with any α_1 subunit (Walker and De Waard, 1998). It is certainly possible, however, that some toxins make micro-contacts with non-α_1 subunits, or interact with intramembrane or even intracellular parts of the α_1 subunit, as suggested for ω-atracotoxin-Hv2a and ω-Aga-IVA (Section 4.4.1). And some data suggest that the anticonvulsant gabapentin, which binds to the $\alpha_2\delta$ subunit (Gee et al., 1996), may inhibit N-type channels with some selectivity over other subtypes (Sutton et al., 2002; Martin et al., 2002).

Given these caveats, it is reassuring that thus far, the location of channel mutations that affect toxin affinity are consistent with studies of the mechanism of toxin block. Broadly, mutations that affect binding of the putative pore-blocking conotoxins ω-CTx-MVIIA and ω-CTx-GVIA are in sequences of the $Ca_v2.2$ channel adjacent to the selectivity filter. And the affinity of the gating modifier ω-Aga-IVA for $Ca_v2.1$ channels is altered dramatically by mutations in the short extracellular S3-S4 loop in domain IV, an ideal position to disrupt movement of the charged transmembrane S4 segment involved in voltage dependent gating. In general, however, surprisingly few experiments have examined the effects of channel mutations on the potency of toxin inhibition of currents, and no experiments have examined the actual binding of toxins to mutant channels.

4.1 ω-Conotoxins: Pore Blockers?

4.1.1. Effects of Mutations at the Pore Domain on ω-Conotoxin Inhibition

The first work to localize the binding site of ω-CTx-GVIA (or of any calcium channel-blocking toxin) was done by R. W. Tsien and colleagues (Ellinor et al., 1994). These authors assayed the effects of calcium channel mutations on the speed of inhibition by ω-CTx-GVIA of N-type channels expressed in *Xenopus* oocytes. The clone expressed was mostly $Ca_v2.2$ sequence from human hippocampal library and IMR-32 cells (a mix of human neuroblasts and fibroblasts), with a small amount of $Ca_v2.1$ sequence inserted near the 5' end to increase expression. Currents resulting from expression of this channel were inhibited completely by 5 μM toxin. When pseudosubunits from the ω-CTx-GVIA-insensitive $Ca_v2.1$ clone were swapped for the corresponding $Ca_v2.2$ pseudo-subunits, many different combinations slowed the on-rate, but the third pseudo-subunit had a somewhat larger effect than others in determining the toxin on-rate (Figure 10). One interpretation is that all domains of the channel contribute to the macro binding site, but the most important micro contacts were made between the toxin and domain III of the channel. Accordingly, point mutations were made in the third domain to convert $Ca_v2.2$ residues to $Ca_v2.1$ residues, many of them introducing positive charges. Mutations in the S5-H5 sequence of domain III, an area preceding the pore domain and thus likely to contribute to the pore vestibule, slowed the ω-CTx-GVIA on-rate by up to 30-fold, whereas mutations in domain III H5-S6 or in domain II had little effect (Ellinor et al., 1994).

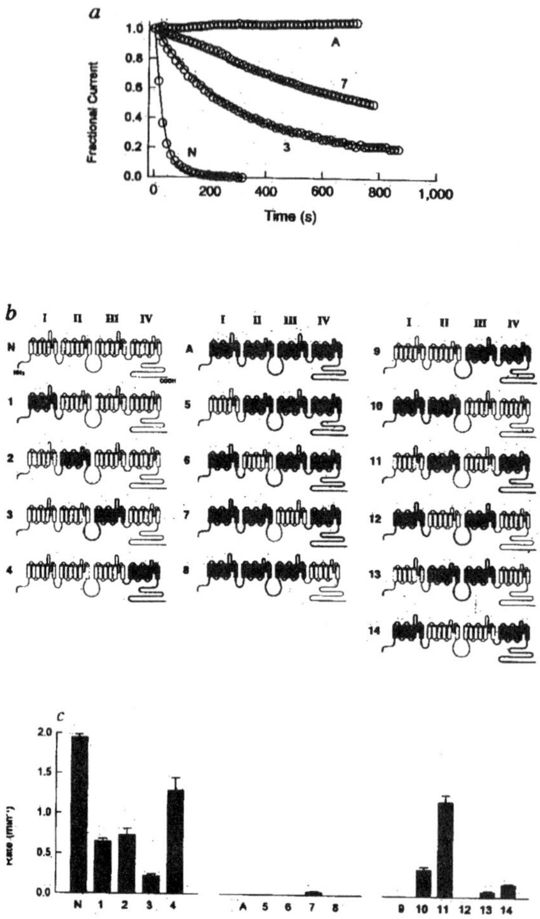

Figure 10. Multiple domains contribute to the speed of ω-CTx-GVIA inhibition of $Ca_V2.2$ channels. (A) onset of inhibition for $Ca_V2.2$ ("N"), $Ca_V2.1$ ("A"), and two chimeric ("3" and "7") channels; (B) composition of chimeras composed of domains from $Ca_V2.1$ (black) and $Ca_V2.2$ (hollow) channels; (C) rate (min^{-1}) of ω-CTx-GVIA inhibition for all chimeras. Reprinted with permission from Ellinor et al., 1994.

This part of the channel was also highlighted by Feng, Zamponi, and colleagues. Point mutations in the $Ca_v2.2$ sequence, this time at residues conserved between $Ca_v2.2$ and $Ca_v2.1$ channels, also slowed the rate of ω-CTx-GVIA block and the rate of ω-CTx-MVIIA block, while preserving 1:1 toxin:channel binding stoichiometry (Feng et al., 2001a). Most interestingly, point mutation G1326P actually speeded the onset of ω-CTx-GVIA inhibition by a factor of ~ 20, but also dramatically increased the off-rate, such that ω-CTx-GVIA block was reversible within ~ 15 minutes (Figure 11, Figure 12). Block of this G1326P variant by ω-CTx-MVIIA was also fully reversible within ~ 2 minutes (Feng et al., 2001a), suggesting a similar microsite for ω-CTx-GVIA and ω-CTx-MVIIA. Effects of the G1326P mutation were specific, as mutations G1326K, G1326E, and E1332P (see Figure 12) did not render toxin block reversible. Thus mutation of residues that likely lie just outside the calcium channel selectivity filter interfere with ω-conotoxin block.

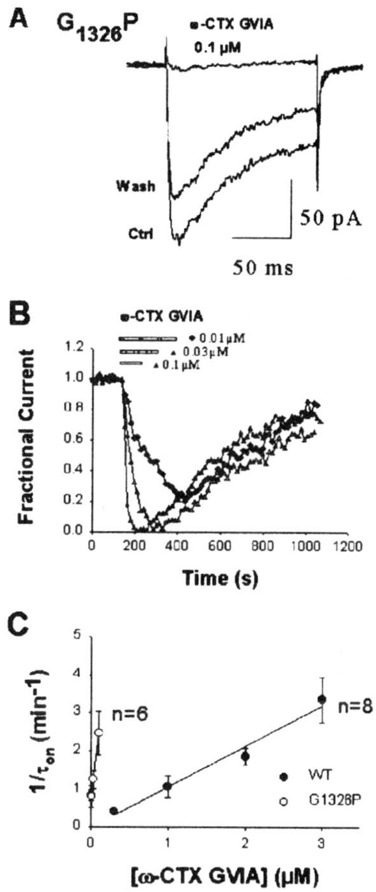

Figure 11. Point mutation G1326P, in the calcium-binding motif of $Ca_V2.2$ domain III near the pore, speeds dramatically the onset and reversal of ω-CTx-GVIA inhibition. (A) currents before and after inhibition by 100 nM toxin; (B) fast, concentration-independent reversal of toxin inhibition; (C) increased toxin on-rate on the mutant. Reprinted with permission from Feng et al., 2001a.

Do these residues in the domain III S5-H5 extracellular linker form the conotoxin binding site? Certainly this part of the channel at least mediates toxin access, and the location of these residues just N-terminal to those that undoubtedly contribute to the selectivity filter is consistent with a pore-blocking toxin. Moreover, these channel residues in domain III form a consensus EF-hand calcium binding motif, and mutations within this sequence (including G1326P) alter single channel conductances (but perhaps not ion selectivity; Feng et al., 2001b). As reviewed above (Section 3.1.1), divalent ion concentration greatly affects ω-CTx-GVIA kinetics and potency, as does reducing the divalent ion concentration to near zero (Carbone and Lux, 1988; Liang and Elmslie, 2002). It seems likely that the effects of divalent ions on ω-conotoxin affinity are caused by interactions between the toxin and divalent ions bound at the EF-hand calcium binding site. Importantly, however, no mutation thus far has actually prevented ω-conotoxin block, rather than affecting on- or off-rates. For example, mutation G1326P, which dramatically increased toxin on- and off-rates, only had a small effect on the calculated toxin equilibrium binding affinity. Perhaps most likely is the model presented by Zamponi and colleagues (Feng et

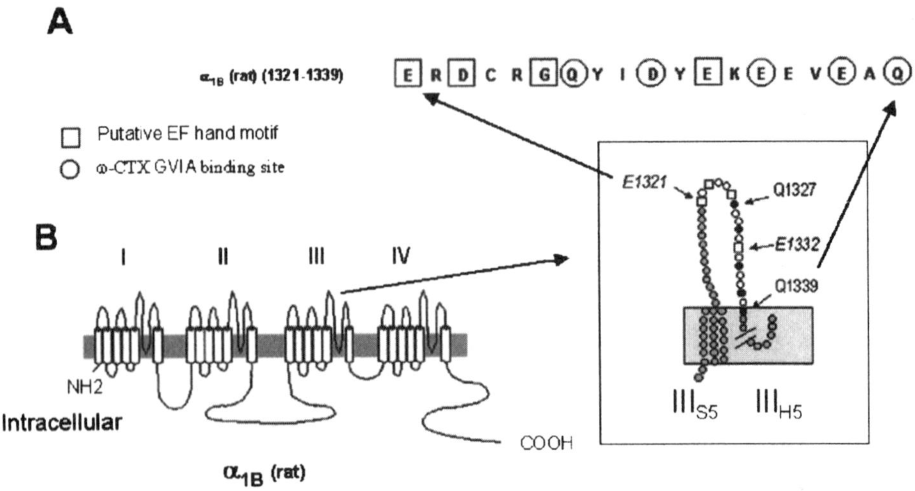

Figure 12. Location of residues whose mutation affects the kinetics of ω-CTx-GVIA inhibition of $Ca_V2.2$ channels. (A) amino acid sequence of the domain III S5-H5 region of the rat $Ca_V2.2$ (α_{1B}) channel, with EF-hand motif and ω-CTx-GVIA-interacting residues as marked. (B) predicted channel topology. Reprinted with permission from Feng et al., 2001a.

al., 2001a), in which divalent ion occupancy of the domain III EF-hand motif governs access to the toxin binding site, but the high-affinity toxin binding site itself lies closer to the pore. Given the high conservation of E1321, D1323, G1326, and Y1328 (Figure 12) across different calcium channel subtypes and species (Feng et al., 2001a), it is hard to imagine how these residues could bind the $Ca_v2.2$-selective conotoxins. Yet the proximity of these "gate" residues to the channel pore, and the biophysical features of ω-conotoxin block, suggest that the toxin does bind within the pore vestibule. The non-negligible contribution from domains I, II, and IV (Figure 10; Ellinor et al., 1994) also argues for micro binding sites to these channel domains in addition to domain III. It is also possible to imagine that the toxin might bind to the outside of the pore vestibule, at a site limited to $Ca_v2.2$ channels, then couple allosterically to the permeation pathway via the EF-hand motif—the standard caveat of site-directed mutagenesis studies.

It is interesting to speculate that if these residues indeed control access to the binding site, rather than forming the binding site themselves, the subtype-specificity of conotoxin block may be determined by access to the binding site rather than by the residues making up the binding site itself. Thus, the fast and slow block and unblock by ω-CTx-MVIIC of currents through N-type and P-type channels respectively (McDonough et al., 1996) could perhaps be understood in terms of a similar binding site at the pore of both channel types, but with easy toxin access to N-type channels and hindered toxin access to P-type channels. The $Ca_v1.2$ L-type channel blocker KP4 loses all potency when extracellular barium is replaced by calcium (Gage et al., 2002), again allowing speculation as to a "guard" role for a calcium-binding domain of the channel.

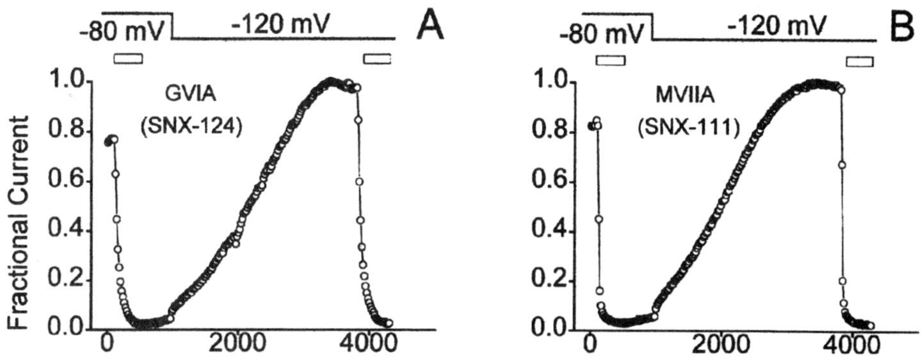

Figure 13. ω-conotoxins have higher affinity for partially inactivated than for noninactivated $Ca_V2.2$ N-type channels. Reversal of (A) ω-CTx-GVIA and (B) ω-CTx-MVIIA inhibition is speeded dramatically by changing the holding potential from −80 mV to −120 mV. Hollow bars indicate toxin application; time axis is in seconds. Reprinted with permission from Stocker et al., 1997; copyright 1997 by the Society for Neuroscience.

4.1.2. Effects of Non-Pore Domains on ω–Conotoxin Inhibition

Several cases are known in which non-pore, even intracellular, domains of the calcium channel govern ω-conotoxin affinity or kinetics. Such observations are at first difficult to reconcile with the pore-blocking mechanism outlined above, but may be explained by some state-dependence to toxin affinity. For conotoxins ω-CTx-GVIA, ω-CTx-MVIIA, ω-CTx-TVIA, and SNX-331 (variant Y13W of ω–CTx-MVIIC), inhibition of $Ca_v2.2$ channels expressed in oocytes was more potent and the toxin off-rate much slower at holding voltages that cause significant inactivation (Figure 13). Quantitatively, results were consistent with a ~ 60-fold higher toxin affinity for inactivated than resting channel states, interpreted as a change in shape of the pore vestibule upon inactivation that favors toxin binding (Stocker et al., 1997). Such changes in channel gating properties may also underlie the difference in ω-CTx-MVIIC inhibition of $Ca_v2.1$ channels expressed in oocytes with alternatively spliced versions of the β_4 subunit (Helton et al., 2002). Channels incorporating splice form β_{4b} inactivated at more negative voltages than form β_{4a} and were inhibited more quickly by ω-CTx-MVIIC (see Figure 12 of Lipscombe and Castiglioni, this volume). Mutation of just four residues in the β_{4b} sequence changed channel voltage dependence *and* toxin sensitivity to match those of β_{4a}. It seems likely that the mutations reduced occupancy of the inactivated state and so made the channels less sensitive to toxin block. Higher conotoxin affinity for inactivated states has not yet been demonstrated on calcium channels from native tissue.

Other mutations at non-pore areas of the channel affect ω-conotoxin block without affecting channel gating, illustrating additional mechanisms by which differences in primary sequence can control toxin affinity. N-type channel variant $Ca_v2.2$-Δ1, missing much of the intracellular loop between the second and third pseudosubunit channel repeats, was blocked by ω-CTx-MVIIA with IC_{50} ~ 15 nM, whereas the full-length channel was blocked with IC_{50} ~ 1 nM (Kaneko et al., 2002). Since $Ca_v2.2$-Δ1 inactivated at voltages ~ 25 mV more positive, one possibility is that the deletion reduced toxin sensitivity by lessening fractional inactivation. However, variant $Ca_v2.2$-Δ2, missing 119 fewer residues than $Ca_v2.2$-Δ1, had similar gating properties but the same ω-conotoxin sensitivity as the full-length channel. As the 119 residues present in $Ca_v2.2$-Δ2 but not in $Ca_v2.2$-Δ1 have a net fixed charge of −12, the authors suggest that these residues help attract the positively charged (+4 and +5) ω-CTx-GVIA and ω-CTx-MVIIA to their binding site and thereby increase toxin affinity (Kaneko et al., 2002). Another structurally similar toxin, ω-CTx-CVID from *C. catus*, inhibited $Ca_v2.2$ expressed in oocytes in the absence of coexpressed β subunits with 1000-fold greater affinity than when coexpressed with β_3 (Lewis et al., 2000). The mechanism is unclear, but again, this illustrates that areas of the channel that are not in the pore domain, and not even extracellular, can strongly modulate ω-conotoxin affinity, even though the macro binding site is at the pore.

In sum, a preponderance of the evidence indicates that ω-CTx-GVIA, related ω-conotoxins, and other toxins sharing this macro binding site block the channel pore by binding to multiple channel domains, with access controlled by calcium-binding residues in an EF-hand motif just outside the pore region in the S5-H5 region of channel domain III. The residues that actually bind ω-conotoxins are not known. It seems likely that the fit of the toxin within the pore vestibule is exact, since so many non-pore modifications of the channel affect toxin affinity strongly. Non-pore domains of the channel from α_1, β, and possibly other subunits may modulate toxin affinity by altering fractional channel inactivation and/ or by changing the conformation of the pore vestibule allosterically. This raises the possibility that many physiological modulations of the calcium channel that alter gating, such as phosphorylation or interaction with G proteins, calmodulin, or syntaxin (Catterall, 2000) might also alter toxin affinity. This makes ω-conotoxins somewhat less than infallibly specific channel blockers, but more valuable tools for probing the shape of the channel itself.

4.2 Toxins Sharing the ω-Conotoxin Macrosite

As detailed above, ω-CTx-GVIA and ω-CTx-MVIIA share a similar mechanism of block and a binding site on the $Ca_v2.2$ N-type channel (Olivera et al., 1987). To what extent can results on the mechanisms and binding sites of these comparatively well-studied conotoxins be extended to infer the mechanism and binding sites of other toxins? Probably quite well, especially among the many toxins sharing structural and sequence homology (Norton et al., this volume). Many binding studies assay the potency of a given toxin by measuring how efficiently it competes for binding with radiolabeled ω-CTx-GVIA, strongly implying a common binding site or interacting binding sites (e.g., Figure 1). In this manner, many toxins have been linked to the ω-CTx-GVIA ($Ca_v2.2$) or ω-CTx-MVIIC ($Ca_v2.1$) binding sites: ω-conotoxins CVIA, CVIB, CVIC, and CVID (Lewis et al., 2000), MVIIA and hybrids (Olivera et al., 1987; Wang et al., 1998; Nielsen et al., 1999), MVIIC hybrids (Hillyard et al., 1992; Gandia et al., 1997; Nielsen et al., 1999), MVIID (Gandia et al., 1997), SVIA and SVIB (Ramilo et al., 1992), TVIA and hybrids (Wang et al., 1998), GVIIA and GVIIB (Olivera et al., 1985; Abe and Saisu, 1987), and the spider toxins ω-PTxIIA (Dos Santos et al., 2002), and ω-Aga-IIA (Adams et al., 1990). ω-Aga-IA (Adams et al., 1990) and ω-Aga-IVA (Hillyard et al., 1992; Adams et al., 1993) do not affect ω-conotoxin binding, SNX-325 displaces ω-CTx-MVIIA binding only weakly (IC_{50} = 75 nM; Newcomb et al., 1995), suggesting a separate site, and the effects of ω-Aga-IIIA on ω-conotoxin binding are complex (Section 4.3).

Functional electrophysiological measurements also link different toxins to the same binding site. The same mutations on $Ca_v2.2$ affect the kinetics of ω-CTx-MVIIA as well as ω-CTx-GVIA inhibition (Ellinor et al., 1994; Feng et al., 2001a). Reversible block of N-type currents by ω-CTx-MVIIC prevents the irreversible ω-CTx-GVIA block (McDonough et al., 2002). Clearly one site on the calcium channel is a common target for many toxins—again suggesting that this common site is the channel pore.

4.3. A Macrosite for ω-Aga-IIIA and DW13.3?

Both binding and electrophysiological experiments show that ω-Aga-IIIA binds to a single site on the channel (Yan and Adams, 2000) that interacts with the ω-conotoxin binding site, but is distinct. ω-Aga-IIIA prevents binding of ω-CTx-GVIA (McEnery et al., 1991; McIntosh et al., 1992; Venema et al., 1992; Adams et al., 1993; Ertel et al., 1994) and ω-CTx-MVIIC (Adams et al., 1993) to various membrane preparations. Similarly, maximal inhibition (giving partial block) of P-type currents by ω-Aga-IIIA prevents the normal complete block of currents by ω-CTx-MVIIC (McDonough et al., 2002). Clearly, ω-Aga-IIIA prevents ω-conotoxin binding. The reverse, however, is not the case—ω-CTx-GVIA and ω-CTx-MVIIC only slightly affect binding of radiolabeled ω-Aga-IIIA (Adams et al., 1993). Moreover, ω-Aga-IIIA dramatically slows the dissociation of ω-CTx-GVIA, but not vice-versa (Yan and Adams, in press). ω-Aga-IIIA has never been tested on mutant channels, and so the residues that bind ω-Aga-IIIA or control toxin access are not known, but these data strongly suggest that ω-Aga-IIIA binds to a different epitope than the conotoxins. Possibly ω-Aga-IIIA binds in the pore vestibule, but away from the selectivity filter and atop the conotoxin binding site. This might explain the reduction of single channel conductance by ω-Aga-IIIA as well as the prevention of ω–conotoxin binding. It has not been tested directly whether ω-Aga-IIIA shows higher potency for inactivated than noninactivated channels. The higher potency for binding to membrane preparations ($K_D \sim 10$ pM; Yan and Adams, 2000), which presumably contain a high proportion of inactivated channels, compared to the concentration for half-maximal inhibition of currents (~ 1 nM; Mintz, 1991; Cohen et al., 1992; Ertel et al., 1994; Mintz, 1994) suggests that this may be the case, although such differences may also arise from the supersaturating concentrations of toxin used in electrophysiological experiments due to the extremely slow toxin on-rate.

The spider toxin DW13.3 shows similarities in selectivity and mechanism to ω-Aga-IIIA that suggest it may bind at the same site. In a detailed report, Sutton and colleagues (Sutton et al., 1998) reported that DW13.3 is a large (74 residue) toxin that potently (~ 4 nM to ~ 100 nM) inhibits expressed $Ca_v1.2$, $Ca_v2.1$, $Ca_v2.2$, and $Ca_v2.3$ channels, but not T-type channels from GH3 cells. The inhibition of currents was only partial when complete, likewise the inhibition of N-type channels from rat sympathetic neurons (85% block) and P-type channels from Purkinje neurons (76% block). DW13.3 slowed but (unlike ω-Aga-IIIA) did not prevent inhibition of the remaining current by ω-CTx-GVIA. By analogy with ω-Aga-IIIA, this suggests high-affinity binding to the outer part of the pore vestibule at a site close to but distinct from the ω-conotoxin macro binding site. Occlusion experiments between DW13.3 and ω-Aga-IIIA that would help resolve this issue have not been reported. While ω-Aga-IIIA and DW13.3 share overall charge and number of cysteine residues, they lack sequence homology. This is no barrier to a common binding site, however, as the mechanistically almost identical ω-CTx-GVIA and ω-CTx-MVIIA share only ~ 30% sequence identity outside the common disulfide structure (Olivera et al., 1991). By contrast, the spider toxin ω-PTx-IIA has some homology to ω-Aga-IIIA, but evidently binds to the conotoxin macro site (Dos Santos et al., 2002).

4.4. Binding Sites of Gating Modifier Toxins

For ω-Aga-IVA and SNX-482, channel mutations that alter toxin affinity have been found, and experiments probing interactions of kurtoxin and grammotoxin with ω-Aga-IVA lead to conclusions about their binding sites. Again, mutations affecting the potency of gating modifier toxins are outside pore-forming regions of the channel, supporting the grouping of calcium channel-targeting toxins by function.

4.4.1. The ω-Aga-IVA Binding Site

Work from three independent groups has found residues in the fourth pseudosubunit repeat of the channel, within the short ~ 10-residue extracellular loop linking the third and fourth transmembrane helices (S3-S4), whose mutation affects ω-Aga-IVA binding. Since expression of $Ca_v2.1$ channels in oocytes produced channels only weakly inhibited by ω-Aga-IVA, Bourinet, Soong, Snutch and colleagues (Bourinet et al., 1999) expressed $Ca_v2.1$ clones from rat brain in mammalian HEK-293 cells. An asparagine-proline insertion in the domain IV S3-S4 loop reduced the ω-Aga-IVA on-rate and increased the off-rate, for an overall reduction in toxin affinity from 16 nM to 146 nM. This same NP sequence was also identified by Hans et al. (1999), using human $Ca_v2.1$ expressed in HEK-293 cells. Deletion of the NP increased inhibition by 1 μM ω-Aga-IVA from 50% to 80%. Channels carrying the four-residue sequence IAET from $Ca_v2.2$ channels at the NP site were inhibited only ~ 20% by 1 μM ω-Aga-IVA (Hans et al., 1999), consistent with the low affinity ω-Aga-IVA inhibition of native N-type channels (Sidach and Mintz, 2000).

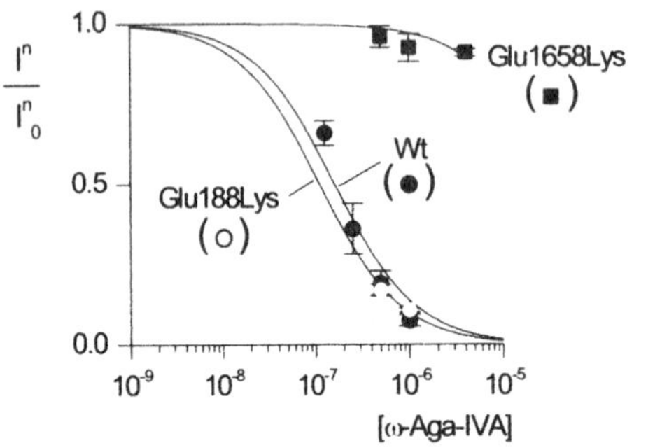

Figure 14. Low-affinity ω-Aga-IVA binding site. Mutation E1658K in the domain IV S3-S4 linker, but not homologous mutation E188K in the domain I S3-S4 linker, abolishes ω-Aga-IVA inhibition of $Ca_V2.1$ channels expressed in oocytes. Reprinted with permission from Winterfield and Swartz, 2000.

Winterfield and Swartz (2000) examined further the low-affinity (IC_{50} = 160 nM) ω-Aga-IVA inhibition of $Ca_v2.1$ channels expressed in oocytes. Mutation of a single glutamate residue to lysine in the domain IV S3-S4 linker (E1658K) (of the clone lacking NP) made the channel completely insensitive to 1 μM ω-Aga-IVA (Figure 14), whereas the analogous mutation E188K in the domain I S3-S4 linker was without effect. Although there is some concern that a low-affinity ω-Aga-IVA binding site might differ from the high-affinity binding site of native channels, clearly the domain IV S3-S4 linker contributes to an ω-Aga-IVA receptor. Experiments with K^+ channels suggest that this linker may continue the S3 α-helix into the aqueous environment, with a kink in the helix at the lipid:aqueous interface (Li-Smerin and Swartz, 2001).

Does this S3-S4 linker region form the toxin binding site itself, or serve as gatekeeper to a binding site, like the calcium-binding EF hand motif of $Ca_v2.2$ channels that governs ω-CTx-GVIA block (Feng et al., 2001a,b)? The question is impossible to answer, but since the S3-S4 loop is so short, it seems probable that it forms the binding site. At the very least it forms a crucial part of the binding site, in conjunction with as yet undetermined residues from other channel domains. The NP insertion and mutation E1658K both affect channel voltage dependence (Bourinet et al., 1999; Hans et al., 1999; Winterfield and Swartz, 2000), supporting the idea that ω-Aga-IVA binding at this site could alter channel gating. The alterations of toxin affinity produced by these mutations are not explainable by a redistribution of channels into gating states that have lower affinity for toxin. Mutation E188K in the homologous position of $Ca_v2.1$ domain I affects gating to the same extent as mutation E1658K in domain IV, but has no effect on toxin affinity. Moreover, the toxin-induced alteration of gating (~ +70 mV shift in half-maximal activation, McDonough et al., 1997b) is larger than the alterations in gating caused by the mutations (~ +30 mV for E1658K and E188K, Winterfield and Swartz, 2000; ~ +8 mV for +NP, Bourinet et al., 1999).

In addition to the sequence of the domain IV S3-S4 linker, cell type-specific post-translational modifications are clearly an important determinant of ω-Aga-IVA binding affinity. When expressed in oocytes, even channels formed by the $Ca_v2.1$ (-NP) clone are relatively insensitive to ω-Aga-IVA, and there is no depolarization-induced knockoff of the toxin (Bourinet et al., 1999). By comparison, knockoff was prominent in all neuronal types in which it was examined (Mintz et al., 1992a). Expression of $Ca_v2.1$ in mammalian cells increased toxin affinity and allowed knockoff, suggesting that processing of the channel is somehow deficient in oocytes (Berrow et al., 1997; Bourinet et al., 1999). Moreover, clones identified by RT-PCR from single mouse Purkinje neurons and expressed in HEK cells were not inhibited by ω-Aga-IVA (IC_{50} ~ 50 nM; Tsunemi et al., 2002) as potently as the channel from Purkinje neurons (IC_{50} = 1 – 2 nM, Mintz et al. 1992b). As the splice variants studied by Tsunemi et al. are by definition the ones expressed in Purkinje neurons, this suggests that post-translational modifications specific to neurons, and not just to mammalian cells, or else unidentified ancillary proteins in Purkinje neurons, contribute to high-affinity ω-Aga-IVA binding. Expression of several different splice forms of $Ca_v2.1$ in neurons cultured from $Ca_v2.1$ -/- knockout mice (Jun et al., 1999) is underway now and will settle this question.

There is also evidence that transmembrane or intracellular domains of the channel contribute to high-affinity ω-Aga-IVA binding, possibly by localizing the toxin at the domain IV S3-S4 linker. The ω-Aga-IVA toxin has an unstructured, lipophilic C-terminal tail that is essential for toxin activity (Kim et al., 1995). The homologous lipophilic C-terminal tail of ω-ACTX-Hv2a (Wang et al., 2001) is also essential for its activity, as is post-translational palmitoylation of the C-terminus of PLTX-II (Branton et al., 1993). The lipophilic C-terminal portion of these toxins may initiate toxin binding by interacting with transmembrane segments of the channel, possibly exposing a cryptic high-affinity toxin binding site (Yu et al., 1993; Wang et al., 2001). Such a step in ω-Aga-IVA binding, with the channel docking site itself possibly produced by posttranslational modification, would be consistent with the difficulty of reproducing high-affinity ω-Aga-IVA binding in heterologous expression systems. This "dock and lock" mechanism, in which one interaction brings the toxin and channel into proximity and another is responsible for high-affinity binding, has been proposed by Olivera and colleagues for α-conotoxin binding to acetylcholine receptors (McIntosh et al., 1999).

4.4.2. P-type, Q-type, and P/Q-type Channels: Splice-Form Specific Affinity of ω-Aga-IVA for $Ca_v2.1$

The NP insertion in domain IV S3-S4 that reduces ω-Aga-IVA affinity elevenfold is not an *in vitro* mutation, but is expressed in some forms of $Ca_v2.1$ *in vivo* due to alternative splicing (Bourinet et al., 1999; Hans et al., 1999). This discovery makes ω-Aga-IVA potentially very useful for uncovering splice form-specific functions of $Ca_v2.1$ channels and was a major contribution towards solving the puzzle of P-, Q-, and P/Q-type channels.

P-type channels were originally defined pharmacologically in cerebellar Purkinje neurons (Llinas et al., 1989) as channels resistant to dihydropyridines and ω-CTx-GVIA, weakly sensitive to micromolar concentrations of ω-CTx-MVIIC (Regan et al., 1991; Hillyard et al., 1992), but inhibited completely by the polyamine toxin FTX (Llinas et al., 1989; but see Norris et al., 1996) and by nanomolar concentrations of ω-Aga-IVA (Mintz et al., 1992b). These P-type channels also showed little or no inactivation during short (up to ~ 100 ms) test pulses (Regan, 1991).

These properties did not match well with the properties of the $Ca_v2.1$ (α_{1A}) clone, the first neuronal calcium channel clone expressed (Mori et al., 1991), although $Ca_v2.1$ localizes strongly to cerebellum (Starr et al., 1991; Westenbroek et al., 1995). Expressed in oocytes, $Ca_v2.1$ inactivated markedly during a test pulse, was blocked well by ω-CTx-MVIIC, and, importantly, was only weakly sensitive to block by ω-Aga-IVA (Sather et al., 1993; Zhang et al., 1993)—properties opposite to those of native P-type channels. Similar to this, hippocampal synaptic transmission between Schaffer collateral fibers and CA1 pyramidal neurons was insensitive to 30 nM ω-Aga-IVA but blocked fully by ω-CTx-MVIIC and almost fully by 1 μM ω-Aga-IVA. This led to the proposal that synaptic transmission in hippocampal neurons was governed by a "Q-type" channel with these properties encoded by the $Ca_v2.1$ gene (Wheeler et al., 1994). An analysis of the kinetics of currents in cerebellar granule neurons concluded that currents with "Q-type" pharmacological properties inactivated rapidly during a test pulse, consistent with the $Ca_v2.1$ channels expressed in oocytes (Randall and Tsien, 1995).

Since then, antisense and antibody experiments (Gillard et al., 1997; Pinto et al., 1998), recordings from knockout mice (Jun et al., 1999), and single cell RT-PCR (Tsunemi et al., 2002) have demonstrated that one or more forms of $Ca_v2.1$ indeed makes up the P-type currents in cerebellar Purkinje neurons. ω-CTx-MVIIC has since been found to inhibit P-type channels potently, with $K_D \sim 50$ nM. The extremely slow toxin on-rate caused the original underestimation of its potency on $Ca_v2.1$ channels (McDonough et al., 1996). At micromolar concentrations, ω-Aga-IVA inhibits N-type channels as well as P-type channels (Sidach and Mintz, 2000), and the very weak inhibition of $Ca_v2.1$ in oocytes by ω-Aga-IVA was not due to the channel itself but to limitations of the oocyte expression system (Bourinet et al., 1999; see Figure 7 of Lipscombe and Castiglioni, this volume). Finally, analysis of the expression patterns of splice variants of $Ca_v2.1$ showed that splice forms of $Ca_v2.1$ with higher sensitivity to ω-Aga-IVA were expressed in cerebellar mRNA and in Purkinje neurons, as opposed to hippocampal neurons (Bourinet et al., 1999; Tsunemi et al., 2002).

Given that many forms of $Ca_v2.1$ (Soong et al., 2002) exist and can combine with different β subunits (Moreno et al., 1997; Mermelstein et al., 1999) to produce channels with different voltage dependence, inactivation speeds, and ω-Aga-IVA sensitivity, clearly the distinction between "P-type" and "Q-type" is invalid (Miller, 1995; Beam, 1999). The compromise has been to refer to $Ca_v2.1$ currents as "P/Q-type." Certainly the best way to describe a channel is by its exact genetic composition. As this is rarely possible when dealing with native channels, however, some terminology is needed to describe currents carried by $Ca_v2.1$ channels in native neurons. Pharmacologically, some forms of $Ca_v2.1$ channels can be inhibited with high affinity and specificity by ω-Aga-IVA (complete and fairly fast inhibition by ~ one hundred nanomolar), and most or all forms by ω–CTx-MVIIC (slow and complete inhibition by several hundred nanomolar). My own feeling is that the "type" in the appellation "P-type" was consciously chosen to incorporate the suggestion that multiple isoforms of channels similar to the channel in Purkinje neurons fit this description, just as "L-type" and "N-type" channels incorporate variants with different sensitivity to dihydropyridines and to ω-CTx-GVIA. The cell bodies of Purkinje neurons express an enormous number of functional $Ca_v2.1$ channels, the most of any known cell type, so "P-type" is appropriate for more than historical reasons.

4.4.3. Binding Sites for Kurtoxin, Grammotoxin, and SNX-482

Binding sites for kurtoxin, grammotoxin, and SNX-482 have not been investigated with such precision as the ω-Aga-IVA binding site. The binding of SNX-482 was tested (Bourinet et al., 2001) on $Ca_v2.3$ channels from rat brain (Soong et al., 1993) expressed in tsA-201 cells with β_{2a} and $\alpha_2\delta$. For this particular isoform of $Ca_v2.3$, inhibition of inward currents by SNX-482 was rapid and complete due to a ~ +70 mV shift in the voltage dependence of channel activation (Bourinet et al., 2001). SNX-482 also shifted the activation of $Ca_v1.2$ L-type channels, but by only ~ +10 mV. Chimeric constructs showed that substitution of multiple channel domains between $Ca_v1.2$ and $Ca_v2.3$ channels was necessary to reconstruct the $Ca_v2.3$-type shift in activation gating. The exact residues are not known, but S1-S2 linkers and S3-S4 linkers are the likely sites, since these are extracellular epitopes adjacent to transmembrane regions S2 and S4, regions that physically move during activation of *shaker* potassium channels (Hille, 2001; Bezanilla, 2002).

No investigations have addressed the molecular nature of the binding sites for kurtoxin or grammotoxin. On native P-type channels from Purkinje neurons, however, each of these two gating modifier toxins has been applied together with ω-Aga-IVA. When applied to channels already bound to ω-Aga-IVA, grammotoxin caused an additional depolarizing shift in the voltage dependence of activation. The shift due to both toxins applied together was almost equal to the sum of the shifts due to each toxin applied individually (Figure 15; McDonough et al., 1997a). This suggests that whatever the grammotoxin binding site (or sites), the domain IV S3-S4 sequence is not involved, discounting the possibility that both toxins can bind to the same short stretch of channel. On potassium channels, the S3-S4 linker forms the high-affinity binding site for hanatoxin (Swartz and MacKinnon, 1995, 1997a,b), a gating modifier toxin closely

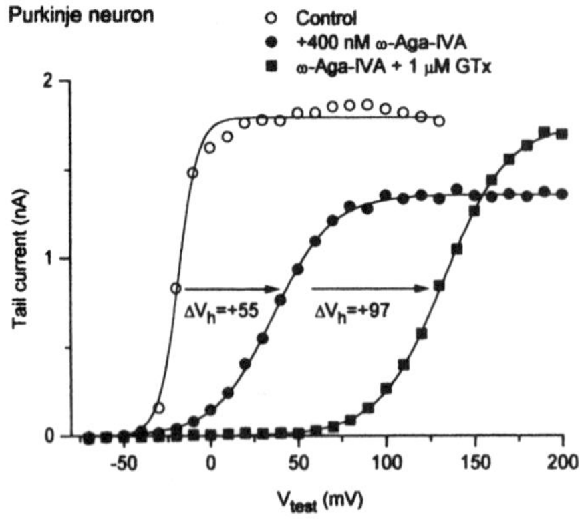

Figure 15. Additive shift in half-maximal activation voltage of P-type channels when ω-Aga-IVA and grammotoxin are applied consecutively. Figure reprinted with permission from McDonough et al., 1997a, copyright 1997 by the Biophysical Society.

homologous to grammotoxin. The K^+ channel S3-S4 linker also forms a low-affinity binding site for grammotoxin (Li-Smerin and Swartz, 1998). By extension, this suggests that grammotoxin may bind to an S3-S4 loop of the calcium channel, one other than domain IV.

Gating modification of P-type channels by kurtoxin, unlike grammotoxin, is not independent of modification by ω-Aga-IVA. Binding of kurtoxin, detected by a dramatic slowing of channel activation speed, greatly slowed the subsequent onset of inhibition of inward currents by 200 nM ω-Aga-IVA, suggesting some common binding determinants between kurtoxin and ω-Aga-IVA (Sidach and Mintz, 2002). Nevertheless, kurtoxin did not prevent inhibition by ω-Aga-IVA. Either the toxin binding sites are allosterically linked, or the slow inhibition by ω-Aga-IVA reflected slow replacement of kurtoxin by ω-Aga-IVA on a common binding site. Since kurtoxin barely affects the steady-state voltage dependence of P-type channel activation, whereas ω-Aga-IVA affects it strongly, separate, allosterically linked binding sites for the two toxins at S1-S2 or S3-S4 loops are perhaps most likely.

4.4. Binding Site of Calcicludine

Swapping domains between $Ca_v1.2$ and $Ca_v2.3$ channels showed evidence for a calcicludine binding site distributed among different channel domains: domains II or III for the altered voltage dependence of activation, and domains I, II, and III (IV was not tested) for inhibition of currents (Stotz et al., 2000). The involvement of different domains in modification of gating and permeation would be consistent with more than one binding site for the toxin, suggested by the poor fit of the dose-response curve to a Hill coefficient of one. Binding (Schweitz et al., 1994) and electrophysiological (Stotz et al., 2000) assays agree that calcicludine does not inhibit the action of ω-CTx-GVIA on $Ca_v2.2$ N-type channels, suggesting a calcicludine binding site different from the ω-conotoxin binding site. Addition of calcicludine to $Ca_v2.2$ channels did not prevent ω-CTx-GVIA block, whereas ω-Aga-IIIA would have prevented this (Venema et al., 1992). These suggest that calcicludine binds away from the conotoxin binding site and the ω-Aga-IIIA binding site, at one or more sites (Stotz et al., 1999) contributed to by many channel domains. It may bind outside the pore and effect pore constriction and alter channel gating allosterically.

5. EXPERIMENTAL USE OF TOXINS

As detailed above, the mechanisms by which toxins inhibit channels are grounds for a few major cautions when using either gating modifier toxins or permeation inhibitor toxins. The composition of the ionic solution can affect powerfully toxin potency and reversibility. In particular, for ω-conotoxins, high concentrations of divalent ions may impede toxin access to the binding site (Sections 3.1.1 and 4.1), and in the absence of divalent ions, toxin inhibition is reversible. The very slow on-rate of many toxins means that equilibrium binding of low toxin concentrations is difficult to achieve in an electrophysiological experiment, when channel rundown and weakening of the gigaseal are often problems. For example, ω-Aga-IIIA binding at concentrations near the K_D takes hours (Yan and Adams, 2000). For gating modifier toxins, care must be taken to ensure that the test depolarizations used to open the calcium channels do not exceed the voltage at which toxin-bound channels open. This is an important consideration when using toxins such as ω-Aga-IVA to inhibit calcium entry evoked by seconds-long depolarizations with high-potassium solutions, a protocol often used to assay the coupling of calcium channels to gene expression. Not only will sufficiently large depolarizations open toxin-bound channels and thus allow calcium entry, but depending on the particular toxin, large depolarizations may put channels into a gating state from which toxin unbinds quickly.

In addition to mechanistic considerations, toxin potency, particularly of the gating modifier toxins that often bind to hypervariable regions of the channel, may depend strongly on the splice variant or gating state of the α_1 subunit under study. As reviewed in detail above, toxin potency may be affected strongly by fractional channel inactivation, by splicing or post-translational modification of the α_1 subunit, by the development stage of channel expression, or by the particular co-expressed ancillary subunits. Toxins reported to be affected by such considerations include ω-Aga-IVA (Moreno et al., 1997; Bourinet et al., 1999; Hans et al., 1999), ω-CTx-MVIIA (Stocker et al., 1997; Lewis et al., 2000), ω-CTx-MVIIC (Helton et al., 2002), SNX-482 (Tottene et al., 2000), ω-CTx-GVIA (White et al., 1997), and ω-CTx-CVID (Lewis et al., 2000).

Especially in electrophysiological experiments requiring acute toxin application, several technical considerations may be important to reproduce the high potency and selectivity for particular α_1 subunits. The peptides can adsorb to tubing used to deliver them to the preparation, meaning that the toxin concentration actually delivered to the sample is less than that of the stock solution. Thus it is standard to include ~ 1 mg/ mL of an inert peptide such as BSA or cytochrome c in solutions, to saturate nonspecific peptide binding sites in the tubing, and/ or to use siliconized pipette tips, recording chamber, and storage vials (Sidach and Mintz, 2002). In my hands, it also seems helpful to run toxin-containing solutions through the tubing for several minutes before beginning the experiment. A corollary is that since toxins can stick to tubing, once a given tube has been used for toxin application, it should either be discarded or checked carefully after washing to ensure that no toxin remained in the tube before using it to perfuse a different compound. Weak, non-specific, and quickly-reversible inhibition by a toxin may be due to trace heavy metals in the water supply or lyophilized toxin powder. In my hands, this can often be prevented by including several hundred micromolar EDTA in the recording solution.

6. FUTURE DIRECTIONS

Gaps remain in the pharmacology of calcium channels that may be filled with new toxins or with molecules derived from toxins. Notably, $Ca_v1.2$ and $Ca_v1.3$ L-type channels are widely coexpressed (Hell et al., 1993), yet have different biophysical properties (Lipscombe, 2002), suggesting different functions. A high-affinity blocker able to distinguish between the two would be valuable for investigating their different functional roles. Within the Ca_v2 family, a high-affinity blocker with selectivity for all splice forms of $Ca_v2.1$ P-type and $Ca_v2.3$ R-type channels would be useful. To block all splice forms, such a molecule is likely to bind within the pore vestibule, an area not known to be subject to alternative splicing (Lipscombe et al., 2002). Conversely, determining functional differences of different splice forms of a given α_1 subunit would be aided greatly by splice form-specific inhibitors. Further development of gating modifier toxins may be key: ω-Aga-IVA and SNX-482 already appear capable of this (Section 2.3 and 4.4). The ability of kurtoxin (Sidach and Mintz, 2002) and grammotoxin (McDonough et al., 1997a) to actually increase peak currents through P-type channels gives some hope for the feasibility of drugs that enhance currents through non-L-type channels, for use in disorders such as migraine that might result from reduced channel function or altered gating (Knight et al., 2002; Kraus and Cohen, this volume). A toxin selective for native T-type channels, or better yet able to discriminate among $Ca_v3.1$, 3.2, and 3.3 T-type channels would be greatly helpful, although paralytic venoms are perhaps unlikely to be as rich a source of new ligands, as T-type channels are generally not coupled to transmitter release.

For each of these cases, existing toxins of known structure, active surfaces, mechanisms, and binding sites may be useful for the development of compounds with improved selectivity. For example, a pore-blocking toxin (ω-CTx-MVIIC) and a gating modifier toxin (grammotoxin) are known that each inhibit both N-type and P-type channels, in addition to the pore-blocking toxins selective for N-type channels (ω-CTx-GVIA and ω-CTx-MVIIA) and gating modifier toxins with good selectivity for P-type channels (ω-Aga-IVA and ω-Aga-IVB). Further mechanistic information on known toxins is also needed. Surprisingly few experiments have assayed the affinity of toxins for mutant channels: biophysically well-characterized toxins such as grammotoxin, ω-Aga-IIIA, and calciseptine have not been assayed on mutant channels, and the residues binding even ω-CTx-GVIA have not yet been found. Mutant cycle analysis of toxins and channels has not been reported, although many mutations already exist for toxins (e.g., Lew et al., 1997; Flinn et al., 1999; Nielsen et al., 1999) and channels (e.g., Ellinor et al., 1994; Feng et al., 2001a). Although no structure is reported for a calcium channel at this writing, the advent of high-resolution structural studies of voltage-gated channels makes this approach perhaps less interesting for study of the channel itself, although still informative for pharmacological and pharmaceutical purposes. Such information is crucial to refining the next generation of calcium channel-targeting drugs.

ACKNOWLEDGMENTS

I am grateful to Dr. Ray Norton, Dr. Gerald Zamponi, and Dr. Charles Cohen for review of the manuscript, and to Dr. Michael Adams for helpful discussions. Supported by NCRR:P41RR01395 and by a grant from the American Heart Association.

REFERENCES

Abe, T., Koyano, K., Saisu, H., Nishiuchi, Y. and Sakakibara, S., 1986, Binding of ω-conotoxin to receptor sites associated with the voltage-sensitive calcium channel, *Neurosci Lett* **71**:203-208.

Abe, T. and Saisu, H., 1987, Identification of the receptor for ω-conotoxin in brain. Probable components of the calcium channel, *J Biol Chem* **262**:9877-9882.

Adams, D. J., Smith, A. B., Schroeder, C. I., Yasuda, T., and Lewis, R. J., 2003, ω-conotoxin CVID inhibits a pharmacologically distinct voltage sensitive calcium channel associated with transmitter release from preganglionic nerve terminals, *J Biol Chem* **278**:4057-4062.

Adams, M. E., Bindokas, V. P., Hasegawa, L. and Venema, V. J., 1990, ω-agatoxins: novel calcium channel antagonists of two subtypes from funnel web spider (Agelenopsis aperta) venom, *J Biol Chem* **265**:861-867.

Adams, M. E., Mintz, I. M., Reily, M. D., Thanabal, V. and Bean, B. P., 1993, Structure and properties of ω-agatoxin IVB, a new antagonist of P-type calcium channels, *Mol Pharmacol* **44**:681-688.

Adams, M. E., Myers, R. A., Imperial, J. S. and Olivera, B. M., 1993, Toxityping rat brain calcium channels with ω-toxins from spider and cone snail venoms, *Biochemistry* **32**:12566-12570.

Adams, M. E. and Olivera, B. M., 1994, Neurotoxins: overview of an emerging research technology, *Trends Neurosci* **17**:151-155.

Aiyar, J., Withka, J. M., Rizzi, J. P., Singleton, D. H., Andrews, G. C., Lin, W., Boyd, J., Hanson, D. C., Simon, M., Dethlefs, B. and et al., 1995, Topology of the pore-region of a K^+ channel revealed by the NMR-derived structures of scorpion toxins, *Neuron* **15**:1169-1181.

Albillos, A., Garcia, A. G., Olivera, B. and Gandia, L., 1996, Re-evaluation of the P/Q Ca^{2+} channel components of Ba^{2+} currents in bovine chromaffin cells superfused with solutions containing low and high Ba^{2+} concentrations, *Pflugers Arch* **432**:1030-1038.

Aosaki, T. and Kasai, H., 1989, Characterization of two kinds of high-voltage-activated Ca-channel currents in chick sensory neurons. Differential sensitivity to dihydropyridines and omega-conotoxin GVIA, *Pflugers Arch* **414**:150-156.

Avery, R. B. and Johnston, D., 1996, Multiple channel types contribute to the low-voltage-activated calcium current in hippocampal CA3 pyramidal neurons, *J Neurosci* **16**:5567-5582.

Bargas, J., Ayala, G. X., Hernandez, E. and Galarraga, E., 1998, Ca^{2+}-channels involved in neostriatal glutamatergic transmission, *Brain Res Bull* **45**:521-524.

Bargas, J., Howe, A., Eberwine, J., Cao, Y. and Surmeier, D. J., 1994, Cellular and molecular characterization of Ca^{2+} currents in acutely isolated, adult rat neostriatal neurons, *J Neurosci* **14**:6667-6686.

Barnes, S. and Hille, B., 1989, Ionic channels of the inner segment of tiger salamander cone photoreceptors, *J Gen Physiol* **94**:719-743.

Beam, K., 1999, Calcium channel splicing: mind your Ps and Qs, *Nat Neurosci* **2**:393-394.

Bean, B. P., 1984, Nitrendipine block of cardiac calcium channels: high-affinity binding to the inactivated state, *Proc Natl Acad Sci U S A* **81**:6388-6392.

Bech-Hansen, N. T., Naylor, M. J., Maybaum, T. A., Pearce, W. G., Koop, B., Fishman, G. A., Mets, M., Musarella, M. A. and Boycott, K. M., 1998, Loss-of-function mutations in a calcium-channel alpha1-subunit gene in Xp11.23 cause incomplete X-linked congenital stationary night blindness, *Nat Genet* **19**:264-267.

Beck, H., Steffens, R., Heinemann, U. and Elger, C. E., 1997, Properties of voltage-activated Ca^{2+} currents in acutely isolated human hippocampal granule cells, *J Neurophysiol* **77**:1526-1537.

Berrow, N. S., Brice, N. L., Tedder, I., Page, K. M. and Dolphin, A. C., 1997, Properties of cloned rat alpha1A calcium channels transiently expressed in the COS-7 cell line, *Eur J Neurosci* **9**:739-748.

Bezanilla, F., 2000, The voltage sensor in voltage-dependent ion channels, *Physiol Rev* **80**:555-592.

Bezanilla, F., 2002, Voltage sensor movements, *J Gen Physiol* **120**:465-473.

Bindokas, V. P. and Adams, M. E., 1989, ω-Aga-I: a presynaptic calcium channel antagonist from venom of the funnel web spider, *Agelenopsis aperta*, *J Neurobiol* **20**:171-188.

Bindokas, V. P., Venema, V. J. and Adams, M. E., 1991, Differential antagonism of transmitter release by subtypes of ω-agatoxins, *J Neurophysiol* **66**:590-601.

Boland, L. M., Morrill, J. A. and Bean, B. P., 1994, ω-Conotoxin block of N-type calcium channels in frog and rat sympathetic neurons, *J Neurosci* **14**:5011-5027.

Bourinet, E., Soong, T. W., Sutton, K., Slaymaker, S., Mathews, E., Monteil, A., Zamponi, G. W., Nargeot, J. and Snutch, T. P., 1999, Splicing of α_{1A} subunit gene generates phenotypic variants of P- and Q-type calcium channels, *Nat Neurosci* **2**:407-415.

Bourinet, E., Stotz, S. C., Spaetgens, R. L., Dayanithi, G., Lemos, J., Nargeot, J. and Zamponi, G. W., 2001, Interaction of SNX482 with domains III and IV inhibits activation gating of α_{1E} ($Ca_V2.3$) calcium channels, *Biophys J* **81**:79-88.

Bowers, C. W., Phillips, H. S., Lee, P., Jan, Y. N. and Jan, L. Y., 1987, Identification and purification of an irreversible presynaptic neurotoxin from the venom of the spider Hololena curta, *Proc Natl Acad Sci U S A* **84**:3506-3510.

Bowersox, S. S. and Luther, R., 1998, Pharmacotherapeutic potential of omega-conotoxin MVIIA (SNX-111), an N-type neuronal calcium channel blocker found in the venom of Conus magus, *Toxicon* **36**:1651-1658.

Bowersox, S. S., Miljanich, G. P., Sugiura, Y., Li, C., Nadasdi, L., Hoffman, B. B., Ramachandran, J. and Ko, C. P., 1995, Differential blockade of voltage-sensitive calcium channels at the mouse neuromuscular junction by novel omega-conopeptides and omega-agatoxin-IVA, *J Pharmacol Exp Ther* **273**:248-256.

Branton, W. D., Kolton, L., Jan, Y. N. and Jan, L. Y., 1987, Neurotoxins from Plectreurys spider venom are potent presynaptic blockers in Drosophila, *J Neurosci* 7:4195-4200.

Branton, W. D., Rudnick, M. S., Zhou, Y., Eccleston, E. D., Fields, G. B. and Bowers, L. D., 1993, Fatty acylated toxin structure, *Nature* **365**:496-497.

Brejc, K., van Dijk, W. J., Smit, A. B. and Sixma, T. K., 2002, The 2.7 A structure of AChBP, homologue of the ligand-binding domain of the nicotinic acetylcholine receptor, *Novartis Found Symp* **245**:22-29; discussion 29-32, 165-168.

Brown, A. M., Sayer, R. J., Schwindt, P. C. and Crill, W. E., 1994, P-type calcium channels in rat neocortical neurones, *J Physiol* **475**:197-205.

Carbone, E. and Lux, H. D., 1988, Omega-conotoxin blockade distinguishes Ca from Na permeable states in neuronal calcium channels, *Pflugers Arch* **413**:14-22.

Carbone, E., Prestipino, G., Spadavecchia, L., Franciolini, F. and Possani, L. D., 1987, Blocking of the squid axon K^+ channel by noxiustoxin: a toxin from the venom of the scorpion Centruroides noxius, *Pflugers Arch* **408**:423-431.

Cartier, G. E., Yoshikami, D., Gray, W. R., Luo, S., Olivera, B. M. and McIntosh, J. M., 1996, A new alpha-conotoxin which targets α3β2 nicotinic acetylcholine receptors, *J Biol Chem* **271**:7522-7528.

Cassola, A. C., Jaffe, H., Fales, H. M., Castro Afeche, S., Magnoli, F. and Cipolla-Neto, J., 1998, omega-phonetoxin-IIA: a calcium channel blocker from the spider phoneutria nigriventer, *Pflugers Arch* **436**:545-552.

Catterall, W. A., 2001, A 3D view of sodium channels, *Nature* **409**:988-989, 991.

Catterall, W. A., 1977, Activation of the action potential Na^+ ionophore by neurotoxins. An allosteric model, *J Biol Chem* **252**:8669-8676.

Catterall, W. A., 1977, Membrane potential-dependent binding of scorpion toxin to the action potential Na^+ ionophore. Studies with a toxin derivative prepared by lactoperoxidase-catalyzed iodination, *J Biol Chem* **252**:8660-8668.

Catterall, W. A., 1998, Structure and function of neuronal Ca^{2+} channels and their role in neurotransmitter release, *Cell Calcium* **24**:307-323.

Catterall, W. A., 2000, Structure and regulation of voltage-gated Ca^{2+} channels, *Annu Rev Cell Dev Biol* **16**:521-555.

Cestele, S. and Catterall, W. A., 2000, Molecular mechanisms of neurotoxin action on voltage-gated sodium channels, *Biochimie* **82**:883-892.

Cestele, S., Qu, Y., Rogers, J. C., Rochat, H., Scheuer, T. and Catterall, W. A., 1998, Voltage sensor-trapping: enhanced activation of sodium channels by beta-scorpion toxin bound to the S3-S4 loop in domain II, *Neuron* **21**:919-931.

Chang, N. S., French, R. J., Lipkind, G. M., Fozzard, H. A. and Dudley, S., Jr., 1998, Predominant interactions between mu-conotoxin Arg-13 and the skeletal muscle Na^+ channel localized by mutant cycle analysis, *Biochemistry* **37**:4407-4419.

Chuang, R. S., Jaffe, H., Cribbs, L., Perez-Reyes, E. and Swartz, K. J., 1998, Inhibition of T-type voltage-gated calcium channels by a new scorpion toxin, *Nat Neurosci* **1**:668-674.

Churchill, D. and Macvicar, B. A., 1998, Biophysical and pharmacological characterization of voltage-dependent Ca^{2+} channels in neurons isolated from rat nucleus accumbens, *J Neurophysiol* **79**:635-647.

Cohen, C. J., Ertel, E. A., Smith, M. M., Venema, V. J., Adams, M. E. and Leibowitz, M. D., 1992, High affinity block of myocardial L-type calcium channels by the spider toxin omega-Aga-toxin IIIA: advantages over 1,4-dihydropyridines, *Mol Pharmacol* **42**:947-951.

Connor, M. and Christie, M. J., 1998, Modulation of $Ca^{2}+$ channel currents of acutely dissociated rat periaqueductal grey neurons, *J Physiol* **509 (Pt 1)**:47-58.

Conticello, S. G., Gilad, Y., Avidan, N., Ben-Asher, E., Levy, Z. and Fainzilber, M., 2001, Mechanisms for evolving hypervariability: the case of conopeptides, *Mol Biol Evol* **18**:120-131.

Cox, D. H. and Dunlap, K., 1992, Pharmacological discrimination of N-type from L-type calcium current and its selective modulation by transmitters, *J Neurosci* **12**:906-914.

Cruz, L. J., Gray, W. R., Olivera, B. M., Zeikus, R. D., Kerr, L., Yoshikami, D. and Moczydlowski, E., 1985, Conus geographus toxins that discriminate between neuronal and muscle sodium channels, *J Biol Chem* **260**:9280-9288.

Cruz, L. J. and Olivera, B. M., 1986, Calcium channel antagonists. Omega-conotoxin defines a new high affinity site, *J Biol Chem* **261**:6230-6233.

de Weille, J. R., Schweitz, H., Maes, P., Tartar, A. and Lazdunski, M., 1991, Calciseptine, a peptide isolated from black mamba venom, is a specific blocker of the L-type calcium channel, *Proc Natl Acad Sci U S A* **88**:2437-2440.

Delmas, P., Abogadie, F. C., Buckley, N. J., and Brown, D. A., 2000, Calcium channel gating and modulation by neurotransmitters depend on cellular compartmentalization, *Nature Neurosci* **3**:670-678.

Desmadryl, G., Chambard, J. M., Valmier, J. and Sans, A., 1997, Multiple voltage-dependent calcium currents in acutely isolated mouse vestibular neurons, *Neuroscience* **78**:511-522.

Dolphin, A. C., 1998, Mechanisms of modulation of voltage-dependent calcium channels by G proteins, *J Physiol* **506 (Pt 1)**:3-11.

Dos Santos, R. G., Van Renterghem, C., Martin-Moutot, N., Mansuelle, P., Cordeiro, M. N., Diniz, C. R., Mori, Y., De Lima, M. E. and Seagar, M., 2002, Phoneutria nigriventer omega-phonetoxin IIA blocks the Ca_V2 family of calcium channels and interacts with ω-conotoxin-binding sites, *J Biol Chem* **277**:13856-13862.

Doyle, D. A., Morais Cabral, J., Pfuetzner, R. A., Kuo, A., Gulbis, J. M., Cohen, S. L., Chait, B. T. and MacKinnon, R., 1998, The structure of the potassium channel: molecular basis of K^+ conduction and selectivity, *Science* **280**:69-77.

Dreijer, A. M. and Kits, K. S., 1995, Multiple second messenger routes enhance two high-voltage-activated calcium currents in molluscan neuroendocrine cells, *Neuroscience* **64**:787-800.

Ellinor, P. T., Zhang, J. F., Horne, W. A. and Tsien, R. W., 1994, Structural determinants of the blockade of N-type calcium channels by a peptide neurotoxin, *Nature* **372**:272-275.

Ellinor, P. T., Zhang, J. F., Randall, A. D., Zhou, M., Schwarz, T. L., Tsien, R. W. and Horne, W. A., 1993, Functional expression of a rapidly inactivating neuronal calcium channel, *Nature* **363**:455-458.

Ertel, E. A., Campbell, K. P., Harpold, M. M., Hofmann, F., Mori, Y., Perez-Reyes, E., Schwartz, A., Snutch, T. P., Tanabe, T., Birnbaumer, L., Tsien, R. W. and Catterall, W. A., 2000, Nomenclature of voltage-gated calcium channels, *Neuron* **25**:533-535.

Ertel, E. A., Cohen, C. J., Leibowitz, M. D., Adams, M., Nguyen, T., Ashley, I. M., and Smith, M. M., 1993, Omega-Aga-IIB: A high affinity peptide blocker of multiple types of voltage-gated Ca channels. *Biophys J* **64**(2 part 2): A117.

Ertel, E. A., Smith, M. M., Leibowitz, M. D. and Cohen, C. J., 1994, Isolation of myocardial L-type calcium channel gating currents with the spider toxin ω-Aga-IIIA, *J Gen Physiol* **103**:731-753.

Ertel, E. A., Warren, V. A., Adams, M. E., Griffin, P. R., Cohen, C. J. and Smith, M. M., 1994, Type III ω-agatoxins: a family of probes for similar binding sites on L- and N-type calcium channels, *Biochemistry* **33**:5098-5108.

Espiritu, D. J., Watkins, M., Dia-Monje, V., Cartier, G. E., Cruz, L. J. and Olivera, B. M., 2001, Venomous cone snails: molecular phylogeny and the generation of toxin diversity, *Toxicon* **39**:1899-1916.

Fainzilber, M., Lodder, J. C., van der Schors, R. C., Li, K. W., Yu, Z., Burlingame, A. L., Geraerts, W. P. and Kits, K. S., 1996, A novel hydrophobic omega-conotoxin blocks molluscan dihydropyridine-sensitive calcium channels, *Biochemistry* **35**:8748-8752.

Feigenbaum, P., Garcia, M. L. and Kaczorowski, G. J., 1988, Evidence for distinct sites coupled to high affinity omega-conotoxin receptors in rat brain synaptic plasma membrane vesicles, *Biochem Biophys Res Commun* **154**:298-305.

Feldman, D. H., Olivera, B. M. and Yoshikami, D., 1987, Omega Conus geographus toxin: a peptide that blocks calcium channels, *FEBS Lett* **214**:295-300.

Feng, Z. P., Hamid, J., Doering, C., Bosey, G. M., Snutch, T. P. and Zamponi, G. W., 2001a, Residue Gly1326 of the N-type calcium channel alpha 1B subunit controls reversibility of omega-conotoxin GVIA and MVIIA block, *J Biol Chem* **276**:15728-15735.

Feng, Z. P., Hamid, J., Doering, C., Jarvis, S. E., Bosey, G. M., Bourinet, E., Snutch, T. P. and Zamponi, G. W., 2001b, Amino acid residues outside of the pore region contribute to N-type calcium channel permeation, *J Biol Chem* **276**:5726-5730.

Fisher, T. E. and Bourque, C. W., 1995, Voltage-gated calcium currents in the magnocellular neurosecretory cells of the rat supraoptic nucleus, *J Physiol* **486 (Pt 3)**:571-580.

Flinn, J. P., Pallaghy, P. K., Lew, M. J., Murphy, R., Angus, J. A. and Norton, R. S., 1999, Roles of key functional groups in omega-conotoxin GVIA synthesis, structure and functional assay of selected peptide analogues, *Eur J Biochem* **262**:447-455.

Foehring, R. C. and Scroggs, R. S., 1994, Multiple high-threshold calcium currents in acutely isolated rat amygdaloid pyramidal cells, *J Neurophysiol* **71**:433-436.

Fujita, Y., Mynlieff, M., Dirksen, R. T., Kim, M. S., Niidome, T., Nakai, J., Friedrich, T., Iwabe, N., Miyata, T., Furuichi, T. and et al., 1993, Primary structure and functional expression of the omega-conotoxin-sensitive N-type calcium channel from rabbit brain, *Neuron* **10**:585-598.

Gage, M. J., Rane, S. G., Hockerman, G. H. and Smith, T. J., 2002, The virally encoded fungal toxin KP4 specifically blocks L-type voltage-gated calcium channels, *Mol Pharmacol* **61**:936-944.

Gandia, L., Lara, B., Imperial, J. S., Villarroya, M., Albillos, A., Maroto, R., Garcia, A. G. and Olivera, B. M., 1997, Analogies and differences between omega-conotoxins MVIIC and MVIID: binding sites and functions in bovine chromaffin cells, *Pflugers Arch* **435**:55-64.

Garcia, M. C., Hernandez-Gallegos, Z., Escamilla, J. and Sanchez, J. A., 2001, Calciseptine, a Ca^{2+} channel blocker, has agonist actions on L-type Ca^{2+} currents of frog and mammalian skeletal muscle, *J Membr Biol* **184**:121-129.

Garcia, M. L., Hanner, M., Knaus, H. G., Slaughter, R. and Kaczorowski, G. J., 1999, Scorpion toxins as tools for studying potassium channels, *Methods Enzymol* **294**:624-639.

Gee, N. S., Brown, J. P., Dissanayake, V. U., Offord, J., Thurlow, R. and Woodruff, G. N., 1996, The novel anticonvulsant drug, gabapentin (Neurontin), binds to the $\alpha 2\delta$ subunit of a calcium channel, *J Biol Chem* **271**:5768-5776.

Gillard, S. E., Volsen, S. G., Smith, W., Beattie, R. E., Bleakman, D. and Lodge, D., 1997, Identification of pore-forming subunit of P-type calcium channels: an antisense study on rat cerebellar Purkinje cells in culture, *Neuropharmacology* **36**:405-409.

Glauner, K. S., Mannuzzu, L. M., Gandhi, C. S. and Isacoff, E. Y., 1999, Spectroscopic mapping of voltage sensor movement in the Shaker potassium channel, *Nature* **402**:813-817.

Goldstein, S. A., Pheasant, D. J. and Miller, C., 1994, The charybdotoxin receptor of a Shaker K^+ channel: peptide and channel residues mediating molecular recognition, *Neuron* **12**:1377-1388.

Guida, S., Trettel, F., Pagnutti, S., Mantuano, E., Tottene, A., Veneziano, L., Fellin, T., Spadaro, M., Stauderman, K., Williams, M., Volsen, S., Ophoff, R., Frants, R., Jodice, C., Frontali, M. and Pietrobon, D., 2001, Complete loss of P/Q calcium channel activity caused by a CACNA1A missense mutation carried by patients with episodic ataxia type 2, *Am J Hum Genet* **68**:759-764.

Gurevitz, M., Gordon, D., Ben-Natan, S., Turkov, M. and Froy, O., 2001, Diversification of neurotoxins by C-tail 'wiggling': a scorpion recipe for survival, *Faseb J* **15**:1201-1205.

Hall, J. C., 1994, The mating of a fly, *Science* **264**:1702-1714.

Hamilton, S. L. and Perez, M., 1987, Toxins that affect voltage-dependent calcium channels, *Biochem Pharmacol* **36**:3325-3329.

Hans, M., Urrutia, A., Deal, C., Brust, P. F., Stauderman, K., Ellis, S. B., Harpold, M. M., Johnson, E. C. and Williams, M. E., 1999, Structural elements in domain IV that influence biophysical and pharmacological properties of human α_{1A}-containing high-voltage-activated calcium channels, *Biophys J* **76**:1384-1400.

Hell, J. W., Westenbroek, R. E., Warner, C., Ahlijanian, M. K., Prystay, W., Gilbert, M. M., Snutch, T. P. and Catterall, W. A., 1993, Identification and differential subcellular localization of the neuronal class C and class D L-type calcium channel alpha 1 subunits, *J Cell Biol* **123**:949-962.

Helton, T. D., Kojetin, D. J., Cavanagh, J. and Horne, W. A., 2002, Alternative splicing of a β4 subunit proline-rich motif regulates voltage-dependent gating and toxin block of $Ca_V2.1$ Ca^{2+} channels, *J Neurosci* **22**:9331-9339.

Hidalgo, P. and MacKinnon, R., 1995, Revealing the architecture of a K^+ channel pore through mutant cycles with a peptide inhibitor, *Science* **268**:307-310.

Hill, J. M., Alewood, P. F. and Craik, D. J., 1996, Three-dimensional solution structure of μ-conotoxin GIIIB, a specific blocker of skeletal muscle sodium channels, *Biochemistry* **35**:8824-8835.

Hille, B., 2001, *Ion Channels of Excitable Membranes*, Sinauer Associates, Sunderland, MA.

Hillyard, D. R., Monje, V. D., Mintz, I. M., Bean, B. P., Nadasdi, L., Ramachandran, J., Miljanich, G., Azimi-Zoonooz, A., McIntosh, J. M., Cruz, L. J. and et al., 1992, A new *Conus* peptide ligand for mammalian presynaptic Ca^{2+} channels, *Neuron* **9**:69-77.

Hirooka, K., Bertolesi, G. E., Kelly, M. E., Denovan-Wright, E. M., Sun, X., Hamid, J., Zamponi, G. W., Juhasz, A. E., Haynes, L. W. and Barnes, S., 2002, T-Type calcium channel α_{1G} and α_{1H} subunits in human retinoblastoma cells and their loss after differentiation, *J Neurophysiol* **88**:196-205.

Hockerman, G. H., Dilmac, N., Scheuer, T. and Catterall, W. A., 2000, Molecular determinants of diltiazem block in domains IIIS6 and IVS6 of L-type Ca^{2+} channels, *Mol Pharmacol* **58**:1264-1270.

Horne, W. A., Hawrot, E. and Tsien, R. W., 1991, ω-Conotoxin GVIA receptors of Discopyge electric organ. Characterization of ω-conotoxin binding to the nicotinic acetylcholine receptor, *J Biol Chem* **266**:13719-13725.

Jackson, H. and Usherwood, P. N., 1988, Spider toxins as tools for dissecting elements of excitatory amino acid transmission, *Trends Neurosci* **11**:278-283.

Jain, K. K., 2000, An evaluation of intrathecal ziconotide for the treatment of chronic pain, *Expert Opin Investig Drugs* **9**:2403-2410.

Jimenez, E. C., Olivera, B. M., Gray, W. R. and Cruz, L. J., 1996, Contryphan is a D-tryptophan-containing Conus peptide, *J Biol Chem* **271**:28002-28005.

Jones, S. W. and Marks, T. N., 1989, Calcium currents in bullfrog sympathetic neurons. I. Activation kinetics and pharmacology, *J Gen Physiol* **94**:151-167.

Jun, K., Piedras-Renteria, E. S., Smith, S. M., Wheeler, D. B., Lee, S. B., Lee, T. G., Chin, H., Adams, M. E., Scheller, R. H., Tsien, R. W. and Shin, H. S., 1999, Ablation of P/Q-type Ca^{2+} channel currents, altered synaptic transmission, and progressive ataxia in mice lacking the α_{1A}-subunit, *Proc Natl Acad Sci U S A* **96**:15245-15250.

Kaneko, S., Cooper, C. B., Nishioka, N., Yamasaki, H., Suzuki, A., Jarvis, S. E., Akaike, A., Satoh, M. and Zamponi, G. W., 2002, Identification and characterization of novel human $Ca_V2.2$ (alpha 1B) calcium channel variants lacking the synaptic protein interaction site, *J Neurosci* **22**:82-92.

Kasai, H., Aosaki, T. and Fukuda, J., 1987, Presynaptic Ca-antagonist omega-conotoxin irreversibly blocks N-type Ca-channels in chick sensory neurons, *Neurosci Res* **4**:228-235.

Kasai, H. and Neher, E., 1992, Dihydropyridine-sensitive and omega-conotoxin-sensitive calcium channels in a mammalian neuroblastoma-glioma cell line, *J Physiol* **448**:161-188.

Kerr, L. M. and Yoshikami, D., 1984, A venom peptide with a novel presynaptic blocking action, *Nature* **308**:282-284.

Kim, J. I., Konishi, S., Iwai, H., Kohno, T., Gouda, H., Shimada, I., Sato, K. and Arata, Y., 1995, Three-dimensional solution structure of the calcium channel antagonist omega-agatoxin IVA: consensus molecular folding of calcium channel blockers, *J Mol Biol* **250**:659-671.

Kits, K. S., Lodder, J. C., van der Schors, R. C., Li, K. W., Geraerts, W. P. and Fainzilber, M., 1996, Novel omega-conotoxins block dihydropyridine-insensitive high voltage-activated calcium channels in molluscan neurons, *J Neurochem* **67**:2155-2163.

Kits, K. S. and Mansvelder, H. D., 1996, Voltage gated calcium channels in molluscs: classification, Ca^{2+} dependent inactivation, modulation and functional roles, *Invert Neurosci* **2**:9-34.

Knight, Y. E., Bartsch, T., Kaube, H. and Goadsby, P. J., 2002, P/Q-type calcium-channel blockade in the periaqueductal gray facilitates trigeminal nociception: a functional genetic link for migraine?, *J Neurosci* **22**:RC213.

Koschak, A., Reimer, D., Huber, I., Grabner, M., Glossmann, H., Engel, J. and Striessnig, J., 2001, alpha 1D ($Ca_V1.3$) subunits can form L-type Ca^{2+} channels activating at negative voltages, *J Biol Chem* **276**:22100-22106.

Lamb, G. D. and Walsh, T., 1987, Calcium currents, charge movement and dihydropyridine binding in fast- and slow-twitch muscles of rat and rabbit, *J Physiol* **393**:595-617.

Lampe, R. A., Defeo, P. A., Davison, M. D., Young, J., Herman, J. L., Spreen, R. C., Horn, M. B., Mangano, T. J. and Keith, R. A., 1993, Isolation and pharmacological characterization of omega-grammotoxin SIA, a novel peptide inhibitor of neuronal voltage-sensitive calcium channel responses, *Mol Pharmacol* **44**:451-460.

Lancelin, J. M., Kohda, D., Tate, S., Yanagawa, Y., Abe, T., Satake, M. and Inagaki, F., 1991, Tertiary structure of conotoxin GIIIA in aqueous solution, *Biochemistry* **30**:6908-6916.

Laurent, G., 2002, Olfactory network dynamics and the coding of multidimensional signals, *Nat Rev Neurosci* **3**:884-895.

Leao, R. M., Cruz, J. S., Diniz, C. R., Cordeiro, M. N. and Beirao, P. S., 2000, Inhibition of neuronal high-voltage activated calcium channels by the ω *Phoneutria nigriventer* Tx3-3 peptide toxin, *Neuropharmacology* **39**:1756-1767.

Ledwell, J. L. and Aldrich, R. W., 1999, Mutations in the S4 region isolate the final voltage-dependent cooperative step in potassium channel activation, *J Gen Physiol* **113**:389-414.

Lennon, V. A., Kryzer, T. J., Griesmann, G. E., O'Suilleabhain, P. E., Windebank, A. J., Woppmann, A., Miljanich, G. P. and Lambert, E. H., 1995, Calcium-channel antibodies in the Lambert-Eaton syndrome and other paraneoplastic syndromes, *N Engl J Med* **332**:1467-1474.

Leung, H. T., Branton, W. D., Phillips, H. S., Jan, L. and Byerly, L., 1989, Spider toxins selectively block calcium currents in Drosophila, *Neuron* **3**:767-772.

Lewis, R. J., Nielsen, K. J., Craik, D. J., Loughnan, M. L., Adams, D. A., Sharpe, I. A., Luchian, T., Adams, D. J., Bond, T., Thomas, L., Jones, A., Matheson, J. L., Drinkwater, R., Andrews, P. R. and Alewood, P. F., 2000, Novel ω-conotoxins from *Conus catus* discriminate among neuronal calcium channel subtypes, *J Biol Chem* **275**:35335-35344.

Li, R. A., Ennis, I. L., French, R. J., Dudley, S. C., Jr., Tomaselli, G. F. and Marban, E., 2001, Clockwise domain arrangement of the sodium channel revealed by μ-conotoxin (GIIIA) docking orientation, *J Biol Chem* **276**:11072-11077.

Li, R. A., Ennis, I. L., Velez, P., Tomaselli, G. F. and Marban, E., 2000, Novel structural determinants of μ-conotoxin (GIIIB) block in rat skeletal muscle (μ1) Na^+ channels, *J Biol Chem* **275**:27551-27558.

Liang, H. and Elmslie, K. S., 2002, Rapid and reversible block of N-type calcium channels (Ca_V 2.2) by ω-conotoxin GVIA in the absence of divalent cations, *J Neurosci* **22**:8884-8890.

Lin, Z., Harris, C. and Lipscombe, D., 1996, The molecular identity of Ca channel α1-subunits expressed in rat sympathetic neurons, *J Mol Neurosci* **7**:257-267.

Lin, Z., Haus, S., Edgerton, J. and Lipscombe, D., 1997, Identification of functionally distinct isoforms of the N-type Ca^{2+} channel in rat sympathetic ganglia and brain, *Neuron* **18**:153-166.

Lin, Z., Lin, Y., Schorge, S., Pan, J. Q., Beierlein, M. and Lipscombe, D., 1999, Alternative splicing of a short cassette exon in α_{1B} generates functionally distinct N-type calcium channels in central and peripheral neurons, *J Neurosci* **19**:5322-5331.

Lipscombe, D., 2002, L-type calcium channels: highs and new lows, *Circ Res* **90**:933-935.

Lipscombe, D., Pan, J. Q. and Gray, A. C., 2002, Functional diversity in neuronal voltage-gated calcium channels by alternative splicing of $Ca_V\alpha_1$, *Mol Neurobiol* **26**:21-44.

Li-Smerin, Y. and Swartz, K. J., 1998, Gating modifier toxins reveal a conserved structural motif in voltage-gated Ca^{2+} and K^+ channels, *Proc Natl Acad Sci U S A* **95**:8585-8589.

Li-Smerin, Y. and Swartz, K. J., 2001, Helical structure of the COOH terminus of S3 and its contribution to the gating modifier toxin receptor in voltage-gated ion channels, *J Gen Physiol* **117**:205-218.

Li-Smerin, Y. and Swartz, K. J., 2000, Localization and molecular determinants of the Hanatoxin receptors on the voltage-sensing domains of a K^+ channel, *J Gen Physiol* **115**:673-684.

Llinas, R., Sugimori, M., Lin, J. W. and Cherksey, B., 1989, Blocking and isolation of a calcium channel from neurons in mammals and cephalopods utilizing a toxin fraction (FTX) from funnel-web spider poison, *Proc Natl Acad Sci U S A* **86**:1689-1693.

Lu, Q., and Dunlap, K., 1999, Cloning and functional expression of novel N-type Ca^{2+} channel variants, *J Biol Chem* **274**:34566-34575.

MacKinnon, R., 1991, Determination of the subunit stoichiometry of a voltage-activated potassium channel, *Nature* **350**:232-235.

MacKinnon, R., Cohen, S. L., Kuo, A., Lee, A. and Chait, B. T., 1998, Structural conservation in prokaryotic and eukaryotic potassium channels, *Science* **280**:106-109.

MacKinnon, R. and Miller, C., 1988, Mechanism of charybdotoxin block of the high-conductance, Ca^{2+}-activated K^+ channel, *J Gen Physiol* **91**:335-349.

Magee, J. C., Avery, R. B., Christie, B. R. and Johnston, D., 1996, Dihydropyridine-sensitive, voltage-gated Ca^{2+} channels contribute to the resting intracellular Ca^{2+} concentration of hippocampal CA1 pyramidal neurons, *J Neurophysiol* **76**:3460-3470.

Mannuzzu, L. M., Moronne, M. M. and Isacoff, E. Y., 1996, Direct physical measure of conformational rearrangement underlying potassium channel gating, *Science* **271**:213-216.

Maricq, A. V. and Korenbrot, J. I., 1988, Calcium and calcium-dependent chloride currents generate action potentials in solitary cone photoreceptors, *Neuron* **1**:503-515.

Martin, D. J., McClelland, D., Herd, M. B., Sutton, K. G., Hall, M. D., Lee, K., Pinnock, R. D. and Scott, R. H., 2002, Gabapentin-mediated inhibition of voltage-activated Ca^{2+} channel currents in cultured sensory neurones is dependent on culture conditions and channel subunit expression, *Neuropharmacology* **42**:353-366.

Matsuyama, Z., Wakamori, M., Mori, Y., Kawakami, H., Nakamura, S. and Imoto, K., 1999, Direct alteration of the P/Q-type Ca^{2+} channel property by polyglutamine expansion in spinocerebellar ataxia 6, *J Neurosci* **19**:RC14.

McCleskey, E. W., Fox, A. P., Feldman, D. H., Cruz, L. J., Olivera, B. M., Tsien, R. W. and Yoshikami, D., 1987, ω-conotoxin: direct and persistent blockade of specific types of calcium channels in neurons but not muscle, *Proc Natl Acad Sci U S A* **84**:4327-4331.

McDonough, S. I., Boland, L. M., Mintz, I. M. and Bean, B. P., 2002, Interactions among toxins that inhibit N-type and P-type calcium channels, *J Gen Physiol* **119**:313-328.

McDonough, S. I., Lampe, R. A., Keith, R. A. and Bean, B. P., 1997a, Voltage-dependent inhibition of N- and P-type calcium channels by the peptide toxin ω-grammotoxin-SIA, *Mol Pharmacol* **52**:1095-1104.

McDonough, S. I., Mintz, I. M. and Bean, B. P., 1997b, Alteration of P-type calcium channel gating by the spider toxin ω-Aga-IVA, *Biophys J* **72**:2117-2128.

McDonough, S. I., Swartz, K. J., Mintz, I. M., Boland, L. M. and Bean, B. P., 1996, Inhibition of calcium channels in rat central and peripheral neurons by ω-conotoxin MVIIC, *J Neurosci* **16**:2612-2623.

McEnery, M. W., Snowman, A. M., Sharp, A. H., Adams, M. E. and Snyder, S. H., 1991, Purified ω-conotoxin GVIA receptor of rat brain resembles a dihydropyridine-sensitive L-type calcium channel, *Proc Natl Acad Sci U S A* **88**:11095-11099.

McFarlane, M. B., 1997, Depolarization-induced slowing of Ca^{2+} channel deactivation in squid neurons, *Biophys J* **72**:1607-1621.

McIntosh, J. M., Hasson, A., Spira, M. E., Gray, W. R., Li, W., Marsh, M., Hillyard, D. R. and Olivera, B. M., 1995, A new family of conotoxins that blocks voltage-gated sodium channels, *J Biol Chem* **270**:16796-16802.

McIntosh, J. M., Santos, A. D. and Olivera, B. M., 1999, Conus peptides targeted to specific nicotinic acetylcholine receptor subtypes, *Annu Rev Biochem* **68**:59-88.

Mermelstein, P. G., Foehring, R. C., Tkatch, T., Song, W. J., Baranauskas, G. and Surmeier, D. J., 1999, Properties of Q-type calcium channels in neostriatal and cortical neurons are correlated with beta subunit expression, *J Neurosci* **19**:7268-7277.

Meunier, F. A., Feng, Z.-P., Molgo, J., Zamponi, G. W., and Schiavo, G., 2002, Glycerotoxin from Glycera convolute stimulates neurosecretion by up-regulating N-type Ca^{2+} channel activity, *EMBO J* **21**:6733-6743.

Miljanich, G. P. and Ramachandran, J., 1995, Antagonists of neuronal calcium channels: structure, function, and therapeutic implications, *Annu Rev Pharmacol Toxicol* **35**:707-734.

Miller, C., 1995, The charybdotoxin family of K^+ channel-blocking peptides, *Neuron* **15**:5-10.

Mintz, I. M., 1994, Block of Ca channels in rat central neurons by the spider toxin ω-Aga-IIIA, *J Neurosci* **14**:2844-2853.

Mintz, I. M., Adams, M. E. and Bean, B. P., 1992a, P-type calcium channels in rat central and peripheral neurons, *Neuron* **9**:85-95.

Mintz, I. M., Sabatini, B. L. and Regehr, W. G., 1995, Calcium control of transmitter release at a cerebellar synapse, *Neuron* **15**:675-688.

Mintz, I. M., Venema, V. J., Adams, M. E. and Bean, B. P., 1991, Inhibition of N- and L-type Ca^{2+} channels by the spider venom toxin ω-Aga-IIIA, *Proc Natl Acad Sci U S A* **88**:6628-6631.

Mintz, I. M., Venema, V. J., Swiderek, K. M., Lee, T. D., Bean, B. P. and Adams, M. E., 1992b, P-type calcium channels blocked by the spider toxin ω-Aga-IVA, *Nature* **355**:827-829.

Mitra, A. K., McCarthy, M. P. and Stroud, R. M., 1989, Three-dimensional structure of the nicotinic acetylcholine receptor and location of the major associated 43-kD cytoskeletal protein, determined at 22 A by low dose electron microscopy and x-ray diffraction to 12.5 A, *J Cell Biol* **109**:755-774.

Moise, L., Zeng, H., Caffery, P., Rogowski, R. S., and Hawrot, E., 2002, Structure and function of α-bungarotoxin, *J Toxicol Toxin Rev* **21**: 293-317.

Moreno, H., Rudy, B. and Llinas, R., 1997, beta subunits influence the biophysical and pharmacological differences between P- and Q-type calcium currents expressed in a mammalian cell line, *Proc Natl Acad Sci U S A* **94**:14042-14047.

Morgans, C. W., Gaughwin, P. and Maleszka, R., 2001, Expression of the α_{1F} calcium channel subunit by photoreceptors in the rat retina, *Mol Vis* **7**:202-209.

Mori, Y., Friedrich, T., Kim, M. S., Mikami, A., Nakai, J., Ruth, P., Bosse, E., Hofmann, F., Flockerzi, V., Furuichi, T., Mikoshiba, K., Imoto, K., Tanabe, T., and Numa, S., 1991, Primary structure and functional expression from complementary DNA of a brain calcium channel, *Nature* **350**:398-402.

Morita, H., Shi, J., Ito, Y. and Inoue, R., 2002, T-channel-like pharmacological properties of highvoltage-activated, nifedipine-insensitive Ca^{2+} currents in the rat terminal mesenteric artery, *Br J Pharmacol* **137**:467-476.

Mozhayeva, G. N., Naumov, A. P., Nosyreva, E. D. and Grishin, E. V., 1980, Potential-dependent interaction of toxin from venom of the scorpion *Buthus eupeus* with sodium channels in myelinated fibre: voltage clamp experiments, *Biochim Biophys Acta* **597**:587-602.

Newcomb, R., Palma, A., Fox, J., Gaur, S., Lau, K., Chung, D., Cong, R., Bell, J. R., Horne, B., Nadasdi, L., and Ramachandran, J., 1995, SNX-325, a novel calcium antagonist from the spider *Segestria florentina*, *Biochemistry* **34**:8341-8347.

Newcomb, R., Szoke, B., Palma, A., Wang, G., Chen, X., Hopkins, W., Cong, R., Miller, J., Urge, L., Tarczy-Hornoch, K., Loo, J. A., Dooley, D. J., Nadasdi, L., Tsien, R. W., Lemos, J. and Miljanich, G., 1998, Selective peptide antagonist of the class E calcium channel from the venom of the tarantula *Hysterocrates gigas*, *Biochemistry* **37**:15353-15362.

Nielsen, K. J., Adams, D., Thomas, L., Bond, T., Alewood, P. F., Craik, D. J. and Lewis, R. J., 1999, Structure-activity relationships of omega-conotoxins MVIIA, MVIIC and 14 loop splice hybrids at N and P/Q-type calcium channels, *J Mol Biol* **289**:1405-1421.

Nobile, M., Noceti, F., Prestipino, G. P. and Possani, L. D., 1996, Helothermine, a lizard venom toxin, inhibits calcium current in cerebellar granules, *Exp Brain Res* **110**:15-20.

Norris, T. M., Moya, E., Blagbrough, I. S. and Adams, M. E., 1996, Block of high-threshold calcium channels by the synthetic polyamines sFTX-3.3 and FTX-3.3, *Mol Pharmacol* **50**:939-946.

Olivera, B. M., 1997, E.E. Just Lecture, 1996. Conus venom peptides, receptor and ion channel targets, and drug design: 50 million years of neuropharmacology, *Mol Biol Cell* **8**:2101-2109.

Olivera, B. M., Cruz, L. J., de Santos, V., LeCheminant, G. W., Griffin, D., Zeikus, R., McIntosh, J. M., Galyean, R., Varga, J., Gray, W. R., and Rivier, J., 1987, Neuronal calcium channel antagonists. Discrimination between calcium channel subtypes using ω-conotoxin from *Conus magus* venom, *Biochemistry* **26**:2086-2090.

Olivera, B. M., Gray, W. R., Zeikus, R., McIntosh, J. M., Varga, J., Rivier, J., de Santos, V. and Cruz, L. J., 1985, Peptide neurotoxins from fish-hunting cone snails, *Science* **230**:1338-1343.

Olivera, B. M., McIntosh, J. M., Cruz, L. J., Luque, F. A. and Gray, W. R., 1984, Purification and sequence of a presynaptic peptide toxin from *Conus geographus* venom, *Biochemistry* **23**:5087-5090.

Olivera, B. M., Miljanich, G. P., Ramachandran, J. and Adams, M. E., 1994, Calcium channel diversity and neurotransmitter release: the ω-conotoxins and ω-agatoxins, *Annu Rev Biochem* **63**:823-867.

Olivera, B. M., Rivier, J., Clark, C., Ramilo, C. A., Corpuz, G. P., Abogadie, F. C., Mena, E. E., Woodward, S. R., Hillyard, D. R. and Cruz, L. J., 1990, Diversity of Conus neuropeptides, *Science* **249**:257-263.

Olivera, B. M., Rivier, J., Scott, J. K., Hillyard, D. R. and Cruz, L. J., 1991, Conotoxins, *J Biol Chem* **266**:22067-22070.

Ophoff, R. A., Terwindt, G. M., Vergouwe, M. N., van Eijk, R., Oefner, P. J., Hoffman, S. M., Lamerdin, J. E., Mohrenweiser, H. W., Bulman, D. E., Ferrari, M., Haan, J., Lindhout, D., van Ommen, G. J., Hofker, M. H., Ferrari, M. D. and Frants, R. R., 1996, Familial hemiplegic migraine and episodic ataxia type-2 are caused by mutations in the Ca^{2+} channel gene CACNL1A4, *Cell* **87**:543-552.

Ousley, A. H. and Froehner, S. C., 1994, An anti-peptide antibody specific for the class A calcium channel $\alpha 1$ subunit labels mammalian neuromuscular junction, *Proc Natl Acad Sci U S A* **91**:12263-12267.

Pan, J. Q. and Lipscombe, D., 2000, Alternative splicing in the cytoplasmic II-III loop of the N-type Ca channel α_{1B} subunit: functional differences are β subunit-specific, *J Neurosci* **20**:4769-4775.

Pardo, B. and Honegger, P., 1999, High sensitivity of immature GABAergic neurons to blockers of voltage-gated calcium channels, *Brain Res Dev Brain Res* **115**:9-16.

Park, C. S. and Miller, C., 1992, Interaction of charybdotoxin with permeant ions inside the pore of a K^+ channel, *Neuron* **9**:307-313.

Peng, K., Chen, X. D. and Liang, S. P., 2001, The effect of Huwentoxin-I on Ca^{2+} channels in differentiated NG108-15 cells, a patch-clamp study, *Toxicon* **39**:491-498.

Perchenet, L., Benardeau, A. and Ertel, E. A., 2000, Pharmacological properties of $Ca_V3.2$, a low voltage-activated Ca^{2+} channel cloned from human heart, *Naunyn Schmiedebergs Arch Pharmacol* **361**:590-599.

Piedras-Renteria, E. S., Watase, K., Harata, N., Zhuchenko, O., Zoghbi, H. Y., Lee, C. C. and Tsien, R. W., 2001, Increased expression of α_{1A} Ca^{2+} channel currents arising from expanded trinucleotide repeats in spinocerebellar ataxia type 6, *J Neurosci* **21**:9185-9193.

Pinto, A., Gillard, S., Moss, F., Whyte, K., Brust, P., Williams, M., Stauderman, K., Harpold, M., Lang, B., Newsom-Davis, J., Bleakman, D., Lodge, D. and Boot, J., 1998, Human autoantibodies specific for the alpha1A calcium channel subunit reduce both P-type and Q-type calcium currents in cerebellar neurons, *Proc Natl Acad Sci U S A* **95**:8328-8333.

Plummer, M. R., Logothetis, D. E. and Hess, P., 1989, Elementary properties and pharmacological sensitivities of calcium channels in mammalian peripheral neurons, *Neuron* **2**:1453-1463.

Possani, L. D., Merino, E., Corona, M., Bolivar, F. and Becerril, B., 2000, Peptides and genes coding for scorpion toxins that affect ion-channels, *Biochimie* **82**:861-868.

Protti, D. A. and Llano, I., 1998, Calcium currents and calcium signaling in rod bipolar cells of rat retinal slices, *J Neurosci* **18**:3715-3724.

Protti, D. A., Reisin, R., Mackinley, T. A. and Uchitel, O. D., 1996, Calcium channel blockers and transmitter release at the normal human neuromuscular junction, *Neurology* **46**:1391-1396.

Ramilo, C. A., Zafaralla, G. C., Nadasdi, L., Hammerland, L. G., Yoshikami, D., Gray, W. R., Kristipati, R., Ramachandran, J., Miljanich, G., Olivera, B. M. and et al., 1992, Novel α- and ω-conotoxins from *Conus striatus* venom, *Biochemistry* **31**:9919-9926.

Randall, A. and Tsien, R. W., 1995, Pharmacological dissection of multiple types of Ca^{2+} channel currents in rat cerebellar granule neurons, *J Neurosci* **15**:2995-3012.

Randall, A. D. and Tsien, R. W., 1997, Contrasting biophysical and pharmacological properties of T-type and R-type calcium channels, *Neuropharmacology* **36**:879-893.

Ranganathan, R., Lewis, J. H. and MacKinnon, R., 1996, Spatial localization of the K^+ channel selectivity filter by mutant cycle-based structure analysis, *Neuron* **16**:131-139.

Rash, L. D. and Hodgson, W. C., 2002, Pharmacology and biochemistry of spider venoms, *Toxicon* **40**:225-254.

Regan, L. J., 1991, Voltage-dependent calcium currents in Purkinje cells from rat cerebellar vermis, *J Neurosci* **11**:2259-2269.

Regan, L. J., Sah, D. W. and Bean, B. P., 1991, Ca^{2+} channels in rat central and peripheral neurons: high-threshold current resistant to dihydropyridine blockers and ω-conotoxin, *Neuron* **6**:269-280.

Regehr, W. G. and Mintz, I. M., 1994, Participation of multiple calcium channel types in transmission at single climbing fiber to Purkinje cell synapses, *Neuron* **12**:605-613.

Regehr, W. G. and Mintz, I. M., 1994, Participation of multiple calcium channel types in transmission at single climbing fiber to Purkinje cell synapses, *Neuron* **12**:605-613.

Rigby, A. C., Baleja, J. D., Furie, B. C. and Furie, B., 1997a, Three-dimensional structure of a gamma-carboxyglutamic acid-containing conotoxin, conantokin G, from the marine snail *Conus geographus*: the metal-free conformer, *Biochemistry* **36**:6906-6914.

Rigby, A. C., Baleja, J. D., Li, L., Pedersen, L. G., Furie, B. C. and Furie, B., 1997b, Role of gamma-carboxyglutamic acid in the calcium-induced structural transition of conantokin G, a conotoxin from the marine snail Conus geographus, *Biochemistry* **36**:15677-15684.

Rios, E. and Brum, G., 1987, Involvement of dihydropyridine receptors in excitation-contraction coupling in skeletal muscle, *Nature* **325**:717-720.

Rogers, J. C., Qu, Y., Tanada, T. N., Scheuer, T. and Catterall, W. A., 1996, Molecular determinants of high affinity binding of α-scorpion toxin and sea anemone toxin in the S3-S4 extracellular loop in domain IV of the Na^+ channel alpha subunit, *J Biol Chem* **271**:15950-15962.

Saegusa, H., Kurihara, T., Zong, S., Kazuno, A., Matsuda, Y., Nonaka, T., Han, W., Toriyama, H. and Tanabe, T., 2001, Suppression of inflammatory and neuropathic pain symptoms in mice lacking the N-type Ca^{2+} channel, *EMBO J* **20**:2349-2356.

Sano, K., Enomoto, K. and Maeno, T., 1987, Effects of synthetic ω-conotoxin, a new type Ca^{2+} antagonist, on frog and mouse neuromuscular transmission, *Eur J Pharmacol* **141**:235-241.

Sasaki, T., Feng, Z. P., Scott, R., Grigoriev, N., Syed, N. I., Fainzilber, M. and Sato, K., 1999, Synthesis, bioactivity, and cloning of the L-type calcium channel blocker ω-conotoxin TxVII, *Biochemistry* **38**:12876-12884.

Sather, W. A., Tanabe, T., Zhang, J. F., Mori, Y., Adams, M. E. and Tsien, R. W., 1993, Distinctive biophysical and pharmacological properties of class A (BI) calcium channel alpha 1 subunits, *Neuron* **11**:291-303.

Sato, C., Ueno, Y., Asai, K., Takahashi, K., Sato, M., Engel, A. and Fujiyoshi, Y., 2001, The voltage-sensitive sodium channel is a bell-shaped molecule with several cavities, *Nature* **409**:1047-1051.

Scholze, A., Plant, T. D., Dolphin, A. C., and Nurnberg, B., 2001, Functional expression and characterization of a voltage-gated $Ca_V1.3$ (α1D) calcium channel subunit from an insulin-secreting cell line, *Mol Endocrinol* **15**:1211-1221.

Schweitz, H., Heurteaux, C., Bois, P., Moinier, D., Romey, G. and Lazdunski, M., 1994, Calcicludine, a venom peptide of the Kunitz-type protease inhibitor family, is a potent blocker of high-threshold Ca^{2+} channels with a high affinity for L-type channels in cerebellar granule neurons, *Proc Natl Acad Sci U S A* **91**:878-882.

Scott, R. H., Dolphin, A. C., Bindokas, V. P. and Adams, M. E., 1990, Inhibition of neuronal Ca^{2+} channel currents by the funnel web spider toxin ω-Aga-IA, *Mol Pharmacol* **38**:711-718.

Scott, R. H., Gorton, V. J., Harding, L., Patel, D., Pacey, S., Kellenberger, C., Hietter, H. and Bermudez, I., 1997, Inhibition of neuronal high voltage-activated calcium channels by insect peptides: a comparison with the actions of ω-conotoxin GVIA, *Neuropharmacology* **36**:195-208.

Sheets, M. F. and Hanck, D. A., 1999, Gating of skeletal and cardiac muscle sodium channels in mammalian cells, *J Physiol* **514**:425-436.

Sheets, M. F. and Hanck, D. A., 1995, Voltage-dependent open-state inactivation of cardiac sodium channels: gating current studies with Anthopleurin-A toxin, *J Gen Physiol* **106**:617-640.

Sidach, S. S. and Mintz, I. M., 2002, Kurtoxin, a gating modifier of neuronal high- and low-threshold Ca channels, *J Neurosci* **22**:2023-2034.

Sidach, S. S. and Mintz, I. M., 2000, Low-affinity blockade of neuronal N-type Ca channels by the spider toxin ω-agatoxin-IVA, *J Neurosci* **20**:7174-7182.

Smit, A. B., Syed, N. I., Schaap, D., van Minnen, J., Klumperman, J., Kits, K. S., Lodder, H., van der Schors, R. C., van Elk, R., Sorgedrager, B., Brejc, K., Sixma, T. K. and Geraerts, W. P., 2001, A glia-derived acetylcholine-binding protein that modulates synaptic transmission, *Nature* **411**:261-268.

Smith-Maxwell, C. J., Ledwell, J. L. and Aldrich, R. W., 1998a, Role of the S4 in cooperativity of voltage-dependent potassium channel activation, *J Gen Physiol* **111**:399-420.

Smith-Maxwell, C. J., Ledwell, J. L. and Aldrich, R. W., 1998b, Uncharged S4 residues and cooperativity in voltage-dependent potassium channel activation, *J Gen Physiol* **111**:421-439.

Soong, T. W., DeMaria, C. D., Alvania, R. S., Zweifel, L. S., Liang, M. C., Mittman, S., Agnew, W. S. and Yue, D. T., 2002, Systematic identification of splice variants in human P/Q-type channel α1(2.1) subunits: implications for current density and Ca^{2+}-dependent inactivation, *J Neurosci* **22**:10142-10152.

Soong, T. W., Stea, A., Hodson, C. D., Dubel, S. J., Vincent, S. R. and Snutch, T. P., 1993, Structure and functional expression of a member of the low voltage-activated calcium channel family, *Science* **260**:1133-1136.

Srinivasan, K. N., Gopalakrishnakone, P., Tan, P. T., Chew, K. C., Cheng, B., Kini, R. M., Koh, J. L., Seah, S. H. and Brusic, V., 2002, SCORPION, a molecular database of scorpion toxins, *Toxicon* **40**:23-31.

Stapleton, A., Blankenship, D. T., Ackermann, B. L., Chen, T. M., Gorder, G. W., Manley, G. D., Palfreyman, M. G., Coutant, J. E. and Cardin, A. D., 1990, Curtatoxins. Neurotoxic insecticidal polypeptides isolated from the funnel-web spider *Hololena curta*, *J Biol Chem* **265**:2054-2059.

Starr, T. V., Prystay, W. and Snutch, T. P., 1991, Primary structure of a calcium channel that is highly expressed in the rat cerebellum, *Proc Natl Acad Sci U S A* **88**:5621-5625.

Stocker, J. W., Nadasdi, L., Aldrich, R. W. and Tsien, R. W., 1997, Preferential interaction of ω-conotoxins with inactivated N-type Ca^{2+} channels, *J Neurosci* **17**:3002-3013.

Stotz, S. C., Spaetgens, R. L. and Zamponi, G. W., 2000, Block of voltage-dependent calcium channel by the green mamba toxin calcicludine, *J Membr Biol* **174**:157-165.

Strom, T. M., Nyakatura, G., Apfelstedt-Sylla, E., Hellebrand, H., Lorenz, B., Weber, B. H., Wutz, K., Gutwillinger, N., Ruther, K., Drescher, B., Sauer, C., Zrenner, E., Meitinger, T., Rosenthal, A. and Meindl, A., 1998, An L-type calcium-channel gene mutated in incomplete X-linked congenital stationary night blindness, *Nat Genet* **19**:260-263.

Sutton, K. G., Martin, D. J., Pinnock, R. D., Lee, K. and Scott, R. H., 2002, Gabapentin inhibits high-threshold calcium channel currents in cultured rat dorsal root ganglion neurones, *Br J Pharmacol* **135**:257-265.

Sutton, K. G., Siok, C., Stea, A., Zamponi, G. W., Heck, S. D., Volkmann, R. A., Ahlijanian, M. K. and Snutch, T. P., 1998, Inhibition of neuronal calcium channels by a novel peptide spider toxin, DW13.3, *Mol Pharmacol* **54**:407-418.

Swartz, K. J. and MacKinnon, R., 1997a, Hanatoxin modifies the gating of a voltage-dependent K^+ channel through multiple binding sites, *Neuron* **18**:665-673.

Swartz, K. J. and MacKinnon, R., 1995, An inhibitor of the Kv2.1 potassium channel isolated from the venom of a Chilean tarantula, *Neuron* **15**:941-949.

Swartz, K. J. and MacKinnon, R., 1997b, Mapping the receptor site for hanatoxin, a gating modifier of voltage-dependent K^+ channels, *Neuron* **18**:675-682.

Tabata, T., Olivera, B. M. and Ishida, A. T., 1996, ω-conotoxin-MVIID blocks an ω-conotoxin-GVIA-sensitive, high-threshold Ca^{2+} current in fish retinal ganglion cells, *Neuropharmacology* **35**:633-636.

Takamori, M., Maruta, T. and Komai, K., 2000, Lambert-Eaton myasthenic syndrome as an autoimmune calcium-channelopathy, *Neurosci Res* **36**:183-191.

Tanabe, T., Takeshima, H., Mikami, A., Flockerzi, V., Takahashi, H., Kangawa, K., Kojima, M., Matsuo, H., Hirose, T. and Numa, S., 1987, Primary structure of the receptor for calcium channel blockers from skeletal muscle, *Nature* **328**:313-318.

Teramoto, N., Ogata, R., Okabe, K., Kameyama, A., Kameyama, M., Watanabe, T. X., Kuriyama, H. and Kitamura, K., 1996, Effects of calciseptine on unitary barium channel currents in guinea-pig portal vein, *Pflugers Arch* **432**:462-470.

Teramoto, T., Kuwada, M., Niidome, T., Sawada, K., Nishizawa, Y. and Katayama, K., 1993, A novel peptide from funnel web spider venom, ω-Aga-TK, selectively blocks, P-type calcium channels, *Biochem Biophys Res Commun* **196**:134-140.

Terlau, H., Stocker, M., Shon, K. J., McIntosh, J. M. and Olivera, B. M., 1996, μO-conotoxin MrVIA inhibits mammalian sodium channels, but not through site I, *J Neurophysiol* **76**:1423-1429.

Tottene, A., Volsen, S. and Pietrobon, D., 2000, α_{1E} subunits form the pore of three cerebellar R-type calcium channels with different pharmacological and permeation properties, *J Neurosci* **20**:171-178.
Toyoshima, C. and Unwin, N., 1988, Ion channel of acetylcholine receptor reconstructed from images of postsynaptic membranes, *Nature* **336**:247-250.
Tsunemi, T., Saegusa, H., Ishikawa, K., Nagayama, S., Murakoshi, T., Mizusawa, H. and Tanabe, T., 2002, Novel Cav2.1 splice variants isolated from Purkinje cells do not generate P-type Ca^{2+} current, *J Biol Chem* **277**:7214-7221.
Unwin, N., 1995, Acetylcholine receptor channel imaged in the open state, *Nature* **373**:37-43.
Unwin, N., 1998, The nicotinic acetylcholine receptor of the Torpedo electric ray, *J Struct Biol* **121**:181-190.
Usowicz, M. M., Sugimori, M., Cherksey, B. and Llinas, R., 1992, P-type calcium channels in the somata and dendrites of adult cerebellar Purkinje cells, *Neuron* **9**:1185-1199.
Venema, V. J., Swiderek, K. M., Lee, T. D., Hathaway, G. M. and Adams, M. E., 1992, Antagonism of synaptosomal calcium channels by subtypes of ω-agatoxins, *J Biol Chem* **267**:2610-2615.
Wagner, J. A., Snowman, A. M., Biswas, A., Olivera, B. M. and Snyder, S. H., 1988, ω-conotoxin GVIA binding to a high-affinity receptor in brain: characterization, calcium sensitivity, and solubilization, *J Neurosci* **8**:3354-3359.
Wakamatsu, K., Kohda, D., Hatanaka, H., Lancelin, J. M., Ishida, Y., Oya, M., Nakamura, H., Inagaki, F. and Sato, K., 1992, Structure-activity relationships of μ-conotoxin GIIIA: structure determination of active and inactive sodium channel blocker peptides by NMR and simulated annealing calculations, *Biochemistry* **31**:12577-12584.
Walker, D. and De Waard, M., 1998, Subunit interaction sites in voltage-dependent Ca^{2+} channels: role in channel function, *Trends Neurosci* **21**:148-154.
Wang, G., Dayanithi, G., Newcomb, R. and Lemos, J. R., 1999, An R-type Ca^{2+} current in neurohypophysial terminals preferentially regulates oxytocin secretion, *J Neurosci* **19**:9235-9241.
Wang, X., Connor, M., Smith, R., Maciejewski, M. W., Howden, M. E., Nicholson, G. M., Christie, M. J. and King, G. F., 2000, Discovery and characterization of a family of insecticidal neurotoxins with a rare vicinal disulfide bridge, *Nat Struct Biol* **7**:505-513.
Wang, X., Smith, R., Fletcher, J. I., Wilson, H., Wood, C. J., Howden, M. E. and King, G. F., 1999, Structure-function studies of ω-atracotoxin, a potent antagonist of insect voltage-gated calcium channels, *Eur J Biochem* **264**:488-494.
Wang, X. H., Connor, M., Wilson, D., Wilson, H. I., Nicholson, G. M., Smith, R., Shaw, D., Mackay, J. P., Alewood, P. F., Christie, M. J. and King, G. F., 2001, Discovery and structure of a potent and highly specific blocker of insect calcium channels, *J Biol Chem* **276**:40306-40312.
Wang, Y. X., Bezprozvannaya, S., Bowersox, S. S., Nadasdi, L., Miljanich, G., Mezo, G., Silva, D., Tarczy-Hornoch, K. and Luther, R. R., 1998, Peripheral versus central potencies of N-type voltage-sensitive calcium channel blockers, *Naunyn Schmiedebergs Arch Pharmacol* **357**:159-168.
Watanabe, T. X., Itahara, Y., Kuroda, H., Chen, Y. N., Kimura, T. and Sakakibara, S., 1995, Smooth muscle relaxing and hypotensive activities of synthetic calciseptine and the homologous snake venom peptide FS2, *Jpn J Pharmacol* **68**:305-313.
Westenbroek, R. E., Sakurai, T., Elliott, E. M., Hell, J. W., Starr, T. V., Snutch, T. P. and Catterall, W. A., 1995, Immunochemical identification and subcellular distribution of the alpha 1A subunits of brain calcium channels, *J Neurosci* **15**:6403-6418.
Wheeler, D. B., Randall, A. and Tsien, R. W., 1994, Roles of N-type and Q-type Ca^{2+} channels in supporting hippocampal synaptic transmission, *Science* **264**:107-111.
White, M. G., Crumling, M. A. and Meriney, S. D., 1997, Developmental changes in calcium current pharmacology and somatostatin inhibition in chick parasympathetic neurons, *J Neurosci* **17**:6302-6313.
Williams, M. E., Brust, P. F., Feldman, D. H., Patthi, S., Simerson, S., Maroufi, A., McCue, A. F., Velicelebi, G., Ellis, S. B. and Harpold, M. M., 1992, Structure and functional expression of an ω-conotoxin-sensitive human N-type calcium channel, *Science* **257**:389-395.
Williams, M. E., Feldman, D. H., McCue, A. F., Brenner, R., Velicelebi, G., Ellis, S. B. and Harpold, M. M., 1992, Structure and functional expression of α1, α2, and β subunits of a novel human neuronal calcium channel subtype, *Neuron* **8**:71-84.
Williams, M. E., Marubio, L. M., Deal, C. R., Hans, M., Brust, P. F., Philipson, L. H., Miller, R. J., Johnson, E. C., Harpold, M. M. and Ellis, S. B., 1994, Structure and functional characterization of neuronal α_{1E} calcium channel subtypes, *J Biol Chem* **269**:22347-22357.
Wilson, G. F., Richardson, F. C., Fisher, T. E., Olivera, B. M. and Kaczmarek, L. K., 1996, Identification and characterization of a Ca^{2+}-sensitive nonspecific cation channel underlying prolonged repetitive firing in Aplysia neurons, *J Neurosci* **16**:3661-3671.

Wilson, S. M., Toth, P. T., Oh, S. B., Gillard, S. E., Volsen, S., Ren, D., Philipson, L. H., Lee, E. C., Fletcher, C. F., Tessarollo, L., Copeland, N. G., Jenkins, N. A. and Miller, R. J., 2000, The status of voltage-dependent calcium channels in α1E knock-out mice, *J Neurosci* **20**:8566-8571.

Winterfield, J. R. and Swartz, K. J., 2000, A hot spot for the interaction of gating modifier toxins with voltage-dependent ion channels, *J Gen Physiol* **116**:637-644.

Witcher, D. R., De Waard, M. and Campbell, K. P., 1993a, Characterization of the purified N-type Ca^{2+} channel and the cation sensitivity of ω-conotoxin GVIA binding, *Neuropharmacology* **32**:1127-1139.

Witcher, D. R., De Waard, M., Sakamoto, J., Franzini-Armstrong, C., Pragnell, M., Kahl, S. D. and Campbell, K. P., 1993b, Subunit identification and reconstitution of the N-type Ca^{2+} channel complex purified from brain, *Science* **261**:486-489.

Woppmann, A., Ramachandran, J. and Miljanich, G. P., 1994, Calcium channel subtypes in rat brain: biochemical characterization of the high-affinity receptors for ω-conopeptides SNX-230 (synthetic MVIIC), SNX-183 (SVIB), and SNX-111 (MVIIA), *Mol Cell Neurosci* **5**:350-357.

Wu, L. G. and Saggau, P., 1994, Pharmacological identification of two types of presynaptic voltage-dependent calcium channels at CA3-CA1 synapses of the hippocampus, *J Neurosci* **14**:5613-5622.

Xu, W. and Lipscombe, D., 2001, Neuronal $Ca_V1.3\alpha_1$ L-type channels activate at relatively hyperpolarized membrane potentials and are incompletely inhibited by dihydropyridines, *J Neurosci* **21**:5944-5951.

Yan, L. and Adams, M. E., 2000, The spider toxin ω-Aga IIIA defines a high affinity site on neuronal high voltage-activated calcium channels, *J Biol Chem* **275**:21309-21316.

Yasuda, O., Morimoto, S., Chen, Y., Jiang, B., Kimura, T., Sakakibara, S., Koh, E., Fukuo, K., Kitano, S. and Ogihara, T., 1993, Calciseptine binding to a 1,4-dihydropyridine recognition site of the L-type calcium channel of rat synaptosomal membranes, *Biochem Biophys Res Commun* **194**:587-594.

Yasuda, O., Morimoto, S., Jiang, B., Kuroda, H., Kimura, T., Sakakibara, S., Fukuo, K., Chen, S., Tamatani, M. and Ogihara, T., 1994, FS2, a mamba venom toxin, is a specific blocker of the L-type calcium channels, *Artery* **21**:287-302.

Yu, H., Rosen, M. K., Saccomano, N. A., Phillips, D., Volkmann, R. A. and Schreiber, S. L., 1993, Sequential assignment and structure determination of spider toxin ω-Aga-IVB, *Biochemistry* **32**:13123-13129.

Zagotta, W. N., Hoshi, T. and Aldrich, R. W., 1989, Gating of single Shaker potassium channels in Drosophila muscle and in Xenopus oocytes injected with Shaker mRNA, *Proc Natl Acad Sci U S A* **86**:7243-7247.

Zengel, J. E., Sosa, M. A. and Poage, R. E., 1993, ω-Conotoxin reduces facilitation of transmitter release at the frog neuromuscular junction, *Brain Res* **611**:25-30.

Zhang, J. F., Randall, A. D., Ellinor, P. T., Horne, W. A., Sather, W. A., Tanabe, T., Schwarz, T. L. and Tsien, R. W., 1993, Distinctive pharmacology and kinetics of cloned neuronal Ca^{2+} channels and their possible counterparts in mammalian CNS neurons, *Neuropharmacology* **32**:1075-1088.

CALCIUM CHANNEL BLOCKING POLYPEPTIDES

Structure, Function, and Molecular Mimicry

Raymond S. Norton[1], Jonathan B. Baell[1], and James A. Angus[2]

1. INTRODUCTION

Venomous creatures produce a wealth of interesting peptide and protein toxins as components of their venoms. Many such peptide and protein toxins are potent and highly selective blockers or modulators of calcium channel function, and as such are valuable pharmacological tools and potentially valuable leads in human therapeutic development. Cone shells and spiders are rich sources of such toxins, although they are also found in scorpions and insects. In this chapter we compare the amino acid sequences of toxins active against calcium channels and describe their three-dimensional structures and structure-function relationships. Certain structural motifs, in particular the inhibitor cystine knot, prove to be quite common amongst this class of toxins. Aspects of the pharmacology and physiology of these toxins in mammalian systems are discussed, with an emphasis on their application to the treatment of chronic pain. We then consider how these naturally occurring toxins might be mimicked in non-peptide (peptidomimetic) compounds that could be useful therapeutically for the treatment of pain and in other indications.

2. PEPTIDE AND PROTEIN BLOCKERS OF CALCIUM CHANNELS

The largest and most extensively characterized group of polypeptide toxins that block calcium channels are the ω-conotoxins, one of many classes of polypeptide toxin derived from the more than 500 species of cone shell found in the Indo-Pacific region (Olivera et al., 1990; 1994; Jones and Bulaj, 2000). The ω-conotoxins block neuronal voltage-gated calcium channels, thereby inhibiting synaptic neurotransmitter release. They were originally dubbed 'shaker' peptides because upon intracerebral injection in mice they

---

[1]Walter and Eliza Hall Institute of Medical Research, Parkville 3052, Australia; [2]Department of Pharmacology, University of Melbourne, Parkville 3052, Australia. Address correspondence to Professor Raymond S. Norton, Walter and Eliza Hall Institute of Medical Research, 1G Royal Parade, Parkville 3052, Australia.

Calcium Channel Pharmacology, Edited by McDonough
Kluwer Academic/Plenum Publishers, 2004

```
                1       5         10        15        20        25
GVIA            C K S O G S S C S O T S Y N C C R - S C N O Y T K R C - - Y
GVIB            C K S O G S S C S O T S Y N C C R - S C N O Y T K R C - - Y G
GVIC            C K S O G S S C S O T S Y N C C R - S C N O Y T K R C
GVIIA           C K S O G T P C S R G M R D C C T - S C L L Y S N K C R R Y
GVIIB           C K S O G T P C S R G M R D C C T - S C L S Y S N K C R R Y

MVIIA           C K G K G A K C S R L M Y D C C T G S C R - - S G K C
MVIIB           C K G K G A S C H R T S Y D C C T G S C N R - - G K C
MVIIC           C K G K G A P C R K T M Y D C C S G S C G R - R G K C
MVIID           C Q G R G A S C R K T M Y N C C S G S C N R - - G R C

SVIA            C R S S G S X C G V T S I - C C - G R C - - Y R G K C T
SVIB            C K L K G Q S C R K T S Y D C C S G S C G R - S G K C
SO3           R C K A A G K P C S R I A Y N C C T G S C R - - S G K C
SO5       S T S C M E A G S Y C G S T T R I C C - G Y C A Y F G K K C I D Y P S N

TVIA            C L S O G S S C S O T S Y N C C R - S C N O Y S R K C

RVIA            C K P O G S O C R V S S Y N C C S - S C K S Y N K K C G

CVIA            C K S T G A S C R R T S Y D C C T G S C R - - S G R C
CVIB            C K G K G A S C R K T M Y D C C R G S C R - - S G R C
CVIC            C K G K G Q S C S K L M Y D C C T G S C S R - R G K C
CVID            C K S K G A K C S K L M Y D C C S G S C S G T V G R C

TxVII           C K Q A D E P C D V F S L D C C T G I C - - - L G V C M W

CnVIIA          C K G K G A O C T R L M Y D C C H G S C S S S K.G R C
```

Figure 1. Sequence alignment of ω-conotoxins (and closely related toxins) that block Ca^{2+} channels. All sequences shown are C-terminally amidated. The half-cystines are highlighted in bold and the horizontal lines denote the disulfide connectivities. O denotes hydroxyproline and X an uncharacterized residue (where the state of hydroxylation of a proline has not been defined). Species, toxins and SwissProt accession numbers are as follows: *Conus geographus* GVIA, GVIB, GVIC, P01522; GVIIA/GVIIB (SNX-178), P05484. *C. magus* MVIIA, MVIIB, MVIIC, and MVIID, MVIIA precursor (SNX-111), P05484; MVIIB (SNX-159), P05485; MVIID precursor Q26350. *C. striatus* SVIA precursor (SNX-157), P28880; SVIB precursor (SNX-183), P28881; SO3 precursor, Q9XZK2; SO5 precursor, Q9XZK4. *C. tulipa* TVIA (SNX-185), P58915. *C. radiatus* RVIA, P58914. *C. catus* CVIA precursor, P58917; CVIB, P58918; CVIC, P58919; CVID precursor, P58920. *C. consors* CnVIIA, P58916. *C. textile* TxVII, P56714. Note that names with SNX prefixes were assigned by Neurex Corporation, now a subsidiary of Elan Corporation. Not all known ω-conotoxin sequences are shown; for example, PnVIA and PnVIB from *C. pennaceus* (P56712 and P56713), which target molluscan channels, and the precursors of PuIA and PuIIA from *C. pulicarius* (Q9XYZ0 and Q9XYZ1) are not included. SO3 and SO5 are shown as representatives of ω-type conotoxins identified in *C. striatus* and *C. textile* (Lu et al., 1999).

induce a persistent tremor that can last for days. They are '4-loop' peptides with the consensus sequence C······C······CC···C······C and a 1-4/2-5/3-6 disulfide bonding pattern. The ω- prefix identifies conotoxins active against voltage-gated calcium channels (Olivera and Cruz, 2001) and has been adopted for other classes of toxin that act on these channels.

Figure 1 summarizes the amino acid sequences of mature ω-conotoxins published to date. A striking feature of this sequence alignment is the sequence diversity, with the only common features of the sequences being the six half-cystine residues and the Gly at

```
                 1   5    10   15   20           25   30   35   40
Aga-IIIA         S C I D I G G D C D G E K D D C Q C C R R N G Y - - - - C S C Y S L F G Y L K S G C K C V
Aga-IIIB         S C I D F G G D C D G E K D D C Q C C R S N G Y - - - - C S C Y N L F G Y L K S G C K C E
Aga-IIIC         N C I D F G G D C D G E K D D C Q C C X R N G Y - - - - C S C Y N L F G Y L K R G C K X E
Aga-IIID         S C I K I G E D C D G D K D D C Q C C R T N G Y - - - - C S X Y X L F G Y L K S G

PtxIIA           S C I N V G D F C D G K K D D C Q C C R D N A F - - - - C S C S V I F G Y - K T N C R C

Tx1      A E L T S C F P V G H E C D G D A S N C N C C G D D V Y - - - - C G C - - - - G W G R W N C K C

Tx3-3            G C A N A Y K S C N G - P H T - - C C W G Y N G Y K K A C I C S - - - G X - N W K

Tx3-4            S C I N V G D F C D G K K D - - - C C Q C D R D N A F - C S C S V I F G Y - K T N C R C E


                   45   50   55   60   65   70   75
Aga-IIIA         V G T S A E F Q G I C R R K A R Q C Y N S D P D K C E S H N K P K R R
Aga-IIIB         V G T S A E F R R I C R R K A K Q C Y N S D P D K C V S V Y K P K R R
Aga-IIIC         V G

PtxIIA           E V G T T A T S Y G I C M A K H K C G R Q T T C T K P C L S K R C K K N H

Tx1              K V A D Q S Y A Y G I C K D K V N C P N R H L W P A K V C K K P C R R E C
```

Figure 2. Sequence alignment of spider toxins from the Tx3 and ω-agatoxin families that block Ca^{2+} channels. Tx3-3 and Tx3-4 are two of the six spider toxins belonging to the Tx3 family. In all of the sequences, the half-cystines are highlighted in bold. Sequences, excluding Tx3-3 and Tx3-4, were aligned using CLUSTAL W (1.82). Tx3-3 and Tx3-4 were aligned based on the location of their cysteine residues. Species, toxins and SwissProt accession numbers are as follows: *Phoneutria nigriventer* ω-PtxIIA; Tx1, P17727; Tx3-3 fragment, P81789; Tx3-4 fragment, P81790. *Agelenopsis aperta* ω-Aga IIIA, P33034; ω-Aga IIIB, P81744; ω-Aga IIIC fragment, P81745; ω-Aga IIID fragment, P81746.

position 5 (the exception being TxVII), although residues 2 and 24 (using the GVIA numbering) are usually Lys or Arg and residue 18 is usually Ser. This dramatic diversity is a characteristic feature of cone shell toxins in general, not just the ω-conotoxins, and makes it all the more remarkable that the three-dimensional structures of the ω-conotoxins are so well conserved, as discussed below. As noted by Olivera et al. (1995), cone shells have successfully exploited combinatorial synthesis long before it became fashionable in the pharmaceutical industry.

Toxins that block or modulate calcium channel activity have been isolated from several other sources, most notably spiders. Toxins from the funnel web spider *Agelenopsis aperta* illustrate two major classes observed to date. ω-agatoxin IIIA is a 76-residue protein that blocks several classes of Ca^{2+} channel, and a number of variants have been isolated from the same spider (Ertel et al., 1994), as shown in Figure 2. ω-Aga IIIA was originally described as blocking L- and N-type channels with equally high potency, although it has been shown to bind to all high voltage-activated channels, including P/Q- and R-types (Yan and Adams, 2000). By contrast, the type IV ω-agatoxins ω-Aga IVA (Mintz et al., 1992) and IVB (Adams et al., 1993), also known as ω-Aga TK, consist of 48 residues and selectively block P-type channels (Figure 3). A fascinating feature of the ω-agatoxins is the presence of a post-translational modification that produces a D-amino acid at position 46. ω-Aga IVB and IVC were found to differ only in the configuration at Ser46. Both were selective for P-type channels but ω-Aga

```
            1       5         10        15        20        25        30
Aga-IVA     K K K C I A K D Y G R C K W G - G T P - - - C C R G R G C I C S I M
Aga-IVB     E D N C I A E D Y G K C T W G - G T K - - - C C R G R P C R C S M I
SNX-325       G S C I E S G K - S C T H S R S M K N G L C C P K S R C N C R Q I
DW 13.3       A E C L M I G D T S C V P R L G R R - - - C C Y G A W C Y C D Q -

                                        35        40        45
Aga-IVA     G T N - - - - - - - - - - C E C K P R L I M E G L G L A
Aga-IVB     G T N - - - - - - - - - - C E C T P R L I M E G L S F A
SNX-325     Q H R H D Y L G K R K Y S C R C S
DW 13.3     Q - - - - - - - - - - L S C R R V G R K R E C G W V E V N C K C G W S W S Q R I D D W R A D Y

              70
DW13.3      S C K C P E D Q

            1       5         10        15            20        25        30              35        40
SNX-482     G V D K A G C R Y M F G G C S V N - - D D C C P R L G C H S L F - - - S Y C A W D L T F S D
SIA                   D C V R F W G K C S Q T - - S D C C P H L A C K S K W P R N - I C V W D G S V
Agelenin            G G C L P H N R F C N A L S G P R C C S G L K C K E L S I W D S R C L
```

Figure 3. Sequence alignment of related spider toxins that block Ca^{2+} channels. In all of the sequences, the half-cystines are highlighted in bold. Species, toxins and SwissProt accession numbers are as follows: *Agelenopsis aperta* ω-Aga IVA, P30288; ω-AgaIVB precursor, P37045. *Segestria florentina* SNX-325, P58605. *Filistata hibernalis* DW 13.3 (this sequence is included here rather than in Figure 2 because comparison of the amino acid sequences in CLUSTAL W indicates greater similarity to the ω-Aga IV sequences than ω-Aga III). *Grammostola spatulata* ω-grammotoxin SIA. *Agelena opulenta* agelenin, P31328. *Hysterocrates gigas* SNX-482, P56854.

IVB (containing D-Ser46) was more potent and more resistant to proteolysis (Heck et al., 1994; Kuwada et al., 1994).

A. aperta also contains two classes of long toxins which have not been as extensively studied. ω-agatoxins IA, IB and IC are heterodimeric proteins of mass about 7.5 kDa and with five disulfide bridges. ω-Aga IA blocks insect Ca^{2+} channels, blocks neurotransmission at the frog neuromuscular junction, and suppresses high threshold Ca^{2+} currents in rat dorsal root ganglion cells (Bindokas and Adams, 1989; Scott et al., 1990) but does not affect N-type channels (Venema et al., 1992). The type II ω-agatoxins (IIA, IIB) have masses around 10 kDa, show limited sequence similarity to the type I ω-agatoxins, and are active against N-type channels (Adams et al., 1990; Venema et al., 1992) and insect Ca^{2+} channels (Bindokas et al., 1991).

A family of six toxins has been isolated from the Brazilian 'armed' spider *Phoneutria nigriventer* (Cordeiro et al., 1993), one of which (Tx3-3, or ω-PnTx3-3) (Figure 2) inhibits a range of high-voltage activated Ca^{2+} channels, but most effectively P/Q- and R-types (Leão et al., 2000). Tx3-4 (Figure 2) had similar activity to Tx3-3 but lower potency (Miranda et al., 1998). Tx3-2 and Tx3-5 did not affect T-type channels but were active against L-type (Kalapothakis et al., 1998; Leão et al., 2000). Tx1 is a 77-residue protein from the same spider (Diniz et al., 1990) which may act on calcium channels (Santos et al., 1999) and shows some sequence similarity to the ω-agatoxins III (Figure 2). ω-phonetoxin-IIA (ω-PtxIIA), also with 76 residues, 14 Cys forming seven disulfide bonds and sequence similarity to the ω-agatoxins III, blocks P/Q- and N-type currents and has limited activity on L-type currents (Cassola et al., 1998). It is only partially displaced by GVIA and MVIIC and defines a unique binding site (dos Santos et al., 2002).

ω-grammotoxin SIA is a 36-residue polypeptide from the tarantula *Grammostola spatulata* (Lampe et al., 1993) which potently inhibits both P/Q- and N-type channels by impeding channel gating and binds to distinct or additional sites on P-type channels compared to ω-Aga-IVA (McDonough et al., 1997). It shares 42% sequence identity with the K^+ channel blocker hanatoxin (isolated from the same spider) and in fact the two toxins cross-react, although grammotoxin is more potent against Ca^{2+} channels and hanatoxin more potent on K^+ channels (Li-Smerin and Swartz, 1998). SNX-482, a 41-residue toxin from the tarantula *Hysterocrates gigas*, is homologous to ω-grammotoxin SIA and hanatoxin (Figure 3) but blocks the α_{1E} ($Ca_V2.3$) subunits of R-type channels (Newcomb et al., 1998). Another toxin with obvious sequence similarity to ω-grammotoxin SIA and SNX-482 is the P-type blocker agelenin, from the funnel web spider *Agelena opulenta* (Hagiwara et al., 1990; Inui et al., 1992).

SNX-325, from *Segestria florentina* (Figure 3), is active on N-type channels (Newcomb et al., 1995). Although equipotent with MVIIA and GVIA in some preparations, it is 4-5 orders of magnitude less potent at displacing radioiodinated MVIIA from rat brain synaptosomal membranes. DW13.3 is a novel 74-residue toxin from *Filistata hibernalis* that potently blocks a broad range of Ca^{2+} channels with the exception of T-type (Sutton et al., 1998). The Chinese bird spider *Selenocosmia huwena* elaborates a series of polypeptide toxins, one of which, huwentoxin-I (Liang et al., 1993), blocks high-voltage-activated Ca^{2+} channels, particularly N-type, in prostaglandin E1 differentiated NG108-15 cells (Peng et al., 2001).

Spiders, not surprisingly, elaborate a number of toxic polypeptides targeted at insects, and many of these are ion channel blockers. In most cases the exact protein target in the insect remains to be established, but some spider toxins have been shown to target calcium channels. Wang et al. (2001) recently isolated the ω-atracotoxins, a novel family of insect-selective neurotoxins from the Australian funnel-web spider *Hadronyche versuta* that appear to be the most potent blockers of insect voltage-gated Ca^{2+} channels reported so far, displaying at least a 10^4-fold preference for insect versus vertebrate channels. They consist of around 45 residues and contain three disulfide bridges. Highly selective toxins of this type have attracted considerable interest because of their potential as novel insecticides. Another insect calcium channel blocker, the 44-residue PLTX-II from the spider *Plectreurys tristis* (Branton et al., 1993), contains a C-terminal palmitoyl group that is essential for biological activity and may perform the same role as the hydrophobic C-terminal tail of the funnel web toxins, as discussed below.

This summary of spider toxins active on Ca^{2+} channels covers most of the major classes characterized to date. A comprehensive review of spider toxins generally, including Ca^{2+} channel toxins, has been published recently by Rash and Hodgson (2002).

Scorpions are another well-studied source of polypeptide toxins, and a large number of blockers or modulators of K^+ and Na^+ channels have been obtained from their venoms. Fewer Ca^{2+} channel toxins have been described. The most important of these in the context of this chapter is kurtoxin, a 60-residue polypeptide with four disulfides, which distinguishes between T-type calcium channels and other classes (Chuang et al., 1998), although not with absolute specificity (Sidach and Mintz, 2002). Like the scorpion α-toxins to which it is closely related, it also slows the inactivation of voltage-gated Na^+ channels. A range of scorpion toxins, eg. ryanotoxin (Morrissette et al., 1996) and imperatoxin I (Zamudio et al., 1997a), also affect ryanodine-sensitive calcium channels in skeletal muscle, although by apparently different mechanisms. Imperatoxin A (Zamudio et al., 1997b) and maurocalcine (Fajloun et al., 2000) are smaller, with 33 residues and three disulfides and with greater sequence similarity to spider toxins such as Tx2-9 and agelenin than to other scorpion toxins.

Relatively few snake toxins have been found with activity against calcium channels. The black mamba elaborates a 60-residue polypeptide called calciseptine (de Weille et al., 1991), which contains four disulfide bonds. Like the 1,4-dihydropyridines, calciseptine selectively blocks L-type Ca^{2+} channels and is totally inactive on other voltage-dependent channels such as N- and T-type channels. FS2 is another mamba venom toxin, consisting of 60 residues, three of which differ from those of calciseptine (Yasuda et al., 1994). It is also an L-type blocker sharing the binding site on the channels with the 1,4-dihydropyridines.

Calcicludine, a 60-residue protein isolated from green mamba venom, blocks most of the high-threshold Ca^{2+} channels (L-, N-, or P-type) in the 10-100 nM range (Schweitz et al., 1994). In rat cerebellar granule neurons in primary culture, calcicludine blocks potently the L-type component of the Ca^{2+} current, with a $K_{0.5}$ of 0.2 nM. This toxin has clear sequence similarity to the Kunitz-type protease inhibitors such as bovine pancreatic trypsin inhibitor, with roughly half the residues being identical or similar, including the locations of the three disulfide bonds.

All of the polypeptides described above were obtained from venomous creatures, but some Ca^{2+} channel blockers have also been found in insects, and we mention a few examples here. Three novel polypeptides were isolated from the brain and fat body of adult *Locusta migratoria* (Nakakura et al., 1992). They range in size from 35 residues (PMP-D2) to 54 and all have three disulfides (Figure 4). Also known as LMCI I and II, the two smaller polypeptides are protease inhibitors that block the prophenoloxidase-phenoloxidase proteolytic activation cascade in hemocyte extracts of the same insect (Boigegrain et al., 1992). Nonetheless, Scott et al. (1997) have demonstrated that they are able to block mammalian N-type channels. Intriguingly, PMP-C is post-translationally modified by a single fucose moiety, and the presence of this sugar abolished the activity of this toxin against Ca^{2+} channels.

Three new polypeptides have also been identified recently in toxic saliva from the predatory insect *Peirates turpis*, known as the assassin bug (Corzo et al., 2001). Ptu1, the shortest peptide, was chemically synthesized and shown to reversibly block N-type channels expressed in BHK cells. Its amino acid sequence is shown in Figure 4.

```
          1       5         10        15         20        25        30
PTU1      A E K D C I A P G A P C F G T D K P C - C N P R - A W - C S S Y A N K C L
PMP-D2    - E E K C T - P G Q V K Q Q D C N T C T C T P T G V W G C T R - - K G C Q P A
```

Figure 4. Sequence alignment of related insect polypeptides that block Ca^{2+} channels. In both of the sequences, the half-cystines are highlighted in bold. The disulfide connectivities of Ptu1 follow the ICK motif, whereas those of PMP-D2 do not. Sequences were aligned using CLUSTAL W (1.82). Species and SwissProt accession numbers are as follows: *Peirates turpis* Ptu1, P58606; *Locusta migratoria* PMP-D2, Protease inhibitors precursor, P80060.

3. THE ω-CONOTOXINS

3.1 Three-Dimensional Structures

The solution structure of ω-conotoxin GVIA was solved independently by four groups in 1993 (Sevilla et al., 1993; Davis et al., 1993; Skalicky et al., 1993; Pallaghy et al., 1993), and since then the structure has been refined to higher resolution (Pallaghy and Norton, 1999). GVIA consists of a triple-stranded, anti-parallel β-sheet stabilized by a disulfide knot and contains four reverse turns (at Ser3-Ser6, Ser9-Ser12, Cys15-Ser18 and Asn20-Lys24, respectively), as shown in Figure 5. Solution structures have been determined since for ω-conotoxins MVIIA, from *C. magus* (Basus et al., 1995; Kohno et al., 1995; Nielsen et al., 1996), which also blocks N-type calcium channels, MVIIC (Farr-Jones et al., 1995; Nemoto et al., 1995; Nielsen et al., 1999b), which blocks P- and Q-type calcium channels, and SVIB (Nielsen et al., 1996), which blocks both N- and P/Q-type, with approximately equal affinity (Woppmann et al., 1994). The structure and dynamics of ω-conotoxin MVIIA-Gly, a precursor to MVIIA, have also been described (Goldenberg et al., 2001). More recently, ω-conotoxin structures have been reported for MVIID (Civera et al., 1999), which has similar although not identical subtype specificity to MVIIC, TxVII (Kobayashi et al., 2000), an L-type channel blocker (Figure 6), and CVID (Lewis et al., 2000), a highly selective N-type blocker (Adams et al., 2003).

Striking amino acid sequence diversity is a feature of conotoxins in general, and this is certainly true for the ω-conotoxins, as noted in Section 2. It is remarkable, therefore, that three-dimensional structures are so highly conserved among toxins. GVIA and MVIIA have only 40% sequence homology, but 71% of their backbone dihedral angles are identical and they are both selective for N-type channels, whereas MVIIA and MVIIC, which have different subtype selectivities, have 81% homology but only 61% of their backbone dihedral angles are identical (Basus et al., 1995). This implies that local structural differences as well as specific amino acid residue differences may contribute to subtype specificity in the ω-conotoxins, even though the global structures are very similar. Essentially the same conclusion was reached from a detailed comparison of the structures of MVIIA and SVIB, an N- and P/Q-type blocker (Nielsen et al., 1996). While the overall structures were similar, individual loops showed some differences in backbone conformation and hydrogen bonding pattern, and these differences, coupled with differences in amino acid sequence, led to differences in surface polarity and charge. A slight variation from the consensus structure of the ω-conotoxins was noted in the recently solved structure of CVID (Lewis et al., 2000), in which two additional hydrogen bonds appear to stabilize loop 2 and loop 4 curves towards loop 2 to create a globular surface. CVID is of particular interest because it appears to be the ω-conotoxin most selective for N-type over P/Q-type channels, a desirable property in avoiding potential side effects associated with peripheral N-type channels (Nielsen et al., 2000; see Section 5). Recently, Adams et al. (2003) attributed this selectivity to interaction with a pharmacologically distinct variant of the N-type ($Ca_V2.2$) α-subunit.

The structure of TxVII (Kobayashi et al., 2000) is also similar to those of the other ω-conotoxins (Figure 6), even though it shows almost no sequence similarity to the others (Figure 1). It maintains Lys at position 2, but Gly5 and Tyr13 are replaced by Asp5 and Leu13 respectively. The absence of Tyr13 may be partly responsible for the specificity of TxVII for L-type channels, as discussed in more detail below. Residues 3-6 in TxVII are described as forming a type II β-turn, just as in the other ω-conotoxins. This type of turn usually requires Gly at position $i + 2$ (Venkatachalam, 1968). When Gly5 in GVIA was replaced with Ala the folding was disrupted (Lew et al., 1997), and D-Ala, which can adopt a positive φ-angle, was required to restore correct folding. It would be interesting to assess the effect on folding of replacing Asp5 in TxVII with Gly.

The ω-conotoxin structures also provide a striking example of a structural scaffold that has been used widely in nature to generate polypeptides with a range of biological activities. This common structural motif, consisting of a small, triple-stranded, anti-parallel β-sheet and a disulfide knot (Narasimhan et al., 1994; Pallaghy et al., 1994), has been designated the inhibitor cystine knot (ICK) motif to distinguish it from other disulfide knot motifs (Isaacs, 1995). Polypeptides containing this motif come from a diverse range of species, including plants and animals, and have a variety of receptors as their biological targets, although ion channels feature prominently. Examples include the plant proteins gurmarin (a sweet taste suppressor), kalata (an inducer of uterine contractions) and a range of protease inhibitors, the AVR9 elicitor protein from a fungus affecting tomatoes, and a number of spider toxins, as well as several newly isolated cone shell toxins (Norton and Pallaghy, 1998). This structural motif appears to provide a robust and versatile scaffold for the presentation of a variety of functional groups, thereby generating a range of polypeptides with diverse biological targets.

Interestingly, the converse scenario is found in blockers of potassium channels, where a functional diad consisting of a Lys and a hydrophobic (usually aromatic) residue that are essential for blockade are presented to the channel on a wide range of different scaffolds. Thus, the scorpion toxins display this diad on a β-sheet scaffold while sea anemone toxins utilize a helical scaffold (Dauplais et al., 1997). And the snake toxin α-dendrotoxin (Gasparini et al., 1998) and κ-conotoxin (Garcia et al., 1999) also display this diad, even though the former adopts a BPTI-like fold and the latter has a similar fold to the ω-conotoxins.

The issue of structural and functional diversity in cone shell toxins was considered in a recent paper by Miles et al (2002), in which the structure of conotoxin gm9a, another ICK motif based structure, was described. Across a range of conotoxin structures that contain six Cys residues linked in the same 1-4/2-5/3-6 disulfide bonding pattern as in the ω-conotoxins, functional diversity is achieved by a combination of sequence variation and structural variation. As a result, structural analysis reveals few clues as to the potential molecular target of a given polypeptide's action. The μ-conotoxins GIIIA and GIIIB and ψ-conotoxin PIIIE all share a common Cys framework (-$CCX_{4\text{-}5}CX_{4\text{-}5}CX_4CC$-, where X denotes residues between half-cystines) and the 1-4/2-5/3-6 bonding pattern (Balaji et al., 2000). These polypeptides are similarly structured but functionally distinct: the μ-conotoxins (Sato et al., 1991; Wakamatsu et al., 1992; Hill et al., 1996) bind selectively to sodium channels, and ψ-PIIIE (Mitchell et al., 1998) inhibits the acetylcholine receptor. An analogous situation exists amongst the ω and κ O-superfamily conotoxins. All of these polypeptides exhibit the ICK-motif within the same Cys framework (-$CX_6CX_6CCX_{2\text{-}3}CX_{4\text{-}6}C$-) and despite significant sequence diversity all have a remarkably similar structure. However, the ω-conotoxins SO3 (Yan and Zhang, 2000), MVIIA, and GVIA all selectively inhibit voltage-gated calcium channels, and PVIIA, a κ-conotoxin (Savarin et al., 1998), selectively inhibits Shaker potassium channels. In contrast, conotoxin GS, which is structurally related to the O-superfamily conotoxins through a common Cys framework and ICK motif (Hill et al., 1997), is functionally related to the μ-conotoxins, as it binds competitively with μ-GIIIA to skeletal muscle sodium channels, even though conotoxin GS has low sequence identity with either GIIIA or GIIB and is 50% larger.

Despite having a common Cys framework, there can be significant structural diversity amongst members of a given superfamily. This can be attributed mainly to the different inter-cysteine spacings across the sequences and is illustrated well by considering the structural variation among the O-superfamily polypeptides. There is significant sequence variation among ω-SO3, ω-MVIIA, ω-GVIA, and κ-PVIIA, although they have very similar structures. Importantly, the number of residues between sequential half-cystines is reasonably consistent across these four polypeptides, i.e., 6 (C_1-C_2), 6 (C_2-C_3), 0 (C_3-C_4), 2-3 (C_4-C_5) and 4-6 (C_5-C_6). By contrast, conotoxin GS has inter-cysteine spacings of 6, 3, 0, 4 and 7 and a structure that differs from the ω- and κ-conotoxins. In TVIIA (Hill et al., 2000) these spacings are 6, 3, 0, 4 and 4, and it has essentially the same structure as conotoxin GS over the disulfide-linked region of the molecule, except that the loop between C_5 and C_6 is about half as large in TVIIA. A similarly significant departure from the classic ω-conotoxin fold is observed in EVIA (Volpon et al., 2000), which has the spacings 6, 9, 0, 3, and 3. Thus, the inter-cysteine spacings are a good indicator of how similar structures are likely to be.

We have seen in this section how a common structural scaffold, the ICK motif, can serve a range of different functions by modifications to its surface chemistry and the size of loops protruding from its surface. It is worth noting that even variants with the same structure and same function can achieve their goals using different surfaces. As will be described below, GVIA and MVIIA, which both target N-type channels with high specificity and potency, have some features in common, such as Tyr13, and some marked differences amongst their key residues. It remains to be determined if these two ω-conotoxins bind to N-type channels in an identical fashion, or if they make slightly different interactions with the channel, achieving blockade in different orientations.

3.2. Flexibility, Folding and Stability

Before considering the question of which residues in the ω-conotoxins determine their calcium channel subtype specificity, it is appropriate to consider how rigid these small polypeptide scaffolds are and what role the disulfides play in their stability.

In our initial structure determination of GVIA (Pallaghy et al., 1993) residues 10-13 (in loop 2) were less well defined than the rest of the molecule, suggesting the possibility of some conformational averaging in this part of the structure. In a subsequent refinement of the structure (Pallaghy and Norton, 1999), two families of structures differing in the vicinity of residues 11-14 were consistent with most of the NMR data, and it was only the presence of some close interresidue contacts in the second family that were not reflected in the NOE data set which distinguished the first as being more representative of the average solution structure of GVIA.

The possibility of conformational averaging in loop 2 was also noted in early studies on MVIIA (Basus et al., 1995; Kohno et al, 1995), although Nielsen et al. (1996)

concluded that MVIIA adopted a rigid structure that was also insensitive to changes in solvent polarity. Since then the dynamics of MVIIA and a putative precursor extended by a Gly at the C-terminus have been investigated in detail by analysis of ^{13}C (Atkinson et al., 2000) and ^{15}N (Goldenberg et al., 2000) NMR relaxation measurements, respectively, in both cases at 10°C and slightly acidic pH. In MVIIA the loop between Cys8 and Cys15 (loop 2) did not adopt a unique structure in the calculated structures, and relaxation data indicated the presence of slow conformational exchange on a timescale of around 4 ms. It was suggested that conformational exchange was associated with isomerization of the Cys8-Cys20 disulfide bridge. A similar conclusion regarding conformational exchange of residues 9-15 was reached for MVIIA-Gly, although here the time constant was estimated to be 35 μs. These observations are highly relevant to consideration of the structure-function relationships in MVIIA because this motion affects key residues on the surface of MVIIA, including Arg10, Leu11 and Tyr13, as discussed below. It is possible that such motions allow MVIIA to interact with a broader range of calcium channel subtypes than its more rigid counterparts and may contribute to the reversibility of MVIIA binding. Although other ω-conotoxins may also experience similar conformational averaging, quantitative analyses using NMR relaxation measurements have not been undertaken. In the family of structures of CVID (Lewis et al., 2000), loop 2 appears to be as well defined as the rest of the molecule, implying that this loop is not undergoing conformational exchange. Perhaps this contributes to the higher subtype selectivity of CVID.

These observations highlight an important point concerning structures determined in solution from NMR data, which is the case for all of the ω-conotoxins considered here. These ω-conotoxin structures span almost a decade, and in that period the quality of NMR data has improved significantly as a result of the use of higher field magnets as well as improvements in probe sensitivity and methods of spectral acquisition. Furthermore, methods for calculating and refining structures determined from NMR data have become more robust and more efficient. As a result, the precision and accuracy of structures determined now is superior to those from a decade ago, and we can be more confident in detecting disordered regions of peptide and protein structures, where a single family of structures may not fit all the available NMR data. Moreover, NMR relaxation measurements have become more common and represent an excellent means of quantifying the rates and amplitudes of internal motions within a protein. Indeed, in the process of characterizing the molecular surface of a toxin and understanding how it might block an ion channel, it is important not to ignore the possibility of side chain and even backbone flexibility. Even well defined structures may change upon channel binding. As long duration (ns) molecular dynamics simulations become more accessible we can expect to see them used to gain a better understanding of the conformational space available to ligand surfaces.

It is a *sine qua non* that the disulfide bonds in small polypeptide toxins are essential for their structure and thus their activity, and ω-conotoxins are no exception. Goldenberg and co-workers (Price-Carter et al., 1998) reported losses of binding affinity of between 70- and 5,200-fold in MVIIA analogues with single disulfides deleted, with removal of the 3-6 disulfide (Cys15-Cys26) causing the greatest loss of affinity. Flinn et al. (1999a) found that [Ser15,26]GVIA had no detectable activity at 10 μM, corresponding to a loss of potency of > 8000-fold. A preliminary study by Sabo et al.

(1992) in which disulfide bridges were deleted from GVIA reported some residual biological activity when the Cys1-Cys16 disulfide bridge was absent but complete loss of activity when either of the other bridges was deleted. However, these results were obtained from a goldfish lethality assay and there are likely to be some differences between the goldfish and mammalian responses; moreover, no disulfide mapping was reported for these analogues.

Goldenberg and co-workers (Price-Carter et al., 1996a,b; 1998; 2002) have also undertaken an extensive study of the refolding of MVIIA and the role of disulfide formation therein. Three MVIIA analogs lacking one of the native disulfides were refolded in three different buffers, one containing I_2, which causes rapid and irreversible disulfide formation, the second glutathione, to permit disulfide shuffling, and the third containing glutathione and 8 M urea to disrupt non-covalent interactions. In all three refolding solutions the species with two native disulfides accumulated to detectable levels, but in no case were they the major product. Moreover, the addition of urea had little effect, implying that formation of the native pairings is not significantly favored by non-covalent interactions. It appears that the oxidative folding of MVIIA *in vitro* is characterized by a broad distribution of disulfide-bonded intermediates, with no particular preference for native pairings among the intermediates containing only two disulfides. Once two native disulfides form, there is a high probability of progression to the native (three-disulfide) fold, which eventually accumulates in preference to non-native three-disulfide species because it is energetically more stable. However, when formation of the third disulfide is blocked, non-native two-disulfide species dominate. The accumulation of non-native two-disulfide forms of GVIA was also observed in analogues with individual pairs of half-cystines replaced by Ala or Ser (Flinn et al., 1999a).

To investigate whether the precursor sequences of ω-conotoxins contain information that helps direct folding of the mature sequences, the disulfide-coupled folding of mature MVIIA was compared with that of two putative precursor forms: pro-MVIIA-Gly, which contains the propeptide and the C-terminal Gly residue, and MVIIA-Gly (Price-Carter et al., 1996b). The three forms folded with similar kinetics, but the folding efficiency of MVIIA-Gly was > 80%, versus approximately 50% for both mature MVIIA and the form containing the propeptide. MVIIA-Gly has a very similar structure to MVIIA, but the terminal Gly stabilizes the structure by around 1 kcal mol^{-1} via hydrogen bonding to residues 17 and 19 (Goldenberg et al., 2001).

3.3. Structure-Function Relationships

The residues essential for calcium channel blocking activity have been mapped for GVIA (Kim et al., 1994; Lew et al., 1997) and MVIIA (Nadasdi et al., 1995; Kim et al., 1995a). The key residue is Tyr13, particularly its phenolic hydroxyl, with Lys2 making a lesser contribution. In GVIA, replacement of Arg17, Tyr22 and Lys24 with Ala resulted in significant reductions in potency in functional assays, but not in rat brain membrane binding (Lew et al., 1997). Analog potencies were strongly correlated between the different functional assays, but not between the functional and binding assays, implying that the binding assay gave a somewhat simplified view of the channel binding surface of GVIA (this issue is also discussed by McDonough, this volume). The location of these residues on the surface of GVIA is shown in Figure 7.

In MVIIA, Arg10, Leu11 and Arg21 were also important for binding to rat brain synaptosomal membranes (Nadasdi et al., 1995). Proteolytic cleavage between residues 13 and 14 in MVIIA decreased the binding affinity by two orders of magnitude, and loss of the 12-13 dipeptide unit led to almost total loss of binding (Nadasdi et al., 1995).

The contributions of key functional groups to the pharmacophore of GVIA have been investigated in finer detail by structural and *in vitro* functional studies of analogs substituted at one or two positions with residues other than Ala (Flinn et al., 1999b). Substitutions of Lys2 with non-standard amino acids and Arg quantified the roles of the length and charge of the Lys side chain, both of which play a role. The orientation of the Tyr13 side chain and its hydroxyl moiety was demonstrated to be important by substitution with D-handed Tyr and the constrained analogue Tic(OH) (7-hydroxy-1,2,3,4-tetrahydroisoquinoline-3-carboxylic acid). The roles of the Hyp10 and Hyp21 hydroxyl groups, investigated by proline substitutions, appear to be more structural (as monitored by NMR) than functional, although small decreases in potency were observed in some assays. The reversibility of channel blockade was also studied, and several analogues with faster wash-out characteristics than native GVIA were identified. Rapid reversibility (as for MVIIA) may be beneficial for therapeutic applications. The di-substituted analogues revealed some interesting cooperative effects not predicted from single- residue substitutions. A di-substituted chimera of GVIA and MVIIA (GVIA containing Arg10 and Leu11) was more potent than either native molecule.

The more detailed picture of the pharmacophore of GVIA obtained by Flinn et al. (1999b) provides a better basis for the development of truncated peptide or completely non-peptidic analogues. Limited progress has been made with truncated peptide analogues (Pallaghy and Norton, 2000; Pallaghy et al., 2003). Strategies for designing (non-peptidic) mimetics are described in Section 6.

More recently the structure-function relationship in MVIIA has been explored in greater detail by Nielsen et al. (1999a,b). It was confirmed that loops 2 and 4 make the greatest contributions to calcium channel subtype specificity and that the chirality at Tyr13 was critical for activity. It is clear that Lys2 is less important for activity in MVIIA than in GVIA, and Nielsen et al. have proposed that there is a salt bridge between Lys2 and Asp14 in MVIIA, giving Lys2 a structural role. Such a salt bridge cannot exist in GVIA, and although the possibility of a hydrogen bond between Lys2 and Asn14 in GVIA has been suggested (Nielsen et al., 2000), there is no evidence for this. Nevertheless, replacing Lys2 in GVIA with Ala did cause some local conformational disturbances detectable by NMR (Lew et al., 1997). Nielsen et al. (2000) also noted that the poor correlation between functional assays and binding assays noted by Lew et al. (1997) was improved if Tyr22 of GVIA were omitted from the analysis, perhaps implying a role for this residue in interactions with peripheral as opposed to central N-type channels.

The key residues in MVIIC have been investigated by Sato and co-workers (Sasaki et al., 2000; Sato et al., 2000a,b; Minami et al., 2001) using various synthetic approaches, including MVIIC/MVIIA chimeras and combinatorial methods. Thr11, Tyr13 and Lys2 are important for binding of this polypeptide to both N- and P/Q-type channels, while Lys4 and Arg22 are important for P/Q-type binding. Mutating Thr11 to Leu (as found in MVIIA, see Figure 1) increased the preference for N-type channels. A series of MVIIC/MVIIA chimeras was also used by Nielsen et al. (1999) to determine which features of the structure and surface chemistry were important for N-type specificity.

The recently described CVID structure (Lewis et al., 2000) appears to be one of the most well-defined observed so far, and the relative orientation of loops 2 and 4 in this structure may resemble the conformation in complex with N-type channels, for which CVID is highly selective. In a recent comparison, Nielsen et al. (2000) showed that CVID was the most selective of the ω-conotoxins for N-type channels (see also Adams et al., 2003). GVIA ranked second, even though the structures shown in that paper suggest that the relative orientations of loops 2 and 4 are rather different in CVID and GVIA. Comparison of the CVID structure with a more recently determined and better defined structure of GVIA (Pallaghy and Norton, 1999) would be more informative, but a detailed comparison is not possible because the coordinates of the CVID structure have not been deposited in the Protein Data Bank and are not available from the authors. The MVIIA structure compared with CVID by Nielsen et al. (2000) is an older structure than the more representative structure described by Atkinson et al. (2000), which is the basis for the comparison made by Lewis et al. (2000).

The L-type blocker TxVII is unusual in several respects: Gly5 and Tyr13 are replaced by Asp5 and Leu13, respectively, and loop 4 contains only three residues (Figure 1). Even though its structure is similar to those of the other ω-conotoxins (Figure 6), its surface is much more hydrophobic. It has been suggested that TxVII may bind to the L-type channel at a similar site to that occupied by low molecular weight L-type blockers such as dihydropyridines, phenylalkylamines, and benzothiazepines, with Phe11 and Trp26 playing key roles (Kobayashi et al., 2000).

Some progress has also been made in mapping the GVIA binding site on the calcium channel, which appears to have contributions from all four domains (I-IV) of the α subunit of the channel, but most significantly from the putative extracellular loop between transmembrane helix S5 and the pore region H5 of domain III (Ellinor et al., 1994). Additional residues within this region were shown recently to be important determinants of GVIA blockade (Feng et al., 2001). Substitution of Glu1332 with Arg slowed the time course of development of block while mutation of Gly1326 to Arg, Glu or Pro significantly decreased the time constant for development of the block. In the G1326P mutant channel activity was almost completely recovered following washout. It was concluded that Gly1326 may form a barrier which controls the access of ω-conotoxins to their blocking site in the outer vestibule of the channel pore and stabilizes the toxin-channel interaction. Modeling of the interaction of GVIA with loop regions of the calcium channel has been carried out (Doughty et al., 1995; 1998).

4. SPIDERS, SNAKES and OTHERS

4.1 Three-Dimensional Structures

Among the calcium channel blockers derived from spiders, solution structures have been determined for ω-Aga IVB (Yu et al., 1993; Adams et al., 1993; Reily et al., 1995), ω-Aga IVA (Reily et al., 1994; Kim et al., 1995b) and ω-grammotoxin SIA (Takeuchi et al., 2002). As noted previously (Pallaghy et al., 1994; Narasimhan et al., 1994; Norton and Pallaghy, 1998), ω-Aga IVA and IVB adopt ICK motif structures (Figure 8), with a

Figure legends for Color Plate 1 (following page)

Figure 5. Stereo view of the structure of GVIA showing the backbone (as a ribbon) and all side chains. The structure closest to the mean over the family of NMR-derived structures (Pallaghy and Norton, 1999) is shown. The disulfide bridges are shown in gold and the positions of the four reverse turns are indicated by Roman numerals I-IV. Residues found to be important for binding or activity in an alanine scan of GVIA (Lew et al., 1997) are shaded as follows: blue, >100-fold potency reduction in at least two assay types; cyan, >10-fold potency reduction in at least two assay types; pink, >10-fold potency reduction in at least one assay type. This diagram was created using MOLMOL (Koradi et al., 1996).

Figure 6. Comparison of structures of GVIA, MVIIA, MVIIC and TxVII. The closest structures to the geometric average are shown for GVIA (2CCO, Pallaghy and Norton, 1999), MVIIA (1DW4, Atkinson et al., 2000; this structure is in close agreement with that for MVIIA-Gly determined by Goldenberg et al., 2001), MVIIC (1CNN, Nielsen et al., 1999b) and TxVII (1F3K, Kobayashi et al., 2000). The structures were aligned over the backbone heavy atoms of the three β-strands and the six Cys residues. This diagram was created using MOLMOL (Koradi et al., 1996).

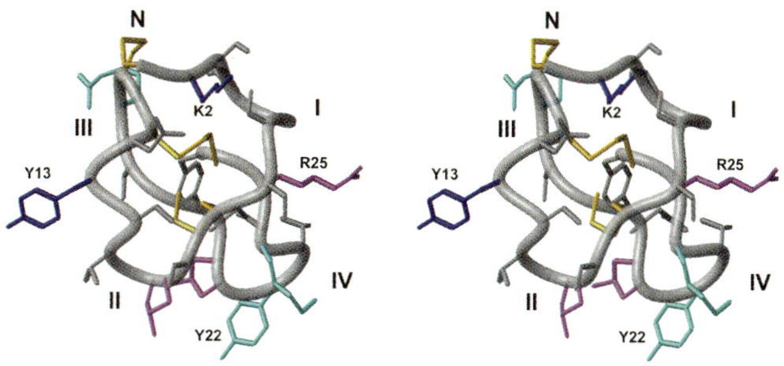

Figure 5.

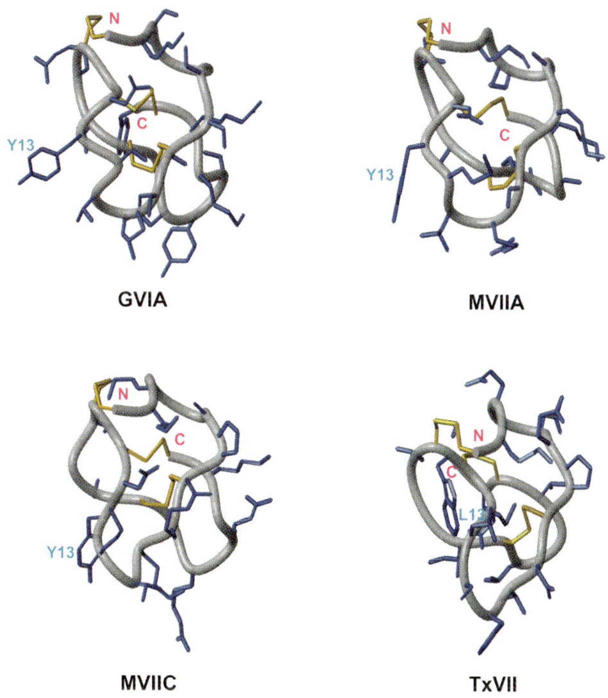

Figure 6.

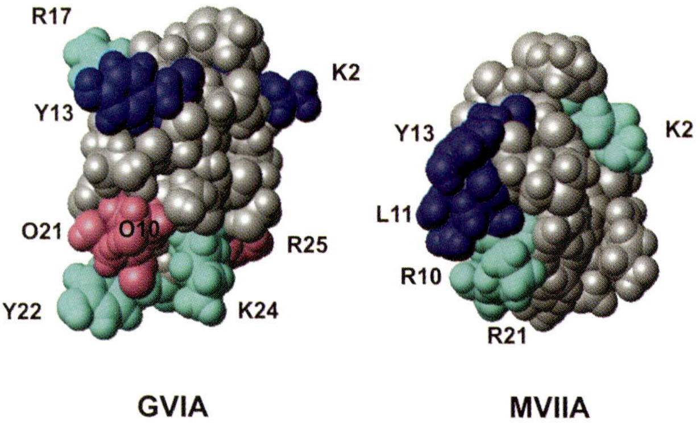

Figure 7.

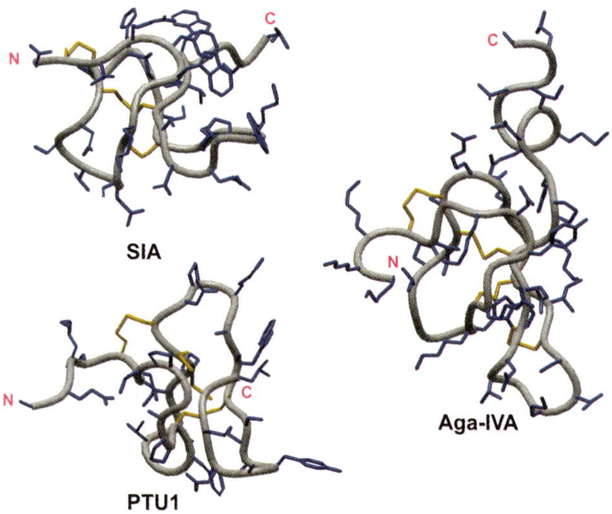

Figure 8.

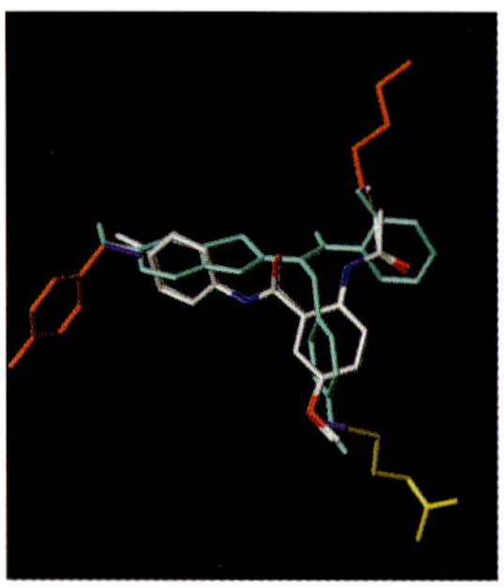

Figure 12a.

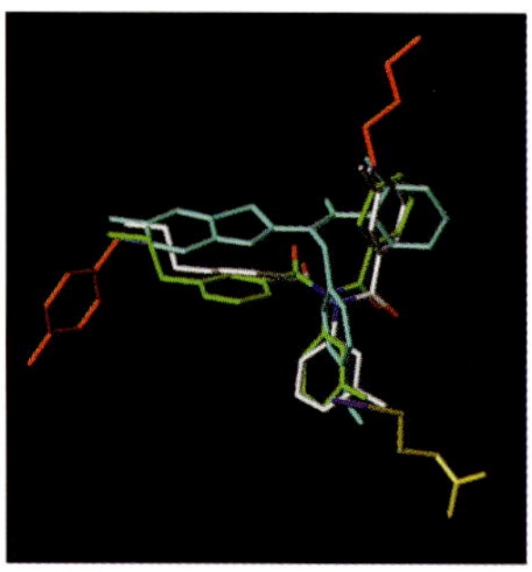

Figure 12b.

Figure legends for Color Plate 2 (preceding page)

Figure 7. CPK representation of the structures of GVIA and MVIIA. The structures, alignment and color coding are as described for GVIA in the captions to Figs. 5 and 6. This diagram was created using MOLMOL (Koradi et al., 1996).

Figure 8. Comparison of structures of ω-grammotoxin SIA, ω-Aga IVA, and Ptu1. The closest structures to the geometric average are shown for SIA (1KOZ, Takeuchi et al., 2002), ω-Aga IVA (1IVA, Reily et al., 1994), and Ptu1 (1I26, Bernard et al., 2001). The structures were approximately aligned over the backbone heavy atoms of the following pairs of Cys residues: SIA 15-16 and ω-Aga IVA and Ptu1 19-20. This diagram was created using MOLMOL (Koradi et al., 1996).

Figure 12. The K2-Y13-R-17 GVIA pharmacophore. K2 and Y13 are colored red, R17 yellow, and the three αC-βC bonds purple. In **(a)**, the designed benzothiazole scaffold of mimetic **1** is shown in light blue, superimposed on the GVIA pharmacophore, as is the designed anthranilamide scaffold of mimetic **2**, colored by atom type. The solid-state structure of the anthranilamide scaffold was determined and is superimposed on the theoretical structure, showing near identical conformation. In **(b)**, the benzothiazole is shown again but this time, the designed imide scaffold of **3** is shown in green, with its subsequently-determined solid-state conformation superimposed, colored by atom-type. Reprinted with permission from Baell et al. (2001).

fourth disulfide stabilizing loop 4, which is longer than in the ω-conotoxins (compare Figures 1 and 2). ω-Aga IVA and IVB also have a C-terminal extension of about 10 residues that it is not stabilized by a disulfide bridge and is disordered in solution. Ser46 in the disordered tail of ω-Aga IVB is in the D-configuration, but this was not known at the time of the structure determinations. Molecular dynamics simulations indicated that the switch from L- to D-Ser46 would affect the local conformation near the C-terminus (Kuwada et al., 1994).

The recently determined structure of ω-grammotoxin SIA shows that it also adopts the ICK motif (Figure 8). Its structure is very similar to that of hanatoxin1, which binds to the voltage sensing domains of K^+ and Ca^{2+} channels, although with higher affinity for K^+ channels. A detailed comparison of the two structures identified a conserved face containing a large hydrophobic patch surrounded by a number of basic residues (Takeuchi et al., 2002). It was suggested that slight differences in the surface shape resulting from the orientation of surface aromatic residues and the distribution of charged residues might explain differences in the binding affinity of these gating modifiers for voltage-gated Ca^{2+} and K^+ channels. Huwentoxin-I also adopts an ICK motif structure (Qu et al., 1997), as predicted from its amino acid sequence (Pallaghy et al., 1994).

Wang et al. (2001) recently described a novel family of insect-selective Ca^{2+} channel blockers. The structure of one of the toxins, ω-ACTX-Hv2a, reveals a highly structured, disulfide-rich core and a structurally disordered C-terminal extension that is essential for channel blocking activity. A synthetic analogue of ω-ACTX-Hv2a lacking residues 33-45 had the same core structure as the full-length toxin but was inactive against insect Ca^{2+} channels and did not competitively inhibit the activity of the native toxin. Weak structural/functional homology with ω-Aga-IVA/B led to the suggestion that these two toxin families might share a similar mechanism of action despite their different phylogenetic specificities.

Snakes elaborate a family of toxins about 60 residues in length that show high sequence similarity to Kunitz-type protease inhibitors such as BPTI. As expected, the solution structure of calcicludine (Gilquin et al., 1999; Nishio et al., 1999) is very similar to those of BPTI and dendrotoxin-I, a K^+ channel blocker. The different channel specificities for calcicludine and dendrotoxin-I must originate from differences in the chemistry and structure of the channel binding surfaces, with differences in the N-terminal region being potentially the most important. The structures of calciseptine or FS2, which are selective for L-type channels, have not yet been determined experimentally.

Among the scorpion toxins active on calcium channels, a structure is available only for maurocalcine (Mosbah et al., 2000). In keeping with its limited sequence similarity to spider toxins and the presence of three disulfides in a 1-4/2-5/3-6 pattern, it adopts an ICK fold. The binding site of its close relative imperatoxin A on the ryanodine receptor has been shown by cryo-electron microscopy to be about 11 nm away from the transmembrane pore (Samso et al., 1999). These toxins are thought to mimic the dihydropyridine receptor II-III loop.

Finally, we consider the structures of two classes of toxins from insects which have been shown to block Ca^{2+} channels. PMP-D2 from *Locusta migratoria* consists of a three-stranded antiparallel β-sheet (Mer et al., 1994) preceded by a reverse turn, inviting comparison with the canonical ω-conotoxin structure. Nevertheless, the disulfide patterns are not the same, nor are the overall topologies. Suggestions made by the authors about similar features of the surface of PMP-D2 and GVIA preceded the availability of alanine scan and other data on key residues in the ω-conotoxins and should be re-evaluated in the light of that substantial body of information. PMP-C has a similar sequence to PMP-D2, with 14 of its 36 residues (including the six half-cystines) identical. It adopts a prolate ellipsoid shape with a tertiary fold not observed previously in the family of small proteinase inhibitors (Mer et al., 1996). The proteinase binding loop is located in the C-terminal part of the molecule, between two half-cystines, and resembles that found in other small canonical proteinase inhibitors. Comparison of the PMP-C and PMP-D2 structures led to identification of two salt bridges in the latter.

Ptu1 is a toxin from the assassin bug *Peirates turpis* which binds reversibly to N-type calcium channels, albeit with lower affinity than MVIIA. As expected from its sequence similarity to the ω-conotoxins (Figures 1 and 4) it adopts an ICK fold, as shown in Figure 8, although only the second and third strands of the β-sheet are present, a result consistent with CD measurements that show a lower content of β-sheet than GVIA or MVIIA (Bernard et al., 2001). Comparison of the structure with those of GVIA and MVIIA suggested that Phe13 in Ptu1 might mimic Tyr13 in the ω-conotoxins. In addition, Ptu1 contains four basic residues, some of which probably mimic functionally important Lys and Arg residues in the ω-conotoxins. However, these putative similarities have not yet been tested by mutational analysis.

Although it is somewhat removed from the main theme of this chapter, it would be remiss not to mention the structure of KP4, a virally encoded toxin isolated from a fungus, which shows activity against calcium channels. The crystal structure of this 105-residue protein from the corn smut fungus, *Ustilago maydis*, shows it to be an α/β-sandwich with some structural features in common with the shorter scorpion α-toxins (Gu et al., 1995). This similarity led to an exploration of its ability to inhibit various cation channels, and thereby to the finding that it can block mammalian calcium channels. Subsequent studies showed that it specifically blocks L-type channels (Gage et al., 2002) although its biological target is most likely fungal calcium channels.

4.2 Structure-Function Relationships

ω-Aga IVA and IVB both have an abundance of positively charged side chains and a markedly hydrophobic tail. Three conserved Arg residues at positions 21, 23 and 39 constitute a large positive potential that might contribute to P-type channel binding (Reily et al., 1994). The N-terminal regions of ω-Aga IVA and IVB differ significantly (Figure 3), the former being quite basic and the latter acidic. These residues clearly cannot form part of the common channel binding surface of these two toxins, but they do appear to influence the kinetics of binding (Reily et al., 1994).

Yu et al. (1993) suggested that the disordered hydrophobic tail in ω-Aga IVB could anchor the toxin in the lipid bilayer, thereby facilitating its interaction with the channel. Consistent with this idea, Wang et al. (2001) showed that a synthetic analogue of ω-ACTX-Hv2a lacking residues 33-45 had the same core structure as the full-length toxin but was inactive against insect Ca^{2+} channels and did not competitively inhibit the activity of the native toxin. Furthermore, functional data on ω-Aga IVA presented by

Kim et al. (1995b) also suggested that the C-terminal tail played an important role in channel blockade. Thus, these two toxin families might share a similar mechanism of action despite their different phylogenetic specificities. It remains to be established whether the C-terminal tail interacts with the lipid bilayer or elements of the Ca^{2+} channel, or indeed both.

Generally speaking, there is far less published information available on the identity of key residues in other spider toxins or the snake and insect toxins described above. Suggestions about key residues or key regions of the molecular surface have been included with the description of their tertiary structures.

5. PHARMACOLOGICAL ASSAYS AND THERAPEUTIC APPLICATIONS

5.1 Introduction

The challenge in any bench-to-bedside therapeutics discovery program is to determine when a new entity becomes a realistic and achievable therapeutic target. Robust basic science with a large dose of disinterested financial support prepares the field for development and commercialization of an authentic new target. Frequently, however, the target, having been robustly defined, eludes the best chemistry because of the location of the target or perhaps because nature's secrets are locked away in a very specific and rigid chemical structure. In this section we review advances in taking highly selective N-type Ca^{2+} channel antagonists through robust in vitro tissue assays containing the Ca^{2+} channels of interest in their natural environment and onto in vivo whole animal tests of cardiovascular changes and the treatment of chronic pain. Preclinical pharmacological testing is the moment of truth for testing the concept of target and drug specificity in a controlled assay. These assays are selected because of their capacity to test certain hypotheses where we control the drug concentration and limit the complexity of the system under test, but in as natural an environment as possible. The inevitable test that follows the successful in vitro assays is administration in vivo. The challenge here is to keep pushing the science to overcome the limitations posed by the distribution of the specific target and the apparent need to stay closely aligned to the natural toxin ligand.

Calcium channels are current or potential targets for therapeutics directed against intractable pain, ischemic brain damage, migraine, and some forms of epilepsy and ataxia (McDonough et al., 2002). The discovery that GVIA was a highly selective N-type Ca^{2+} channel antagonist allowed the physiological role of these channels to be fully explored. Isolated tissue assays have been a robust and sensitive tool to determine the location and action of N-type channels on sympathetic nerves. GVIA at 1-10 nM can completely inhibit the twitch contraction response to the nerve-muscle electrical stimulation of the rat and mouse vas deferens, the contraction of rat mesenteric resistance artery, and tachycardia of mouse and rat right atria (Wright and Angus, 1996). Evidence that GVIA binds to N-type channels located only pre-junctionally was that GVIA at 1000 times higher concentration (10 μM) than necessary to completely block the release of transmitter (10 nM) was without any effect on the response to exogenously applied noradrenaline (Whorlow et al, 1996a). Our work indicates that in some sympathetic-nerve innervated tissues in some species, channels in addition to the

N-type may contribute to the influx of Ca^{2+} and subsequent release of transmitter. For example, in the mouse and rat vas deferens, when stimulated with high frequency (high intensity) a GVIA insensitive component (< 20%) of the response is blocked by ω-Aga IVA, a P-type channel blocker (≤ 100 nM), and by the non-selective N, P/Q-type channel blocker ω-conotoxin MVIIC. In the motor nerve-muscle assay of the mouse phrenic nerve hemidiaphragm, GVIA was completely inactive but ω-Aga IVA and MVIIC reduced the nerve response more than 80%, indicating a role for P/Q- but not N-type channels at the motor end plate (Wright and Angus, 1996).

Parasympathetic nerves such as the cardiac vagus are sensitive to GVIA. GVIA blocks vagal-nerve mediated bradycardia and negative inotropic responses in guinea pig atria (Serone and Angus, 1999) but less so in the rabbit atria (Wright and Angus, 1995). If N-type channels play an important role in peripheral autonomic nerve transmission demonstrated in isolated tissues, then GVIA would be expected to cause autonomic dysfunction if given into the systemic circulation.

Thus, intravenous injections of GVIA (3 or 10 μg/kg) in conscious rabbits caused regional bed vasodilatation and hypotension with tachycardia as the cardiac vagus is withdrawn and the sympathetic reflexes blocked (Pruneau and Angus, 1990). This sympatholytic action of GVIA, while sparing the vagus to some extent in the rabbit, was supported by the findings that the nasopharyngeal reflex, an acute vagal bradycardia in response to a noxious stimulus such as ammonia or cigarette smoke passed across the nares, and Bezold-Jarisch-like vagal bradycardia reflex, induced by bolus intravenous injections of serotonin, were completely unaffected by GVIA (Wright and Angus, 1995). These studies suggest that N-type channels may be important for only some autonomic reflexes (i.e., the baroreflex) with little effect on the peripheral vagal varicosities.

GVIA i.v. at 10 μg/kg caused a long lasting hypotension and blunting of the baroreceptor heart rate reflex for over 48 h. Repeated daily doses continued to block the reflex with little sign of tachyphylaxis. The blood pressure recovered somewhat towards pre-GVIA levels prior to the next dose, probably from major fluid shifts and activation of the renin-angiotensin system (Wright and Angus, 1995).

The administration of the angiotension-converting enzyme inhibitor enalaprilat or the AT_1-receptor antagonist losartan to conscious rabbits caused the blood pressure to fall further and faster after GVIA (10 μg/kg) than in the absence of the antagonists. This finding suggests that N-type channels are not involved in the action of converting enzyme or in neurohumoral secretion of angiotensin II. Importantly, this study highlights the powerful hypotensive action of GVIA and N-type channel inhibition when neurohumoral buffering is obviated. Postural hypotension is observed with ganglion blocking drugs and during early treatment with α_1-adrenoceptor antagonists such as prazosin, when the patient would faint on standing as the normal venoconstriction is impaired. In rabbits, an abrupt 90° tilt causes rapid recovery of blood pressure in response to the postural adaptation. After GVIA (10 μg/kg) the tilt response is impaired to the same degree as prazosin, suggesting that N-type channels on sympathetic venoconstrictor nerves are impaired (Hawkes et al, 1995).

Central (intracerebroventricular septum, icv) administration of GVIA (30 pmol/kg) caused a small progressive fall in blood pressure of ~20 mmHg over 48 h and only a slight attenuation of the baroreceptor-heart rate reflex. Perhaps the peptide does not readily access cardiovascular control centres given the modest effects and lack of

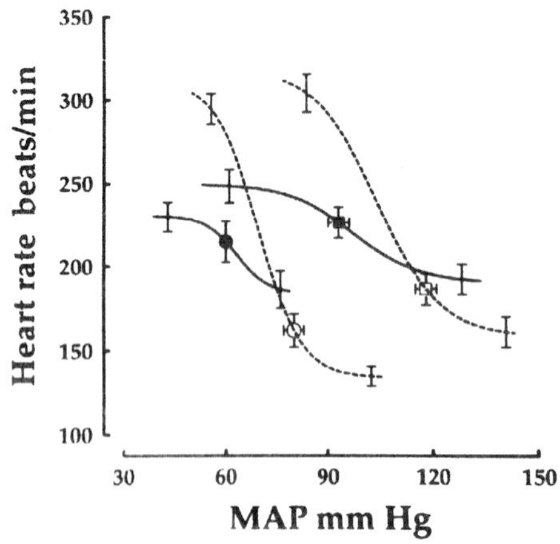

Figure 9. Average baroreceptor-heart rate reflex curves. Curves relate mean arterial pressure (MAP) to heart rate before (open symbols, dashed lines) and 2 h after (solid symbols, solid lines) administration of ω-conotoxin GVIA (10 μg/kg i.v. bolus) in normotensive (circles; n=4) and hypertensive (squares; n=6) conscious rabbits. The symbols on the curves represent the average resting values for MAP and heart rate. Error bars on the symbols are SEM (those not shown are within the symbol), and those on the curves represent the SEM of the lower plateau (right) and the heart rate range (left). Reprinted with permission from Wright and Angus (1995).

vagolytic activity (Whorlow et al, 1994; 1996b). Thus, in intact conscious rabbits, GVIA is a potent sympatholytic on the cardiovascular system when injected i.v., but has relatively modest and slow onset actions when given centrally.

5.2 Structure-Activity Relationships

When comparing the potencies of N-type channel blocking toxins it is important to allow sufficient time for penetration of the peptide to its site of action and equilibration of the concentration, which is usually indicated by a steady plateau response in in vitro tissue assays. We found that the GVIA Ala-scan analogues Gly5→D-Ala and Ser6→Ala, both at 10 μg/kg i.v., caused similar falls in blood pressure and tachycardia to those of GVIA (Lew et al., 1997). More recently, we compared the N-type channel inhibitors GVIA, MVIIA and CVID in three isolated tissue assays of atria, vas deferens and mesenteric artery. All three peptides were equipotent in the three assays with one exception: CVID was 10-fold less potent in the artery neuromuscular preparation compared with the vas deferens or atria assays. In addition, both CVID and MVIIA have faster onset and offset rates than GVIA, and both CVID and MVIIA can be reapplied with no loss of potency after washout of 30 minutes (Wright et al., 2000b). Structural factors may underlie the long onset/offset action of GVIA as it binds to the channel α_1-subunit. Indeed, when Tyr13 in GVIA was replaced by D-Tyr, the analogue could be rapidly washed out of the tissue despite the 17-fold loss of potency as an N-type antagonist (Flinn et al., 1999b).

Returning to the comparison of the three peptides CVID, GVIA and MVIIA in conscious rabbits, all three cause similar falls in blood pressure, increases in heart rate, and attenuation of some autonomic reflexes following i.v. but not intrathecal administration. CVID and MVIA would appear to offer some advantage over GVIA in having faster kinetics, which would aid adjustment of dose, especially if side effects were encountered. In addition, the apparent sparing of the sympathetic vasoconstrictor action by CVID in vitro was also observed in vivo by the short-lived (30 min) blockade of the orthostatic (tilt) reflex compared with the prolonged action (>90 min) by MVIA after the same dose (100 μg/kg i.v) (Wright et al., 2000a). These observations may point to subtle differences in N-type channel properties in different regions where CVID may have some selectivity not shared with MVIIA or GVIA (*cf* Adams et al., 2003). These cardiovascular studies highlight the dramatic sympatholytic actions of N-type blocking agents given systemically (Figure 9).

5.3 Analgesia

Chronic pain is the most common reason for seeking medical care. The world market in 2001 for analgesics was US$22 billion. In the USA alone, nearly 7 million people per annum are affected by neuropathic pain, with 10% of those suffering intractable pain. Patients with pain associated with cancer number 1.4 million. As a clinical problem, neuropathic pain arises from the nervous system itself when there is disease or injury affecting the central or peripheral nerves. Neuropathies may result from widespread disease (e.g., multiple sclerosis), localized trauma (stroke), systemic toxins (vinca alkaloids), systemic disorders (diabetes mellitus), or peripheral nerve damage (injury, trauma). A symptom of neuropathic pain is allodynia, described commonly as 'burning', 'tingling' or 'electric shocks'. Hyperalgesia follows nerve injury where the

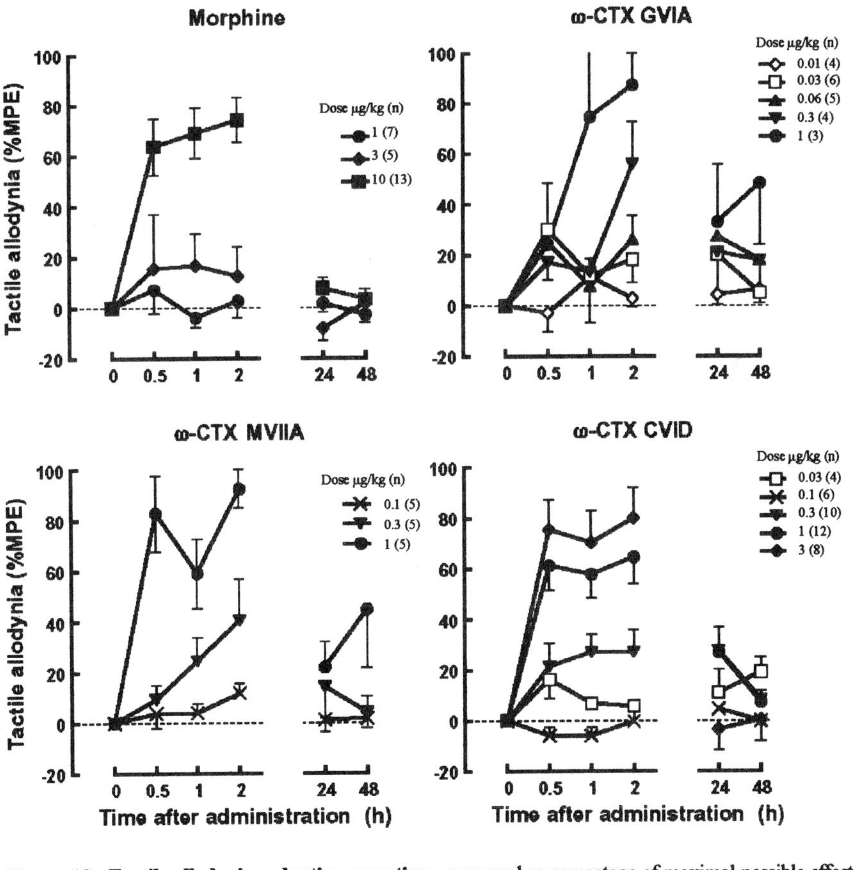

Figure 10. Tactile allodynia reduction over time, expressed as percentage of maximal possible effect (%MPE) (µg/kg IT). Data are shown only for treatment groups containing more than two animals. ω-ctx = ω-conotoxin. Reprinted with permission from Scott et al. (2002).

affected area becomes super-sensitive to a weak touch stimulus. N-type Ca^{2+} channels have been found by using specific antibodies to the α_{1B} subunit to be evenly distributed throughout all laminae of the rat dorsal horn, with diffuse fluorescence on the neuronal soma and punctate location in nerve terminals indicating synapses (Westenbroek et al., 1998). Using double labeling techniques, immunocytochemical evidence suggests that N-type channels control the release of substance P and glutamate (Westenbroek et al., 1998).

Opioid agonists (morphine) are effective for low intensity acute pain but are ineffective in many non-cancer states and neuropathic pain. A decrease in transmitter release in nociceptive neurons by blocking N-type channels is an attractive proposition, especially in combination with other analgesics. On this basis, Neurex Corporation (Menlo Park, CA) has conducted extensive clinical trials with MVIIA (SNX-111, ziconotide) for the treatment of intractable neuropathic pain. A case study described by Brose et al. (1997) illustrates the value of this therapy. 'A 43-year old man with left upper extremity pain due to brachial plexus avulsion was admitted to the Stanford Pain Management Service. This patient had a 23-year history of intractable pain, described as a constant burning and aching of his upper left extremity, with episodic lancinating pains. Brachial plexus avulsion had been confirmed at the time of injury by myelography. One year after the injury, above-the-elbow, left arm amputation was performed because of recurrent injury to the flail limb. This amputation did not alter the nature or severity of his pain in the phantom limb or the remaining stump." Continuous intrathecal infusion of MVIIA at 2 ng/kg/h resulted in complete loss of pain with no side effects.

Early trials in acute postoperative pain showed that higher doses (7 μg/h) of MVIIA intrathecally were associated with dizziness, blurred vision, nystagmus and sedation causing the drug to be discontinued after 24 h in 4 out of 6 patients. The remainder were given 0.7 μg/h and showed no adverse effects and required significantly less patient-controlled morphine than placebo (Atanassof et al., 2000). In a rat model of post-operative pain on the left hindpaw plantar region, intravenous ziconotide (3 μg/kg) had no effect on thermal pain threshold on the injured or normal hindpaw. Intrathecal ziconotide, however, was more potent and longer acting than intrathecal morphine (Wang et al., 2000a). In combination experiments, chronic (7 day) intrathecal morphine infusion caused tolerance that did not cross over to ziconotide (Wang et al, 2000b).

In rats with chronic neuropathic pain in a unilateral spinal nerve ligation model, all three ω-conotoxins (CVID, GVIA, MVIIA) were potent at attenuating tactile allodynia after intrathecal bolus injection. These peptides did not attenuate acute nociceptive responses to a heat stimulus in the tail flick assay. In contrast, morphine was effective in the tail flick assay and tactile allodynia assay, but 12-36 times less potent than the peptides in controlling the neuropathic pain (Table 1, Figure 10; Scott et al., 2002). These findings show that CVID had similar potency to MVIIA, but showed less toxicity in the therapeutic range in rats. In a recent Phase 1 trial, CVID (AM336, Amrad Corporation Ltd., Victoria, Australia) 0.6-6 μg/h intrathecal infusion was generally well tolerated at or below 3 μg/h and produced a reduction in severe pain from cancer in patients refractory to standard analgesics (Cousins et al., 2002).

Table 1. Summary of potencies of intrathecal administration of ω-conotoxins and morphine to attenuate tactile allodynia and to demonstrate moderate or severe toxicity.

	GVIA (μg/kg)	MVIIA (μg/kg)	CVID (μg/kg)	Morphine (μg/kg)
ED_{50} (μg/kg) (n)	0.12 (22)	0.32 (17)	0.36 (41)	4.36 (28)
C.I.	(0.06 – 0.24)	(0.23 – 0.45)	(0.27 – 0.48)	(2.9 – 6.5)
Potency relative to				
a) Morphine	36	13.6	12.1	1
b) CVID	3.0	1.1	1	
TD_{50} (μg/kg) (n)	0.60 (33)	0.68 (21)	3.5 (48)	
C.I.	(0.56 – 0.65)	(0.26 – 1.8)	(3.1 – 4.0)	
Relative toxicity	5.8	5.1	1	
Therapeutic index (TD_{50}: ED_{50})	5.0	2.1	9.7	

Values expressed with 95% C.I. (confidence interval).
ED_{50} is dose of agent (μg/kg) causing 50% MPE in tactile allodynia attenuation. Relative potency: a) is the ED_{50} of the peptides compared to morphine, b) is the ED_{50} of GVIA and MVIIA compared with CVID. TD_{50} is dose of agent causing moderate to severe behavioral toxicity. Relative toxicity is the TD_{50} of GVIA and MVIIA compared to CVID. (n) is number of rats – note that for TD_{50}, n includes sham neuropathy animals. See Scott et al. (2002).

The future therapy of neuropathic pain and some cancer pain may well be through the development of peptide or, preferably, non-peptide small organic N-type channel antagonists that may not need to be given by cassette and the intrathecal route. The key will be analgesic efficacy without cardiovascular side effects. There are other potential applications for N-type channel antagonists apart from analgesia. Ischemic injury, e.g., from stroke, causes an excessive influx of calcium into depolarized neurons that contributes to cerebral tissue damage. MVIIA given as a single i.v. bolus up to 24 h after global cerebral ischemia in rats, or by infusion for 24 h commencing 1 h after re-perfusion, protected the rat significantly. Neuroprotection appears to be long lasting and perhaps more efficacious if delayed by 6 h rather than 1 h after re-perfusion (Bowersox et al., 1997). Similarly, spinal cord ischaemia and resulting paraplegia, a major complication from surgical repair of the thoraco-abdominal aorta, can be prevented by MVIIA i.t. infusion (Burns et al., 1999).

6. MOLECULAR MIMICRY OF CALCIUM CHANNEL BLOCKING POLYPEPTIDES

6.1. Background

Structurally defined, pharmacologically active toxins such as the Ca^{2+} channel-blocking polypeptides have many advantages in therapeutic research. The relative ease with which analogs can be synthesized means that toxins can serve as useful biochemical probes of channel topology and function, as well as being therapeutically useful in their own right, as described above. Peptides can be radiolabeled and used in the screening of compound libraries for new ligands that are then optimized by medicinal chemistry. Numerous small molecule blockers of the N-type channel have been discovered in this way. These have been reviewed previously (Norton et al., 1999) and since then, more neuronal calcium channel antagonists have been discovered by similar means (Asakura et al., 1999; de los Rios et al., 2002; Dooley et al., 2002; Hainsworth et al., 2000; Hu et al., 2000a,b; Lukyanetz et al., 2002; Rhim et al., 2002; Ryder et al., 1999a,b; Ryder et al., 2000; Schelkum et al., 1999; Seko et al., 2002; Song et al., 2000; Sutton and Snutch 2002; Yan et al., 2002). The major advantages of small molecule ion channel blockers over their polypeptide counterparts are seen to be cheaper production costs and the potential for oral availability.

For the same reasons, the pharmacophores elucidated for these toxins present attractive targets for the rational design of small molecule mimetics. Wider interest in the successful mimicry of structurally defined polypeptides by rationally designed, smaller molecules stems from parallels to protein binding epitopes. The discovery of small molecule antagonists of protein-protein interactions represents a large unmet demand. This may be alleviated through the rational design of non-peptidic binding epitope mimetics, since much of the binding energy in protein-protein interactions may be provided by only a handful of residues (Clackson and Wells, 1995, Shapiro et al., 2000). The use of structural information on pharmacologically active toxins and protein-protein interfaces should in principle allow for the rational discovery of organic templates with minimal synthetic effort, but this has not been realized. Few examples of a successful design outcome exist because it is a deceptively difficult task to create synthesizable and drug-like scaffolds of discontinuous binding epitopes that are conformationally appropriate and biologically active. Serendipity has triumphed over design (Chen et al., 2002; Gadek et al., 2002; Guo et al., 2000; Hauel et al., 2002; Menzler et al., 2000; Naerum et al., 2002; Nakayama et al., 2001; Perrier et al., 2000; Roumelioti et al., 2002; Sarabu et al., 1997; Tilley et al., 1997; Yang et al., 2001). It is important to make the distinction between this type of peptidomimetic research and the highly successful – though usually synthetically intensive - “traditional” peptidomimetic research. The latter typically starts with a short, biologically active sequence, with or without structural information. We include beta-turn mimetics in this category (Ball and Alewood 1990, Burgess, 2001).

Amongst scaffold design results, the work of the Hamilton group (Ernst et al., 2002; Kutzki et al., 2002; Orner et al., 2001) and of Gadek et al. (2002) deserve particular comment. The terphenyl scaffold designed by the Hamilton group stands out in terms of simplicity of design and synthesis of non-peptide, 3-residue mimetics of the discontinuous binding epitope representing one face of an α-helix. From models it was deduced that an appropriately substituted terphenyl system could project mimetic side chains appropriately, based on knowledge of relative side chain orientations in α-helices. Synthesis and x-ray crystallographic analysis of a conformational representative supported this, before fully functionalised terphenyl mimetics of targeted α-helices were synthesized and tested. This scaffold has thrice produced potent mimetics of a hydrophobic helical face, and toxins containing helical segments important for binding are potential targets.

The potent LFA-1 antagonist reported by Gadek et al. (2002) provides hope that a nonlinear, discontinuous but contiguous, protein epitope (encompassing five residues spanning three different β-strands across the face of a protein surface, this being ICAM-1) could be mimicked structurally and functionally by a small molecule. Serendipity did, however, play an important part in the discovery of this mimetic.

The only successes with any pharmacologically active toxin mimetics since this time appear to be our own (Baell et al., 2001; Baell et al., 2002). We set out to synthesize molecular scaffolds designed specifically to target certain residues, to synthesize conformational representatives in order to verify conformation experimentally, and to finally obtain a biological "readout" in a fully functionalized mimetic. To our knowledge, this has not been achieved systematically for any discontinuous binding epitope of a polypeptide or protein that does not involve a well understood secondary structure such as the α-helix.

6.2. Rational Design of GVIA Mimetics

Residues K2 and Y13 were targeted for mimicry as these consistently showed up in alanine scanning experiments as most important for binding to the N-type calcium channel. In order to not be too simplistic, it was decided to target a third residue for mimicry, and R17 was chosen, also on the basis of alanine scanning data. Since the backbone conformation was better defined than the side chain conformation, scaffolds were designed to project side chains appropriately through mimicry of the αC-βC bond vector. Through an iterative process of interactive building, minimization and reference to the Cambridge Crystallographic Data Files, the benzothiazole derivative **1** was designed (Figure 11) as a K2-Y13-R17 GVIA mimetic. In order to see whether it was possible to discover unrelated scaffolds that might also be suitable, this interactive design process was repeated twice more to give the anthranilamide derivative **2** and the imide derivative **3** in Figure 11 as synthetic targets. The details of these design principles and processes can be found in Baell et al. (2001, 2002).

At the time of design, no suitable examples of the solid-state conformation of these systems had been lodged in the Cambridge Crystallographic Data Files. It was anticipated that the unfunctionalized scaffolds in mimetics **1-3** would be relatively simple to synthesize in reasonable amounts, would have a reasonable chance of being crystalline, and so would be suited to crystallization attempts to yield x-ray crystallographic-quality crystals. On the other hand, synthesis of the fully functionalized mimetics was anticipated to require substantially more effort, involving smaller quantities of material, which might or might not be suited to crystallization. The importance of verifying conformation before initiating potentially lengthy syntheses of the fully functionalised scaffolds lies in the fact that if the conformation is inappropriate, the chances of observing a binding event are likely to be minimal, and

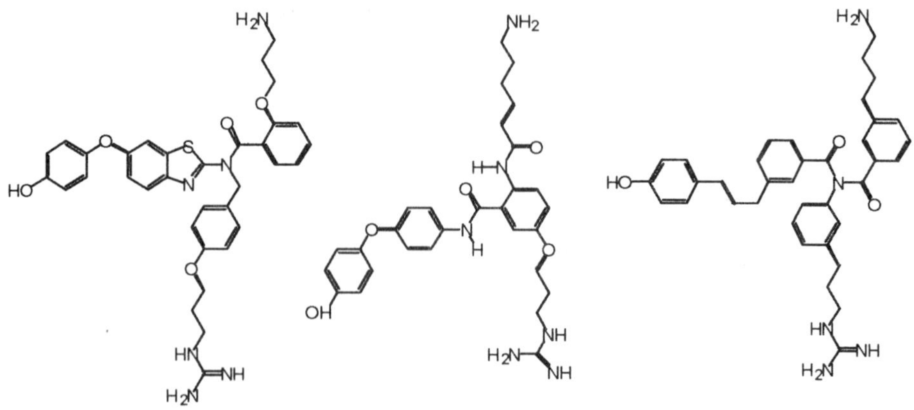

1 2 3

Figure 11. Putative mimetics of the K2-Y13-R17 GVIA pharmacophore, containing lysine-like, tyrosine-like and arginine-like side chains. The mimetic αC-βC bonds are emboldened.

synthesis might be a complete waste of time. Therefore, simpler conformational representatives were synthesized, crystallized, and subject to x-ray crystallographic analysis. These results are shown in Figure 12 for the anthranílamide system (a) and imide system (b). In both depictions, the benzothiazole system is light blue, K2 and Y13 are red, R17 is yellow, and K2-Y13-R17 α-β bonds are purple. The theoretical and solid-state conformations of the anthranilamide system in (a) are superimposed on each other, colored by atom type, and are essentially identical. The theoretical and solid-state conformations of the imide system in (b) are superimposed on each other, colored green or by atom type respectively, and differ only in the degree of twist about partially conjugated bonds. This result was seen as a highly successful design outcome. The experimental conformational behaviour of **1** is in line with theoretical predictions, but is more complex and has been discussed in detail elsewhere (Baell et al., 2001).

Having ascertained that the designed scaffolds would be conformationally appropriate, the fully functionalized mimetics **1** and **2** were synthesized and tested for biological activity. Radiolabeled ^{125}I-GVIA displacement studies on rat brain membrane preparations have recently yielded a Kd of around 1 μM for mimetic **1**, with mimetic **2** yet to be tested (Baell, Duggan, Lewis, Lok, Schroeder, unpublished results). Although less active in functional assays, these mimetics still inhibited N-type channel activity in stimulated rat vas deferens with IC_{50} values of approximately 100 μM and 70 μM respectively (Baell et al., 2002).

6.3. GVIA Mimetics as Potential Therapeutics

At this point, several important questions are raised. Firstly, what might be expected in terms of a dissociation constant of a K2-Y13-R17 GVIA mimetic based on alanine scanning data? The free energy of binding for a compound with a dissociation constant of 1 μM is around –34 kJ mol^{-1} at 295K. Upon binding, an energetic penalty of around

60 kJ mol^{-1} is incurred in loss of rotational and translational energy (Williams and Westwell, 1998), so that in mimetic **1**, it may be proposed that the three mimetic side chains provide around 95 kJ mol^{-1} worth of energy to the binding process. On the other hand, alanine scanning data suggest that these side chains should contribute in sum no more than approximately 25 kJ mol^{-1} to the binding process (Baell et al., 2001), in which case, mimetic **1** would not be observed to bind at all to the N-type calcium channel. Rudimentary though this thermodynamic treatment is, it seems likely that one or more of the mimetic side chains contribute energetically more than their counterparts in GVIA to the binding process. It can be envisaged, for example, that upon replacement of R17 in GVIA with an alanine residue, loss of a polar interaction could be partially compensated for through strengthened interaction of the ion channel with other nearby polar residues on GVIA. There is evidence that K24 and R25 in GVIA support each other in this manner, because the K24A-R25A double mutant is much less active than either one of the single point mutants (Flinn et al., 1999b). In this case, a positively charged side chain in a K24 mimetic, for example, could contribute much more to the binding process than suggested from the K24A alanine scanning data. The same phenomenon could occur with a hydrophobic interaction, and hydrophobic burial of the phenol ring in mimetic **1** could be driven more than Y23 in GVIA, being more solvent exposed.

A 3-point mimetic of ShK toxin was similarly more active than might be expected based on alanine scanning data and superficial thermodynamic analysis (Baell et al., 2002). Thus, at least for mimetics of pharmacologically active toxins, side chains may contribute more to the binding process than their counterparts in the native toxin, although extrapolation to mimetics of other polypeptide or protein binding epitopes is unwarranted. It would be quite possible for a polar residue on a polypeptide or protein surface, for example, to contribute substantially more than its mimetic counterpart in a small molecule if, in the polypeptide or protein, this residue is protected from solvation by flanking residues, akin to a protein "hot spot" (Bogan and Thorn, 1998), while the mimetic side chain would be expected to be highly exposed to solvent. Thus, positively and negatively cooperative effects could be rife on polypeptide and protein surfaces, and it is difficult to predict how active a mimetic might be based on alanine scanning data alone. This has been identified as a problem in the interpretation of site-directed mutagenesis experiments in protein-protein complexes (Vaughan et al., 1999; Shapiro et al., 2000).

Are these compounds really mimicking GVIA in binding to the N-type calcium channel? Apart from the observation that they not only displace GVIA binding but also inhibit the N-type calcium channel, several other lines of evidence suggest that this is indeed the case. First, complete removal of any one of the side chains in mimetic **1** leads to a large drop in activity (Baell, Duggan, Lewis, Lok, and Schroeder, unpublished results), showing that each side chain interacts with the N-type calcium channel as expected for a GVIA 3-residue mimetic. Second, the large difference between binding and functional data in mimetic **1** mirrors GVIA behavior and in fact the typical behavior of pharmacologically active toxins (Nielsen et al., 2000). Third, the pharmacodynamics of mimetic **1** mirrors GVIA in that the inhibitory effect on the N-type calcium channel does not reverse even with a prolonged wash (Figure 13). This is quite remarkable in such a simple GVIA mimetic, though is not necessarily a desirable property to have in a

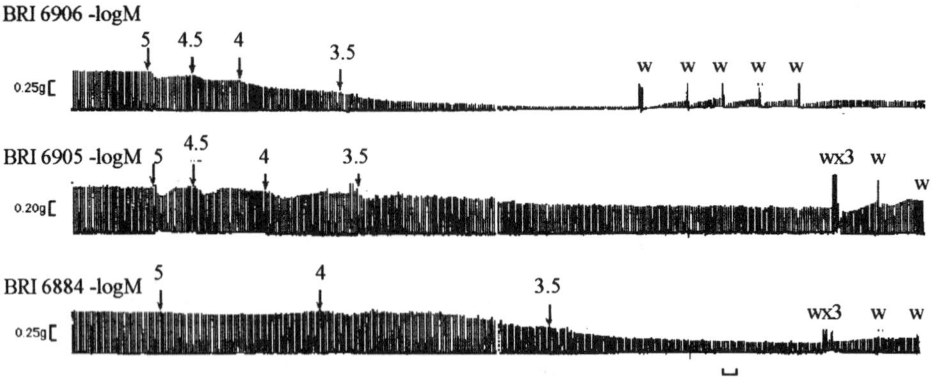

Figure 13. Inhibitory effect of increasing concentrations of mimetic 1 (=BRI6906) on median nerve-stimulated rat vas deferens, showing the pseudo-irreversible behaviour. BRI6905 is an analogue of BRI6906 where the guanidine group has been replaced by an amine, while in BRI6884 the entire R17 mimetic side chain has been removed. w indicates wash and wx3 a triple wash.

drug, as it makes dosing studies potentially difficult (Nielsen et al., 2000; Smith et al., 2002). Parenthetically we note a recent report (Liang and Elmslie, 2002) that N-type channel blockade by GVIA is rapid and reversible in the absence of divalent cations, suggesting that the issue of reversibility for mimetics such as **1** should be investigated under a range of conditions.

Could better results be obtained using more conventional structure-based approaches such as database mining? To test this, databases of available compounds were searched for those that contained the K2-Y13-R17 GVIA pharmacophore, but no hits were found (Baell et al., 2001). Database diversity was simply insufficient.

The fourth question that arises is whether these are indeed lead compounds. While falling outside some generally accepted criteria for lead-likeness in terms of being slightly too large, for example (Hann et al., 2001, Oprea et al., 2001), these mimetics are highly optimizable, which is a valued feature in a lead compound. By design they contain synthetic junctions so that the scaffold could be constructed and adorned with optimized side chains in a combinatorial sense. Not being restricted to native amino acid side chains, new interactions with the N-type channel may be created that are not utilized by GVIA. Side chain building blocks, once constructed, could be used on alternative scaffolds, saving time. It is to be remembered that 17 kJ mol^{-1} of free energy is sufficient to increase binding 10^3-fold. This could be supplied by a small degree of conformational constraint and a few extra favorable electrostatic and/or hydrophobic interactions.

One reason to aspire to lead-likeness criteria is to maximize the chances of obtaining orally active therapeutics after optimization. Systemically acting N-type blockers could cause peripherally-mediated hypotensive side effects (McGuire et al., 1997, Nielsen et al., 2000, Norton et al., 1999, Smith et al., 2002, Wright et al., 2000a,b). That this apparently has not hampered a substantial effort by the pharmaceutical industry to find orally active blockers of the N-type calcium is probably more a testament to the market size for antinociceptives and the uncertainty as to how a pharmacological profile in vitro may manifest itself in vivo. The N-type channel blocker reported by Pfizer constitutes such an example (Song et al., 2000). Nevertheless, it may be an advantage for antagonism of the N-type channel to remain purely central. This might be realized in a dicationic mimetic such as **1** if it were combined with emerging brain-targeting delivery methods (Bodor and Buchwald. 2002; Pardridge. 2002).

Are these mimetics selective, or could they be made selective? This important question is perhaps the most difficult to answer, and is linked to the question of optimization. Mimetics **1** and **2** have yet to be assessed for selectivity, though unless they mimic residues responsible for selectivity in GVIA - which they don't - it is probably unreasonable to expect them to be selective for N-type channels over, for example, P/Q-type channels. This selectivity would be therapeutically necessary. N-type selectivity in the ω-conotoxins is thought to have its basis in loops 2 and 4 (Nielsen et al., 2000; see Section 3.3). Loop 4 is near loop 2, which includes Y13. One way of obtaining selective mimetics might be to extend mimicry to include those residues in loops 2 and 4 that are believed to be important for interaction with the N-type channel and not the P/Q-type channel. This would, however, make any resulting mimetic less "drug-like" in the small molecule sense, if mimicry is to already include the disparate residues K2, Y13 and R17. Alternatively, selectivity may arise in the optimization step, since modified mimetics may utilize interactions not possible in GVIA or analogs, which are restricted in choice of side chain. GVIA presumably contacts a much greater surface area of the N-type channel than would a small molecule mimetic. A mimetic should in principle be able to access many more alternative molecular surfaces of the channel vestibule as potential inhibitory epitopes than would GVIA.

6.4 Prospects for Toxin Mimetics

Interactive molecular design has allowed for the relatively easy discovery of compounds that mimic three binding residues in the pharmacologically active toxins GVIA and ShK toxin. Most toxins are likely to have several relevant 3-point pharmacophores, providing some versatility in choice of mimetic size and physicochemical properties. These characteristics relate to how far apart the targeted residues are and the nature of their side chains. Three residues presenting a contiguous surface and containing not more than one charged side chain are likely to give rise to smaller mimetics that have a good chance of oral availability. By contrast, the GVIA mimetics discussed herein target three distant residues, two of which are positively charged at physiological pH.

The focus on mimicking bond projection of the side chain junction alleviates the need for detailed information on the side chain conformation in the bound state. By targeting only three residues, first generation mimetics can be kept simpler and more synthesizable, but necessarily would be expected to bind relatively weakly. Building in optimizability, therefore, is essential in this type of mimetic. While initial scaffold discovery was previously the rate-limiting step in mimetic development, in our approach the rate-limiting step is now the optimization of potency and selectivity to therapeutically relevant levels. This has yet to be attempted, and success in this endeavor will be necessary for this type of toxin mimetic to progress from representing academically interesting moieties to a competitive approach to drug discovery.

7. CONCLUDING REMARKS

Naturally occurring polypeptide and protein blockers of voltage-gated Ca^{2+} channels are found throughout Nature. They come in a variety of shapes and sizes, although for the most part are in the size range more appropriately described as peptides or polypeptides rather than proteins. This presumably reflects their highly specific function of channel blockade or channel modulation, which can be effected by a small cluster of residues presented on a rather rigid scaffold. Multi-domain proteins capable of allosteric interactions are not required. A limited range of scaffolds has been found to date, with the 'inhibitor cystine knot' motif being one of the more common structures utilized.

This chapter has summarized the sequences and three-dimensional structures of polypeptides and proteins active on voltage-gated calcium channels. The pharmacological consequences and potential therapeutic benefits of channel blockade are described in detail for one particularly abundant and active class of toxins, the ω-conotoxins. One representative of this class, ziconotide, has undergone extensive clinical evaluation for the treatment of chronic intractable pain, and others are at earlier stages of clinical development. The importance of these therapeutic indications and the potential limitations of polypeptides as drugs has stimulated attempts to mimic the activity of these polypeptides in small non-peptidic analogues. Progress towards this goal is documented, and this example is used to illustrate some of the broader issues that have to be dealt with in such attempts. These approaches, in conjunction with emerging computational tools such as *in silico* screening, will hopefully lead to small molecule mimetics of toxins that are useful therapeutically. Currently *in silico* approaches are restricted to mimicking the toxin ligands, but we look forward with much anticipation to a high-resolution crystal structure of a voltage-gated Ca^{2+} channel, not only as a target for *in silico* screening but also as a major step forward in our understanding of the molecular basis for channel function and toxin action.

ACKNOWLEDGMENTS

We thank Jennifer Sabo for assistance in preparation of the sequence and structure figures and Pierre Escoubas for helpful comments.

REFERENCES

Adams, D. J., Smith, A. B., Schroeder, C. I., Yasuda T., Lewis R. J., 2003, ω-conotoxin CVID inhibits a pharmacologically distinct voltage-sensitive calcium channel associated with transmitter release from preganglionic nerve terminals, *J Biol Chem* **278**:4057-4062.

Adams, M. E., Bindokas, V. P., Hasegawa, L. and Venema, V. J., 1990, ω-Agatoxins: novel calcium channel antagonists of two subtypes from funnel web spider (*Agelenopsis aperta*) venom, *J Biol Chem* **265**:861-867.

Adams, M. E., Mintz, I. M., Reily, M. D., Thanabal, V. and Bean, B. P., 1993, Structure and properties of ω-agatoxin IVB, a new antagonist of P-type calcium channels, *Mol Pharmacol* **44**:681-688.

Asakura, K., Kanemasa, T., Minagawa, K., Kagawa, K. and Ninomiya, M., 1999, The nonpeptide alpha-eudexp6l from *Juniperus virginiana* Linn. (*Cupressaceae*) inhibits ω-agatoxin IVA-sensitive Ca^{2+} currents and synaptosomal $^{45}Ca^{2+}$ uptake, *Brain Res* **823**:169-176.

Atanassoff, P. G., Hartmannsgruber, M. W., Thrasher, J., Wermeling, D., Longton, W., Gaeta, R., Singh, T., Mayo, M., McGuire, D. and Luther, R. R., 2000, Ziconotide, a new N-type calcium channel blocker, administered intrathecally for acute postoperative pain, *Reg Anesth Pain Med* **25**:274-278.

Atkinson RA, Kieffer B, Dejaegere A, Sirockin F, and Lefevre JF, 2000, Structural and dynamic characterization of ω-conotoxin MVIIA: the binding loop exhibits slow conformational exchange, *Biochemistry* **39**:3908-19.

Baell, J. B., Forsyth, S. A., Gable, R. W., Norton, R. S. and Mulder, R. J., 2001, Design and synthesis of type-III mimetics of ω-conotoxin GVIA, *J Comput Aided Mol Des* **15**:1119-1136.

Baell, J. B., Harvey, A. J. and Norton, R. S., 2002, Design and synthesis of type-III mimetics of ShK toxin, *J Comput Aided Mol Des* **16**:245-262.

Balaji, R. A., Ohtake, A., Sato, K., Gopalakrishnakone, P., Kini, R. M., Seow, K. T. and Bay, B. H., 2000, λ-conotoxins, a new family of conotoxins with unique disulfide pattern and protein folding. Isolation and characterization from the venom of *Conus marmoreus*, *J Biol Chem* **275**:39516-39522.

Ball, J. B. and Alewood, P. F., 1990, Conformational constraints: nonpeptide beta-turn mimics, *J Mol Recognit* **3**:55-64.

Basus, V. J., Nadasdi, L., Ramachandran, J. and Miljanich, G. P., 1995, Solution structure of ω-conotoxin MVIIA using 2D NMR spectroscopy, *FEBS Lett* **370**:163-169.

Bernard, C., Corzo, G., Mosbah, A., Nakajima, T. and Darbon, H., 2001, Solution structure of Ptu1, a toxin from the assassin bug *Peirates turpis* that blocks the voltage-sensitive calcium channel N-type, *Biochemistry* **40**:12795-12800.

Bindokas, V. P. and Adams, M. E., 1989, ω-Aga-I: a presynaptic calcium channel antagonist from venom of the funnel web spider, Agelenopsis aperta, *J Neurobiol* **20**:171-188.

Bindokas, V. P., Venema, V. J. and Adams, M. E., 1991, Differential antagonism of transmitter release by subtypes of ω-agatoxins, *J Neurophysiol* **66**:590-601.

Blaber, M., Baase, W. A., Gassner, N. and Matthews, B. W., 1995, Alanine scanning mutagenesis of the α-helix 115-123 of phage T4 lysozyme: effects on structure, stability and the binding of solvent, *J Mol Biol* **246**:317-330.

Bodor, N., and Buchwald, P., 2002, Barriers to remember: brain-targeting chemical delivery systems and Alzheimer's disease, *Drug Discov Today* **7**:766-774.

Bogan, A. A. and Thorn, K. S., 1998, Anatomy of hot spots in protein interfaces, *J Mol Biol* **280**:1-9.

Boigegrain, R. A., Mattras, H., Brehelin, M., Paroutaud, P. and Coletti-Previero, M. A., 1992, Insect immunity: two proteinase inhibitors from hemolymph of *Locusta migratoria*, *Biochem Biophys Res Commun* **189**:790-793.

Bowersox, S. S., Singh, T., and Luther, R. R., 1997, Selective blockade of N-type voltage-sensitive calcium channels protects against brain injury after transient focal cerebral ischemia in rats, *Brain Res* **747**:343-347.

Branton, W. D., Rudnick, M. S., Zhou, Y., Eccleston, E. D., Fields, G. B. and Bowers, L. D., 1993, Fatty acylated toxin structure, *Nature* **365**:496-497.

Brose, W. G., Gutlove, D. P., Luther, R. R., Bowersox, S. S., and McGuire, D., 1997, Use of intrathecal SNX-111, a novel, N-type, voltage-sensitive, calcium channel blocker, in the management of intractable brachial plexus avulsion pain, *Clin J Pain* **13**:256-259.

Burgess, K., 2001, Solid-phase syntheses of β-turn analogues to mimic or disrupt protein-protein interactions, *Acc Chem Res* **34**:826-835.

Burns, L. H., Jin, Z. and Bowersox, S. S., 1999, The neuroprotective effects of intrathecal administration of the selective N-type calcium channel blocker ziconotide in a rat model of spinal ischemia, *J Vasc Surg* **30**:334-343.

Cassola, A. C., Jaffe, H., Fales, H. M., Castro Afeche, S., Magnoli, F. and Cipolla-Neto, J., 1998, ω-phonetoxin-IIA: a calcium channel blocker from the spider *Phoneutria nigriventer*, *Pflugers Arch* **436**:545-552.

Chen, L., Tilley, J., Trilles, R. V., Yun, W., Fry, D., Cook, C., Rowan, K., Schwinge, V. and Campbell, R., 2002, N-acyl-L-phenylalanine derivatives as potent VLA-4 antagonists that mimic a cyclic peptide conformation, *Bioorg Med Chem Lett* **12**:137-140.

Chuang, R. S., Jaffe, H., Cribbs, L., Perez-Reyes, E. and Swartz, K. J., 1998, Inhibition of T-type voltage-gated calcium channels by a new scorpion toxin, *Nat Neurosci* **1**:668-674.

Civera, C., Vazquez, A., Sevilla, J. M., Bruix, M., Gago, F., Garcia, A. G. and Sevilla, P., 1999, Solution structure determination by two-dimensional ^{1}H NMR of ω-conotoxin MVIID, a calcium channel blocker peptide, *Biochem Biophys Res Commun* **254**:32-35.

Clackson, T. and Wells, J. A., 1995, A hot spot of binding energy in a hormone-receptor interface, *Science* **267**:383-6.

Cordeiro Mdo, N., de Figueiredo, S. G., Valentim Ado, C., Diniz, C. R., von Eickstedt, V. R., Gilroy, J. and Richardson, M., 1993, Purification and amino acid sequences of six Tx3 type neurotoxins from the venom of the Brazilian 'armed' spider *Phoneutria nigriventer*, *Toxicon* **31**:35-42.

Corzo, G., Adachi-Akahane, S., Nagao, T., Kusui, Y. and Nakajima, T., 2001, Novel peptides from assassin bugs (Hemiptera: Reduviidae): isolation, chemical and biological characterization, *FEBS Lett* **499**:256-261.

Cousins, M. J., Goucke, C., Cher, L. M., Brooker, C. D., Amor, P. E., and Crump, D. E., A phase 1 clinical trial of AM336, a novel N-type calcium channel blocker. In, *Abstracts of 10th World Congress on Pain*, August 17-22, 2002, San Diego, CA, USA, IASP Press, Seattle, p. 201.

Dauplais, M., Lecoq, A., Song, J., Cotton, J., Jamin, N., Gilquin, B., Roumestand, C., Vita, C., de Medeiros, C. L., Rowan, E. G., Harvey, A. L. and Menez, A., 1997, On the convergent evolution of animal toxins. Conservation of a diad of functional residues in potassium channel-blocking toxins with unrelated structures, *J Biol Chem* **272**:4302-4309.

Davis, J. H., Bradley, E. K., Miljanich, G. P., Nadasdi, L., Ramachandran, J. and Basus, V. J., 1993, Solution structure of ω-conotoxin GVIA using 2-D NMR spectroscopy and relaxation matrix analysis, *Biochemistry* **32**:7396-7405.

de los Rios, C., Marco, J. L., Carreiras, M. D., Chinchon, P. M., Garcia, A. G. and Villarroya, M., 2002, Novel tacrine derivatives that block neuronal calcium channels, *Bioorg Med Chem* **10**:2077-2088.

de Weille, J. R., Schweitz, H., Maes, P., Tartar, A. and Lazdunski, M., 1991, Calciseptine, a peptide isolated from black mamba venom, is a specific blocker of the L-type calcium channel, *Proc Natl Acad Sci U S A* **88**:2437-2440.

Diniz, C. R., Cordeiro Mdo, N., Junor, L. R., Kelly, P., Fischer, S., Reimann, F., Oliveira, E. B. and Richardson, M., 1990, The purification and amino acid sequence of the lethal neurotoxin Tx1 from the venom of the Brazilian 'armed' spider *Phoneutria nigriventer*, *FEBS Lett* **263**:251-253.

Dooley, D. J., Donovan, C. M., Meder, W. P. and Whetzel, S. Z., 2002, Preferential action of gabapentin and pregabalin at P/Q-type voltage-sensitive calcium channels: inhibition of K^+-evoked [^{3}H]-norepinephrine release from rat neocortical slices, *Synapse* **45**:171-190.

Dos Santos, R. G., Van Renterghem, C., Martin-Moutot, N., Mansuelle, P., Cordeiro, M. N., Diniz, C. R., Mori, Y., De Lima, M. E. and Seagar, M., 2002, *Phoneutria nigriventer* ω-phonetoxin IIA blocks the Ca_V2 family of calcium channels and interacts with ω-conotoxin-binding sites, *J Biol Chem* **277**:13856-13862.

Doughty, S. W., Blaney, F. E., Orlek, B. S. and Richards, W. G., 1998, A molecular mechanism for toxin block in N-type calcium channels, *Protein Eng* **11**:95-99.

Doughty, S. W., Blaney, F. E. and Richards, W. G., 1995, Models of ion pores in N-type voltage-gated calcium channels, *J Mol Graph* **13**:342-348.

Ellinor, P. T., Zhang, J. F., Horne, W. A. and Tsien, R. W., 1994, Structural determinants of the blockade of N-type calcium channels by a peptide neurotoxin, *Nature* **372**:272-275.

Ernst, J. T., Kutzki, O., Debnath, A. K., Jiang, S., Lu, H. and Hamilton, A. D., 2002, Design of a protein surface antagonist based on alpha-helix mimicry: inhibition of gp41 assembly and viral fusion, *Angew Chem Int Ed Engl* **41**:278-281.

Ertel, E. A., Warren, V. A., Adams, M. E., Griffin, P. R., Cohen, C. J. and Smith, M. M., 1994, Type III ω-agatoxins: a family of probes for similar binding sites on L- and N-type calcium channels, *Biochemistry* **33**:5098-5108.

Fajloun, Z., Kharrat, R., Chen, L., Lecomte, C., Di Luccio, E., Bichet, D., El Ayeb, M., Rochat, H., Allen, P. D., Pessah, I. N., De Waard, M. and Sabatier, J. M., 2000, Chemical synthesis and characterization of maurocalcine, a scorpion toxin that activates Ca^{2+} release channel/ryanodine receptors, *FEBS Lett* **469**:179-185.

Farr-Jones, S., Miljanich, G. P., Nadasdi, L., Ramachandran, J. and Basus, V. J., 1995, Solution structure of ω-conotoxin MVIIC, a high affinity ligand of P-type calcium channels, using ^{1}H NMR spectroscopy and complete relaxation matrix analysis, *J Mol Biol* **248**:106-124.

Feng, Z. P., Hamid, J., Doering, C., Bosey, G. M., Snutch, T. P. and Zamponi, G. W., 2001, Residue Gly^{1326} of the N-type calcium channel α_{1B} subunit controls reversibility of ω-conotoxin GVIA and MVIIA block, *J Biol Chem* **276**:15728-15735.

Flinn, J. P., Pallaghy, P. K., Lew, M. J., Murphy, R., Angus, J. A. and Norton, R. S., 1999a, Role of disulfide bridges in the folding, structure and biological activity of ω-conotoxin GVIA, *Biochim Biophys Acta* **1434**:177-190.

Flinn, J. P., Pallaghy, P. K., Lew, M. J., Murphy, R., Angus, J. A. and Norton, R. S., 1999b, Roles of key functional groups in ω-conotoxin GVIA synthesis, structure and functional assay of selected peptide analogues, *Eur J Biochem* **262**:447-455.

Gadek, T. R., Burdick, D. J., McDowell, R. S., Stanley, M. S., Marsters, J. C., Jr., Paris, K. J., Oare, D. A., Reynolds, M. E., Ladner, C., Zioncheck, K. A., Lee, W. P., Gribling, P., Dennis, M. S., Skelton, N. J., Tumas, D. B., Clark, K. R., Keating, S. M., Beresini, M. H., Tilley, J. W., Presta, L. G. and Bodary, S. C., 2002, Generation of an LFA-1 antagonist by the transfer of the ICAM-1 immunoregulatory epitope to a small molecule, *Science* **295**:1086-1089.

Gage, M. J., Rane, S. G., Hockerman, G. H. and Smith, T. J., 2002, The virally encoded fungal toxin KP4 specifically blocks L-type voltage-gated calcium channels, *Mol Pharmacol* **61**:936-944.

Garcia, E., Scanlon, M. and Naranjo, D., 1999, A marine snail neurotoxin shares with scorpion toxins a convergent mechanism of blockade on the pore of voltage-gated K channels, *J Gen Physiol* **114**:141-157.

Gasparini, S., Danse, J. M., Lecoq, A., Pinkasfeld, S., Zinn-Justin, S., Young, L. C., de Medeiros, C. C., Rowan, E. G., Harvey, A. L. and Menez, A., 1998, Delineation of the functional site of α-dendrotoxin. The functional topographies of dendrotoxins are different but share a conserved core with those of other Kv1 potassium channel-blocking toxins, *J Biol Chem* **273**:25393-25403.

Gilquin, B., Lecoq, A., Desne, F., Guenneugues, M., Zinn-Justin, S. and Menez, A., 1999, Conformational and functional variability supported by the BPTI fold: solution structure of the Ca^{2+} channel blocker calcicludine, *Proteins* **34**:520-532.

Goldenberg, D. P., Koehn, R. E., Gilbert, D. E. and Wagner, G., 2001, Solution structure and backbone dynamics of an ω-conotoxin precursor, *Protein Sci* **10**:538-550.

Gu, F., Khimani, A., Rane, S. G., Flurkey, W. H., Bozarth, R. F. and Smith, T. J., 1995, Structure and function of a virally encoded fungal toxin from *Ustilago maydis*: a fungal and mammalian Ca^{2+} channel inhibitor, *Structure* **3**:805-814.

Guo, Z.-X., Cammidge, A. N., and Horwell, D. C., 2000. Dendroid peptide structural mimetics of ω-conotoxin-MVIIA based on a 2(1H)-quinolinone core, *Tetrahedron* **56**:5169-5175.

Hagiwara, K., Sakai, T., Miwa, K., Kawai, N., and Nakajima, T., 1991, Complete amino-acid sequence of a new type of neurotoxin from the venom of the spider, *Agelena opulenta*, *Biomed Res* **11**:181-186.

Hainsworth, A. H., Stefani., A., Calabresi, P., Smith, T. W., and Leach, M. J., 2000, Sipatrigine (BW 619C89) is a neuroprotective agent and a sodium channel and calcium channel inhibitor, *CNS Drug Reviews* **6**:111-134.

Hann, M. M., Leach, A. R. and Harper, G., 2001, Molecular complexity and its impact on the probability of finding leads for drug discovery, *J Chem Inf Comput Sci* **41**:856-864.

Hauel, N. H., Nar, H., Priepke, H., Ries, U., Stassen, J. M. and Wienen, W., 2002, Structure-based design of novel potent nonpeptide thrombin inhibitors, *J Med Chem* **45**:1757-1766.

Hawkes, A. L., Angus, J. A. and Wright, C. E., 1995, ω-Conotoxin GVIA and prazosin, but not felodipine, cause postural hypotension in rabbits, *Clin Exp Pharmacol Physiol* **22**:711-716.

Heck, S. D., Siok, C. J., Krapcho, K. J., Kelbaugh, P. R., Thadeio, P. F., Welch, M. J., Williams, R. D., Ganong, A. H., Kelly, M. E., Lanzetti, A. J., Gray, W. R., Phillips, D., Parks, T. N., Jackson, H., Ahlijanian, M. K., Saccomano, N. A., and Volkmann, R. A., 1994, Functional consequences of posttranslational isomerization of Ser^{46} in a calcium channel toxin, *Science* **266**:1065-1068.

Hill, J. M., Alewood, P. F. and Craik, D. J., 1996, Three-dimensional solution structure of μ-conotoxin GIIIB, a specific blocker of skeletal muscle sodium channels, *Biochemistry* **35**:8824-8835.

Hill, J. M., Alewood, P. F. and Craik, D. J., 1997, Solution structure of the sodium channel antagonist conotoxin GS: a new molecular caliper for probing sodium channel geometry, *Structure* **5**:571-583.

Hill, J. M., Alewood, P. F. and Craik, D. J., 2000, Conotoxin TVIIA, a novel peptide from the venom of *Conus tulipa* 2. Three-dimensional solution structure, *Eur J Biochem* **267**:4649-4657.

Hu, L. Y., Ryder, T. R., Rafferty, M. F., Siebers, K. M., Malone, T., Chatterjee, A., Feng, M. R., Lotarski, S. M., Rock, D. M., Stoehr, S. J., Taylor, C. P., Weber, M. L., Miljanich, G. P., Millerman, E. and Szoke, B. G., 2000a, Neuronal N-type calcium channel blockers: a series of 4-piperidinylaniline analogs with analgesic activity, *Drug Des Discov* **17**:85-93.

Hu, L. Y., Ryder, T. R., Rafferty, M. F., Taylor, C. P., Feng, M. R., Kuo, B. S., Lotarski, S. M., Miljanich, G. P., Millerman, E., Siebers, K. M. and Szoke, B. G., 2000b, The discovery of [1-(4-dimethylamino-benzyl)-piperidin-4-yl]-[4-(3,3-dimethylbutyl)-phenyl]-(3-methyl-but-2-enyl)-amine, an N-type Ca^{2+} channel blocker with oral activity for analgesia, *Bioorg Med Chem* **8**:1203-1212.

Inui, T., Hagiwara, K., Nakajima, K., Kimura, T., Nakajima, T. and Sakakibara, S., 1992, Synthesis and disulfide structure determination of agelenin: identification of the carboxy-terminus as an amide form, *Pept Res* **5**:140-144.

Isaacs, N. W., 1995, Cystine knots, *Curr Opin Struct Biol* **5**:391-395.

Jones, R. M. and Bulaj, G., 2000, Conotoxins - new vistas for peptide therapeutics, *Curr Pharm Des* **6**:1249-1285.

Kalapothakis, E., Penaforte, C. L., Leao, R. M., Cruz, J. S., Prado, V. F., Cordeiro, M. N., Diniz, C. R., Romano-Silva, M. A., Prado, M. A., Gomez, M. V. and Beirao, P. S., 1998, Cloning, cDNA sequence analysis and patch clamp studies of a toxin from the venom of the armed spider (*Phoneutria nigriventer*), *Toxicon* **36**:1971-1980.

Kim, J. I., Konishi, S., Iwai, H., Kohno, T., Gouda, H., Shimada, I., Sato, K. and Arata, Y., 1995a, Three-dimensional solution structure of the calcium channel antagonist ω-agatoxin IVA: consensus molecular folding of calcium channel blockers, *J Mol Biol* **250**:659-671.

Kim, J. I., Takahashi, M., Ohtake, A., Wakamiya, A. and Sato, K., 1995b, Tyr13 is essential for the activity of ω-conotoxin MVIIA and GVIA, specific N-type calcium channel blockers, *Biochem Biophys Res Commun* **206**:449-454.

Kim, J. I., Takahashi, M., Ogura, A., Kohno, T., Kudo, Y. and Sato, K., 1994, Hydroxyl group of Tyr13 is essential for the activity of ω-conotoxin GVIA, a peptide toxin for N-type calcium channel, *J Biol Chem* **269**:23876-23878.

Kobayashi, K., Sasaki, T., Sato, K. and Kohno, T., 2000, Three-dimensional solution structure of ω-conotoxin TxVII, an L-type calcium channel blocker, *Biochemistry* **39**:14761-14767.

Kohno, T., Kim, J. I., Kobayashi, K., Kodera, Y., Maeda, T. and Sato, K., 1995, Three-dimensional structure in solution of the calcium channel blocker ω-conotoxin MVIIA, *Biochemistry* **34**:10256-10265.

Koradi, R., Billeter, M., and Wüthrich, K., 1996, MOLMOL: a program for display and analysis of macromolecular structures, *J Mol Graph* **14**:51-55, 29-32.

Kutzki, O., Park, H. S., Ernst, J. T., Orner, B. P., Yin, H. and Hamilton, A. D., 2002, Development of a potent Bcl-x(L) antagonist based on α-helix mimicry, *J Am Chem Soc* **124**:11838-11839.

Kuwada, M., Teramoto, T., Kumagaye, K. Y., Nakajima, K., Watanabe, T., Kawai, T., Kawakami, Y., Niidome, T., Sawada, K., Nishizawa, Y. and et al., 1994, ω-agatoxin-TK containing D-serine at position 46, but not synthetic ω-[L-Ser^{46}]agatoxin-TK, exerts blockade of P-type calcium channels in cerebellar Purkinje neurons, *Mol Pharmacol* **46**:587-593.

Lampe, R. A., Defeo, P. A., Davison, M. D., Young, J., Herman, J. L., Spreen, R. C., Horn, M. B., Mangano, T. J. and Keith, R. A., 1993, Isolation and pharmacological characterization of ω-grammotoxin SIA, a novel peptide inhibitor of neuronal voltage-sensitive calcium channel responses, *Mol Pharmacol* **44**:451-460.

Leão, R. M., Cruz, J. S., Diniz, C. R., Cordeiro, M. N. and Beirao, P. S., 2000, Inhibition of neuronal high-voltage activated calcium channels by the ω-*Phoneutria nigriventer* Tx3-3 peptide toxin, *Neuropharmacology* **39**:1756-1767.

Lew, M. J., Flinn, J. P., Pallaghy, P. K., Murphy, R., Whorlow, S. L., Wright, C. E., Norton, R. S. and Angus, J. A., 1997, Structure-function relationships of ω-conotoxin GVIA. Synthesis, structure, calcium channel binding, and functional assay of alanine-substituted analogues, *J Biol Chem* **272**:12014-12023.

Lewis, R. J., Nielsen, K. J., Craik, D. J., Loughnan, M. L., Adams, D. A., Sharpe, I. A., Luchian, T., Adams, D. J., Bond, T., Thomas, L., Jones, A., Matheson, J. L., Drinkwater, R., Andrews, P. R. and Alewood, P. F., 2000, Novel ω-conotoxins from *Conus catus* discriminate among neuronal calcium channel subtypes, *J Biol Chem* **275**:35335-35344.

Liang H., and Elmslie, K.S., 2002, Rapid and reversible block of N-type calcium channels (Ca_v 2.2) by ω-conotoxin GVIA in the absence of divalent cations. *J Neurosci* 15:8884-8890.

Liang, S. P., Zhang, D. Y., Pan, X., Chen, Q. and Zhou, P. A., 1993, Properties and amino acid sequence of huwentoxin-I, a neurotoxin purified from the venom of the Chinese bird spider *Selenocosmia huwena*, *Toxicon* **31**:969-978.

Li-Smerin, Y. and Swartz, K. J., 1998, Gating modifier toxins reveal a conserved structural motif in voltage-gated Ca^{2+} and K^+ channels, *Proc Natl Acad Sci U S A* **95**:8585-8589.

Lu, B.-S., Yu, F., Zhao, D., Huang, P.-T., Huang, C.-F., 1999, Conopeptides from *Conus striatus* and *Conus textile* by cDNA cloning. *Peptides* **20**:1139-1144.

Lukyanetz, E. A., Shkryl, V. M. and Kostyuk, P. G., 2002, Selective blockade of N-type calcium channels by levetiracetam, *Epilepsia* **43**:9-18.

McDonough, S. I., Boland, L. M., Mintz, I. M. and Bean, B. P., 2002, Interactions among toxins that inhibit N-type and P-type calcium channels, *J Gen Physiol* **119**:313-328.

McDonough, S. I., Lampe, R. A., Keith, R. A. and Bean, B. P., 1997, Voltage-dependent inhibition of N- and P-type calcium channels by the peptide toxin ω-grammotoxin-SIA, *Mol Pharmacol* **52**:1095-1104.

McGuire, D., Bowersox, S., Fellmann, J. D. and Luther, R. R., 1997, Sympatholysis after neuron-specific, N-type, voltage-sensitive calcium channel blockade: first demonstration of N-channel function in humans, *J Cardiovasc Pharmacol* **30**:400-403.

Menzler, S., Bikker, J. A., Suman-Chauhan, N. and Horwell, D. C., 2000, Design and biological evaluation of non-peptide analogues of ω-conotoxin MVIIA, *Bioorg Med Chem Lett* **10**:345-347.

Mer, G., Hietter, H., Kellenberger, C., Renatus, M., Luu, B. and Lefevre, J. F., 1996, Solution structure of PMP-C: a new fold in the group of small serine proteinase inhibitors, *J Mol Biol* **258**:158-171.

Mer, G., Kellenberger, C., Koehl, P., Stote, R., Sorokine, O., Van Dorsselaer, A., Luu, B., Hietter, H. and Lefevre, J. F., 1994, Solution structure of PMP-D2, a 35-residue peptide isolated from the insect *Locusta migratoria*, *Biochemistry* **33**:15397-15407.

Miles, L. A., Dy, C. Y., Nielsen, J., Barnham, K. J., Hinds, M. G., Olivera, B. M., Bulaj, G. and Norton, R. S., 2002, Structure of a novel P-superfamily spasmodic conotoxin reveals an inhibitory cystine knot motif, *J Biol Chem* **277**:43033-43040.

Minami, K., Raymond, C., Martin-Moutot, N., Ohtake, A., Van Renterghem, C., Takahashi, M., Seagar, M.J., Mori, Y., and Sato, K., 2001, Role of Thr(11) in the binding of ω-conotoxin MVIIC to N-type Ca^{2+} channels. *FEBS Lett* **491**:127-30

Mintz, I. M., Venema, V. J., Swiderek, K. M., Lee, T. D., Bean, B. P. and Adams, M. E., 1992, P-type calcium channels blocked by the spider toxin ω-Aga-IVA, *Nature* **355**:827-829.

Miranda, D. M., Romano-Silva, M. A., Kalapothakis, E., Diniz, C. R., Cordeiro, M. N., Santos, T. M., Prado, M. A. and Gomez, M. V., 1998, *Phoneutria nigriventer* toxins block tityustoxin-induced calcium influx in synaptosomes, *Neuroreport* **9**:1371-1373.

Mitchell, S. S., Shon, K. J., Foster, M. P., Davis, D. R., Olivera, B. M. and Ireland, C. M., 1998, Three-dimensional solution structure of conotoxin ψ-PIIIE, an acetylcholine gated ion channel antagonist, *Biochemistry* **37**:1215-1220.

Morrissette, J., Beurg, M., Sukhareva, M. and Coronado, R., 1996, Purification and characterization of ryanotoxin, a peptide with actions similar to those of ryanodine, *Biophys J* **71**:707-721.

Mosbah, A., Kharrat, R., Fajloun, Z., Renisio, J. G., Blanc, E., Sabatier, J. M., El Ayeb, M. and Darbon, H., 2000, A new fold in the scorpion toxin family, associated with an activity on a ryanodine-sensitive calcium channel, *Proteins* **40**:436-442.

Nadasdi, L., Yamashiro, D., Chung, D., Tarczy-Hornoch, K., Adriaenssens, P. and Ramachandran, J., 1995, Structure-activity analysis of a *Conus* peptide blocker of N-type neuronal calcium channels, *Biochemistry* **34**:8076-8081.

Naerum, L., Norskov-Lauritsen, L., and Oleson, P. H., 2002, Scaffold hopping and optimization towards libraries of glycogen synthase kinase-3 inhibitors, *Bioorg Med Chem Lett* **12**: 1525-1528.

Nakakura, N., Hietter, H., Van Dorsselaer, A. and Luu, B., 1992, Isolation and structural determination of three peptides from the insect *Locusta migratoria*. Identification of a deoxyhexose-linked peptide, *Eur J Biochem* **204**:147-153.

Nakayama, K., Kawato, H. C., Inagaki, H. and Ohta, T., 2001, Novel peptidomimetics of the antifungal cyclic peptide rhodopeptin: design of mimetics utilizing scaffolding methodology, *Org Lett* **3**:3447-3450.

Narasimhan, L., Singh, J., Humblet, C., Guruprasad, K. and Blundell, T., 1994, Snail and spider toxins share a similar tertiary structure and 'cystine motif', *Nat Struct Biol* **1**:850-852.

Nemoto, N., Kubo, S., Yoshida, T., Chino, N., Kimura, T., Sakakibara, S., Kyogoku, Y. and Kobayashi, Y., 1995, Solution structure of ω-conotoxin MVIIC determined by NMR, *Biochem Biophys Res Commun* **207**:695-700.

Newcomb, R., Palma, A., Fox, J., Gaur, S., Lau, K., Chung, D., Cong, R., Bell, J. R., Horne, B., Nadasdi, L. and et al., 1995, SNX-325, a novel calcium antagonist from the spider *Segestria florentina*, *Biochemistry* **34**:8341-8347.

Newcomb, R., Szoke, B., Palma, A., Wang, G., Chen, X., Hopkins, W., Cong, R., Miller, J., Urge, L., Tarczy-Hornoch, K., Loo, J. A., Dooley, D. J., Nadasdi, L., Tsien, R. W., Lemos, J. and Miljanich, G., 1998, Selective peptide antagonist of the class E calcium channel from the venom of the tarantula *Hysterocrates gigas*, *Biochemistry* **37**:15353-15362.

Nielsen, K. J., Adams, D. A., Alewood, P. F., Lewis, R. J., Thomas, L., Schroeder, T. and Craik, D. J., 1999a, Effects of chirality at Tyr13 on the structure-activity relationships of ω-conotoxins from *Conus magus*, *Biochemistry* **38**:6741-6751.

Nielsen, K. J., Adams, D., Thomas, L., Bond, T., Alewood, P. F., Craik, D. J. and Lewis, R. J., 1999b, Structure-activity relationships of ω-conotoxins MVIIA, MVIIC and 14 loop splice hybrids at N and P/Q-type calcium channels, *J Mol Biol* **289**:1405-1421.

Nielsen, K. J., Schroeder, T., and Lewis, R., 2000, Structure-activity relationships of ω-conotoxins at N-type voltage-sensitive calcium channels, *J Mol Recognit* **13**:55-70.

Nielsen, K. J., Thomas, L., Lewis, R. J., Alewood, P. F., and Craik, D. J., 1996, A consensus structure for ω-conotoxins with different selectivities for voltage-sensitive calcium channel subtypes: comparison of MVIIA, SVIB and SNX-202, *J Mol Biol* **263**:297-310.

Nishio, H., Katoh, E., Yamazaki, T., Inui, T., Nishiuchi, Y. and Kimura, T., 1999, Structure-activity relationships of calcicludine and dendrotoxin-I, homologous peptides acting on different targets, calcium and potassium channels, *Biochem Biophys Res Commun* **262**:319-321.

Norton, R. S. and Pallaghy, P. K., 1998, The cystine knot structure of ion channel toxins and related polypeptides, *Toxicon* **36**:1573-1583.

Norton, R. S., Pallaghy, P. K., Baell, J. B., Wright, C. E., Lew, M. J., and Angus, J. A., 1999, Polypeptide ω-conotoxin GVIA as a basis for new analgesic and neuroprotective agents, *Drug Develop Res* **46**:206-218.

Olivera B. M., and Cruz, L. J., 2001, Conotoxins, in retrospect, *Toxicon* **39**: 7-14.

Olivera, B. M., Hillyard, D. R., Marsh, M. and Yoshikami, D., 1995, Combinatorial peptide libraries in drug design: lessons from venomous cone snails, *Trends Biotechnol* **13**:422-426.

Olivera, B. M., Miljanich, G. P., Ramachandran, J. and Adams, M. E., 1994, Calcium channel diversity and neurotransmitter release: the ω-conotoxins and ω-agatoxins, *Annu Rev Biochem* **63**:823-867.

Olivera, B. M., Rivier, J., Clark, C., Ramilo, C. A., Corpuz, G. P., Abogadie, F. C., Mena, E. E., Woodward, S. R., Hillyard, D. R. and Cruz, L. J., 1990, Diversity of *Conus* neuropeptides, *Science* **249**:257-263.

Oprea, T. I., Davis, A. M., Teague, S. J. and Leeson, P. D., 2001, Is there a difference between leads and drugs? A historical perspective, *J Chem Inf Comput Sci* **41**:1308-1315.

Orner, B. P., Ernst, J. T. and Hamilton, A. D., 2001, Toward proteomimetics: terphenyl derivatives as structural and functional mimics of extended regions of an α-helix, *J Am Chem Soc* **123**:5382-5383.

Pallaghy, P. K., Duggan, B. M., Pennington, M. W. and Norton, R. S., 1993, Three-dimensional structure in solution of the calcium channel blocker ω-conotoxin, *J Mol Biol* **234**:405-420.

Pallaghy, P.K., Kumar, J.R., Miles, L.A., Pennington, M.W., and Norton, R.S., 2003, Non-linear epitope mimicry on complete chiral families of cyclic peptides. *Submitted*

Pallaghy, P. K., Nielsen, K. J., Craik, D. J. and Norton, R. S., 1994, A common structural motif incorporating a cystine knot and a triple-stranded β-sheet in toxic and inhibitory polypeptides, *Protein Sci* **3**:1833-1839.

Pallaghy, P. K. and Norton, R. S., 1999, Refined solution structure of ω-conotoxin GVIA: implications for calcium channel binding, *J Pept Res* **53**:343-351.

Pallaghy, P. K. and Norton, R. S., 2000, The cyclic contryphan motif CPxXPXC, a robust scaffold potentially useful as an ω-conotoxin mimic, *Biopolymers* **54**:173-179.

Pardridge, W. M., 2002, Targeting neurotherapeutic agents through the blood-brain barrier, *Arch Neurol* **59**:35-40.

Peng, K., Chen, X. D. and Liang, S. P., 2001, The effect of huwentoxin-I on Ca^{2+} channels in differentiated NG108-15 cells, a patch-clamp study, *Toxicon* **39**:491-498.

Perrier, V., Wallace, A. C., Kaneko, K., Safar, J., Prusiner, S. B. and Cohen, F. E., 2000, Mimicking dominant negative inhibition of prion replication through structure-based drug design, *Proc Natl Acad Sci U S A* **97**:6073-6078.

Price-Carter, M., Bulaj, G. and Goldenberg, D. P., 2002, Initial disulfide formation steps in the folding of an ω-conotoxin, *Biochemistry* **41**:3507-3519.

Price-Carter, M., Gray, W. R. and Goldenberg, D. P., 1996a, Folding of ω-conotoxins. 1. Efficient disulfide-coupled folding of mature sequences in vitro, *Biochemistry* **35**:15537-15546.

Price-Carter, M., Gray, W. R. and Goldenberg, D. P., 1996b, Folding of ω-conotoxins. 2. Influence of precursor sequences and protein disulfide isomerase, *Biochemistry* **35**:15547-15557.

Price-Carter, M., Hull, M. S. and Goldenberg, D. P., 1998, Roles of individual disulfide bonds in the stability and folding of an ω-conotoxin, *Biochemistry* **37**:9851-9861.

Pruneau, D. and Angus, J. A., 1990, ω-conotoxin GVIA, the N-type calcium channel inhibitor, is sympatholytic but not vagolytic: consequences for hemodynamics and autonomic reflexes in conscious rabbits, *J Cardiovasc Pharmacol* **16**:675-680.

Qu, Y., Liang, S., Ding, J., Liu, X., Zhang, R. and Gu, X., 1997, Proton nuclear magnetic resonance studies on huwentoxin-I from the venom of the spider *Selenocosmia huwena*: 2. Three-dimensional structure in solution, *J Protein Chem* **16**:565-574.

Rash, L. D. and Hodgson, W. C., 2002, Pharmacology and biochemistry of spider venoms, *Toxicon* **40**:225-254.

Reily, M. D., Holub, K. E., Gray, W. R., Norris, T. M. and Adams, M. E., 1994, Structure-activity relationships for P-type calcium channel-selective ω-agatoxins, *Nat Struct Biol* **1**:853-856.

Reily, M. D., Thanabal, V. and Adams, M. E., 1995, The solution structure of ω-Aga-IVB, a P-type calcium channel antagonist from venom of the funnel web spider, *Agelenopsis aperta*, *J Biomol NMR* **5**:122-132.

Rhim, H., Kim, H., Lee, D. Y., Oh, T. H. and Nah, S. Y., 2002, Ginseng and ginsenoside Rg3, a newly identified active ingredient of ginseng, modulate Ca^{2+} channel currents in rat sensory neurons, *Eur J Pharmacol* **436**:151-158.

Roumelioti, P., Polevaya, L., Zoumpoulakis, P., Giatus, N., Mutule, I., Keivish, T., Zoga, A., Vlahakos, D., Iliodromitis, E., Kremastinos, D., Grdadolnik, S. G., Mavromoustakos, T., and Matsoukas, J., 2002, Design, synthesis and biological evaluation of cyclic angiotensin II analogues with 3,5 side-chain bridges: role of C-terminal aromatic residue and ring cluster for activity and implications in the drug design of AT1 non-peptide antagonists, *Bioorg Med Chem Lett* **12**:2627-2633.

Ryder, T. R., Hu, L. Y., Rafferty, M. F., Lotarski, S. M., Rock, D. M., Stoehr, S. J., Taylor, C. P., Weber, M. L., Miljanich, G. P., Millerman, E. and Szoke, B. G., 2000, Structure-activity relationship at the leucine side chain in a series of N,N-dialkyldipeptidyl-amines as N-type calcium channel blockers, *Drug Des Discov* **16**:317-322.

Ryder, T., Hu, L.-Y., Rafferty, M. F., Lotarski, S. M., Rock, D. M., Stoehr, S. J., Taylor, C. P., Weber, M. L., Miljanich, G. P., Millerman, E., and Szoke, B. G., 1999a, Structure-activity relationship at the proximal phenyl group in a series of non-peptidyl N-type calcium channel antagonists, *Bioorg Med Chem Lett* **9**:2453-2458.

Ryder, T., Hu, L.-Y., Rafferty, M. F., Lotarski, S. M., Rock, D. M., Stoehr, S. J., Taylor, C. P., Weber, M. L., Miljanich, G. P., Millerman, E., and Szoke, B. G., 1999b, Multiple parallel synthesis of N,N-dialkyldipeptidylamines as N-type calcium channel blockers, *Bioorg Med Chem Lett* **9**:1813-1818.

Sabo, T., Gilon, C., Shafferman, A., and Elhanaty, E., 1992, in *Peptides: Chemistry and Biology. Proceedings of the Twelfth American Peptide Symposium*, Smith, J. A., and Rivier, J. E., eds., Escom: Leiden, pp 159-160.

Samso, M., Trujillo, R., Gurrola, G. B., Valdivia, H. H. and Wagenknecht, T., 1999, Three-dimensional location of the imperatoxin A binding site on the ryanodine receptor, *J Cell Biol* **146**:493-499.

Santos, R. G., Diniz, C. R., Cordeiro, M. N. and De Lima, M. E., 1999, Binding sites and actions of Tx1, a neurotoxin from the venom of the spider *Phoneutria nigriventer*, in guinea pig ileum, *Braz J Med Biol Res* **32**:1565-1569.

Sarabu, R., Cooper, J. P., Cook, C. M., Gillespie, P., Perrotta, A. V. and Olson, G. L., 1997, Design and synthesis of small molecule interleukin-1 receptor antagonists based on a benzene template, *Drug Des Discov* **15**:191-198.

Sasaki, T., Kobayashi, K., Kohno, T., and Sato, K., 2000, Combinatorial synthesis of ω-conotoxin MVIIC analogues and their binding with N- and P/Q-type calcium channels. *FEBS Lett* **466**:125-9.

Sato, K., Ishida, Y., Wakamatsu, K., Kato, R., Honda, H., Ohizumi, Y., Nakamura, H., Ohya, M., Lancelin, J. M., and Kohda, D., 1991, Active site of μ-conotoxin GIIIA, a peptide blocker of muscle sodium channels, *J Biol Chem* **266**:16989-16991.

Sato, K., Raymond, C., Martin-Moutot, N., Sasaki, T., Ohtake, A., Minami, K., Van Renterghem, C., Kim, J.I., Takahashi, M., and Seagar, M.J., 2000a, Binding of Ala-scanning analogs of ω-conotoxin MVIIC to N- and P/Q-type calcium channels. FEBS Lett **469**:147-150

Sato K., Raymond, C., Martin-Moutot, N., Sasaki, T., Ohtake, A., Minami, K., Van Renterghem, C., Takahashi, M., and Seagar, M.J., 2000b, Binding of six chimeric analogs of ω-conotoxin MVIIA and MVIIC to N- and P/Q-type calcium channels, *Biophys Res Commun* **269**:254-256.

Savarin, P., Guenneugues, M., Gilquin, B., Lamthanh, H., Gasparini, S., Zinn-Justin, S. and Menez, A., 1998, Three-dimensional structure of κ-conotoxin PVIIA, a novel potassium channel-blocking toxin from cone snails, *Biochemistry* **37**:5407-5416.

Schelkum, R. M., yuen, P., Malone, T. C., Rock, D. M., Stoehr, S., Szoke, B., and Tarczy-Hornoch, K., 1999, Synthesis and biological activity of substituted bis-(4-hydroxyphenyl)methanes as N-type calcium channel blockers, *Bioorg Med Chem Lett* **9**:2447-2452.

Schweitz, H., Heurteaux, C., Bois, P., Moinier, D., Romey, G. and Lazdunski, M., 1994, Calcicludine, a venom peptide of the Kunitz-type protease inhibitor family, is a potent blocker of high-threshold Ca^{2+} channels with a high affinity for L-type channels in cerebellar granule neurons, *Proc Natl Acad Sci U S A* **91**:878-882.

Scott, D. A., Wright, C. E. and Angus, J. A., 2002, Actions of intrathecal ω-conotoxins CVID, GVIA, MVIIA, and morphine in acute and neuropathic pain in the rat, *Eur J Pharmacol* **451**:279-286.

Scott, R. H., Dolphin, A. C., Bindokas, V. P. and Adams, M. E., 1990, Inhibition of neuronal Ca^{2+} channel currents by the funnel web spider toxin ω-Aga-IA, *Mol Pharmacol* **38**:711-718.

Scott, R. H., Gorton, V. J., Harding, L., Patel, D., Pacey, S., Kellenberger, C., Hietter, H. and Bermudez, I., 1997, Inhibition of neuronal high voltage-activated calcium channels by insect peptides: a comparison with the actions of ω-conotoxin GVIA, *Neuropharmacology* **36**:195-208.

Seko, T., Kato, M., Kohno, H., Ono, S., Hashimura, K., Takenobu, Y., Takimizu, H., Nakai, K., Maegawa, H., Katsube, N. and Toda, M., 2002, L-Cysteine based N-type calcium channel blockers: structure-activity relationships of the C-terminal lipophilic moiety, and oral analgesic efficacy in rat pain models, *Bioorg Med Chem Lett* **12**:2267-2269.

Serone, A. P. and Angus, J. A., 1999, Role of N-type calcium channels in autonomic neurotransmission in guinea-pig isolated left atria, *Br J Pharmacol* **127**:927-934.

Sevilla, P., Bruix, M., Santoro, J., Gago, F., Garcia, A. G. and Rico, M., 1993, Three-dimensional structure of ω-conotoxin GVIA determined by ^{1}H NMR, *Biochem Biophys Res Commun* **192**:1238-1244.

Shapiro, R., Ruiz-Gutierrez, M. and Chen, C. Z., 2000, Analysis of the interactions of human ribonuclease inhibitor with angiogenin and ribonuclease A by mutagenesis: importance of inhibitor residues inside versus outside the C-terminal "hot spot", *J Mol Biol* **302**:497-519.

Sidach, S. S. and Mintz, I. M., 2002, Kurtoxin, a gating modifier of neuronal high- and low-threshold Ca channels, *J Neurosci* **22**:2023-2034.

Skalicky, J. J., Metzler, W. J., Ciesla, D. J., Galdes, A. and Pardi, A., 1993, Solution structure of the calcium channel antagonist ω-conotoxin GVIA, *Protein Sci* **2**:1591-1603.

Smith, M. T., Cabot, P. J., Ross, F. B., Robertson, A. D. and Lewis, R. J., 2002, The novel N-type calcium channel blocker, AM336, produces potent dose-dependent antinociception after intrathecal dosing in rats and inhibits substance P release in rat spinal cord slices, *Pain* **96**:119-127.

Song, Y., Bowersox, S. S., Connor, D. T., Dooley, D. J., Lotarski, S. M., Malone, T., Miljanich, G., Millerman, E., Rafferty, M. F., Rock, D., Roth, B. D., Schmidt, J., Stoehr, S., Szoke, B. G., Taylor, C., Vartanian, M. and Wang, Y. X., 2000, (S)-4-Methyl-2-(methylamino)pentanoic acid [4, 4-bis(4-fluorophenyl)butyl]amide hydrochloride, a novel calcium channel antagonist, is efficacious in several animal models of pain, *J Med Chem* **43**:3474-3477.

Sutton, K. G., and Snutch, T. P., 2002, Gabapentin: A novel analgesic targeting voltage-gated calcium channels, *Drug Develop Res* **54**:167-172.

Sutton, K. G., Siok, C., Stea, A., Zamponi, G. W., Heck, S. D., Volkmann, R. A., Ahlijanian, M. K. and Snutch, T. P., 1998, Inhibition of neuronal calcium channels by a novel peptide spider toxin, DW13.3, *Mol Pharmacol* **54**:407-418.

Takeuchi, K., Park, E., Lee, C., Kim, J., Takahashi, H., Swartz, K. and Shimada, I., 2002, Solution structure of ω-grammotoxin SIA, a gating modifier of P/Q and N-type Ca^{2+} channel, *J Mol Biol* **321**:517-526.

Tilley, J. W., Chen, L., Fry, D. C., Emerson, S. D., Powers, G. D., Biondi, D., Varnell, T., Trilles, R., Guthrie, R., Mennona, F., Kaplan, G., LeMahieu, R. A., Carson, M., Han, R.-J., Liu, C.-M., Palermo, R., and Ju, G., 1997, Identification of a small molecule inhibitor of the IL-2/IL-2R receptor interaction which binds to IL-2, *J Amer Chem Soc* **119**:7589-7690.

Vaughan, C. K., Buckle, A. M. and Fersht, A. R., 1999, Structural response to mutation at a protein-protein interface, *J Mol Biol* **286**:1487-1506.

Venema, V. J., Swiderek, K. M., Lee, T. D., Hathaway, G. M. and Adams, M. E., 1992, Antagonism of synaptosomal calcium channels by subtypes of ω-agatoxins, *J Biol Chem* **267**:2610-2615.

Venkatachalam, C. M., 1968, Stereochemical criteria for polypeptides and proteins. V. Conformation of a system of three linked peptide units, *Biopolymers* **6**:1425-1436.

Volpon, L., Lamthanh, H., Le Gall, F., Menez, A., Lancelin, J. M., 2000, *Protein Data Bank submission*, accession no. 1GIP/1GIZ.

Wakamatsu, K., Kohda, D., Hatanaka, H., Lancelin, J. M., Ishida, Y., Oya, M., Nakamura, H., Inagaki, F. and Sato, K., 1992, Structure-activity relationships of μ-conotoxin GIIIA: structure determination of active and inactive sodium channel blocker peptides by NMR and simulated annealing calculations, *Biochemistry* **31**:12577-12684.

Wang, X. H., Connor, M., Wilson, D., Wilson, H. I., Nicholson, G. M., Smith, R., Shaw, D., Mackay, J. P., Alewood, P. F., Christie, M. J. and King, G. F., 2001, Discovery and structure of a potent and highly specific blocker of insect calcium channels, *J Biol Chem* **276**:40306-40312.

Wang, Y. X., Gao, D., Pettus, M., Phillips, C. and Bowersox, S. S., 2000a, Interactions of intrathecally administered ziconotide, a selective blocker of neuronal N-type voltage-sensitive calcium channels, with morphine on nociception in rats, *Pain* **84**:271-281.

Wang, Y. X., Pettus, M., Gao, D., Phillips, C. and Bowersox, S. S., 2000b, Effects of intrathecal administration of ziconotide, a selective neuronal N-type calcium channel blocker, on mechanical allodynia and heat hyperalgesia in a rat model of postoperative pain, *Pain* **84**:151-158.

Westenbroek, R. E., Hoskins, L. and Catterall, W. A., 1998, Localization of Ca^{2+} channel subtypes on rat spinal motor neurons, interneurons, and nerve terminals, *J Neurosci* **18**:6319-6330.

Whorlow, S. L., Angus, J. A. and Wright, C. E., 1994, The effects of central administration of ω-conotoxin GVIA on cardiovascular parameters and autonomic reflexes in conscious rabbits, *Clin Exp Pharmacol Physiol* **21**:865-873.

Whorlow, S. L., Angus, J. A. and Wright, C. E., 1996a, Selectivity of ω-conotoxin GVIA for N-type calcium channels in rat isolated small mesenteric arteries, *Clin Exp Pharmacol Physiol* **23**:16-21.

Whorlow, S. L., Loiacono, R. E., Angus, J. A. and Wright, C. E., 1996b, Distribution of N-type Ca^{2+} channel binding sites in rabbit brain following central administration of ω-conotoxin GVIA, *Eur J Pharmacol* **315**:11-18.

Williams, D. H., and Westwell, M. S., 1998, Aspects of weak interactions, *Chem Soc Rev* **27**:57.

Woppmann, A., Ramachandran, J. and Miljanich, G. P., 1994, Calcium channel subtypes in rat brain: biochemical characterization of the high-affinity receptors for ω-conopeptides SNX-230 (synthetic MVIIC), SNX-183 (SVIB), and SNX-111 (MVIIA), *Mol Cell Neurosci* **5**:350-357.

Wright, C. E. and Angus, J. A., 1995, Hemodynamic and autonomic reflex effects of chronic N-type Ca^{2+} channel blockade with ω-conotoxin GVIA in conscious normotensive and hypertensive rabbits, *J Cardiovasc Pharmacol* **25**:459-468.

Wright, C. E. and Angus, J. A., 1996, Effects of N-, P- and Q-type neuronal calcium channel antagonists on mammalian peripheral neurotransmission, *Br J Pharmacol* **119**:49-56.

Wright, C. E., Hawkes, A. L. and Angus, J. A., 2000a, Postural hypotension following N-type Ca^{2+} channel blockade is amplified in experimental hypertension, *J Hypertens* **18**:65-73.

Wright, C. E., Robertson, A. D., Whorlow, S. L. and Angus, J. A., 2000b, Cardiovascular and autonomic effects of ω-conotoxins MVIIA and CVID in conscious rabbits and isolated tissue assays, *Br J Pharmacol* **131**:1325-1336.

Yan, L. and Adams, M. E., 2000, The spider toxin ω-Aga IIIA defines a high affinity site on neuronal high voltage-activated calcium channels, *J Biol Chem* **275**:21309-21316.

Yan, Y. B., and Zhang, R. Q., 2000, *Protein Data Bank submission*, accession no. 1FYG.

Yan, Z., Chi, P., Bibb, J. A., Ryan, T. A. and Greengard, P., 2002, Roscovitine: a novel regulator of P/Q-type calcium channels and transmitter release in central neurons, *J Physiol* **540**:761-770.

Yang, W., He, H., and Drueckhammer, D. G., 2001, Computer-guided design in molecular recognition: design and synthesis of a glucopyranose receptor, *Angew Chem Int Ed* **40**:1714-1718.

Yasuda, O., Morimoto, S., Jiang, B., Kuroda, H., Kimura, T., Sakakibara, S., Fukuo, K., Chen, S., Tamatani, M. and Ogihara, T., 1994, FS2, a mamba venom toxin, is a specific blocker of the L-type calcium channels, *Artery* **21**:287-302.

Yu, H., Rosen, M. K., Saccomano, N. A., Phillips, D., Volkmann, R. A. and Schreiber, S. L., 1993, Sequential assignment and structure determination of spider toxin ω-Aga-IVB, *Biochemistry* **32**:13123-13129.

Zamudio, F. Z., Conde, R., Arevalo, C., Becerril, B., Martin, B. M., Valdivia, H. H. and Possani, L. D., 1997a, The mechanism of inhibition of ryanodine receptor channels by imperatoxin I, a heterodimeric protein from the scorpion *Pandinus imperator*, *J Biol Chem* **272**:11886-11894.

Zamudio, F. Z., Gurrola, G. B., Arevalo, C., Sreekumar, R., Walker, J. W., Valdivia, H. H. and Possani, L. D., 1997b, Primary structure and synthesis of imperatoxin A ($IpTx_a$), a peptide activator of Ca^{2+} release channels/ryanodine receptors, *FEBS Lett* **405**:385-389.

PHARMACOLOGY OF Ca_V3 (T-TYPE) CHANNELS

Eric A. Ertel[1]

1. INTRODUCTION

Voltage-operated Ca^{2+} channels are complex membrane proteins that are responsible for mediating Ca^{2+} influx in response to membrane depolarization (Hille 1984). Ca^{2+} entry through these channels serves to regulate intracellular processes such as contraction, secretion, neurotransmission, and gene expression, in response to action potentials and other membrane potential changes. Although all voltage-operated Ca^{2+} channels are activated by a conformational modification of protein structure induced by a change in membrane potential, these channels can vary considerably in their electrophysiological properties such as voltage-dependences, activation and inactivation kinetics, sensitivity to extracellular and intracellular ligands, ion selectivity, and single channel properties. Historically, the characterization of voltage-operated Ca^{2+} channels has progressed on two separate fronts: the electrophysiological and pharmacological characterization of voltage-operated Ca^{2+} channel currents and the biochemical and molecular characterization of voltage-operated Ca^{2+} channel proteins. This has led to the emergence of two separate nomenclatures. The correspondence between these two nomenclatures has been difficult to define because many of the features used to associate cloned and native Ca^{2+} channels depend on many factors, such as cellular environment, co-expression of accessory subunits, and modulation by second messengers and G-proteins.

The different electrophysiological and pharmacological properties of native voltage-operated Ca^{2+} channels has led to the description of a number of distinct voltage-operated Ca^{2+} currents. So far, six main categories have been described, each with many subtypes: L (Long-lasting), T (Transient), N (Neither T nor L, or Neuronal), P (Purkinje cells), Q (after P), and R (Remaining, or Resistant, or after Q) (Kenakin et al., 1992; Spedding et al., 1992a,b; Spedding et al., 1993).

[1] PRNS, 73/103b, F. Hoffmann-La Roche Ltd., Postfach, Basel, CH-4070, Switzerland.

Calcium Channel Pharmacology, Edited by McDonough
Kluwer Academic/Plenum Publishers, 2004

2. T-TYPE Ca^{2+} CURRENTS AND ASSOCIATED CHANNEL PROTEINS

2.1. Electrophysiological Characterization

The existence of T-type Ca^{2+} channels was first suggested after the identification of two distinct Ca^{2+} currents in starfish eggs in 1975 (Hagiwara et al., 1975). In vertebrate cells, T-type Ca^{2+} currents were originally described in sensory neurons (Carbone et al., 1984a; Carbone et al., 1984b), atrial (Bean, 1985) and ventricular (Nilius et al., 1985) myocytes, and pituitary cells (Armstrong et al., 1985) (reviewed in Huguenard 1996; Ertel et al., 1997a; Ertel et al., 1997b). T-type Ca^{2+} currents are often referred to as transient Ca^{2+} currents, as low-voltage-activated (LVA) Ca^{2+} currents, or as slowly-deactivating Ca^{2+} currents. Initially, Ca^{2+} currents were called T-type when their time course was Transient, *i.e.*, when they inactivated faster ($\tau \approx$ 5-50 ms) than L-type Ca^{2+} currents (Bean, 1989; Nowycky et al., 1985). However, the inactivation rate of L-type Ca^{2+} currents is dependent on the experimental conditions (Bean, 1989; Hagiwara et al., 1988), and Ca^{2+}-dependent inactivation, in particular, can accelerate the time course of L-type Ca^{2+} currents to resemble that of T-type Ca^{2+} currents. Therefore, the inactivation rate is nowadays mostly used as a corroborating identifier of T-type Ca^{2+} currents. More commonly used are the voltage dependences of activation and inactivation, which occur at fairly negative voltages for T-type Ca^{2+} currents compared to most other Ca^{2+} currents (Figure 1). For T-type Ca^{2+} currents, activation starts between -70 mV and -50 mV, and the current reaches its maximal amplitude around -40 mV to -10 mV (Bonvallet, 1987; Bean, 1989; Carbone et al., 1984b; Hagiwara et al., 1988; Zhou et al., 1994): for this reason, T-type Ca^{2+} currents are often called "low-voltage-activated Ca^{2+} currents." Similarly, the voltage for half-maximal inactivation of T-type Ca^{2+} currents ranges from -75 mV to -55 mV, compared to -30 mV to -10 mV for L-type Ca^{2+} currents (Hirano et al., 1989; Hagiwara et al., 1988; Bonvallet et al., 1989). However, these voltage characteristics are not unique to T-type Ca^{2+} currents and are shared, for example, with

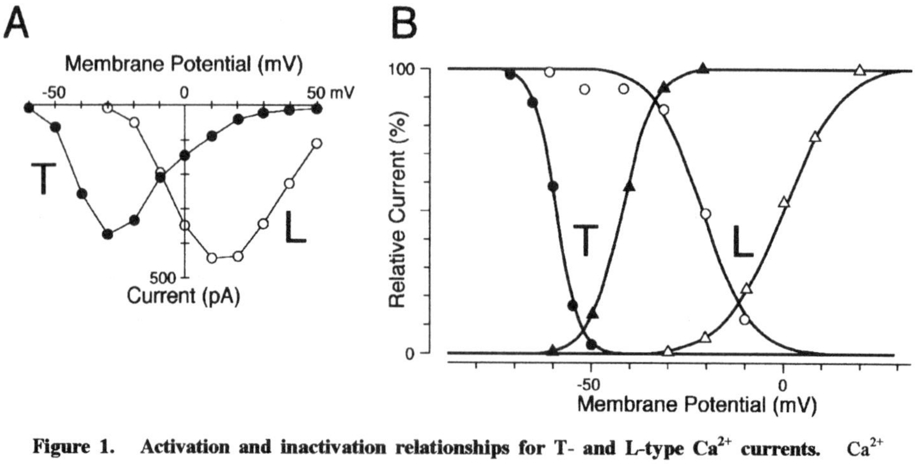

Figure 1. Activation and inactivation relationships for T- and L-type Ca^{2+} currents. Ca^{2+} currents were recorded from canine cardiac Purkinje cells. (A) peak current-voltage relationships. (B) Boltzmann fits to data for activation (triangles) and inactivation (circles) for T-type (filled symbols) and L-type (hollow symbols) Ca^{2+} currents. For both types of current these relationships overlap, suggesting the presence of a steady-state component (window current). $[Ca^{2+}]_o$ was 5 mM. Adapted from Hirano et al., 1989.

some R-type Ca^{2+} currents (Zhang et al., 1993a). Also, recent results indicate that L-type Ca^{2+} currents can sometimes activate at atypical negative voltages (Rossier et al., 1998; Johnston et al., 1998). In addition, under various normal or pathological circumstances, the voltage-dependences of many Ca^{2+} current types can be shifted significantly, adding to the unreliability of distinctions based on these properties. In summary, voltage dependence alone is not a sufficient biophysical determinant of T-type Ca^{2+} currents.

What could be an appropriate defining characteristic? At present, there are four candidates (Table 1): single-channel conductance, ionic selectivity, deactivation rate, and voltage-dependence of inactivation rate. First, the underlying conductance of T-type Ca^{2+} currents described to date is small (5-9 pS) compared to other Ca^{2+} currents (> 10 pS) (Rampe et al., 1993; Nilius et al., 1985; Dirksen et al., 1995; Carbone et al., 1987). Unfortunately, measuring single-channel conductance requires single-channel analysis, which is technically more demanding than classical microelectrode or patch-clamp analysis and has thus been performed only in a restricted number of physiological systems. It remains to be seen whether this specificity of T-type Ca^{2+} currents will hold as more data is published.

Second, T-type Ca^{2+} channels conduct Ca^{2+} as well as Ba^{2+} (Fedulova et al., 1985; Carbone et al., 1984b; Bean, 1985) or better than Ba^{2+} (Friedman et al., 1986; Bossu et al., 1985). In contrast, other Ca^{2+} channels have significantly larger conductance for Ba^{2+} than Ca^{2+} (Randall et al., 1993; Zhang et al., 1993a; Rampe et al., 1993; Fox et al., 1987; Bean, 1985). A problem with this property is that it can be easily tested only in systems where a single type of Ca^{2+} channel is present because it is difficult to separate simultaneous changes of permeability of multiple Ca^{2+} channels. Furthermore, ionic replacement of Ca^{2+} for Ba^{2+} modifies the voltage dependences of Ca^{2+} channels, which confuses characterization even further.

Third, T-type Ca^{2+} currents deactivate much more slowly than other Ca^{2+} currents, *i.e.* the channels return slowly from the open state to the resting state (Armstrong et al., 1985; Bean, 1985). T-type Ca^{2+} currents have been described with deactivation rates in the range 1-10 ms, while other Ca^{2+} currents have deactivation rates faster than 0.5 ms (Cohen et al., 1992a; Ertel et al., 1994; Cohen et al., 1988; Carbone et al., 1987; McCarthy et al., 1989). This parameter has been found to be relatively independent of ionic conditions and can be measured both with classical patch-clamp and single-channel measurements. However, dihydropyridine agonists modify L-type Ca^{2+} currents so they deactivate slowly, and it is not known whether physiological circumstances could do likewise and confuse the separation of T-type Ca^{2+} currents from this or other types of Ca^{2+} currents.

Fourth, the voltage-dependent inactivation of T-type Ca^{2+} currents proceeds at a rate that is itself voltage-dependent, so that T-type Ca^{2+} currents inactivate faster at more depolarized voltages (Beam et al., 1988a; Monteil et al., 2000b). In contrast, the inactivation rate of other Ca^{2+} currents is voltage-independent, if one eliminates Ca^{2+}-dependent inactivation, which depends on current and consequently on voltage.

In summary, the electrophysiological characteristics of T-type Ca^{2+} currents most useful and reliable in distinguishing them from other voltage-gated Ca^{2+} currents are the slow rate of deactivation and the voltage-dependent rate of inactivation. Pharmacological tools are also useful to distinguish Ca^{2+} current types from one another. Selective blockers include dihydropyridine antagonists for L-type Ca^{2+} currents, ω-CTx-GVIA for

N-type Ca^{2+} currents, and ω-Aga-IVA for P-type Ca^{2+} currents. T-type Ca^{2+} currents are less sensitive than L-type to classical Ca^{2+} antagonists and not at all to ω-CTx-GVIA: in fact, no specific blocker of this current is known. However, various pharmacological compounds show some selectivity for T-type Ca^{2+} currents under defined conditions, as detailed in section 3.

Table 1. Electrophysiological characteristics of T-type Ca^{2+} currents. The underlined characteristics are most specific to T-type Ca^{2+} currents.

	T-type	non T-type
Activation voltage	low	high (L,N,P,Q), low-medium (R)
Inactivation voltage	low	high (L,P,Q), medium (N), low (R)
Inactivation rate	fast	fast (P), slow (L,N,Q,R)
<u>Deactivation rate</u>	slow	fast
Ca inactivation	no	yes (L,P,R), yes/no (N)
<u>V-dep inactivation rate</u>	yes	no
<u>Conductance (pS)</u>	5-9	>10
<u>Permeation</u>	Ca^{2+}=Ba^{2+}	Ca^{2+} < Ba^{2+}

2.2. Molecular Characterization

In parallel, the biochemical and molecular characterization of voltage-operated Ca^{2+} channels revealed complex proteins. Presently, the best-characterized channel is the one responsible for L-type Ca^{2+} currents (Seagar et al., 1988; Campbell et al., 1988; Abernethy et al., 2002; Catterall, 2000; Striessnig, 1999; Takahashi et al., 1987). It is a protein composed of four distinct subunits (α_1, β, $\alpha_2\delta$, and γ) encoded by four distinct genes (Ellis et al., 1988; Ruth et al., 1989; Tanabe et al., 1987; Jay et al., 1990) (α_2 and δ are derived from the same gene with the product cleaved into two peptides linked by a disulfide bond (De Jongh et al., 1990)). This structure appears to be conserved for L-type Ca^{2+} channels throughout the body. A possible exception may be the γ-subunit, which may be present only in muscle L-type Ca^{2+} channels (Takahashi et al., 1987; Ahlijanian et al., 1990; McEnery et al., 1991; Witcher et al., 1993), although recently described subunits may be associated with neuronal channels as well (Letts et al., 1998). At present, the subunit composition of T-type Ca^{2+} channels remains very controversial, although their α_1-subunit is certainly homologous to that of L-type Ca^{2+} channels (Perez-Reyes et al., 1998; Perez-Reyes 2003). There is no convincing evidence for a β-subunit, a γ-subunit may be associated with functional channels and modify their electrophysiological properties slightly (Green et al., 2001; Klugbauer et al., 2000), and an $\alpha_2\delta$-subunit may be involved in channel expression and/or membrane insertion but may or may not be associated with functional channels (Gao et al., 2000; Dolphin et al., 1999).

The most recent nomenclature of voltage-operated Ca^{2+} channels recognizes 3 families, each including 3 or 4 subtypes (Ertel et al., 2000). Table 2 shows the correspondence between the voltage-operated Ca^{2+} channels and the voltage-operated Ca^{2+} currents, which is nowadays well established with a few exceptions. The third Ca^{2+} channel family includes $Ca_v3.1$, $Ca_v3.2$, and $Ca_v3.3$ (α_{1G}, α_{1H}, and α_{1I}), and functional expression studies have demonstrated that these genes encode T-type Ca^{2+} channels. $Ca_v3.1$ (α_{1G}) was originally cloned from rat brain, and homologues were found in human and mouse (Perez-Reyes et al., 1998; Klugbauer et al., 1999; McRory et al., 2001; Mittman et al., 1999a); $Ca_v3.2$ (α_{1H}) was cloned from human heart (Cribbs et al., 1998), from a human medullary thyroid carcinoma cell line (Williams et al., 1999), from rat brain (McRory et al., 2001), and from a rat insulin-secreting cell line (Zhuang et al., 2000); and $Ca_v3.3$ (α_{1I}) was cloned from rat (Lee et al., 1999; McRory et al., 2001) and human brain (Gomora et al., 2002; Monteil et al., 2000a; Mittman et al., 1999b). Alternative splicing has been observed for all three α_1-subunits, and the structural differences found in the splice variants result in altered biophysical, physiological, and pharmacological properties of $Ca_v3.1$ (α_{1G}, Chemin et al., 2001a), $Ca_v3.2$ (α_{1H}, Jagannathan et al., 2002), and $Ca_v3.3$ (α_{1I}, McRory et al., 2001).

Table 2. Nomenclature of voltage-operated Ca^{2+} channels.

Family	Type	Former name	Ca^{2+} current
Ca_V1	$Ca_V1.1$	α_{1S}	L
	$Ca_V1.2$	α_{1C}	
	$Ca_V1.3$	α_{1D}	
	$Ca_V1.4$	α_{1F}	
Ca_V2	$Ca_V2.1$	α_{1A}	P/Q
	$Ca_V2.2$	α_{1B}	N
	$Ca_V2.3$	α_{1E}	R
Ca_V3	$Ca_V3.1$	α_{1G}	T
	$Ca_V3.2$	α_{1H}	
	$Ca_V3.3$	α_{1I}	

2.3. R-type Ca^{2+} Currents and Associated Channel Proteins

It is sometimes difficult to discriminate R-type Ca^{2+} currents from T-type Ca^{2+} currents because the two currents can bear significant similarities. Therefore, in some cases, properties assigned to T-type Ca^{2+} currents may in fact relate to R-type Ca^{2+} currents and vice-versa. For this reason, a short description of R-type Ca^{2+} currents is given here.

R-type Ca^{2+} currents were originally described in cerebellar granule cells (Randall et al., 1995; Zhang et al., 1993a) as the Remaining or Resistant current after blocking other Ca^{2+} current types with drugs and toxins. Based on this definition by default, it is not surprising that R-type Ca^{2+} currents in different cells possess a wide variety of distinct electrophysiological properties (Tottene et al., 2000). It is quite likely that multiple channel types are responsible for R-type Ca^{2+} currents in different cells or even in the same cell (Tottene et al., 1996). It is even possible that R-type Ca^{2+} currents may sometimes be carried by channel proteins that normally underlie L-, T-, N-, P-, or Q-type Ca^{2+} currents but have been unusually processed or regulated.

In many (but not all) cases, R-type Ca^{2+} currents are low- to moderate-voltage-activated and inactivate faster than L- or N-type Ca^{2+} currents (Miljanich et al., 1995). Therefore, R-type Ca^{2+} currents often resemble T-type Ca^{2+} currents and distinguishing the two current types can be difficult (Bourinet et al., 1996), although not impossible (Randall et al., 1997). R-type Ca^{2+} currents have been mainly described in neuronal cells, but some reports have also suggested that they exist in non-neuronal cells as well. However, in these cases, it remains difficult to ascertain whether the recorded currents are definitely R-type or rather T-type Ca^{2+} currents. In rat atrial myocytes, it is likely that part of the described T-type Ca^{2+} currents may really be R-type Ca^{2+} currents (Piedras-Renteria et al., 1997). In contrast, in vascular smooth muscle, part or most of the suggested R-type Ca^{2+} currents may really be T-type Ca^{2+} currents (Bkaily et al., 1998; Morita et al., 1999).

The second voltage-operated Ca^{2+} channel family includes $Ca_v2.1$, $Ca_v2.2$, and $Ca_v2.3$ (α_{1A}, α_{1B}, and α_{1E}, Table 2). So far, these genes appear to encode Ca^{2+} channels that are localized mostly in the nervous system and lack the characteristic DHP-sensitivity of L-type Ca^{2+} channels. The $Ca_v2.3$ (α_{1E}) gene is the most recent addition to the "neuronal" (non-L) channel family (Niidome et al., 1992). Its expression is widespread in mammalian neurons (Williams et al., 1994; Soong et al., 1993), but its assignment to one of the native Ca^{2+} currents has been problematic. $Ca_v2.3$ (α_{1E}) has been proposed to encode a T-type Ca^{2+} channel because of its activation and inactivation at negative potentials, its sensitivity to Ni^{2+} (Soong et al., 1993), and its permeation characteristics (Bourinet et al., 1996; Dolphin 1998; Stephens et al., 1997). However, the current produced by expression of $Ca_v2.3$ (α_{1E}) differs from the "classical" T-type Ca^{2+} current in having a slower inactivation rate, a larger single-channel conductance (5-9 pS for T-type *vs.* 14 pS for $Ca_v2.3$), a high sensitivity to Cd^{2+}, and a low sensitivity to octanol and amiloride. On the other hand, the human and rat $Ca_v2.3$ currents share a number of characteristics with R-type Ca^{2+} currents (Ellinor et al., 1993; Randall et al., 1995; Williams et al., 1994), which are sensitive to both Ni^{2+} and Cd^{2+} but not to DHPs or to ω-CTx-GVIA. There is a striking resemblance between currents through $Ca_v2.3$ (α_{1E}) expressed in HEK293 cells and R-type Ca^{2+} currents with respect to voltage-dependence, rates of activation and inactivation, and the relative size of Ba^{2+} and Ca^{2+} currents (Randall et al., 1995; Williams et al., 1994).

3. MOLECULAR PHARMACOLOGY

Pharmacological tools are also useful to distinguish Ca^{2+} channels types from one another. Selective blockers include dihydropyridine antagonists for L-type Ca^{2+} channels, ω-CTx-GVIA for N-type Ca^{2+} channels, ω-Aga-IVA for P-type Ca^{2+} channels. No satisfactory tool is currently known for Q-, R-, or T-type Ca^{2+} channels (Table 3). Diverse compounds have been reported, at one time or another, to block T-type Ca^{2+} channels selectively. Unfortunately, this selectivity has generally turned out to be limited to specific tissues and/or test conditions (for a review, see Heady et al., 2001).

3.1. Inorganic Ions

Ni^{2+} is the most commonly used "selective" blocker of T-type Ca^{2+} channels. Indeed, it blocks many T-type Ca^{2+} currents at concentrations ranging from 5 to 40 μM, which is 10-100-fold more potent than most other Ca^{2+} currents (Bonvallet, 1987; Gu et al., 1993). However, a few T-type Ca^{2+} currents appear less sensitive to Ni^{2+} (Akaike et al., 1989a; Bossu et al., 1998) while some non-T-type Ca^{2+} currents can be quite sensitive, notably R-type Ca^{2+} currents (Zhang et al., 1993a). In fact, Ni^{2+} block of Ca^{2+} channels depends on the subunit composition of the channel and the ion used as charge carrier (Zamponi et al., 1996; Lacinova et al., 2000). Ni^{2+} block is a combination of four events: a reduction in peak current, a slowing of activation, a slowing of inactivation, and a shift of the voltage dependence towards more positive values. As each of these events is affected differently by the molecular composition of the channels and the experimental conditions, the potency of Ni^{2+} block cannot always be reliably used to classify an expressed channel as T-type. Conversely, whereas Cd^{2+} was initially considered a weaker blocker of T-type Ca^{2+} channels than Ni^{2+} (Bonvallet, 1987; Fox et al., 1987; Hagiwara et al., 1988) the opposite has also been observed (Akaike et al., 1989b; Akaike et al., 1989a; Takahashi et al., 1990; Lacinova et al., 2000). An IC_{50} as low as 15 μM has even been reported for Cd^{2+} on T-type Ca^{2+} channels in rat pyramidal neurons (Ye et al., 1993).

Table 3. Pharmacological characteristics of T-type and HVA Ca^{2+} currents.

	T-type sensitivity	HVA sensitivity
Ni^{2+}	high, sometimes low	high (R), low (L,N,P)
Cd^{2+}	low, sometimes high	high (L,N,P,R)
ω-CTx-GVIA	no	high (N), no/low (L,P,Q)
ω-CTx-MVIIC	no	high (N,P,Q), no/low (L)
ω-Aga-IIIA	no	high (L,N,R), partial (P)
ω-Aga-IVA	no	high (P), medium (Q), no/low (N,R), no (L)
kurtoxin	medium/high	medium (L,N,P,R)
ω-Aga-IIB	high	high (L), unknown (N,P,Q,R)
DHP agonists	no	high (L), no (N,P,Q,R)
DHP antagonists	medium	high (L), low (N,P,Q,R)
mibefradil	high	medium (L,N,R)

3.2. Peptide Toxins

As mentioned previously, snail and spider toxins have proven very useful tools to block selectively high voltage-activated Ca^{2+} channels. However, T-type Ca^{2+} currents are insensitive to classical snail and spider toxins (Ertel et al., 1994; Fox et al., 1987). In the last few years, two peptide toxins have been reported to block T-type Ca^{2+} currents: kurtoxin from the scorpion *Parabuthus transvaalicus* and ω-Aga-IIB from the spider *Agelenopsis aperta*. Block by ω-Aga-IIB has only been reported in an abstract (Ertel et al., 1993): in guinea pig cardiac myocytes, the toxin is a highly potent blocker of T-type Ca^{2+} currents (IC_{50} ~ 40 nM) but it also blocks L-type Ca^{2+} currents at similar concentrations. Block by kurtoxin was initially reported for recombinant Ca^{2+} channels expressed in *Xenopus* oocytes (Chuang et al., 1998): the toxin blocked currents through $Ca_v3.1$ (α_{1G}) channels potently ($IC_{50} \sim 15$ nM) and selectively when compared to $Ca_v1.2$, $Ca_v2.1$, $Ca_v2.2$, and $Ca_v2.3$ (α_{1C}, α_{1A}, α_{1B}, and α_{1E}), although it also affected Na^+ channels. Kurtoxin was shown to slow the gating of $Ca_v3.1$ (α_{1G}) channels. Although a later study using native T-type Ca^{2+} currents confirmed this modification of gating, it failed to confirm the selectivity of kurtoxin: the toxin affected with high affinity the gating of T-type as well as L-, N-, and P-type Ca^{2+} currents in rat neurons (Sidach et al., 2002).

3.3. Classical Ca^{2+} Channel Blockers (Anti-Hypertensive Agents)

Most of the so-called "classical" Ca^{2+} channel blockers, which target primarily L-type Ca^{2+} currents, also block T-type Ca^{2+} currents with lower potency. Although this secondary activity is probably not critical to the main therapeutic action of these compounds, differences in blocking potency of T-type Ca^{2+} current may actually explain minor differences in therapeutic benefit (*e.g.*, additional anti-arrhythmic properties) or side-effect profiles (*e.g.*, reduced neuro-hormonal reflexes or increased conduction block).

The classical dihydropyridine antagonists are generally able to block T-type in addition to L-type Ca^{2+} currents with a selectivity ratio ranging from ~ 2 (oxodipine; Galan et al., 1998) to ~500 (nifedipine; Stengel et al., 1998). Interestingly, the new DHP analog PPK-5 may actually possess selectivity for T-type over L-type Ca^{2+} currents (Kumar et al., 2002); this property, however, was reported based on experiments using recombinant $Ca_v3.1$ (α_{1G}) and $Ca_v1.2$ (α_{1C}) and needs to be confirmed using additional preparations. Indeed, other dihydropyridines have occasionally been shown to block T-type Ca^{2+} currents more potently than L-type Ca^{2+} currents in specific preparations: nicardipine in rat hypothalamic (Akaike et al., 1989b) and hippocampal (Takahashi et al., 1989) neurons (Figure 2), as well as in rat aortic smooth muscle cells (Kuga et al., 1990), and felodipine in guinea pig atrial cells (Cohen et al., 1992b) (see also Romanin et al., 1992 for niguldipine). These tissue-specific results suggest that dihydropyridine block of T-type Ca^{2+} currents depends on the molecular composition of the channel subtype found in the tissue.

3.4. Mibefradil

Presently, mibefradil (Figure 2) may be the most selective small molecule blocking T-type Ca^{2+} currents. It has 10-30 fold selectivity for T-type over L-type Ca^{2+} currents in cardiac (Mangoni et al., 1997; Bénardeau et al., 1998; but see Protas et al., 2000) and vascular (Mishra et al., 1994b) myocytes as well as in endocrine cells (Rossier et al., 1998). Mibefradil is also selective for T-type over HVA Ca^{2+} currents in some (Pan et al., 2001; McDonough et al., 1998) but not all neurons (Viana et al., 1997; Randall et al., 1997). Furthermore, in side-by-side comparisons, mibefradil appears to have similar potency on all three types of cloned T-type Ca^{2+} channels ($Ca_v3.1$, $Ca_v3.2$, $Ca_v3.3$ or α_{1G}, α_{1H}, α_{1I}) (Martin et al., 2000) and on cloned and native channels ($Ca_v3.2$ or α_{1H} and guinea pig atrial myocytes; Perchenet et al., 2000).

The IC_{50} values reported for the block of T-type Ca^{2+} currents by mibefradil range from 14 nM in cerebellar Purkinje neurons (McDonough et al., 1998) to 5000 nM in spermatogenic cells (Arnoult et al., 1998). This wide span is quite likely due in part to differences in voltage and ionic conditions when the block was measured. For example, the potency of block by mibefradil decreases up to 10-fold when the charge carrier is switched from 2 mM Ca^{2+} (close to physiological conditions) to 10 mM Ba^{2+} (a very commonly used condition in pharmacology studies) (Martin et al., 2000). Similarly, the potency of block by mibefradil decreases over 10-fold when the holding potential is changed from slightly depolarized (-70 mV to –50 mV) to slightly hyperpolarized (-90 mV to -110 mV) (McDonough et al., 1998; Bénardeau et al., 1998). Therefore, due to the lack of consistency in testing conditions among laboratories, it is generally not possible to make a precise comparison of block potencies in different preparations based on published data.

Although, in the majority of cases, mibefradil was found to be selective for T-type Ca^{2+} channels over high-voltage-activated channels, there are notable exceptions. In spinal motoneurons, mibefradil blocks L-, N-, P/Q-, and T-type Ca^{2+} currents with similar potency (Viana et al., 1997). In cerebellar granule cells, mibefradil blocks R-type Ca^{2+} currents with IC_{50} ~ 0.5-1.0 μM, which is its "usual" potency on T-type Ca^{2+} currents (Randall et al., 1997). In rabbit sinus node cells, mibefradil was also reported to block L- and T-type Ca^{2+} currents with similar potency but this is likely due to the difference in holding voltage used to measure each channel (Protas et al., 2000). In addition to this lack of selectivity between voltage-operated Ca^{2+} channels in some preparations, mibefradil also blocks other ion channels with similar potency: delayed-rectifier K^+ channels (Chouabe et al., 1998; Liu et al., 1999), ATP-activated K^+ channels (Gomora et al., 1999), Na^+ channels (Eller et al., 2000), tetrodotoxin-sensitive Ca^{2+} current (Heubach et al., 2000; but see Guatimosim et al., 2001; and Cole et al., 1997 for arguments that this current is carried by Na^+ channels), as well as Ca^{2+}- and volume-activated Cl^- channels (Nilius et al., 1997).

3.5. Antiepileptics

A number of antiepileptic drugs block T-type Ca^{2+} currents with significant potency. For some of these drugs, block of T-type Ca^{2+} currents could actually be the primary mechanism of therapeutic activity (see section 6).

Ethosuximide, the preferred drug for the treatment of petit-mal epilepsy, blocks T-type Ca^{2+} currents in thalamic neurons with little effect on HVA Ca^{2+} currents (Coulter et al., 1989b). Block of T-type Ca^{2+} currents occurs in the therapeutic concentration range (0.3-0.7 mM). T-type Ca^{2+} currents are also blocked (with somewhat lower selectivity) by the therapeutically active ethosuximide analog methyl-phenyl-succinimide (MPS) but they are not blocked by the inactive analog succinimide or the pro-convulsant analog tetra-methyl-succinimide (Coulter et al., 1990). As with mibefradil, block by ethosuximide or MPS is not selective among $Ca_V3.1$, $Ca_V3.2$, and $Ca_V3.3$. Block is also voltage-dependent: IC_{50}s are ~ 10 mM for a holding voltage of –100 mV and ~ 1 mM for –75 mV (Gomora et al., 2001). The selectivity of ethosuximide between recombinant Ca^{2+} channels is unknown.

Hydantoin antiepileptic drugs (*e.g.*, phenytoin and phenobarbital) are believed to act principally through the block of Na^+ channels. Nevertheless, these drugs also block T-type Ca^{2+} currents (Twombly et al., 1988; Todorovic et al., 1998), completely or partially (~60% with IC_{50} ~ 7 μM for phenytoin, 100% with IC_{50} ~ 1.7 mM for phenobarbital). Phenytoin is slightly selective for T-type over HVA Ca^{2+} currents and for

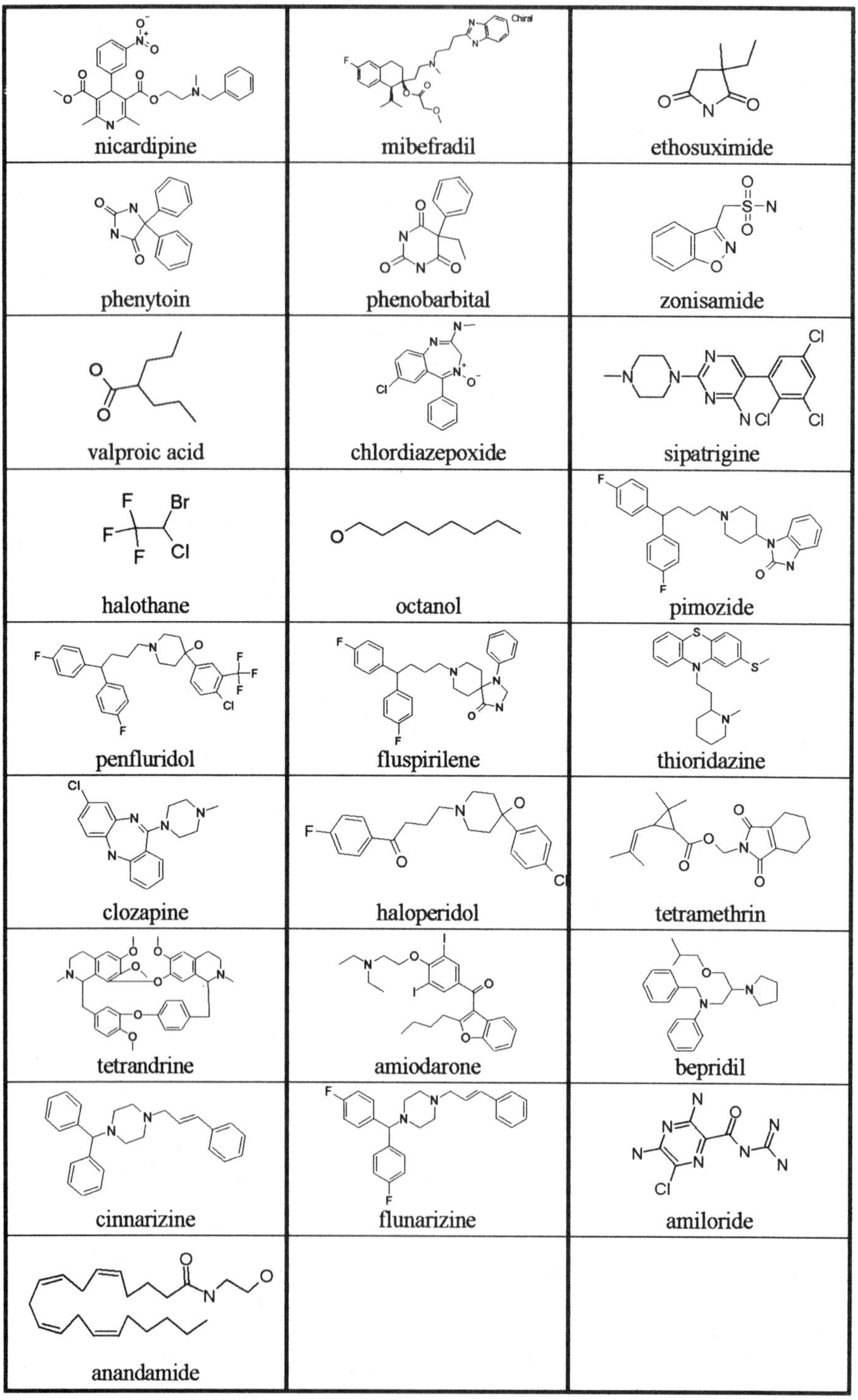

Figure 2. Chemical structures of drugs blocking T-type Ca^{2+} channels.

$Ca_v3.2$ over $Ca_v3.1$ and $Ca_v3.3$, and it also exhibits a higher affinity at depolarized voltages (Twombly et al., 1988; Todorovic et al., 2000).

Finally, zonisamide, valproic acid, chlordiazepoxide and sipatrigine have also been reported to block T-type Ca^{2+} currents, completely or partially: ~ 60% for 0.5 mM zonisamide (Suzuki et al., 1992), ~ 20% for 0.3 mM valproic acid (Todorovic et al., 1998; Todorovic et al., 2000), 100% with IC_{50} ~ 0.3 mM for chlordiazepoxide (Reuveny et al., 1993), and 100% with IC_{50} ~ 14 μM for sipatrigine (McNaughton et al., 2000). The chemical structures of some antiepileptic drugs blocking T-type Ca^{2+} channels are shown in Figure 2.

3.6. Anesthetics

The mechanism of action of general anesthetics is not well understood. Many such compounds block or enhance the activity of a variety of ion channels, which may form a component of their therapeutic activity. In particular, they often block T-type Ca^{2+} currents, which may help reduce neuronal activity.

Isoflurane, sevoflurane, enflurane, halothane, and nitrous oxide (N_2O) block native and cloned T-type Ca^{2+} currents with (sevoflurane, isoflurane, and N_2O) or without (halothane) selectivity over HVA Ca^{2+} currents (Eskinder et al., 1991; Ries et al., 1999; McDowell et al., 1999; Camara et al., 2001; Yamakage et al., 2001; Todorovic et al., 1998; Todorovic et al., 2001a; Herrington et al., 1991). For isoflurane and halothane at least, block occurs at therapeutically relevant concentrations. For most compounds, selectivity between $Ca_v3.1$, $Ca_v3.2$, and $Ca_v3.3$ and voltage dependence of block are not well characterized; an exception is N_2O, which is a selective blocker of $Ca_v3.2$ over $Ca_v3.1$ channels and exhibits no voltage-dependence of block (Todorovic et al., 2001a).

The aliphatic alcohol octanol also blocks native and cloned T-type Ca^{2+} currents with IC_{50} ~ 100-300 μM (Llinas et al., 1986; Todorovic et al., 1998; Herrington et al., 1992; Todorovic et al., 2000; Takahashi et al., 1989) but without selectivity among subtypes (Todorovic et al., 2000) or even over HVA Ca^{2+} currents (Takahashi et al., 1989).

A few other anesthetics and barbiturates have also been shown to block T-type Ca^{2+} currents (pentobarbital, thiopental, etomidate, propofol, ketamine); however, for these compounds, only 1-2 reports are available and more detailed work is clearly needed. The chemical structures of some anesthetic agents blocking T-type Ca^{2+} channels are shown in Figure 2.

3.7. Antipsychotics

The predominant view is that antipsychotic agents act primarily by blocking D2 dopaminergic receptors in the forebrain. Additional block of D1 dopaminergic, 5-HT2 serotonergic and α_2-adrenergic receptors may contribute to the therapeutic activity of some of these drugs. In addition, like anesthetics, many antipsychotics block T-type Ca^{2+} currents, an effect that may contribute to therapeutic activity. Diphenylbutylpiperidines, such as pimozide, penfluridol, and fluspirilene, are particularly potent. Unfortunately, antipsychotics are often very potent blockers of K^+ channels as well, particularly the rapid delayed-rectifier K^+ channel (herg gene product): this explains their common side-effect of prolonging the QT-interval of the ECG.

Pimozide is a second-line antipsychotic agent used in the treatment of tics in Tourette's syndrome. In addition to its anti-dopaminergic activity, pimozide is also a Ca^{2+} antagonist (Qar et al., 1987), and it blocks native and cloned T-type Ca^{2+} currents fairly potently. The compound is remarkably potent on cloned channels, with IC_{50}s in the range 0.030-0.055 μM for $Ca_v3.1$, $Ca_v3.2$, and $Ca_v3.3$ (Santi et al., 2002). It is somewhat less potent on native channels, with IC_{50}s in the range 0.3-1 μM (Enyeart et al., 1993; Bertolesi et al., 2002; Arnoult et al., 1998). Pimozide is slightly selective for T-type over L-type Ca^{2+} currents (Enyeart et al., 1990) and blocks N- and P-type Ca^{2+} currents even less potently (Sah et al., 1994).

Penfluridol was the most effective diphenylbutylpiperidine at blocking T-type Ca^{2+} currents of human thyroid cancer cells (Enyeart et al., 1992). Penfluridol is ~ 10-fold selective for T-type over L-type Ca^{2+} currents (Enyeart et al., 1993). Fluspirilene is 5-10 fold less potent than pimozide or penfluridol, whereas thioridazine, clozapine and haloperidol are ~ 100-fold less active ($IC_{50}s \geq 10$ μM). Like pimozide, haloperidol appears somewhat more potent on cloned channels, with IC_{50}s in the range 1.2-1.4 μM for $Ca_v3.1$, $Ca_v3.2$, and $Ca_v3.3$ (Santi et al., 2002). All antipsychotics appear to block T-type Ca^{2+} currents in a voltage-dependent manner (Santi et al., 2002; Arnoult et al., 1998; Enyeart et al., 1992). The chemical structures of some antipsychotic agents blocking T-type Ca^{2+} channels are shown in Figure 2.

3.8. Other Agents

Among the various non-dihydropyridine Ca^{2+} channel antagonists that block T-type Ca^{2+} channels—tetramethrin (Tsunoo et al., 1985; Hagiwara et al., 1988), tetrandrine (Liu et al., 1992; Rossier et al., 1993; Yao et al., 2002; Kwan et al., 2002), amiodarone (Cohen et al., 1992b), bepridil (Cohen et al., 1992b), cinnarizine (Cohen et al., 1992b), or flunarizine (Akaike et al., 1989a; Tytgat et al., 1988)—none is selective or even effective for T-type Ca^{2+} channels from different cells or species (see also Hirano et al., 1989; Huguenard 1996).

The mechanism of action of the potassium-sparing diuretic amiloride appears to be the block of a non-voltage-activated Na^+ channel in the kidney (ENaC). Nevertheless, high concentrations of amiloride block T-type Ca^{2+} currents selectively over other Ca^{2+} currents in mouse neuroblastoma and chick dorsal root ganglion neurons, with $IC_{50} \sim$ 30-80 μM (Tang et al., 1988). However, the extent of block varies considerably with the system used (Hirano et al., 1989; Tytgat et al., 1990b; Takahashi et al., 1989), which could be due in part to selectivity of the drug among different T-type Ca^{2+} channel isoforms (Monteil et al., 2000a; Williams et al., 1999). In any case, block occurs at relatively high concentrations, and amiloride blocks Na^+ currents with higher potency than Ca^{2+} currents.

Finally, the endogenous cannabinoid receptor ligand anandamide has been proposed recently as a potential endogenous blocker of T-type Ca^{2+} currents (Chemin et al., 2001b).

4. TISSUE LOCALIZATION

In 1984, T-type Ca^{2+} currents were first described in rat and chick sensory neurons (Carbone et al., 1984a; Carbone et al., 1984b). Since then, they have been observed in a variety of excitable and non-excitable cells (Bean 1989; Fox et al., 1987; Hess 1990).

4.1. Cardiac Muscle

T-type Ca^{2+} currents have not been consistently identified in cardiac myocytes, and significant species differences are reported. Generally speaking, T-type Ca^{2+} currents are found more frequently, and they exhibit higher current densities, in the spontaneously-beating cardiac tissues (e.g., sinoatrial node and atrial muscle) than in normally quiescent tissues (e.g., ventricular muscle). Furthermore, T-type Ca^{2+} currents tend to decrease in magnitude during pre- and post-natal development of the heart, as the ventricular muscle loses its spontaneous activity. Conversely, T-type Ca^{2+} currents become re-expressed or increase in magnitude in pathological conditions that produce de-differentiation or new growth of tissue. Table 4 shows the distribution of T-type Ca^{2+} currents in cardiac muscle of diverse species. In the various studies, the T-type Ca^{2+} currents were recorded using different voltage pulse protocols and ionic conditions, which renders difficult the comparison of current magnitudes. However, the effect of $[Ca^{2+}]_0$ on T-type Ca^{2+} current has been investigated for rabbit sinoatrial cells (Hagiwara et al., 1988), where an increase in $[Ca^{2+}]_0$ from 2.5 to 5 mM results in a 1.3-fold increase in T-type Ca^{2+} current; thus, after correction for $[Ca^{2+}]_0$ differences, one can expect that the current magnitudes will remain generally small.

4.2. Smooth Muscle Cells

T-type Ca^{2+} currents have been observed in vascular smooth muscle cells from rat aorta (Friedman et al., 1986; Akaike et al., 1989a; but see Quignard et al., 2001 for a dissenting view) and in smaller arteries from many species: coronary from human (Quignard et al., 1997), guinea pig (Ganitkevich et al., 1990), and dog (Mishra et al., 1994a); mesenteric from human (Smirnov et al., 1992a) and rat (Bean et al., 1986); cerebral from rat, rabbit, and guinea pig (Alborch et al., 1995); rabbit ear (Benham et al., 1987); and rat tail (Wang et al., 1989). They were also recorded in veins: dog saphenous (Yatani et al., 1987), rat portal (Loirand et al., 1989), and rat azygos (Mishra et al., 1994b).

T-type Ca^{2+} channels are also found in smooth muscle cells from human uterus (Young et al., 1993), sheep mesenteric lymphatic vessels (Hollywood et al., 1997), dog bronchi (Janssen 1997), toad stomach (Vivaudou et al., 1988), rat ileum (Smirnov et al., 1992b), as well as human (Xiong et al., 1995) and canine (Huizinga et al., 1991) colon.

4.3. Secretory Cells

In most secretory cells, voltage-gated Ca^{2+} channels couple electrical activity to the secretion of transmitters and hormones. Because the voltage dependences of activation and inactivation of T-type Ca^{2+} channels overlap (Figure 1), significant steady state T-type Ca^{2+} current can occur. This "window current" generated over a restricted range of potentials, where a small fraction of channels is constantly activated, could be a sufficient signal for the stimulation of hormone secretion (Lledo et al., 1990; Cohen et al., 1987; Barrett et al., 1991b). Currents with T-type properties have been described in numerous secretory cells and cell lines including rat clonal pituitary cells (Cohen et al., 1985; Armstrong et al., 1985), sheep growth-hormone-secreting pituitary somatotrophs (Chen et al., 2000), human (Payet et al., 1994) and bovine (Cohen et al., 1988) aldosterone-secreting adrenal glomerulosa cells, rat (Barbara et al., 1995) and bovine (Mlinar et al., 1993) cortisol-secreting adrenal zona fasciculata cells, human calcitonin-secreting thyroid C-cells (Berger et al., 1984; Biagi et al., 1991; Biagi et al., 1992), rat (Hollins et al., 1996; Bournaud et al., 2001) and bovine (Artalejo et al., 1994) catecholamine-secreting chromaffin cells, human (Barnett et al., 1995) and rat (Sala et al., 1990; Ashcroft et al., 1990) insulin-secreting pancreatic β-cells, and human (Barnett et al., 1995; Davalli et al., 1996) and guinea pig (Rorsman, 1988) glucagon-secreting pancreatic α-cells.

4.4. Neurons

T-type Ca^{2+} channels have been reported in sensory and central neurons and to a smaller extent in sympathetic neurons of a variety of species including human (Barnes et al., 1992; Beck et al., 1997; Carbone et al., 1990; Sah 1995; Sayer et al., 1993), rat, mouse, guinea pig, and chick. In neurons, T-type Ca^{2+} channels appear to be preferentially localized in dendrites, which may explain the occasional difficulties encountered in recording them from isolated neurons (Christie et al., 1995; Gauck et al., 2001; Karst et al., 1993; Kavalali et al., 1997; Pouille et al., 2000). The reader is referred to excellent reviews on neuronal T-type Ca^{2+} currents (Bean, 1989; Huguenard, 1996; Perez-Reyes, 2003). Table 5 (from Perez-Reyes, 2003) recapitulates our present knowledge on their localization in the nervous system.

4.5. Other Cells

T-type Ca^{2+} channels are also found in rat and mouse skeletal muscle cells (Beam et al., 1988b; Beam et al., 1988a; Dirksen et al., 1995; Garcia et al., 1994), cultured endothelial cells from human (Yokokawa et al., 1990; Yazawa et al., 2002) and cow (Bossu et al., 1992), rat osteoblasts (Chesnoy-Marchais et al., 1988), rat bone marrow cells (Publicover et al., 1995), bovine ciliary epithelial cells (Jacob, 1991), mouse fibroblasts (Chen et al., 1988), mouse (Akopian et al., 1996) and rabbit (Bringmann et al., 2000) glial cells, human spermatozoa (Shiomi et al., 1996), and human and mouse spermatogenic cells (Arnoult et al., 1996; Liévano et al., 1996; Jagannathan et al., 2002). Their function in these cells remains very speculative.

4.6. Human Tissues

T-type Ca^{2+} channels have been studied much more extensively in animals than in humans, due principally to the low availability of human intact or diseased tissues. Table 6 summarizes our present knowledge on the distribution of these channels in human cells. It is clear that T-type Ca^{2+} channels have been found in much the same locations in humans and animals, with the notable exception of cardiac myocytes. Since the presence and density of these channels in contractile heart myocytes is highly species-dependent, this is not a surprising result. In pacemaking cells from animals, T-type Ca^{2+} channels are more consistently found; unfortunately, as could be expected, human Purkinje or nodal tissues have never been studied in that respect. Although T-type Ca^{2+} channels were never observed in adult ventricular tissue, they have been observed in ventricular myocytes from human fetus, consistent with the observation that the levels of $Ca_v3.2$ (α_{1H}) transcripts in human ventricular cells decrease with heart development (Qu et al., 2001).

Table 4. Distribution of T-type Ca^{2+} currents in cardiac muscle.

Tissue	State	Species	Density pA/pF	$[Ca]_o$ mM	References
Ventricle	Adult healthy	Human	No	2	(Beuckelmann 1991)
		Dog	1	5	(Tseng 1989)
		Cat	No	5	(Nuss 1993)
		Calf	No		(Bean 1989)
		Rabbit	Yes	10	(Wetzel 1991)
		Hamster	6	20Ba	(Bkaily 1997; Sen 1994)
		Guinea pig	1	2-5	(Nilius 1985; Balke 1992; Zygmunt 1990)
		Rat	No	5	(Richard 1990; Tytgat 1990a)
		Frog	No	1.8	(Argibay 1988; Bean 1989)
		Shark	10	5	(Maylie 1995)
		Squid	Yes	10	(Odblom 2000)
		Lymnaea	Yes	3.5	(Yeoman 1999)
	Fetal or neonatal	Human	Yes	2-10	(Bkaily 1995; Bkaily 1992)
		Rat	3	5-10	(Furukawa 1992; Ferron 2002)
		Mouse	3	2.5	(Cribbs 2001; Nuss 1994)
	Hyper-trophy	Cat	1.8	5	(Nuss 1993)
		Rat	4.8	5	(Martinez 1999)
	Infarct	Rat	1.1	1.8	(Huang 2000)
	Myopathy	Hamster	12	20	(Bkaily 1997; Sen 1994)
	Dedifferen-tiated cells	Rat	1.2	1.8	(Fares 1996)
Atrium	Adult healthy	Human	No	2	(Ouadid 1991; Li 1997a)
		Dog	0.3	5	(Bean 1985)
		Cat	0.7	2.7	(Zhou 1994)
		Calf	No		(Bean 1989)
		Rabbit	No		(Bean 1989)
		Guinea pig	Yes	5	(Bénardeau 1998)
		Rat	0.4	2	(Xu 1990; Leuranguer 2000)
		Frog	Yes	1.8	(Bonvallet 1987)
Purkinje	Adult healthy	Dog	1.7-2.9	2-5	(Hirano 1989; Tseng 1989)
Sinoatrial node	Adult healthy	Rabbit	2.6	2.5	(Hagiwara 1988)
Atrial latent pacemaker	Adult healthy	Cat	3.3	2.7	(Zhou 1994)

Table 5. Properties of T-type Ca^{2+} currents in isolated neurons.

Tissue	Dival mM	Thres mV	IV_{peak} mV	I_{max} pA	$V_{0.5}$ mV	h∞ mV	τ_{inact} ms	τ_{recov} ms	Nickel µM	Ref.
Basal forebrain										
Striatum	5 Ca	-60	-40	-90	-53[M]	-88		288		1
Accumbens	5 Ca	-64	-33	-51	-43[C]	-80		397		2
Septum	2.5 Ca	-54		-1320	-40[M]	-49	16		5	3
Septum	2 Ba	-70	-40	-500	-59[M]	-84				4
Amygdala	10 Ca	-60	-20	-20	-38[C]	-70			300	5
Cerebral cortex										
Temporal	2 Ca	-60	-30	-47		-86		400	50(28%)	6
Piriform	5 Ca	-60	-30	-250	-44[T]	-86	22			7
Hippocampus										
Dentate granule	2 Ca	-50	-30	-224		-65	25			8
LM interneurons	2 Ca	-60	-30	-100	-48[M]	-84	12	100	400	9
CA1 pyramidal	10 Ca	-60	-30	-150	-46[C]	-79	15	430 7000		10
CA1,3 pyramidal	2 Ba	-57		-100		-76	25	820		11
Thalamus										
Ventrobasal	3 Ca	-60	-25	-350	-52[T]	-84	20	251		12
Ventrobasal	3 Ca	-60	-30	-280	-59[T]	-81	30	350	200(66%)	13
Ventrobasal	5 Ca	-70		-1400		-96	15	353		14
LGN	10 Ca	-48	-15	-186		-67	36	299	500(83%)	15
LGN	5 Ca	-60	-23	-200	-45[M]	-64	20	615	50(100%)	16
LGN-interneuron	2 Ca	-70		-110	-55[M]	-81	20	218		17
LGN-relay	2 Ca	-70		-289	-64[M]	-86	20	206		18
Reticular	3 Ca	-50	-10	-131	-49[T]	-78	80	570	200(54%)	19
Reticular	2 Ca	-70	-50	-128	-64[T]	-85	25	275	50(38%)	20
Lateral habenula	1 Ca	-65	-35	-150	-62[T]	-81	58[W]	507		21
Hypothalamus										
Hypothalamus	10 Ca	-65	-30			-93	40	940	600	22
Hypothalamus	10 Ca	-60	-30	-150	-46[C]	-79	15	215		23
Hypothalamus	5 Ca	-50	-20	-400			20		50(50%)	24
Supraoptic	2 Ca	-64	-60	-250		-65			200	25
Supraoptic	2 Ca	-60	-30	-400			42		50(84%)	26
Paraventricular	10 Ca	-59	-33	-402		-67	12		100(67%)	27
Midbrain & pons										
Area postrema	5 Ca	-65	-10	-36		-72			10(100%)	28
Cochlear nucleus	2 Ca	-75	-40	-400	-59[M]	-89			500(50%)	29
Cerebellum										
Purkinje	10 Ca	-60	-20	-400	-40[C]	-79	23	3300	110	30
DRG										
-	2.5 Ca	-65	-35	-2000		-82			100(95%)	31
-	10 Ba	-60	-25		-41[C]	-60	25	330		32
Nodose ganglion										
-	5 Ca	-60	-20	-400		-48	30	400		33
-	10 Ca	-55	-15	-400	-29[C]	-44	30			34
Pelvic ganglion	10 Ca	-50	-25	-500	-38[T]	-57			10	35
Olfactory										
-	2.6 Ca	-57	-36	-50		-82				36
-	3 Ca	-70	-40	-60	-45[C]	-66			100(71%)	37
Retina										
Rod bipolar	10 Ca	-60	-30	-35	-36[T]	-59	30		1000(80%)	38
Cone bipolar	10 Ca	-60	-20	-66	-35[T]	-60	55		1000(85%)	39
Oligodendrocytes	20 Ba	-50	-25	-150		-63	10			40

Table 5, continued. The divalent cation concentration used to measure the currents is shown. Properties of the current-voltage relationship are described in terms of the voltage threshold where currents become detectable, the voltage where they peak (IV_{peak}), and the maximum amplitude observed (I_{max}). Methods used to calculate midpoints of the activation curves ($V_{0.5}$) are denoted with the following superscripts: C, chord conductance; T, tail currents; M, I/I_{max}. The midpoint of the steady-state inactivation curve is reported ($h\infty$). Kinetics of inactivation (τ_{inact}) can vary with test potentials; therefore, the values shown represent the voltage-independent rate, which was typically measured at -10 mV. W superscript indicates that currents decayed in a biexponential manner; to simplify comparisons, a weighted τ was calculated according to the following equation $\tau_W = A_1\tau_1 + A_2\tau_2$, where A represents the fraction of current decaying with the respective time constant. The values shown for recovery from inactivation represent the voltage-independent rate, which was typically measured at -90 mV. Tail current kinetics vary continuously with repolarization potential, so when possible the values shown represent those obtained at -90 mV. In some cases recovery from inactivation was found to be biphasic, so the values represent the tau of the fast component, the fraction of channels that recover fast (in parentheses), and the tau of the slow component. Nickel responses represent either the IC_{50} value obtained or the concentration tested and the percent block observed (in parentheses). LGN, lateral geniculate nucleus; MDN, magnocellular dorsal nucleus; LM, lacunosum-moleculare of hippocampus; DRG, dorsal root ganglion; inact., inactivation; recov., recovery; dival., divalent; thres., threshold. References: 1 Hoehn et al., 1993; 2 Churchill et al., 1998; 3 Allen et al., 1993b; 4 Griffith et al., 1994; 5 Kaneda et al., 1989; 6 Sayer et al., 1993; 7 Magistretti et al., 1998; 8 Zhang et al., 1993b; 9 Fraser et al., 1991; 10 Takahashi et al., 1991; 11 Thompson et al., 1991; 12 Coulter et al., 1989a; 13 Huguenard et al., 1992; 14 Kuo et al., 2001; 15 Suzuki et al., 1989; 16 Hernandez-Cruz et al., 1989; 17 Pape et al., 1994; 18 Pape et al., 1994; 19 Huguenard et al., 1992; 20 Tsakiridou et al., 1995; 21 Huguenard et al., 1993; 22 Akaike et al., 1989b; 23 Takahashi et al., 1991; 24 Muller et al., 1992; 25 Erickson et al., 1993; 26 Fisher et al., 1995; 27 Luther et al., 2000; 28 Hay et al., 1996; 29 Doughty et al., 1998; 30 Kaneda et al., 1990; 31 White et al., 1989; 32 Nakashima et al., 1998; 33 Bossu et al., 1985; 34 Lambert et al., 1997; 35 Lee et al., 2002; 36 Wang et al., 1996b; 37 Kawai 1999, Kawai et al., 2001; 38 Pan, 2000; 39 Pan, 2000; 40 Blankenfeld et al., 1992. Adapted from Perez-Reyes, 2003.

5. MODULATION

In many tissues, neurotransmitters or hormones modulate the activity of T-type Ca^{2+} currents. In the studies reviewed below (summarized in Table 7), divergent or conflicting results are sometimes reported. As suggested by Alvarez et al. (1992), this may be due to species or tissue specificity, or even to alteration of the Ca^{2+} channels following enzymatic dissociation. For example, in frog atrial cells, these authors observe a T-type Ca^{2+} current in 91% of the cells dissociated with collagenase and trypsin but only in 30% of the cells dissociated with pronase. In many cases, there is still a need to link more strongly the relevance of results from studies in isolated cells to the physiology of whole organs or animals.

T-type Ca^{2+} currents are usually not affected by β-adrenergic stimulation, which tends to increase the open probability of L-type Ca^{2+} channels via a cAMP-dependent protein kinase (PKA) (Bean, 1985; Bonvallet, 1987; Hirano et al., 1989; Maylie et al., 1995). Indeed, in most patch-clamp experiments, the magnitude of T-type Ca^{2+} currents does not decrease spontaneously (no "run-down") and their properties do not change much with time: this suggests that they are not PKA-phosphorylated or that such phosphorylation does not modify their electrophysiological properties very much (Ohya et al., 1989). Nevertheless, isoproterenol can increase T-type Ca^{2+} currents in frog atrial cardiomyocytes (Alvarez et al., 1996) and guinea pig ventricular myocytes (Mitra et al., 1986; Zhang et al., 2002a).

Table 6. T-type Ca^{2+} channels in human cells.

	Cell type	Origin	(a)	Size	(b)	References
Myocytes	ventricle	heart failure	fresh	0		(Beuckelmann 1991)
	atrium	coronary disease	fresh	0		(Ouadid 1991)
	mesenteric artery	icolec-/gastrectomy	fresh	+	Most	(Smirnov 1992a)
	coronary artery	heart transplant	fresh	0		(Quignard 1997)
			culture	+++++	Most	
	colon	colonic disease	fresh	+++++	50%	(Xiong 1995)
	uterus	cesarean delivery	fresh	+++++	Most	(Young 1993)
	unspecified skeletal	normal & dystrophy	culture	+++++	Most	(Rivet 1990)
	vastus lateralis	normal & hyperthermia	culture	+++++	57%	(Lehmann-Horn 1995)
Endocrine	glomerulosa cells	renal transplant donors	fresh	+++++	Most	(Payet 1994)
	chromaffin cells	carcinoma	fresh	0		(Gandia 1998)
	pancreatic b-cells	life-supported cadavers	culture	++	75%	(Barnett 1995)
	pancreatic b-cells	renal transplant donors	fresh	+++	25%	(Davalli 1996)
	medullary thyroid	carcinoma	culture	+++++	Most	(Biagi 1992)
	pituitary, GH	acromegalic tumor	culture	++++	Most	(Takei 1996)
	pituitary, ACTH	Cushing's disease	culture	+++++	Most	(Guerineau 1991)
Neurons	fetal central	9-16 week fetus	culture	++	80%	(Sah 1995)
	neocortex pyramidal	epileptics	fresh	++++	90%	(Sayer 1993)
	hippocampal granule	epileptics	fresh	+++++	Most	(Beck 1997)
	neuroblastoma	cell line	culture	+++	Most	(Carbone 1990)
	neuroblastoma	cell line	culture	++	73%	(Kito 1996)
	neuroblastoma	cell line	culture	+++	43%	(Urbano 1997)
	retinoblastoma	cell line	culture	+++++	Most	(Barnes 1992)
	spermatogenic cells	healthy men	fresh	+	Few	(Shiomi 1996)
	endothelial cells	umbilical vein / aorta	culture	(c)	Most	(Yokokawa 1990)

(a) Cells were either studied shortly (<48h) after isolation ("fresh") or after multiple platings ("culture"). (b) Number of studied cells exhibiting T-type Ca^{2+} channels. (c) The presence of T-type Ca channels is suggested by the fact that Ni^{2+} decreases the influx of Ca^{2+} induced by endothelin-1.

In addition, Growth Hormone Releasing Hormone (GHRH) enhances T-type Ca^{2+} currents in sheep pituitary somatotrophs (Chen et al., 2000), most likely through an increase in cAMP followed by PKA-mediated phosphorylation. In contrast, the enhancements of T-type Ca^{2+} currents cells by insulin in rat GH_3 (Meza et al., 1994) or by ACTH or Vasoactive Intestinal Peptide (VIP) in rat adrenal fasciculata (Barbara et al., 1995) are probably due to the insertion of new channels in the plasma membrane. Some more conflicting data are also available: in GH_3 cells, glucocorticoids are reported to increase (Fomina et al., 1993) or decrease (Meza et al., 1994) T-type Ca^{2+} currents, and

Table 7 Modulation of T-type Ca^{2+} currents. Changes in parentheses occur less frequently.

Modulator	⇑/⇓/ no change	Receptor	Mechanism
Isoproterenol	= (⇑ ⇓)	?	?
GHRH	⇑	GHRH receptor	PKA
ACTH / VIP / insulin	⇑	?	Insertion of new channels
Glucocorticoid / estrogen	⇑ ⇓	?	?
Norepinephrine	⇑ ⇓ =	?	?
Carbachol	⇑ ⇓ =	m_3, m_5, (m_1)	PKA, PKC
Substance P	⇑ ⇓	?	?
Serotonin	⇑ ⇓	5-HT_{1A}, 5-HT_{1D}	G-protein independent
GABA	⇓ (⇑ =)	$GABA_B$	?
Angiotensin II	⇑	AT_1	CaMKII and/or G_i
	⇓	AT_2	PTX-insensitive G-protein
Endothelin-1	⇑ (=)	?	PKC
ANP	⇓	?	?
Somatostatin	⇓ (=)	?	PTX-sensitive G-protein
Bradykinin	⇓	?	?
Morphine	⇓ =	μ-opioid	PTX-sensitive G-protein
Nociceptin	⇓ =	ORL_1	G-protein independent
Dopamine	⇓	D_1	$G_{s,\beta\gamma}$ and cAMP
	⇓	D_2	PTX-sensitive G-protein
Arachidonic acid	⇓	Direct effect	
Strong depolarization	⇑ (=)		Voltage or Tyr kinase or Gi-protein or Ca^{2+}/Na^+ interaction

the estrogen 17 beta-estradiol was found to increase T-type Ca^{2+} currents in GH_3 cells (Ritchie 1993) but decrease them in mouse spermatogenic cells (Espinosa et al., 2000).

Conflicting results have also been reported on α-adrenergic modulation of T-type Ca^{2+} channels. Thus, norepinephrine enhances T-type Ca^{2+} currents in rat portal vein smooth muscle cells (Pacaud et al., 1987), but it decreases them in chick DRG and sympathetic ganglion (SG) neurons (Marchetti et al., 1986). Moreover, norepinephrine increases T-type Ca^{2+} currents in 70% of canine Purkinje cells but decreases it in 30% (Tseng et al., 1989). Finally, in a few preparations, norepinephrine selectively reduced the activity of HVA Ca^{2+} currents while having no effects on T-type Ca^{2+} currents (Song et al., 1991; Li et al., 1998).

Muscarinic agonists can also enhance or decrease T-type Ca^{2+} currents depending on the preparation. For example, carbachol, acting through muscarinic receptors, increases T-type Ca^{2+} currents (~55%) in rat hippocampal interneurons (Fraser et al., 1991) but decreases them (~90%) in hen granulosa cells (Wan et al., 1996). In addition, muscarinic agonists have no effect on T-type Ca^{2+} currents in rat magnocellular neurons (Allen et al., 1993a), dog bronchi (Janssen, 1997), or shark cardiac myocytes (Maylie et al., 1995). A dissection of the actions of the various muscarinic receptors was performed using recombinant receptors expressed in 3T3 cells (Pemberton et al., 2000). These cells express a native T-type Ca^{2+} current, which becomes up-regulated (~25%) by external acetylcholine when cells are transfected with either m3 or m5 muscarinic receptors.

Pharmacological dissection of the mechanism suggests that the increase in current is due to the activation of PKA following an increase in intracellular cAMP. In contrast, the native T-type Ca^{2+} current is unaffected by acetylcholine when cells are transfected with either m2, or m4 muscarinic receptors, unless the cells are "pre-activated" by forskolin. In that case, acetylcholine decreases both intracellular cAMP and the T-type Ca^{2+} current. Finally, the native T-type Ca^{2+} current is unaffected by acetylcholine when cells are transfected with m1 muscarinic receptors (despite the fact that intracellular cAMP is increased), unless PKC is "pre-inhibited" in the cells. In that case, acetylcholine increases both intracellular cAMP and the T-type Ca^{2+} current. Thus, muscarinic regulation of T-type Ca^{2+} currents may depend simultaneously on the PKA and PKC pathways.

Divergent results with the same neurotransmitter are also observed with substance P. This compound (as well as neurotensin) reduces T-type Ca^{2+} currents slightly (~25%) in rat cholinergic neurons (Margeta-Mitrovic et al., 1997) but it stimulates the currents (~45%) in most neurons in a slice preparation of dorsal horn (although, in some of the neurons, the currents were inhibited) (Ryu et al., 1990).

Similarly, serotonin increases T-type Ca^{2+} currents (~60%) in rat hippocampal interneurons (Fraser et al., 1991) but reduces them (~25%) in sensory neurons from *Xenopus* larvae (Sun et al., 1997; Sun et al., 1999). In *Xenopus*, serotonin acts through both 5-HT_{1A} and 5-HT_{1D} receptors followed by a membrane-delimited mechanism that does not seem to involve G-proteins. Serotonin did not produce an $h\infty$ shift, but effects on open probability Po were not evaluated. In rat adrenal glomerulosa cells, serotonin acts through 5-HT_7 receptors to increase T-type Ca^{2+} currents (~110%) and stimulate aldosterone secretion; this effect appears to depend on the PKA pathway (Lenglet et al., 2002). Serotonin also stimulates aldosterone secretion in humans, acting through 5-HT_4 receptors, but the actual mechanism is not known.

In the ND7–23 cells, T-type Ca^{2+} currents were reduced (~55%) by the $GABA_B$ agonist baclofen (Kobrinsky et al., 1994); a similar inhibition was reported in hippocampal interneurons (Fraser et al., 1991). However, in mouse brain stem neurons, baclofen increased the currents ~30% (Zhang et al., 1999) and, in rat DRG neurons, it increased them at low concentrations but decreased them at higher concentrations (Scott et al., 1990). No effect was seen in rat globus pallidus neurons (Stefani et al., 1999).

Angiotensin II (AII) can also either enhance or reduce T-type Ca^{2+} currents although, in this case, the mechanism is better understood. In adrenal glomerulosa cells, an enhancement of T-type Ca^{2+} currents is mediated through the binding to AT_1-receptors (McCarthy et al., 1993; Cohen et al., 1988; Bonvallet et al., 1989) followed by the activation of two distinct pathways in parallel: the calmodulin-dependent protein kinase II (CaMKIIγc) (Lu et al., 1994; Wolfe et al., 2002), and the GTP-binding protein G_i (Lu et al., 1996). In non-differentiated NG108-15 cells, a small reduction of T-type Ca^{2+} currents (~25%) is mediated through the binding to AT_2-receptors, followed by the activation of a pertussis toxin-insensitive G-protein (Buisson et al., 1992; Buisson et al., 1995).

Endothelin-1 enhances T-type Ca^{2+} currents more consistently. This hormone increases T-type Ca^{2+} currents in human endothelial cells (Yokokawa et al., 1990), in smooth muscle cells from guinea pig (Inoue et al., 1990) but not rat (Gordienko et al., 1994), and in ventricular myocytes from human fetus (Bkaily et al., 1995), chick (Bkaily et al., 1995), and neonatal rat (Furukawa et al., 1992). The modulation of T-type Ca^{2+} channels by endothelin-1 probably involves the activation of protein kinase C (PKC) (Furukawa et al., 1992). However, the effects of PKC on T-type Ca^{2+} currents are still controversial since both enhanced currents (*e.g.*, in rat ventricular myocytes; Furukawa et al., 1992) and decreased currents (*e.g.*, in rat GH3 cells, Marchetti et al., 1988; or bovine adrenal glomerulosa cells, Tseng et al., 1991) have been observed.

Conversely, many hormones and neurotransmitters reduce T-type Ca^{2+} currents. This is generally accomplished through one or both of two mechanisms: 1) a decrease in the opening probability of the channels (Po decrease) reducing the macroscopic current through an effect on single channels, or 2) a shift of the voltage dependence of inactivation to more negative potentials ($h\infty$ shift) decreasing the availability of functional T-type Ca^{2+} channels.

Atrial natriuretic peptide (ANP or ANF) decreases the magnitude of T-type Ca^{2+} currents in bovine adrenal glomerulosa cells ($h\infty$ shift) (Barrett et al., 1991a; Barrett et al., 1991b; McCarthy et al., 1990). Because the shift in the voltage dependence of inactivation is not accompanied by a change in the voltage dependence of activation, the Ca^{2+} window current decreases in amplitude and aldosterone secretion is reduced. In contrast, L-type Ca^{2+} currents are enhanced by ANP, which could provide a means of fine-tuning the inhibition of total Ca^{2+} currents.

Somatostatin reduces T-type Ca^{2+} currents in rat somatotrophs ($h\infty$ shift) via a pertussis-sensitive G-protein (Chen et al., 1990). However, somatostatin does not affect T-type Ca^{2+} currents in neuroblastoma N1E-115 cells (Bean, 1989).

Bradykinin inhibits T-type Ca^{2+} currents in ND7–23 and F-11 cells (20-50%), two cell lines derived from DRG (Kobrinsky et al., 1994; Kusano et al., 1993). This effect was pertussis-toxin-sensitive in F-11 cells but not in ND7–23 cells.

Opioids also appear to down-regulate T-type Ca^{2+} currents, although the effect is small and variable and it remains somewhat controversial. In one study in rat DRG neurons, the enkephalin analog DAGO, morphine, and morphiceptin inhibited T-type Ca^{2+} currents (0-100%) by activating μ-opioid receptors (Schroeder et al., 1991). However, morphine was found to have no effects on T-type Ca^{2+} currents in another study with rat DRG neurons (Moran et al., 2000; Abdulla et al., 1997) as well as in mouse nociceptive trigeminal ganglion neurons (Borgland et al., 2001). In a neuroblastoma cell line, an enkephalin analog inhibits T-type Ca^{2+} currents (20-40%) via a pertussis-sensitive G-protein (Kasai, 1992). In these cells, the activation or the down-regulation of protein kinase C, the addition of arachidonic acid, and the activation of PKA had no effects.

Nociceptin also inhibits T-type Ca^{2+} currents in rat DRG neurons (>90%) by a G-protein-independent mechanism (Abdulla et al., 1997). In this preparation, nociceptin did not produce an $h\infty$ shift, while effects on Po were not evaluated. On the other hand, nociceptin has no effect on T-type Ca^{2+} currents from mouse nociceptive trigeminal ganglion neurons (Borgland et al., 2001).

Dopamine effects on T-type Ca^{2+} currents probably represent the best-characterized modulation of this channel type by a neurotransmitter. Dopamine, or a selective dopamine receptor agonist, inhibits T-type Ca^{2+} currents in chick DRG and SG neurons (~60%, Marchetti et al., 1986), in bass retinal cone-type horizontal cells (~55%, Pfeiffer-Linn et al., 1993), and in rat endocrine cells (adrenal glomerulosa, 30-60%, Osipenko et al., 1994; Drolet et al., 1997), melanotrophs (30-50%, Nussinovitch et al., 1992; Keja et al., 1992), and lactotrophs (~25%, Lledo et al., 1990; Lledo et al., 1992). The mechanism of action of dopamine remains controversial and may be different in the different tissues: dopamine produces an h∞ shift in lactotrophs but it may only reduce Po in the other cell types. In most cases, dopamine appears to act through activation of D_2 receptors followed by a pertussis-sensitive G-protein, but D_1 receptors, the $G_{\beta\gamma}$-subunit of the G_s-protein, and a rise in intracellular cAMP may be involved in adrenal glomerulosa cells (Drolet et al., 1997; but not Osipenko et al., 1994) and in bass horizontal cells (Pfeiffer-Linn et al., 1993).

Arachidonic acid (AA) decreases native T-type Ca^{2+} currents in NG108-15 neuroblastoma x glioma hybrid cells (Schmitt et al., 1995a) as well as currents through recombinant $Ca_V3.2$ (α_{1H}) (Zhang et al., 2000). The effect is slow, partly reversible, and it is accompanied by a small h∞ shift. It is likely that AA directly affects channel activity without intermediate messengers.

Strong depolarizing voltages can also modulate the activity of T-type Ca^{2+} currents, in addition to the "normal" voltage-dependent gating of the channels. This phenomenon, called "facilitation", has been observed for both native and recombinant T-type Ca^{2+} channels. With native channels, facilitation was observed in rat bone marrow cells (Publicover et al., 1995), mouse spermatogenic cells (Arnoult et al., 1997), and myocytes from guinea pig coronary (Ganitkevich et al., 1990; Mishra et al., 1994a) and bullfrog atria (Alvarez et al., 1996) but not guinea pig ventricle (Mitra et al., 1986; Zygmunt et al., 1990). Consistent with this "tissue distribution," facilitation was only observed with recombinant $Ca_v3.3$ (α_{1I}) channels but not with $Ca_v3.1$ or $Ca_v3.2$ (α_{1G}, α_{1H}) (Chemin et al., 2002a; Gomora et al., 2002; Klockner et al., 1999). Facilitation increases and saturates with the magnitude and the duration of the strong depolarizing pulses, and it is larger at physiological than at room temperature. Facilitation is slowly reversible upon cessation of the depolarizing pulses. Facilitated currents are larger (50-200%) and they generally inactivate faster than basal currents.

The mechanism of facilitation of T-type Ca^{2+} channels remains uncertain, and it may be different in different cell types. In rat bone marrow cells, facilitation did not require intracellular ATP or GTP and was not affected by protein kinase inhibitors, suggesting a direct effect of voltage on channel activity (Publicover et al., 1995). In mouse spermatogenic cells, tyrosine kinase inhibitors increased T-type Ca^{2+} currents under basal conditions but not once facilitation had occurred, whereas tyrosine phosphatase inhibitors reduced T-type Ca^{2+} currents under basal conditions and prevented facilitation, suggesting a role for tyrosine phosphorylation in facilitation (Arnoult et al., 1997). In bullfrog atrial cells, facilitation apparently occurs through a voltage-dependent relief of a Gi-protein-mediated inhibitory tone (Alvarez et al., 1996) as well as a voltage-dependent interaction of Ca^{2+} and Na^+ ions at the permeation sites (Alvarez et al., 2000).

6. IN VIVO PHARMACOLOGY AND FUNCTION

The lack of highly specific blockers has hindered the elucidation of the function of T-type Ca^{2+} currents. Indeed, a remarkable variety of physiological roles have been proposed for these currents, but few, if any, have been unequivocally demonstrated at this time.

6.1. T-type Ca^{2+} Channels in Growth and Development

Several lines of evidence suggest an important contribution of T-type Ca^{2+} channels to the signal transduction pathways regulating growth, differentiation, tissue remodeling and repair. Indeed, the expression of the various T-type Ca^{2+} channels is highly regulated during embryogenesis and fetal development as well as during the perinatal period. Overall the following results suggest an important regulatory role for T-type Ca^{2+} channels in cellular proliferation, whether the growth is associated with normal development, pathological evolution, or injury repair.

6.1.1. Growth and Differentiation

T-type Ca^{2+} channels are predominantly expressed in the early stages of differentiation of many embryonic and neonatal systems, whereas upon maturation, they decrease in density and can even disappear completely. Such differential expression of T-type Ca^{2+} channels has been reported in neuronal cells from chick, mouse, and rat (Gottman et al., 1988; Gottmann et al., 1991; Kostyuk et al., 1993; Yaari et al., 1987; McCobb et al., 1989; Desmadryl et al., 1998; Martin-Caraballo et al., 2001; reviewed in Perrier et al., 2000), in human retinoblastoma cells (Hirooka et al., 2002), in glial cells from rabbit (Bringmann et al., 2000), in skeletal muscle cells from rat and mouse (Rohwedel et al., 1994; Beam et al., 1988a; Berthier et al., 2002), and in atrial and ventricular myocytes from chick and rat (Furukawa et al., 1992; Kawano et al., 1989; Leuranguer et al., 2000; Ferron et al., 2002). As an example, in chick embryonic (Kawano et al., 1989) and rat neonatal (Furukawa et al., 1992) ventricular myocytes, T-type Ca^{2+} currents predominate over L-type, with densities of 3 and 4.2 pA/pF, but they are small or absent in adult cells. Although a developmental study of T-type Ca^{2+} currents in human cardiac ventricle has not been published, a similar pattern exists for $Ca_v3.2$ (α_{1H}) transcripts in that tissue (Qu et al., 2001).

Before disappearing upon cellular maturation, T-type Ca^{2+} currents may be important in providing the cytoplasmic calcium needed for diverse cellular functions, such as control of gene expression, synapse formation, and neurotransmitter phenotype. In *Xenopus* neurons, for instance, the T-type Ca^{2+} currents are responsible for the spontaneous Ca^{2+} spikes that are necessary for K^+ channel modulation (Spitzer et al., 1993): thus they play an important role in specific aspects of the differentiation process (Gu et al., 1993). A remarkable exception to this pattern of T-type Ca^{2+} channel expression occurs in the rabbit sinoatrial node, where the expression of T-type Ca^{2+} channels remains unchanged with development, while that of L-type Ca^{2+} channels increases (Protas et al., 2001). This exception may be linked to the fact that T-type Ca^{2+} currents are critical in the generation of rhythmic activity in mature sinoatrial node cells (see below).

In human skeletal muscle; recent studies have shown that T-type Ca^{2+} currents are important for myoblast fusion (Bijlenga et al., 2000). In particular, T-type Ca^{2+} currents can be large enough to increase intracellular Ca^{2+} and thereby trigger fusion, both of which are prevented by nickel, amiloride, or antisense nucleotides against $Ca_v3.2$. Remarkably, the opening of T-type Ca^{2+} channels in myoblasts requires the previous activation of K^+ channels in order to hyperpolarize the cells into the voltage range of T-type Ca^{2+} window currents (Bijlenga et al., 2000; Bernheim et al., 2002).

In several instances, a positive correlation has been observed between T-type Ca^{2+} channel expression and cell proliferation. For example, in rat atrial myocytes, a sharp increase in T-type Ca^{2+} current density (up to 1.3-1.4 pA/pF) occurs during the period of rapid growth that characterizes normal postnatal development of this tissue (Xu et al., 1992). In this model, the current density correlates directly with the rate of cardiac growth (Figure 3). Cellular proliferation has also been closely associated with the

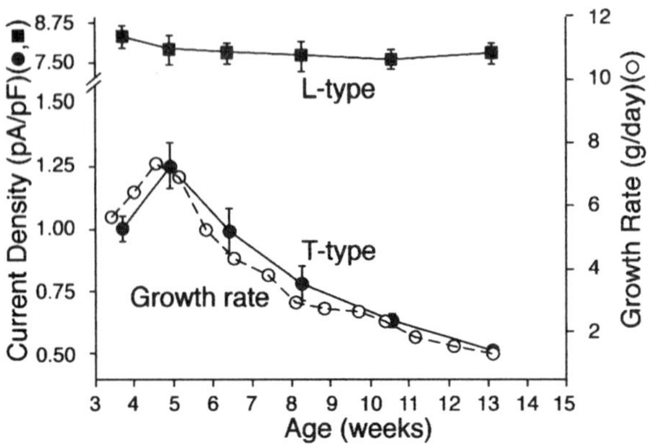

Figure 3. T-type Ca^{2+} currents in growth and development. Cardiac growth rate and T- and L-type Ca^{2+} current densities in atrial myocytes are plotted as a function of the postnatal age of female rats. The T-type Ca^{2+} current density correlates well with growth rate (correlation coefficient = 0.97 for female; 0.83 for male, not shown). Reproduced from Xu et al., 1992.

presence of T-type Ca^{2+} currents in various cultured cells including bovine capillary endothelial cells (Bossu et al., 1992) and rat aortic smooth muscle cells (Richard et al., 1992). In the latter, the expression of T-type Ca^{2+} channels coincides with the loss of contractility associated with growth, whereas the expression of L-type channels is predominant in contracting, non-proliferating cells. At confluence, both the mitotic processes and the expression of T-type Ca^{2+} channels are stopped, leading to the disappearance of T-type Ca^{2+} currents. Furthermore, the expression of T-type Ca^{2+} channels depends on the cell cycle (Kuga et al., 1996). Thus, in rat aortic smooth muscle cells, T-type Ca^{2+} channels are expressed in the G_1-phase, most abundantly in the proliferation S-phase, but not in the G_o-phase, a result consistent with the preferential presence of T-type Ca^{2+} currents in proliferating cells. Recently, more direct evidence for a role of T-type Ca^{2+} channels in cell growth came form experiments showing that over-expression of recombinant $Ca_v3.2$ channels in HEK293 cells significantly increases their proliferation rate in cell cultures (Wang et al., 2002; but for a contradictory result, see Chemin et al., 2000).

Further evidence for the growth-regulating role of T-type Ca^{2+} channels comes from their involvement in the sustained intracellular Ca^{2+} increase induced by a variety of mitogens including angiotensin II (Ganten et al., 1975), endothelin (Hirata et al., 1989), and platelet-derived growth factor (PDGF) (Estacion et al., 1993; Wang et al., 1993). Hence, the progression of density-arrested mouse fibroblasts to the S-phase after PDGF stimulation correlates with the presence of T-type Ca^{2+} channels. Furthermore, the T-type Ca^{2+} current density increases 4- to 10-fold in GH_3 anterior pituitary (Ritchie 1993) cells grown in the presence of estrogen, whereas the L-type Ca^{2+} current remains unchanged; this process is blocked by cycloheximide and consequently requires protein synthesis. In prostate tumors, neuroendocrine differentiation of epithelial cells is associated with increased tumor aggressivity and invasiveness; in culture, this differentiation is associated with increased levels of T-type Ca^{2+} currents as well as $Ca_v3.2$ (α_{1H}) transcripts (Mariot et al., 2002). Similarly, a threefold increase in T-type Ca^{2+} current density is observed in rat atrial myocytes following treatment with growth hormone-secreting tumors (Xu et al., 1990). This increase, which precedes those of heart weight and myocyte volume, could be a cause rather than a consequence of anatomical changes. Because insulin-like growth factor 1 (IGF_1) can directly increase the density of T-type Ca^{2+} channels in cultured, non-proliferating cells, Best et al. (1998) suggest that the cascade of events resulting in T-type Ca^{2+} channel modulation involves the growth hormone-induced secretion of IGF_1 and subsequent stimulation of the synthesis of new Ca^{2+} channels.

The prominent role of T-type Ca^{2+} channel activation in the regulation of protein synthesis and cell growth is corroborated by the inhibitory action of T-type Ca^{2+} channel blockers on steroidogenesis and cell proliferation. Mibefradil has an antiproliferative effect on rat smooth muscle cells *in vitro* and *in vivo*, an effect not observed with L-type Ca^{2+} antagonists such as verapamil or amlodipine (Schmitt et al., 1995b). Likewise, the reduction of T-type Ca^{2+} currents by the lipoxygenase inhibitor nordihydroguariaretic acid prevents PDGF-induced mouse fibroblast proliferation (Wang et al., 1993), whereas their reduction by angiotensin II bound to AT_2 receptors prevents the growth of rat coronary endothelial cells (Stoll et al., 1995). Block of T-type Ca^{2+} currents by ethosuximide prevents EPO- or VEGF-induced neurite outgrowth in rat retinal explants (Bocker-Meffert et al., 2002), and block by mibefradil, nickel, or anandamide prevents neuritogenesis in the NG108–15 cell line (Chemin et al., 2002b). Finally, mibefradil reduces the proliferation rate of human lymphocytes *in vitro* (Lijnen 2002; Lijnen et al., 1999) and of rat leucocytes *in vivo* (Bilici et al., 2001); mibefradil also reduces the adhesion of human leucocytes *in vitro* (Nebe et al., 2002), apparently by reducing the surface expression of β_2-integrins and L-selectin.

6.1.2. Tissue Remodeling and Repair

In addition to their potential role in the normal growth of various tissues, T-type Ca^{2+} channels may also play a major role in tissue remodeling during disease or tissue repair after injury.

6.1.2a. Cardiac Hypertrophy and Cardiomyopathy. The myocardium comprises different cell types in addition to the cardiac myocytes (see Weber et al., 1991 for a review). In particular, there are many cells present in the cardiac interstitium including endothelial cells, vascular smooth muscle cells, cardiac fibroblasts, macrophages, and mast cells. The hypertrophied myocardium can be the site of myocyte growth, which accounts for left ventricular hypertrophy (expressed as increased myocardial mass or wall thickness), and/or fibroblast growth responsible for collagen synthesis and subsequent interstitial and perivascular fibrosis. Endothelial and vascular smooth muscle cells can also resume mitosis, resulting in intimal and medial thickening of the coronary vasculature and ensuing reduction of the luminal area.

In adults, ventricular overload is known to stimulate cardiac myocytes to enter a hypertrophic growth phase (Mann et al., 1989): in fact, an increase in hemodynamic load triggers a general re-expression of fetal or neonatal phenotypes (Marban et al., 1990). In feline myocytes, hypertrophic growth induced by pressure-overload is associated with altered ionic currents including the repolarizing K^+ current (Kleiman et al., 1989), the L-type Ca^{2+} current (Nuss et al., 1991), and the T-type Ca^{2+} current (Nuss et al., 1993): in normal adult feline myocytes, T-type Ca^{2+} currents are scarce or absent but they are reexpressed in the hypertrophied ventricle. T-type Ca^{2+} currents also reappear in cardiac myocytes from rats with cardiac hypertrophy induced by pressure-overload (Martinez et al., 1999) or myocardial infarction (Huang et al., 2000), and in hamsters with cardiomyopathy (Sen et al., 1994). The reappearance of T-type Ca^{2+} currents in rat ventricle is associated with increased levels of $Ca_v3.1$ mRNA (Huang et al., 2000). Increased T-type Ca^{2+} currents are also observed in rat atrial myocytes, when volume overload is induced by growth hormone-secreting tumors (Xu et al., 1990). Finally, *in vitro*, endothelin-1 simultaneously induces the hypertrophy of cultured rat ventricular myocytes, enhances the expression of muscle-specific genes, stimulates protein synthesis, and increases T-type Ca^{2+} currents (Furukawa et al., 1992; Ito et al., 1991) (Figure 4). Thus, in addition to their involvement in healthy development and growth, T-type Ca^{2+} currents could also contribute to hypertrophic proliferation, alter the basic electrophysiological properties of myocardial cells, and increase the incidence of spontaneous action potentials and subsequent arrhythmias.

Nevertheless, there are models of cardiac hypertrophy and failure where T-type Ca^{2+} currents are not upregulated. For example, in dogs with congestive heart failure induced by rapid pacing, myocytes from atria or Purkinje fibers are hypertrophied but they do not

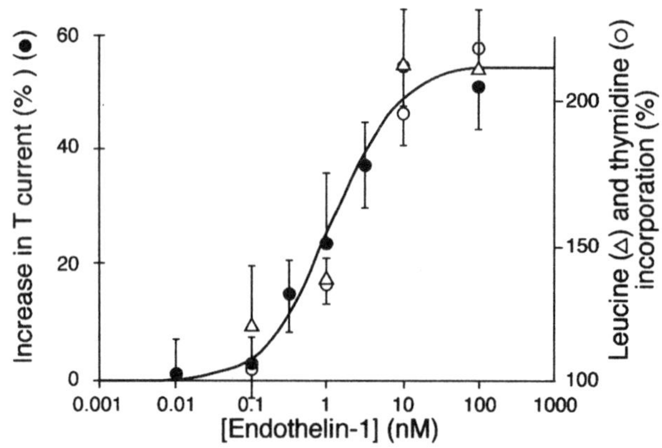

Figure 4. T-type Ca^{2+} currents in cardiac hypertrophy. Simultaneous increase of T-type Ca^{2+} current, protein and DNA synthesis in neonatal cultured rat ventricular myocytes exposed to endothelin-1. Enhanced T-type Ca^{2+} currents were measured after 20-min exposure and increased amino-acid incorporation after 24 hours (muscle-specific gene expression was measured after 6 hours and cell size after 48 hours; not shown on figure). Adapted from Furukawa et al., 1992 and Ito et al., 1991.

exhibit increased T-type Ca^{2+} currents (Li et al., 2000; Han et al., 2001). Although this could reflect species differences, it seems more likely due to a difference in tissues: contrary to rat or cat ventricular myocytes, myocytes from dog atria or Purkinje fibers possess T-type Ca^{2+} currents under normal conditions and they may not require additional currents for the hypertrophic response.

Whereas growth of cardiac myocytes is governed by mechanical overload, the cardiac fibroblast growth and collagen deposition associated with myocardial fibrosis is rather associated with an elevation in circulating hormones such as angiotensin II (Brilla et al., 1992; Villarreal et al., 1993) or in substances produced locally such as endothelins (Guarda et al., 1993). Because T-type Ca^{2+} channels are up-regulated by these peptides and are over-expressed in rapidly proliferating cells including mouse fibroblasts (Estacion et al., 1993; Wang et al., 1993), they could participate in the development of fibrosis.

Because hypertension is often associated with cardiovascular remodeling (cardiac hypertrophy, decrease of coronary vascular reserve, interstitial and perivascular fibrosis), the effects of mibefradil on remodeling have been well characterized. Mibefradil alleviates these symptoms in various models of hypertension by effectively lowering blood pressure (Karam et al., 1996; Veniant et al., 1995; Vacher et al., 1996) but it also provides additional benefits through a direct antiproliferative action. In the DOCA hypertensive rat treated with mibefradil, resorption of hypertrophy and fibrosis is observed and appears related to a direct action on cardiomyocytes and fibroblast growth. In a rat model of vascular injury, mibefradil has an antiproliferative effect on smooth muscle cells (Schmitt et al., 1995b; Schmitt et al., 1996). Since L-type Ca^{2+} channel antagonists such as verapamil, amlodipine, and felodipine do not prevent neointima formation (Nyborg et al., 1985), the antiproliferative effects of mibefradil are likely to be derived from the block of T-type Ca^{2+} channels.

Similar observations can be made regarding the effects of mibefradil on renal function and histology in animal models of renal failure. Despite similar hemodynamic improvements, mibefradil is markedly superior to amlodipine in protecting against kidney injury caused by deoxycorticosterone-salt hypertension in rats (Baylis et al., 2001).

In animal models of cardiac hypertrophy and cardiomyopathy, mibefradil was often found to reduce remodeling or improve function: in cardiomyopathic hamsters (Villame et al., 2001), in aortic-banded rats (Min et al., 2002), in myocardial infarction-induced heart failure rats (Sandmann et al., 1998; Sandmann et al., 2001; Sandmann et al., 2000), and in ischemia-induced myocardial damage in rat hearts (Arh et al., 2000).

6.1.2b. Atrial Tachycardia-Induced Electrical Remodeling. Sustained atrial tachycardia can cause electrophysiological alterations (shortening of atrial effective refractory period and increases in heterogeneity of atrial refractoriness) that promote the inducibility and maintenance of atrial fibrillation (Fareh et al., 2001; Fareh et al., 1999; Daoud et al., 1997; Yu et al., 1998; Tieleman et al., 1997). These modifications appear to result from intracellular Ca^{2+} overload followed by changes in the expression of multiple genes (Nattel, 1999). The exact mechanisms leading to Ca^{2+} overload are not well understood, but Ca^{2+} entry through T-type Ca^{2+} channels has been implicated after it was demonstrated that such electrical remodeling in dogs can be prevented by chronic treatment with mibefradil but not diltiazem (Fareh et al., 2001) or verapamil (Lee et al., 2000).

6.1.2c. Other Pathologies. T-type Ca^{2+} currents are also up-regulated by acute injury in very different settings such as in rat thalamic neurons after corticotomy (Chung et al., 1993) and in rat smooth muscle cells following vascular injury (Schmitt et al., 1995b). However, they are not increased by direct damage of the sciatic nerve by axotomy or constriction (Hogan et al., 2000; Baccei et al., 2000; Abdulla et al., 2001).

6.2. T-type Ca^{2+} Channels in Cardiovascular Physiology and Pathology

In the last few years, due to availability of mibefradil, the role of T-type Ca^{2+} channels has been most extensively studied in cardiovascular pathophysiology. Many roles have been suggested for these channels in health and disease, including a contribution to the control of vascular tone and to the pacemaker depolarization of cardiac cells.

6.2.1. Vascular Tone and Hypertension

The contribution of T-type Ca^{2+} currents to the maintenance of normal vascular tone is still controversial (Leuranguer et al., 2001), although suggestive data exist regarding specific vascular beds. For example, in rat mesenteric arterioles, mibefradil reduces local and conducted vasoconstriction, in contrast to nifedipine or nimodipine (Gustafsson et al., 2001). Consistent with this pharmacology, this tissue contains mRNA for $Ca_v3.1$ and $Ca_v3.2$ (α_{1G}, α_{1H}) but not for $Ca_v1.2$ (α_{1C}). T-type Ca^{2+} currents also appear to play a major role in the maintenance of efferent arteriolar tone in dog (Honda et al., 2001) or rat (Ozawa et al., 2001) kidney, of the myogenic tone of skeletal muscle resistance arteries (VanBavel et al., 2002), and of the vasoconstriction induced by NO-synthase inhibition in hypertensive lungs (Muramatsu et al., 1997).

It is generally agreed that hypertension is associated with abnormal activity of the Ca^{2+} channels of vascular smooth muscle, which result in intracellular Ca^{2+} overload (Self et al., 1994). Indeed, Ca^{2+} channel activity is increased in various models of hypertension (Self et al., 1994; Wilde et al., 1994). However, the respective contribution of L- and T-type Ca^{2+} channels to this pathological Ca^{2+} overload is still debated. In venous and arterial smooth muscle cells of moderately hypertensive rats such as the SHR, earlier studies reported increased L-type Ca^{2+} currents without appreciable changes in T-type Ca^{2+} currents (Hermsmeyer 1991; Ohya et al., 1993). Evidence for increased expression of L-type Ca^{2+} channels in SHR has also been provided by radiolabeled dihydropyridine binding studies (Aoki et al., 1986; Bruner et al., 1990; Ebata et al., 1991). However, data from extremely hypertensive rats (stroke-prone spontaneously hypertensive rats or SP-SHR) indicate an important contribution of T-type Ca^{2+} currents (Self et al., 1994). In these animals, T-type Ca^{2+} currents are significantly increased relative to normotensives, due conjointly to an increase in the number of T-type Ca^{2+} channels and a shift of the activation and inactivation curves towards more negative potentials. These alterations observed in newborn SP-SHR suggest that T-type Ca^{2+} channel dysfunction could be a mechanism of genetic hypertension.

6.2.2. Cardiac Pacemaking and Abnormal Electrical Activity

Under normal conditions, the T-type Ca^{2+} current is unlikely to contribute significantly to the electrical activity of atrial and ventricular myocytes. When T-type Ca^{2+} currents are present in these cells, they are very weak and their activation voltage range overlaps with that of the much larger Na^+ current (see section 4.1). The main function of these cells is contraction, which requires a large Ca^{2+} influx to initiate Ca^{2+} release from the sarcoplasmic reticulum. The slowly-inactivating L-type Ca^{2+} channels, which sustain the action potential plateau, serve this purpose more appropriately than T-type Ca^{2+} channels. Conversely, T-type Ca^{2+} channels may contribute to impulse conduction when Na^+ channels are partially or totally inactivated by membrane depolarization such as in pathologically depolarized ischemic cells or normally depolarized nodal or Purkinje cells.

Indeed, in cardiac pacemaker cells, T-type Ca^{2+} currents are much larger than in working myocardial cells, suggesting that T-type Ca^{2+} channels are important for impulse conduction and pacemaking in these tissues (Noble 1984; Bean 1985; Armstrong et al., 1985; Hagiwara et al., 1988). Generally, the activation and inactivation curves of T-type Ca^{2+} channels overlap at the diastolic voltage of pacemaker cells (Hirano et al., 1989; Tseng et al., 1989; Schneider et al., 1994), and the resulting Ca^{2+} window current is well suited to participate in the late phase of diastolic depolarization.

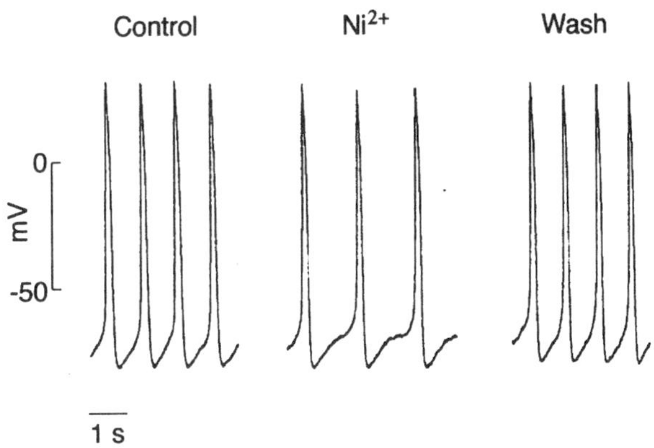

Figure 5. T-type Ca^{2+} currents in the maintenance of heart rate. The effect of Ni^{2+} is shown on spontaneous action potentials in cat atrial latent pacemaker cells. The pacemaker cycle length is prolonged 67% by 40 μM Ni^{2+}. Reproduced from Zhou et al., 1994.

Pharmacological studies confirm this view. Block of T-type Ca^{2+} channels by Ni^{2+} increases pacemaker cycle length in cat atrial latent pacemaker cells (Zhou et al., 1994) (Figure 5). This is also seen to a lesser extent in rabbit sinoatrial node cells with Ni^{2+} (Hagiwara et al., 1988; Satoh 1995), mibefradil (Masumiya et al., 1999), tetramethrin (Satoh 1995), efonidipine (Masumiya et al., 1997), or ryanodine (Li et al., 1997b; Li et al., 1997c). Finally, mibefradil reduces heart rate without decreasing cardiac contractility in animals as well as in normal human volunteers (Schmitt et al., 1989), hypertensive patients (Schmitt et al., 1992), patients with angina pectoris (Portegies et al., 1991; Rousseau et al., 1996) or heart failure (Muntinga et al., 1996). In a detailed electrophysiological study in patients, sinus node automaticity and heart rate were reduced more than atrioventricular node conduction. These observations strongly support the idea that T-type Ca^{2+} channels participate in the regulation of heart rate.

Abnormal rhythmicity has also been linked to T-type Ca^{2+} channels. In the cardiomyopathic Syrian hamster, a model of congestive heart failure, the density of T-type Ca^{2+} currents in ventricular myocytes is increased despite a shift of the activation and inactivation curves towards more negative potentials (Sen et al., 1994). In fact, this shift brings the voltage range of the T-type Ca^{2+} window current closer to the resting potential (Hirano et al., 1989), which could result in the increased incidence of spontaneous beating and arrhythmias observed in this genetically determined cardiomyopathy.

6.2.3. Vascular Pacemaking

In vascular smooth muscle cells, T-type Ca^{2+} channels could be responsible for the spontaneous firing of Ca-dependent action potentials (Friedman et al., 1986). Because smooth muscle cells, in contrast to cardiac myocytes, lack abundant Na^{+} channels, T-type Ca^{2+} channels could also play a role in activating contraction. Thus, vascular tone would be increased by pathological enhancement of T-type Ca^{2+} currents or decreased by their pharmacological block.

T-type Ca^{2+} currents are found in sheep mesenteric lymphatic smooth muscle cells (Hollywood et al., 1997), where they are probably involved in controlling the pumping frequency of the lymphatic vessels (Thornbury 1999). Indeed, nickel slows the contraction frequency without affecting the stroke volume, whereas nifedipine reduces the stroke volume without affecting the frequency (Convery et al., 1997).

In canine colonic smooth muscle cells, their density increases with development to be maximal in mature cells (Xiong et al., 1993). This unusual particularity could serve a very specific purpose by contributing to the slow action potential involved in colon motility, a function increasingly needed after birth.

6.2.4. Hypercholesterolemia

An increase in membrane cholesterol content has been suggested to alter the contractility of canine vascular smooth muscle, possibly by increasing the permeability of the cell membrane to calcium (Yokoyama et al., 1979). More recently Ca^{2+} channels have been implicated in this phenomenon (Sen et al., 1992): in rat arterial smooth muscle cells, L-type and to a lesser extent T-type Ca^{2+} currents increase upon exposure to cholesterol-rich liposomes. The modulation of Ca^{2+} channels by cholesterol could be important in the pathophysiology of atherosclerosis.

6.3. T-type Ca^{2+} Channels in Nervous System Physiology and Pathology

In neuronal cells as in cardiac myocytes, the electrophysiological properties of the T-type Ca^{2+} channels are well suited for a role in pacemaking and burst-firing processes, including epilepsies. Indeed, T-type Ca^{2+} channels are predominant in the soma and dendrites of neurons with strong Ca^{2+}-dependent burst-firing behaviors such as thalamic relay neurons and cerebellar Purkinje cells. Other potential roles include contributions to the initiation of long-term potentiation or depression (LTP/LTD) and to tolerance to and dependence on morphine.

6.3.1. Repetitive Neuronal Firing

T-type Ca^{2+} channels are involved in the regulation of thalamocortical neuron firing. During wakefulness, these neurons are in the tonic-firing mode (characterized by a desynchronized EEG) but, in the transition towards quiescent sleep or altered state of consciousness (characterized by a synchronized EEG), they shift to the oscillatory firing mode, which progressively entrains the whole thalamocortical system into rhythmic activity. This transition between rhythmic and oscillatory firing mode is controlled by T-type Ca^{2+} channels, which generate the low-threshold spikes that depolarize the cell membrane and initiate the oscillatory firing (Crunelli et al., 1991; Huguenard et al., 1992; Steriade et al., 1988; Suzuki et al., 1989; Huguenard, 1999).

In mouse cerebellar Purkinje neurons, mibefradil slows but does not stop spontaneous firing, similar to its effects on cardiac sinoatrial node cells (Raman et al., 1999). Similarly, in rat or hamster spinal laminae neurons, mibefradil, Ni^{2+}, or La^{3+} (but not antagonists of HVA Ca^{2+} channels) decrease the frequency of spontaneous miniature excitatory synaptic currents (Bao et al., 1998).

Epileptic disorders are characterized by neuronal hyper-excitability, and generalized absence epilepsies reflect synchronized oscillations of cortical, reticular thalamic, and cortico-thalamic neurons (Gloor et al., 1988). Antiepileptic drugs are thought to act by reducing neuronal activity via two main mechanisms: the block of excitatory Na^{+} or Ca^{2+} channels or the enhancement of inhibitory GABA-mediated responses. The importance of T-type Ca^{2+} channels in the development of epileptic rhythms is suggested by the observation that the most effective antiepileptic drugs, *i.e.*, the succinimides (Kostyuk et al., 1992), the benzodiazepines (Huguenard et al., 1994; Reuveny et al., 1993), valproic acid (Kelly et al., 1990), flunarizine (Janis et al., 1991), and zonisamide (Suzuki et al., 1992), seem to exert their effect, at least in part, on the thalamic T-type Ca^{2+} current. Furthermore, in a rat model of absence epilepsy, T-type Ca^{2+} currents increase in density in the neurons of the nucleus reticularis compared to non-epileptic control animals and also contribute to the pathological synchronization of spike and wave discharges observed in this genetic model of epilepsy (Tsakiridou et al., 1995). In these animals, the expression of $Ca_v3.2$ (α_{1H}) in the reticular thalamic nuclei is enhanced and expression of $Ca_v3.1$ (α_{1G}) in the ventrobasal thalamic nucleus is enhanced (Talley et al., 2000). Expression of T-type Ca^{2+} currents in hippocampal CA_1 pyramidal neurons also increases in three different models of epilepsy: focal (Faas et al., 1996) or temporal lobe (Su et al., 2002) epilepsy in rats, and bursting activity in hippocampal slices (Niesen et al., 1999).

Genetic data also supports the view that that T-type Ca^{2+} currents contribute to the maintenance of repetitive or burst firing. Thalamocortical relay neurons of $Ca_v3.1$-deficient transgenic mice do not exhibit burst firing, and the mice are resistant to absence seizures (Kim et al., 2001; see Saegusa and Tanabe, this volume). Conversely, mutations in several genes for subunits of HVA Ca^{2+} channels are linked to spike-wave seizure phenotypes in mice, despite the fact that changes in HVA Ca^{2+} currents should not increase intrinsic membrane excitability. Remarkably, these mutations are associated with increased densities not only of HVA Ca^{2+} currents but also of T-type Ca^{2+} currents; this is accompanied by an $h\infty$ shift to more positive voltages, which increases the availability of T-type Ca^{2+} currents. Yet, no alterations were found in the mRNA levels of $Ca_v3.1$, $Ca_v3.2$, or $Ca_v3.3$ (α_{1G}, α_{1H}, α_{1I}) (Zhang et al., 2002b).

Results in mouse dopaminergic midbrain neurons provide an interesting exception to the dogma that T-type Ca^{2+} currents contribute to the maintenance of repetitive or burst firing (Wolfart et al., 2002). In these cells, inhibition of T-type Ca^{2+} currents actually initiates burst firing, apparently due to functional coupling between Ca^{2+} entry through T-type Ca^{2+} channels and activation of SK_3 K^+ channels.

6.3.2. Nociception

Spinal delivery of HVA Ca^{2+} channel antagonists can attenuate or prevent pain as well as hyperalgesia (Vanegas et al., 2000; see Norton et al., this volume). The role of T-type Ca^{2+} channels is less clear but recent results have brought new understanding and suggest very promising potential applications for the treatment of neuropathic pain.

Neuronal hyperexcitability is a common underlying mechanism for neuropathic pain and epilepsy, and some of the classical antiepileptic drugs are effective in the treatment of neuropathic pain. In particular, spinal delivery of ethosuximide produces a dose-related inhibition of neuronal responses evoked by electrical, mechanical, or thermal stimuli in a rodent model of neuropathy (Matthews et al., 2001). Similarly, systemic administration of mibefradil induces mechanical and thermal antinociception in adult rats (Todorovic et al., 2002) and local injection of mibefradil blocks the thermal hyperalgesia induced by injected reducing agents (Todorovic et al., 2001b). Indeed, in rat sensory neurons, redox agents increase T-type Ca^{2+} currents but do not affect other voltage- and ligand-gated channels thought to mediate pain sensitivity; redox agents also increase currents through recombinant $Ca_v3.2$ channels (Todorovic et al., 2001b). Finally, as mentioned previously, nociceptin inhibits T-type Ca^{2+} currents in rat DRG neurons (Abdulla et al., 1997).

T-type Ca^{2+} channels also appear to play a role in morphine antinociception as well as in the development of tolerance and dependence to morphine. In mice, mibefradil enhances the antinociceptive effects of acute morphine, and it actually blocks the development of tolerance to these effects but not the development of morphine dependence (Dogrul et al., 2002).

These data strongly suggest an important role for T-type Ca^{2+} channels in peripheral nociception and that T-type Ca^{2+} channels blockers could be useful non-central antinociceptive agents.

6.3.3. Other Roles in the Nervous System

Recently, limited but promising data have suggested additional roles for T-type Ca^{2+} channels, such in the initiation of long-term potentiation or depression (LTP/LTD) (Wasling et al., 2002; Salami et al., 2000) or in smell transduction (Kawai 1999; Kawai et al., 2001).

6.4. T-type Ca^{2+} Channels in the Endocrine System

In endocrine cells, the electrophysiological properties of the T-type Ca^{2+} channels are well suited for a role in mediating secretion. Indeed, under normal circumstances, most endocrine cells exhibit small membrane depolarizations upon stimulation, and the low-voltage-activated T-type Ca^{2+} channels are more likely to support Ca^{2+} entry than the HVA Ca^{2+} channels.

In animals and humans, treatment with dihydropyridine Ca^{2+} antagonists tends to be associated with reflex stimulation of the sympathetic system, resulting in increased levels of plasma catecholamines, renin, and aldosterone (Bellet et al., 1987). Yet, the clinical antihypertensive effect of mibefradil is not associated with neurohormonal stimulation (Schmitt et al., 1992). Moreover, in the DOCA-salt rat model of hypertension, where the renin-angiotensin system is depressed, mibefradil (in contrast to enalapril) can still decrease plasma renin activity, possibly by a direct effect on renin secretion by juxtaglomerular cells (Karam et al., 1996). This absence of neurohormonal stimulation is consistent with the role of T-type Ca^{2+} channels in many neurohormonal secretory systems.

6.4.1. Aldosterone

Serum levels of potassium are regulated by the secretion of aldosterone from adrenal glomerulosa cells. These cells have a remarkably negative resting membrane voltage, and they can depolarize and release aldosterone upon very small increases in serum K^+. Hormones can also stimulate aldosterone secretion: circulating angiotensin II (A_{II}) acts synergistically with potassium (Chen et al., 1999; Lotshaw 2001), and serotonin (5-HT) also may act as a paracrine factor (Lefebvre et al., 2001).

Adrenal glomerulosa cells express both T- and L-type Ca^{2+} channels, but physiological increases in extracellular K^+ stimulate aldosterone secretion primarily by enhancing Ca^{2+} entry through T-type Ca^{2+} channels (Barrett et al., 1995). In bovine, A_{II} activates calmodulin-dependent protein kinase II (CaMKII), thereby shifting the activation of T-type Ca^{2+} channels to more negative voltages and increasing Ca^{2+} entry at lower extracellular K^+ (Chen et al., 1999; Barrett et al., 2000). In rats, however, A_{II} slightly depolarizes the cells, thereby opening T-type Ca^{2+} channels directly (Lotshaw, 2001). Mibefradil inhibits K^+-, A_{II}-, and 5-HT-stimulated aldosterone secretion. On the other hand, 5-HT appears to increase Ca^{2+} entry through T-type Ca^{2+} channels by activating the cAMP/PKA pathway (Lenglet et al., 2002).

6.4.2. ANP

The secretion of atrial natriuretic peptide (ANP) by the atrium is increased by atrial workload. However, diltiazem decreases atrial dynamics but increases ANP secretion, and the L-type Ca^{2+} channel agonist BayK-8644 increases atrial dynamics but decreases ANP secretion. In contrast, mibefradil decreases both atrial dynamics and ANP secretion. These results suggest that Ca^{2+} entry through L- and T-type Ca^{2+} channels elicit opposite effects on the release of ANP from atrial myocytes (Wen et al., 2000): release of ANP is inhibited by Ca^{2+} entry through L-type Ca^{2+} channels whereas it is increased by Ca^{2+} entry through T-type Ca^{2+} channels. This is somewhat similar to aldosterone secretion from bovine adrenal glomerulosa (Barrett et al., 1995).

Similarly, in neonatal rats, Ca^{2+} entry through T-type Ca^{2+} channels appears to be a major mechanism of depolarization-induced ANP secretion by the atria (Leuranguer et al., 2000).

6.4.3. Cortisol

Rat and bovine adrenal fasciculata cells express almost exclusively T-type Ca^{2+} channels, which are thus the main pathway for Ca^{2+} entry (Mlinar et al., 1993; Barbara et al., 1995). Indeed, Ni^{2+} and mibefradil each inhibit cortisol secretion stimulated by adrenocorticotropic hormone (ACTH) or angiotensin II (Enyeart et al., 1993; Mlinar et al., 1993; Gomora et al., 2000), suggesting an important role for T-type Ca^{2+} currents in cortisol synthesis and secretion. In contrast, pharmacological block of L-type Ca^{2+} channels does not alter secretion substantially (Enyeart et al., 1993). Adrenal fasciculata cells can generate Ca^{2+}-dependent action potentials but it is not known if this property plays a role in regulating secretion (Barbara et al., 1995). Finally, ACTH may also regulate the expression of T-type Ca^{2+} channels (Enyeart et al., 1993; Mlinar et al., 1993).

ACTH appears to stimulate cortisol secretion by depolarizing the cells, thereby activating T-type Ca^{2+} channels and raising intracellular Ca^{2+}. Two possible mechanisms have been proposed: the inhibition of a K^+ conductance, probably TREK-1 (Enyeart et al., 2001; Enyeart et al., 2002), and/or the activation of a Cl- conductance (Dupre-Aucouturier et al., 2002). The two mechanisms are not mutually incompatible, but more data are needed to resolve this question. Endothelin also stimulates cortisol secretion from human and rat adrenal fasciculata cells, but in this case the role of Ca^{2+} entry remains unknown (Rebuffat et al., 2001).

6.4.4. Insulin

T-type Ca^{2+} currents have been described in human and rat pancreatic β-cells (Barnett et al., 1995; Sala et al., 1990; Ashcroft et al., 1990), and they are augmented in diabetic rats (Kato et al., 1996). Interestingly, T-type Ca^{2+} currents were found in β-cells from diabetes-prone NOD mice but not in normal animals (Wang et al., 1996a). The most widely-accepted model of glucose-stimulated insulin secretion asserts that: 1) increases in plasma glucose raises intracellular ATP in β-cells 2) intracellular ATP inhibits K_{ATP} channels, 3) K_{ATP} channel closure depolarizes the β-cells slightly, 4) T-type Ca^{2+} channel opening initiates a slow plateau depolarization, 5) this depolarization triggers rapid voltage spikes dependent on Ca^{2+} entry through R-, P/Q-, and L-type Ca^{2+} channels (reviewed in Kitasato et al., 1996). Due to the complexity of this cascade of events, a variety of compounds can inhibit insulin secretion, and the pharmacological dissection of the role of a single ion channel type has been difficult.

Consistent with this model, nickel reduces insulin secretion and delays the onset of the rapid voltage spikes (Bhattacharjee et al., 1997). Also, long-term treatment with mibefradil both prevents and reverses the development of hyperinsulinemia, hypertriglyceridemia, and hypertension in fructose-hypertensive rats (Verma et al., 1997).

6.5. Other Roles for T-type Ca^{2+} Channels

A further role for T-type Ca^{2+} currents has been suggested in the capacitation and acrosome reaction of sperm (Bai et al., 2002; Jagannathan et al., 2002). In mice, the only Ca^{2+} currents in spermatocytes are T-type Ca^{2+} currents and, in man, the predominant isoform in testis is $Ca_v3.2$ (α_{1H}) (Son et al., 2000; Jagannathan et al., 2002). However, there are still significant uncertainties regarding the exact identity of the low-voltage-activated channel in spermatocytes (T-type or R-type) and their relevance in human sperm (Garcia et al., 1999; Wennemuth et al., 2000; Bonaccorsi et al., 2001; Sakata et al., 2001; see Saegusa and Tanabe, this volume).

7. CONCLUSION

In the last 10 years, our understanding of the biophysics, pharmacology and physiopathology of T-type Ca^{2+} channels has blossomed, due to the availability of a few semi-selective blockers and to the molecular identification of the channels. These channels are expressed in many types of spontaneously active cells, including cardiac pacemaker cells, thalamic neurons, and hormone secreting cells, where they are involved in physiological firing pattern regulation. Conversely, T-type channels may be associated with pathophysiological conditions involving abnormal impulse generation. A second major role for T-type Ca^{2+} channels can be found in the regulation of growth, differentiation and tissue remodeling. Finally, T-type channels are also clearly important in hormone secretion by some endocrine cells that do not necessarily fire action potentials.

T-type Ca^{2+} channels represent the most likely molecular target of some Ca^{2+} antagonists, such as ethosuximide for epilepsy or mibefradil for cardiovascular disorders. More selective blockers of T-type Ca^{2+} channels are needed to confirm their role in complex processes involving many Ca^{2+} channel types and to demonstrate definitively the value of T-type Ca^{2+} channel antagonism as a therapeutic modality.

ACKNOWLEDGMENTS

I am grateful to Dr. Agnès Bénardeau for valuable comments on the manuscript.

REFERENCES

Abdulla, F. A. and Smith, P. A. (1997). Nociceptin inhibits T-type Ca^{2+} channel current in rat sensory neurons by a G-protein-independent mechanism. *J Neurosci* **17**(22): 8721-8.

Abdulla, F. A. and Smith, P. A. (2001). Axotomy- and autotomy-induced changes in Ca^{2+} and K^+ channel currents of rat dorsal root ganglion neurons. *J Neurophysiol* **85**(2): 644-58.

Abernethy, D. R. and Soldatov, N. M. (2002). Structure-functional diversity of human L-type Ca^{2+} channel: perspectives for new pharmacological targets. *J Pharmacol Exp Ther* **300**(3): 724-8.

Ahlijanian, M. K., Westenbroek, R. E., et al., (1990). Subunit structure and localization of dihydropyridine-sensitive calcium channels in mammalian brain, spinal cord, and retina. *Neuron* **4**(6): 819-32.

Akaike, N., Kanaide, H., et al., (1989a). Low-voltage-activated calcium current in rat aorta smooth muscle cells in primary culture. *J Physiol (Lond)* **416**: 141-160.

Akaike, N., Kostyuk, P. G., et al., (1989b). Dihydropyridine-sensitive low-threshold calcium channels in isolated rat hypothalamic neurones. *J Physiol* **412**: 181-95.

Akopian, G., Kressin, K., et al., (1996). Identified glial cells in the early postnatal mouse hippocampus display different types of Ca^{2+} currents. *Glia* **17**(3): 181-94.

Alborch, E., Salom, J. B., et al., (1995). Calcium channels in cerebral arteries. *Pharmacol Ther* **68**(1): 1-34.

Allen, T. G. and Brown, D. A. (1993a). M2 muscarinic receptor-mediated inhibition of the Ca^{2+} current in rat magnocellular cholinergic basal forebrain neurones. *J Physiol* **466**: 173-89.

Allen, T. G., Sim, J. A., et al., (1993b). The whole-cell calcium current in acutely dissociated magnocellular cholinergic basal forebrain neurones of the rat. *J Physiol* **460**: 91-116.

Alvarez, J. L., Artiles, A., et al., (2000). Modulation of voltage-dependent facilitation of the T-type calcium current by sodium ion in isolated frog atrial cells. *Pflugers Arch* **441**(1): 39-48.

Alvarez, J. L., Rubio, L. S., et al., (1996). Facilitation of T-type calcium current in bullfrog atrial cells: voltage-dependent relief of a G protein inhibitory tone. *J Physiol* **491 (Pt 2)**: 321-34.

Alvarez, J. L. and Vassort, G. (1992). Properties of the low threshold Ca current in single frog atrial cardiomyocytes. A comparison with the high threshold Ca current. *J Gen Physiol* **100**(3): 519-45.

Aoki, K. and Asano, M. (1986). Effects of Bay K 8644 and nifedipine on femoral arteries of spontaneously hypertensive rats. *Br J Pharmacol* **88**(1): 221-30.

Argibay, J. A., Fischmeister, R., et al., (1988). Inactivation, reactivation and pacing dependence of calcium current in frog cardiocytes: correlation with current density. *J Physiol* **401**: 201-26.

Arh, M. and Budihna, M. V. (2000). Comparison of effects of nitrendipine, lacidipine and mibefradil on postischaemic myocardial damage in isolated rat hearts. *Pflugers Arch* **440**(5 Suppl): R149-50.

Armstrong, C. M. and Matteson, D. R. (1985). Two distinct populations of calcium channels in a clonal line of pituitary cells. *Science* **227**(4682): 65-7.

Arnoult, C., Cardullo, R. A., et al., (1996). Activation of mouse sperm T-type Ca^{2+} channels by adhesion to the egg zona pellucida. *Proceedings of the National Academy of Sciences U S A* **93**(23): 13004-13009.

Arnoult, C., Lemos, J. R., et al., (1997). Voltage-dependent modulation of T-type calcium channels by protein tyrosine phosphorylation. *Embo J* **16**(7): 1593-9.

Arnoult, C., Villaz, M., et al., (1998). Pharmacological properties of the T-type Ca^{2+} current of mouse spermatogenic cells. *Mol Pharmacol* **53**(6): 1104-11.

Artalejo, C. R., Adams, M. E., et al., (1994). Three types of Ca^{2+} channel trigger secretion with different efficacies in chromaffin cells. *Nature* **367**(6458): 72-6.

Ashcroft, F. M., Kelly, R. P., et al., (1990). Two types of Ca channel in rat pancreatic beta-cells. *Pflugers Arch* **415**(4): 504-6.

Baccei, M. L. and Kocsis, J. D. (2000). Voltage-gated calcium currents in axotomized adult rat cutaneous afferent neurons. *J Neurophysiol* **83**(4): 2227-38.

Bai, J. and Shi, Y. (2002). Inhibition of T-type Ca^{2+} currents in mouse spermatogenic cells by gossypol, an antifertility compound. *Eur J Pharmacol* **440**(1): 1-6.

Balke, C. W., Rose, W. C., et al., (1992). Macroscopic and unitary properties of physiological ion flux through T-type Ca^{2+} channels in guinea-pig heart cells. *J Physiol* **456**: 247-65.

Bao, J., Li, J. J., et al., (1998). Differences in Ca^{2+} channels governing generation of miniature and evoked excitatory synaptic currents in spinal laminae I and II. *J Neurosci* **18**(21): 8740-50.

Barbara, J. G. and Takeda, K. (1995). Voltage-dependent currents and modulation of calcium channel expression in zona fasciculata cells from rat adrenal gland. *J Physiol* **488 (Pt 3)**: 609-22.

Barnes, S. and Haynes, L. W. (1992). Low-voltage-activated calcium channels in human retinoblastoma cells. *Brain Research* **598**(1-2): 19-22.

Barnett, D. W., Pressel, D. M., et al., (1995). Voltage-dependent Na^{+} and Ca^{2+} currents in human pancreatic islet beta- cells: evidence for roles in the generation of action potentials and insulin secretion. *Pflugers Archiv* **431**(2): 272-82.

Barrett, P. Q., Ertel, E. A., et al., (1995). Voltage-gated calcium currents have two opposing effects on the secretion of aldosterone. *Am J Physiol* **268**(4 Pt 1): C985-92.

Barrett, P. Q., Isales, C. M., et al., (1991a). Ca^{2+} channels and aldosterone secretion: modulation by K+ and atrial natriuretic peptide. *Am J Physiol* **261**(4 Pt 2): F706-19.

Barrett, P. Q., Isales, C. M., et al., (1991b). Modulation of Ca^{2+} channels by atrial natriuretic peptide in the bovine adrenal glomerulosa cell. *Can J Physiol Pharmacol* **69**(10): 1553-60.

Barrett, P. Q., Lu, H. K., et al., (2000). Stimulation of unitary T-type Ca^{2+} channel currents by calmodulin-dependent protein kinase II. *Am J Physiol Cell Physiol* **279**(6): C1694-703.

Baylis, C., Qiu, C., et al., (2001). Comparison of L-type and mixed L- and T-type calcium channel blockers on kidney injury caused by deoxycorticosterone-salt hypertension in rats. *Am J Kidney Dis* **38**(6): 1292-7.

Beam, K. G. and Knudson, C. M. (1988a). Calcium currents in embryonic and neonatal mammalian skeletal muscle. *J Gen Physiol* **91**(6): 781-98.

Beam, K. G. and Knudson, C. M. (1988b). Effect of postnatal development on calcium currents and slow charge movement in mammalian skeletal muscle. *J Gen Physiol* **91**(6): 799-815.

Bean, B. P. (1985). Two kinds of calcium channels in canine atrial cells. Differences in kinetics, selectivity, and pharmacology. *J Gen Physiol* **86**(1): 1-30.

Bean, B. P. (1989). Classes of calcium channels in vertebrate cells. *Annu Rev Physiol* **51**: 367-84.

Bean, B. P., Sturek, M., et al., (1986). Calcium channels in muscle cells isolated from rat mesenteric arteries: modulation by dihydropyridine drugs. *Circ Res* **59**(2): 229-35.

Beck, H., Steffens, R., et al., (1997). Properties of voltage-activated Ca^{2+} currents in acutely isolated human hippocampal granule cells. *J Neurophysiol* 77(3): 1526-37.

Bellet, M., Sassano, P., et al., (1987). Converting-enzyme inhibition buffers the counter-regulatory response to acute administration of nicardipine. *Br J Clin Pharmacol* **24**(4): 465-72.

Bénardeau, A. and Ertel, E. A. (1998). Selective block of myocardial T-type calcium channels by mibefradil: A comparison with the 1,4-dihydropyridine amlodipine. *Low-voltage-activated T-type Ca channels*. R. W. Tsien, J.-P. Clozel and J. Nargeot. Chester, UK, Adis International: 386-394.

Benham, C. D., Hess, P., et al., (1987). Two types of calcium channels in single smooth muscle cells from rabbit ear artery studied with whole-cell and single-channel recordings. *Circulation Research* **61**(4 Pt 2): I10-I16.

Berger, C. L., de Bustros, A., et al., (1984). Human medullary thyroid carcinoma in culture provides a model relating growth dynamics, endocrine cell differentiation, and tumor progression. *J Clin Endocrinol Metab* **59**(2): 338-43.

Bernheim, L. and Bader, C. R. (2002). Human myoblast differentiation: Ca^{2+} channels are activated by K^+ channels. *News Physiol Sci* **17**: 22-6.

Berthier, C., Monteil, A., et al., (2002). α_{1H} mRNA in single skeletal muscle fibres accounts for T-type calcium current transient expression during fetal development in mice. *J Physiol* **539**(Pt 3): 681-91.

Bertolesi, G. E., Shi, C., et al., (2002). The Ca^{2+} channel antagonists mibefradil and pimozide inhibit cell growth via different cytotoxic mechanisms. *Mol Pharmacol* **62**(2): 210-9.

Best, P. M., Chen, C. C., et al., (1998). Regulation of the density of T-type calcium current by growth hormone and insulin-like growth factor 1 in atrial myocytes. *Low-voltage-activated T-type Ca channels.* R. W. Tsien, J.-P. Clozel and J. Nargeot. Chester, UK, Adis International: 343-350.

Beuckelmann, D. J., Näbauer, M., et al., (1991). Characteristics of calcium-current in isolated human ventricular myocytes from patients with terminal heart failure. *J Mol Cell Cardiol* **23**(8): 929-937.

Bhattacharjee, A., Whitehurst, R. M., Jr., et al., (1997). T-type calcium channels facilitate insulin secretion by enhancing general excitability in the insulin-secreting beta-cell line, INS-1. *Endocrinology* **138**(9): 3735-3740.

Biagi, B. A. and Enyeart, J. J. (1991). Multiple calcium currents in a thyroid C-cell line: biophysical properties and pharmacology. *Am J Physiol* **260**(6 Pt 1): C1253-63.

Biagi, B. A., Mlinar, B., et al., (1992). Membrane currents in a calcitonin-secreting human C cell line. *American Journal of Physiology* **263**(5 Pt 1): C986-C994.

Bijlenga, P., Liu, J. H., et al., (2000). T-type alpha 1H Ca2+ channels are involved in Ca^{2+} signaling during terminal differentiation (fusion) of human myoblasts. *Proc Natl Acad Sci U S A* **97**(13): 7627-32.

Bilici, D., Akpinar, E., et al., (2001). Protective effect of T-type calcium channel blocker in histamine-induced paw inflammation in rat. *Pharmacol Res* **44**(6): 527-31.

Bkaily, G., Naik, R., et al., (1998). Endothelin-1 and insulin activate the steady-state voltage dependent R-type Ca2+ channel in aortic smooth muscle cells via a pertussis toxin and cholera toxin sensitive G-protein. *Mol Cell Biochem* **183**(1-2): 39-47.

Bkaily, G., Sculptoreanu, A., et al., (1992). Apamin, a highly potent fetal L-type Ca^{2+} current blocker in single heart cells. *Am J Physiol* **262**(2 Pt 2): H463-71.

Bkaily, G., Sculptoreanu, A., et al., (1997). Increases of T-type Ca^{2+} current in heart cells of the cardiomyopathic hamster. *Mol Cell Biochem* **176**(1-2): 199-204.

Bkaily, G., Wang, S., et al., (1995). ET-1 stimulates Ca^{2+} currents in cardiac cells. *J Cardiovasc Pharmacol* **26 Suppl 3**: S293-6.

Blankenfeld, G. G., Verkhratsky, A. N., et al., (1992). Ca^{2+} Channel Expression in the Oligodendrocyte Lineage. *Eur J Neurosci* **4**(11): 1035-1048.

Bocker-Meffert, S., Rosenstiel, P., et al., (2002). Erythropoietin and VEGF promote neural outgrowth from retinal explants in postnatal rats. *Invest Ophthalmol Vis Sci* **43**(6): 2021-6.

Bonaccorsi, L., Forti, G., et al., (2001). Low-voltage-activated calcium channels are not involved in capacitation and biological response to progesterone in human sperm. *Int J Androl* **24**(6): 341-51.

Bonvallet, R. (1987). A low threshold calcium current recorded at physiological Ca concentrations in single frog atrial cells. *Pflugers Arch* **408**(5): 540-2.

Bonvallet, R. and Rougier, O. (1989). Existence of two calcium currents recorded at normal calcium concentrations in single frog atrial cells. *Cell Calcium* **10**(7): 499-508.

Borgland, S. L., Connor, M., et al., (2001). Nociceptin inhibits calcium channel currents in a subpopulation of small nociceptive trigeminal ganglion neurons in mouse. *J Physiol* **536**(Pt 1): 35-47.

Bossu, J.-L., Mouginot, D., et al., (1998). Low threshold fast-inactivating Ca^{2+} channels in dendrites of cerebellar Purkinje cells. *Low-voltage-activated T-type Ca channels.* R. W. Tsien, J.-P. Clozel and J. Nargeot. Chester, UK, Adis International: 63-71.

Bossu, J. L., Elhamdani, A., et al., (1992). Voltage-dependent calcium entry in confluent bovine capillary endothelial cells. *FEBS Lett* **299**(3): 239-42.

Bossu, J. L., Feltz, A., et al., (1985). Depolarization elicits two distinct calcium currents in vertebrate sensory neurones. *Pflugers Arch* **403**(4): 360-8.

Bourinet, E., Zamponi, G. W., et al., (1996). The alpha 1E calcium channel exhibits permeation properties similar to low-voltage-activated calcium channels. *J Neurosci* **16**(16): 4983-93.

Bournaud, R., Hidalgo, J., et al., (2001). Low threshold T-type calcium current in rat embryonic chromaffin cells. *J Physiol* **537**(Pt 1): 35-44.

Brilla, C. G., Zhou, G., et al., (1992). Aldosterone-mediated stimulation of collagen synthesis in cultured cardiac fibroblasts. *Proc 14th Sci Meeting, Intl Soc Hypertens*: P23.

Bringmann, A., Schopf, S., et al., (2000). Developmental regulation of calcium channel-mediated currents in retinal glial (Muller) cells. *J Neurophysiol* **84**(6): 2975-83.

Bruner, C. A. and Webb, R. C. (1990). Increased vascular reactivity to Bay K 8644 in genetic hypertension. *Pharmacology* **41**(1): 24-35.

Buisson, B., Bottari, S. P., et al., (1992). The angiotensin AT2 receptor modulates T-type calcium current in non-differentiated NG108-15 cells. *FEBS Lett* **309**(2): 161-4.

Buisson, B., Laflamme, L., et al., (1995). A G protein is involved in the angiotensin AT2 receptor inhibition of the T-type calcium current in non-differentiated NG108-15 cells. *J Biol Chem* **270**(4): 1670-4.

Camara, A. K., Begic, Z., et al., (2001). Differential modulation of the cardiac L- and T-type calcium channel currents by isoflurane. *Anesthesiology* **95**(2): 515-24.

Campbell, K. P., Leung, A. T., et al., (1988). The biochemistry and molecular biology of the dihydropyridine-sensitive calcium channel. *Trends Neurosci* **11**(10): 425-30.

Carbone, E. and Lux, H. D. (1984a). A low voltage-activated calcium conductance in embryonic chick sensory neurons. *Biophys J* **46**(3): 413-8.

Carbone, E. and Lux, H. D. (1984b). A low voltage-activated, fully inactivating Ca channel in vertebrate sensory neurones. *Nature* **310**(5977): 501-2.

Carbone, E. and Lux, H. D. (1987). Kinetics and selectivity of a low-voltage-activated calcium current in chick and rat sensory neurones. *J Physiol (Lond)* **386**: 547-570.

Carbone, E., Sher, E., et al., (1990). Ca currents in human neuroblastoma IMR32 cells: kinetics, permeability and pharmacology. *Pflugers Archiv* **416**(1-2): 170-9.

Catterall, W. A. (2000). Structure and regulation of voltage-gated Ca^{2+} channels. *Annu Rev Cell Dev Biol* **16**: 521-55.

Chemin, J., Monteil, A., et al., (2001a). Alternatively spliced α_{1G} ($Ca_V3.1$) intracellular loops promote specific T-type Ca^{2+} channel gating properties. *Biophys J* **80**(3): 1238-50.

Chemin, J., Monteil, A., et al., (2000). Overexpression of T-type calcium channels in HEK-293 cells increases intracellular calcium without affecting cellular proliferation. *FEBS Lett* **478**(1-2): 166-72.

Chemin, J., Monteil, A., et al., (2002a). Specific contribution of human T-type calcium channel isotypes (α_{1G}, α_{1H} and α_{1I}) to neuronal excitability. *J Physiol* **540**(Pt 1): 3-14.

Chemin, J., Monteil, A., et al., (2001b). Direct inhibition of T-type calcium channels by the endogenous cannabinoid anandamide. *Embo J* **20**(24): 7033-40.

Chemin, J., Nargeot, J., et al., (2002b). Neuronal T-type alpha 1H calcium channels induce neuritogenesis and expression of high-voltage-activated calcium channels in the NG108-15 cell line. *J Neurosci* **22**(16): 6856-62.

Chen, C., Xu, R., et al., (2000). Diverse intracellular signalling systems used by growth hormone-releasing hormone in regulating voltage-gated Ca^{2+} or K channels in pituitary somatotropes. *Immunol Cell Biol* **78**(4): 356-68.

Chen, C., Zhang, J., et al., (1990). Two types of voltage-dependent calcium current in rat somatotrophs are reduced by somatostatin. *Journal of Physiology (London)* **425**(29): 29-42.

Chen, C. F., Corbley, M. J., et al., (1988). Voltage-sensitive calcium channels in normal and transformed 3T3 fibroblasts. *Science* **239**(4843): 1024-6.

Chen, X. L., Bayliss, D. A., et al., (1999). A role for T-type Ca^{2+} channels in the synergistic control of aldosterone production by ANG II and K^+. *Am J Physiol* **276**(5 Pt 2): F674-83.

Chesnoy-Marchais, D. and Fritsch, J. (1988). Voltage-gated sodium and calcium currents in rat osteoblasts. *J Physiol* **398**: 291-311.

Chouabe, C., Drici, M. D., et al., (1998). HERG and KvLQT1/IsK, the cardiac K^+ channels involved in long QT syndromes, are targets for calcium channel blockers. *Mol Pharmacol* **54**(4): 695-703.

Christie, B. R., Eliot, L. S., et al., (1995). Different Ca^{2+} channels in soma and dendrites of hippocampal pyramidal neurons mediate spike-induced Ca^{2+} influx. *J Neurophysiol* **73**(6): 2553-7.

Chuang, R. S., Jaffe, H., et al., (1998). Inhibition of T-type voltage-gated calcium channels by a new scorpion toxin. *Nat Neurosci* **1**(8): 668-74.

Chung, J. M., Huguenard, J. R., et al., (1993). Transient enhancement of low-threshold calcium current in thalamic relay neurons after corticectomy. *J Neurophysiol* **70**(1): 20-27.

Churchill, D. and Macvicar, B. A. (1998). Biophysical and pharmacological characterization of voltage-dependent Ca^{2+} channels in neurons isolated from rat nucleus accumbens. *J Neurophysiol* **79**(2): 635-47.

Cohen, C. J., Ertel, E. A., et al., (1992a). High affinity block of myocardial L-type calcium channels by the spider toxin ω-Aga-toxin IIIA: advantages over 1,4-dihydropyridines. *Molecular Pharmacology* **42**(6): 947-951.

Cohen, C. J. and McCarthy, R. T. (1985). Differential effects of dihydropyridines on two populations of Ca channels in anterior pituitary cells. *Biophysical Journal* **47**: 513.

Cohen, C. J. and McCarthy, R. T. (1987). Nimodipine block of calcium channels in rat anterior pituitary cells. *J Physiol* **387**: 195-225.

Cohen, C. J., McCarthy, R. T., et al., (1988). Ca channels in adrenal glomerulosa cells: K^+ and angiotensin II increase T-type Ca channel current. *Proc Natl Acad Sci U S A* **85**(7): 2412-6.

Cohen, C. J., Spires, S., et al., (1992b). Block of T-type Ca channels in guinea pig atrial cells by antiarrhythmic agents and Ca channel antagonists. *J Gen Physiol* **100**(4): 703-28.

Cole, W. C., Chartier, D., et al., (1997). Ca^{2+} permeation through Na+ channels in guinea pig ventricular myocytes. *Am J Physiol* **273**(1 Pt 2): H128-37.

Convery, M. C., Hollywood, M. A., et al., (1997). Role of inward currents in pumping activity of isolated sheep lymphatics. *J Physiol* **501**: 110-111P.

Coulter, D. A., Huguenard, J. R., et al., (1989a). Calcium currents in rat thalamocortical relay neurones: kinetic properties of the transient, low-threshold current. *J Physiol* **414**: 587-604.

Coulter, D. A., Huguenard, J. R., et al., (1989b). Characterization of ethosuximide reduction of low-threshold calcium current in thalamic neurons. *Ann Neurol* **25**(6): 582-93.

Coulter, D. A., Huguenard, J. R., et al., (1990). Differential effects of petit mal anticonvulsants and convulsants on thalamic neurones: calcium current reduction. *Br J Pharmacol* **100**(4): 800-6.

Cribbs, L. L., Lee, J. H., et al., (1998). Cloning and characterization of α1H from human heart, a member of the T-type Ca^{2+} channel gene family. *Circ Res* **83**(1): 103-9.

Cribbs, L. L., Martin, B. L., et al., (2001). Identification of the t-type calcium channel ($Ca_v3.1d$) in developing mouse heart. *Circ Res* **88**(4): 403-7.

Crunelli, V. and Leresche, N. (1991). A role for GABAB receptors in excitation and inhibition of thalamocortical cells. *Trends Neurosci* **14**(1): 16-21.

Daoud, E. G., Knight, B. P., et al., (1997). Effect of verapamil and procainamide on atrial fibrillation-induced electrical remodeling in humans. *Circulation* **96**(5): 1542-50.

Davalli, A. M., Biancardi, E., et al., (1996). Dihydropyridine-sensitive and -insensitive voltage-operated calcium channels participate in the control of glucose-induced insulin release from human pancreatic beta cells. *J Endocrinol* **150**(2): 195-203.

De Jongh, K. S., Warner, C., et al., (1990). Subunits of purified calcium channels. Alpha 2 and delta are encoded by the same gene. *J Biol Chem* **265**(25): 14738-41.

Desmadryl, G., Hilaire, C., et al., (1998). Developmental regulation of T-, N- and L-type calcium currents in mouse embryonic sensory neurones. *Eur J Neurosci* **10**(2): 545-52.

Dirksen, R. T. and Beam, K. G. (1995). Single calcium channel behavior in native skeletal muscle. *J Gen Physiol* **105**(2): 227-47.

Dogrul, A., Zagli, U., et al., (2002). The role of T-type calcium channels in morphine analgesia, development of antinociceptive tolerance and dependence to morphine, and morphine abstinence syndrome. *Life Sci* **71**(6): 725-34.

Dolphin, A. C. (1998). Properties and modulation of T currents in dorsal root ganglia and ND7-23 cells: comparison with α_{1E} currents expressed in COS-7 cells. *Low-voltage-activated T-type Ca channels*. R. W. Tsien, J.-P. Clozel and J. Nargeot. Chester, UK, Adis International: 269-278.

Dolphin, A. C., Wyatt, C. N., et al., (1999). The effect of α2-δ and other accessory subunits on expression and properties of the calcium channel α1G. *J Physiol* **519 Pt 1**: 35-45.

Doughty, J. M., Barnes-Davies, M., et al., (1998). Contrasting Ca^{2+} channel subtypes at cell bodies and synaptic terminals of rat anterioventral cochlear bushy neurones. *J Physiol* **512 (Pt 2)**: 365-76.

Drolet, P., Bilodeau, L., et al., (1997). Inhibition of the T-type Ca^{2+} current by the dopamine D1 receptor in rat adrenal glomerulosa cells: requirement of the combined action of the G βγ protein subunit and cyclic adenosine 3',5'-monophosphate. *Mol Endocrinol* **11**(4): 503-14.

Dupre-Aucouturier, S., Penhoat, A., et al., (2002). ACTH-induced Cl^- current in bovine adrenocortical cells: correlation with cortisol secretion. *Am J Physiol Endocrinol Metab* **282**(2): E355-65.

Ebata, H., Natsume, T., et al., (1991). Reduced calcium sensitivity of dihydropyridine binding to calcium channels in spontaneously hypertensive rats. *Hypertension* **17**(2): 234-41.

Eller, P., Berjukov, S., et al., (2000). High affinity interaction of mibefradil with voltage-gated calcium and sodium channels. *Br J Pharmacol* **130**(3): 669-77.

Ellinor, P. T., Zhang, J. F., et al., (1993). Functional expression of a rapidly inactivating neuronal calcium channel. *Nature* **363**(6428): 455-8.

Ellis, S. B., Williams, M. E., et al., (1988). Sequence and expression of mRNAs encoding the alpha 1 and alpha 2 subunits of a DHP-sensitive calcium channel. *Science* **241**(4873): 1661-4.

Enyeart, J. J., Biagi, B. A., et al., (1990). Blockade of low and high threshold Ca^{2+} channels by diphenylbutylpiperidine antipsychotics linked to inhibition of prolactin gene expression. *J Biol Chem* **265**(27): 16373-9.

Enyeart, J. J., Biagi, B. A., et al., (1992). Preferential block of T-type calcium channels by neuroleptics in neural crest-derived rat and human C cell lines. *Mol Pharmacol* **42**(2): 364-72.

Enyeart, J. J., Mlinar, B., et al., (1993). T-type Ca^{2+} channels are required for adrenocorticotropin-stimulated cortisol production by bovine adrenal zona fasciculata cells. *Mol Endocrinol* **7**(8): 1031-40.

Enyeart, J. J., Xu, L., et al., (2002). An ACTH- and ATP-regulated Background K^+ Channel in Adrenocortical Cells Is TREK-1. *J Biol Chem* **277**(51): 49186-99.

Enyeart, J. J., Xu, L., et al., (2001). Reciprocal modulation of voltage-gated and background K^+ channels mediated by nucleotides and corticotropin. *Mol Pharmacol* **60**(1): 114-23.

Erickson, K. R., Ronnekleiv, O. K., et al., (1993). Role of a T-type calcium current in supporting a depolarizing potential, damped oscillations, and phasic firing in vasopressinergic guinea pig supraoptic neurons. *Neuroendocrinology* **57**(5): 789-800.

Ertel, E. A., Campbell, K. P., et al., (2000). Nomenclature of voltage-gated calcium channels. *Neuron* **25**(3): 533-5.

Ertel, E. A., Cohen, C. J., et al., (1993). ω-Aga-IIB: A high affinity peptide blocker of multiple types of voltage-gated Ca channels. *Biophys J* **64**(2 part 2): A117.

Ertel, E. A., Smith, M. M., et al., (1994). Isolation of myocardial L-type calcium channel gating currents with the spider toxin ω-Aga-IIIA. *J Gen Physiol* **103**(5): 731-53.

Ertel, S. I. and Ertel, E. A. (1997a). Low-voltage-activated T-type Ca^{2+} channels. *Trends Pharmacol Sci* **18**(2): 37-42.

Ertel, S. I., Ertel, E. A., et al., (1997b). T-type Ca^{2+} channels and pharmacological blockade: potential pathophysiological relevance. *Cardiovasc Drugs Ther* **11**(6): 723-39.

Eskinder, H., Rusch, N. J., et al., (1991). The effects of volatile anesthetics on L- and T-type calcium channel currents in canine cardiac Purkinje cells. *Anesthesiology* **74**(5): 919-26.

Espinosa, F., Lopez-Gonzalez, I., et al., (2000). Dual regulation of the T-type Ca^{2+} current by serum albumin and beta-estradiol in mammalian spermatogenic cells. *FEBS Lett* **475**(3): 251-6.

Estacion, M. and Mordan, L. J. (1993). Expression of voltage-gated calcium channels correlates with PDGF-stimulated calcium influx and depends upon cell density in C3H 10T1/2 mouse fibroblasts. *Cell Calcium* **14**(2): 161-71.

Faas, G. C., Vreugdenhil, M., et al., (1996). Calcium currents in pyramidal CA1 neurons in vitro after kindling epileptogenesis in the hippocampus of the rat. *Neuroscience* **75**(1): 57-67.

Fareh, S., Benardeau, A., et al., (2001). Differential efficacy of L- and T-type calcium channel blockers in preventing tachycardia-induced atrial remodeling in dogs. *Cardiovasc Res* **49**(4): 762-70.

Fareh, S., Benardeau, A., et al., (1999). The T-type Ca^{2+} channel blocker mibefradil prevents the development of a substrate for atrial fibrillation by tachycardia-induced atrial remodeling in dogs. *Circulation* **100**(21): 2191-7.

Fares, N., Gomez, J. P., et al., (1996). T-type calcium current is expressed in dedifferentiated adult rat ventricular cells in primary culture. *C R Acad Sci III* **319**(7): 569-76.

Fedulova, S. A., Kostyuk, P. G., et al., (1985). Two types of calcium channels in the somatic membrane of new-born rat dorsal root ganglion neurones. *J Physiol* **359**: 431-46.

Ferron, L., Capuano, V., et al., (2002). Functional and Molecular Characterization of a T-type Ca^{2+} Channel During Fetal and Postnatal Rat Heart Development. *J Mol Cell Cardiol* **34**(5): 533-46.

Fisher, T. E. and Bourque, C. W. (1995). Voltage-gated calcium currents in the magnocellular neurosecretory cells of the rat supraoptic nucleus. *J Physiol* **486 (Pt 3)**: 571-80.

Fomina, A. F., Kostyuk, P. G., et al., (1993). Glucocorticoid modulation of calcium currents in growth hormone 3 cells. *Neuroscience* **55**(3): 721-5.

Fox, A. P., Nowycky, M. C., et al., (1987). Kinetic and pharmacological properties distinguishing three types of calcium currents in chick sensory neurones. *J Physiol* **394**: 149-72.

Fraser, D. D. and MacVicar, B. A. (1991). Low-threshold transient calcium current in rat hippocampal lacunosum-moleculare interneurons: kinetics and modulation by neurotransmitters. *J Neurosci* **11**(9): 2812-20.

Friedman, M. E., Suarez-Kurtz, G., et al., (1986). Two calcium currents in a smooth muscle cell line. *American Journal of Physiology* **250**(4 Pt 2): H699-H703.

Furukawa, T., Ito, H., et al., (1992). Endothelin-1 enhances calcium entry through T-type calcium channels in cultured neonatal rat ventricular myocytes. *Circ Res* **71**(5): 1242-53.

Galan, L., Talavera, K., et al., (1998). Characteristics of Ca^{2+} channel blockade by oxodipine and elgodipine in rat cardiomyocytes. *Eur J Pharmacol* **357**(1): 93-105.

Gandia, L., Mayorgas, I., et al., (1998). Human adrenal chromaffin cell calcium channels: drastic current facilitation in cell clusters, but not in isolated cells. *Pflugers Arch* **436**(5): 696-704.

Ganitkevich, V. and Isenberg, G. (1990). Contribution of two types of calcium channels to membrane conductance of single myocytes from guinea-pig coronary artery. *J Physiol* **426**: 19-42.

Ganten, D., Schelling, P., et al., (1975). Effect of angiotensin and the angiotensin antagonist P113 on iso-renin and cell growth in 3T3 mouse cells. *IRCS J Med Sci* **3**: 327-332.

Gao, B., Sekido, Y., et al., (2000). Functional properties of a new voltage-dependent calcium channel $\alpha 2\delta$ auxiliary subunit gene (CACNA2D2). *J Biol Chem* **275**(16): 12237-42.

Garcia, J. and Beam, K. G. (1994). Calcium transients associated with the T type calcium current in myotubes. *J Gen Physiol* **104**(6): 1113-28.

Garcia, M. A. and Meizel, S. (1999). Progesterone-mediated calcium influx and acrosome reaction of human spermatozoa: pharmacological investigation of T-type calcium channels. *Biol Reprod* **60**(1): 102-9.

Gauck, V., Thomann, M., et al., (2001). Spatial distribution of low- and high-voltage-activated calcium currents in neurons of the deep cerebellar nuclei. *J Neurosci* **21**(15): RC158.

Gloor, P. and Fariello, R. G. (1988). Generalized epilepsy: some of its cellular mechanisms differ from those of focal epilepsy. *Trends Neurosci* **11**(2): 63-8.

Gomora, J. C., Daud, A. N., et al., (2001). Block of cloned human T-type calcium channels by succinimide antiepileptic drugs. *Mol Pharmacol* **60**(5): 1121-32.

Gomora, J. C., Enyeart, J. A., et al., (1999). Mibefradil potently blocks ATP-activated K^+ channels in adrenal cells. *Mol Pharmacol* **56**(6): 1192-7.

Gomora, J. C., Murbartian, J., et al., (2002). Cloning and expression of the human T-type channel $Ca_v3.3$: Insights into prepulse facilitation. *Biophys J* **83**(1): 229-41.

Gomora, J. C., Xu, L., et al., (2000). Effect of mibefradil on voltage-dependent gating and kinetics of T-type Ca^{2+} channels in cortisol-secreting cells. *J Pharmacol Exp Ther* **292**(1): 96-103.

Gordienko, D. V., Clausen, C., et al., (1994). Ionic currents and endothelin signaling in smooth muscle cells from rat renal resistance arteries. *Am J Physiol* **266**(2 Pt 2): F325-41.

Gottman, K., Dietzel, I. D., et al., (1988). Development of inward currents in chick sensory and automatic neuronal precursor cells in culture. *J Neurosci* **8**(10): 3722-32.

Gottmann, K., Rohrer, H., et al., (1991). Distribution of Ca^{2+} and Na^+ conductances during neuronal differentiation of chick DRG precursor cells. *J Neurosci* **11**(11): 3371-8.

Green, P. J., Warre, R., et al., (2001). Kinetic modification of the α_{1I} subunit-mediated T-type Ca^{2+} channel by a human neuronal Ca^{2+} channel gamma subunit. *J Physiol* **533**(Pt 2): 467-78.

Griffith, W. H., Taylor, L., et al., (1994). Whole-cell and single-channel calcium currents in guinea pig basal forebrain neurons. *J Neurophysiol* **71**(6): 2359-76.

Gu, X. and Spitzer, N. C. (1993). Low-threshold Ca^{2+} current and its role in spontaneous elevations of intracellular Ca^{2+} in developing Xenopus neurons. *J Neurosci* **13**(11): 4936-48.

Guarda, E., Katwa, L. C., et al., (1993). Effects of endothelins on collagen turnover in cardiac fibroblasts. *Cardiovasc Res* **27**(12): 2130-4.

Guatimosim, S., Sobie, E. A., et al., (2001). Molecular identification of a TTX-sensitive Ca^{2+} current. *Am J Physiol Cell Physiol* **280**(5): C1327-39.

Guerineau, N., Corcuff, J. B., et al., (1991). Spontaneous and corticotropin-releasing factor-induced cytosolic calcium transients in corticotrophs. *Endocrinology* **129**(1): 409-20.

Gustafsson, F., Andreasen, D., et al., (2001). Conducted vasoconstriction in rat mesenteric arterioles: role for dihydropyridine-insensitive Ca^{2+} channels. *Am J Physiol Heart Circ Physiol* **280**(2): H582-90.

Hagiwara, N., Irisawa, H., et al., (1988). Contribution of two types of calcium currents to the pacemaker potentials of rabbit sino-atrial node cells. *J Physiol* **395**: 233-53.

Hagiwara, S., Ozawa, S., et al., (1975). Voltage clamp analysis of two inward current mechanisms in the egg cell membrane of a starfish. *J Gen Physiol* **65**(5): 617-44.

Han, W., Chartier, D., et al., (2001). Ionic remodeling of cardiac Purkinje cells by congestive heart failure. *Circulation* **104**(17): 2095-100.

Hay, M., Hasser, E. M., et al., (1996). Area postrema voltage-activated calcium currents. *J Neurophysiol* **75**(1): 133-41.

Heady, T. N., Gomora, J. C., et al., (2001). Molecular pharmacology of T-type Ca^{2+} channels. *Jpn J Pharmacol* **85**(4): 339-50.

Hermsmeyer, K. (1991). Differences of calcium channels in vascular muscle in hypertension. *Am J Hypertens* **4**(7 Pt 2): 412S-415S.

Hernandez-Cruz, A. and Pape, H. C. (1989). Identification of two calcium currents in acutely dissociated neurons from the rat lateral geniculate nucleus. *J Neurophysiol* **61**(6): 1270-83.

Herrington, J. and Lingle, C. J. (1992). Kinetic and pharmacological properties of low voltage-activated Ca^{2+} current in rat clonal (GH3) pituitary cells. *J Neurophysiol* **68**(1): 213-32.

Herrington, J., Stern, R. C., et al., (1991). Halothane inhibits two components of calcium current in clonal (GH3) pituitary cells. *J Neurosci* **11**(7): 2226-40.

Hess, P. (1990). Calcium channels in vertebrate cells. *Annu Rev Neurosci* **13**: 337-56.

Heubach, J. F., Kohler, A., et al., (2000). T-Type and tetrodotoxin-sensitive Ca^{2+} currents coexist in guinea pig ventricular myocytes and are both blocked by mibefradil. *Circ Res* **86**(6): 628-35.

Hille, B. (1984). *Ionic channels of excitable membranes.* Sunderland, Mass., Sinauer Associates.

Hirano, Y., Fozzard, H. A., et al., (1989). Characteristics of L- and T-type currents in canine cardiac Purkinje cells. *Am J Physiol* **256**(5 Pt 2): H1478-92.

Hirata, Y., Takagi, Y., et al., (1989). Endothelin is a potent mitogen for rat vascular smooth muscle cells. *Atherosclerosis* **78**(2-3): 225-8.

Hirooka, K., Bertolesi, G. E., et al., (2002). T-Type Calcium Channel α_{1G} and α_{1H} Subunits in Human Retinoblastoma Cells and Their Loss After Differentiation. *J Neurophysiol* **88**(1): 196-205.

Hoehn, K., Watson, T. W., et al., (1993). Multiple types of calcium channels in acutely isolated rat neostriatal neurons. *J Neurosci* **13**(3): 1244-57.

Hogan, Q. H., McCallum, J. B., et al., (2000). Painful neuropathy decreases membrane calcium current in mammalian primary afferent neurons. *Pain* **86**(1-2): 43-53.

Hollins, B. and Ikeda, S. R. (1996). Inward currents underlying action potentials in rat adrenal chromaffin cells. *J Neurophysiol* **76**(2): 1195-211.

Hollywood, M. A., Cotton, K. D., et al., (1997). Isolated sheep mesenteric lymphatic smooth muscle cells possess both T- and L-type calcium currents. *J Physiol* **501**: 109P.

Honda, M., Hayashi, K., et al., (2001). Divergent renal vasodilator action of L- and T-type calcium antagonists in vivo. *J Hypertens* **19**(11): 2031-7.

Huang, B., Qin, D., et al., (2000). Reexpression of T-type Ca^{2+} channel gene and current in post-infarction remodeled rat left ventricle. *Cardiovasc Res* **46**(3): 442-9.

Huguenard, J. R. (1996). Low-threshold calcium currents in central nervous system neurons. *Annu Rev Physiol* **58**: 329-48.

Huguenard, J. R. (1999). Neuronal circuitry of thalamocortical epilepsy and mechanisms of antiabsence drug action. *Adv Neurol* **79**: 991-9.

Huguenard, J. R., Gutnick, M. J., et al., (1993). Transient Ca^{2+} currents in neurons isolated from rat lateral habenula. *J Neurophysiol* **70**(1): 158-66.

Huguenard, J. R. and Prince, D. A. (1992). A novel T-type current underlies prolonged Ca^{2+}-dependent burst firing in GABAergic neurons of rat thalamic reticular nucleus. *J Neurosci* **12**(10): 3804-17.

Huguenard, J. R. and Prince, D. A. (1994). Clonazepam suppresses GABAB-mediated inhibition in thalamic relay neurons through effects in nucleus reticularis. *J Neurophysiol* **71**(6): 2576-81.

Huizinga, J. D., Farraway, L., et al., (1991). Generation of slow-wave-type action potentials in canine colon smooth muscle involves a non-L-type Ca^{2+} conductance. *J Physiol* **442**: 15-29.

Inoue, Y., Oike, M., et al., (1990). Endothelin augments unitary calcium channel currents on the smooth muscle cell membrane of guinea-pig portal vein. *J Physiol* **423**: 171-91.

Ito, H., Hirata, Y., et al., (1991). Endothelin-1 induces hypertrophy with enhanced expression of muscle-specific genes in cultured neonatal rat cardiomyocytes. *Circ Res* **69**(1): 209-15.

Jacob, T. J. (1991). Identification of a low-threshold T-type calcium channel in bovine ciliary epithelial cells. *Am J Physiol* **261**(5 Pt 1): C808-13.

Jagannathan, S., Punt, E. L., et al., (2002). Identification and localization of T-type voltage-operated calcium channel subunits in human male germ cells. Expression of multiple isoforms. *J Biol Chem* **277**(10): 8449-56.

Janis, R. A. and Triggle, D. J. (1991). Drugs acting on calcium channels. *Calcium channels: their properties, functions, regulation and clinical relevance.* L. H. Hurwitz, L. D. Partridge and J. K. Leach. Boca Raton, CRC Press: 195-249.

Janssen, L. J. (1997). T-type and L-type Ca^{2+} currents in canine bronchial smooth muscle: characterization and physiological roles. *Am J Physiol* **272**(6 Pt 1): C1757-65.

Jay, S. D., Ellis, S. B., et al., (1990). Primary structure of the gamma subunit of the DHP-sensitive calcium channel from skeletal muscle. *Science* **248**(4954): 490-2.

Johnston, D., Avery, R. B., et al., (1998). Physiology, pharmacology and function of low voltage-activated Ca^{2+} channels in hippocampal pyramidal neurons. *Low-voltage-activated T-type Ca channels.* R. W. Tsien, J.-P. Clozel and J. Nargeot. Chester, UK, Adis International: 133-142.

Kaneda, M. and Akaike, N. (1989). The low-threshold Ca current in isolated amygdaloid neurons in the rat. *Brain Res* **497**(1): 187-90.

Kaneda, M., Wakamori, M., et al., (1990). Low-threshold calcium current in isolated Purkinje cell bodies of rat cerebellum. *J Neurophysiol* **63**(5): 1046-51.

Karam, H., Heudes, D., et al., (1996). Respective role of humoral factors and blood pressure in cardiac remodeling of DOCA hypertensive rats. *Cardiovascular Research* **31**(2): 287-295.

Karst, H., Joels, M., et al., (1993). Low-threshold calcium current in dendrites of the adult rat hippocampus. *Neurosci Lett* **164**(1-2): 154-8.

Kasai, H. (1992). Voltage- and time-dependent inhibition of neuronal calcium channels by a GTP-binding protein in a mammalian cell line. *J Physiol* **448**: 189-209.

Kato, S., Ishida, H., et al., (1996). Alterations in basal and glucose-stimulated voltage-dependent Ca^{2+} channel activities in pancreatic beta cells of non-insulin-dependent diabetes mellitus GK rats. *J Clin Invest* **97**(11): 2417-25.

Kavalali, E. T., Zhuo, M., et al., (1997). Dendritic Ca^{2+} channels characterized by recordings from isolated hippocampal dendritic segments. *Neuron* **18**(4): 651-63.

Kawai, F. (1999). Odorants suppress T- and L-type Ca^{2+} currents in olfactory receptor cells by shifting their inactivation curves to a negative voltage. *Neurosci Res* **35**(4): 253-63.

Kawai, F. and Miyachi, E. (2001). Enhancement by T-type Ca^{2+} currents of odor sensitivity in olfactory receptor cells. *J Neurosci* **21**(10): RC144.

Kawano, S. and DeHaan, R. L. (1989). Low-threshold current is major calcium current in chick ventricle cells. *Am J Physiol* **256**(5 Pt 2): H1505-8.

Keja, J. A., Stoof, J. C., et al., (1992). Dopamine D2 receptor stimulation differentially affects voltage-activated calcium channels in rat pituitary melanotropic cells. *J Physiol* **450**: 409-35.

Kelly, K. M., Gross, R. A., et al., (1990). Valproic acid selectively reduces the low-threshold (T) calcium current in rat nodose neurons. *Neurosci Lett* **116**(1-2): 233-8.

Kenakin, T. P., Bond, R. A., et al., (1992). Definition of pharmacological receptors. *Pharmacol Rev* **44**(3): 351-62.

Kim, D., Song, I., et al., (2001). Lack of the burst firing of thalamocortical relay neurons and resistance to absence seizures in mice lacking α_{1G} T-type Ca^{2+} channels. *Neuron* **31**(1): 35-45.

Kitasato, H., Kai, R., et al., (1996). The intrinsic rhythmicity of spike-burst generation in pancreatic beta-cells and intercellular interaction within an islet. *Jpn J Physiol* **46**(5): 363-73.

Kito, M., Maehara, M., et al., (1996). Mechanisms of T-type calcium channel blockade by zonisamide. *Seizure* **5**(2): 115-9.

Kleiman, R. B. and Houser, S. R. (1989). Outward currents in normal and hypertrophied feline ventricular myocytes. *Am J Physiol* **256**(5 Pt 2): H1450-61.

Klockner, U., Lee, J. H., et al., (1999). Comparison of the Ca^{2+} currents induced by expression of three cloned alpha1 subunits, α_{1G}, α_{1H} and α_{1I}, of low-voltage-activated T-type Ca^{2+} channels. *Eur J Neurosci* **11**(12): 4171-8.

Klugbauer, N., Dai, S., et al., (2000). A family of gamma-like calcium channel subunits. *FEBS Lett* **470**(2): 189-97.

Klugbauer, N., Marais, E., et al., (1999). A T-type calcium channel from mouse brain. *Pflugers Arch* **437**(5): 710-5.

Kobrinsky, E. M., Pearson, H. A., et al., (1994). Low- and high-voltage-activated calcium channel currents and their modulation in the dorsal root ganglion cell line ND7-23. *Neuroscience* **58**(3): 539-52.

Kostyuk, P., Pronchuk, N., et al., (1993). Calcium currents in aged rat dorsal root ganglion neurones. *J Physiol* **461**: 467-83.

Kostyuk, P. G., Molokanova, E. A., et al., (1992). Different action of ethosuximide on low- and high-threshold calcium currents in rat sensory neurons. *Neuroscience* **51**(4): 755-8.

Kuga, T., Kobayashi, S., et al., (1996). Cell cycle--dependent expression of L- and T-type Ca^{2+} currents in rat aortic smooth muscle cells in primary culture. *Circ Res* **79**(1): 14-9.

Kuga, T., Sadoshima, J., et al., (1990). Actions of Ca^{2+} antagonists on two types of Ca^{2+} channels in rat aorta smooth muscle cells in primary culture. *Circ Res* **67**(2): 469-80.

Kumar, P. P., Stotz, S. C., et al., (2002). Synthesis and evaluation of a new class of nifedipine analogs with T-type calcium channel blocking activity. *Mol Pharmacol* **61**(3): 649-58.

Kuo, C. C. and Yang, S. (2001). Recovery from inactivation of T-type Ca^{2+} channels in rat thalamic neurons. *J Neurosci* **21**(6): 1884-92.

Kusano, K. and Gainer, H. (1993). Modulation of voltage-activated Ca currents by pain-inducing agents in a dorsal root ganglion neuronal line, F-11. *J Neurosci Res* **34**(2): 158-69.

Kwan, C. Y. and Achike, F. (2002). Tetrandrine and related bis-benzylisoquinoline alkaloids from medicinal herbs: cardiovascular effects and mechanisms of action. *Acta Pharmacol Sin* **23**(12): 1057-1068.

Lacinova, L., Klugbauer, N., et al., (2000). Regulation of the calcium channel α_{1G} subunit by divalent cations and organic blockers. *Neuropharmacology* **39**(7): 1254-66.

Lambert, R. C., Maulet, Y., et al., (1997). T-type Ca^{2+} current properties are not modified by Ca^{2+} channel beta subunit depletion in nodosus ganglion neurons. *J Neurosci* **17**(17): 6621-8.

Lee, J. H., Daud, A. N., et al., (1999). Cloning and expression of a novel member of the low voltage-activated T-type calcium channel family. *J Neurosci* **19**(6): 1912-21.

Lee, J. H., Kim, E. G., et al., (2002). Identification of T-type α_{1H} Ca^{2+} channels ($Ca_v3.2$) in major pelvic ganglion neurons. *J Neurophysiol* **87**(6): 2844-50.

Lee, S. H., Yu, W. C., et al., (2000). Effect of verapamil on long-term tachycardia-induced atrial electrical remodeling. *Circulation* **101**(2): 200-6.

Lefebvre, H., Compagnon, P., et al., (2001). Production and metabolism of serotonin (5-HT) by the human adrenal cortex: paracrine stimulation of aldosterone secretion by 5-HT. *J Clin Endocrinol Metab* **86**(10): 5001-7.

Lehmann-Horn, F., Sipos, I., et al., (1995). Altered calcium currents in human hypokalemic periodic paralysis myotubes expressing mutant L-type calcium channels. *Soc Gen Physiol Ser* **50**: 101-13.

Lenglet, S., Louiset, E., et al., (2002). Activation of 5-HT(7) receptor in rat glomerulosa cells is associated with an increase in adenylyl cyclase activity and calcium influx through T-type calcium channels. *Endocrinology* **143**(5): 1748-60.

Letts, V. A., Felix, R., et al., (1998). The mouse stargazer gene encodes a neuronal Ca^{2+}-channel gamma subunit. *Nat Genet* **19**(4): 340-7.

Leuranguer, V., Mangoni, M. E., et al., (2001). Inhibition of T-type and L-type calcium channels by mibefradil: physiologic and pharmacologic bases of cardiovascular effects. *J Cardiovasc Pharmacol* **37**(6): 649-61.

Leuranguer, V., Monteil, A., et al., (2000). T-type calcium currents in rat cardiomyocytes during postnatal development: contribution to hormone secretion. *Am J Physiol Heart Circ Physiol* **279**(5): H2540-8.

Li, D., Melnyk, P., et al., (2000). Effects of experimental heart failure on atrial cellular and ionic electrophysiology. *Circulation* **101**(22): 2631-8.

Li, G. R. and Nattel, S. (1997a). Properties of human atrial ICa at physiological temperatures and relevance to action potential. *American Journal of Physiology* **272**(1 Pt 2): H227-35.

Li, J. and Nathan, R. D. (1997b). Ionic basis of ryanodine's negative chronotropic effect on S-A node pacemaker cells. *Biophysical Journal* **72**(2): A49.

Li, J., Qu, J., et al., (1997c). Ionic basis of ryanodine's negative chronotropic effect on pacemaker cells isolated from the sinoatrial node. *Am J Physiol* **273**(5 Pt 2): H2481-9.

Li, Y. W., Guyenet, P. G., et al., (1998). Voltage-dependent calcium currents in bulbospinal neurons of neonatal rat rostral ventrolateral medulla: modulation by α2-adrenergic receptors. *J Neurophysiol* **79**(2): 583-94.

Liévano, A., Santi, C. M., et al., (1996). T-type Ca^{2+} channels and α_{1E} expression in spermatogenic cells, and their possible relevance to the sperm acrosome reaction. *FEBS Lett* **388**(2-3): 150-154.

Lijnen, P., Fagard, R., et al., (1999). Mibefradil-induced inhibition of proliferation of human peripheral blood mononuclear cells. *J Cardiovasc Pharmacol* **33**(4): 595-604.

Lijnen, P. J. (2002). Impaired mechanisms of leukocyte adhesion in vitro by the calcium channel antagonist mibefradil. *Cardiovasc Drugs Ther* **16**(3): 179-80.

Liu, J. H., Bijlenga, P., et al., (1999). Mibefradil (Ro 40-5967) inhibits several Ca^{2+} and K^+ currents in human fusion-competent myoblasts. *Br J Pharmacol* **126**(1): 245-50.

Liu, Q. Y., Karpinski, E., et al., (1992). Tetrandrine inhibits both T and L calcium channel currents in ventricular cells. *J Cardiovasc Pharmacol* **20**(4): 513-9.

Lledo, P. M., Homburger, V., et al., (1992). Differential G protein-mediated coupling of D2 dopamine receptors to K^+ and Ca^{2+} currents in rat anterior pituitary cells. *Neuron* **8**(3): 455-63.

Lledo, P. M., Legendre, P., et al., (1990). Dopamine inhibits two characterized voltage-dependent calcium currents in identified rat lactotroph cells. *Endocrinology* **127**(3): 990-1001.

Llinas, R. and Yarom, Y. (1986). Specific blockade of the low threshold calcium channel by high molecular weight alcohols. *Soc Neurosci Abstr* **12**: 174.

Loirand, G., Mironneau, C., et al., (1989). Two types of calcium currents in single smooth muscle cells from rat portal vein. *J Physiol (Lond)* **412**: 333-349.

Lotshaw, D. P. (2001). Role of membrane depolarization and T-type Ca^{2+} channels in angiotensin II and K^+ stimulated aldosterone secretion. *Mol Cell Endocrinol* **175**(1-2): 157-71.

Lu, H.-K., Fern, R. J., et al., (1996). Angiotensin II stimulates T-type Ca^{2+} channel currents via activation of a G protein, Gi. *American Journal of Physiology* **271**: C1340-C1349.

Lu, H. K., Fern, R. J., et al., (1994). Ca^{2+}-dependent activation of T-type Ca^{2+} channels by calmodulin-dependent protein-kinase-II. *American Journal of Physiology* **267**(1): F183-189.

Luther, J. A. and Tasker, J. G. (2000). Voltage-gated currents distinguish parvocellular from magnocellular neurones in the rat hypothalamic paraventricular nucleus. *J Physiol* **523 Pt 1**: 193-209.

Magistretti, J. and de Curtis, M. (1998). Low-voltage activated T-type calcium currents are differently expressed in superficial and deep layers of guinea pig piriform cortex. *J Neurophysiol* **79**(2): 808-16.

Mangoni, M., Leuranguer, V., et al., (1997). Block of cardiac Ca^{2+} channels by the new Ca^{2+} antagonist Ro 40-5967: consequences on heart rate and cardiac output. *Biophysical Journal* **72**(2): A256.

Mann, D. L., Kent, R. L., et al., (1989). Load regulation of the properties of adult feline cardiocytes: growth induction by cellular deformation. *Circulation Research* **64**(6): 1079-1090.

Marban, E. and Koretsune, Y. (1990). Cell calcium, oncogenes, and hypertrophy. *Hypertension* **15**(6 Pt 1): 652-8.

Marchetti, C. and Brown, A. M. (1988). Protein kinase activator 1-oleoyl-2-acetyl-sn-glycerol inhibits two types of calcium currents in GH3 cells. *Am J Physiol* **254**(1 Pt 1): C206-10.

Marchetti, C., Carbone, E., et al., (1986). Effects of dopamine and noradrenaline on Ca channels of cultured sensory and sympathetic neurons of chick. *Pflugers Arch* **406**(2): 104-11.

Margeta-Mitrovic, M., Grigg, J. J., et al., (1997). Neurotensin and substance P inhibit low- and high-voltage-activated Ca^{2+} channels in cultured newborn rat nucleus basalis neurons. *J Neurophysiol* **78**(3): 1341-52.

Mariot, P., Vanoverberghe, K., et al., (2002). Overexpression of an α1H (Ca_v3.2) T-type calcium channel during neuroendocrine differentiation of human prostate cancer cells. *J Biol Chem* **277**(13): 10824-33.

Martin, R. L., Lee, J. H., et al., (2000). Mibefradil block of cloned T-type calcium channels. *J Pharmacol Exp Ther* **295**(1): 302-8.

Martin-Caraballo, M. and Greer, J. J. (2001). Voltage-sensitive calcium currents and their role in regulating phrenic motoneuron electrical excitability during the perinatal period. *J Neurobiol* **46**(4): 231-48.

Martinez, M. L., Heredia, M. P., et al., (1999). Expression of T-type Ca^{2+} channels in ventricular cells from hypertrophied rat hearts. *J Mol Cell Cardiol* **31**(9): 1617-25.

Masumiya, H., Kase, J., et al., (1999). Effects of mibefradil, a selective T-type Ca^{2+} channel antagonist, on sino-atrial node and ventricular myocardia. *Res Commun Mol Pathol Pharmacol* **104**(3): 321-9.

Masumiya, H., Tanaka, H., et al., (1997). Effects of Ca^{2+} channel antagonists on sinus node: prolongation of late phase 4 depolarization by efonidipine. *Eur J Pharmacol* **335**(1): 15-21.

Matthews, E. A. and Dickenson, A. H. (2001). Effects of ethosuximide, a T-type Ca^{2+} channel blocker, on dorsal horn neuronal responses in rats. *Eur J Pharmacol* **415**(2-3): 141-9.

Maylie, J. and Morad, M. (1995). Evaluation of T- and L-type Ca^{2+} currents in shark ventricular myocytes. *Am J Physiol* **269**(5 Pt 2): H1695-703.

McCarthy, R. T. and Cohen, C. J. (1989). Nimodipine block of calcium channels in rat vascular smooth muscle cell lines. Exceptionally high-affinity binding in A7r5 and A10 cells. *J Gen Physiol* **94**(4): 669-92.

McCarthy, R. T., Isales, C., et al., (1993). T-type calcium channels in adrenal glomerulosa cells: GTP-dependent modulation by angiotensin II. *Proc Natl Acad Sci U S A* **90**(8): 3260-4.

McCarthy, R. T., Isales, C. M., et al., (1990). Atrial natriuretic peptide differentially modulates T- and L-type calcium channels. *Am J Physiol* **258**(3 Pt 2): F473-8.

McCobb, D. P., Best, P. M., et al., (1989). Development alters the expression of calcium currents in chick limb motoneurons. *Neuron* **2**(6): 1633-43.

McDonough, S. I. and Bean, B. P. (1998). Mibefradil inhibition of T-type calcium channels in cerebellar purkinje neurons. *Mol Pharmacol* **54**(6): 1080-7.

McDowell, T. S., Pancrazio, J. J., et al., (1999). Volatile anesthetic sensitivity of T-type calcium currents in various cell types. *Anesth Analg* **88**(1): 168-73.

McEnery, M. W., Snowman, A. M., et al., (1991). Purified ω-conotoxin GVIA receptor of rat brain resembles a dihydropyridine-sensitive L-type calcium channel. *Proc Natl Acad Sci U S A* **88**(24): 11095-9.

McNaughton, N. C., Hainsworth, A. H., et al., (2000). Inhibition of recombinant low-voltage-activated Ca^{2+} channels by the neuroprotective agent BW619C89 (Sipatrigine). *Neuropharmacology* **39**(7): 1247-53.

McRory, J. E., Santi, C. M., et al., (2001). Molecular and functional characterization of a family of rat brain T-type calcium channels. *J Biol Chem* **276**(6): 3999-4011.

Meza, U., Avila, G., et al., (1994). Long-term regulation of calcium channels in clonal pituitary cells by epidermal growth factor, insulin, and glucocorticoids. *J Gen Physiol* **104**(6): 1019-38.

Miljanich, G. P. and Ramachandran, J. (1995). Antagonists of neuronal calcium channels: structure, function, and therapeutic implications. *Annu Rev Pharmacol Toxicol* **35**: 707-34.

Min, J. Y., Meissner, A., et al., (2002). Mibefradil improves beta-adrenergic responsiveness and intracellular Ca^{2+} handling in hypertrophied rat myocardium. *Exp Biol Med (Maywood)* **227**(5): 336-44.

Mishra, S. K. and Hermsmeyer, K. (1994a). Inhibition of signal Ca^{2+} in dog coronary arterial vascular muscle cells by Ro 40-5967 [published erratum appears in J Cardiovasc Pharmacol 1995 Apr;25(4):679]. *J Cardiovasc Pharmacol* **24**(1): 1-7.

Mishra, S. K. and Hermsmeyer, K. (1994b). Selective inhibition of T-type Ca^{2+} channels by Ro 40-5967. *Circ Res* **75**(1): 144-8.

Mitra, R. and Morad, M. (1986). Two types of calcium channels in guinea pig ventricular myocytes. *Proc Natl Acad Sci U S A* **83**(14): 5340-4.

Mittman, S., Guo, J., et al., (1999a). Structure and alternative splicing of the gene encoding α1G, a human brain T calcium channel alpha1 subunit. *Neurosci Lett* **274**(3): 143-6.

Mittman, S., Guo, J., et al., (1999b). Structure and alternative splicing of the gene encoding alpha1I, a human brain T calcium channel alpha1 subunit. *Neurosci Lett* **269**(3): 121-4.

Mlinar, B., Biagi, B. A., et al., (1993). Voltage-gated transient currents in bovine adrenal fasciculata cells. I. T-type Ca2+ current. *J Gen Physiol* **102**(2): 217-37.

Monteil, A., Chemin, J., et al., (2000a). Molecular and functional properties of the human α_{1G} subunit that forms T-type calcium channels. *J Biol Chem* **275**(9): 6090-100.

Monteil, A., Chemin, J., et al., (2000b). Specific properties of T-type calcium channels generated by the human alpha 1I subunit. *J Biol Chem* **275**(22): 16530-5.

Moran, T. D., Abdulla, F. A., et al., (2000). Cellular neurophysiological actions of nociceptin/orphanin FQ. *Peptides* **21**(7): 969-76.

Morita, H., Cousins, H., et al., (1999). Predominant distribution of nifedipine-insensitive, high voltage-activated Ca^{2+} channels in the terminal mesenteric artery of guinea pig. *Circ Res* **85**(7): 596-605.

Muller, T. H., Misgeld, U., et al., (1992). Ionic currents in cultured rat hypothalamic neurones. *J Physiol* **450**: 341-62.

Muntinga, H. J., van der Vring, J. A., et al., (1996). Effect of mibefradil on left ventricular diastolic function in patients with congestive heart failure. *J Cardiovasc Pharmacol* **27**(5): 652-6.

Muramatsu, M., Tyler, R. C., et al., (1997). Possible role of T-type Ca^{2+} channels in L-NNA vasoconstriction of hypertensive rat lungs. *Am J Physiol* **272**(6 Pt 2): H2616-21.

Nakashima, Y. M., Todorovic, S. M., et al., (1998). Properties of Ba^{2+} currents arising from human α1E and α1Eβ3 constructs expressed in HEK293 cells: physiology, pharmacology, and comparison to native T-type Ba^{2+} currents. *Neuropharmacology* **37**(8): 957-72.

Nattel, S. (1999). Atrial electrophysiological remodeling caused by rapid atrial activation: underlying mechanisms and clinical relevance to atrial fibrillation. *Cardiovasc Res* **42**(2): 298-308.

Nebe, B., Holzhausen, C., et al., (2002). Impaired mechanisms of leukocyte adhesion in vitro by the calcium channel antagonist mibefradil. *Cardiovasc Drugs Ther* **16**(3): 183-93.

Niesen, C. E. and Ge, S. (1999). Chronic epilepsy in developing hippocampal neurons: electrophysiologic and morphologic features. *Dev Neurosci* **21**(3-5): 328-38.

Niidome, T., Kim, M. S., et al., (1992). Molecular cloning and characterization of a novel calcium channel from rabbit brain. *FEBS Lett* **308**(1): 7-13.

Nilius, B., Hess, P., et al., (1985). A novel type of cardiac calcium channel in ventricular cells. *Nature* **316**(6027): 443-6.

Nilius, B., Prenen, J., et al., (1997). Inhibition by mibefradil, a novel calcium channel antagonist, of Ca^{2+} and volume-activated Cl^- channels in macrovascular endothelial cells. *Br J Pharmacol* **121**(3): 547-55.

Noble, D. (1984). The surprising heart: a review of recent progress in cardiac electrophysiology. *J Physiol* **353**: 1-50.

Nowycky, M. C., Fox, A. P., et al., (1985). Three types of neuronal calcium channel with different calcium agonist sensitivity. *Nature* **316**(6027): 440-3.

Nuss, H. B. and Houser, S. R. (1991). Voltage dependence of contraction and calcium current in severely hypertrophied feline ventricular myocytes. *J Mol Cell Cardiol* **23**(6): 717-26.

Nuss, H. B. and Houser, S. R. (1993). T-type Ca^{2+} current is expressed in hypertrophied adult feline left ventricular myocytes. *Circulation Research* **73**(4): 777-782.

Nuss, H. B. and Marban, E. (1994). Electrophysiological properties of neonatal mouse cardiac myocytes in primary culture. *J Physiol* **479** (**Pt 2**): 265-79.

Nussinovitch, I. and Kleinhaus, A. L. (1992). Dopamine inhibits voltage-activated calcium channel currents in rat pars intermedia pituitary cells. *Brain Res* **574**(1-2): 49-55.

Nyborg, N. C. and Mulvany, M. J. (1985). Lack of effect of anti-hypertensive treatment with felodipine on cardiovascular structure of young spontaneously hypertensive rats. *Cardiovasc Res* **19**(9): 528-36.

Odblom, M. P., Williamson, R., et al., (2000). Ionic currents in cardiac myocytes of squid, Alloteuthis subulata. *J Comp Physiol [B]* **170**(1): 11-20.

Ohya, Y., Abe, I., et al., (1993). Voltage-dependent Ca^{2+} channels in resistance arteries from spontaneously hypertensive rats. *Circ Res* **73**(6): 1090-9.

Ohya, Y. and Sperelakis, N. (1989). ATP regulation of the slow calcium channels in vascular smooth muscle cells of guinea pig mesenteric artery. *Circ Res* **64**(1): 145-54.

Osipenko, O. N., Varnai, P., et al., (1994). Dopamine blocks T-type calcium channels in cultured rat adrenal glomerulosa cells. *Endocrinology* **134**(1): 511-4.

Ouadid, H., Seguin, J., et al., (1991). Properties and Modulation of Ca channels in adult human atrial cells. *J Mol Cell Cardiol* **23**(1): 41-54.

Ozawa, Y., Hayashi, K., et al., (2001). Effect of T-type selective calcium antagonist on renal microcirculation: studies in the isolated perfused hydronephrotic kidney. *Hypertension* **38**(3): 343-7.

Pacaud, P., Loirand, G., et al., (1987). Opposing effects of noradrenaline on the two classes of voltage-dependent calcium channels of single vascular smooth muscle cells in short-term primary culture. *Pflugers Arch* **410**(4-5): 557-9.

Pan, Z. H. (2000). Differential expression of high- and two types of low-voltage-activated calcium currents in rod and cone bipolar cells of the rat retina. *J Neurophysiol* **83**(1): 513-27.

Pan, Z. H., Hu, H. J., et al., (2001). T-type Ca^{2+} channels mediate neurotransmitter release in retinal bipolar cells. *Neuron* **32**(1): 89-98.

Pape, H. C., Budde, T., et al., (1994). Prevention of Ca^{2+} -mediated action potentials in GABAergic local circuit neurones of rat thalamus by a transient K^+ current. *J Physiol* **478 Pt 3**: 403-22.

Payet, M. D., Durroux, T., et al., (1994). Characterization of K^+ and Ca^{2+} ionic currents in glomerulosa cells from human adrenal glands. *Endocrinology* **134**(6): 2589-98.

Pemberton, K. E., Hill-Eubanks, L. J., et al., (2000). Modulation of low-threshold T-type calcium channels by the five muscarinic receptor subtypes in NIH 3T3 cells. *Pflugers Arch* **440**(3): 452-61.

Perchenet, L., Bénardeau, A., et al., (2000). Pharmacological properties of $Ca_V3.2$, a low voltage-activated Ca2+ channel cloned from human heart. *Naunyn Schmiedebergs Arch Pharmacol* **361**(6): 590-9.

Perez-Reyes, E. (2003). Molecular Physiology of Low-Voltage-Activated T-type Calcium Channels. *Physiol Rev* **83**(1): 117-61.

Perez-Reyes, E., Cribbs, L. L., et al., (1998). Molecular characterization of a neuronal low-voltage-activated T-type calcium channel. *Nature* **391**(6670): 896-900.

Perrier, J. F. and Hounsgaard, J. (2000). Development and regulation of response properties in spinal cord motoneurons. *Brain Res Bull* **53**(5): 529-35.

Pfeiffer-Linn, C. and Lasater, E. M. (1993). Dopamine modulates in a differential fashion T- and L-type calcium currents in bass retinal horizontal cells. *J Gen Physiol* **102**(2): 277-94.

Piedras-Renteria, E. S., Chen, C. C., et al., (1997). Antisense oligonucleotides against rat brain α1E DNA and its atrial homologue decrease T-type calcium current in atrial myocytes. *Proc Natl Acad Sci U S A* **94**(26): 14936-41.

Portegies, M. C., Schmitt, R., et al., (1991). Lack of negative inotropic effects of the new calcium antagonist Ro 40-5967 in patients with stable angina pectoris. *J Cardiovasc Pharmacol* **18**(5): 746-751.

Pouille, F., Cavelier, P., et al., (2000). Dendro-somatic distribution of calcium-mediated electrogenesis in purkinje cells from rat cerebellar slice cultures. *J Physiol* **527 Pt 2**: 265-82.

Protas, L., DiFrancesco, D., et al., (2001). L-type but not T-type calcium current changes during postnatal development in rabbit sinoatrial node. *Am J Physiol Heart Circ Physiol* **281**(3): H1252-9.

Protas, L. and Robinson, R. B. (2000). Mibefradil, an $I_{Ca,T}$ blocker, effectively blocks $I_{Ca,L}$ in rabbit sinus node cells. *Eur J Pharmacol* **401**(1): 27-30.

Publicover, S. J., Preston, M. R., et al., (1995). Voltage-dependent potentiation of low-voltage-activated Ca^{2+} channel currents in cultured rat bone marrow cells. *J Physiol* **489 (Pt 3)**: 649-61.

Qar, J., Galizzi, J. P., et al., (1987). Receptors for diphenylbutylpiperidine neuroleptics in brain, cardiac, and smooth muscle membranes. Relationship with receptors for 1,4-dihydropyridines and phenylalkylamines and with Ca2+ channel blockade. *Eur J Pharmacol* **141**(2): 261-8.

Qu, Y. and Boutjdir, M. (2001). Gene expression of SERCA2a and L- and T-type Ca channels during human heart development. *Pediatr Res* **50**(5): 569-74.

Quignard, J. F., Frapier, J. M., et al., (1997). Voltage-gated calcium channel currents in human coronary myocytes. Regulation by cyclic GMP and nitric oxide. *J Clin Invest* **99**(2): 185-93.

Quignard, J. F., Harricane, M. C., et al., (2001). Transient down-regulation of L-type Ca^{2+} channel and dystrophin expression after balloon injury in rat aortic cells. *Cardiovasc Res* **49**(1): 177-88.

Raman, I. M. and Bean, B. P. (1999). Ionic currents underlying spontaneous action potentials in isolated cerebellar Purkinje neurons. *J Neurosci* **19**(5): 1663-74.

Rampe, D. and Triggle, D. J. (1993). New synthetic ligands for L-type voltage-gated calcium channels. *Prog Drug Res* **40**: 191-238.

Randall, A. and Tsien, R. W. (1995). Pharmacological dissection of multiple types of Ca^{2+} channel currents in rat cerebellar granule neurons. *J Neurosci* **15**(4): 2995-3012.

Randall, A. D. and Tsien, R. W. (1997). Contrasting biophysical and pharmacological properties of T-type and R-type calcium channels. *Neuropharmacology* **36**(7): 879-93.

Randall, A. D., Wendland, B., et al., (1993). Five pharmacologically distinct high voltage-activated Ca^{2+} channels in cerebellar granule cells. *Soc Neurosci Abstr* **19**: 1478.

Rebuffat, P., Aragona, F., et al., (2001). Signaling pathways involved in the A and B receptor-mediated cortisol secretagogue effect of endothelins in the human adrenal cortex. *Int J Mol Med* **7**(3): 301-5.

Reuveny, E., Twombly, D. A., et al., (1993). Chlordiazepoxide block of two types of calcium channels in neuroblastoma cells. *J Pharmacol Exp Ther* **264**(1): 22-8.

Richard, S., Neveu, D., et al., (1992). Differential expression of voltage-gated Ca^{2+} currents in cultivated aortic myocytes. *Biochim Biophys Acta* **1160**(1): 95-104.

Richard, S., Tiaho, F., et al., (1990). Two pathways for Ca^{2+} channel gating differentially modulated by physiological stimuli. *Am J Physiol* **258**(6 Pt 2): H1872-81.

Ries, C. R. and Puil, E. (1999). Mechanism of anesthesia revealed by shunting actions of isoflurane on thalamocortical neurons. *J Neurophysiol* **81**(4): 1795-801.

Ritchie, A. K. (1993). Estrogen increases low voltage-activated calcium current density in GH3 anterior pituitary cells. *Endocrinology* **132**(4): 1621-9.

Rivet, M., Cognard, C., et al., (1990). Calcium currents in normal and dystrophic human skeletal muscle cells in culture. *Cell Calcium* **11**(8): 507-14.

Rohwedel, J., Maltsev, V., et al., (1994). Muscle cell differentiation of embryonic stem cells reflects myogenesis in vivo: developmentally regulated expression of myogenic determination genes and functional expression of ionic currents. *Dev Biol* **164**(1): 87-101.

Romanin, C., Seydl, K., et al., (1992). The dihydropyridine niguldipine inhibits T-type Ca^{2+} currents in atrial myocytes. *Pflugers Arch* **420**(3-4): 410-2.

Rorsman, P. (1988). Two types of Ca^{2+} currents with different sensitivities to organic Ca^{2+} channel antagonists in guinea pig pancreatic alpha 2 cells. *J Gen Physiol* **91**(2): 243-54.

Rossier, M. F., Burnay, M. M., et al., (1998). Distinct functions of T- and L-type calcium channels during activation of aldosterone production in adrenal glomerulosa cells. *Low-voltage-activated T-type Ca channels*. R. W. Tsien, J.-P. Clozel and J. Nargeot. Chester, UK, Adis International: 176-185.

Rossier, M. F., Python, C. P., et al., (1993). Blocking T-type calcium channels with tetrandrine inhibits steroidogenesis in bovine adrenal glomerulosa cells. *Endocrinology* **132**(3): 1035-43.

Rousseau, M. F., Hayashida, W., et al., (1996). Hemodynamic and cardiac effects of the selective T-type and L-type calcium channel blocking agent mibefradil in patients with varying degrees of left ventricular systolic dysfunction. *J Am Coll Cardiol* **28**(4): 972-979.

Ruth, P., Rohrkasten, A., et al., (1989). Primary structure of the beta subunit of the DHP-sensitive calcium channel from skeletal muscle. *Science* **245**(4922): 1115-8.

Ryu, P. D. and Randic, M. (1990). Low- and high-voltage-activated calcium currents in rat spinal dorsal horn neurons. *J Neurophysiol* **63**(2): 273-85.

Sah, D. W. (1995). Human fetal central neurons in culture: voltage- and ligand-gated currents. *J Neurophysiol* **74**(5): 1889-99.

Sah, D. W. and Bean, B. P. (1994). Inhibition of P-type and N-type calcium channels by dopamine receptor antagonists. *Mol Pharmacol* **45**(1): 84-92.

Sakata, Y., Saegusa, H., et al., (2001). Analysis of Ca^{2+} currents in spermatocytes from mice lacking $Ca_v2.3$ (α_{1E}) Ca^{2+} channel. *Biochem Biophys Res Commun* **288**(4): 1032-6.

Sala, S. and Matteson, D. R. (1990). Single-channel recordings of two types of calcium channels in rat pancreatic beta-cells. *Biophys J* **58**(2): 567-71.

Salami, M. and Fathollahi, Y. (2000). Do Ca^{2+} channels share NMDA receptors in plasticity of synaptic transmission in the rat visual cortex? *Neuroreport* **11**(17): 3887-91.

Sandmann, S., Bohle, R. M., et al., (2000). The T-type calcium channel blocker mibefradil reduced interstitial and perivascular fibrosis and improved hemodynamic parameters in myocardial infarction-induced cardiac failure in rats. *Virchows Arch* **436**(2): 147-57.

Sandmann, S., Claas, R., et al., (2001). Calcium channel blockade limits cardiac remodeling and improves cardiac function in myocardial infarction-induced heart failure in rats. *J Cardiovasc Pharmacol* **37**(1): 64-77.

Sandmann, S., Spitznagel, H., et al., (1998). Effects of the calcium channel antagonist mibefradil on haemodynamic and morphological parameters in myocardial infarction-induced cardiac failure in rats. *Cardiovasc Res* **39**(2): 339-50.

Santi, C. M., Cayabyab, F. S., et al., (2002). Differential inhibition of T-type calcium channels by neuroleptics. *J Neurosci* **22**(2): 396-403.

Satoh, H. (1995). Role of T-type Ca^{2+} channel inhibitors in the pacemaker depolarization in rabbit sino-atrial nodal cells. *Gen Pharmacol* **26**(3): 581-7.

Sayer, R. J., Brown, A. M., et al., (1993). Calcium currents in acutely isolated human neocortical neurons. *J Neurophysiol* **69**(5): 1596-606.

Schmitt, H. and Meves, H. (1995a). Modulation of neuronal calcium channels by arachidonic acid and related substances. *J Membr Biol* **145**(3): 233-44.

Schmitt, R., Clozel, J. P., et al., (1995b). Mibefradil prevents neointima formation after vascular injury in rats. Possible role of the blockade of the T-type voltage-operated calcium channel. *Arterioscler Thromb Vasc Biol* **15**(8): 1161-1165.

Schmitt, R., Clozel, J. P., et al., (1996). Prevention of neointima formation by mibefradil after vascular injury in rats: comparison with ACE inhibition. *Cardiovasc Drugs Ther* **10**(2): 101-5.

Schmitt, R., Kleinbloesem, C. H., et al., (1992). Hemodynamic and humoral effects of the novel calcium antagonist Ro 40-5967 in patients with hypertension. *Clin Pharmacol Ther* **52**(3): 314-23.

Schmitt, R., Kleinbloesem, C. H., et al., (1989). Ro 40-5967 a calcium antagonist of a new generation? *Eur J Clin Pharmacol* **36 (Suppl)**: A336.

Schneider, T., Wei, X., et al., (1994). Molecular analysis and functional expression of the human type E neuronal Ca2+ channel alpha 1 subunit. *Receptors Channels* **2**(4): 255-70.

Schroeder, J. E., Fischbach, P. S., et al., (1991). Activation of mu opioid receptors inhibits transient high- and low-threshold Ca2+ currents, but spares a sustained current. *Neuron* **6**(1): 13-20.

Scott, R. H., Wootton, J. F., et al., (1990). Modulation of neuronal T-type calcium channel currents by photoactivation of intracellular guanosine 5'-O(3-thio) triphosphate. *Neuroscience* **38**(2): 285-94.

Seagar, M. J., Takahashi, M., et al., (1988). Molecular properties of dihydropyridine-sensitive calcium channels. *Ann N Y Acad Sci* **522**: 162-75.

Self, D. A., Bian, K., et al., (1994). Stroke-prone SHR vascular muscle Ca^{2+} current amplitudes correlate with lethal increases in blood pressure. *J Vasc Res* **31**(6): 359-66.

Sen, L., Bialecki, R. A., et al., (1992). Cholesterol increases the L-type voltage-sensitive calcium channel current in arterial smooth muscle cells. *Circ Res* **71**(4): 1008-14.

Sen, L. and Smith, T. W. (1994). T-type Ca^{2+} channels are abnormal in genetically determined cardiomyopathic hamster hearts. *Circulation Research* **75**(1): 149-155.

Shiomi, H., Yamano, S., et al., (1996). Characteristics of calcium ion influx induced by human follicular fluid in individual human sperm. *Archives of Andrology* **37**(2): 79-86.

Sidach, S. S. and Mintz, I. M. (2002). Kurtoxin, a gating modifier of neuronal high- and low-threshold ca channels. *J Neurosci* **22**(6): 2023-34.

Smirnov, S. V. and Aaronson, P. I. (1992a). Ca^{2+} currents in single myocytes from human mesenteric arteries: evidence for a physiological role of L-type channels. *J Physiol (Lond)* **457**: 455-75.

Smirnov, S. V., Zholos, A. V., et al., (1992b). Potential-dependent inward currents in single isolated smooth muscle cells of the rat ileum. *J Physiol* **454**: 549-71.

Son, W. Y., Lee, J. H., et al., (2000). Acrosome reaction of human spermatozoa is mainly mediated by alpha1H T-type calcium channels. *Mol Hum Reprod* **6**(10): 893-7.

Song, S. Y., Saito, K., et al., (1991). Adrenergic and cholinergic inhibition of Ca^{2+} channels mediated by different GTP-binding proteins in rat sympathetic neurones. *Pflugers Arch* **418**(6): 592-600.

Soong, T. W., Stea, A., et al., (1993). Structure and functional expression of a member of the low voltage-activated calcium channel family. *Science* **260**(5111): 1133-6.

Spedding, M. and Paoletti, R. (1992a). Classification of calcium channels and calcium antagonists: progress report. *Cardiovascular Drugs and Therapy* **6**(1): 35-9.

Spedding, M. and Paoletti, R. (1992b). Classification of calcium channels and the sites of action of drugs modifying channel function. *Pharmacol Rev* **44**(3): 363-76.

Spedding, M. and Vanhoutte, P. M. (1993). Nomenclature of calcium channels and channel modulators: IUPHAR classification. *J Cardiovasc Pharmacol* **22**(6): 906.

Spitzer, N. C., Debaca, R. C., et al., (1993). Calcium dependence of differentiation of GABA immunoreactivity in spinal neurons. *J Comp Neurol* **337**(1): 168-75.

Stefani, A., Spadoni, F., et al., (1999). The modulation of calcium current by GABA metabotropic receptors in a sub-population of pallidal neurons. *Eur J Neurosci* **11**(11): 3995-4005.

Stengel, W., Jainz, M., et al., (1998). Different potencies of dihydropyridine derivatives in blocking T-type but not L-type Ca^{2+} channels in neuroblastoma-glioma hybrid cells. *Eur J Pharmacol* **342**(2-3): 339-45.

Stephens, G. J., Page, K. M., et al., (1997). Functional expression of rat brain cloned α_{1E} calcium channels in COS-7 cells. *Pflugers Arch* **433**(4): 523-32.

Steriade, M. and Llinas, R. R. (1988). The functional states of the thalamus and the associated neuronal interplay. *Physiol Rev* **68**(3): 649-742.

Stoll, M., Steckelings, U. M., et al., (1995). The angiotensin AT2-receptor mediates inhibition of cell proliferation in coronary endothelial cells. *J Clin Invest* **95**(2): 651-657.

Striessnig, J. (1999). Pharmacology, structure and function of cardiac L-type Ca^{2+} channels. *Cell Physiol Biochem* **9**(4-5): 242-69.

Su, H., Sochivko, D., et al., (2002). Upregulation of a T-type Ca^{2+} channel causes a long-lasting modification of neuronal firing mode after status epilepticus. *J Neurosci* **22**(9): 3645-55.

Sun, Q. Q. and Dale, N. (1997). Serotonergic inhibition of the T-type and high voltage-activated Ca^{2+} currents in the primary sensory neurons of Xenopus larvae. *J Neurosci* **17**(18): 6839-49.

Sun, Q. Q. and Dale, N. (1999). G-proteins are involved in 5-HT receptor-mediated modulation of N- and P/Q- but not T-type Ca^{2+} channels. *J Neurosci* **19**(3): 890-9.

Suzuki, S., Kawakami, K., et al., (1992). Zonisamide blocks T-type calcium channel in cultured neurons of rat cerebral cortex. *Epilepsy Res* **12**(1): 21-7.

Suzuki, S. and Rogawski, M. A. (1989). T-type calcium channels mediate the transition between tonic and phasic firing in thalamic neurons. *Proc Natl Acad Sci U S A* **86**(18): 7228-32.

Takahashi, K. and Akaike, N. (1990). Nicergoline inhibits T-type Ca^{2+} channels in rat isolated hippocampal CA1 pyramidal neurones. *British Journal of Pharmacology* **100**(4): 705-10.

Takahashi, K., Ueno, S., et al., (1991). Kinetic properties of T-type Ca^{2+} currents in isolated rat hippocampal CA1 pyramidal neurons. *J Neurophysiol* **65**(1): 148-55.

Takahashi, K., Wakamori, M., et al., (1989). Hippocampal CA1 pyramidal cells of rats have four voltage-dependent calcium conductances. *Neurosci Lett* **104**(1-2): 229-234.

Takahashi, M., Seagar, M. J., et al., (1987). Subunit structure of dihydropyridine-sensitive calcium channels from skeletal muscle. *Proc Natl Acad Sci U S A* **84**(15): 5478-82.

Takei, T., Takano, K., et al., (1996). Enhancement of Ca^{2+} currents by GHRH and its relation to PKA and [Ca2+]i in human GH-secreting adenoma cells. *Am J Physiol* **271**(5 Pt 1): E801-7.

Talley, E. M., Solorzano, G., et al., (2000). Low-voltage-activated calcium channel subunit expression in a genetic model of absence epilepsy in the rat. *Brain Res Mol Brain Res* **75**(1): 159-65.

Tanabe, T., Takeshima, H., et al., (1987). Primary structure of the receptor for calcium channel blockers from skeletal muscle. *Nature* **328**(6128): 313-8.

Tang, C. M., Presser, F., et al., (1988). Amiloride selectively blocks the low threshold (T) calcium channel. *Science* **240**(4849): 213-215.

Thompson, S. M. and Wong, R. K. (1991). Development of calcium current subtypes in isolated rat hippocampal pyramidal cells. *J Physiol* **439**: 671-89.

Thornbury, K. D. (1999). Tonic and phasic activity in smooth muscle. *Ir J Med Sci* **168**(3): 201-7.

Tieleman, R. G., De Langen, C., et al., (1997). Verapamil reduces tachycardia-induced electrical remodeling of the atria. *Circulation* **95**(7): 1945-53.

Todorovic, S. M., Jevtovic-Todorovic, V., et al., (2001a). $Ca_v3.2$ channel is a molecular substrate for inhibition of T-type calcium currents in rat sensory neurons by nitrous oxide. *Mol Pharmacol* **60**(3): 603-10.

Todorovic, S. M., Jevtovic-Todorovic, V., et al., (2001b). Redox modulation of T-type calcium channels in rat peripheral nociceptors. *Neuron* **31**(1): 75-85.

Todorovic, S. M. and Lingle, C. J. (1998). Pharmacological properties of T-type Ca^{2+} current in adult rat sensory neurons: effects of anticonvulsant and anesthetic agents. *J Neurophysiol* **79**(1): 240-52.

Todorovic, S. M., Meyenburg, A., et al., (2002). Mechanical and thermal antinociception in rats following systemic administration of mibefradil, a T-type calcium channel blocker. *Brain Res* **951**(2): 336-40.

Todorovic, S. M., Perez-Reyes, E., et al., (2000). Anticonvulsants but not general anesthetics have differential blocking effects on different T-type current variants. *Mol Pharmacol* **58**(1): 98-108.

Tottene, A., Moretti, A., et al., (1996). Functional diversity of P-type and R-type calcium channels in rat cerebellar neurons. *J Neurosci* **16**(20): 6353-63.

Tottene, A., Volsen, S., et al., (2000). α_{1E} subunits form the pore of three cerebellar R-type calcium channels with different pharmacological and permeation properties. *J Neurosci* **20**(1): 171-8.

Tsakiridou, E., Bertollini, L., et al., (1995). Selective increase in T-type calcium conductance of reticular thalamic neurons in a rat model of absence epilepsy. *J Neurosci* **15**(4): 3110-7.

Tseng, G. N. and Boyden, P. A. (1989). Multiple types of Ca^{2+} currents in single canine Purkinje cells. *Circ Res* **65**(6): 1735-50.

Tseng, G. N. and Boyden, P. A. (1991). Different effects of intracellular Ca and protein kinase C on cardiac T and L Ca currents. *Am J Physiol* **261**(2 Pt 2): H364-79.

Tsunoo, A., Yoshii, M., et al., (1985). Differential block of two calcium channels in neuroblastoma cells. *Biophysical Journal* **47**: 433a.

Twombly, D. A., Yoshii, M., et al., (1988). Mechanisms of calcium channel block by phenytoin. *J Pharmacol Exp Ther* **246**(1): 189-95.

Tytgat, J., Vereecke, J., et al., (1988). Differential effects of verapamil and flunarizine on cardiac L-type and T-type Ca channels. *Naunyn Schmiedebergs Arch Pharmacol* **337**(6): 690-2.

Tytgat, J., Vereecke, J., et al., (1990a). A combined study of sodium current and T-type calcium current in isolated cardiac cells. *Pflugers Arch* **417**(2): 142-8.

Tytgat, J., Vereecke, J., et al., (1990b). Mechanism of cardiac T-type Ca channel blockade by amiloride. *J Pharmacol Exp Ther* **254**(2): 546-51.

Urbano, F. J., Sierra, F., et al., (1997). Differential expression of voltage-gated Ca^{2+} conductances in human neuroblastoma NB69 cells cultured in defined serum-free and astrocyte- conditioned media. *Glia* **20**(1): 70-78.

Vacher, E., Richer, C., et al., (1996). Mibefradil, a selective calcium T-channel blocker, in stroke-prone spontaneously hypertensive rats. *J Cardiovasc Pharmacol* **27**(5): 686-94.

VanBavel, E., Sorop, O., et al., (2002). Role of T-type calcium channels in myogenic tone of skeletal muscle resistance arteries. *Am J Physiol Heart Circ Physiol* **283**(6): H2239-43.

Vanegas, H. and Schaible, H. (2000). Effects of antagonists to high-threshold calcium channels upon spinal mechanisms of pain, hyperalgesia and allodynia. *Pain* **85**(1-2): 9-18.

Veniant, M., Gray, G. A., et al., (1995). Structural changes and cyclic GMP content of the aorta after calcium antagonism or angiotensin converting enzyme inhibition in renovascular hypertensive rats. *J Hypertens* **13**(7): 731-737.

Verma, S., Bhanot, S., et al., (1997). Chronic T-type Ca^{2+} channel blockade with mibefradil in hyperinsulinemic, insulin-resistant and hypertensive rats. *Cardiovasc Res* **34**(1): 121-8.

Viana, F., Van den Bosch, L., et al., (1997). Mibefradil (Ro 40-5967) blocks multiple types of voltage-gated calcium channels in cultured rat spinal motoneurones. *Cell Calcium* **22**(4): 299-311.

Villame, J., Massicotte, J., et al., (2001). Effects of mibefradil, a T- and L-type calcium channel blocker, on cardiac remodeling in the UM-X7.1 cardiomyopathic hamster. *Cardiovasc Drugs Ther* **15**(1): 41-8.

Villarreal, F. J., Kim, N. N., et al., (1993). Identification of functional angiotensin II receptors on rat cardiac fibroblasts. *Circulation* **88**(6): 2849-61.

Vivaudou, M. B., Clapp, L. H., et al., (1988). Regulation of one type of Ca^{2+} current in smooth muscle cells by diacylglycerol and acetylcholine. *Faseb J* **2**(9): 2497-504.

Wan, X., Desilets, M., et al., (1996). Muscarinic activation inhibits T-type Ca^{2+} current in hen granulosa cells. *Endocrinology* **137**(6): 2514-21.

Wang, L., Bhattacharjee, A., et al., (1996a). Abnormally expressed low-voltage-activated calcium channels in beta-cells from NOD mice and a related clonal cell line. *Diabetes* **45**(12): 1678-83.

Wang, R., Karpinski, E., et al., (1989). Two types of calcium channels in isolated smooth muscle cells from rat tail artery. *American Journal of Physiology* **256**(5 Pt 2): H1361-H1368.

Wang, X., McKenzie, J. S., et al., (1996b). Whole cell calcium currents in acutely isolated olfactory bulb output neurons of the rat. *J Neurophysiol* **75**(3): 1138-51.

Wang, Y. Q., Brooks, G., et al., (2002). [Functional analysis of the human T-type calcium channel α_{1H} subunit gene in cellular proliferation]. *Yi Chuan Xue Bao* **29**(8): 659-65.

Wang, Z., Estacion, M., et al., (1993). Ca^{2+} influx via T-type channels modulates PDGF-induced replication of mouse fibroblasts. *Am J Physiol* **265**(5 Pt 1): C1239-46.

Wasling, P., Hanse, E., et al., (2002). Long-term depression in the developing hippocampus: low induction threshold and synapse nonspecificity. *J Neurosci* **22**(5): 1823-30.

Weber, K. T. and Brilla, C. G. (1991). Pathological hypertrophy and cardiac interstitium. Fibrosis and renin-angiotensin-aldosterone system. *Circulation* **83**(6): 1849-65.

Wen, J. F., Cui, X., et al., (2000). Distinct roles for L- and T-type Ca^{2+} channels in regulation of atrial ANP release. *Am J Physiol Heart Circ Physiol* **279**(6): H2879-88.

Wennemuth, G., Westenbroek, R. E., et al., (2000). CaV2.2 and CaV2.3 (N- and R-type) Ca^{2+} channels in depolarization-evoked entry of Ca^{2+} into mouse sperm. *J Biol Chem* **275**(28): 21210-7.

Wetzel, G. T., Chen, F., et al., (1991). Calcium current measurements in acutely isolated neonatal cardiac myocytes. *Pediatr Res* **30**(1): 83-8.

White, G., Lovinger, D. M., et al., (1989). Transient low-threshold Ca^{2+} current triggers burst firing through an afterdepolarizing potential in an adult mammalian neuron. *Proc Natl Acad Sci U S A* **86**(17): 6802-6.

Wilde, D. W., Furspan, P. B., et al., (1994). Calcium current in smooth muscle cells from normotensive and genetically hypertensive rats. *Hypertension* **24**(6): 739-746.

Williams, M. E., Marubio, L. M., et al., (1994). Structure and functional characterization of neuronal alpha 1E calcium channel subtypes. *J Biol Chem* **269**(35): 22347-57.

Williams, M. E., Washburn, M. S., et al., (1999). Structure and functional characterization of a novel human low-voltage activated calcium channel. *J Neurochem* **72**(2): 791-9.

Witcher, D. R., De Waard, M., et al., (1993). Subunit identification and reconstitution of the N-type Ca^{2+} channel complex purified from brain. *Science* **261**(5120): 486-9.

Wolfart, J. and Roeper, J. (2002). Selective coupling of T-type calcium channels to SK potassium channels prevents intrinsic bursting in dopaminergic midbrain neurons. *J Neurosci* **22**(9): 3404-13.

Wolfe, J. T., Wang, H., et al., (2002). Stimulation of recombinant $Ca_v3.2$, T-type, Ca^{2+} channel currents by CaMKIIgamma(C). *J Physiol* **538**(Pt 2): 343-55.

Xiong, Z., Sperelakis, N., et al., (1993). Changes in calcium channel current densities in rat colonic smooth muscle cells during development and aging. *Am J Physiol* **265**(3 Pt 1): C617-25.

Xiong, Z., Sperelakis, N., et al., (1995). Ca^{2+} currents in human colonic smooth muscle cells. *American Journal of Physiology* **269**(3 Pt 1): G378-85.

Xu, X. and Best, P. M. (1992). Postnatal changes in T-type calcium current density in rat atrial myocytes. *J Physiol* **454**: 657-72.

Xu, X. P. and Best, P. M. (1990). Increase in T-type calcium current in atrial myocytes from adult rats with growth hormone-secreting tumors. *Proceedings of the National Academy of Sciences U S A* **87**(12): 4655-4659.

Yaari, Y., Hamon, B., et al., (1987). Development of two types of calcium channels in cultured mammalian hippocampal neurons. *Science* **235**(4789): 680-2.

Yamakage, M., Chen, X., et al., (2001). Different inhibitory effects of volatile anesthetics on T- and L-type voltage-dependent Ca2+ channels in porcine tracheal and bronchial smooth muscles. *Anesthesiology* **94**(4): 683-93.

Yao, W. X. and Jiang, M. X. (2002). Effects of tetrandrine on cardiovascular electrophysiologic properties. *Acta Pharmacol Sin* **23**(12): 1069-74.

Yatani, A., Seidel, C. L., et al., (1987). Whole-cell and single-channel calcium currents of isolated smooth muscle cells from saphenous vein. *Circulation Research* **60**(4): 523-533.

Yazawa, K., Ono, K., et al., (2002). Modulation by Mibefradil of the Histamine-Induced Ca^{2+} Entry in Human Aortic Endothelial Cells. *Jpn J Pharmacol* **90**(2): 125-130.

Ye, J. H. and Akaike, N. (1993). Calcium currents in pyramidal neurons acutely dissociated from the rat frontal cortex: a study by the nystatin perforated patch technique. *Brain Res* **606**(1): 111-7.

Yeoman, M. S., Brezden, B. L., et al., (1999). LVA and HVA Ca^{2+} currents in ventricular muscle cells of the Lymnaea heart. *J Neurophysiol* **82**(5): 2428-40.

Yokokawa, K., Kohno, M., et al., (1990). Effect of endothelin-1 on cytosolic calcium ions in cultured human endothelial cells. *J Hypertens* **8**(9): 843-9.

Yokoyama, M. and Henry, P. D. (1979). Sensitization of isolated canine coronary arteries to calcium ions after exposure to cholesterol. *Circ Res* **45**(4): 479-86.

Young, R. C., Smith, L. H., et al., (1993). T-type and L-type calcium currents in freshly dispersed human uterine smooth muscle cells. *American Journal of Obstetrics and Gynecology* **169**(4): 785-92.

Yu, W. C., Chen, S. A., et al., (1998). Tachycardia-induced change of atrial refractory period in humans: rate dependency and effects of antiarrhythmic drugs. *Circulation* **97**(23): 2331-7.

Zamponi, G. W., Bourinet, E., et al., (1996). Nickel block of a family of neuronal calcium channels: subtype- and subunit-dependent action at multiple sites. *J Membr Biol* **151**(1): 77-90.

Zhang, J. F., Randall, A. D., et al., (1993a). Distinctive pharmacology and kinetics of cloned neuronal Ca^{2+} channels and their possible counterparts in mammalian CNS neurons. *Neuropharmacology* **32**(11): 1075-88.

Zhang, L., Valiante, T. A., et al., (1993b). Contribution of the low-threshold T-type calcium current in generating the post-spike depolarizing afterpotential in dentate granule neurons of immature rats. *J Neurophysiol* **70**(1): 223-31.

Zhang, L. M., Wang, Z., et al., (2002a). Effects of sustained beta-adrenergic stimulation on ionic currents of cultured adult guinea pig cardiomyocytes. *Am J Physiol Heart Circ Physiol* **282**(3): H880-9.

Zhang, W., Elsen, F., et al., (1999). Postnatal development of GABAB receptor-mediated modulation of voltage-activated Ca^{2+} currents in mouse brain-stem neurons. *Eur J Neurosci* **11**(7): 2332-42.

Zhang, Y., Cribbs, L. L., et al., (2000). Arachidonic acid modulation of α_{1H}, a cloned human T-type calcium channel. *Am J Physiol Heart Circ Physiol* **278**(1): H184-93.

Zhang, Y., Mori, M., et al., (2002b). Mutations in high-voltage-activated calcium channel genes stimulate low-voltage-activated currents in mouse thalamic relay neurons. *J Neurosci* **22**(15): 6362-71.

Zhou, Z. and Lipsius, S. L. (1994). T-type calcium current in latent pacemaker cells isolated from cat right atrium. *J Mol Cell Cardiol* **26**(9): 1211-9.

Zhuang, H., Bhattacharjee, A., et al., (2000). Cloning of a T-type Ca^{2+} channel isoform in insulin-secreting cells. *Diabetes* **49**(1): 59-64.

Zygmunt, A. C. and Maylie, J. (1990). Stimulation-dependent facilitation of the high threshold calcium current in guinea-pig ventricular myocytes. *J Physiol* **428**: 653-71.

CELLULAR FUNCTIONS OF CALCIUM CHANNEL SUBTYPES

Erika S. Piedras-Rentería[1], Paul G. Mermelstein[2], and Geoffrey S. Pitt[3]

1. INTRODUCTION

Intracellular calcium regulation is critically important for numerous cellular functions, such as muscle contraction, secretion, neurotransmitter release and cell migration. Voltage-activated calcium channels play a crucial role in several of these processes (see Figure 1). Although the functional diversity of calcium channels can be explained by the rich molecular variety of this channel superfamily, a common denominator of calcium channel function is that they all couple incoming excitatory electrical signals to fundamental processes within the cell. Different calcium channel subtypes can play a variety of physiological roles within cells, depending on the stage of development, cell type and other factors. A couple of notable exceptions are the skeletal L-type calcium channel α_{1S}, which is the only calcium channel that coordinates muscle contraction in skeletal muscle, and the L-type channel subtype α_{1F}, whose expression seems to be restricted to retinal tissue.

The various cellular functions of different calcium channel subtypes are summarized in Table 1. The role of calcium channels in muscle contraction is discussed by Triggle (this volume; see also Shirokova et al., 1999; Bers, 2001). The role of T-type calcium channels in cell growth and proliferation is discussed by Ertel (this volume). In this chapter, we will examine the role of calcium channels in the coupling of calcium influx to functions such as neurotransmitter release, neuronal plasticity, and gene expression. In addition, we will review the functional association of calcium channels with calcium-activated potassium channels, as this family of potassium channels plays a vital role in the feedback modulation of calcium influx in excitable cells. We will end this chapter by reviewing the current knowledge of intracellular interactions of calcium channels with

[1]Department of Physiology, Loyola University Chicago, Maywood, IL 60153; [2]Department of Neuroscience, University of Minnesota, Minneapolis, MN 55455; [3]Department of Pharmacology, Columbia University, New York, NY 10032. Correspondence to Dr. Erika S. Piedras-Rentería.

Calcium Channel Pharmacology, Edited by McDonough
Kluwer Academic/Plenum Publishers, 2004

the calcium-binding protein calmodulin and with CaMKII, which are intrinsic regulators of calcium channel function.

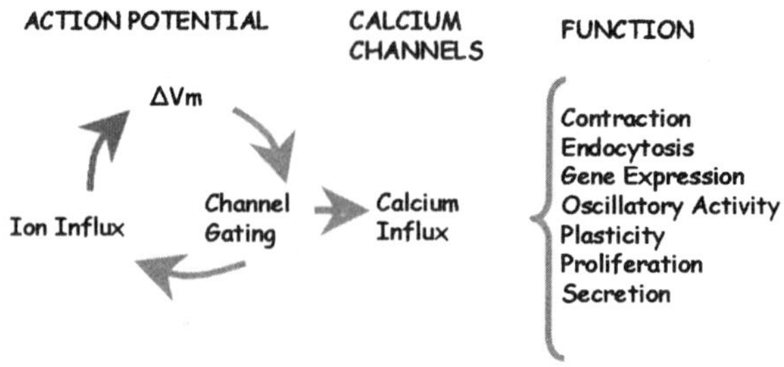

Figure 1. Calcium channels link electrical signals to various cellular functions.

Table 1. Cellular functions of calcium channel subtypes

Ca_V	α_1 Subunit	Type	Cellular Function
1.1	S	L	Excitation-Contraction
1.2	C		Excitation-Contraction
1.3	D		Excitation-Secretion Excitation-Transcription
1.4	F		Excitation-Secretion
2.1	A	P/Q	Excitation- Secretion Excitation-Transcription (?)
2.2	B	N	
2.3	E	R	
3.1	G	T	Oscillatory activity Cell Proliferation (?)
3.2	H		
3.3	I		

2. EXCITATION-SECRETION: Ca^{2+} CHANNELS AND NEUROTRANSMISSION

Fast release of neurotransmitter is a fundamental neuronal function that requires calcium ions to trigger the fusion of neurotransmitter-filled vesicles into the plasma membrane. This is a rapid process that takes less than 100 μs to occur (Sabatini and Regehr, 1999) and requires the participation of multiple synaptic proteins (Bennett, 1997; Burgoyne and Morgan, 1998; Sudhof and Scheller, 2001). The need for calcium entry for neurotransmitter release has been recognized for quite some time. Seminal research by Katz and collaborators demonstrated the control of neurotransmitter release by Ca^{2+} ions (Katz, 1969), which act in a cooperative manner (3-4 calcium ions) to induce neurotransmitter release (Dodge and Rahamimoff, 1967). Simultaneous recordings from presynaptic terminals and postsynaptic sites demonstrated the presence of presynaptic calcium channels that triggered neurotransmitter release (Llinas et al., 1981; Augustine et al., 1985). Moreover, measurement of local increases of calcium within 100 nm of the calcium binding sites (Adler et al., 1991) and the requirement of very high local calcium concentrations (~100 μM) for triggering vesicle release (Llinas et al., 1992) suggested a close association between calcium channels and the release machinery.

At the molecular level, the SNARE complex (soluble NSF attachment protein receptors), which consists of the vesicle-associated protein VAMP (synaptobrevin) and the plasma membrane proteins syntaxin and SNAP-25 (synaptosome-associated protein-25) plays an essential in the fusion of synaptic vesicles. The interaction of the SNARE complex with the calcium sensor synaptotagmin initiates the vesicle fusion process, but the assembly and disassembly of the core complex is modulated by interactions with several synaptic proteins (reviewed by Lin and Scheller, 2000; Mochida, 2000).

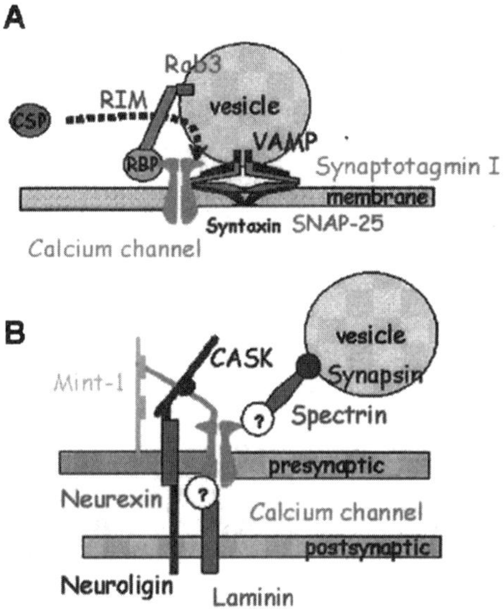

Figure 2. Synaptic proteins associated with calcium channels. Calcium channels constitute part of the functional secretory complex (A) and the structural complex that localizes channels and vesicle clusters to the active zone (B). Syntaxin, synaptotagmin, SNAP25, RIM-binding protein, CSP, Mint-1 and CASK are among the presynaptic proteins that interact with calcium channels.

The rapid release of neurotransmitter by calcium is tightly synchronized thanks (in part) to the physical association of the synaptic release machinery with voltage-gated calcium channels. The calcium channels involved in transmitter release include the members of the neuronal ABE channel family of P/Q-type, N-type and R-type calcium channels (α_{1A} α_{1B} and α_{1E}), which mediate neurotransmitter release in the peripheral and central nervous system (for a review see Reuter, 1995; Tsien and Wheeler, 1999). L-type calcium channel types (α_{1C}, α_{1D}) also mediate secretion in neuroendocrine cells and in the retina (α_{1F}). Members of this group of channels can interact with several proteins of synaptic origin, some of which are components of the vesicle release machinery, modulators of vesicle release, adaptor proteins or extracellular matrix proteins present in active zones (see Figure 2).

2.1. Calcium Channel Interactions With Synaptic Proteins

Co-immunoprecipitation experiments have demonstrated that members of the L (Ca_v1) and ABE (Ca_v2) channel families directly interact with two SNARE proteins that play a major role in the vesicle fusion process: syntaxin and SNAP-25 (Saisu et al., 1991; Bennett et al., 1992; Leveque et al., 1994; el Far et al., 1995; Martin-Moutot et al., 1996; Kim and Catterall, 1997). Calcium channels also interact with the calcium-sensor synaptotagmin and with CSP (cysteine string protein), a synaptic protein with homology to the molecular chaperones DnaJ proteins (Leveque et al., 1992; Yoshida et al., 1992; Leveque et al., 1994; Charvin et al., 1997; Leveque et al., 1998). The intracellular region of the calcium channel responsible for these synaptic proteins interactions is referred to as the synprint site and resides in the cytosolic region between the II and III transmembrane domains of the channel.

The overall effect of single SNARE protein interaction with calcium channels is inhibitory. Coexpression of syntaxin 1A/B with either N-type, Q-type or L-type calcium channels in the oocyte expression system leads to changes in channel kinetic properties, changes in the steady state inactivation and activation of the channels, and decreases in N-type and L-type calcium current amplitude (Wiser et al., 1996, 1997; Tobi et al., 1998; Wiser et al., 1999; Zhong et al., 1999; Bezprozvanny et al., 2000; Degtiar et al., 2000). The modulation of calcium channels by syntaxin 1A has recently been reported to depend on the permeant ion, suggesting that the channel-syntaxin interaction might respond to channel conformational changes during the depolarization process (Wiser et al., 2002).

SNAP-25 moderately increased the rate of inactivation of L-type channels with no effect on the current amplitude (Wiser et al., 1996; Atlas, 2001). N-type channel amplitude was slightly decreased by SNAP25 without effect on the current kinetics (Wiser et al., 1996), while the steady state inactivation of the P/Q-type channels was altered (Zhong et al., 1999).

Interestingly, co-expression of both SNARE proteins led to the reversal of the inhibitory effect by syntaxin 1A on N-type currents (Wiser et al., 1996; Tobi et al., 1998; Wiser et al., 1999). The reversal of the inhibitory effect by the SNARE protein pair had a different effect on L-type channels from that observed in N-type channels: it affected the current amplitude but not the rate of inactivation (Atlas, 2001). Despite the differences between neuronal and neuroendocrine channels, both experiments suggested that calcium channels activity may be regulated by the SNARE complex.

Synaptotagmins are also known to interact with voltage dependent calcium channels. These calcium sensors provide calcium-dependent regulation of the core fusion machinery complex (Jahn and Sudhof, 1999). The synaptotagmin effect on calcium channels appears to be mediated via its association with the syntaxin/SNAP-25 complex. Synaptotagmin assembly with the ternary SNARE complex occurs via interactions with the SH3/membraine domain of syntaxin (Davis et al., 1999). In line with this observation, functional assays showed that both the inhibitory effects of syntaxin 1A on N- and L-type currents and the effect of SNAP25 on P/Q-type channels were reversed by co-expression of each SNARE protein with synaptotagmin (Wiser et al., 1996; Zhong et al., 1999; Atlas, 2001). Indeed, it seems that perturbations in the subtle balance of syntaxin/synaptotagmin or SNAP-25/syntaxin resulted in abnormal neurotransmitter release: overexpression of syntaxin 1A diminished neurotransmitter release, while lowering of syntaxin concentration led to increased levels of neurotransmitter secretion (Wu et al., 1998; Watanabe et al., 1999; Atlas, 2001).

Despite the subtle differences in the effect of synaptic protein co-expression with particular calcium channel subtypes, a general mechanism of channel regulation by synaptic proteins can be considered (Atlas, 2001):

1) A basal interaction at the plasma membrane between calcium channels and syntaxin that results in channel inhibition during "rest" conditions. This stage would presumably reduce random, non-regulated calcium channel openings;

2) A channel/ syntaxin/ SNAP-25 or a channel/ syntaxin/ synaptotagmin complex interaction at the active zone. During this stage syntaxin inhibition is removed, thus "priming" the channel for fusion;

3) A "ready" state, consisting of channel/ syntaxin/ SNAP-25/ synaptotagmin complexes where calcium channels are ready to activate at fast rates and maximal amplitude upon membrane depolarization resulting in vesicle fusion (Atlas, 2001).

CSP proteins, on the other hand, are synaptic components thought to regulate synaptic activity upon docking of the synaptic vesicle to the release sites by positive regulation of presynaptic calcium channels (Umbach et al., 1994). Earlier coimmunoprecipitation experiments have shown CSP interaction with P/Q-type calcium channels at the synprint site of the channels (Leveque et al., 1998), although recent data suggest the interaction occurs via CSP binding to the calcium effector domain of syntaxin, thus effectively acting as a competitor of the syntaxin-synprint interaction (Wu et al., 1999). In support of this hypothesis, injection of CSP in chick ciliary neuron calyx synapses enhanced presynaptic N-type currents by recruitment of dormant (syntaxin-bound?) channels (Chen et al., 2002).

RIM binding proteins (RBPs) are molecules that interact with the GTP-binding protein Rab3. Rab3 is a negative modulator of vesicle release (reviewed by Geppert and Sudhof, 1998). It has been recently shown that RBPs associate with N- and P/Q-type channels (Coppola et al., 2001), and with the $Ca_v1.3$ L-type calcium channel (α_{1D}) at the RQLPQTP motif located at the C terminus of the channel (Hibino et al., 2002). Notably, RBP also associates with α_{1B} and can presumably interact with other channels, given the fact that the RXXPXXP motif is tightly conserved within the ABE family, between the L-type channels α_{1C} and α_{1F}, and even in the more distant LVA channel α_{1G}. Functionally, disruption of the RIM-channel interaction produced increased hormone release in neuroendocrine cells (Hibino et al., 2002). RBPs are suggested to play a regulatory role in synaptic transmission strength through the functional link between the tethering apparatus and the fusion machinery.

2.2. Calcium Channels, Adaptor Proteins, and Structural Components of the Active Zone

Molecular adaptors are multidomain-containing proteins that can interact with intracellular targets to maintain a proper localization of such components within subcellular regions. For instance, PSD95 proteins are adaptors that sustain the postsynaptic colocalization of NMDA receptors, potassium channels and neuroligin via interactions with their PDZ domains. The presynaptic side of the active zone contains an adaptor triptych composed of CASK, Mint1 and Veli (Butz et al., 1998), all of which contain one or more PDZ cassettes. CASK is a MAGUK protein (membrane-associated guanylate kinase domain-containing protein) that binds to the cytosolic C-terminal tail of the neuron-specific cell surface protein neurexin, which in turn interacts with the postsynaptic cell surface protein neuroligin (see Figure 2). Mint-1 also interacts with neurexin, and together with Veli and CASK forms a ternary complex. This presynaptic multi-PDZ complex allows interaction with other synaptic targets such as syndecans, SynCAM, Munc-18, amyloid precursor protein and calcium channels (Biederer and Sudhof, 2000; Biederer et al., 2002).

Indeed, presynaptic calcium channels have been shown to interact with Mint-1 and CASK (Maximov et al., 1999). The carboxyl termini of N-type channels ending with the

(E/D)XW(C/S)-COOH motif were able to associate with one of the PDZ domains of Mint-1, while a proline-rich region near the channel's C tail was responsible for the interaction with the SH3 domain of CASK. Interestingly, only the "long" carboxy terminus variants of N-type and P/Q-type channels exhibit the C-terminal domain (Maximov et al., 1999). Another region within the C-terminus located between aa 2177 to 2339 of the N-type channel sequence contains three regions homologous to PSD-95/disc large/zona occludens-1 and Src motifs that are necessary and sufficient to target the channel to the presynaptic site (Maximov and Bezprozvanny, 2002).

Additional experiments locate calcium channels as an integral part of pre-post synaptic transmembrane complexes in the neuromuscular junction and electric organ synapses from the marine ray *Torpedo californica* (Sunderland et al., 2000). Coimmunoprecipitation experiments revealed the association of calcium channels with laminin $\alpha_4\beta_2\gamma_1$ complexes (laminin-9) and with the cytoskeletal protein spectrin. $\alpha_4\beta_2$ laminin is greatly enriched in postsynaptic membranes, while spectrin is localized and associates with synapsin 1 at presynaptic vesicles. Although calcium channels and these transmembrane complexes were found together, it is not clear whether the channel interacts directly or indirectly with laminin (Sunderland et al., 2000).

Thus the data presented in this section indicate a possible second function of presynaptic calcium channels as components of the transmembrane complexes that define the apposition of pre- and post-synaptic membranes at the active zone.

3. CALCIUM CHANNELS AND NEURONAL PLASTICITY

Calcium also plays a vital role in guiding long-term changes in neuronal excitability. Neuronal plasticity allows for behavioral adaptation, helping to promote an organism's continued growth and development, as well as increasing its chances of survival in times of crisis. Calcium entry through voltage-gated calcium channels triggers many of these changes through activation of the intracellular signaling cascades that alter neuronal excitability. Over the next several pages we will discuss the role of voltage-gated calcium channels in mediating stimulus-induced changes in synaptic efficacy, the resulting activation of several different transcription factors, and how these changes modify cell function.

3.1. L-type Calcium Channels Influencing Long-Term Potentiation/ Depression and Memory

With brief periods of heightened synaptic activity, neurons from a multitude of brain regions exhibit prolonged changes in excitability. Two examples of such plasticity are long-term potentiation (LTP) and long-term depression (LTD). LTP is an activity-dependent increase in synaptic efficacy, while LTD results in a decrease in synapse excitability. Generation of LTP versus LTD is dependent upon the frequency and duration of the stimulus. In sites such as the CA1 region of the hippocampus, LTP is induced by high frequency, short duration stimulation parameters (~25-100 Hz for only a few seconds), whereas LTD is achieved through lower frequency yet longer duration protocols (~1-5 Hz for several minutes). Both are believed to be major components of the cellular underpinnings regulating learning and memory, and thus have been under intense study since the mid 1970s (Bliss and Lømo, 1973). For further details of LTP and

LTD outside the scope of this chapter, readers are directed to recent reviews on the topic (e.g. Milner et al., 1998; Kandel, 2001).

Classical models of LTP, particularly at the widely studied hippocampal CA3-CA1 synapse, denote calcium entry through *N*-methyl-D-aspartate (NMDA) receptors as the trigger for plasticity. And while NMDA receptor-mediated cellular modulation has received the most attention, in recent years, other routes of calcium entry have also been shown to affect activity-induced potentiation (Ániksztejn and Ben, 1991). NMDA receptor-independent LTP most often requires the activation of L-type calcium channels (Grover and Teyler, 1990).

Within the hippocampal CA1 region, varying the stimuli that trigger changes in synaptic efficacy determines whether NMDA receptor and/or L-type calcium channel-mediated LTP is achieved. More specifically, following electrical stimulation to the Schaffer Collaterals, NMDA receptor-mediated LTP occurs across a variety of tetanus frequencies. Conversely, LTP triggered via L-type calcium channels only arises following very high frequency stimulations, such as theta-bursts (Grover and Teyler, 1990; Morgan and Teyler, 1999; Morgan and Teyler, 2001a; Dudek and Fields, 2002; Evers et al., 2002). Another mechanism that triggers L-type calcium channel-mediated LTP is synaptic stimulation following blockade of voltage-gated potassium channels with tetraethylammonium (TEA), resulting in prolonged depolarizations (Huang and Malenka, 1993; Huber et al., 1995a; Evers et al., 2002). This can also be achieved by direct activation of L-type calcium channels with repetitive depolarizing current pulses (Huang and Malenka, 1993). Epileptiform activity has also been found to induce L-type calcium channel LTP (Morgan and Teyler, 2001b). The requirement for more tightly clustered or broader stimulation parameters to induce L-type calcium channel-mediated plasticity may be due to unique biophysical properties of the channel and frequency-dependent modulation by calcium. These features are described in greater detail below.

NMDA receptor- and L-type calcium channel-dependent LTP may also be regulated through distinct second messenger systems. The initial increase in synaptic efficacy triggered by activation of NMDA receptors is dependent upon serine/threonine protein kinases. On the other hand, several studies have found L-type calcium channels target tyrosine kinases to stimulate LTP (Grover and Teyler, 1995; Cavus and Teyler, 1996; c.f. Huber et al., 1995b). The cellular localization of NMDA receptors and L-type calcium channels differ, which may be at least partially responsible for their ability to elicit distinct signal transduction cascades.

L-type calcium channel regulation of LTP extends beyond that measured at the CA3-CA1 hippocampal synapse. In dissociated hippocampal cultures, LTP was found dependent upon L-type calcium channels (Bi and Poo, 1998). Further, at least one form of mossy fiber LTP also requires L-type calcium channel activation (Kapur et al., 1998). Within the dentate gyrus, L-type calcium channels are responsible for LTP following high frequency electrical stimulation (Manahan-Vaughan et al., 1998). Outside the hippocampus, L-type calcium channels have also been demonstrated to be critical for LTP at the thalamoamygdala synapse, and are believed to play a critical role in fear memory (Weisskopf et al., 1999; Bauer et al., 2002; Lee et al., 2002). Recently, a component of retinocollicular LTP was also found to be dependent upon L-type calcium channels (Lo and Mize, 2002).

Within experimental settings, L-type calcium channel and NMDA receptor-mediated LTP can be independently activated (e.g. Evers et al., 2002; Hardingham et al., 2002). However, *in vivo* these two pathways must work in unison to ensure proper cellular function. They are invariably linked, as the stimuli that open L-type calcium channels typically work through activation of glutamate receptors. Some recent data suggest that NMDA- and L-type calcium channel-mediated LTP are responsible for separate stages of learning and memory. In a spatial memory task, Borroni et al. (2000) found blockade of NMDA receptors disrupts memory acquisition, while inhibition of L-type calcium channels results in a decrease in memory retention. Interestingly, hippocampal slices taken from older rats exhibit predominantly L-type calcium channel-mediated LTP and may mirror age-dependent changes in learning and memory (Izumi and Zorumski, 1998; Shankar et al., 1998). Somewhat surprisingly, however, several studies have found enhancement of memory following blockade of L-type calcium channels (McMonagle-Strucko and Fanelli, 1993; Quevedo et al., 1998; Quartermain et al., 2001). It is somewhat difficult to reconcile these data, as they employ different routes of drug administration, dose of antagonists, and utilize different memory tasks. However, at frequencies that border between optimal LTP and LTD stimulations, L-type calcium channels may normally oppose LTP in favor of LTD (Norris et al., 1998). Thus, with blockade of L-type calcium channels, LTP induced by "lower" frequency stimulations may be enhanced, and in theory, so would be the results of some learning and memory tests.

The role of L-type calcium channels in LTD has also been well established. Originally characterized in cerebellar Purkinje neurons, LTD resulted following calcium entry through voltage-gated calcium channels (Sakurai, 1990; Daniel et al., 1992). Shortly thereafter, hippocampal CA1 LTD was also found to be dependent upon activation of L-type calcium channels (Wickens and Abraham, 1991; Bolshakov and Siegelbaum, 1994; Christie et al., 1997; Norris et al., 1998). LTD in the dentate gyrus after stimulation of the perforant pathway also exhibits sensitivity to L-type calcium channel blockers (Christie and Abraham, 1994; Wu et al., 2001). L-type calcium channel-mediated LTD is also observed in hippocampal cell cultures (Bi and Poo, 1998) and retinocollicular synapses (Lo and Mize, 2002).

Regardless of whether the initial stimulus leads to increased or decreased cell excitability, neurons respond in a similar fashion. The immediate observance of LTP and LTD requires the alteration of preexisting proteins. However, to maintain these changes in cell function, the synthesis of new protein is essential (Milner et al., 1998). Production of new protein can be highly localized, as translation of some mRNA species occurs in post-synaptic densities (Wells et al., 2000; Steward and Schuman, 2001; Steward and Worley, 2001), altering the excitability of that particular synapse. However, synaptic activity can also have global effects on neuronal excitability, as information transmitted from the synapse to the nucleus leads to the activation of transcription factors. These transcription factors regulate cell excitability through alterations in gene expression. Many of the intracellular signaling pathways leading to gene expression and neuronal plasticity are again tightly regulated by calcium. Thus, considerable attention has focused upon the mechanisms by which calcium modifies gene expression. As discussed next, L-type calcium channels, which are required for induction of LTP and LTD, are also vital for its maintenance through the regulation of gene expression.

4. L-TYPE CALCIUM CHANNELS AND GENE EXPRESSION

Of the ten α_1 proteins forming the primary subunit of a voltage-gated calcium channel, nine are expressed in nervous tissue, the exception being the skeletal muscle α_{1S} ($Ca_v1.1$). With localization apparently restricted to the soma and proximal dendrites in a variety of different cell types, L-type calcium channels generated from α_{1C} ($Ca_v1.2$) and α_{1D} ($Ca_v1.3$) subunits were always assumed to be the primary regulators of gene expression (Ahlijanian et al., 1990; Westenbroek et al., 1990; Hell et al., 1993; Westenbroek et al., 1998). Indeed, research over the last dozen or so years has sustained this hypothesis. However, as will be discussed later in the chapter, the mechanisms underlying L-type calcium channel-mediated gene expression are far more complex than initially appreciated.

The most prominent transcription factor implicated in neuronal plasticity is cyclic AMP response element binding protein (CREB). CREB, a member of the bZIP superfamily of transcription factors, has been shown to play a critical role in many forms of activity-dependent plasticity. In a variety of brain regions, CREB-mediated gene expression is critical for cell survival (Walton and Dragunow, 2000). CREB signaling is also required for behaviors pertaining to learning and memory (Silva et al., 1998). It plays a crucial role in the regulation of circadian rhythms (Kako and Ishida, 1998), regulates emotional responses (Barrot et al., 2002) and is responsible for many of the cellular events underlying drug addiction (Berke and Hyman, 2000; Nestler, 2001). Furthermore, CREB becomes activated during injury, and while CREB-dependent gene expression may prevent cell death following neuronal insult (Mabuchi et al., 2001; Tanaka, 2001), it has also been implicated in the induction of hyperalgesia (Anderson and Seybold, 2000; Melzack et al., 2001). Outside of animal models, genetic disorders that result in a disruption of CREB signaling have been linked to Rubenstein-Taybi syndrome (Petrij et al., 1995) and Huntington's disease (McCampbell et al., 2000; Steffan et al., 2000; Nucifora et al., 2001; Wyttenbach et al., 2001; Mantamadiotis et al., 2002).

A critical step in CREB activation is its phosphorylation on Ser^{133}. Recent advances have also determined that phosphorylation at Ser^{142} and Ser^{143} regulate CREB-dependent gene expression, and readers are encouraged to familiarize themselves with these data as well (Sun et al., 1994; Parker et al., 1998; Gau et al., 2002; Kornhauser et al., 2002). Although not obvious given its name (protein kinase A was first shown to phosphorylate CREB following increases in cAMP), CREB phosphorylation is highly sensitive to changes in intracellular calcium. Calcium entry through L-type calcium channels leads to CREB phosphorylation in a variety of neuronal tissue including the hippocampus (Sheng et al., 1991; Bading et al., 1993; Deisseroth et al., 1996; Mermelstein et al., 2000), cerebral cortex (Murphy et al., 1991; Shieh et al., 1998; Tao et al., 1998), striatum (Liu and Graybiel, 1996; Rajadhyaksha et al., 1999), cerebellum (Bito et al., 1999), olfactory bulb (Cigola et al., 1998), and retina (Yoshida et al., 1995). Because of their influential role in learning and memory, the majority of experiments examining this mode of CREB phosphorylation have utilized hippocampal and cortical neurons. Although the signaling mechanisms utilized by L-type calcium channels are generally believed to be consistent across brain regions (Mermelstein et al., 2001), some differences may exist (see Deisseroth and Tsien, 2002). It is generally accepted that with brief periods of heightened synaptic activity, calcium entry through L-type calcium channels triggers phosphorylation of CREB through activation of the calmodulin (CaM) kinases, CaM kinase kinase (CaMKK) and CaM kinase IV (CaMKIV) (Enslen et al., 1995; Tokumitsu et al., 1995; Bito et al., 1996; Ahn et al., 1999; Mermelstein et al., 2001). With longer depolarizations, calcium entry through L-type calcium channels will phosphorylate CREB via ras/mitogen activated protein kinase (MAPK) signaling (Xing et al., 1996; Impey et al., 1998). Further, calcium-sensitive adenylate cyclases may also play a role in CREB phosphorylation, and thus L-type calcium channels may utilize yet another mechanism to activate this transcription factor (Xia and Storm, 1997; Smit and Iyengar, 1998). The reasons behind L-type calcium channels employing multiple kinases remain poorly understood. However, robust CREB-dependent gene expression requires sustained CREB phosphorylation (Bito et al., 1997). Thus, the recruitment of several different kinases throughout a prolonged stimulus period may ensure maximal responses in gene expression.

CREB-dependent gene expression is initiated by a CREB dimer binding to a CREB responsive element (CRE) consisting of the conserved palindromic sequence (TGACGTCA) and the recruitment of a coactivator, such as CREB binding protein (CBP) or the closely related P300 (Montminy, 1997). Interestingly, nuclear calcium concentrations (to which L-type calcium channels contribute) have also been shown to regulate the activity of CBP through phosphorylation of Ser^{301} (Hardingham et al., 1999; Hu et al., 1999; Impey et al., 2002). With activation of CREB and CBP, the transcriptional complex (including RNA polymerase II and other general transcriptional machinery) is believed to become active.

Considering that CREB-dependent gene expression has been implicated in the regulation of a wide variety of cellular processes, it is not surprising that many different genes contain a CRE element. These include genes that encode for proteins regulating neurotransmission, cell structure, growth factors, cell metabolism, signal transduction and gene expression (for a detailed list of genes, see Lonze and Ginty, 2002). Thus, through the regulation of just a single transcription factor, L-type calcium channels can potentially regulate a host of genes responsible for a wide range of cellular functions.

With an important role in a diversity of cell behaviors, it is not surprising that several signaling pathways, independent of calcium entry though L-type calcium channels, also lead to CREB activation. For example, CRE-dependent transcription follows the opening of NMDA receptors, and is required for this separate form of LTP. A variety of other stimuli, including growth factors and hormones, will also activate this transcription factor. How then, are specific changes in cell behavior and function achieved if each of so many different stimuli employ such a common endpoint? Possibly, by differentially influencing other mediators of gene expression, L-type calcium channels generate a unique 'fingerprint' of cell activation, which in turn underlies their unique contributions to neuronal plasticity.

Nuclear factor of activated T-cells (NFATc) is one class of transcription factor that in at least several cell types appears directly responsive to calcium entry specifically through L-type calcium channels. Originally characterized in immune tissue, a recent study by Graef et al. (1999) discovered that NFATc4 (also termed NF-ATc4 and NFAT3) was also expressed in brain. Within hippocampal neurons, NFAT-dependent transcription was triggered following heightened synaptic stimulation; opening of NMDA receptors under these conditions was required for NFATc4 activation. However, Graef and colleagues found that the primary function of NMDA receptors was to sufficiently depolarize these cells such that the L-type calcium channels became activated, as blockade of L-type calcium channels completely inhibited NFAT-dependent transcription, regardless of how much calcium entered the cell via other ligand- or voltage-gated channels.

Results from Graef et al. (1999) suggest that L-type calcium channels regulate NFAT-dependent gene expression via three distinct mechanisms. A primary requirement for transcription initiation is the translocation of NFATc4 from the cytosol (off-state) into the nucleus (on-state). Translocation follows calcineurin-mediated dephosphorylation of NFATc4, resulting in the unmasking of multiple nuclear localization signals. Activation of L-type calcium channels triggers this calcineurin-mediated dephosphorylation and subsequent nuclear translocation. For sustained NFAT-dependent transcription, NFATc4 must remain dephosphorylated, as phosphorylation would promote nuclear export. As such, L-type calcium channels were also found to modulate the export of NFATc4, possibly through inhibition of glycogen synthase kinase 3β (GSK3β). For NFAT-dependent transcription to occur, NFATc proteins require the assistance of a nuclear partner (NFATn) in order to stabilize their interactions at an NFAT binding site (GGAAA). The nuclear binding partners of NFATc4 require phosphorylation for activation. As mentioned above, calcium entry through L-type calcium channels stimulates several signaling pathways that lead to the activation of candidate NFATn protein kinases.

The genes regulated by NFAT-dependent transcription are best characterized outside of the nervous system, where they are known to encode for cytokines, cell surface proteins, and other transcription factors critical for immune responses. Within neurons, evidence suggests NFAT regulates the gene encoding the inositol 1,2,5-trisphosphate receptor type 1 (IP_3R1) (Graef et al., 1999), thereby implementing a mechanism by which L-type calcium channels can alter calcium signaling and cell excitability (Inoue et al., 1998). Undoubtedly, more neuronal genes regulated by NFATc4 will soon be discovered.

Although L-type calcium channels activate both CREB and NFATc4, it is interesting to note that the transcription factors respond somewhat differently to external stimulation. Specifically, CREB phosphorylation is more pronounced with LTP- versus LTD-like stimulations, while the opposite is true for NFATc4. Recent data suggest that the calmodulin-sensitive proteins downstream of the L-type calcium channel are principally responsible for the variations seen in CREB and NFAT activity. Calcineurin exhibits a higher affinity for the calcium bound form of calmodulin than the CaM-dependent kinases (Van Eldik and Watterson, 1998). This may be one reason why NFAT-dependent gene expression is preferentially activated with low frequency stimulation that gives rise to more modest increases in intracellular calcium. Furthermore, calcineurin promotes dephosphorylation of CREB through disinhibition of protein phosphatase 1, signaling mediated via inhibitor 1 (Bito et al., 1996). Thus, even with some CaMKIV-mediated CREB phosphorylation following low frequency stimulation, calcineurin would continue to oppose CREB dependent transcription.

With higher frequency stimulation, calcineurin becomes inactivated via oxidation by free radicals (Stemmer et al., 1995; Bito et al., 1996; Wang et al., 1996). This would prolong CREB phosphorylation while attenuating NFAT signaling. Moreover, increased calcium entry generated by high frequency stimulation further activates CaMKK and CaMKIV. Robust activation of CaMKII, mediated by high frequency stimulation (Miller and Kennedy, 1986; Lisman, 1989; Lisman, 1994) results in the facilitation of L-type channel openings (Zühlke et al., 1999; Dzhura et al., 2000). The functional consequence is to further inhibit calcineurin, while concurrently increasing CaMKK- and CaMKIV-mediated CREB phosphorylation. Here again is an example whereby differences in gene expression can be achieved through signaling to various second messenger systems, even though the initial calcium signal may arise from the same source.

4.1. Significance of the Link Between L-type Calcium Channels and Gene Expression

Since L-type channels are privileged in signaling to CREB and NFAT, considerable interest has been generated in understanding the types of stimuli that maximally activate these channels. Interestingly enough, L-type calcium channels have the ability to distinguish between different physiological stimulations, allowing for gene expression only following particular patterns of activation. Since neurons are constantly being depolarized, this feature is critical in ensuring that long-term changes in plasticity only occur under very specific circumstances.

Action potentials (APs) and excitatory post-synaptic potentials (EPSPs) are two examples of physiological stimuli that a neuron can discriminate between. Dendritic APs are a result of back propagation of spikes originating in the cell soma: EPSPs are the result of synaptic neurotransmission. There are several reports indicating that L-type calcium channel mediated CREB phosphorylation occurs following EPSPs, but not simple spike stimulation (Bito et al., 1996; Deisseroth et al., 1996; Mermelstein et al., 2000). The observance of L-type calcium channel-mediated CREB phosphorylation following a series of EPSPs and not APs makes sense, as it provides a means by which 'routine' intercellular communication can be distinguished between stimulation events that are destined to bring upon long-term changes in cell excitability.

The ability of neurons to discriminate between APs and EPSPs is due to the unique biophysical properties of L-type calcium channels. As such, experiments performed in hippocampal pyramidal neurons suggest that L-type calcium channels act as a kinetic filter. Depolarizations that mimic a train of EPSPs are adequate stimuli to trigger CREB phosphorylation since L-type calcium channels activate at relatively hyperpolarized potentials. Conversely, APs are poor activators of L-type calcium channels because these channels exhibit relatively slow activation kinetics (Mermelstein et al., 2000). These distinctive properties of L-type calcium channels have also been observed in cortical, cerebellar and striatal neurons as well as neuroblastoma cells (Kasai and Neher, 1992; Marchetti et al., 1995; Randall and Tsien, 1995; Song and Surmeier, 1996; Nakazawa and Murphy, 1999), suggesting a shared mechanism across neurons to discriminate between different physiological stimuli.

The inability of APs to induce CREB phosphorylation can be overcome with the L-type calcium channel agonist, Bay K8644. Application of Bay K8644 acts in a sense to "broaden" the AP stimulus. Recently, Dudek and Fields (2002) found that theta burst stimulation will also induce MAPK activation and CREB phosphorylation via opening of L-type calcium channels. These findings are in very good agreement with the data previously described: L-type calcium channel-mediated LTP occurs only following high frequency tetanus, or with stimulations that result in broadening the duration of synaptic activity. As will be discussed later in this chapter, calcium-dependent inactivation and facilitation of L-type calcium channels may also play a critical role in 'tuning' L-type calcium channels to respond favorably to particular patterns of stimulation.

4.2. Linking L-type Calcium Channels to Transcription Factors

A fascinating twist emerged upon examination of the details surrounding CREB and NFATc4 activation. Following synaptic stimulation, increases in intracellular calcium are produced by calcium entering the neuron via a variety of voltage-gated channels and ligand-gated receptors, as well as by release of calcium from intracellular stores (Magee and Johnston, 1995; Spruston et al., 1995; Eilers et al., 1996; Finch and Augustine, 1998). Surprisingly, under many circumstances, L-type calcium channels contribute only a small percentage to the overall rise in intracellular calcium, yet blocking calcium entry through these channels will dramatically attenuate gene expression (Deisseroth et al., 1998; Dolmetsch et al., 2001). Calcium entry through L-type channels is therefore somehow privileged in regulating CREB and NFATc4. The question remained: what are the underlying mechanisms by which L-type calcium channels can play such a disproportionate role in the activation of gene expression?

The means by which L-type calcium channels regulate CREB activity has received the most attention. In 1996, Deisseroth and colleagues preloaded hippocampal neurons with one of three separate calcium chelators, each exhibiting different buffering characteristics: BAPTA, a calcium chelator with high affinity and capacity; EGTA, exhibiting a lower affinity for calcium than BAPTA, but equal capacity; and Br_2-BAPTA, having an equal affinity for calcium as BAPTA, but a lower capacity. Following field stimulation of cultured hippocampal neurons, CREB phosphorylation was inhibited only in cells preloaded with BAPTA, suggesting a "functional window" in which calcium must rise (i.e. micromolar concentrations within ~1 μm from the L-type channel) in order for gene expression to occur.

This concept of localized calcium signaling has been further substantiated. Recently, Dolmetsch et al. (2001) found that removal of the calmodulin binding IQ motif from the carboxy terminus of the calcium channel α_{1C} subunit results in L-type calcium channels incapable of signaling to CREB. Their data demonstrate an inherent requirement of the calcium channel in order for it to signal to the calcium-sensitive kinases that phosphorylate CREB. However, other calcium channel α_1 subunits contain similar calmodulin binding IQ motifs (Lee et al., 1999; Peterson et al., 1999), indicating that while interactions between calmodulin and the calcium channel α_1 subunit are required for CREB activation, they are not sufficient to explain the privileged role of L-type calcium channels in nuclear signaling. Interestingly, the α_{1C} subunit also contains a PDZ interaction sequence (Koch et al., 1990; Snutch et al., 1991), known to bind at least two PDZ domain proteins, NIL-16 and CIPP (Kurschner et al., 1998; Kurschner and Yuzaki, 1999). Since other PDZ domain proteins are critical for linking channels and receptors to specific second messenger systems (Craven and Bredt, 1998; Sheng and Pak, 2000), it is possible that the normal subcellular localization of α_{1C}, promoted by its interaction with PDZ domain proteins, is required for L-type calcium channels to activate CREB. Indeed, overexpression of the PDZ interaction sequence in hippocampal neurons disrupts both the normal localization of α_{1C}-comprised L-type calcium channels, as well as L-type calcium channel-mediated CREB phosphorylation and CRE-dependent transcription, without influencing calcium entry through the channel (Weick et al., 2002). Hypothetically, anchoring α_{1C} to regions where signaling to CREB is initiated would not only promote the ability of these particular L-type calcium channels to activate gene expression, but would also help to exclude non-α_{1C}-comprised L-type calcium channels from localizing to these same regions, and thus signaling to the nucleus themselves. The data also suggest that calcium entry through α_{1C}-comprised L-type calcium channels, and not those containing the α_{1D} subunit, are responsible for CREB activation.

Less is known about the mechanism by which L-type calcium channels are privileged in signaling to NFATc4. Preliminary evidence from Weick et al. (2002) suggests that interactions of α_{1C} with PDZ domain proteins are not required. Interestingly, there are several other differences between activation of CREB and NFATc4 following calcium entry through L-type calcium channels. While calcineurin is required to activate NFATc4 (Rao et al., 1997), this phosphatase has been shown to limit CREB phosphorylation (Bito et al., 1996). CREB and NFATc4 also exhibit differential sensitivities to the patterns of synaptic stimulation used to trigger their responses (Deisseroth et al., 1996; Graef et al., 1999). Further, calcium entry through NMDA receptors, but not release from intracellular stores, will lead to rapid CREB activation, while the opposite is true for NFATc4. What can account for these differences in CREB and NFATc4 activation? One possibility is that NFAT-dependent transcription is being regulated by calcium entry through α_{1D} ($Ca_V1.3$) L-type calcium channels. This is an intriguing possibility, especially since in several tissues, potential α_{1D} L-type calcium channels activate at very hyperpolarized potentials, rivaling the classical T-type low-voltage activated channels (Avery and Johnston, 1996; Magee et al., 1996; Xu and Lipscombe, 2001). These L-type calcium channels may also show enhanced sensitivity to stimuli destined to produces changes in cell excitability (Schjott and Plummer, 2000). If, in fact, these channels were conferred with the ability to signal changes in gene expression and protein synthesis, it would have profound implications as to the types of stimulations that affect neuronal plasticity.

4.3. Functional Consequences of L-type Calcium Channel-Mediated Changes in Gene Expression

The ability of L-type calcium channels to signal to the nucleus allows these channels to induce a variety of changes to the nervous system. CREB and NFAT-dependent transcription have been implicated in regulating the expression of a wide variety of proteins involved with the many aspects of cell function. Immediate-early genes, neurotransmitters, growth factors, ion channels, intracellular signaling enzymes, and structural proteins are all regulated by these changes in gene expression. Along with changes in cell excitability, calcium entry through L-type calcium channels regulates cell fate (Brosenitsch et al., 1998). During development and regeneration, these same channels help guide axons and dendrites to their appropriate targets (McAllister et al., 1996; Bishop and Milton, 1998; Ohbayashi et al., 1998). Thus, based upon specific patterns of depolarization, calcium entry through voltage-gated calcium channels can activate the second messenger systems that are not only required for triggering immediate changes in synaptic efficacy, but also those responsible for long-term changes in cell excitability and function.

4.4. Regulation of Gene Expression and Protein Synthesis by other Calcium Channels

While historically, the majority of studies examining the role of voltage-gated calcium channels on gene expression have focused upon L-type calcium channels, recent evidence from several laboratories have found non L-type calcium channels also influence gene expression and protein synthesis. For example, Sutton et al. (1999) found that P/Q-type calcium channels in cerebellar neurons regulate the expression of syntaxin-1A. These results are particularly intriguing, as syntaxin-1A interacts not only with P/Q-type calcium channels, but also with other presynaptic proteins that regulate synaptic vesicle release. Syntaxin-1A binds P/Q-type calcium channels to promote inactivation. Thus, an activity-dependent feedback mechanism may exist whereby calcium entry through P/Q-type calcium channels regulates future synaptic efficacy.

Along with P/Q-type calcium channels, N-type calcium channels may also regulate protein synthesis as a means of negative feedback. Raghib et al. (2001) found that truncated isoforms of α_{1B} ($Ca_V2.2$), which are potentially generated from alternative splicing, inhibit the functional expression of full-length N-type calcium channels. The authors believe inhibition results from blockade of protein synthesis (and not from α_{1B} gene expression). Furthermore, patients diagnosed with episodic ataxia type-2 (EA-2) exhibit mutations in the α_{1A} gene that predict the expression of truncated protein (Mori et al., this volume). Thus, this form of negative feedback may play an important role in the symptoms of this disease. Undoubtedly, future research will uncover other mechanisms by which voltage-gated calcium channels regulate cell excitability and function.

5. CALCIUM-ACTIVATED POTASSIUM CHANNELS

Calcium-activated potassium channels function as endogenous calcium sensors that can finely regulate calcium influx and cellular excitability. Activation of these channels following intracellular calcium concentration increases leads to cell hyperpolarization, which in turn modulates the calcium signaling process, depending on the calcium entry route. There are three types of calcium-activated potassium channels, named BK, IK and SK. These potassium channel families are encoded for by different genes and thus give rise to three biophysically and pharmacologically different channel types (for review see Sah, 1996; Vergara et al., 1998; Gribkoff et al., 2001; Jensen et al., 2001). The activation of these potassium channels in response to calcium entry indirectly makes voltage-gated calcium channels responsible for their cellular effects.

BK (big $K_{(Ca)}$ or maxi-K) channels are large-conductance potassium channels found in a variety of cells, including excitable cells (Marty, 1981; Pallotta et al., 1981). BK channels are regulated both by voltage and calcium, which renders them ideal for regulating calcium influx and neuronal excitability (Gribkoff et al., 2001). Intermediate conductance or IK channels are expressed in non-excitable cells and regulated by calcium but not by voltage; their main functions include the stabilization of the resting potential, regulation of cell volume, and fluid secretion. Small conductance calcium activated channels (SK) are not voltage-dependent, but they are regulated by calcium. They exist in non-excitable and excitable cells, where they contribute to afterhyperpolarization and to spike frequency adaptation. Table 2 summarizes the properties of these three calcium-activated potassium channel families.

5.1. BK or MaxiK Channels

Maxi K or BK channels are proteins with 7 predicted membrane-spanning domains, encoded by the KCNMA1 gene, which generates multiple splice variants (formerly named slo channels) of the pore-forming α subunit (Lagrutta et al., 1994; Tseng-Crank et al., 1994). An auxiliary β subunit ($\beta 1$ to $\beta 4$) modifies the channel gating properties and

Table 2. Properties of calcium-activated potassium channels.

Property	Channel		
	BK (maxiK, I_C)	**IK**	**SK (mAHP, I_{AHP})**
Calcium sensitivity	Low		
$[Ca^{2+}]_i$ range required for activation	1-10 μM	~50-900 nM	~50-900 nM
Calmodulin binding	No	Yes	Yes
Voltage dependence	Yes (*e* fold/9-15 mV)	No	No
Single-channel conductance	100-250 pS	20-100 pS	4-20 pS
Blockers	Iberiotoxin Charybdotoxin Paxilline TEA	Charybdotoxin Clotrimazole Tram-34	Apamin Scyllatoxin Dequalinium Curare
Openers	NS 1619 NS-004 (benz-imidazolones) DHS-1	1-EBIO DC-EBIO CZ ZOX Theophylline Caffeine, IBMX	Riluzole 1-EBIO
Function	Action Potential repolarization Neurotransmitter release regulation	Resting potential Volume regulation Fluid secretion	After-hyperpolarization Spike frequency adaptation
Localization	Neurons Skeletal muscle Smooth muscle Epithelia	Epithelia Endothelia Erythrocytes Lymphocytes (B and T) Monocytes	Neurons Skeletal muscle
Gene Name	KCNMA1	KCNN4	KCNN1 to 3
Clone name (mammalian)	slo1,2,3	IK1/SK4	SK1,2,3
Topology	7 TM	6 TM	6 TM
Association with voltage-gated calcium channels	Yes L-type, N-type, R-type	No	Yes L-type, N-type, P/Q-type T-type

Abbreviations: 1-EBIO = 1-ethyl-benzimidazolinone; CZ = Chlorozoxazone; DHS-1= dehydrosoyasaponin-1; TM = transmembrane; TRAM-34 = (1-[2-chlorophenyl) diphenylmethyl]-1H-pyrazole; ZOX= Zoxazolamine.

pharmacology (Dworetzky et al., 1994; Knaus et al., 1994; Tseng-Crank et al., 1996; Brenner et al., 2000; Meera et al., 2000). The native channel is thought to exist as an α subunit alone (Shen et al., 1994) or co-assembled with 1-3 β subunits (Adelman et al., 1992; Garcia-Calvo et al., 1994; McManus et al., 1995). The topology of BK channel contains an "extra" transmembrane domain, termed S0, which precedes the traditional S1-S6 segments analogous to the 6TM domains from the voltage-gated channel family (Meera et al., 1997). The single channel conductance of BK channels is greater than 100 pS when measured in symmetrical K^+ solutions.

These channels are steeply voltage-dependent, with the potassium conductance increasing with membrane depolarization. In addition, the open probability and gating rates of BK channels are also functions of intracellular calcium levels: channels open rapidly in the presence of Ca^{2+} and close in a similar fashion in its absence, as predicted for a fast, reversible calcium binding reaction. The calcium sensitivity of BK channels arises from domains termed "calcium bowls" containing a number of negatively charged amino acids that presumably bind calcium ions directly, in combination with other regions in the C-terminus (Schreiber and Salkoff, 1997; Xia et al., 2002). The channel K_D for calcium is also sharply voltage-dependent, being relatively insensitive to calcium (in the μM level) at resting membrane potential, but more sensitive during physiological depolarization levels (in the nanomolar range at 20-40 mV) (Latorre et al., 1989).

BK channels are blocked by low concentrations of tetraethylammonium (TEA, 0.5-1 μM) (Blatz and Magleby, 1984), paxilline (Sanchez and McManus, 1996), and the scorpion toxins iberiotoxin and charybdotoxin (Galvez et al., 1990), although the latter is also known to block voltage-dependent potassium channels and IK channels. Some BK channels are specifically activated by dehydrosoyasaponin-1 (DHS-1), a compound isolated from a medicinal herb (McManus et al., 1993).

5.2. IK Channels

The first calcium-activated potassium channel type studied, in the late 1950's, were IK channels from human red cells (Gardos, 1958). IK currents are expressed in various non-excitable tissues, primarily blood and epithelial cells. Their activation plays an important role in the immune responses of B- and T-cells and in secretion in epithelial tissues. These channels are not voltage-sensitive, but rather are activated only by increases in intracellular calcium concentration. Single channel conductance of these channels in human red cells are ~ 20-100 pS in symmetric high potassium conditions (for review see Jensen et al., 2001).

IK channels are encoded by the KCNN4 gene, a gene with higher homology to SK channels than to other potassium channels. The protein encoded by KCNN4 is a 6 transmembrane domain protein (S1-S6), with an intracellular amino-terminal region similar to that of *shaker* channels, and a long carboxyl terminal tail. Calmodulin interacts with the C-terminal region of IK channels in a non calcium-dependent manner (Xia et al., 1998; Fanger et al., 1999), and it is thought to be a physiological modulator of the channel, along with PKA and CaMKII phosphorylation among others (Jensen et al., 2001).

IK channels are blocked by the scorpion toxin charybdotoxin and with less potency by maurotoxin, margatoxin and kaliotoxin (Garcia-Calvo et al., 1993; Grissmer et al., 1994; Aiyar et al., 1995). The imidazole antimycotic compound clotrimazole is a potent blocker of IK channels (IC_{50}=50 nM) (Alvarez et al., 1992). Positive modulators of these channels include 1-EBIO, chlorzoxazone and others (Table 2), although these drugs also affect SK channels.

The main physiological function of these channels is in volume regulation and secretion. IK channels are also linked genetically to the rare disorder Diamond-Blackfan Anaemia (DBA) (Gustavsson et al., 1997).

5.3. SK Channels

SK (small conductance) channels (Romey and Lazdunski, 1984; Blatz and Magleby, 1986; Lang and Ritchie, 1987) are similar to IK channels in that they also are activated by cytosolic calcium in the 400-800 nM range and are completely voltage-independent (Hille, 2001) (for recent review, see Sah and Louise Faber, 2002). However, SK channels manifest small single channel conductance levels, from 5-20 pS. They are specifically blocked by apamin (Romey and Lazdunski, 1984; Blatz and Magleby, 1986), and they are also inhibited by curare and bicuculline (Seutin and Johnson, 1999), dequallinium (Dunn, 1994), scyllatoxin, and high concentrations of TEA (10-20 mM). IK channels are not sensitive to BK-channel blocking scorpion toxins charybdotoxin and iberiotoxin. SK channel openers include 1-EBIO, riluzole and chlorzoxazone.

SK channels, like BK channels were first discovered in skeletal muscle, but they are also widely distributed among tissues. There are three SK channel subtypes (SK1-3), which show different sensitivity to the peptide blocker apamin. These channel subtypes are encoded for by the KCNN1-3 genes, a gene structurally similar to the KCNN4 gene of IK channels. Accordingly, these genes encode for 6 TM proteins with ~ 60% homology amongst them, and similarity to voltage gated channels limited to the pore region. Similar to IK channels, the proximal region of the C-terminal tail of SK channels binds constitutively to calmodulin, which confers allosteric calcium sensitivity and leads to channel opening upon calcium binding (Xia et al., 1998; Keen et al., 1999; Schumacher et al., 2001).

5.4. Functions of Calcium-Activated Potassium Channels in Neurons

The physiological functions of BK channels are defined by their voltage and calcium sensitivity. Increases in intracellular calcium to micromolar concentrations due to the opening of voltage-dependent calcium channels induces a shift in the activation voltage curve of BK channels towards more physiological voltages (more negative potentials). The steep voltage dependence of BK channels contributes to their activation during the action potential (AP) upstroke and quick closure upon membrane potential repolarization.

SK channels, unlike BK channels, are not activated by voltage. Their higher calcium sensitivity and cellular localization define their physiological role in the AP repolarization (aftherhyperpolarization) and in spike frequency adaptation. In the soma of the neuron, the AP repolarization process has three afterhyperpolarization components, classified as fast, medium and slow hyperpolarizations (fAHP, mAHP, sAHP, respectively) according to their kinetics (Sah and Louise Faber, 2002). These three components are mediated by calcium-activated potassium channel openings that result from voltage-dependent calcium channel activity evoked by the action potential (Lancaster and Nicoll, 1987; Storm, 1987; Storm, 1990). The fast component (fAHP) activates immediately after the AP and lasts tens of milliseconds. BK channel activation is responsible for this component (Adams et al., 1982; Lancaster and Nicoll, 1987; Shao et al., 1999). The medium component (mAHP, I_{AHP}, $gK_{Ca.2}$) also activates quickly upon AP termination, but its decay lasts several hundred of milliseconds. The slow component (sAHP or sI_{AHP}) peaks over several hundreds of milliseconds and lasts up to 5 seconds after APs end. Both mAHP and sAHP (sI_{AHP}) are the result of SK channel opening. The slower kinetics of opening of this component are due to the rate limiting diffusion of calcium following calcium channel opening upon membrane depolarization. To date, the molecular identity of the channels responsible for the slow AHP component is not known, although their conductance levels are compatible with those of SK channels (Sah and Louise Faber, 2002).

Spike frequency adaptation is the result of mAHP and sI_{AHP} component activity. Neurons that express SK channels exhibit tonic AP firing, and the activation of I_{AHP} slows the maximal firing frequency (Engel et al., 1999). This effect is explained by the relatively rapid activation but slow inactivation of SK channels. The slow decay of the sI_{AHP} leads to a prolonged AHP immediately after firing. As sI_{AHP} inactivates, the neuron will slowly reach threshold and can again fire APs.

On the other hand, the slow activation of sI_{AHP} after AP-induced calcium influx allows for repetitive neuronal spiking, before the summation of sI_{AHP} progressively slows down neuron discharge frequency and terminates the AP (Cassell et al., 1986; Sah, 1995). This process is further modulated by neurotransmitter-mediated decrease of sI_{AHP} current, which can finely tune the firing properties of neurons.

The presynaptic role of BK channels in neurotransmitter release is well documented. Invasion of the AP in the presynaptic terminal leads to a significant increase in the local intracellular calcium concentration (>100 μM) near the mouth of the channel which, as reviewed earlier, induces vesicle fusion to the presynaptic membrane and subsequent NT release (Gribkoff et al., 2001). Blockade of BK channels results in increases of NT release, suggesting close positioning between the two channel types (Petersen and Maruyama, 1984; Robitaille et al., 1993; Liu et al., 1999).

5.5. Association between Calcium Channels and Calcium-Activated Potassium Channels

A physical association between voltage-gated calcium channels and calcium-activated potassium channels has been postulated, similar to that occurring between calcium channels and presynaptic proteins. Although there is no evidence for a direct interaction to date, several experimental approaches support this idea. Afterhyperpolarization events mediated by SK channels in the somata of several types of neurons are sensitive to N-type channel blockers (Sah and Davies, 2000). Similarly, P-type channel inhibition caused blockade of the SK-type channels responsible for the sI_{AHP} in cortical and neostriatal neurons (Pineda et al., 1992; Williams et al., 1997). Furthermore, in isolated patches from hippocampal pyramidal neurons containing both BK and SK channels, L-type channel activation preferentially activated SK channels over BK channels (Marrion and Tavalin, 1998). R-type and L-type channels also play a role in the activation of SK conductances. L-type channels, however, may activate either BK or SK channels, depending on the neuronal type (Sah and Davies, 2000).

Close proximity between the two channel entities has been described in several systems, in which the BK channel activity is difficult to dissect from that of calcium channels (Robitaille et al., 1993; Issa and Hudspeth, 1994; Sun et al., 1999). Blockade of BK channels results in increased NT release, suggesting close positioning between the two channel types (Petersen and Maruyama, 1984; Robitaille et al., 1993; Issa and Hudspeth, 1994; Liu et al., 1999). Indeed, BK channel activity has been used to monitor presynaptic calcium dynamics during NT release at *Xenopus* neuromuscular junction (Yazejian et al., 2000). Like the postsynaptic functions of calcium channels, the presynaptic association between calcium and BK channels is often preferential to a particular calcium channel subtype, irrespective of the calcium channel diversity in the neuronal type (Sah and Davies, 2002). It should be remarked that this presynaptic association seems to closely correlate with the dominant channel responsible for the neurotransmitter release process. For example, N-type channel blockers inhibit the activation of BK channels in experimental preparations where N-type channels mediate neurotransmitter release, such as sympathetic neurons and the frog neuromuscular junction (Wisgirda and Dryer, 1994; Brock and Cunnane, 1995; Ireland et al., 1998). Blockade of R-type channels in the NMJ from the P-type channel knockout animals inhibited BK currents, even though both N- and R-type channels take part in neurotransmitter release. Interestingly, only the R-type channels were found closely located to the vesicle release machinery, while N-type channels were located farther from the release sites (Urbano et al., 2001). Thus, the possibility that BK channels associate with presynaptic proteins rather than with calcium channels cannot be excluded.

6. CALCIUM CHANNELS AND CALMODULIN

6.1. Calcium-Dependent Gating

The voltage-gated L-type Ca^{2+} channel is unique among ion channels in displaying two gating properties that are regulated by the permeating ion: calcium-dependent inactivation (CDI) and calcium-dependent facilitation (CDF). These feedback mechanisms are of critical importance for regulation of the electromechanical activity of the heart and other important physiological processes. CDI helps determine the length of the cardiac action potential plateau (Beeler and Reuter, 1970), and CDF contributes to the positive force-frequency relationship of cardiac contraction (Lemaire et al., 1998).

CDI of the L-type channel refers to the accelerated rate of channel inactivation observed when currents are recorded in the presence of Ca^{2+}. For experimental purposes, the amount and rate of inactivation in the presence of Ca^{2+} is compared to the amount and rate of inactivation recorded in the presence of Ba^{2+}, another divalent cation that permeates the channel well. This is clearly demonstrated during the 400 ms test pulse to +20 mV from a holding potential of –90 mV shown in Figure 3. Inactivation can be quantified by measuring the residual current at an arbitrary time after initiation of the test pulse. The residual current at the end of the 400 ms test pulse, r_{400}, is shown for recordings in Ca^{2+} and Ba^{2+} to quantitate this measure of the accelerated, Ca^{2+}-dependent inactivation.

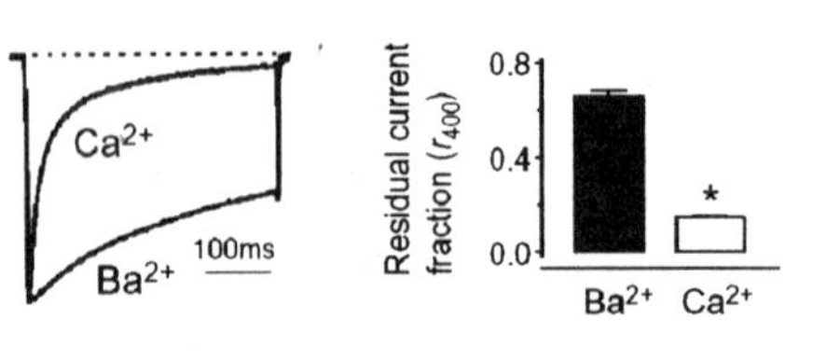

Figure 3. Calcium-dependent inactivation (CDI) of L-type channels. *Left,* representative currents recorded from *Xenopus* oocytes during test pulses to +20 mV from a holding voltage of –90 mV. I_{Ca} was scaled to peak I_{Ba}. *Right,* residual fractions of peak currents remaining at the end of a 400-ms test-pulse (r_{400}), plotted for I_{Ba} and I_{Ca}. *, p<0.001 vs. I_{Ba}, paired t-test. Adapted from Zühlke et al., 1999. Reprinted with permission from Nature publishing group.

CDF is a frequency- and Ca^{2+}-dependent potentiation of the L-type channel current (Figure 4), a process that may be tightly correlated with CDI;

Zygmunt and Maylie proposed that CDF results from *less inactivation* rather than from increased activation (Zygmunt and Maylie, 1990). Further, both of these Ca^{2+}-gating properties appear to be triggered by Ca^{2+} at or near the mouth of the channel rather than by a change in the level of cytosolic Ca^{2+}. This implies that the Ca^{2+} sensor lies within or near the pore-forming α_{1C} subunit. Channels reconstituted in lipid bilayers display CDI, even when the concentration of Ca^{2+} on the cytosolic face of the bilayer is buffered to ~ 20 nM (Haack and Rosenberg, 1994), lower than the concentration of Ca^{2+} in resting cells, and several orders of magnitude less than the calculated Ca^{2+} concentration at the cytoplasmic mouth of an open channel. Single channel recordings from ventricular myocytes also implicate Ca^{2+} entry through the channel or from a neighboring channel, rather than a change in bulk intracellular Ca^{2+} concentration, as responsible for CDI (Imredy and Yue, 1992). Intracellular Ca^{2+} buffering with EGTA—which cannot eliminate the Ca^{2+} gradient around the mouth of the channel—diminishes neither CDI nor CDF, although high concentrations of the faster Ca^{2+} chelator, BAPTA, will inhibit CDF (Hryshko and Bers, 1990). Together, these results suggest that the Ca^{2+}-sensing apparatus triggering both of these gating modes resides within or near the channel pore.

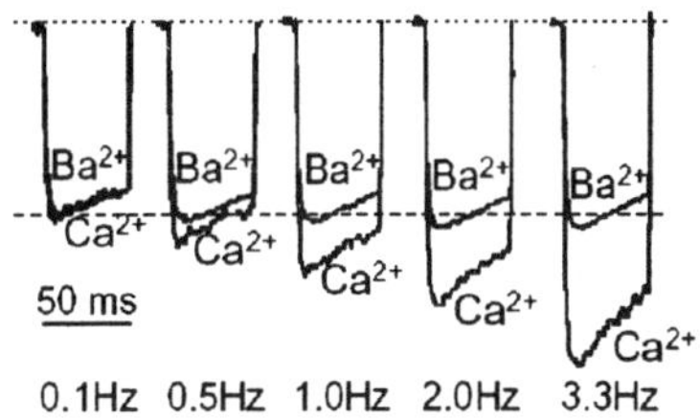

Figure 4. Calcium-dependent facilitation of the L-type channel. Frequency-dependent facilitation of I_{Ca} and I_{Ba} currents recorded in *Xenopus* oocytes, 48 hours after injection of mRNAs for the LTC α_{1C} (I1654A mutant), β_1, and $\alpha_2\delta$ subunits. Shown is the 40th of a train of 40 test pulses to +20 mV from V_h of –90 mV at 0.1 to 3.3 Hz. The I_{Ca} and I_{Ba} currents have been normalized to the current amplitude at the beginning of the train (dashed line). Recordings from the α_{1C} I1654A mutant, which displays enhanced facilitation, are shown for illustrative purposes. Adapted from Zühlke et al., 1999. Reprinted with permission from Nature publishing group.

6.2. Calmodulin Interactions With α_{1C}

Several lines of evidence from recent work have demonstrated that the Ca^{2+}-binding protein calmodulin (CaM) and a CaM binding motif (IQ motif) within the cytoplasmic C-terminus of α_{1C}, the pore-forming subunit, form the Ca^{2+}-sensor and effector domains, respectively, for both CDI and CDF. First, the IQ motif in α_{1C} is a Ca^{2+}-dependent CaM binding sequence (Peterson et al., 1999; Qin et al., 1999; Zühlke et al., 1999), in which mutations that render the channel subunit unable to bind CaM also disrupt CDI and CDF (Zühlke et al., 1999; Zühlke et al., 2000). Second, CDI can be blocked in a dominant negative fashion by CaM mutants that lack Ca^{2+}-binding in their C-terminal EF-hand domains (Peterson et al., 1999; Zühlke et al., 1999). Dominant negative calmodulins and certain mutations within the IQ motif also eliminate CDF, while other mutations within the IQ motif enhance CDF, underlining the common Ca^{2+}-sensing apparatus for both gating properties.

6.2.1. The Determinants of CDI and CDF in α_{1C}

The critical determinants for CDI and CDF lie within α_{1C}. While expression of the β subunits β1, β3, or β4 (but not β2) accelerates CDI (Noceti et al., 1998; Cens et al., 1999), CDI is observed when α_{1C} is expressed alone (Noceti et al., 1998). The critical determinants have been mapped to the large cytoplasmic tail (Figure 5; de Leon et al.,

1995; Soldatov et al., 1997; Soldatov et al., 1998; Zühlke and Reuter, 1998). Replacement of a 53 amino acid domain containing a consensus EF-hand (Babitch, 1990) with the homologous region of α_{1E} ablates CDI (de Leon et al., 1995). Downstream of the EF-hand, an 80 amino acid region has also been recognized as essential. The $\alpha_{1C,86}$ splice variant, in which a non-identical 81 amino acid segment replaces those 80 amino acids in the conventional α_{1C} subunit (labeled $\alpha_{1C,77}$), inactivates 10-fold faster when Ba^{2+} permeates the channel, yet displays no additional increase in its rate of inactivation when Ca^{2+} goes through the channel (Soldatov et al., 1997; see Lipscombe and Castiglioni, this volume). Deletion of this segment in $\alpha_{1C,77}$ also accelerates inactivation in Ba^{2+} and abolishes any increased rate of inactivation in Ca^{2+} (Zühlke and Reuter, 1998). This 80 amino acid segment has been extensively dissected to reveal several domains that contribute to CDI (Soldatov et al., 1998; Zühlke and Reuter, 1998). Prominent among these is the CaM binding IQ motif (Zühlke and Reuter, 1998). Analysis of the role of mutations in the IQ motif then revealed a role for CDF as well (Zühlke et al., 1999).

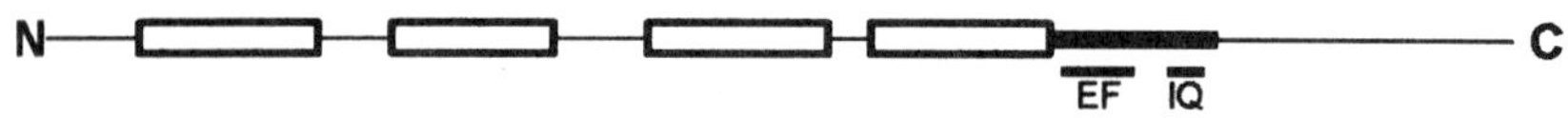

Figure 5. Determinants of calcium-dependent inactivation in α_{1C}. Schematic of α_{1C}. Open boxes represent transmembrane domain segments I-IV. Thick line represents the 80 amino acid segment that differs between the $\alpha_{1C,77}$ and $\alpha_{1C,86}$ splice variants. The relative positions of the EF-hand and IQ domain are indicated.

6.2.2. Constitutive CaM-α_{1C} Interactions

A constitutive interaction between CaM and the channel had been postulated to keep CaM strategically positioned to sense Ca^{2+} entry and to induce inactivation rapidly, and recent work has confirmed this hypothesis. Initial characterization of the IQ motif suggested that it might form a type of consensus calmodulin (CaM) binding motif that interacts preferentially with apoCaM rather than Ca^{2+}/CaM (Rhoads and Friedberg, 1997), and therefore serve also as a constitutive CaM tether,. However, the IQ motif could not support apoCaM interactions nor binding to the dominant negative CaMs *in vitro*, effectively ruling out its role as a tethering site (Pitt et al., 2001). Another consensus CaM binding site conforming to a 1-8-14 Type A motif (Rhoads and Friedberg, 1997), was identified within the α_{1C} C-terminal tail *in silico* (Pitt et al., 2001). Interestingly, this site contains the short sequence ^{1572}IKTEG^{1576}, previously identified as essential for CDI by analysis of deletions or domain swaps (de Leon et al., 1995; Soldatov et al., 1997; Zühlke and Reuter, 1998; Peterson et al., 2000). Using the dominant negative CaM_{1234}, GST-pull down experiments confirmed that the proximal portion of the C-terminal tail contained a CaM tethering site (Pitt et al., 2001). Subsequently, fluorescence resonance energy transfer (FRET) measurements between CFP-labeled CaM and YFP-labeled α_{1C} confirmed this finding (Erickson et al., 2001). An interesting aspect to this tethering of CaM to α_{1C} is that there appears to be a Ca^{2+}-dependence to the interaction that reflects a structural requirement for Ca^{2+} binding at low nanomolar concentrations to a site within this proximal portion of the cytoplasmic tail of α_{1C} (Romanin et al., 2000; Pitt et al., 2001). These results are consistent with the bilayer experiments performed by Haack and Rosenberg (Haack and Rosenberg, 1994), in which they observed the loss of CDI in some experiments after buffering $[Ca^{2+}]_i$ to ~ 20 nM;

working close to the calculated K_d for this structural requirement for Ca^{2+}, they may have lost the tethered CaM in those experiments which failed to display CDI.

6.2.3. Non-Contiguous Sequences Involved in CaM Tethering

Using a series of peptides to search for CaM binding within this identified tethering region, the constitutive CaM binding sites within α_{1C} have been mapped to two non-contiguous sequences between the recognized EF hand and the IQ motif. One region contained the signature sequence IKTEG, and the other included another sequence, LLDQV, also previously implicated as essential for CDI (Zühlke and Reuter, 1998). Fulfilling their requirements to serve as CaM tethering sites, both of these regions were demonstrated to be able to interact with CaM at Ca^{2+} levels below the basal levels found in the cell, clearly distinguishing these sites from the IQ motif, for which a higher Ca^{2+} level is required for CaM interaction.

6.2.4. Lobe-Specific Interactions of CaM with α_{1C}

The finding that multiple, non-contiguous sequences in the C-terminal region interact with CaM, and do so with different Ca^{2+} requirements, suggested that certain sequences might accelerate inactivation by binding CaM, and so tether CaM in readiness for such action. A clear criterion for the accelerator site was established by previous experiments with CaM mutants in which various Ca^{2+}-binding EF hands were disabled (Peterson et al., 1999). CDI was spared following expression of CaM_{12}, a CaM mutant in which the two N-terminal EF hands were mutated to Ca^{2+}-insensitive forms. In contrast CaM_{34}, a CaM mutant in which the two C-terminal EF hands are disabled, ablated CDI in a dominant negative manner. These results predicted that the accelerator site should interact preferentially with the C-terminal lobe of CaM, the region where wild-type Ca^{2+} binding is critical, rather than the N-terminal lobe, where loss of Ca^{2+} binding can be tolerated. On the other hand, the findings with dominant negative CaMs made no specific prediction about which domain(s) of CaM should interact with the tethering site, so long as this binding occurs at resting Ca^{2+} levels (Peterson et al., 1999).

Using either the amino- or carboxyl-lobes of CaM generated by limited proteolysis (Newton et al., 1984) it was determined that CaM binding to the IQ motif showed a CaM carboxyl-lobe preference, providing a satisfying explanation for the earlier experiments in which co-expression of CaM_{34} but not CaM_{12} blocked CDI (Peterson et al., 1999). The tethering site containing the other consensus CaM binding motif and the IKTEG signature sequence demonstrated a preference for interaction with one lobe of CaM over the other, but in this case it was the N-terminal lobe of CaM rather than the C-terminal lobe. Lobe-specific regulation by CaM finds precedent in the control of the Ca^{2+}-activated K^+ current in *Paramecium* (Kink et al., 1990), the Ca^{2+}-activated K^+ current generated by SK channels (Keen et al., 1999), and more recently in the differential regulation of CDI and CDF in P/Q-type Ca^{2+} channels (DeMaria et al., 2001).

While CDI and CDF of L-type Ca^{2+} channels share many similarities, there are some essential differences. For example, certain mutations within the IQ motif that do not abolish Ca^{2+}/CaM binding will still support (or even augment) CDF, while abolishing CDI (Zühlke et al., 1999; Zühlke et al., 2000). Second, CDF is a frequency-dependent process, requiring multiple depolarizations within a range of permissible frequencies, while CDI can be revealed within a single depolarization. The frequency detector mechanism for CDF has not yet been identified. Further, CDI appears independent of cytoplasmic components, as demonstrated by bilayer experiments, while single channel recordings from excised cardiac myocyte patches showed that CDF was only prominent after the addition of another component, activated Ca^{2+}/CaM-dependent kinase II (CaMKII) (Dzhura et al., 2000).

6.3. Role of CaM Kinase II: the Frequency Detector for Calcium-Dependent Facilitation?

Several additional lines of evidence suggest that activation of CaMKII plays an essential role in CDF. In cardiac myocytes, pharmacological inhibition of CaMKII abolishes CDF (Xiao et al., 1994; Yuan and Bers, 1994), and immunocytochemistry suggests that the channel and the kinase are co-localized (Xiao et al., 1994). Facilitation was prevented when a non-hydrolyzable ATP analog was added in addition to activated CaMKII to the cytoplasmic face of cardiac myocyte membranes, implicating a phosphorylation event.

The CaMKII holoenzyme is perfectly poised to serve as the calcium spike frequency detector due to its activation and autoregulation. Ca^{2+}/CaM binding disinhibits a pseudo-substrate domain to produce kinase activation, a process that decays in low calcium after CaM dissociates from the enzyme. However, autophosphorylation of Thr^{286} increases the kinase's affinity for CaM by over 10,000-fold ("CaM trapping") and renders subsequent kinase activity independent (autonomous) of Ca^{2+} and CaM after the initial stimulus (Meyer et al., 1992). The autophosphorylation process requires coincident binding of Ca^{2+}/CaM molecules to the subunit acting as kinase as well as the subunit acting as substrate (Hanson et al., 1994; Rich and Schulman, 1998), making this process an intraholoenzyme autophosphorylation mechanism (Bradshaw et al., 2002) on Thr^{286} that is highly cooperative. *In vitro* experiments have shown that CaMKII increases its activity in response to more rapid Ca^{2+} oscillations (De Koninck and Schulman, 1998) and that the isoforms of CaMKII possess different calcium spike frequency sensitivities (Bayer et al., 2002).

There is a family of closely related CaMKII isoforms that are expressed throughout the body. In the brain, where CaMKII is particularly enriched in the postsynaptic density, it comprises 1-2% of the total protein (Erondu and Kennedy, 1985) and the α and β isoforms predominate. The persistent activity afforded by autophosphorylation and CaM trapping has been suggested as an essential underpinning of activity-dependent changes in synaptic strength, and therefore has made CaMKII a candidate for the molecular basis of memory (Lisman et al., 2002). The closely related δ isoform is most abundant in the heart (Edman and Schulman, 1994), where, in addition to its role in regulating CDF, CaMKII has also been implicated as able to induce cardiac hypertrophy and dilated cardiomyopathy (Zhang et al., 2002). Recent studies of CDF of single channels in cardiomyocyte membranes, have utilized exogenously added CaMKII-α (Dzhura et al., 2000; Wu et al., 2001), underlying the similarity in terms of structure and regulation of these two isoforms.

As a kinase, CaMKII is an attractive candidate for the CDF effector. Phosphorylation of the L-type channel, such as after β-agonist stimulation of PKA, produces increases in L-type Ca^{2+} current (Reuter and Scholz, 1977; Yue et al., 1990), similar to that seen after the addition of CaMKII to cardiomyocyte membranes (Dzhura et al., 2000). Src phosphorylation of the neuronal LTC isoform, in response to IGF-1 stimulation, also leads to potentiation of the LTC current (Bence-Hanulec et al., 2000). While the mechanism(s) by which PKA and Src phosphorylation lead to potentiation of current have not been determined, it is interesting to note that the targets of both kinases are residues within the cytoplasmic C-terminal tail of α_{1C}. Phosphorylation of the cytoplasmic C-terminal tail may be a general mechanism by which the LTC current can become potentiated. How the initial Ca^{2+} sensing event, by the constitutively bound CaM and the IQ motif within the α_{1C} subunit, activates a second Ca^{2+}/CaM regulated event, namely activation of CaMKII, to trigger CDF remains unclear.

6.3.1. The α_{1C} Subunit is a CaMKII Substrate

While others have proposed that CaMKII phosphorylation of the α_{1C} subunit underlies CDF (Xiao et al., 1994; Yuan and Bers, 1994; Dzhura et al., 2000), demonstration that α_{1C} is a CaMKII substrate has been shown only recently. Hudmon et al. (submitted) tested whether CaMKII could phosphorylate any of the intracellular domains of α_{1C} *in vitro* by generating GST fusion proteins for each of the intracellular domains and two fragments of the large cytoplasmic tail of the α_{1C} subunit and using them in an *in vitro* kinase assay. Significant incorporation of ^{32}P was observed for both the N-terminal fusion protein of α_{1C} and the C-terminal fusion protein containing amino acids 1669-2171, but not for any of the intracellular loops connecting the transmembrane domains. These data indicate that the loci for CaMKII action via phosphorylation of α_{1C} likely reside within these regions, both of which have previously been shown to be targets of kinases that modulate channel function (Rotman et al., 1995; Bence-Hanulec et al., 2000; McHugh et al., 2000). In addition, the C-terminus contains critical molecular determinants involved with both CDI and CDF (Zühlke et al., 1999; Zühlke et al., 2000).

6.3.2. CaMKII Associates with the L-type Channel

Since the onset of CDF is rapid (less than several hundred milliseconds are necessary for the manifestation of CDF), the pre-association of CaMKII with the LTC might be expected in order to assure this rapid response. In addition to the previous data suggesting that the kinase and the LTC are co-localized at the sarcolemmal membrane in cardiac myocytes (Xiao et al., 1994), recent data has suggested that α-CaMKII can directly associate with the NMDA receptor, presumably resulting in bi-directional regulation of the channel and the kinase (Bayer et al., 2001). Using both confocal microscopy and immunoprecipitation, Hudmon et al. (submitted) demonstrated that the α_{1C} subunit and CaMKII are associated in cells. In further studies, a direct association between multiple intracellular domains of α_{1C} and CaMKII was demonstrated. Depending on the state of activation of CaMKII (native vs. Ca^{2+}/CaM-activated vs. autophosphorylated), different intracellular domains of α_{1C} supported binding, implicating a dynamic regulation between CaMKII activation and CaMKII-α_{1C} interaction.

6.3.3. Calcium-Dependent Facilitation Depends Upon CaMKII interaction with α_{1C}

Analysis of the binding determinants within one of these α_{1C} domains, the C-terminus, is particularly interesting for several reasons. The C-terminus is one of the targets for CaMKII phosphorylation; it binds the autophosphorylated form CaMKII, the form that was able to support facilitation of single channels (Dzhura et al., 2000); and several distinct functions, including CDF, have been mapped to different domains within the tail (Hell et al., 1995; Hell et al., 1996; Gao et al., 1997; Gao et al., 2000; Zühlke et al., 2000;

Gao et al., 2001; Pitt et al., 2001). Mapping this site revealed that the major binding determinants for CaMKII resided just proximal to, but separate from, the IQ motif (Hudmon et al., submitted). Mutation of 6 amino acids within this CaMKII binding site blocked CDF in L-type Ca^{2+} channels expressed in *Xenopus* oocytes (Figure 6).

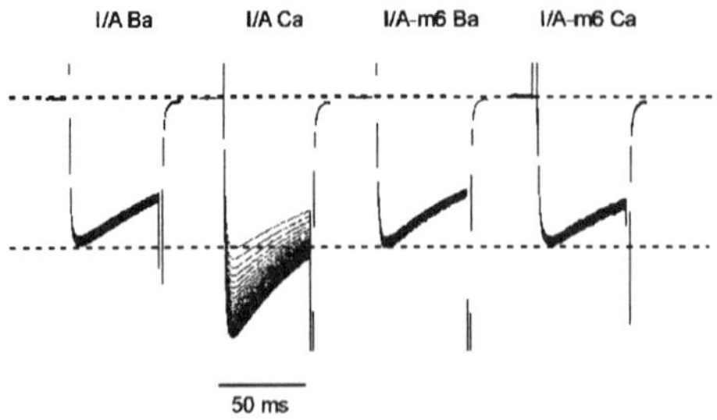

Figure 6. Mutation of the CaMKII interaction site within α_{1C} abolishes calcium-dependent facilitation. I_{Ba} and scaled I_{Ca} traces during a train of 40 test pulses to +20 mV from V_h of –90 mV at 3.3 Hz, recorded from oocytes expressing α_{1C} I1654A (I/A) or α_{1C} harboring a 6 amino acid mutation within the CaMKII binding site in the α_{1C} C-terminus. I/A represents the I1654A mutation that reveals robust CDF in *Xenopus* oocytes.

These results imply that CaMKII might form a stable (irreversible) interaction with α_{1C}, localizing the kinase near the site of the Ca^{2+} stimulus and near its substrate, similar to the way in which an AKAP situates PKA near the L-type channel for β-adrenergic mediated effects (Gao et al., 1997). Also, the cytoplasmic tail of α_{1A}, the pore-forming subunit of the P/Q-type Ca^{2+} channel, which has also recently been shown to display CDF (Lee et al., 1999; Lee et al., 2000; DeMaria et al., 2001), displays a high degree of homology with α_{1C} in this CaMKII binding region. Indeed, 6 of 7 amino acids comprising the CaMKII interaction site on α_{1C} are identical between the two α_1 subunits, suggesting that CaMKII interaction might regulate CDF in P/Q-type channels, too. As a first proof of this hypothesis, Hudmon et al. (submitted) demonstrated that this same region of α_{1A} binds to CaMKII in a fashion similar to how CaMKII binds to α_{1C}.

ACKNOWLEDGMENTS

Supported by Excellence in Academic Medicine Grant, Loyola University Medical Center (E.S.P.R.) and NIH NS41302 and Whitehall Foundation grants (P.G.M.).

REFERENCES

Adams, P. R., Constanti, A., Brown, D. A. and Clark, R. B., 1982, Intracellular Ca^{2+} activates a fast voltage-sensitive K^+ current in vertebrate sympathetic neurones, *Nature* **296**:746-749.

Adelman, J. P., Shen, K. Z., Kavanaugh, M. P., Warren, R. A., Wu, Y. N., Lagrutta, A., Bond, C. T. and North, R. A., 1992, Calcium-activated potassium channels expressed from cloned complementary DNAs, *Neuron* **9**:209-216.

Adler, E. M., Augustine, G. J., Duffy, S. N. and Charlton, M. P., 1991, Alien intracellular calcium chelators attenuate neurotransmitter release at the squid giant synapse, *J Neurosci* **11**:1496-1507.

Ahlijanian, M. K., Westenbroek, R. E. and Catterall, W. A., 1990, Subunit structure and localization of dihydropyridine-sensitive calcium channels in mammalian brain, spinal cord, and retina, *Neuron* **4**:819-832.

Ahn, S., Ginty, D. D. and Linden, D. J., 1999, A late phase of cerebellar long-term depression requires activation of CaMKIV and CREB, *Neuron* **23**:559-568.

Aiyar, J., Withka, J. M., Rizzi, J. P., Singleton, D. H., Andrews, G. C., Lin, W., Boyd, J., Hanson, D. C., Simon, M., Dethlefs, B. and et al., 1995, Topology of the pore-region of a K^+ channel revealed by the NMR-derived structures of scorpion toxins, *Neuron* **15**:1169-1181.

Alvarez, J., Montero, M. and Garcia-Sancho, J., 1992, High affinity inhibition of Ca^{2+}-dependent K^+ channels by cytochrome P-450 inhibitors, *J Biol Chem* **267**:11789-11793.

Anderson, L. E. and Seybold, V. S., 2000, Phosphorylated cAMP response element binding protein increases in neurokinin-1 receptor-immunoreactive neurons in rat spinal cord in response to formalin-induced nociception, *Neurosci Lett* **283**:29-32.

Aniksztejn, L. and Ben, A. Y., 1991, Novel form of long-term potentiation produced by a K^+ channel blocker in the hippocampus, *Nature* **349**:67-69.

Atlas, D., 2001, Functional and physical coupling of voltage-sensitive calcium channels with exocytotic proteins: ramifications for the secretion mechanism, *J Neurochem* 77:972-985.

Avery, R. B. and Johnston, D., 1996, Multiple channel types contribute to the low-voltage-activated calcium current in hippocampal CA3 pyramidal neurons, *J Neurosci* **16**:5567-5582.

Babitch, J., 1990, Channel hands [letter], *Nature* **346**:321-322.

Bading, H., Ginty, D. D. and Greenberg, M. E., 1993, Regulation of gene expression in hippocampal neurons by distinct calcium signaling pathways, *Science* **260**:181-186.

Barrot, M., Olivier, J. D., Perrotti, L. I., DiLeone, R. J., Berton, O., Eisch, A. J., Impey, S., Storm, D. R., Neve, R. L., Yin, J. C., Zachariou, V. and Nestler, E. J., 2002, CREB activity in the nucleus accumbens shell controls gating of behavioral responses to emotional stimuli, *Proc Natl Acad Sci U S A* **99**:11435-11440.

Bauer, E. P., Schafe, G. E. and LeDoux, J. E., 2002, NMDA receptors and L-type voltage-gated calcium channels contribute to long-term potentiation and different components of fear memory formation in the lateral amygdala, *J Neurosci* **22**:5239-5249.

Bayer, K. U., De Koninck, P., Leonard, A. S., Hell, J. W. and Schulman, H., 2001, Interaction with the NMDA receptor locks CaMKII in an active conformation, *Nature* **411**:801-805.

Bayer, K. U., De Koninck, P. and Schulman, H., 2002, Alternative splicing modulates the frequency-dependent response of CaMKII to Ca^{2+} oscillations, *Embo J* **21**:3590-3597.

Beeler, G. W. and Reuter, H., 1970, Membrane calcium current in ventricular myocardial fibres, *J Physiol* **207**:191-209.

Bence-Hanulec, K. K., Marshall, J. and Blair, L. A., 2000, Potentiation of neuronal L calcium channels by IGF-1 requires phosphorylation of the $\alpha 1$ subunit on a specific tyrosine residue, *Neuron* **27**:121-131.

Bennett, M. K., Calakos, N., Kreiner, T. and Scheller, R. H., 1992, Synaptic vesicle membrane proteins interact to form a multimeric complex, *J Cell Biol* **116**:761-775.

Bennett, M. K., 1997, Ca^{2+} and the regulation of neurotransmitter secretion, *Curr Opin Neurobiol* 7:316-322.

Berke, J. D. and Hyman, S. E., 2000, Addiction, dopamine, and the molecular mechanisms of memory, *Neuron* **25**:515-532.

Bers, D. M., 2001, *Excitation-contraction coupling and cardiac contractile force*, Kluwer Academic Publishers, Dordrecht.

Bezprozvanny, I. B., Zhong, P., Scheller, R. H. and Tsien, R. W., 2000, Molecular determinants of the functional interaction between syntaxin and N-type Ca^{2+} channel gating, *In preparation*

Bi, G. Q. and Poo, M. M., 1998, Synaptic modifications in cultured hippocampal neurons: dependence on spike timing, synaptic strength, and postsynaptic cell type, *J Neurosci* **18**:10464-10472.

Biederer, T. and Sudhof, T. C., 2000, Mints as adaptors. Direct binding to neurexins and recruitment of munc18, *J Biol Chem* **275**:39803-39806.

Biederer, T., Sara, Y., Mozhayeva, M., Atasoy, D., Liu, X., Kavalali, E. T. and Sudhof, T. C., 2002, SynCAM, a synaptic adhesion molecule that drives synapse assembly, *Science* **297**:1525-1531.

Bishop, D. L. and Milton, R. L., 1998, Nimodipine suppresses preferential reinnervation of mouse soleus muscles by slow alpha-motoneurons, *Exp Neurol* **154**:366-370.

Bito, H., Deisseroth, K. and Tsien, R. W., 1996, CREB phosphorylation and dephosphorylation: a Ca^{2+}- and stimulus duration-dependent switch for hippocampal gene expression, *Cell* **87**:1203-1214.

Bito, H., Deisseroth, K. and Tsien, R. W., 1997, Ca2+-dependent regulation in neuronal gene expression, *Curr Opin Neurobiol* 7:419-429.

Bito, H., Deisseroth, K. and Tsien, R. W., 1999, Activity-dependent regulation of communication from synapse to nucleus, Proceedings In: 22nd International Foundation, Taniguchi Symposium.

Blatz, A. L. and Magleby, K. L., 1984, Ion conductance and selectivity of single calcium-activated potassium channels in cultured rat muscle, *J Gen Physiol* **84**:1-23.

Blatz, A. L. and Magleby, K. L., 1986, Single apamin-blocked Ca-activated K^+ channels of small conductance in cultured rat skeletal muscle, *Nature* **323**:718-720.

Bliss, T. V. and Lømo, T., 1973, Long-lasting potentiation of synaptic transmission in the dentate area of the anaesthetized rabbit following stimulation of the perforant path, *J Physiol* **232**:331-356.

Bolshakov, V. Y. and Siegelbaum, S. A., 1994, Postsynaptic induction and presynaptic expression of hippocampal long-term depression, *Science* **264**:1148-1152.

Borroni, A. M., Fichtenholtz, H., Woodside, B. L. and Teyler, T. J., 2000, Role of voltage-dependent calcium channel long-term potentiation (LTP) and NMDA LTP in spatial memory, *J Neurosci* **20**:9272-9276.

Bradshaw, J. M., Hudmon, A. and Schulman, H., 2002, Chemical quenched flow kinetic studies indicate an intraholoenzyme autophosphorylation mechanism for Ca^{2+}/calmodulin-dependent protein kinase II, *J Biol Chem* **277**:20991-20998.

Brenner, R., Jegla, T. J., Wickenden, A., Liu, Y. and Aldrich, R. W., 2000, Cloning and functional characterization of novel large conductance calcium-activated potassium channel beta subunits, hKCNMB3 and hKCNMB4, *J Biol Chem* **275**:6453-6461.

Brock, J. A. and Cunnane, T. C., 1995, Effects of Ca^{2+} and K^+ channel blockers on nerve impulses recorded from guinea-pig postganglionic sympathetic nerve terminals, *J Physiol* **489** (**Pt 2**):389-402.

Brosenitsch, T. A., Salgado-Commissariat, D., Kunze, D. L. and Katz, D. M., 1998, A role for L-type calcium channels in developmental regulation of transmitter phenotype in primary sensory neurons, *J Neurosci* **18**:1047-1055.

Burgoyne, R. D. and Morgan, A., 1998, Calcium sensors in regulated exocytosis, *Cell Calcium* **24**:367-376.

Butz, S., Okamoto, M. and Sudhof, T. C., 1998, A tripartite protein complex with the potential to couple synaptic vesicle exocytosis to cell adhesion in brain, *Cell* **94**:773-782.

Cassell, J. F., Clark, A. L. and McLachlan, E. M., 1986, Characteristics of phasic and tonic sympathetic ganglion cells of the guinea-pig, *J Physiol* **372**:457-483.

Cavus, I. and Teyler, T., 1996, Two forms of long-term potentiation in area CA1 activate different signal transduction cascades, *J Neurophysiol* **76**:3038-3047.

Cens, T., Restituito, S., Galas, S. and Charnet, P., 1999, Voltage and calcium use the same molecular determinants to inactivate calcium channels, *J Biol Chem* **274**:5483-5490.

Charvin, N., L'Eveque, C., Walker, D., Berton, F., Raymond, C., Kataoka, M., Shoji-Kasai, Y., Takahashi, M., De Waard, M. and Seagar, M. J., 1997, Direct interaction of the calcium sensor protein synaptotagmin I with a cytoplasmic domain of the α1A subunit of the P/Q-type calcium channel, *Embo J* **16**:4591-4596.

Chen, S., Zheng, X., Schulze, K. L., Morris, T., Bellen, H. and Stanley, E. F., 2002, Enhancement of presynaptic calcium current by cysteine string protein, *J Physiol* **538**:383-389.

Christie, B. R. and Abraham, W. C., 1994, L-type voltage-sensitive calcium channel antagonists block heterosynaptic long-term depression in the dentate gyrus of anaesthetized rats, *Neurosci Lett* **167**:41-45.

Christie, B. R., Schexnayder, L. K. and Johnston, D., 1997, Contribution of voltage-gated Ca^{2+} channels to homosynaptic long-term depression in the CA1 region in vitro, *J Neurophysiol* 77:1651-1655.

Cigola, E., Volpe, B. T., Lee, J. W., Franzen, L. and Baker, H., 1998, Tyrosine hydroxylase expression in primary cultures of olfactory bulb: role of L-type calcium channels, *J Neurosci* **18**:7638-7649.

Coppola, T., Magnin-Luthi, S., Perret-Menoud, V., Gattesco, S., Schiavo, G. and Regazzi, R., 2001, Direct interaction of the Rab3 effector RIM with Ca^{2+} channels, SNAP-25, and synaptotagmin, *J Biol Chem* **276**:32756-32762.

Craven, S. E. and Bredt, D. S., 1998, PDZ proteins organize synaptic signaling pathways, *Cell* **93**:495-498.

Daniel, H., Hemart, N., Jaillard, D. and Crepel, F., 1992, Coactivation of metabotropic glutamate receptors and of voltage-gated calcium channels induces long-term depression in cerebellar Purkinje cells in vitro, *Exp Brain Res* **90**:327-331.

Davis, A. F., Bai, J., Fasshauer, D., Wolowick, M. J., Lewis, J. L. and Chapman, E. R., 1999, Kinetics of synaptotagmin responses to Ca^{2+} and assembly with the core SNARE complex onto membranes, *Neuron* **24**:363-376.

De Koninck, P. and Schulman, H., 1998, Sensitivity of CaM kinase II to the frequency of Ca^{2+} oscillations, *Science* **279**:227-230.

de Leon, M., Wang, Y., Jones, L., Perez-Reyes, E., Wei, X., Soong, T. W., Snutch, T. P. and Yue, D. T., 1995, Essential Ca^{2+}-binding motif for Ca^{2+}-sensitive inactivation of L-type Ca^{2+} channels, *Science* **270**:1502-1506.

Degtiar, V. E., Scheller, R. H. and Tsien, R. W., 2000, Syntaxin Modulation of Slow Inactivation Of N-type Calcium Channels, *Submitted*

Deisseroth, K., Bito, H. and Tsien, R. W., 1996, Signaling from synapse to nucleus: postsynaptic CREB phosphorylation during multiple forms of hippocampal synaptic plasticity, *Neuron* **16**:89-101.

Deisseroth, K., Heist, E. K. and Tsien, R. W., 1998, Translocation of calmodulin to the nucleus supports CREB phosphorylation in hippocampal neurons, *Nature* **392**:198-202.

Deisseroth, K. and Tsien, R. W., 2002, Dynamic multiphosphorylation passwords for activity-dependent gene expression, *Neuron* **34**:179-182.

DeMaria, C. D., Soong, T. W., Alseikhan, B. A., Alvania, R. S. and Yue, D. T., 2001, Calmodulin bifurcates the local Ca2+ signal that modulates P/Q-type Ca^{2+} channels, *Nature* **411**:484-489.

Dodge, F., Jr. and Rahamimoff, R., 1967, Co-operative action of calcium ions in transmitter release at the neuromuscular junction, *J Physiol* **193**:419-432.

Dolmetsch, R. E., Pajvani, V., Fife, K., Spotts, J. and Greenberg, M. E., 2001, Signaling to the nucleus by and L-type calcium channel-calmodulin complex through the MAP kinase pathway, *Science* **294**:333-339.

Dudek, S. M. and Fields, R. D., 2002, Somatic action potentials are sufficient for late-phase LTP-related cell signaling, *Proc Natl Acad Sci U S A* **99**:3962-3967.

Dunn, P. M., 1994, Dequalinium, a selective blocker of the slow afterhyperpolarization in rat sympathetic neurones in culture, *Eur J Pharmacol* **252**:189-194.

Dworetzky, S. I., Trojnacki, J. T. and Gribkoff, V. K., 1994, Cloning and expression of a human large-conductance calcium-activated potassium channel, *Brain Res Mol Brain Res* **27**:189-193.

Dzhura, I., Wu, Y., Colbran, R. J., Balser, J. R. and Anderson, M. E., 2000, Calmodulin kinase determines calcium-dependent facilitation of L-type calcium channels, *Nat Cell Biol* **2**:173-177.

Edman, C. F. and Schulman, H., 1994, Identification and characterization of delta B-CaM kinase and delta C-CaM kinase from rat heart, two new multifunctional Ca2+/calmodulin- dependent protein kinase isoforms, *Biochim Biophys Acta* **1221**:89-101.

Eilers, J., Plant, T. and Konnerth, A., 1996, Localized calcium signalling and neuronal integration in cerebellar Purkinje neurones, *Cell Calcium* **20**:215-226.

el Far, O., Charvin, N., Leveque, C., Martin-Moutot, N., Takahashi, M. and Seagar, M. J., 1995, Interaction of a synaptobrevin (VAMP)-syntaxin complex with presynaptic calcium channels, *Febs Lett* **361**:101-105.

Engel, J., Schultens, H. A. and Schild, D., 1999, Small conductance potassium channels cause an activity-dependent spike frequency adaptation and make the transfer function of neurons logarithmic, *Biophys J* **76**:1310-1319.

Enslen, H., Tokumitsu, H. and Soderling, T. R., 1995, Phosphorylation of CREB by CaM-kinase IV activated by CaM-kinase IV kinase, *Biochem Biophys Res Commun* **207**:1038-1043.

Erickson, M. G., Alseikhan, B. A., Peterson, B. Z. and Yue, D. T., 2001, Preassociation of calmodulin with voltage-gated Ca^{2+} channels revealed by FRET in single living cells, *Neuron* **31**:973-985.

Erondu, N. E. and Kennedy, M. B., 1985, Regional distribution of type II Ca^{2+}/calmodulin-dependent protein kinase in rat brain, *J Neurosci* **5**:3270-3277.

Evers, M. R., Salmen, B., Bukalo, O., Rollenhagen, A., Bosl, M. R., Morellini, F., Bartsch, U., Dityatev, A. and Schachner, M., 2002, Impairment of L-type Ca^{2+} channel-dependent forms of hippocampal synaptic plasticity in mice deficient in the extracellular matrix glycoprotein tenascin-C, *Journal of Neuroscience* **22**:7177-7194.

Fanger, C. M., Ghanshani, S., Logsdon, N. J., Rauer, H., Kalman, K., Zhou, J., Beckingham, K., Chandy, K. G., Cahalan, M. D. and Aiyar, J., 1999, Calmodulin mediates calcium-dependent activation of the intermediate conductance KCa channel, IKCa1, *J Biol Chem* **274**:5746-5754.

Finch, E. A. and Augustine, G. J., 1998, Local calcium signalling by inositol-1,4,5-trisphosphate in Purkinje cell dendrites, *Nature* **396**:753-756.

Galvez, A., Gimenez-Gallego, G., Reuben, J. P., Roy-Contancin, L., Feigenbaum, P., Kaczorowski, G. J. and Garcia, M. L., 1990, Purification and characterization of a unique, potent, peptidyl probe for the high conductance calcium-activated potassium channel from venom of the scorpion Buthus tamulus, *J Biol Chem* **265**:11083-11090.

Gao, T., Yatani, A., Dell'Acqua, M. L., Sako, H., Green, S. A., Dascal, N., Scott, J. D. and Hosey, M. M., 1997, cAMP-dependent regulation of cardiac L-type Ca^{2+} channels requires membrane targeting of PKA and phosphorylation of channel subunits, *Neuron* **19**:185-196.

Gao, T., Bunemann, M., Gerhardstein, B. L., Ma, H. and Hosey, M. M., 2000, Role of the C terminus of the alpha 1C ($Ca_V1.2$) subunit in membrane targeting of cardiac L-type calcium channels, *J Biol Chem* **275**:25436-25444.

Gao, T., Cuadra, A. E., Ma, H., Bunemann, M., Gerhardstein, B. L., Cheng, T., Eick, R. T. and Hosey, M. M., 2001, C-terminal fragments of the alpha 1C ($Ca_V1.2$) subunit associate with and regulate L-type calcium channels containing C-terminal-truncated alpha 1C subunits, *J Biol Chem* **276**:21089-21097.

Garcia-Calvo, M., Leonard, R. J., Novick, J., Stevens, S. P., Schmalhofer, W., Kaczorowski, G. J. and Garcia, M. L., 1993, Purification, characterization, and biosynthesis of margatoxin, a component of *Centruroides margaritatus* venom that selectively inhibits voltage-dependent potassium channels, *J Biol Chem* **268**:18866-18874.

Garcia-Calvo, M., Knaus, H. G., McManus, O. B., Giangiacomo, K. M., Kaczorowski, G. J. and Garcia, M. L., 1994, Purification and reconstitution of the high-conductance, calcium-activated potassium channel from tracheal smooth muscle, *J Biol Chem* **269**:676-682.

Gardos, G., 1958, The function of calcium in the potassium permeability of human erythrocytes, *Biochem Biophys Acta* **30**:653-654.

Gau, D., Lemberger, T., von Gall, C., Kretz, O., Le Minh, N., Gass, P., Schmid, W., Schibler, U., Korf, H. W. and Schutz, G., 2002, Phosphorylation of CREB Ser142 regulates light-induced phase shifts of the circadian clock, *Neuron* **34**:245-253.

Geppert, M. and Sudhof, T. C., 1998, RAB3 and synaptotagmin: the yin and yang of synaptic membrane fusion, *Annu Rev Neurosci* **21**:75-95.

Graef, I. A., Mermelstein, P. G., Stankunas, K., Neilson, J. R., Deisseroth, K., Tsien, R. W. and Crabtree, G. R., 1999, L-type calcium channels and GSK-3 regulate the activity of NF-ATc4 in hippocampal neurons, *Nature* **401**:703-708.

Gribkoff, V. K., Starrett, J. E., Jr. and Dworetzky, S. I., 2001, Maxi-K potassium channels: form, function, and modulation of a class of endogenous regulators of intracellular calcium, *Neuroscientist* **7**:166-177.

Grissmer, S., Nguyen, A. N., Aiyar, J., Hanson, D. C., Mather, R. J., Gutman, G. A., Karmilowicz, M. J., Auperin, D. D. and Chandy, K. G., 1994, Pharmacological characterization of five cloned voltage-gated K^+ channels, types Kv1.1, 1.2, 1.3, 1.5, and 3.1, stably expressed in mammalian cell lines, *Mol Pharmacol* **45**:1227-1234.

Grover, L. M. and Teyler, T. J., 1990, Two components of long-term potentiation induced by different patterns of afferent activation, *Nature* **347**:477-479.

Grover, L. M. and Teyler, T. J., 1995, Different mechanisms may be required for maintenance of NMDA receptor-dependent and independent forms of long-term potentiation, *Synapse* **19**:121-133.

Gustavsson, P., Willing, T. N., van Haeringen, A., Tchernia, G., Dianzani, I., Donner, M., Elinder, G., Henter, J. I., Nilsson, P. G., Gordon, L., Skeppner, G., van't Veer-Korthof, L., Kreuger, A. and Dahl, N., 1997, Diamond-Blackfan anaemia: genetic homogeneity for a gene on chromosome 19q13 restricted to 1.8 Mb, *Nat Genet* **16**:368-371.

Haack, J. A. and Rosenberg, R. L., 1994, Calcium-dependent inactivation of L-type calcium channels in planar lipid bilayers, *Biophys J* **66**:1051-1060.

Hanson, P. I., Meyer, T., Stryer, L. and Schulman, H., 1994, Dual role of calmodulin in autophosphorylation of multifunctional CaM kinase may underlie decoding of calcium signals, *Neuron* **12**:943-956.

Hardingham, G. E., Chawla, S., Cruzalegui, F. H. and Bading, H., 1999, Control of recruitment and transcription-activating function of CBP determines gene regulation by NMDA receptors and L-type calcium channels, *Neuron* **22**:789-798.

Hardingham, G. E., Fukunaga, Y. and Bading, H., 2002, Extrasynaptic NMDARs oppose synaptic NMDARs by triggering CREB shut-off and cell death pathways, *Nat Neurosci* **5**:405-414.

Hell, J. W., Westenbroek, R. E., Warner, C., Ahlijanian, M. K., Prystay, W., Gilbert, M. M., Snutch, T. P. and Catterall, W. A., 1993, Identification and differential subcellular localization of the neuronal class C and class D L-type calcium channel α_1 subunits, *J Cell Biol* **123**:949-962.

Hell, J. W., Yokoyama, C. T., Breeze, L. J., Chavkin, C. and Catterall, W. A., 1995, Phosphorylation of presynaptic and postsynaptic calcium channels by cAMP-dependent protein kinase in hippocampal neurons, *Embo J* **14**:3036-3044.

Hell, J. W., Westenbroek, R. E., Breeze, L. J., Wang, K. K., Chavkin, C. and Catterall, W. A., 1996, N-methyl-D-aspartate receptor-induced proteolytic conversion of postsynaptic class C L-type calcium channels in hippocampal neurons, *Proc Natl Acad Sci U S A* **93**:3362-3367.

Hibino, H., Pironkova, R., Onwumere, O., Vologodskaia, M., Hudspeth, A. J. and Lesage, F., 2002, RIM binding proteins (RBPs) couple Rab3-interacting molecules (RIMs) to voltage-gated Ca^{2+} channels, *Neuron* **34**:411-423.

Hille, B., 2001, *Ion channels of excitable membranes*, Sinauer Associates, Sunderland.

Hryshko, L. V. and Bers, D. M., 1990, Ca current facilitation during postrest recovery depends on Ca entry, *Am J Physiol* **259**:H951-961.

Hu, S. C., Chrivia, J. and Ghosh, A., 1999, Regulation of CBP-mediated transcription by neuronal calcium signaling, *Neuron* **22**:799-808.

Huang, Y. Y. and Malenka, R. C., 1993, Examination of TEA-induced synaptic enhancement in area CA1 of the hippocampus: the role of voltage-dependent Ca^{2+} channels in the induction of LTP, *J Neurosci* **13**:568-576.

Huber, K. M., Mauk, M. D. and Kelly, P. T., 1995, Distinct LTP induction mechanisms: contribution of NMDA receptors and voltage-dependent calcium channels, *J Neurophysiol* **73**:270-279.

Huber, K. M., Mauk, M. D. and Kelly, P. T., 1995, LTP induced by activation of voltage-dependent Ca^{2+} channels requires protein kinase activity, *Neuroreport* **6**:1281-1284.

Impey, S., Obrietan, K., Wong, S. T., Poser, S., Yano, S., Wayman, G., Deloulme, J. C., Chan, G. and Storm, D. R., 1998, Cross talk between ERK and PKA is required for Ca^{2+} stimulation of CREB- dependent transcription and ERK nuclear translocation, *Neuron* **21**:869-883.

Impey, S., Fong, A. L., Wang, Y., Cardinaux, J. R., Fass, D. M., Obrietan, K., Wayman, G. A., Storm, D. R., Soderling, T. R. and Goodman, R. H., 2002, Phosphorylation of CBP mediates transcriptional activation by neural activity and CaM kinase IV, *Neuron* **34**:235-244.

Imredy, J. P. and Yue, D. T., 1992, Submicroscopic Ca^{2+} diffusion mediates inhibitory coupling between individual Ca^{2+} channels, *Neuron* **9**:197-207.

Inoue, T., Kato, K., Kohda, K. and Mikoshiba, K., 1998, Type 1 inositol 1,4,5-trisphosphate receptor is required for induction of long-term depression in cerebellar Purkinje neurons, *J Neurosci* **18**:5366-5373.

Ireland, D. R., Davies, P. J. and McLachlan, E. M., 1998, The role of N-type Ca^{2+} channels in regulating excitability of guinea-pig sympathetic neurones, *J Auton Nerv Syst* **73**:109-114.

Issa, N. P. and Hudspeth, A. J., 1994, Clustering of Ca^{2+} channels and Ca^{2+}-activated K^+ channels at fluorescently labeled presynaptic active zones of hair cells, *Proc Natl Acad Sci U S A* **91**:7578-7582.

Izumi, Y. and Zorumski, C. F., 1998, LTP in CA1 of the adult rat hippocampus and voltage-activated calcium channels, *Neuroreport* **9**:3689-3691.

Jahn, R. and Sudhof, T. C., 1999, Membrane fusion and exocytosis, *Annu Rev Biochem* **68**:863-911.

Jensen, B. S., Strobaek, D., Olesen, S. P. and Christophersen, P., 2001, The Ca^{2+}-activated K^+ channel of intermediate conductance: a molecular target for novel treatments?, *Curr Drug Targets* **2**:401-422.

Kako, K. and Ishida, N., 1998, The role of transcription factors in circadian gene expression, *Neurosci Res* **31**:257-264.

Kandel, E. R., 2001, The molecular biology of memory storage: a dialogue between genes and synapses, *Science* **294**:1030-1038.

Kapur, A., Yeckel, M. F., Gray, R. and Johnston, D., 1998, L-Type calcium channels are required for one form of hippocampal mossy fiber LTP, *J Neurophysiol* **79**:2181-2190.

Kasai, H. and Neher, E., 1992, Dihydropyridine-sensitive and ω-conotoxin-sensitive calcium channels in a mammalian neuroblastoma-glioma cell line, *J Physiol (Lond)* **448**:161-188.

Katz, B., 1969, *The Release of Neural Transmitter Substances*, Liverpool University Press, Liverpool.

Keen, J. E., Khawaled, R., Farrens, D. L., Neelands, T., Rivard, A., Bond, C. T., Janowsky, A., Fakler, B., Adelman, J. P. and Maylie, J., 1999, Domains responsible for constitutive and Ca^{2+}-dependent interactions between calmodulin and small conductance Ca^{2+}-activated potassium channels, *J Neurosci* **19**:8830-8838.

Kim, D. K. and Catterall, W. A., 1997, Ca^{2+}-dependent and -independent interactions of the isoforms of the alpha1A subunit of brain Ca^{2+} channels with presynaptic SNARE proteins, *Proceedings of the National Academy of Sciences of the United States of America* **94**:14782-14786.

Kink, J. A., Maley, M. E., Preston, R. R., Ling, K. Y., Wallen-Friedman, M. A., Saimi, Y. and Kung, C., 1990, Mutations in paramecium calmodulin indicate functional differences between the C-terminal and N-terminal lobes in vivo, *Cell* **62**:165-174.

Knaus, H. G., Folander, K., Garcia-Calvo, M., Garcia, M. L., Kaczorowski, G. J., Smith, M. and Swanson, R., 1994, Primary sequence and immunological characterization of beta-subunit of high conductance Ca^{2+}-activated K^+ channel from smooth muscle, *J Biol Chem* **269**:17274-17278.

Koch, W. J., Ellinor, P. T. and Schwartz, A., 1990, cDNA cloning of a dihydropyridine-sensitive calcium channel from rat aorta. Evidence for the existence of alternatively spliced forms, *J Biol Chem* **265**:17786-17791.

Kornhauser, J. M., Cowan, C. W., Shaywitz, A. J., Dolmetsch, R. E., Griffith, E. C., Hu, L. S., Haddad, C., Xia, Z. and Greenberg, M. E., 2002, CREB transcriptional activity in neurons is regulated by multiple, calcium-specific phosphorylation events, *Neuron* **34**:221-233.

Kurschner, C., Mermelstein, P. G., Holden, W. T. and Surmeier, D. J., 1998, CIPP, a novel multivalent PDZ domain protein, selectively interacts with Kir4.0 family members, NMDA receptor subunits, neurexins, and neuroligins, *Mol Cell Neurosci* **11**:161-172.

Kurschner, C. and Yuzaki, M., 1999, Neuronal interleukin-16 (NIL-16): a dual function PDZ domain protein, *J Neurosci* **19**:7770-7780.

Lagrutta, A., Shen, K. Z., North, R. A. and Adelman, J. P., 1994, Functional differences among alternatively spliced variants of Slowpoke, a *Drosophila* calcium-activated potassium channel, *J Biol Chem* **269**:20347-20351.

Lancaster, B. and Nicoll, R. A., 1987, Properties of two calcium-activated hyperpolarizations in rat hippocampal neurones, *J Physiol* **389**:187-203.

Lang, D. G. and Ritchie, A. K., 1987, Large and small conductance calcium-activated potassium channels in the GH3 anterior pituitary cell line, *Pflugers Arch* **410**:614-622.

Latorre, R., Oberhauser, A., Labarca, P. and Alvarez, O., 1989, Varieties of calcium-activated potassium channels, *Annu Rev Physiol* **51**:385-399.

Lee, A., Wong, S. T., Gallagher, D., Li, B., Storm, D. R., Scheuer, T. and Catterall, W. A., 1999, Ca^{2+}/calmodulin binds to and modulates P/Q-type calcium channels, *Nature* **399**:155-159.

Lee, A., Wong, S. T., Gallagher, D., Li, B., Storm, D. R., Scheuer, T. and Catterall, W. A., 1999, Ca^{2+}/calmodulin binds to and modulates P/Q-type calcium channels, *Nature* **399**:155-159.

Lee, A., Scheuer, T. and Catterall, W. A., 2000, Ca2+/calmodulin-dependent facilitation and inactivation of P/Q-type Ca^{2+} channels, *J Neurosci* **20**:6830-6838.

Lee, O., Lee, C. J. and Choi, S., 2002, Induction mechanisms for L-LTP at thalamic input synapses to the lateral amygdala: requirement of mGluR5 activation, *Neuroreport* **13**:685-691.

Lemaire, S., Piot, C., Leclercq, F., Leuranguer, V., Nargeot, J. and Richard, S., 1998, Heart rate as a determinant of L-type Ca^{2+} channel activity: mechanisms and implication in force-frequency relation, *Basic Res Cardiol* **93 Suppl 1**:51-59.

Leveque, C., Hoshino, T., David, P., Shoji, K. Y., Leys, K., Omori, A., Lang, B., el, F. O., Sato, K., Martin, M. N. and et, a. l., 1992, The synaptic vesicle protein synaptotagmin associates with calcium channels and is a putative Lambert-Eaton myasthenic syndrome antigen, *Proc Natl Acad Sci U S A* **89**:3625-3629.

Leveque, C., el Far, O., Martin-Moutot, N., Sato, K., Kato, R., Takahashi, M. and Seagar, M. J., 1994, Purification of the N-type calcium channel associated with syntaxin and synaptotagmin. A complex implicated in synaptic vesicle exocytosis, *J Biol Chem* **269**:6306-6312.

Leveque, C., Pupier, S., Marqueze, B., Geslin, L., Kataoka, M., Takahashi, M., De Waard, M. and Seagar, M., 1998, Interaction of cysteine string proteins with the α1A subunit of the P/Q-type calcium channel, *J Biol Chem* **273**:13488-13492.

Lin, R. C. and Scheller, R. H., 2000, Mechanisms of synaptic vesicle exocytosis, *Annu Rev Cell Dev Biol* **16**:19-49.

Lisman, J., 1989, A mechanism for the Hebb and the anti-Hebb processes underlying learning and memory, *PNAS* **86**:9574-9578.

Lisman, J., 1994, The CaM kinase II hypothesis for the storage of synaptic memory, *Trends Neurosci* **17**:406-412.

Lisman, J., Schulman, H. and Cline, H., 2002, The molecular basis of CaMKII function in synaptic and behavioural memory, *Nat Rev Neurosci* **3**:175-190.

Liu, F. C. and Graybiel, A. M., 1996, Spatiotemporal dynamics of CREB phosphorylation: transient versus sustained phosphorylation in the developing striatum, *Neuron* **17**:1133-1144.

Liu, Y. C., Patel, H. J., Khawaja, A. M., Belvisi, M. G. and Rogers, D. F., 1999, Neuroregulation by vasoactive intestinal peptide (VIP) of mucus secretion in ferret trachea: activation of BK_{Ca} channels and inhibition of neurotransmitter release, *Br J Pharmacol* **126**:147-158.

Llinas, R., Steinberg, I. Z. and Walton, K., 1981, Relationship between presynaptic calcium current and postsynaptic potential in squid giant synapse, *Biophys J* **33**:323-351.

Llinas, R., Sugimori, M. and Silver, R. B., 1992, Microdomains of high calcium concentration in a presynaptic terminal, *Science* **256**:677-679.

Lo, F. S. and Mize, R. R., 2002, Properties of LTD and LTP of retinocollicular synaptic transmission in the developing rat superior colliculus, *Eur J Neurosci* **15**:1421-1432.

Lonze, B. E. and Ginty, D. D., 2002, Function and regulation of CREB family transcription factors in the nervous system, *Neuron* **35**:605-623.

Mabuchi, T., Kitagawa, K., Kuwabara, K., Takasawa, K., Ohtsuki, T., Xia, Z., Storm, D., Yanagihara, T., Hori, M. and Matsumoto, M., 2001, Phosphorylation of cAMP response element-binding protein in hippocampal neurons as a protective response after exposure to glutamate in vitro and ischemia in vivo, *J Neurosci* **21**:9204-9213.

Magee, J. C. and Johnston, D., 1995, Synaptic activation of voltage-gated channels in the dendrites of hippocampal pyramidal neurons *Science* **268**:301-304.

Magee, J. C., Avery, R. B., Christie, B. R. and Johnston, D., 1996, Dihydropyridine-sensitive, voltage-gated Ca^{2+} channels contribute to the resting intracellular Ca^{2+} concentration of hippocampal CA1 pyramidal neurons, *J Neurophysiol* **76**:3460-3470.

Manahan-Vaughan, D., Braunewell, K. H. and Reymann, K. G., 1998, Subtype-specific involvement of metabotropic glutamate receptors in two forms of long-term potentiation in the dentate gyrus of freely moving rats, *Neuroscience* **86**:709-721.

Mantamadiotis, T., Lemberger, T., Bleckmann, S. C., Kern, H., Kretz, O., Martin Villalba, A., Tronche, F., Kellendonk, C., Gau, D., Kapfhammer, J., Otto, C., Schmid, W. and Schutz, G., 2002, Disruption of CREB function in brain leads to neurodegeneration, *Nat Genet* **31**:47-54.

Marchetti, C., Amico, C. and Usai, C., 1995, Functional characterization of the effect of nimodipine on the calcium current in rat cerebellar granule cells, *J Neurophysiol* **73**:1169-1180.

Marrion, N. V. and Tavalin, S. J., 1998, Selective activation of Ca^{2+}-activated K^{+} channels by co-localized Ca^{2+} channels in hippocampal neurons, *Nature* **395**:900-905.

Martin-Moutot, N., Charvin, N., Leveque, C., Sato, K., Nishiki, T., Kozaki, S., Takahashi, M. and Seagar, M., 1996, Interaction of SNARE complexes with P/Q-type calcium channels in rat cerebellar synaptosomes, *J Biol Chem* **271**:6567-6570.

Marty, A., 1981, Ca-dependent K channels with large unitary conductance in chromaffin cell membranes, *Nature* **291**:497-500.

Maximov, A., Sudhof, T. C. and Bezprozvanny, I., 1999, Association of neuronal calcium channels with modular adaptor proteins, *J Biol Chem* **274**:24453-24456.

Maximov, A. and Bezprozvanny, I., 2002, Synaptic targeting of N-type calcium channels in hippocampal neurons, *J Neurosci* **22**:6939-6952.

McAllister, A. K., Katz, L. C. and Lo, D. C., 1996, Neurotrophin regulation of cortical dendritic growth requires activity, *Neuron* **17**:1057-1064.

McCampbell, A., Taylor, J. P., Taye, A. A., Robitschek, J., Li, M., Walcott, J., Merry, D., Chai, Y., Paulson, H., Sobue, G. and Fischbeck, K. H., 2000, CREB-binding protein sequestration by expanded polyglutamine, *Hum Mol Genet* **9**:2197-2202.

McHugh, D., Sharp, E. M., Scheuer, T. and Catterall, W. A., 2000, Inhibition of cardiac L-type calcium channels by protein kinase C phosphorylation of two sites in the N-terminal domain, *PNAS* **97**:12334-12338.

McManus, O. B., Harris, G. H., Giangiacomo, K. M., Feigenbaum, P., Reuben, J. P., Addy, M. E., Burka, J. F., Kaczorowski, G. J. and Garcia, M. L., 1993, An activator of calcium-dependent potassium channels isolated from a medicinal herb, *Biochemistry* **32**:6128-6133.

McManus, O. B., Helms, L. M., Pallanck, L., Ganetzky, B., Swanson, R. and Leonard, R. J., 1995, Functional role of the beta subunit of high conductance calcium-activated potassium channels, *Neuron* **14**:645-650.

McMonagle-Strucko, K. and Fanelli, R. J., 1993, Enhanced acquisition of reversal training in a spatial learning task in rats treated with chronic nimodipine, *Pharmacol Biochem Behav* **44**:827-835.

Meera, P., Wallner, M., Song, M. and Toro, L., 1997, Large conductance voltage- and calcium-dependent K^+ channel, a distinct member of voltage-dependent ion channels with seven N-terminal transmembrane segments (S0-S6), an extracellular N terminus, and an intracellular (S9-S10) C terminus, *Proc Natl Acad Sci U S A* **94**:14066-14071.

Meera, P., Wallner, M. and Toro, L., 2000, A neuronal beta subunit (KCNMB4) makes the large conductance, voltage- and Ca^{2+}-activated K^+ channel resistant to charybdotoxin and iberiotoxin, *Proc Natl Acad Sci U S A* **97**:5562-5567.

Melzack, R., Coderre, T. J., Katz, J. and Vaccarino, A. L., 2001, Central neuroplasticity and pathological pain, *Ann N Y Acad Sci* **933**:157-174.

Mermelstein, P. G., Bito, H., Deisseroth, K. and Tsien, R. W., 2000, Critical dependence of cAMP response element-binding protein phosphorylation on L-type calcium channels supports a selective response to EPSPs in preference to action potentials, *J Neurosci* **20**:266-273.

Mermelstein, P. G., Deisseroth, K., Dasgupta, N., Isaksen, A. L. and Tsien, R. W., 2001, Calmodulin priming: nuclear translocation of a calmodulin complex and the memory of prior neuronal activity, *Proc Natl Acad Sci U S A* **98**:15342-15347.

Meyer, T., Hanson, P. I., Stryer, L. and Schulman, H., 1992, Calmodulin trapping by calcium-calmodulin-dependent protein kinase, *Science* **256**:1199-1202.

Miller, S. G. and Kennedy, M. B., 1986, Regulation of brain type II Ca^{2+}/calmodulin-dependent protein kinase by autophosphorylation: a Ca^{2+}-triggered switch, *Cell* **44**:861-870.

Milner, B., Squire, L. R. and Kandel, E. R., 1998, Cognitive neuroscience and the study of memory, *Neuron* **20**:445-468.

Mochida, S., 2000, Protein-protein interactions in neurotransmitter release, *Neurosci Res* **36**:175-182.

Montminy, M. R., 1997, Transcriptional regulation by cyclic AMP., *Annual Review of Biochemistry* **66**:in press.

Morgan, S. L. and Teyler, T. J., 1999, VDCCs and NMDARs underlie two forms of LTP in CA1 hippocampus in vivo, *J Neurophysiol* **82**:736-740.

Morgan, S. L. and Teyler, T. J., 2001, Epileptic-like activity induces multiple forms of plasticity in hippocampal area CA1, *Brain Res* **917**:90-96.

Morgan, S. L. and Teyler, T. J., 2001, Electrical stimuli patterned after the theta-rhythm induce multiple forms of LTP, *J Neurophysiol* **86**:1289-1296.

Murphy, T. H., Worley, P. F. and Baraban, J. M., 1991, L-type voltage-sensitive calcium channels mediate synaptic activation of immediate early genes, *Neuron* **7**:625-635.

Nakazawa, H. and Murphy, T. H., 1999, Activation of Nuclear Calcium Dynamics by Synaptic Stimulation in Cultured Cortical Neurons, *J Neurochem* **73**:1075-1083.

Nestler, E. J., 2001, Molecular basis of long-term plasticity underlying addiction, *Nat Rev Neurosci* **2**:119-128.

Newton, D. L., Oldewurtel, M. D., Krinks, M. H., Shiloach, J. and Klee, C. B., 1984, Agonist and antagonist properties of calmodulin fragments, *J Biol Chem* **259**:4419-4426.

Noceti, F., Olcese, R., Qin, N., Zhou, J. and Stefani, E., 1998, Effect of Bay K 8644 (-) and the β2a subunit on Ca^{2+}-dependent inactivation in α1C Ca^{2+} channels, *J Gen Physiol* **111**:463-475.

Norris, C. M., Halpain, S. and Foster, T. C., 1998, Reversal of age-related alterations in synaptic plasticity by blockade of L-type Ca^{2+} channels, *J Neurosci* **18**:3171-3179.

Nucifora, F. C., Jr., Sasaki, M., Peters, M. F., Huang, H., Cooper, J. K., Yamada, M., Takahashi, H., Tsuji, S., Troncoso, J., Dawson, V. L., Dawson, T. M. and Ross, C. A., 2001, Interference by huntingtin and atrophin-1 with cbp-mediated transcription leading to cellular toxicity, *Science* **291**:2423-2428.

Ohbayashi, K., Fukura, H., Inoue, H. K., Komiya, Y. and Igarashi, M., 1998, Stimulation of L-type Ca^{2+} channel in growth cones activates two independent signaling pathways, *J Neurosci Res* **51**:682-696.

Pallotta, B. S., Magleby, K. L. and Barrett, J. N., 1981, Single channel recordings of Ca^{2+}-activated K^+ currents in rat muscle cell culture, *Nature* **293**:471-474.

Parker, D., Jhala, U. S., Radhakrishnan, I., Yaffe, M. B., Reyes, C., Shulman, A. I., Cantley, L. C., Wright, P. E. and Montminy, M., 1998, Analysis of an activator:coactivator complex reveals an essential role for secondary structure in transcriptional activation, *Mol Cell* **2**:353-359.

Petersen, O. H. and Maruyama, Y., 1984, Calcium-activated potassium channels and their role in secretion, *Nature* **307**:693-696.

Peterson, B. Z., DeMaria, C. D., Adelman, J. P. and Yue, D. T., 1999, Calmodulin is the Ca^{2+} sensor for Ca^{2+} -dependent inactivation of L-type calcium channels [published erratum appears in Neuron 1999 Apr;22(4):following 893], *Neuron* **22**:549-558.

Peterson, B. Z., DeMaria, C. D., Adelman, J. P. and Yue, D. T., 1999, Calmodulin is the Ca^{2+} sensor for Ca^{2+} -dependent inactivation of L- type calcium channels, *Neuron* **22**:549-558.

Peterson, B. Z., Lee, J. S., Mulle, J. G., Wang, Y., de Leon, M. and Yue, D. T., 2000, Critical determinants of Ca^{2+}-dependent inactivation within an EF-hand motif of L-type Ca^{2+} channels, *Biophysical Journal* **78**:1906-1920.

Petrij, F., Giles, R. H., Dauwerse, H. G., Saris, J. J., Hennekam, R. C., Masuno, M., Tommerup, N., van Ommen, G. J., Goodman, R. H., Peters, D. J. and et al., 1995, Rubinstein-Taybi syndrome caused by mutations in the transcriptional co-activator CBP *Nature* **376**:348-351.

Pineda, J. C., Galarraga, E., Bargas, J., Cristancho, M. and Aceves, J., 1992, Charybdotoxin and apamin sensitivity of the calcium-dependent repolarization and the afterhyperpolarization in neostriatal neurons, *J Neurophysiol* **68**:287-294.

Pitt, G. S., Zuhlke, R. D., Hudmon, A., Schulman, H., Reuter, H. and Tsien, R. W., 2001, Molecular basis of calmodulin tethering and Ca^{2+}-dependent inactivation of L-type Ca^{2+} channels, *J Biol Chem* **276**:30794-30802.

Qin, N., Olcese, R., Bransby, M., Lin, T. and Birnbaumer, L., 1999, Ca^{2+}-induced inhibition of the cardiac Ca^{2+} channel depends on calmodulin, *Proc Natl Acad Sci U S A* **96**:2435-2438.

Quartermain, D., deSoria, V. G. and Kwan, A., 2001, Calcium channel antagonists enhance retention of passive avoidance and maze learning in mice, *Neurobiol Learn Mem* **75**:77-90.

Quevedo, J., Vianna, M., Daroit, D., Born, A. G., Kuyven, C. R., Roesler, R. and Quillfeldt, J. A., 1998, L-type voltage-dependent calcium channel blocker nifedipine enhances memory retention when infused into the hippocampus, *Neurobiol Learn Mem* **69**:320-325.

Raghib, A., Bertaso, F., Davies, A., Page, K. M., Meir, A., Bogdanov, Y. and Dolphin, A. C., 2001, Dominant-negative synthesis suppression of voltage-gated calcium channel $Ca_v2.2$ induced by truncated constructs, *J Neurosci* **21**:8495-8504.

Rajadhyaksha, A., Barczak, A., Macías, W., Leveque, J. C., Lewis, S. E. and Konradi, C., 1999, L-Type Ca^{2+} channels are essential for glutamate-mediated CREB phosphorylation and *c-fos* gene expression in striatal neurons, *Journal of Neuroscience* **19**:6348-6359.

Randall, A. and Tsien, R. W., 1995, Pharmacological dissection of multiple types of Ca^{2+} channel currents in rat cerebellar granule neurons, *J Neurosci* **15**:2995-3012.

Rao, A., Luo, C. and Hogan, P. G., 1997, Transcription factors of the NFAT family: regulation and function, *Annu Rev Immunol* **15**:707-747.

Reuter, H. and Scholz, H., 1977, The regulation of the calcium conductance of cardiac muscle by adrenaline, *J Physiol* **264**:49-62.

Reuter, H., 1995, Measurements of exocytosis from single presynaptic nerve terminals reveal heterogeneous inhibition by Ca^{2+} channel blockers, *Neuron* **14**:773-779.

Rhoads, A. R. and Friedberg, F., 1997, Sequence motifs for calmodulin recognition, *Faseb J* **11**:331-340.

Rich, R. C. and Schulman, H., 1998, Substrate-directed function of calmodulin in autophosphorylation of Ca^{2+}/calmodulin-dependent protein kinase II, *J Biol Chem* **273**:28424-28429.

Robitaille, R., Garcia, M. L., Kaczorowski, G. J. and Charlton, M. P., 1993, Functional colocalization of calcium and calcium-gated potassium channels in control of transmitter release, *Neuron* **11**:645-655.

Romanin, C., Gamsjaeger, R., Kahr, H., Schaufler, D., Carlson, O., Abernethy, D. R. and Soldatov, N. M., 2000, Ca^{2+} sensors of L-type Ca^{2+} channel, *FEBS Lett* **487**:301-306.

Romey, G. and Lazdunski, M., 1984, The coexistence in rat muscle cells of two distinct classes of Ca^{2+}-dependent K^+ channels with different pharmacological properties and different physiological functions, *Biochem Biophys Res Commun* **118**:669-674.

Rotman, E. I., Murphy, B. J. and Catterall, W. A., 1995, Sites of selective cAMP-dependent phosphorylation of the L-type calcium channel alpha 1 subunit from intact rabbit skeletal muscle myotubes, *J Biol Chem* **270**:16371-16377.

Sabatini, B. L. and Regehr, W. G., 1999, Timing of synaptic transmission, *Annu Rev Physiol* **61**:521-542.

Sah, P., 1995, Properties of channels mediating the apamin-insensitive afterhyperpolarization in vagal motoneurons, *J Neurophysiol* **74**:1772-1776.

Sah, P., 1996, Ca^{2+}-activated K^+ currents in neurones: types, physiological roles and modulation, *Trends Neurosci* **19**:150-154.

Sah, P. and Davies, P., 2000, Calcium-activated potassium currents in mammalian neurons, *Clin Exp Pharmacol Physiol* **27**:657-663.

Sah, P. and Louise Faber, E. S., 2002, Channels underlying neuronal calcium-activated potassium currents, *Prog Neurobiol* **66**:345-353.

Saisu, H., Ibaraki, K., Yamaguchi, T., Sekine, Y. and Abe, T., 1991, Monoclonal antibodies immunoprecipitating ω-conotoxin-sensitive calcium channel molecules recognize two novel proteins localized in the nervous system, *Biochem Biophys Res Commun* **181**:59-66.

Sakurai, M., 1990, Calcium is an intracellular mediator of the climbing fiber in induction of cerebellar long-term depression, *Proc Natl Acad Sci U S A* **87**:3383-3385.

Sanchez, M. and McManus, O. B., 1996, Paxilline inhibition of the α-subunit of the high-conductance calcium-activated potassium channel, *Neuropharmacology* **35**:963-968.

Schjott, J. M. and Plummer, M. R., 2000, Sustained activation of hippocampal L_p-type voltage-gated calcium channels by tetanic stimulation, *J Neurosci* **20**:4786-4797.

Schreiber, M. and Salkoff, L., 1997, A novel calcium-sensing domain in the BK channel, *Biophys J* **73**:1355-1363.

Schumacher, M. A., Rivard, A. F., Bachinger, H. P. and Adelman, J. P., 2001, Structure of the gating domain of a Ca^{2+}-activated K^+ channel complexed with Ca^{2+}/calmodulin, *Nature* **410**:1120-1124.

Seutin, V. and Johnson, S. W., 1999, Recent advances in the pharmacology of quaternary salts of bicuculline, *Trends Pharmacol Sci* **20**:268-270.

Shankar, S., Teyler, T. J. and Robbins, N., 1998, Aging differentially alters forms of long-term potentiation in rat hippocampal area CA1, *J Neurophysiol* **79**:334-341.

Shao, L. R., Halvorsrud, R., Borg-Graham, L. and Storm, J. F., 1999, The role of BK-type Ca^{2+}-dependent K^+ channels in spike broadening during repetitive firing in rat hippocampal pyramidal cells, *J Physiol* **521 Pt 1**:135-146.

Shen, K. Z., Lagrutta, A., Davies, N. W., Standen, N. B., Adelman, J. P. and North, R. A., 1994, Tetraethylammonium block of Slowpoke calcium-activated potassium channels expressed in Xenopus oocytes: evidence for tetrameric channel formation, *Pflugers Arch* **426**:440-445.

Sheng, M., Thompson, M. A. and Greenberg, M. E., 1991, CREB: A Ca^{2+}-regulated transcription factor phosphorylated by calmodulin-dependent kinases, *Science* **252**:1427-1430.

Sheng, M. and Pak, D. T. S., 2000, Ligand-gated ion channel interactions with cytoskeletal and signaling proteins, *Annu Rev Physiol* **62**:755-778.

Shieh, P. B., Hu, S. C., Bobb, K., Timmusk, T. and Ghosh, A., 1998, Identification of a signaling pathway involved in calcium regulation of BDNF expression, *Neuron* **20**:727-740.

Shirokova, N., Gonzalez, A., Kirsch, W. G., Rios, E., Pizarro, G., Stern, M. D. and Cheng, H., 1999, Calcium sparks: release packets of uncertain origin and fundamental role, *J Gen Physiol* **113**:377-384.

Silva, A. J., Kogan, J. H., Frankland, P. W. and Kida, S., 1998, CREB and memory, *Annu Rev Neurosci* **21**:127-148.

Smit, M. J. and Iyengar, R., 1998, Mammalian adenylyl cyclases, *Adv Second Messenger Phosphoprotein Res* **32**:1-21.

Snutch, T. P., Tomlinson, W. J., Leonard, J. P. and Gilbert, M. M., 1991, Distinct calcium channels are generated by alternative splicing and are differentially expressed in the mammalian CNS, *Neuron* **7**:45-57.

Soldatov, N. M., Zühlke, R. D., Bouron, A. and Reuter, H., 1997, Molecular structures involved in L-type calcium channel inactivation. Role of the carboxyl-terminal region encoded by exons 40-42 in α1C subunit in the kinetics and Ca^{2+} dependence of inactivation, *J Biol Chem* **272**:3560-3566.

Soldatov, N. M., Oz, M., O'Brien, K. A., Abernethy, D. R. and Morad, M., 1998, Molecular determinants of L-type Ca^{2+} channel inactivation. Segment exchange analysis of the carboxyl-terminal cytoplasmic motif encoded by exons 40-42 of the human α1C subunit gene, *J Biol Chem* **273**:957-963.

Song, W. J. and Surmeier, D. J., 1996, Voltage-dependent facilitation of calcium channels in rat neostriatal neurons, *J Neurophysiol* **76**:2290-2306.

Spruston, N., Schiller, Y., Stuart, G. and Sakmann, B., 1995, Activity-dependent action potential invasion and calcium influx into hippocampal CA1 dendrites, *Science* **268**:297-300.

Steffan, J. S., Kazantsev, A., Spasic-Boskovic, O., Greenwald, M., Zhu, Y. Z., Gohler, H., Wanker, E. E., Bates, G. P., Housman, D. E. and Thompson, L. M., 2000, The Huntington's disease protein interacts with p53 and CREB-binding protein and represses transcription, *Proc Natl Acad Sci U S A* **97**:6763-6768.

Stemmer, P. M., Wang, X., Krinks, M. H. and Klee, C. B., 1995, Factors responsible for the Ca^{2+}-dependent inactivation of calcineurin in brain, *FEBS Lett* **374**:237-240.

Steward, O. and Schuman, E. M., 2001, Protein synthesis at synaptic sites on dendrites, *Annu Rev Neurosci* **24**:299-325.

Steward, O. and Worley, P. F., 2001, A cellular mechanism for targeting newly synthesized mRNAs to synaptic sites on dendrites, *Proc Natl Acad Sci U S A* **98**:7062-7068.

Storm, J. F., 1987, Action potential repolarization and a fast after-hyperpolarization in rat hippocampal pyramidal cells, *J Physiol* **385**:733-759.

Storm, J. F., 1990, Potassium currents in hippocampal pyramidal cells, *Prog Brain Res* **83**:161-187.

Sudhof, T. C. and Scheller, R. H., 2001, Mechanisms and regulation of neurotransmitter release. Synapses. W. M. Cowall, T. C. Sudhoff and C. F. Stevens. Baltimore, MD, The Johns Hopkins University Press: 177-215.

Sun, P., Enslen, H., Myung, P. S. and Maurer, R. A., 1994, Differential activation of CREB by Ca^{2+}/calmodulin-dependent protein kinases type II and type IV involves phosphorylation of a site that negatively regulates activity, *Genes Dev* **8**:2527-2539.

Sun, X. P., Schlichter, L. C. and Stanley, E. F., 1999, Single-channel properties of BK-type calcium-activated potassium channels at a cholinergic presynaptic nerve terminal, *J Physiol* **518 (Pt 3)**:639-651.

Sunderland, W. J., Son, Y. J., Miner, J. H., Sanes, J. R. and Carlson, S. S., 2000, The presynaptic calcium channel is part of a transmembrane complex linking a synaptic laminin ($\alpha4\beta2\gamma1$) with non-erythroid spectrin, *J Neurosci* **20**:1009-1019.

Sutton, K. G., McRory, J. E., Guthrie, H., Murphy, T. H. and Snutch, T. P., 1999, P/Q-type calcium channels mediate the activity-dependent feedback of syntaxin-1A, *Nature* **401**:800-804.

Tanaka, K., 2001, Alteration of second messengers during acute cerebral ischemia - adenylate cyclase, cyclic AMP-dependent protein kinase, and cyclic AMP response element binding protein, *Prog Neurobiol* **65**:173-207.

Tao, X., Finkbeiner, S., Arnold, D. B., Shaywitz, A. J. and Greenberg, M. E., 1998, Ca^{2+} influx regulates BDNF transcription by a CREB family transcription factor-dependent mechanism, *Neuron* **20**:709-726.

Tobi, D., Wiser, O., Trus, M. and Atlas, D., 1998, N-type voltage-sensitive calcium channel interacts with syntaxin, synaptotagmin and SNAP-25 in a multiprotein complex, *Receptors Channels* **6**:89-98.

Tokumitsu, H., Enslen, H. and Soderling, T. R., 1995, Characterization of a Ca^{2+}/calmodulin-dependent protein kinase cascade. Molecular cloning and expression of calcium/calmodulin-dependent protein kinase kinase, *J Biol Chem* **270**:19320-19324.

Tseng-Crank, J., Foster, C. D., Krause, J. D., Mertz, R., Godinot, N., DiChiara, T. J. and Reinhart, P. H., 1994, Cloning, expression, and distribution of functionally distinct Ca^{2+}-activated K^+ channel isoforms from human brain, *Neuron* **13**:1315-1330.

Tseng-Crank, J., Godinot, N., Johansen, T. E., Ahring, P. K., Strobaek, D., Mertz, R., Foster, C. D., Olesen, S. P. and Reinhart, P. H., 1996, Cloning, expression, and distribution of a Ca^{2+}-activated K^+ channel β-subunit from human brain, *Proc Natl Acad Sci U S A* **93**:9200-9205.

Tsien, R. W. and D. B. Wheeler (1999). Voltage-gated calcium channels. Calcium as a Cellular Regulator. E. Carafoli and C. B. Klee. New York, Oxford University Press: 171-199.

Umbach, J. A., Zinsmaier, K. E., Eberle, K. K., Buchner, E., Benzer, S. and Gundersen, C. B., 1994, Presynaptic dysfunction in *Drosophila* csp mutants, *Neuron* **13**:899-907.

Urbano, F. J., Piedras-Renteria, E. S., Jun, K., Shin, H.-S., Tsien, R. W. and Uchitel, O. D., 2002, Altered properties of quantal neurotransmitter release at endplates of mice lacking P/Q-type Ca^{2+} channels, *Proc Natl Acad Sci U S A, in press.*

Van Eldik, L. and Watterson, D. M., 1998, *Calmodulin and Signal Transduction*, Academic Press, New York.

Vergara, C., Latorre, R., Marrion, N. V. and Adelman, J. P., 1998, Calcium-activated potassium channels, *Curr Opin Neurobiol* **8**:321-329.

Walton, M. R. and Dragunow, I., 2000, Is CREB a key to neuronal survival?, *Trends Neurosci* **23**:48-53.

Wang, X., Culotta, V. C. and Klee, C. B., 1996, Superoxide dismutase protects calcineurin from inactivation, *Nature* **383**:434-437.

Watanabe, T., Fujiwara, T., Komazaki, S., Yamaguchi, K., Tajima, O. and Akagawa, K., 1999, HPC-1/syntaxin 1A suppresses exocytosis of PC12 cells, *J Biochem (Tokyo)* **125**:685-689.

Weick, J. P., Groth, R. D., Isaksen, A. L. and Mermelstein, P. G., 2002, Interactions with PDZ domain proteins are required for L-type calcium channels to activate CREB-dependent gene expression, *J Neurosci, in press.*

Weisskopf, M. G., Bauer, E. P. and LeDoux, J. E., 1999, L-type voltage-gated calcium channels mediate NMDA-independent associative long-term potentiation at thalamic input synapses to the amygdala, *J Neurosci* **19**:10512-10519.

Wells, D. G., Richter, J. D. and Fallon, J. R., 2000, Molecular mechanisms for activity-regulated protein synthesis in the synapto-dendritic compartment, *Curr Opin Neurobiol* **10**:132-137.

Westenbroek, R. E., Ahlijanian, M. K. and Catterall, W. A., 1990, Clustering of L-type Ca^{2+} channels at the base of major dendrites in hippocampal pyramidal neurons, *Nature* **347**:281-284.

Westenbroek, R. E., Hoskins, L. and Catterall, W. A., 1998, Localization of Ca^{2+} channel subtypes on rat spinal motor neurons, interneurons, and nerve terminals, *J Neurosci* **18**:6319-6330.

Wickens, J. R. and Abraham, W. C., 1991, The involvement of L-type calcium channels in heterosynaptic long-term depression in the hippocampus, *Neurosci Lett* **130**:128-132.

Williams, S., Serafin, M., Muhlethaler, M. and Bernheim, L., 1997, Distinct contributions of high- and low-voltage-activated calcium currents to afterhyperpolarizations in cholinergic nucleus basalis neurons of the guinea pig, *J Neurosci* **17**:7307-7315.

Wiser, O., Bennett, M. K. and Atlas, D., 1996, Functional interaction of syntaxin and SNAP-25 with voltage-sensitive L- and N-type Ca^{2+} channels, *Embo J* **15**:4100-4110.

Wiser, O., Tobi, D., Trus, M. and Atlas, D., 1997, Synaptotagmin restores kinetic properties of a syntaxin-associated N-type voltage sensitive calcium channel, *FEBS Lett* **404**:203-207.

Wiser, O., Trus, M., Hernandez, A., Renstrom, E., Barg, S., Rorsman, P. and Atlas, D., 1999, The voltage sensitive L_c-type Ca^{2+} channel is functionally coupled to the exocytotic machinery, *Proc Natl Acad Sci U S A* **96**:248-253.

Wisgirda, M. E. and Dryer, S. E., 1994, Functional dependence of Ca^{2+}-activated K^+ current on L- and N-type Ca^{2+} channels: differences between chicken sympathetic and parasympathetic neurons suggest different regulatory mechanisms, *Proc Natl Acad Sci U S A* **91**:2858-2862.

Wu, J., Rush, A., Rowan, M. J. and Anwyl, R., 2001, NMDA receptor- and metabotropic glutamate receptor-dependent synaptic plasticity induced by high frequency stimulation in the rat dentate gyrus in vitro, *J Physiol* **533**:745-755.

Wu, M. N., Littleton, J. T., Bhat, M. A., Prokop, A. and Bellen, H. J., 1998, ROP, the Drosophila Sec1 homolog, interacts with syntaxin and regulates neurotransmitter release in a dosage-dependent manner, *Embo Journal* **17**:127-139.

Wu, M. N., Fergestad, T., Lloyd, T. E., He, Y., Broadie, K. and Bellen, H. J., 1999, Syntaxin 1A interacts with multiple exocytic proteins to regulate neurotransmitter release in vivo, *Neuron* **23**:593-605.

Wu, Y., Colbran, R. J. and Anderson, M. E., 2001, Calmodulin kinase is a molecular switch for cardiac excitation -contraction coupling, *PNAS* **98**:2877-2881.

Wyttenbach, A., Swartz, J., Kita, H., Thykjaer, T., Carmichael, J., Bradley, J., Brown, R., Maxwell, M., Schapira, A., Orntoft, T. F., Kato, K. and Rubinsztein, D. C., 2001, Polyglutamine expansions cause decreased CRE-mediated transcription and early gene expression changes prior to cell death in an inducible cell model of Huntington's disease, *Hum Mol Genet* **10**:1829-1845.

Xia, X. M., Fakler, B., Rivard, A., Wayman, G., Johnson-Pais, T., Keen, J. E., Ishii, T., Hirschberg, B., Bond, C. T., Lutsenko, S., Maylie, J. and Adelman, J. P., 1998, Mechanism of calcium gating in small-conductance calcium-activated potassium channels, *Nature* **395**:503-507.

Xia, X. M., Zeng, X. and Lingle, C. J., 2002, Multiple regulatory sites in large-conductance calcium-activated potassium channels, *Nature* **418**:880-884.

Xia, Z. and Storm, D. R., 1997, Calmodulin-regulated adenylyl cyclases and neuromodulation, *Curr Opin Neurobiol* **7**:391-396.

Xiao, R. P., Cheng, H., Lederer, W. J., Suzuki, T. and Lakatta, E. G., 1994, Dual regulation of Ca^{2+}/calmodulin-dependent kinase II activity by membrane voltage and by calcium influx, *Proc Natl Acad Sci U S A* **91**:9659-9663.

Xing, J., Ginty, D. D. and Greenberg, M. E., 1996, Coupling of the RAS-MAPK pathway to gene activation by RSK2, a growth factor-regulated CREB kinase, *Science* **273**:959-963.

Xu, W. and Lipscombe, D., 2001, Neuronal $Ca_V1.3\alpha_1$ L-type channels activate at relatively hyperpolarized membrane potentials and are incompletely inhibited by dihydropyridines, *J Neurosci* **21**:5944-5951.

Yazejian, B., Sun, X. P. and Grinnell, A. D., 2000, Tracking presynaptic Ca^{2+} dynamics during neurotransmitter release with Ca^{2+}-activated K^+ channels, *Nat Neurosci* **3**:566-571.

Yoshida, A., Oho, C., Omori, A., Kuwahara, R., Ito, T. and Takahashi, M., 1992, HPC-1 is associated with synaptotagmin and ω-Conotoxin receptor, *J Biol Chem* **267**:24925-24928.

Yoshida, K., Imaki, J., Matsuda, H. and Hagiwara, M., 1995, Light-induced CREB phosphorylation and gene expression in rat retinal cells, *J Neurochem* **65**:1499-1504.

Yuan, W. and Bers, D. M., 1994, Ca-dependent facilitation of cardiac Ca current is due to Ca-calmodulin-dependent protein kinase, *Am J Physiol* **267**:H982-993.

Yue, D., Herzig, S. and Marban, E., 1990, β-Adrenergic stimulation of calcium channels occurs by potentiation of high-activity gating modes, *PNAS* **87**:753-757.

Zhang, T., Johnson, E. N., Gu, Y., Morissette, M. R., Sah, V. P., Gigena, M. S., Belke, D. D., Dillmann, W. H., Rogers, T. B., Schulman, H., Ross, J., Jr. and Brown, J. H., 2002, The cardiac-specific nuclear delta B isoform of Ca^{2+}/Calmodulin-dependent protein kinase II induces hypertrophy and dilated cardiomyopathy associated with increased protein phosphatase 2A activity, *J Biol Chem* **277**:1261-1267.

Zhong, H., Yohoyama, C. T., Scheuer, T. and Catterall, W. A., 1999, Reciprocal regulation of P/Q-type Ca^{2+} channels by SNAP-25, syntaxin and synaptotagmin, *Nature* **2**:939-941.

Zühlke, R. D. and Reuter, H., 1998, Ca^{2+}-sensitive inactivation of L-type Ca^{2+} channels depends on multiple cytoplasmic amino acid sequences of the α1C subunit, *Proceedings of the National Academy of Sciences of the United States of America* **95**:3287-3294.

Zühlke, R. D., Pitt, G. S., Deisseroth, K., Tsien, R. W. and Reuter, H., 1999, Calmodulin supports both inactivation and facilitation of L-type calcium channels, *Nature* **399**:159-162.

Zühlke, R. D., Pitt, G. S., Tsien, R. W. and Reuter, H., 2000, Ca^{2+}-sensitive inactivation and facilitation of L-type Ca^{2+} channels both depend on specific amino acid residues in a consensus calmodulin-binding motif in the α_{1C} subunit, *Journal of Biological Chemistry* **275**:21121-21129.

Zygmunt, A. C. and Maylie, J., 1990, Stimulation-dependent facilitation of the high threshold calcium current in guinea-pig ventricular myocytes, *J Physiol* **428**:653-671.

GENETIC APPROACHES TO THE ELUCIDATION OF CALCIUM CHANNEL FUNCTIONS IN VIVO

Hironao Saegusa[1] and Tsutomu Tanabe[1]

1. INTRODUCTION

Voltage-dependent Ca^{2+} channels (VDCCs) play essential roles in various aspects of cellular activities including muscle contraction, neurotransmitter release, control of neuronal excitability, and gene expression (Catterall, 1998; Hofmann et al., 1999). To date, several kinds of VDCCs have been identified by electrophysiological and pharmacological means, and molecular biological studies have identified ten mammalian genes coding for the pore-forming α_1 subunit (Ertel et al., 2000). Although many studies using pharmacological tools have revealed the physiological functions of VDCC subtypes, such studies have some limitations. The pharmacological specificity of the tools used to distinguish among different channels is not absolute, and the administration of these drugs and toxins to all the cells of interest is sometimes difficult. Genetic manipulation of VDCCs is an important alternative, since it can be used to eliminate completely a specific VDCC in all the cells of an animal. An increasing number of knockout animals carrying mutations in VDCC genes have been reported, to study the *in vivo* functions of different VDCCs. This chapter reviews the abnormalities found in VDCC knockout mice and the deduced *in vivo* functions of the VDCC.

2. KNOCKING OUT CALCIUM CHANNEL ALPHA1 SUBUNITS

2.1. The Ca_v1 Family

Ca_v1 family members constitute the L-type VDCCs, which are blocked by dihydropyridines (DHPs), phenylalkylamines, and benzothiazepines (Triggle, this volume). However, blockers that distinguish among the four members of the Ca_v1 family are not known. A genetic approach seems useful to clarify the physiological functions of each, as

[1] Department of Pharmacology and Neurobiology, Graduate School of Medicine, Tokyo Medical and Dental University, 1-5-45 Yushima, Bunkyo-ku, Tokyo 113-8519, Japan.

it is possible to eliminate selectively a single Ca_v1 channel. At present, knockout mice for $Ca_v1.2$ and $Ca_v1.3$ channels have been reported, and an additional natural mutant for $Ca_v1.1$, the muscular dysgenesis mouse (Tanabe, 1994), is known. In this section, only the targeted mutant mice will be discussed.

2.1.1. The $Ca_v1.2$ Channel

Seisenberger, Hofmann, and colleagues have reported two different lines of mutant mice carrying a targeted allele for the $Ca_v1.2$ channel (Seisenberger et al., 2000). In mouse line A, a selection marker gene for neomycin resistance (*neo*^{R}) was inserted in exon 3 of the *Cacna1c* gene that encodes the $Ca_v1.2$ channel. In mouse line B, exons 14 and 15 were deleted by using the Cre recombinase-loxP system (Rajewsky et al., 1996). The *neo*^{R} selection marker was then deleted with this recombination system, thereby avoiding the possible secondary effects of the insertion of *neo*^{R}. Mutant mice in both lines showed identical phenotypes, and it is highly possible that these two independent targeted alleles lead to null phenotypes. Heterozygous intercrosses did not yield homozygous mutant pups that were born alive, raising the possibility that these mutations resulted in embryonic lethality. Indeed, viable homozygous mutant embryos were obtained in a Mendelian ratio at embryonic day 12.5 (E12.5), but after E14.5 no viable homozygous mutants were recovered.

The $Ca_v1.2$ channel is known to be involved in the control of cardiac functions in adult animals, and, therefore, the heart contraction in the homozygous mutant embryo was analyzed. At E12.5, the beating frequency in homozygous mutants was not significantly different from wild type controls. Although the homozygous mutant embryos did not survive past E14.5, cardiac cells from the E12.5 mutants could be maintained in culture for over a week. In this culture system, cardiac myocytes from homozygous mutants contracted normally, suggesting that either the $Ca_v1.2$ channel is not indispensable for embryonic heart contraction, or that the loss of $Ca_v1.2$ was somehow compensated, possibly by up-regulation of another calcium channel. This result also suggests that the cause of the embryonic lethality is independent of defects in myocyte function; the exact reason for the lethality, however, is not known.

The embryonic heart contraction and the L-type Ca^{2+} current in embryonic cardiac myocytes were further characterized in detail. When the cardiac myocytes cultured *in vitro* from wild type and homozygous mutant embryos were perfused with Ca^{2+}-free medium, contraction stopped immediately. Thus, the embryonic heart contraction is dependent upon influx of extracellular Ca^{2+}. This is in contrast to the contraction of embryonic stem cell-derived cardiac myocytes, in which intracellular Ca^{2+} oscillations are required for contraction (Viatchenko-Karpinski et al., 1999). The embryonic heart contraction was also blocked by application of the dihydropyridine nisoldipine in both wild type and homozygous mutants, suggesting that DHP-sensitive L-type Ca^{2+} channel(s) other than the $Ca_v1.2$ channel were present in the $Ca_v1.2$ homozygous mutant and played an important role in embryonic heart contraction.

Seisenberger et al. further characterized this cryptic L-type channel. First, they performed electrophysiological studies using the cardiac myocytes from embryos with each of the three genotypes. The L-type current in the wild type cardiac myocytes activated at –40 mV and reached its peak between 0 and 10 mV. On the other hand, the L-type current in myocytes from hetero- or homozygous mutant embryos activated at –50 mV and peaked at –10 mV, suggesting the presence of an L-type channel with slightly different properties. The existence of a different L-type component was further supported by experiments to test the sensitivities of the L-type currents to DHP. The sensitivity to nisoldipine was higher in wild type and heterozygous mutant cells than in cells from the homozygous mutant. Nisoldipine (1 μM) reduced barium current (with a holding potential of –80 mV) to 23% and 27% of control in wild type and the heterozygous mutant, respectively, whereas barium current was reduced only to 65% of control in the homozygous mutant. By fitting the nisoldipine dose-inhibition curves to the Hill equation, it was revealed that there were two L-type components with different affinities for nisoldipine in wild type cardiomyocytes: one with low affinity ($IC_{50} \sim 3$ μM) and one with high affinity ($IC_{50} \sim 0.1$ μM). In the homozygous mutant, however, only one component was detected ($IC_{50} \sim 1.1$ μM), thought to correspond to the low affinity component in wild type. Thus, embryonic cardiac myocytes express the $Ca_v1.2$ channel and also an L-type like channel with low affinity for nisoldipine that is also present in the *Cacna1c* knockout embryos.

RT-PCR analysis revealed the expression of $Ca_v1.1$, $Ca_v1.2$, and $Ca_v1.3$ in the wild type cardiac myocytes. However, neither $Ca_v1.1$ nor $Ca_v1.3$ is likely to be the L-type channel with low affinity for nisoldipine: 1) the properties of the $Ca_v1.1$-based channel are not compatible with those of the L-type like channel in embryonic cardiomyocyte; and 2) in $Ca_v1.3$ knockout embryos, a similar L-type like channel with low affinity for nisoldipine was also observed. The L-type like channel in the embryonic cardiomyocytes is likely to be a novel member of the Ca_v1 family.

2.1.2. The Ca$_v$1.3 Channel

So far there have been two reports describing the establishment of $Ca_v1.3$ knockout mice (Platzer et al., 2000; Namkung et al., 2001). Platzer, Striessnig, and colleagues generated the targeted allele by inserting *neo*^{R} into exon 2 of the *Cacna1d* gene encoding $Ca_v1.3$, whereas Namkung, Shin, and colleagues disrupted this gene by inserting an expression marker gene derived from *lacZ-neo*^{R} into exon 1. Both of these targeted disruptions were thought to result in null mutations for *Cacna1d* gene, as demonstrated by Western blot analyses. The mice displayed a variety of phenotypes, but not all of the phenotypes were the same in the two independently generated knockout mice: both lines were deaf and showed cardiac arrhythmia (Platzer et al., 2000; Namkung et al., 2001; Zhang et al., 2002), but a deficit in insulin secretion was observed in only one of them (Namkung et al., 2001).

2.1.2a. Ca$_v$1.3 Knockout Mice are Deaf. $Ca_v1.3$ knockout mice are viable and show almost normal appearance, except for the slightly smaller size in one of the two mutant lines (Namkung et al., 2001). As $Ca_v1.3$ knockout mice displayed no signs of Prayer reflex, normally elicited by auditory stimuli such as hand-clapping in wild type and heterozygous mutants, they were thought to be deaf. Indeed, the hearing threshold as determined by click-evoked auditory brain stem responses was significantly higher in $Ca_v1.3$ knockout mice than in control.

As it is suggested that VDCCs play a role in neurotransmitter release in auditory hair cells, Platzer et al. performed electrophysiological studies using the inner ear hair cells (IHCs) from their knockout mice. In IHCs from wild type pups (P3-P9), an L-type current with unusual kinetics was observed: a relatively negative activation voltage threshold, rapid activation, and slow inactivation. In the $Ca_v1.3$ knockout IHCs, however, the current density was greatly reduced (~ 90% reduction) and the remaining current was not modulated by DHPs. Thus, the $Ca_v1.3$ channel is responsible for most of the Ca^{2+} current in the auditory IHCs.

Platzer et al. next studied the inner ear of the $Ca_v1.3$ knockout by histological means. The histological analyses revealed an almost normal appearance of IHCs and ganglion cells at P14, a time when the knockout mice already exhibited deafness. At this stage, however, degeneration of outer hair cells (OHCs) was evident in the knockout mice. At P35, both OHCs and IHCs were absent, and degeneration of afferent fibers and ganglion cells was also detected. These results suggest that the morphological changes are secondary effects of the loss of the $Ca_v1.3$-based current in the IHCs, since the degeneration of IHCs and OHCs occurs after the onset of deafness.

2.1.2b. The $Ca_v1.3$ Channel and Cardiac Pacemaker Activity. Since expression of $Ca_v1.3$ channels was detected in atrial cardiomyocytes, cardiac functions were examined in the $Ca_v1.3$ knockout mice (Platzer et al., 2000; Zhang et al., 2002). Resting electrocardiograms (ECG) from $Ca_v1.3$ knockout mice showed bradycardia, with prolonged R-R and P-R intervals. Cardiac arrhythmia was also detected, as indicated by a high standard deviation of R-R interval time. These results point to sinoatrial (SA) and atrioventricular (AV) nodal dysfunctions in the $Ca_v1.3$ knockout mice. To examine the possibility that the bradycardia and arrhythmia were caused not by intrinsic cardiac defects but by altered autonomic input, the ECG was recorded in the presence of atropine and propranolol to block parasympathetic and sympathetic activities, respectively. Bradycardia and arrhythmia were still observed in the presence of these drugs, suggesting that the abnormalities are due to altered intrinsic pacemaker activity.

To analyze further the abnormal pacemaker activity in the $Ca_v1.3$ knockout mice, Zhang et al. performed electrophysiological studies. First, they recorded spontaneous action potentials from intact SA nodes. In $Ca_v1.3$ knockout mice, spontaneous action potentials were slowed significantly compared to control littermates. Comparison of the shapes of the action potentials between wild type and knockout mice showed a decrease in the diastolic depolarization at -40 ~ -45 mV in the knockout mice. This result suggests that the $Ca_v1.3$ channel is essential for diastolic depolarization in the SA node.

Zhang et al. further analyzed the L-type current in single SA node cells with the whole-cell patch clamp. In the knockout SA node cells, the L-type current activated at ~ 5 mV more depolarized potentials compared to wild type control. For wild type and knockout mice, the time course of inactivation of inward currents was fitted with a sum of two exponential functions. In $Ca_v1.3$ knockout mice, both time constants were significantly smaller, indicating that inactivation of the Ca^{2+} current in $Ca_v1.3$ knockout mice was faster.

2.1.2c. Postnatal Generation of Pancreatic β Cells Requires the $Ca_v1.3$ Channel. It has been suggested that L-type Ca^{2+} channels play an important role in insulin secretion in pancreatic β cells. From the expression profiles, it has also been suggested that $Ca_v1.3$ and $Ca_v1.2$ are the most likely candidates for the L-type channels in β cells. Therefore, Namkung et al. addressed whether the $Ca_v1.3$ channel is involved in blood glucose homeostasis via the control of insulin secretion (Namkung et al., 2001).

First, serum insulin levels of knockout mice in non-fasting states were examined. The insulin level was significantly lower in the knockout mice than in control littermates, although the blood glucose level was not significantly different. The glucose tolerance test was performed to examine overall pancreatic function. Glucose was injected intraperitoneally, and changes in the blood glucose level were monitored. Blood glucose levels were significantly higher in the knockout mice two hours after glucose injection, whereas in wild type mice the glucose level returned to basal 30 minutes after the injection. In addition, serum insulin levels increased 30 minutes after the glucose injection in the wild type mice, whereas in the knockout mice insulin levels were unchanged in response to glucose injection. These results suggest that the $Ca_v1.3$ channel governs one or more functions in β cells that control insulin secretion. However, the $Ca_v1.3$ knockout mice did not develop diabetes, possibly due to their increased sensitivity to insulin, as suggested by an insulin tolerance test.

L-type currents in the pancreatic islet cells from $Ca_v1.3$ knockout mice were examined directly with the whole-cell patch clamp. Unexpectedly, neither the total current density nor L-type current density was decreased in the mutant pancreatic islet cells, suggesting that compensatory up-regulation of the $Ca_v1.2$ channel occurred in mice lacking $Ca_v1.3$. Confirming this, immunohistochemical analyses revealed overexpression of $Ca_v1.2$ protein in the $Ca_v1.3$ knockout β cells. Despite the apparently normal current density, the current-voltage curve in the knockout islet cells showed a slight shift toward depolarizing potentials. The net result was that currents at moderate depolarizations, between –30 mV and –10 mV, were significantly decreased in the knockout compared to wild type. Since action potential depolarizations elicited by glucose stimulation generally do not overshoot 0 mV, lowered channel activity between –30 mV and –10 mV may reduce insulin secretion dramatically. To test this, insulin secretion from the isolated islets was examined *in vitro*. Indeed, insulin secretion was significantly lowered when the knockout islets were stimulated with low concentration of glucose (3 mM). However, there were no significant differences in insulin secretion between knockout and wild type mice when higher concentrations (6-10 mM) of glucose were used. This suggests that the majority of glucose-sensing and insulin-secretion machinery is unaffected by knockout of $Ca_v1.3$.

Finally, Namkung et al. examined the number of β cells in pancreatic islets, since a decrease in the number of the β cells could contribute to the hypoinsulinemia and glucose intolerance found in the $Ca_v1.3$ knockout mice. Histological examination revealed that the number of islets and the number of larger islets (containing more than 150 cells) were each significantly decreased in the $Ca_v1.3$ knockout mice. Moreover, the percentage of β cells in the islets was significantly lower in $Ca_v1.3$ knockout mice. These results suggest that loss of β cells could be the main reason for the reduced islet mass in the knockout. However, the TUNEL assay to detect apoptosis did not reveal any difference between β cells from wild type and $Ca_v1.3$ knockout mice, and bromodeoxyuridine labeling studies revealed a significantly lower labeling efficiency in the $Ca_v1.3$ knockout mice. Therefore, the reduction in the number of the β cells is due at least partially to a lowered proliferation rate of β cells in the knockout mice.

2.2. The Ca_v2 Family

All of the Ca_v2 family members are expressed almost exclusively in the nervous system and constitute high-voltage-activated channels resistant to DHPs. It is well known that this family of channels, especially $Ca_v2.1$ and $Ca_v2.2$ channels, play a major role in the control of neurotransmitter release. Their different roles *in vivo*, however, are not understood fully. Physiological functions of the $Ca_v2.3$ channel are particularly poorly understood, since a high-affinity inhibitor of least some forms of this channel has been discovered only recently (Newcomb et al., 1998). Genetic studies have been particularly fruitful for the study of $Ca_v2.3$. Here we review the phenotypes of the $Ca_v2.1$, $Ca_v2.2$, and $Ca_v2.3$ knockout mice.

2.2.1. The $Ca_v2.1$ Channel

The $Ca_v2.1$ channel is abundantly expressed in the central nervous system (CNS), and the importance of this channel in the CNS functions has been well recognized. The gene encoding $Ca_v2.1$ (*CACNA1A* in human and *Cacna1a* in mouse) was itself identified to be a causative gene for some neurological disorders in human, including familial hemiplegic migraine, episodic ataxia type 2, and spinocerebellar ataxia type 6, as well as in mice (*tottering*, *leaner*, *rolling Nagoya*, and *rocker*; see Mori et al., this volume). All of these mutations are thought to affect P/Q-type channel activity, but do not seem to result in null phenotypes. Therefore, comparison between these disease-causing mutants and null mutants (knockouts) will provide insight into the mechanisms underlying each of the mutant phenotypes.

2.2.1a. Ataxia in $Ca_v2.1$ Knockout Mice. So far, there are two reports of the generation of *Cacna1a* knockout mice (Jun et al., 1999; Fletcher et al., 2001). Jun, Shin, and colleagues disrupted the *Cacna1a* gene by deleting part of exon 4 and intron 4 and inserting the *neo*^{R} cassette therein, making the $\alpha_12.1$ protein truncate after the first of four internal repeats. Fletcher and colleagues generated their targeted allele by replacing exons 14-17, which encode three of the six transmembrane segments of the second internal repeat, by the *neo*^{R} expression cassette. Both of these targeted mutations are thought to represent null alleles for the *Cacna1a* gene, as determined by Western blot analysis.

Both groups report that their mice are born alive and grow almost normally up to about 10 days, when they begin to show signs of ataxia and dystonia. The knockout mice tend to fall easily and have difficulty in righting themselves. This ataxic phenotype becomes more and more severe as the mice grow, until finally the mice are unable to walk. Most knockout mice died due to this progressive ataxia at about 3 weeks of age. These mice also exhibited episodes of absence epilepsy, during which all movement stopped, like spontaneous *Cacna1a* mutants such as *tottering* and *leaner* (Fletcher et al., 1996). Unaided, knockout animals never survived past 4 weeks of age. When nursed and treated properly, a very small number of knockout mice were able to survive to 100 days in one of the mutant lines (Fletcher et al., 2001).

Histological analyses revealed some abnormalities in cerebellar development in the knockout mice. Jun et al. reported that at P21 (postnatal day 21) the external granule layer (EGL) still persisted in the cerebellum of their knockout mice, and that Purkinje cell axons possessed focal swellings, which were rare in the wild type mice. Although the physiological and/or pathophysiological meanings of the axonal swellings are not clear, the persistent EGL would suggest developmental delay of the cerebellum in *Cacna1a* knockout mice. Consistent with this idea, ectopic expression of tyrosine hydroxylase was detected in the cerebellum from the knockout mice. Another abnormality was reported by Fletcher et al.: sections through the cerebellum of the rare survivors (15 weeks after birth) revealed severe cerebellar atrophy, with loss of some granule cells and Purkinje cells. The loss of granule cells was more severe in the anterior lobes than in posterior lobes of the cerebellum. Interestingly, loss of Purkinje cells exhibited a striped pattern, identical to the situation reported in *leaner* mice. This suggests that there are two populations of Purkinje cells, one whose survival depends on $Ca_v2.1$ channel activity and one whose survival is independent of $Ca_v2.1$. It would be very interesting to explore the mechanisms underlying the dependence of the different populations of cerebellar Purkinje cells on the $Ca_v2.1$ channel.

2.2.1b. The P/Q-type Channel is Encoded Solely by $Ca_v2.1$. Results of the electrophysiological analyses were not completely the same in the two lines of the mutant mice. P/Q-type currents were completely eliminated in both of the knockout mice. A compensatory increase in current density of N- and L-type channels was apparent in cultured cerebellar granule neurons from the knockout mice generated by Fletcher et al. On the other hand, a reduction in the R-type component, instead of up-regulation of N- and L-type channels, was observed in the granule neurons from the knockout mice generated by Jun et al. The situations for heterozygous mutant mice were also different: cerebellar granule cells from heterozygous mutant mice generated by Jun et al. showed a total Ca^{2+} current and P/Q-type current density almost comparable to wild type. Granule neurons from the heterozygous mutant generated by Fletcher et al., however, showed a ~ 50% decrease in P/Q-type current density and in total current density.

In acutely dissociated Purkinje cells, where normally almost all the VDCC activity is derived from the P-type channel (Mintz et al., 1992), VDCC activities were analyzed in the knockout using Ba^{2+} as charge carrier (Jun et al., 1999). The P-type component was completely lacking in the knockout Purkinje cells, and an increase in the current density of L- and N-type channels was apparent. In Purkinje cells from heterozygous mutant mice, however, no decrease in the P-type current or in total current density was observed.

These results clearly demonstrate that P- and Q-type currents are derived from $Ca_v2.1$ channels (see McDonough, this volume). Moreover, using the ^{125}I labeled ω-conotoxin MVIIC, a toxin that blocks P/Q-type channels, Fletcher et al. demonstrated that the ω-CTx-MVIIC receptor was completely lacking in brain membranes from knockout mice. Therefore, the $Ca_v2.1$ channel is thought to be the sole origin of P- and Q-type currents.

2.2.1c. Synaptic Transmission in $Ca_v2.1$ Knockout Mice. The effects of $Ca_v2.1$ channel ablation on synaptic transmission were assessed in slice preparations from the hippocampal CA1 region (Jun et al., 1999). Field excitatory postsynaptic potentials (fEPSPs) were evoked by stimulating the Schaffer collateral and commissural pathways. Despite the presumed importance of the $Ca_v2.1$ channel in the control of neurotransmitter release, synaptic transmission in the knockout preparations was not greatly impaired. However, ω-conotoxin GVIA treatment halved the EPSP slope in the wild-type slice, whereas the same treatment almost eliminated the synaptic transmission in the knockout. Thus, the contribution of N-type channel to excitatory synaptic transmission at CA3-CA1 synapses in knockout mice was enhanced compared to wild type.

2.2.1d. Relationship between P/Q Current Density and Ataxic Phenotype. P/Q-type current was reduced to approximately half normal level in granule cell bodies from heterozygous mutants generated by Fletcher et al. It is not known if this is the case for all neurons in the heterozygous mutant; this fact, however, provides an intriguing insight into the phenotypes of the various *Cacna1a* mutants. For example, it is known that the P/Q-type current density is reduced by 40% in Purkinje neurons from mice homozygous for *tottering* allele, which show an ataxic phenotype (Wakamori et al., 1998). But the heterozygous mutant reported by Fletcher et al. showed no abnormalities in rotarod performance, despite the 50% reduction in the density of functional P/Q-type channels.

Therefore, it is suggested that reduction of the P/Q current to half the normal level is not itself sufficient to cause abnormal neurological phenotypes.

2.2.2. The $Ca_v2.2$ Channel

Like the $Ca_v2.1$ channel, the $Ca_v2.2$ channel is known to be important in the control of neurotransmitter release. It was thus expected that loss of $Ca_v2.2$ would severely affect survival. Unexpectedly, however, all three independent lines of $Ca_v2.2$ knockout mice reported thus far present a generally normal appearance, with just a partial lethality in one of the lines (Ino et al., 2001; Kim et al., 2001a; Saegusa et al., 2001; Saegusa et al., 2002).

Ino and collaborators generated a targeted allele for *Cacna1b* encoding the $Ca_v2.2$ channel such that the cytoplasmic linker between internal repeat domains II and III of $Ca_v2.2$ was disrupted by insertion of the neo^R expression cassette. Saegusa et al. generated a targeted allele for this gene by deleting a part of exon 1 and intron 1, and then inserting a gene for *E. coli* β-galactosidase with a nuclear localization signal at its amino terminus (the *nlacZ* gene). The targeted allele made by Kim et al. was produced by replacing the genomic region corresponding to the S3 region of the first internal repeat domain with $GFP\text{-}neo^R$ cassette. Hereafter we designate the targeted allele made by Ino et al. as $Cacna1b^{tm1Ksw}$, that by Saegusa et al. as $Cacna1b^{tm1Ttan}$, and that by Kim et al. as $Cacna1b^{tm1Hssh}$, according to the accepted nomenclature for mouse mutant alleles (www.informatics.jax.org/mgihome/nomen/gene.shtml). Homozygous mutant mice from each line were shown to be null mutants for this gene by Northern blot and/or immunoblot analyses.

2.2.2a. N-type Current is Missing in $Ca_v2.2$ Knockout Mice. Electrophysiological analyses of mice homozygous for the $Cacna1b^{tm1Ksw}$ allele using neurons isolated from superior cervical ganglion (SCG) revealed that N-type current, which is sensitive to ω-conotoxin GVIA, was selectively abolished without compensatory increases in the activities of other types of VDCCs (Ino et al., 2001). Similarly, in dorsal root ganglion (DRG) neurons, selective loss of N-type currents with no apparent compensatory up-regulation of other VDCCs was observed in mice homozygous for the $Cacna1b^{tm1Ttan}$ and the $Cacna1b^{tm1Hssh}$ allele (Kim et al., 2001a; Saegusa et al., 2001).

2.2.2b. Altered Sympathetic Functions in $Ca_v2.2$ Knockout Mice. Heart rate and blood pressure are under the control of the autonomic nervous system. $Ca_v2.2$ channel expression is abundant in the autonomic system, suggesting that these parameters might be altered in $Ca_v2.2$ knockout mice. Ino et al. measured arterial blood pressure and heart rate through a catheter placed in the carotid artery under anesthetized conditions. Mean blood pressure and heart rate were significantly higher in mice homozygous for the $Cacna1b^{tm1Ksw}$ allele. Saegusa et al., however, reported no significant differences from wild type in the blood pressure and heart rate of mice homozygous for the $Cacna1b^{tm1Ttan}$ allele. Measurements were made in somewhat different ways, and the reason for the discrepancy is not yet clear.

Ino et al. also examined the carotid baroreflex functions in mice homozygous for the $Cacna1b^{tm1Ksw}$ allele. Mean blood pressure was monitored in response to occlusion of the carotid artery. Normally, occlusion of the carotid artery results in a decrease in the blood pressure, which then triggers activation of sympathetic nerves to increase the blood pressure. This baroreflex was assessed in wild type and knockout mice by monitoring mean blood pressure. There were almost no changes in mean blood pressure of knockout mice after carotid artery occlusion, whereas the occlusion caused a significant increase in the blood pressure of wild type mice. Thus, the N-type channel plays an essential role in the carotid baroreflex function.

Ino et al. also examined cardiac inotropic responses, using an isolated left atrial preparation, to observe directly the effects of N-type channel ablation on sympathetic and parasympathetic nerves. Neuronal stimulation of atria from the wild type mouse led to positive inotropic responses in the presence of a blocker for the muscarinic acetylcholine receptor (atropine) and to negative inotropic responses in the presence of an adrenergic receptor blocker (propranolol). When their knockout mice were used for these tests, impaired positive but normal negative inotropic responses were observed. This suggests that the N-type channel governs neurotransmission at sympathetic nerve endings, and that other VDCCs can compensate for loss of the N-type channel in the parasympathetic nerve endings. This result is seemingly inconsistent with the increased basal arterial pressure, since a deficit in transmission at sympathetic nerve endings is likely to lead to a lower circulating level of norepinephrine. However, an almost normal level of circulating norepinephrine was detected in the blood of the knockout mice. Thus, the deficit in the sympathetic nerve function itself cannot explain the abnormal basal blood pressure and heart rate in the knockout mice.

2.2.2c. Abnormal Emotional State in $Ca_v2.2$ Knockout Mice. Although the gross physiology of $Ca_v2.2$ knockout mice was apparently normal, several behavioral tests were performed to search for more subtle abnormalities in neuronal functions. Saegusa et al. performed elevated plus-maze, open-field, and acoustic startle response tests (Tarantino and Bucan, 2000). In the elevated plus-maze test, the number of entrances into open and closed arms of the maze and the time spent in each of the arms was recorded. These parameters are thought to reflect in some measure the level of anxiety of the animal. Usually, normal mice do not like to be exposed in the open arms and tend to enter closed arms more frequently and spend more time there. In contrast, mice homozygous for $Cacna1b^{tm1Ttan}$ tended to enter open arms more frequently and to spend longer time in the open arms, suggesting a lowered level of anxiety in this knockout. In the open-field test, the total length of time and total distance traveled were recorded. These are also thought to indicate the level of anxiety of the animal. The knockout mice showed significantly longer time and distance of locomotor activity than wild type mice, again suggesting an overall lower level of anxiety in the knockout. Finally, startle responses were elicited by acoustic stimuli with various sound intensities, and the knockout mice showed a significantly smaller degree of response, again consistent with an overall less anxious or calmer phenotype.

2.2.2d. Abnormal Pain Responses in $Ca_v2.2$ Knockout Mice. Recently the importance of VDCCs, particularly the N-type VDCC, in pain transmission has been recognized, since several VDCC blockers have analgesic properties. It is thus of great interest to search for pain-related abnormal phenotypes in $Ca_v2.2$ knockout mice, to define further the contribution of N-type channels to pain transmission. To date, there have been three reports describing abnormal pain-related behaviors in $Ca_v2.2$ knockout mice (Hatakeyama et al., 2001; Kim et al., 2001a; Saegusa et al., 2001). Curiously, however, in some experiments different results were obtained from different lines of $Ca_v2.2$ knockout mice, although the experiments were performed in almost the same way. Here we review the

Table 1. Comparison of phenotypes of $Ca_v2.2$ knockout mice in several pain-related behavioral tests. N.D. = not determined.

Test	$Ca_v2.2$ Knockout Line		
	Cacna1b^{tm1Ttan}	*Cacna1b*^{tm1Ksw}	*Cacna1b*^{tm1Hssh}
von Frey	Normal	N.D.	Increased threshold
Tail pressure or tail pinch	Normal	Normal	Increased threshold
Paw-flick	Normal	N.D.	Prolonged latency
Tail-flick	Prolonged latency	N.D.	Prolonged latency
Hot-plate	Normal	Prolonged latency	Normal
Formalin	Reduced licking time in the early half of phase 2	Reduced licking time in phase 2	Reduced licking time in phase 2
Acetic acid writhing	Normal	N.D.	Reduced amount of writhing
Spinal nerve ligation	Reduced thermal hyperalgesia Reduced mechanical allodynia	N.D.	N.D.

pain-related abnormal phenotypes in these mutant mice.

Table 1 shows a comparison among these knockout mice of responses to stimuli such as mechanical force and heat. As stated, inconsistent results were obtained in results concerning acute nociception in these three lines of knockouts. Threshold for mechanical stimuli can be assessed by a von Frey test (Chaplan et al., 1994); a significantly higher threshold was seen only in the *Cacna1b*^{tm1Hssh} line. Tail pinch and tail pressure tests measure thresholds for noxious stimuli. Again, in these assays, only the *Cacna1b*^{tm1Hssh} line showed an increased threshold. Responses to thermal stimuli can be determined by several tests, including paw-flick, tail-flick and hot-plate tests. Paw- and tail-flick tests represent spinal reflexes at the lumbar and sacral levels respectively, and the hot-plate test assesses both spinal and supraspinal involvement in nociception (Chapman et al., 1985). In the *Cacna1b*^{tm1Hssh} line, a prolonged latency to paw withdrawal was apparent in the paw-flick test, but in the *Cacna1b*^{tm1Ttan} line paw-flick latency was unchanged. In the tail-flick test, prolonged latency, corresponding to impaired nociception, was observed in both *Cacna1b*^{tm1Hssh} and *Cacna1b*^{tm1Ttan} lines. However, in the hot plate test, only one knockout line, *Cacna1b*^{tm1Ksw}, showed signs of impaired nociception.

The formalin test and the acetic acid writhing test are often used to assess an animal's responses to chemical stimuli. When formalin is injected into a hindpaw of a mouse, it normally induces licking and/or biting behavior. These responses are divided into two types based on the mechanism by which they are induced: phase 1 responses, which usually occur within 10 minutes of formalin injection, are elicited by direct stimulation of peripheral nociceptors, and phase 2 responses, which begin 10 minutes after injection, are induced by inflammation caused by the formalin. The acetic acid writhing test is a model to assess an animal's response to visceral pain associated with inflammation. The results of the formalin test were almost the same among the different mutant lines: phase 1 responses are not significantly different from control and phase 2 responses are significantly reduced. Phase 2 responses are subdivided into two subphases (Malmberg and Yaksh, 1992; Dirig et al., 1997). Homozygous mutant mice showed reduced responses in the earlier half of phase 2 (Kim et al., 2001a; Saegusa et al., 2001). Results of the acetic acid writhing test were not consistent among the different mutant lines. Only the *Cacna1b*^{tm1Hssh} line showed a reduced amount of writhing, indicating a nociceptive defect, whereas knockout mice from the other lines showed normal responses. These inconsistencies may be related to the different genetic backgrounds of the mutant mice and to the general instability of behavioral studies, as discussed previously (Saegusa et al., 2002). At present it may be safe to conclude that the Ca$_v$2.2 channel is involved in at least the transmission of inflammatory pain.

In addition to these tests, a model of neuropathic pain was used to test one of the Ca$_v$2.2 mutant lines, *Cacna1b*^{tm1Ttan} (Saegusa et al., 2001). Neuropathic pain is caused by nerve injury and is known to be difficult to cure (Devor and Seltzer, 1999; Woolf and Mannion, 1999). When the spinal nerve ligation model was applied to wild-type mice (Kim and Chung, 1992), a robust thermal hyperalgesia and mechanical allodynia was observed. However, in the homozygous mutant, the same treatment yielded markedly reduced signs of both thermal hyperalgesia and mechanical allodynia. These results clearly suggest that Ca$_v$2.2 channel is essential for the development of neuropathic pain, and that controlling N-type channel activity may be useful in its treatment.

2.2.3. The Ca$_v$2.3 Channel

Although originally identified as a component of the low voltage-activated channel, the Ca$_v$2.3 channel clone likely corresponds to at least some of the R-type current in neurons that is resistant to dihydropyridines and to most peptide toxins (Soong et al., 1993). The peptide SNX-482, which blocks at least some forms of Ca$_v$2.3 channel (Newcomb et al., 1998), only recently became commercially available; thus Ca$_v$2.3 channels in native tissue are comparatively uninvestigated. Knockout mice lacking this channel are an ideal resource to delineate the functions of this channel.

So far four independent lines of Ca$_v$2.3 knockout mice have been generated, in four different laboratories (Saegusa et al., 2000; Wilson et al., 2000; Lee et al., 2002; Pereverzev et al., 2002). The targeted allele made by Saegusa et al. was a knock-in allele in which the *nlacZ* gene was inserted in the exon 1 of the *Cacna1e* gene encoding Ca$_v$2.3, designated as *Cacna1e*^{tm1Ttan}. The insertion of *nlacZ* disrupted the *Cacna1e* gene and simultaneously labeled the cells expressing *Cacna1e* with β-galactosidase activity. Wilson et al. replaced the sequence encoding the S4-S6 regions of domain II of the Ca$_v$2.3 protein with a selection marker gene. This allele was designated as *Cacna1e*^{tm1Rjml}. Lee et al. made a targeted allele by inserting a *GFP-neo*^{R} cassette into exon 1, thereby marking the *Cacna1e* expressing cells with GFP. This targeted allele was designated as *Cacna1e*^{tm1Hssh}. Pereverzev et al. first made a floxed (flanked by loxP sequences) allele for the *Cacna1e* gene, thereby making it possible to generate conditional targeted mutant mice. Exon 2 was flanked by two loxP sequences, each of which was located in a different intron. This allele, *Cacna1e*^{tm1Tsch}, behaved just like a wild type allele. A germ-line deleter strain, which expresses Cre recombinase constitutively under the control of the CMV promoter, was used to create a null allele for *Cacna1e*. Cre-mediated recombination between two loxP sequences led to excision of exon 2 of *Cacna1e*, thereby creating a null allele. We refer to this allele created by Cre-mediated recombination as *Cacna1e*$^{tm1.1Tsch}$. All of these mutated alleles (except for the floxed *Cacna1e*^{tm1Tsch} allele made by Pereverzev et al.) were confirmed to be null alleles by Western blot analysis. Mice carrying each allele were analyzed with various tests.

2.2.3a. R-type Currents in Ca$_v$2.3 Knockout Mice. The relationship between R-type currents and the Ca$_v$2.3 channel was analyzed in two of the mutant lines, *Cacna1e*^{tm1Rjml} and *Cacna1e*^{tm1Hssh} (Wilson et al., 2000; Lee et al., 2002). Previous studies using antisense oligonucleotides suggested that the Ca$_v$2.3 channel was responsible for at least a part of the R-type current (Piedras-Renteria and Tsien, 1998). However, it is difficult to conclude unambiguously from antisense experiments whether the Ca$_v$2.3 channel is the sole origin of R-type current, since the antisense strategy generally cannot eliminate completely the expression of a target gene. Thus the additional information obtained from knockout studies is important.

Wilson and colleagues examined the R-type currents in cerebellar granule neurons and neurons from dorsal root ganglia (DRG) by the patch clamp technique. R-type currents were isolated pharmacologically by the application of nimodipine (0.5 or 2 μM), ω-agatoxin IVA (200 nM), and ω-conotoxin GVIA (0.5 or 1 μM), to block L-, P/Q-, and N-type channels respectively. Effects of the peptide inhibitor SNX-482 (1 μM) from tarantula venom were assayed on these resistant currents in cells from wild type mice and from mice homozygous for the *Cacna1e*^{tm1Rjml} allele. R-type currents in granule neurons from homozygous *Cacna1e*^{tm1Rjml} mice were only slightly smaller than in granule neurons from wild type, and no SNX-482 sensitive component was observed in the *Cacna1e*^{tm1Rjml} homozygous mice. In neurons isolated from DRG, two populations with different sensitivity to SNX-482 were found: the resistant current of one population was partially sensitive to SNX-482, whereas the resistant current of the other population was insensitive to SNX-482. In the former, SNX-482 inhibited the current only slightly. Currents from DRG neurons isolated from mice homozygous for the *Cacna1e*^{tm1Rjml} allele were insensitive to SNX-482. These results suggest that Ca$_v$2.3 channels contribute to resistant current in DRG neurons and granule neurons, but that the contribution is small. In other words, there should be other channels that underlie the most of the resistant current.

A different conclusion was reached by Lee et al. on the relation between the Ca$_v$2.3 channel and the resistant current in central amygdala (CeA) neurons. Using 10 μM nifedipine (Wilson et al. used nimodipine), ω-agatoxin IVA, and ω-conotoxin GVIA, they studied the R-type current in CeA neurons from wild type mice and from mice homozygous for the *Cacna1e*^{tm1Hssh} allele. In *Cacna1e*^{tm1Hssh} homozygous mice, most of the R-type current was eliminated, suggesting that most of the R-type current in CeA neurons arises from Ca$_v$2.3 channels. The discrepancy may arise from the different types of neurons studied or from the non-saturating nimodipine concentrations used by Wilson et al, concentrations likely too low to block the Ca$_v$1.3 L-type channel (Koschak et al., 2001; Xu and Lipscombe, 2001).

A recent study examined R-type currents in central neurons from mice homozygous for the *Cacna1e*$^{tm1.1Tsch}$ allele (Sochivko et al., 2002). Using saturating concentrations of L, N, and P/Q-type channel blockers (nifedipine, ω-conotoxin GVIA, ω-conotoxin MVIIC, ω-agatoxin IVA), these authors concluded that $Ca_v2.3$ channels underlie a major part of the R-type currents in neocortical neurons and hippocampal CA1 pyramidal neurons. However, only half of the R-type current in hippocampal dentate gyrus neurons was attributable to the $Ca_v2.3$ channel. It is likely that currents resistant to the spectrum of L, N, and P/Q-type blockers are not made up of a single channel type and that the contribution of the $Ca_v2.3$ channels varies among different types of neuron.

2.2.3b. Function of the $Ca_v2.3$ Channel in Pain Transmission. By assessing the β-galactosidase activity in mice carrying the *nlacZ* knock-in allele of the *Cacna1e* gene (*Cacna1e*^{tm1Ttan}), this gene was found to be expressed in the dorsal horn of the spinal cord and in some DRG neurons (Saegusa et al., 2000). As mentioned above, the $Ca_v2.3$ channel supports at least part of the R-type current in a subset of DRG neurons. This raised the possibility that the $Ca_v2.3$ channel participates in the control of pain transmission, since the primary afferent neurons are known to transmit "pain signals" to dorsal horn neurons in the spinal cord (Basbaum and Jessell, 2000). Pain-related behavioral tests were performed using mice homozygous for the *Cacna1e*^{tm1Ttan} allele to search for possible abnormalities (Saegusa et al., 2000).

Acute nociceptive perception was examined by von Frey, tail-flick, paw-flick, and hot-plate tests using wild type, heterozygous, and homozygous mutant mice. However, none of these tests detected abnormalities in the mutant mice. The responses in phase 1 of the formalin test, which represents the acute responses to chemical stimuli, were also normal in the mutant mice. Responses in phase 2 of the formalin test, indicative of inflammatory pain, were significantly reduced in hetero- and homozygous mutant mice compared to wild type. This suggests that the $Ca_v2.3$ channel is involved in the transmission of inflammatory pain. Curiously, however, only the heterozygous mutant mice showed reduced responses in the acetic acid writhing test, a model for visceral pain associated with inflammation. The reason for the normal writhing responses in homozygous mutant mice is unlikely to reside in the compensatory up-regulation of other VDCCs. Saegusa et al. did not detect any changes in the expression levels of VDCC α_1 subunit mRNAs in thoracic spinal cord and DRG. Moreover, all electrophysiological analyses on $Ca_v2.3$ knockout mice show specific reduction in R-type current, with no significant up-regulation of other subtypes. The exact reason must await further studies, but one possible mechanism to explain the phenotype is a deficit in an antinociceptive mechanism in the mice homozygous for the *Cacna1e*^{tm1Ttan} allele.

2.2.3c. Increased Anxiety in $Ca_v2.3$ Knockout Mice. $Ca_v2.3$ knockout mice (*Cacna1e*^{tm1Ttan} and *Cacna1e*^{tm1Hssh} lines) seemed to show increased level of anxiety, based on their vigorous attempts to escape cages upon handling (Saegusa et al., 2000; Lee et al., 2002). Consistent with this, the knockout mice showed abnormalities in some anxiety-related behavioral tests. Although the knockout mice showed almost normal behaviors in the elevated plus-maze (Saegusa et al., 2000; Lee et al., 2002) and startle response tests (Saegusa et al., 2000), they showed signs of elevated anxiety in the open-field test (Saegusa et al., 2000; Lee et al., 2002). The knockout mice traveled shorter distance than wild-type control (Saegusa et al., 2000). In particular, they tended to stay near the walls and spent less time in the center area of the field (Saegusa et al., 2000; Lee et al., 2002). This suggests an elevated level of anxiety in the $Ca_v2.3$ knockout mice (*Cacna1e*^{tm1Ttan} and *Cacna1e*^{tm1Hssh} lines) upon exposure to a novel environment.

2.2.3d. The $Ca_v2.3$ Channel and Hippocampal Function. The hippocampus is involved in several important functions, including some forms of learning and memory. Underlying mechanisms are thought to include forms of synaptic plasticity including long-term potentiation (LTP). It is known that many mutant mice carrying a targeted deletion in genes involved in Ca^{2+} signaling show impaired hippocampal synaptic plasticity and learning/memory (Chen and Tonegawa, 1997). It is also suggested that VDCCs participate in this process, since under some conditions VDCC-dependent LTP has been detected (Ito et al., 1995; Cavus and Teyler, 1996). The identitiy of the VDCC subtypes involved in this synaptic plasticity are still controversial. Since the $Ca_v2.3$ channel is abundantly expressed in the hippocampus, Kubota, Tanabe, and colleagues addressed the involvement of the $Ca_v2.3$ channel in plasticity (Kubota et al., 2001).

The contribution of $Ca_v2.3$ channels to synaptic transmission was investigated by recording the field EPSP (fEPSP) from the stratum radiatum in CA1 in response to stimulation of the Schaffer collateral pathway. fEPSP shape and the input-output relation were not significantly different between wild type mice and mice homozygous for the *Cacna1e*^{tm1Ttan} allele ($Ca_v2.3$ knockout mice). Paired pulse facilitation, an index for characterizing the presynaptic terminals, was not significantly different between the two genotypes. Thus, basic synaptic functions in the CA1 region are not affected by the deletion of the $Ca_v2.3$ channel. Involvement of $Ca_v2.3$ channels in LTP induced by the theta-burst stimulation (TBS) of the Schaffer collateral-CA1 synapses was also examined. The sensitivity of this form of LTP to Ni^{2+} (25 μM), which blocks preferentially R- and T-type VDCCs, suggested involvement of $Ca_v2.3$ channels. In mice homozygous for the *Cacna1e*^{tm1Ttan} allele, however, TBS still evoked LTP similar to wild type, as indicated by an increase in the slope of the fEPSP.

Kubota et al. also examined the behavioral consequences of $Ca_v2.3$ channel ablation with two types of fear conditioning tests, contextual and tone-dependent fear conditioning, related to hippocampal and non-hippocampal functions, respectively. The contextual fear conditioning test consists of two-day sessions. On day 1, individual mice are put in a conditioning box, where they receive noxious stimuli (i.e., electrical shock) three times. On day 2, they are put in the same box, but with no noxious stimuli. If mice remember the context, they take a specific posture called freezing, which is a sign of fearfulness. Tone-dependent fear conditioning is similar, but tone stimuli are given before each noxious stimulus. On day 2, mice are transferred to a novel environment and receive the tone stimulus. If mice remember the tone, they show the freezing response. In both fear conditioning tests, essentially normal responses were observed in mice homozygous for the *Cacna1e*^{tm1Ttan} allele. Thus fear memory seems intact in these mice.

Finally, Kubota et al. performed the Morris water-maze test (Morris et al., 1982; Tsien et al., 1996) using the mice homozygous for the *Cacna1e*^{tm1Ttan} allele, to assess spatial learning and memory. The water-maze test uses a hidden platform as a training session, a probe test to assess spatial memory, and finally a visible platform test to check for the normal motor and vision functions. In each test, mice are forced to swim in a circular pool from which they can escape if they find the platform. Normal mice can learn the position of the platform and with increased training show a shorter latency time to find it.

In the training session, $Ca_v2.3$ knockout mice tended to find the platform with slightly longer latencies than control mice, but their performance did improve as the test progressed. Overall, their performances were not significantly different from those of wild type mice. However, the probe test revealed that the time spent in the target quadrant by mice homozygous for the *Cacna1e*^{tm1Ttan} allele was significantly decreased. Performance in the visible platform test was normal in mice homozygous for the *Cacna1e*^{tm1Ttan} allele. One peculiar feature observed in the knockout mice was a thigmotaxic swimming—the Ca$_v$2.3 knockout mice tended to swim along the wall of the circular pool. The thigmotaxic swimming may be related to an abnormal emotional state; thus, the apparent subtle deficit in spatial memory in mice homozygous for the *Cacna1e*^{tm1Ttan} allele may be related instead to an abnormal emotional state.

2.2.3e. The Ca$_v$2.3 Channel and Cerebral Ischemia. Since increases in intracellular Ca^{2+} concentration ($[Ca^{2+}]_i$) are critical for ischemic neuronal injury, VDCCs may play important roles in the excitotoxic cascade not only as a pathway of Ca^{2+} influx, but also as a crucial part of the neurotransmitter release machinery. Indeed, L-, N-, and P/Q-type VDCCs are thought to play aggravating functions in ischemic neuronal injury, because blockers for these channels can reduce the ischemic damage (Kobayashi and Mori, 1998). Abundant expression of the Ca$_v$2.3 channel in the CNS suggests that this channel is also involved in the pathophysiology of the neuronal ischemic injury. Toriyama, Tanabe, and colleagues examined the role of the Ca$_v$2.3 channel in the development of neuronal ischemic damage by using Ca$_v$2.3 knockout mice (Toriyama et al., 2002).

Mice homozygous for the *Cacna1e*^{tm1Ttan} allele were subjected to a focal ischemia model. Twenty four hours after occlusion of middle cerebral artery, infarct was localized to the temporo-parietal cerebral cortex, as visualized by staining brain slices with 2,3,5,-triphenyltetrazolium chloride. When the infarct rate, defined as the ratio of infarct volume to contralateral hemispheric volume, was compared between wild type and knockout mice, a significant increase in the infarct rate was detected in the knockout mice. To understand the physiological basis of this result, $[Ca^{2+}]_i$ transients in hippocampal slices were examined under an ischemia-like condition, oxygen-glucose deprivation (OGD). When hippocampal slices from wild type mice and from mice homozygous for the *Cacna1e*^{tm1Ttan} allele were exposed to the OGD condition, a transient increase in $[Ca^{2+}]_i$ was observed in the CA1 pyramidal cell layer and in the dentate gyrus (DG) granular layer. Consistent with the faster infarct rate, the rate of $[Ca^{2+}]_i$ increase was significantly greater in mice homozygous for the *Cacna1e*^{tm1Ttan} allele than in wild type, for both CA1 and DG regions. To examine whether the difference in response to OGD was related to neuronal activity, slices were OGD-treated, with sodium-dependent action potentials blocked by the presence of tetrodotoxin (TTX). The genotype-dependent difference in $[Ca^{2+}]_i$ increase was eliminated by TTX, suggesting that action potential generation was involved in the faster $[Ca^{2+}]_i$ increases in the mice without functional Ca$_v$2.3 channels. This assay was also employed with the γ-aminobutyric acid receptor type A (GABA$_A$) antagonist bicuculline, since the faster $[Ca^{2+}]_i$ increases in mice lacking the Ca$_v$2.3 channel suggested a coupling of the Ca$_v$2.3 channel to GABA release in wild type mice. As with TTX, bicuculline eliminated the genotype-dependent differences in the rate of $[Ca^{2+}]_i$ increase. This suggests that the Ca$_v$2.3 channel and the GABA system are related to the faster $[Ca^{2+}]_i$ increase in mice homozygous for the *Cacna1e*^{tm1Ttan} allele, and that the Ca$_v$2.3 channel exerts some protective functions in ischemic neuronal injury.

2.2.3f. The Ca$_v$2.3 Channel and Sperm Function. Ca^{2+} influx into sperm plays a crucial role in successful fertilization, and VDCCs are thought to be a pathway for this Ca^{2+} influx. Low-voltage-activated (LVA, T-type-like) current has been detected in spermatogenic cells, and this T-type-like current has been suggested to be related to induction of the acrosomal reaction. However, the molecular counterpart of this LVA current has been controversial. The $Ca_v2.3$ channel is known to be expressed in sperm and its precursor cells, as revealed by RT-PCR and immunocytochemistry (Lievano et al., 1996; Westenbroek and Babcock, 1999). The $Ca_v2.3$ channel activates at low voltages when expressed in cultured cells without auxiliary subunits (Meir and Dolphin, 1998), suggesting that $Ca_v2.3$ channels may underlie the LVA current in spermatogenic cells (Lievano et al., 1996).

Sakata, Tanabe, and colleagues addressed this question by using mice homozygous for the $Cacna1e^{tm1Ttan}$ allele. First, LVA current in the pachytene spermatocyte was examined with the whole cell patch clamp, to investigate a contribution of the $Ca_v2.3$ channel (Sakata et al., 2001). The properties of this current were almost identical in wild type and $Ca_v2.3$ knockout mice, leading to the conclusion that the $Ca_v2.3$ channel does not contribute to the T-type-like current in male germ-line cells. Further experiments examined sperm movement and the $[Ca^{2+}]_i$ transient in the sperm head in response to several agonists to search for abnormalities in mature sperm function in $Cacna1e^{tm1Ttan}$ homozygous mice (Sakata et al., 2002). When sperm were stimulated with solubilized zona pellucida (ZP), the physiological agonist for the acrosome reaction, the speed of the increase in $[Ca^{2+}]_i$ tended to be lower in the knockout (although the difference in rate was not statistically significant). When sperm were stimulated with α-D-mannose- bovine serum albumin, an agonist used to elicit the acrosome reaction in human sperm, $[Ca^{2+}]_i$ rose at a significantly lower rate in sperm from mice homozygous for the $Cacna1e^{tm1Ttan}$ allele. Finally, high-potassium medium was applied to elicit the $[Ca^{2+}]_i$ transient by direct depolarization. By this stimulation, however, the rate of $[Ca^{2+}]_i$ increase was faster in the knockout mice. Sperm movement were analyzed *in vitro* using a computer-assisted tracking assay. Under some conditions, the sperm from the mice homozygous for the $Cacna1e^{tm1Ttan}$ allele showed more linear movement. In other words, wild type sperm tended to move in a more randomized fashion.

At present, the physiological meaning of these abnormal phenotypes is not clear. However, it seems that the $Ca_v2.3$ channel may at least contribute to some parts of the mechanisms underlying the acrosome reaction and the control of flagellar movement.

2.2.3g. The $Ca_v2.3$ Channel and Glucose Homeostasis. As discussed in section 2.1.2c on the $Ca_v1.3$ channel knockouts, L-type channels play an important role in glucose-stimulated insulin secretion (GSIS). However, non L-type VDCCs are also reported to be involved in insulin secretion (Ligon et al., 1998). Recently, a specific splice variant of $Ca_v2.3$ has been identified in the INS-1 insulinoma cell line, suggesting its involvement in the control of insulin secretion (Vajna et al., 2001). $Ca_v2.3$ knockout mice were used to test this hypothesis. So far two groups have reported the involvement of the $Ca_v2.3$ channel in glucose homeostasis *in vivo*, obtaining somewhat different results (Matsuda et al., 2001; Pereverzev et al., 2002).

Pereverzev et al. conducted a glucose tolerance test using mice homozygous for the $Cacna1e^{tm1.1Tsch}$ allele. Mice were starved for 14 hours before the start of the test; at this time basal blood glucose and serum insulin levels from the knockout were not different from wild type. After intraperitoneal injection of glucose, mice homozygous for the $Cacna1e^{tm1.1Tsch}$ allele had higher blood glucose levels than wild type. In wild type mice, serum insulin levels increased rapidly after glucose injection, but almost no changes in insulin levels were detected in mice homozygous for the *Cacna1e*$^{tm1.1Tsch}$ allele (at least in young mice). The inability to secrete insulin in response to glucose stimulation in the knockout mice was confirmed using *in vitro* studies. GSIS was not observed from islets of Langerhans isolated from the knockout mice. Thus, the glucose intolerance observed in the $Ca_v2.3$ knockout mice was attributable to a deficit in GSIS. Finally, stress-induced hyperglycemia was examined. Stress activates the hypothalamus-pituitary-adrenal axis, leading to a higher plasma glucose level via enhanced adrenaline release from the adrenal medulla. Mice were restrained individually in narrow cages and the blood glucose level was determined before and after this immobilization stress. In wild type mice, the blood glucose level was markedly elevated 15-30 minutes after the stress administration, whereas in the mice homozygous for the *Cacna1e*$^{tm1.1Tsch}$ allele, this increase in the blood glucose level was greatly diminished.

Matsuda et al. also performed the glucose tolerance test, using mice homozygous for the *Cacna1e*^{tm1Ttan} allele. Similarly, they observed glucose intolerance in the knockout mice. They also detected insulin intolerance in mice homozygous for the *Cacna1e*^{tm1Ttan} allele: an intraperitoneal injection of insulin caused less decrease in blood glucose levels in the knockout compared to wild type. It is possible, therefore, that the glucose intolerance may be explained by insulin intolerance, although serum insulin levels after glucose injection were not titered. These authors found no significant differences in stress-induced hyperglycemia between wild type mice and mice homozygous for the *Cacna1e*^{tm1Ttan} allele.

2.3. The Ca_v3 Family

The Ca_v3 family is the most recently identified group of VDCCs (Perez-Reyes, 2003). So far, three genes that encode T-type channels, $Ca_v3.1$- $Ca_v3.3$, have been discovered, each encoding a T-type channel with distinctive biophysical features. To date there are no drugs that can discriminate among members of this family, and pharmacological separation of T-type and high voltage-activated channels is also somewhat difficult (see Ertel, this volume). A genetic approach would be most fruitful to clarify the *in vivo* functions of each member. Thus far, however, there is only one report of targeted mutant mice for the Ca_v3 family, whose phenotypes we will discuss here.

2.3.1. The $Ca_v3.1$ Channel

To create a null allele for *Cacna1g* encoding the $Ca_v3.1$ channel, Kim, Shin, and colleagues made a targeting vector in which most of the exon coding for amino-terminal residues 82-118 was replaced with a *neo*^{R} cassette (Kim et al., 2001b). No $Ca_v3.1$ channel protein was detected by Western blot analysis of brain membrane proteins from mice homozygous for this allele, confirming that the targeted allele was a null one. Moreoever, whole-cell patch clamp recordings showed that T-type currents were completely absent in thalamocortical (TC) neurons from knockout mice.

2.3.1a. Burst Mode Firing of Action Potentials is Lacking in Neurons From $Ca_v3.1$ Knockout Mice. Kim et al. examined the excitability of TC neurons from the $Ca_v3.1$ knockout mouse by a current clamp method. In wild type TC neurons, hyperpolarizing current injection followed by return to the resting potential of –70 mV elicited a rebound low threshold calcium potential with a burst of action potentials at 200-500 Hz. This rebound burst firing was not detected in TC neurons from knockout mice. Injection of depolarizing current to neurons at –80 mV also caused burst firing in wild type but not knockout mice. On the other hand, tonic action potential firing (100-200 Hz) induced by injection of depolarizing current to a membrane held at –60 mV, a resting voltage sufficiently positive to inactivate T-type channels, was not significantly different between the two genotypes. Thus, $Ca_v3.1$ channels govern the rebound burst mode of action potential firing in TC neurons.

2.3.1b. The $Ca_v3.1$ Channel and Spike-Wave Discharges Associated with Absence Seizures. To examine the involvement of the $Ca_v3.1$ channel in the generation of spike-and-wave discharges (SWDs) associated with absence seizures, $Ca_v3.1$ knockout mice were treated with drugs such as γ-butyrolactone or (RS)-baclofen that induce such seizures. When γ-butyrolactone (70 mg/kg) was injected into wild type mice, electroencephalogram (EEG) recordings showed 3-5 Hz paroxysmal SWDs. In the knockout, such SWDs were not obvious upon drug injection. Similar results were obtained with administration of (RS)-baclofen (20 mg/kg). The conclusion is that the $Ca_v3.1$ channel is required for generation of the SWDs. Field recordings of thalamic nuclei activity confirmed that the SWDs were related to functioning thalamocortical networks.

Involvement of $Ca_v3.1$ channels in the generation of SWDs originating from the cortex was also examined. It is known that an intraperitoneal injection of bicucullinemethobromide (BMB), a $GABA_A$ antagonist, induces SWDs originating primarily from the cortex. When the $Ca_v3.1$ knockout mice were injected with BMB, the EEG pattern showed SWDs similar to wild type control. The potassium channel blocker 4-aminopyridine (4-AP) was also used to test involvement of $Ca_v3.1$ channels in convulsive seizures. Injection of 4-AP induced convulsive seizures in wild type and in $Ca_v3.1$ knockout mice. Therefore, the lack of the $Ca_v3.1$ channel results in the specific loss of SWDs that originate from thalamocortical networks.

3. KNOCKOUTS OF AUXILIARY SUBUNITS

Compared to the pore-forming α_1 subunits, the physiological functions of auxiliary subunits in neurons are much less well characterized, since no known reagents specifically inhibit the functions of auxiliary subunits. Genetic ablation of auxiliary subunits, however, has provided valuable information. To date, knockout mice for β_1, β_2, β_3, and γ_1 subunits have been established, and here we review their phenotypes and deduced functions. *lethargic* mice, which lack the full β_4 subunit, are discussed by Mori et al. (this volume).

3.1. The Beta Subunit

3.1.1. β_1 Subunit

The β_1 subunit was first identified as an auxiliary subunit for the skeletal muscle dihydropyridine receptor. This subunit, when expressed in heterologous systems, modulates channel functions. To assess the role of the β_1 subunit in DHP receptor kinetics and excitation-contraction (E-C) coupling, Gregg et al. established a targeted mutant for the β_1 subunit by constructing a targeting vector in which exons 6, 7a, and 7b were replaced with the *neo*^{R} cassette (Gregg et al., 1996). As exon 6 is used in all β_1 transcripts (exon 7a is skeletal muscle-specific and exon 7b is brain-specific), this targeting strategy likely eliminated all splice variants of the β_1 subunit. The mutation was lethal: mice homozygous for the targeted allele died near the perinatal period. Histological analyses of the β_1 knockout fetuses showed greatly reduced skeletal muscle mass and disorganized myofibrils. Although at E18 the β_1 knockout fetuses were obtained at a Mendelian ratio, they failed to show any movement, suggesting impaired E-C coupling.

Electrical stimulation of the forelimb from wild type E16 fetuses induced twitches, but similar stimulation of β_1 knockout fetuses produced no twitches. When these preparations were treated with caffeine, to increase cytoplasmic [Ca^{2+}] directly by evoking the release of Ca^{2+} from intracellular stores, muscle contraction was observed in both control and β_1 knockout mice. TTX-sensitive action potentials could be induced in both control and knockout myotubes, but action potentials in the knockout myotubes failed to induce a Ca^{2+} transient. These results clearly showed a deficiency in the E-C coupling in the β_1 knockout myotube.

Gregg et al. then examined the Ca^{2+} current from the β_1 knockout myotubes. Whole-cell patch clamp recording revealed a greatly reduced Ca^{2+} current in the β_1 knockout myotubes. There are two possible mechanisms to explain this phenotype: 1) the β_1 subunit is an essential modulator directly controlling the activity of the $Ca_v1.1$ channel; or 2) the β_1 subunit is essential for the transport and/or insertion of the $\alpha_1 1.1$ subunit into the membrane. The former possibility is unlikely because heterologous expression experiments showed $Ca_v1.1$ channel activity in the absence of the β_1 subunit. Moreover, immunohistochemical studies supported the latter possibility. When the control myotubes were stained with anti-β_1 or anti-$\alpha_1 1.1$ antibodies, in both cases a similar punctate staining pattern was observed. Conversely, staining β_1 knockout myotubes with anti-β1 or anti-$\alpha_1 1.1$ antibodies produced no signal. This suggests strongly that the β_1 subunit is essential for proper expression of the $\alpha_1 1.1$ subunit.

3.1.2. β_2 Subunit

The β_2 subunit was originally cloned from the heart and is known to be expressed predominantly in heart, aorta, and brain (Hullin et al., 1992). As a part of a study to clarify β subunit functions in the CNS, a knockout mouse for the β_2 subunit was generated (Ball et al., 2002). The strategy for the gene targeting was to delete exons 4-13 and to insert neo^R therein. Homozygous mutant pups were never obtained, suggesting embryonic lethality. Cardiac dysfunction is thought to produce the lethality, since the β_2 subunit is normally expressed at high levels in the heart, and since homozygous mutant embryos showed no cardiac contraction. To study the functions of the β_2 subunit in the adult CNS, Ball et al. performed transgenic-rescue experiments. A transgenic mouse line was generated expressing the rat β_2 subunit under the control of the α-myosin heavy chain promoter, which drives gene expression in the heart. The lethal phenotype of the β_2 knockout mice was successfully rescued with this transgene. The mice homozygous for the targeted β_2 mutation and carrying the rat β_2 transgene were viable and fertile. These mice were designated as CNS-β_2-null mice.

Ball et al. examined visual system function in CNS-β_2-null mice. Western blot analysis confirmed the absence of the β_2 subunit in retina from the CNS-β_2-null mice, and electroretinograms (ERGs) were performed in response to flash stimuli under dark-adapted conditions, in which responses originated mostly at rod cells. ERGs recorded from control mice contained two major components, termed the a-wave and the b-wave. The former represents light-induced closure of cation channels along the rod photoreceptor outer segments, and the latter represents summed activity of rod bipolar cells and other inner retinal neurons. In the CNS-β_2-null mice, the a-wave was almost normal, but the amplitude of the b-wave was greatly reduced. Under light-adapted conditions, to observe cone cell-dependent functions, b-waves were obvious in the control mice. However, in CNS-β_2-null mice, the b-wave was again greatly diminished. These results suggest that photoreceptor functions are normal in CNS-β_2-null mice, but transmission of the signal to the inner retina is abnormal.

Behavioral consequences of visual functions in the CNS-β_2-null mice were also examined in this study. In a two-way active avoidance test, in which light was used as the conditioning stimulus, CNS-β_2-null mice learned to avoid a foot shock almost as well as control mice. When the intensity of the conditioning stimuli was lowered, however, the avoidance performance declined, suggesting a reduced visual sensitivity in CNS-β_2-null mice. Histological analysis of the retina also showed that the outer plexiform layer, where ribbon synapses transmit the visual signals generated in the photoreceptor cells, was thinner in null mice than in wild type mice. These phenotypes are remarkably close to the symptoms observed in human X-linked congenital stationary night blindness, which is caused by loss-of-function mutations in *CACNA1F* gene encoding the $Ca_v1.4$ L-type channel (Bech-Hansen et al., 1998; Strom et al., 1998). Accordingly, $Ca_v1.4$ channel expression was examined in the retina of the CNS-β_2-null mice. A marked reduction in the intensity of immunostaining was apparent in the OPL. Evidently the β_2 subunit is essential for proper expression and/ or membrane insertion of the $Ca_v1.4$ channel.

3.1.3. β_3 Subunit

Of all four β subunits identified, the β_3 subunit is the most abundant in the brain. In particular, more than half of the N-type channel complex is associated with this subunit. Therefore the consequence of the ablation of the β_3 subunit was expected to be interesting, since the N-type channel plays an important role in neurotransmitter release. Also, the β_3 subunit is known to enhance channel modulation by G-proteins. Namkung, Tsien, Shin and colleagues have generated β_3 knockout mice and examined the Ca^{2+} current in SCG neurons, including modulation by G proteins (Namkung et al., 1998).

The *Cacnb3* gene encoding the β_3 subunit was disrupted by deleting 7 out of the 8 coding exons and inserting the *neo*^{R} gene. Northern blot and RT-PCR analysis using the homozygous mutant brain RNA confirmed successful generation of a null allele for the β_3 subunit. The homozygous mutant mice were viable and, surprisingly, showed no overtly abnormal behavior or tissue morphology.

To examine directly the density of Ca^{2+} current arising from different subtypes, ω-conotoxin GVIA, ω-conotoxin MVIIC, and nimodipine (10 μM) were applied to SCG neurons from wild type and β_3 knockout mice. The Ca^{2+} current density through nimodipine-sensitive L-type channels was significantly lower in the knockout mice. Although the density of ω-CTx-GVIA-sensitive N-type current varied greatly among different neurons, a population of neurons expressing smaller overall amounts of current expressed proportionally less N-type currents in neurons from β_3 knockout than from wild type. Another abnormal phenotype was related to the ω-CTx-MVIIC-sensitive P/Q-type channel. Significant differences were observed in the activation curve for P/Q currents: a second component of channel activation, possibly corresponding to channels in a "reluctant" gating mode, activated at more hyperpolarized voltages in the β_3 knockout SCG cells. Thus, a deficit in a single β subunit can affect multiple subtypes of VDCCs.

To examine the G protein-mediated inhibition of Ca^{2+} currents, norepinephrine was used as a modulatory agent. Norepinephrine inhibited both N- and P/Q-type currents to a similar degree in β_3 knockout and wild type control SCG cells. The voltage-dependent component of the inhibition by norepinephrine was greater for N-type than P/Q-type channels, but this difference was observed in cells from both β_3 knockout and wild-type control mice.

3.1.3a. Visual Pathway Development in β_3 Subunit Knockout Mice

It is known that the refinement of axonal pathways between retina, superior colliculus, and dorsal geniculate nucleus is dependent upon nitric oxide (NO), since mice lacking both endothelial and neuronal NO synthase show delays in this process (Wu et al., 2000a; Wu et al., 2000b). The details of the NO-dependent mechanism are not yet clear, but long-term synaptic depression (LTD) may be involved. Cork, Mize, and colleagues examined axonal refinement in the β_3 knockout mice, mice in which L-type current density is significantly reduced (Namkung et al., 1998; Cork et al., 2001).

After unilateral injection of anterograde tracer (cholera toxin subunit B conjugated with horseradish peroxidase) into the eye, retinal pathways were visualized by tetramethylbenzidine histochemistry. In the β_3 knockout mice, which received the tracer injection on P13, at P15 the ipsilateral retinocollicular pathway was more exuberant than control, with axonal fibers extending to appropriate and to inappropriate sites. The contralateral retinogeniculate projection was also more widespread in the β_3 knockout mice compared to wild type, again suggesting a defect in fiber retraction. Cork et al. speculate that the abnormal projections were related to a deficit in LTD. This would suggest that L-type VDCC-dependent LTD normally underlies NO-dependent axonal refinement and is disrupted in β_3 knockout mice.

3.2. The γ Subunit

3.2.1. γ_1 Subunit

The gamma subunit of the calcium channel was first identified as an accessory subunit for the skeletal muscle DHP receptor and was long thought to be specific for this channel. γ-like subunits expressed in the brain, however, were later identified by positional cloning of the gene responsible for the neurological mutant mouse *stargazer* (Letts et al., 1998). To date eight members of the γ subunit family have been cloned, designated γ_1 through γ_8 (Klugbauer et al., 2000; Burgess et al., 2001). Gene targeting experiments have been reported only for γ_1; two groups have independently established targeted mutant mouse lines for this gene (Freise et al., 2000; Ahern et al., 2001). Freise et al. replaced a 3' part of exon 1 and the adjacent splice donor site with *neo*^{R}, and Ahern et al. deleted all four exons and inserted *neo*^{R} therein. Both of the targeted mutations resulted in null alleles, as confirmed by Western blot analyses.

The γ_1 subunit is a component of the skeletal muscle DHP receptor, which has a dual function: it functions not only as an L-type VDCC through which extracellular Ca^{2+} enters the cytoplasm, but also as a voltage sensor for E-C coupling (Tanabe et al., 1988; Beam et al., 1992). Knockout mice were used to determine the role of the γ_1 subunit in both of these functions.

Whole-cell patch clamp recording of myotubes isolated from the γ_1 knockout mice revealed that L-type Ca^{2+} current density was significantly *greater* than in myotubes from wild type mice, whereas Ca^{2+} selectivity, gating, and activation kinetics were largely unchanged. The increase in current density can be explained by three possible mechanisms: 1) increased single channel conductance; 2) increased channel open probability; 3) increased number of functional channels in the membrane. Since no significant differences were observed in the single-channel conductance and in the number of functional channels, as indicated by isradipine binding, it is most probable that the γ_1 ablation resulted in an increase in the open probability (Freise et al., 2000). Studied with long depolarizations, L-type currents in myotubes from γ_1 knockout mice inactivated more slowly than currents from wild type myotubes. Steady state inactivation in the absence of γ_1 subunit was also shifted to more depolarizing potentials. These results suggest that the γ_1 subunit affects the Ca^{2+} channel function of the skeletal muscle DHP receptor in a manner so as to decrease the amount of Ca^{2+} entry. Overall, no differences in intracellular Ca^{2+} release or E-C force production were found between wild type and γ_1 knockout mice (Ursu et al., 2001). Ahern et al. reported almost the same conclusions: γ_1 ablation increased L-type current density, shifted the steady-state inactivation to more positive potentials, and exerted no effects on Ca^{2+} release (Ahern et al., 2001).

4. CONCLUDING REMARKS

As stated above, many targeted mutant mice for VDCC subunits have been generated, clarifying many previously unknown VDCC functions. A genetic approach encompassing gene-targeting technology is a powerful strategy to understand various biological processes. The most-cited limitation to the gene targeting strategy is that loss of function of the targeted gene can result in compensatory up-regulation of a related gene, complicating the interpretation of mutant phenotypes. The next generation of mutant mice will incorporate a conditional approach, with time- and/or tissue-specific gene inactivation or expression, and thereby overcome some of the problems inherent to conventional gene targeting strategies. In the future, the combination of pharmacological methodology and refined genetic technology should extend greatly our knowledge on the functions of different calcium channel subtypes *in vivo*.

REFERENCES

Ahern, C. A., Powers, P. A., Biddlecome, G. H., Roethe, L., Vallejo, P., Mortenson, L., Strube, C., Campbell, K. P., Coronado, R. and Gregg, R. G., 2001, Modulation of L-type Ca^{2+} current but not activation of Ca^{2+} release by the gamma1 subunit of the dihydropyridine receptor of skeletal muscle, *BMC Physiol* **1**:8.

Ball, S. L., Powers, P. A., Shin, H. S., Morgans, C. W., Peachey, N. S. and Gregg, R. G., 2002, Role of the β_2 subunit of voltage-dependent calcium channels in the retinal outer plexiform layer, *Invest Ophthalmol Vis Sci* **43**:1595-1603.

Basbaum, A. and Jessell, T., 2000, The perception of pain. in *Principles of Neural Science*, E. Kandel, J. Schwartz and T. Jessell, ed., McGraw-Hill, New York, p. 472-491.

Beam, K. G., Adams, B. A., Niidome, T., Numa, S. and Tanabe, T., 1992, Function of a truncated dihydropyridine receptor as both voltage sensor and calcium channel, *Nature* **360**:169-171.

Bech-Hansen, N. T., Naylor, M. J., Maybaum, T. A., Pearce, W. G., Koop, B., Fishman, G. A., Mets, M., Musarella, M. A. and Boycott, K. M., 1998, Loss-of-function mutations in a calcium-channel α_1-subunit gene in Xp11.23 cause incomplete X-linked congenital stationary night blindness, *Nat Genet* **19**:264-267.

Burgess, D. L., Gefrides, L. A., Foreman, P. J. and Noebels, J. L., 2001, A cluster of three novel Ca^{2+} channel gamma subunit genes on chromosome 19q13.4: evolution and expression profile of the gamma subunit gene family, *Genomics* **71**:339-350.

Catterall, W. A., 1998, Structure and function of neuronal Ca^{2+} channels and their role in neurotransmitter release, *Cell Calcium* **24**:307-323.

Cavus, I. and Teyler, T., 1996, Two forms of long-term potentiation in area CA1 activate different signal transduction cascades, *J Neurophysiol* **76**:3038-3047.

Chaplan, S. R., Bach, F. W., Pogrel, J. W., Chung, J. M. and Yaksh, T. L., 1994, Quantitative assessment of tactile allodynia in the rat paw, *J Neurosci Methods* **53**:55-63.

Chapman, C. R., Casey, K. L., Dubner, R., Foley, K. M., Gracely, R. H. and Reading, A. E., 1985, Pain measurement: an overview, *Pain* **22**:1-31.

Chen, C. and Tonegawa, S., 1997, Molecular genetic analysis of synaptic plasticity, activity-dependent neural development, learning, and memory in the mammalian brain, *Annu Rev Neurosci* **20**:157-184.

Cork, R. J., Namkung, Y., Shin, H. S. and Mize, R. R., 2001, Development of the visual pathway is disrupted in mice with a targeted disruption of the calcium channel β_3-subunit gene, *J Comp Neurol* **440**:177-191.

Devor, M. and Seltzer, Z., 1999, Pathophysiology of damaged nerves in relation to chronic pain. in *Textbook of Pain*, P. D. Wall and R. Melzak, ed., Churchill Livingstone, London, p. 129-164.

Dirig, D. M., Konin, G. P., Isakson, P. C. and Yaksh, T. L., 1997, Effect of spinal cyclooxygenase inhibitors in rat using the formalin test and in vitro prostaglandin E2 release, *Eur J Pharmacol* **331**:155-160.

Ertel, E. A., Campbell, K. P., Harpold, M. M., Hofmann, F., Mori, Y., Perez-Reyes, E., Schwartz, A., Snutch, T. P., Tanabe, T., Birnbaumer, L., Tsien, R. W. and Catterall, W. A., 2000, Nomenclature of voltage-gated calcium channels, *Neuron* **25**:533-535.

Fletcher, C. F., Lutz, C. M., O'Sullivan, T. N., Shaughnessy, J. D., Jr., Hawkes, R., Frankel, W. N., Copeland, N. G. and Jenkins, N. A., 1996, Absence epilepsy in *tottering* mutant mice is associated with calcium channel defects, *Cell* **87**:607-617.

Fletcher, C. F., Tottene, A., Lennon, V. A., Wilson, S. M., Dubel, S. J., Paylor, R., Hosford, D. A., Tessarollo, L., McEnery, M. W., Pietrobon, D., Copeland, N. G. and Jenkins, N. A., 2001, Dystonia and cerebellar atrophy in *Cacna1a* null mice lacking P/Q calcium channel activity, *Faseb J* **15**:1288-1290.

Freise, D., Held, B., Wissenbach, U., Pfeifer, A., Trost, C., Himmerkus, N., Schweig, U., Freichel, M., Biel, M., Hofmann, F., Hoth, M. and Flockerzi, V., 2000, Absence of the gamma subunit of the skeletal muscle dihydropyridine receptor increases L-type Ca^{2+} currents and alters channel inactivation properties, *J Biol Chem* **275**:14476-14481.

Gregg, R. G., Messing, A., Strube, C., Beurg, M., Moss, R., Behan, M., Sukhareva, M., Haynes, S., Powell, J. A., Coronado, R. and Powers, P. A., 1996, Absence of the beta subunit (cchb1) of the skeletal muscle dihydropyridine receptor alters expression of the alpha 1 subunit and eliminates excitation-contraction coupling, *Proc Natl Acad Sci U S A* **93**:13961-13966.

Hatakeyama, S., Wakamori, M., Ino, M., Miyamoto, N., Takahashi, E., Yoshinaga, T., Sawada, K., Imoto, K., Tanaka, I., Yoshizawa, T., Nishizawa, Y., Mori, Y., Niidome, T. and Shoji, S., 2001, Differential nociceptive responses in mice lacking the α_{1B} subunit of N-type Ca^{2+} channels, *Neuroreport* **12**:2423-2427.

Hofmann, F., Lacinova, L. and Klugbauer, N., 1999, Voltage-dependent calcium channels: from structure to function, *Rev Physiol Biochem Pharmacol* **139**:33-87.

Hullin, R., Singer-Lahat, D., Freichel, M., Biel, M., Dascal, N., Hofmann, F. and Flockerzi, V., 1992, Calcium channel beta subunit heterogeneity: functional expression of cloned cDNA from heart, aorta and brain, *EMBO J* **11**:885-890.

Ino, M., Yoshinaga, T., Wakamori, M., Miyamoto, N., Takahashi, E., Sonoda, J., Kagaya, T., Oki, T., Nagasu, T., Nishizawa, Y., Tanaka, I., Imoto, K., Aizawa, S., Koch, S., Schwartz, A., Niidome, T., Sawada, K. and Mori, Y., 2001, Functional disorders of the sympathetic nervous system in mice lacking the α_{1B} subunit (Ca_v 2.2) of N-type calcium channels, *Proc Natl Acad Sci U S A* **98**:5323-5328.

Ito, K., Miura, M., Furuse, H., Zhixiong, C., Kato, H., Yasutomi, D., Inoue, T., Mikoshiba, K., Kimura, T., Sakakibara, S. and et al., 1995, Voltage-gated Ca^{2+} channel blockers, ω-AgaIVA and Ni^{2+}, suppress the induction of theta-burst induced long-term potentiation in guinea-pig hippocampal CA1 neurons, *Neurosci Lett* **183**:112-115.

Jun, K., Piedras-Renteria, E. S., Smith, S. M., Wheeler, D. B., Lee, S. B., Lee, T. G., Chin, H., Adams, M. E., Scheller, R. H., Tsien, R. W. and Shin, H. S., 1999, Ablation of P/Q-type Ca^{2+} channel currents, altered synaptic transmission, and progressive ataxia in mice lacking the α_{1A}-subunit, *Proc Natl Acad Sci U S A* **96**:15245-15250.

Kim, S. H. and Chung, J. M., 1992, An experimental model for peripheral neuropathy produced by segmental spinal nerve ligation in the rat, *Pain* **50**:355-363.

Kim, C., Jun, K., Lee, T., Kim, S. S., McEnery, M. W., Chin, H., Kim, H. L., Park, J. M., Kim, D. K., Jung, S. J., Kim, J. and Shin, H. S., 2001a, Altered nociceptive response in mice deficient in the α_{1B} subunit of the voltage-dependent calcium channel, *Mol Cell Neurosci* **18**:235-245.

Kim, D., Song, I., Keum, S., Lee, T., Jeong, M. J., Kim, S. S., McEnery, M. W. and Shin, H. S., 2001b, Lack of the burst firing of thalamocortical relay neurons and resistance to absence seizures in mice lacking α_{1G} T-type Ca^{2+} channels, *Neuron* **31**:35-45.

Klugbauer, N., Dai, S., Specht, V., Lacinova, L., Marais, E., Bohn, G. and Hofmann, F., 2000, A family of gamma-like calcium channel subunits, *FEBS Lett* **470**:189-197.

Kobayashi, T. and Mori, Y., 1998, Ca^{2+} channel antagonists and neuroprotection from cerebral ischemia, *Eur J Pharmacol* **363**:1-15.

Koschak, A., Reimer, D., Huber, I., Grabner, M., Glossmann, H., Engel, J. and Striessnig, J., 2001, alpha 1D ($Ca_v1.3$) subunits can form L-type Ca^{2+} channels activating at negative voltages, *J Biol Chem* **276**:22100-22106.

Kubota, M., Murakoshi, T., Saegusa, H., Kazuno, A., Zong, S., Hu, Q., Noda, T. and Tanabe, T., 2001, Intact LTP and fear memory but impaired spatial memory in mice lacking $Ca_v2.3$ (α_{IE}) channel, *Biochem Biophys Res Commun* **282**:242-248.

Lee, S. C., Choi, S., Lee, T., Kim, H. L., Chin, H. and Shin, H. S., 2002, Molecular basis of R-type calcium channels in central amygdala neurons of the mouse, *Proc Natl Acad Sci U S A* **99**:3276-3281.

Letts, V. A., Felix, R., Biddlecome, G. H., Arikkath, J., Mahaffey, C. L., Valenzuela, A., Bartlett, F. S., 2nd, Mori, Y., Campbell, K. P. and Frankel, W. N., 1998, The mouse stargazer gene encodes a neuronal Ca^{2+}-channel gamma subunit, *Nat Genet* **19**:340-347.

Lievano, A., Santi, C. M., Serrano, C. J., Trevino, C. L., Bellve, A. R., Hernandez-Cruz, A. and Darszon, A., 1996, T-type Ca^{2+} channels and α_{1E} expression in spermatogenic cells, and their possible relevance to the sperm acrosome reaction, *FEBS Lett* **388**:150-154.

Ligon, B., Boyd, A. E., 3rd and Dunlap, K., 1998, Class A calcium channel variants in pancreatic islets and their role in insulin secretion, *J Biol Chem* **273**:13905-13911.

Malmberg, A. B. and Yaksh, T. L., 1992, Antinociceptive actions of spinal nonsteroidal anti-inflammatory agents on the formalin test in the rat, *J Pharmacol Exp Ther* **263**:136-146.

Matsuda, Y., Saegusa, H., Zong, S., Noda, T. and Tanabe, T., 2001, Mice lacking $Ca_v2.3$ α_{1E} calcium channel exhibit hyperglycemia, *Biochem Biophys Res Commun* **289**:791-795.

Meir, A. and Dolphin, A. C., 1998, Known calcium channel α_1 subunits can form low threshold small conductance channels with similarities to native T-type channels, *Neuron* **20**:341-351.

Mintz, I. M., Adams, M. E. and Bean, B. P., 1992, P-type calcium channels in rat central and peripheral neurons, *Neuron* **9**:85-95.

Morris, R. G., Garrud, P., Rawlins, J. N. and O'Keefe, J., 1982, Place navigation impaired in rats with hippocampal lesions, *Nature* **297**:681-683.

Namkung, Y., Smith, S. M., Lee, S. B., Skrypnyk, N. V., Kim, H. L., Chin, H., Scheller, R. H., Tsien, R. W. and Shin, H. S., 1998, Targeted disruption of the Ca^{2+} channel $\beta3$ subunit reduces N- and L-type Ca^{2+} channel activity and alters the voltage-dependent activation of P/Q-type Ca^{2+} channels in neurons, *Proc Natl Acad Sci U S A* **95**:12010-12015.

Namkung, Y., Skrypnyk, N., Jeong, M. J., Lee, T., Lee, M. S., Kim, H. L., Chin, H., Suh, P. G., Kim, S. S. and Shin, H. S., 2001, Requirement for the L-type Ca^{2+} channel α_{1D} subunit in postnatal pancreatic beta cell generation, *J Clin Invest* **108**:1015-1022.

Newcomb, R., Szoke, B., Palma, A., Wang, G., Chen, X., Hopkins, W., Cong, R., Miller, J., Urge, L., Tarczy-Hornoch, K., Loo, J. A., Dooley, D. J., Nadasdi, L., Tsien, R. W., Lemos, J. and Miljanich, G., 1998, Selective peptide antagonist of the class E calcium channel from the venom of the tarantula *Hysterocrates gigas*, *Biochemistry* **37**:15353-15362.

Pereverzev, A., Mikhna, M., Vajna, R., Gissel, C., Henry, M., Weiergraber, M., Hescheler, J., Smyth, N. and Schneider, T., 2002, Disturbances in glucose-tolerance, insulin-release, and stress-induced hyperglycemia upon disruption of the $Ca_v2.3$ α_{1E} subunit of voltage-gated Ca^{2+} channels, *Mol Endocrinol* **16**:884-895.

Perez-Reyes, E., 2003, Molecular Physiology of Low-Voltage-Activated T-type Calcium Channels, *Physiol Rev* **83**:117-161.

Piedras-Renteria, E. S. and Tsien, R. W., 1998, Antisense oligonucleotides against α_{1E} reduce R-type calcium currents in cerebellar granule cells, *Proc Natl Acad Sci U S A* **95**:7760-7765.

Platzer, J., Engel, J., Schrott-Fischer, A., Stephan, K., Bova, S., Chen, H., Zheng, H. and Striessnig, J., 2000, Congenital deafness and sinoatrial node dysfunction in mice lacking class D L-type Ca^{2+} channels, *Cell* **102**:89-97.

Rajewsky, K., Gu, H., Kuhn, R., Betz, U. A., Muller, W., Roes, J. and Schwenk, F., 1996, Conditional gene targeting, *J Clin Invest* **98**:600-603.

Saegusa, H., Kurihara, T., Zong, S., Minowa, O., Kazuno, A., Han, W., Matsuda, Y., Yamanaka, H., Osanai, M., Noda, T. and Tanabe, T., 2000, Altered pain responses in mice lacking α_{1E} subunit of the voltage-dependent Ca^{2+} channel, *Proc Natl Acad Sci U S A* **97**:6132-6137.

Saegusa, H., Kurihara, T., Zong, S., Kazuno, A., Matsuda, Y., Nonaka, T., Han, W., Toriyama, H. and Tanabe, T., 2001, Suppression of inflammatory and neuropathic pain symptoms in mice lacking the N-type Ca^{2+} channel, *EMBO J* **20**:2349-2356.

Saegusa, H., Matsuda, Y. and Tanabe, T., 2002, Effects of ablation of N- and R-type Ca^{2+} channels on pain transmission, *Neurosci Res* **43**:1-7.

Sakata, Y., Saegusa, H., Zong, S., Osanai, M., Murakoshi, T., Shimizu, Y., Noda, T., Aso, T. and Tanabe, T., 2001, Analysis of Ca^{2+} currents in spermatocytes from mice lacking $Ca_v2.3$ (α_{1E}) Ca^{2+} channel, *Biochem Biophys Res Commun* **288**:1032-1036.

Sakata, Y., Saegusa, H., Zong, S., Osanai, M., Murakoshi, T., Shimizu, Y., Noda, T., Aso, T. and Tanabe, T., 2002, $Ca_v2.3$ (α_{1E}) Ca^{2+} channel participates in the control of sperm function, *FEBS Lett* **516**:229-233.

Seisenberger, C., Specht, V., Welling, A., Platzer, J., Pfeifer, A., Kuhbandner, S., Striessnig, J., Klugbauer, N., Feil, R. and Hofmann, F., 2000, Functional embryonic cardiomyocytes after disruption of the L-type α_{1C} ($Ca_v1.2$) calcium channel gene in the mouse, *J Biol Chem* **275**:39193-39199.

Sochivko, D., Pereverzev, A., Smyth, N., Gissel, C., Schneider, T. and Beck, H., 2002, The $Ca_V2.3$ Ca^{2+} channel subunit contributes to R-type Ca^{2+} currents in murine hippocampal and neocortical neurones, *J Physiol* **542**:699-710.

Soong, T. W., Stea, A., Hodson, C. D., Dubel, S. J., Vincent, S. R. and Snutch, T. P., 1993, Structure and functional expression of a member of the low voltage-activated calcium channel family, *Science* **260**:1133-1136.

Strom, T. M., Nyakatura, G., Apfelstedt-Sylla, E., Hellebrand, H., Lorenz, B., Weber, B. H., Wutz, K., Gutwillinger, N., Ruther, K., Drescher, B., Sauer, C., Zrenner, E., Meitinger, T., Rosenthal, A. and Meindl, A., 1998, An L-type calcium-channel gene mutated in incomplete X-linked congenital stationary night blindness, *Nat Genet* **19**:260-263.

Tanabe, T., Beam, K. G., Powell, J. A. and Numa, S., 1988, Restoration of excitation-contraction coupling and slow calcium current in dysgenic muscle by dihydropyridine receptor complementary DNA, *Nature* **336**:134-139.

Tanabe, T., 1994, Structure and function of skeletal muscle and cardiac dihydropyridine receptors. in *Handbook of membrane channels: Molecular and cellular physiology*, C. Perrachia, ed., Academic Press, San Diego, p. 177-186.

Tarantino, L. M. and Bucan, M., 2000, Dissection of behavior and psychiatric disorders using the mouse as a model, *Hum Mol Genet* **9**:953-965.

Toriyama, H., Wang, L., Saegusa, H., Zong, S., Osanai, M., Murakoshi, T., Noda, T., Ohno, K. and Tanabe, T., 2002, Role of $Ca_v2.3$ (α_{1E}) Ca^{2+} channel in ischemic neuronal injury, *Neuroreport* **13**:261-265.

Tsien, J. Z., Huerta, P. T. and Tonegawa, S., 1996, The essential role of hippocampal CA1 NMDA receptor-dependent synaptic plasticity in spatial memory, *Cell* **87**:1327-1338.

Ursu, D., Sebille, S., Dietze, B., Freise, D., Flockerzi, V. and Melzer, W., 2001, Excitation-contraction coupling in skeletal muscle of a mouse lacking the dihydropyridine receptor subunit γ_1, *J Physiol* **533**:367-377.

Vajna, R., Klockner, U., Pereverzev, A., Weiergraber, M., Chen, X., Miljanich, G., Klugbauer, N., Hescheler, J., Perez-Reyes, E. and Schneider, T., 2001, Functional coupling between 'R-type' Ca^{2+} channels and insulin secretion in the insulinoma cell line INS-1, *Eur J Biochem* **268**:1066-1075.

Viatchenko-Karpinski, S., Fleischmann, B. K., Liu, Q., Sauer, H., Gryshchenko, O., Ji, G. J. and Hescheler, J., 1999, Intracellular Ca^{2+} oscillations drive spontaneous contractions in cardiomyocytes during early development, *Proc Natl Acad Sci U S A* **96**:8259-8264.

Wakamori, M., Yamazaki, K., Matsunodaira, H., Teramoto, T., Tanaka, I., Niidome, T., Sawada, K., Nishizawa, Y., Sekiguchi, N., Mori, E., Mori, Y. and Imoto, K., 1998, Single *tottering* mutations responsible for the neuropathic phenotype of the P-type calcium channel, *J Biol Chem* **273**:34857-34867.

Westenbroek, R. E. and Babcock, D. F., 1999, Discrete regional distributions suggest diverse functional roles of calcium channel α_1 subunits in sperm, *Dev Biol* **207**:457-469.

Wilson, S. M., Toth, P. T., Oh, S. B., Gillard, S. E., Volsen, S., Ren, D., Philipson, L. H., Lee, E. C., Fletcher, C. F., Tessarollo, L., Copeland, N. G., Jenkins, N. A. and Miller, R. J., 2000, The status of voltage-dependent calcium channels in alpha 1E knock-out mice, *J Neurosci* **20**:8566-8571.

Woolf, C .J. and Mannion, R. J., 1999, Neuropathic pain: aetiology, symptoms, mechanisms, and management, *Lancet* **353**:1959-1964.

Wu, H. H., Cork, R. J., Huang, P. L., Shuman, D. L. and Mize, R. R., 2000a, Refinement of the ipsilateral retinocollicular projection is disrupted in double endothelial and neuronal nitric oxide synthase gene knockout mice, *Brain Res Dev Brain Res* **120**:105-111.

Wu, H. H., Cork, R. J. and Mize, R. R., 2000b, Normal development of the ipsilateral retinocollicular pathway and its disruption in double endothelial and neuronal nitric oxide synthase gene knockout mice, *J Comp Neurol* **426**:651-665.

Xu, W. and Lipscombe, D., 2001, Neuronal $Ca_v1.3\alpha_1$ L-type channels activate at relatively hyperpolarized membrane potentials and are incompletely inhibited by dihydropyridines, *J Neurosci* **21**:5944-5951.

Zhang, Z., Xu, Y., Song, H., Rodriguez, J., Tuteja, D., Namkung, Y., Shin, H. S. and Chiamvimonvat, N., 2002,

Functional Roles of $Ca_v1.3$ (α_{1D}) calcium channel in sinoatrial nodes: insight gained using gene-targeted null mutant mice, *Circ Res* **90**:981-987.

CALCIUM CHANNEL MUTATIONS AND ASSOCIATED DISEASES

Yasuo Mori[1,2], Yuko Itsukaichi[1,2], Motohiro Nishida[1], and Hiroaki Oka[3]

1. INTRODUCTION

Ca^{2+} influx across the plasma membrane makes a major contribution to augmenting the cytosolic free Ca^{2+} concentration and evokes diverse cellular responses (Clapham, 1995). Voltage-dependent Ca^{2+} channels are composed of the main pore-forming α_1 subunit, encoded by a family of genes: α_{1A} ($Ca_v2.1$), α_{1B} ($Ca_v2.2$), α_{1C} ($Ca_v1.2$), α_{1D} ($Ca_v1.3$), α_{1E} ($Ca_v2.3$), α_{1F} ($Ca_v1.4$), α_{1G} ($Ca_v3.1$), α_{1H} ($Ca_v3.2$), α_{1I} ($Ca_v3.3$), and α_{1S} ($Ca_v1.1$) (Kobayashi and Mori, 1998; Lee et al., 1999; Ertel et al., 2000), plus the accessory $\alpha_2\delta$, β, and γ subunits that regulate expression and function of the α_1 subunit (Campbell et al., 1988; Ahlijanian et al., 1990; Glossmann and Striessnig, 1990; Witcher et al., 1993; Letts et al., 1998). Multiple voltage-dependent Ca^{2+} current subtypes, including the high-threshold L-, N-, P/Q-, R-types and the low-threshold T-type, are major Ca^{2+} entry pathways in neurons (Bean, 1989; Tsien et al., 1991; Llinas et al., 1992; Kobayashi and Mori, 1998). The critical role of Ca^{2+} channels, particularly the P- and N-types, in neurotransmitter release in the synaptic terminals has been well established (Hirning et al., 1988; Turner et al., 1992; Takahashi and Momiyama, 1993; Artalejo et al., 1994; Regehr and Mintz, 1994), whereas the contributions of Ca^{2+} channels in signal integration or synaptic plasticity have been less well understood. Several types of Ca^{2+} channels are co-localized in single neurons and are believed to contribute to the fine-tuning of neuronal activity, since each type is differently modulated.

Recent genetic and molecular biological analyses have revealed many forms of inherited channelopathies. Important association with hereditary neurological disorders has been demonstrated for the genes encoding voltage-dependent Ca^{2+} channel subunits. In addition, Ca^{2+} channel subunit mutations are found widely in lower vertebrates,

[1]Division of Molecular and Cellular Physiology, Center for Integrative Bioscience, Okazaki National Research Institutes, Okazaki 444-8585, Japan; [2]School of Life Science, The Graduate University for Advanced Studies, Okazaki 444-8585, Japan; [3]Nanotechnology Research Laboratory, Advanced Technology Research Laboratories, Matsushita Electric Industrial Co., Ltd., Seika-cho, Soraku-gun, Kyoto 6190237, Japan. Address correspondence to Dr. Mori.

Calcium Channel Pharmacology, Edited by McDonough
Kluwer Academic/Plenum Publishers, 2004

invertebrates, and yeast. These mutants have served as important models for human neurological diseases. In this chapter, we survey the molecular physiology of genetic Ca^{2+} channel diseases and mutations, in an attempt to unveil general mechanisms that rule the expression of phenotypes elicited by Ca^{2+} channel mutations.

2. HUMAN CALCIUM CHANNEL MUTATIONS

2.1. α_{1S} ($Ca_v1.1$: *Cacnla3*) Mutations and Hypokalemic Periodic Paralysis

Hypokalemic periodic paralysis (HypoPP) is a genetic muscle disorder of autosomal dominant inheritance characterized by acute attacks of muscle weakness and a decrease in the blood potassium level. Linkage studies have shown that the HypoPP gene maps to chromosome 1a31-32 and colocalizes with the gene encoding the skeletal muscle DHP receptor L-type Ca^{2+} channel α_{1S} subunit (*Cacnla3*) (Fontaine et al., 1994). Three point mutations resulting in non-conservative changes were found within the coding sequence of the α_{1S} gene, supporting it as the gene responsible for HypoPP (Figure 1) (Jurkat-Rott et al., 1994; Ptacek et al., 1994). These mutations lead to arginine-to-histidine (R528H, R1239H) and arginine-to-glycine (R1239G) substitutions that occur within the IIS4 and IVS4 voltage sensor regions of the channel.

The functional properties of ionic currents conducted by L-type α_{1S} channels containing HypoPP mutations have been studied in acutely biopsied muscle fibers (Jurkat-Rott et al., 1998), myotubes cultured from muscle biopsies (Morrill et al., 1998), and in heterologous expression systems (Table 1) (Lapie et al., 1996; Jurkat-Rott et al., 1998; Morrill and Cannon, 1999). A consistent finding among all the studies is that the Ca^{2+} current density is reduced by about 50% compared to controls, suggesting a loss-of-function defect for HypoPP mutations. The location of the missense mutations within voltage sensors S4 of the L-type channel predicted an alteration of the voltage dependence of channel gating, but recent studies have given conflicting results. R528H mutant channels revealed a hyperpolarizing shift in voltage dependence of activation (Jurkat-Rott et al., 1998; Morrill and Cannon, 1999) and inactivation (Sipos et al., 1995; Jurkat-Rott et al., 1998), and R1239 mutant channels revealed a depolarizing shift of activation (Morrill and Cannon, 1999). Other studies, however, failed to confirm the change (Lapie et al., 1996; Lerche et al., 1996; Morrill et al., 1998).

Despite the studies revealing functional alteration of the L-type α_{1S} channel by HypoPP-causing mutations, the pathophysiology of the disease remains unclear. In HypoPP, hypokalemia is induced by the physiological effect of glucose intake and the release of insulin, which stimulates the Na^+-K^+ pump and shifts K^+ from the extracellular space into the intracellular compartment (Minaker et al., 1988). The hypokalemia causes sustained depolarization in HypoPP fibers, in contrast to the membrane hyperpolarization observed in normal muscle, and initiates the attack (Rudel et al., 1984). The mechanism underlying this mutant depolarization remains unclear: no clear evidence has been presented that the L-type Ca^{2+} channel makes a significant contribution to the muscle resting potential through its voltage sensing function. Recently, it has been proposed that this hypokalemia-induced depolarization is caused by reduction of outward K^+ current. Patch-clamp recordings from fibers carrying the R528H mutant channels demonstrated an impairment of the ATP-sensitive K^+ current (Tricarico et al., 1999). Another study using

Table 1. Effects of hypokalemic periodic paralysis (HypoPP) mutations in the skeletal muscle α_{1S} channel

Mutation	Location	Biophysical properties	Predicted effect on Ca^{2+} influx	References
R528H	II S4	Current density ↓ or → $V_{1/2}$ activation ↓ or → $V_{1/2}$ inactivation ↓ or → Activation time constant ↑ or ↓ Inactivation time constant →	↓	a, b, c, d, e
R1239H	IV S4	Current density ↓ $V_{1/2}$ activation ↑ or → $V_{1/2}$ inactivation → Activation time constant ↑ Inactivation time constant →	↓	c, e
R1239G	IV S4	Current density ↓ $V_{1/2}$ activation ↑ $V_{1/2}$ inactivation → Activation time constant ↑ Inactivation time constant →	↓	c

Symbols: ↑ = Increased, ↓ = Decreased, → = Unchanged, ? = Unknown. References: a, Jurkat-Rott et al., 1998; b, Morrill et al., 1998; c, Morrill and Cannon, 1999; d, Lapie et al., 1996; e, Sipos et al., 1995.

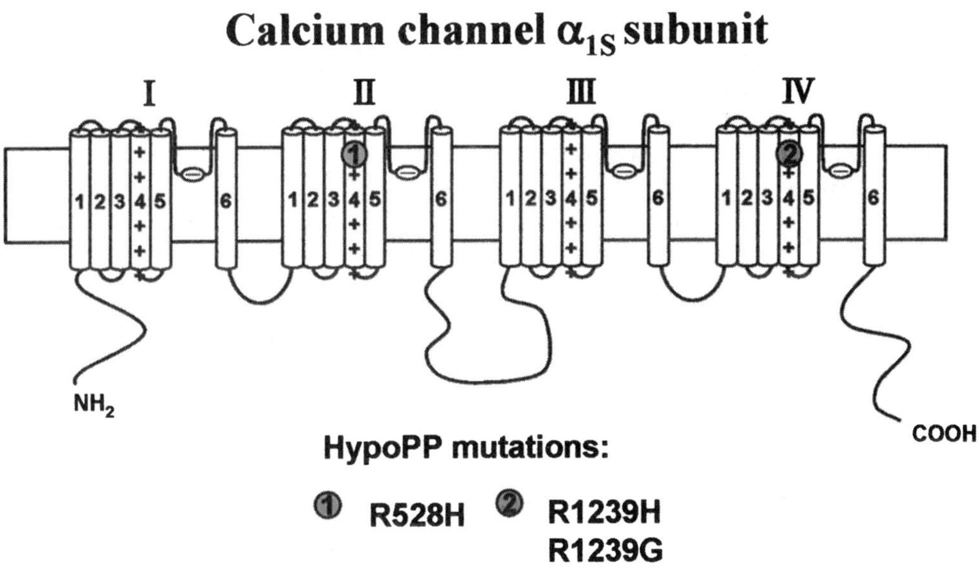

Figure 1. Schematic representation of HypoPP mutations in the CACNLA3 gene

the three-electrode voltage clamp showed a reduction in the membrane K^+ conductance of resting HypoPP fibers, suggesting a defect in an inward rectifier K^+ channel (Ruff, 1999). A chronic reduction in Ca^{2+} influx through L-type channels might alter the expression, distribution, or regulation of ion channels that control the membrane potential, thereby indirectly predisposing the fiber to depolarization. Puzzlingly, the R528H mutation showed no significant effect on excitation-contraction (E-C) coupling or on the electrical behavior of skeletal muscles (Jurkat-Rott et al., 1998), although the most straightforward expectation about the pathophysiology of HypoPP is abnormal E-C coupling. It is, therefore, possible that an unknown function of the Ca^{2+} channel α_{1S} subunit mediates expression of pathophysiological phenotypes of HypoPP.

2.2. α_{1A} (Ca_v2.1: *Cacna1a*) Mutations: Familial Hemiplegic Migraine, Episodic Ataxia Type 2, and Spinocerebellar Ataxia Type 6

The α_{1A} subunit was originally characterized as a voltage-sensing, pore-forming component of a brain high-voltage activated Ca^{2+} channel resistant to blockade by the N-type antagonist ω-conotoxin-GVIA and the dihydropyridines that inhibit L-type channels (Mori et al., 1991). It is now accepted that P- and Q-type channels, which differ in their sensitivity to ω-agatoxin-IVA and in their inactivation kinetics (Llinas et al., 1989; Regan et al., 1991; Mintz et al., 1992; Zhang et al., 1993), are produced from the single α_{1A} gene by alternative splicing (Mori et al., 1991; Sather et al., 1993; Bourinet et al., 1999) and/or through association with different isoforms of accessory subunits (Stea et al., 1994), although the mechanism by which different phenotypes are produced has not yet been fully explained (Bourinet et al., 1999) (see McDonough, this volume, for discussion of P- and Q-type channels). Molecular genetic analyses have identified mutations of the α_{1A} gene that cause migraine, cerebellar ataxia, and other neurological disorders. In the human α_{1A} gene, missense mutations, nonsense mutations, splice site mutations, and CAG expansions have all been shown to underlie neurological disorders.

Migraine is a frequent neurological disorder, characterized by recurrent attacks of disabling headache, vomiting, photo- and phonophobia, and malaise (migraine without an aura). In 20% of patients, attacks are accompanied by transient focal neurological symptoms (migraine with aura). The frequency, duration, and severity of the attacks vary substantially among patients (Russell et al., 1995). Familial hemiplegic migraine (FHM) is a rare autosomal dominant subtype of migraine with aura, characterized by unilateral weakness suggestive of brainstem, subcortical, or cortical involvement. Progressive ataxia suggestive of cerebellar atrophy is developed in some patients. Episodic ataxia type 2 (EA-2) is another autosomal dominant paroxysmal cerebral disorder, characterized by interictal nystagmus, episodes of ataxia, migraine-like symptoms, and cerebellar atrophy. Great intra- and inter-familial variability exist in the symptoms of EA-2 patients (Pietrobon, 2002), suggesting a genetic variation for EA-2 as well as FHM, as indicated above. A locus for FHM was mapped to chromosome 19p13, together with that for EA-2 (Ophoff et al., 1996). Importantly, the more common forms of migraine were mapped to the same gene locus (May et al., 1995). Considering this together with the similarity between the symptoms of the headache and aura phase of FHM and the attacks of normal migraine, it should be stressed that FHM can be used as a model to study the complex genetics of the more common forms of migraine (Ophoff et al., 2001; Pietrobon, 2002).

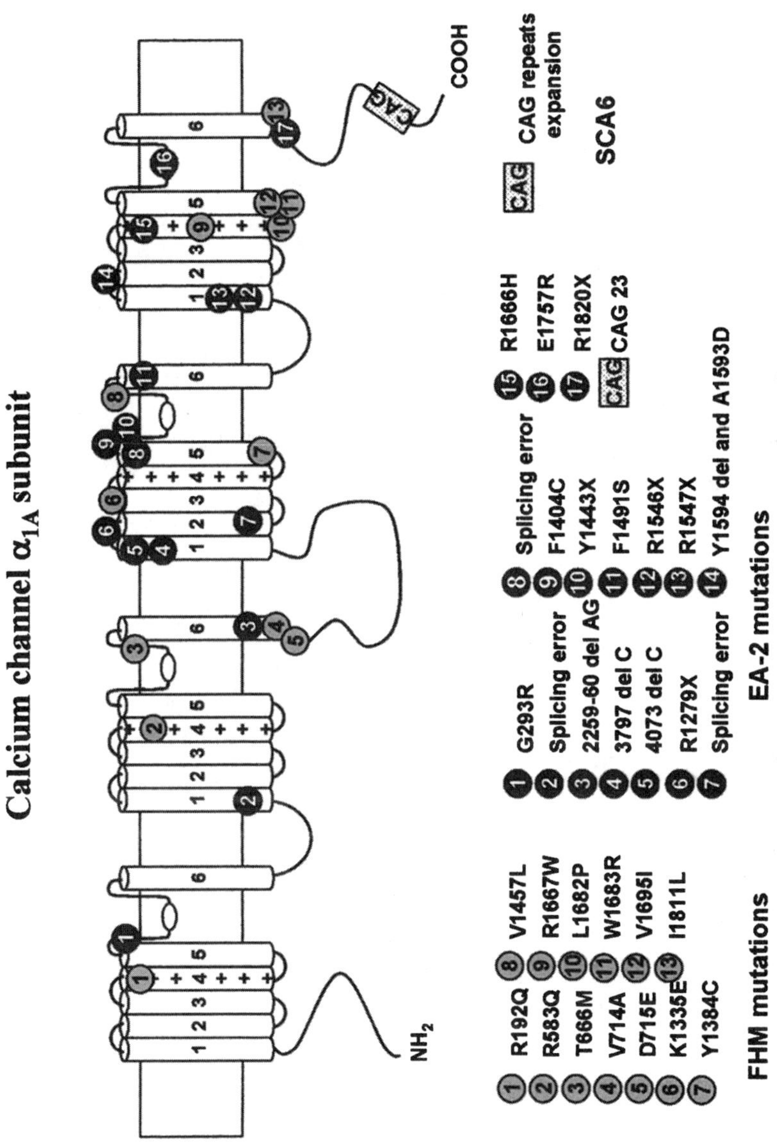

Figure 2. Schematic representation of FHM, EA-2 and SCA6 mutations in the CACNA1A gene.

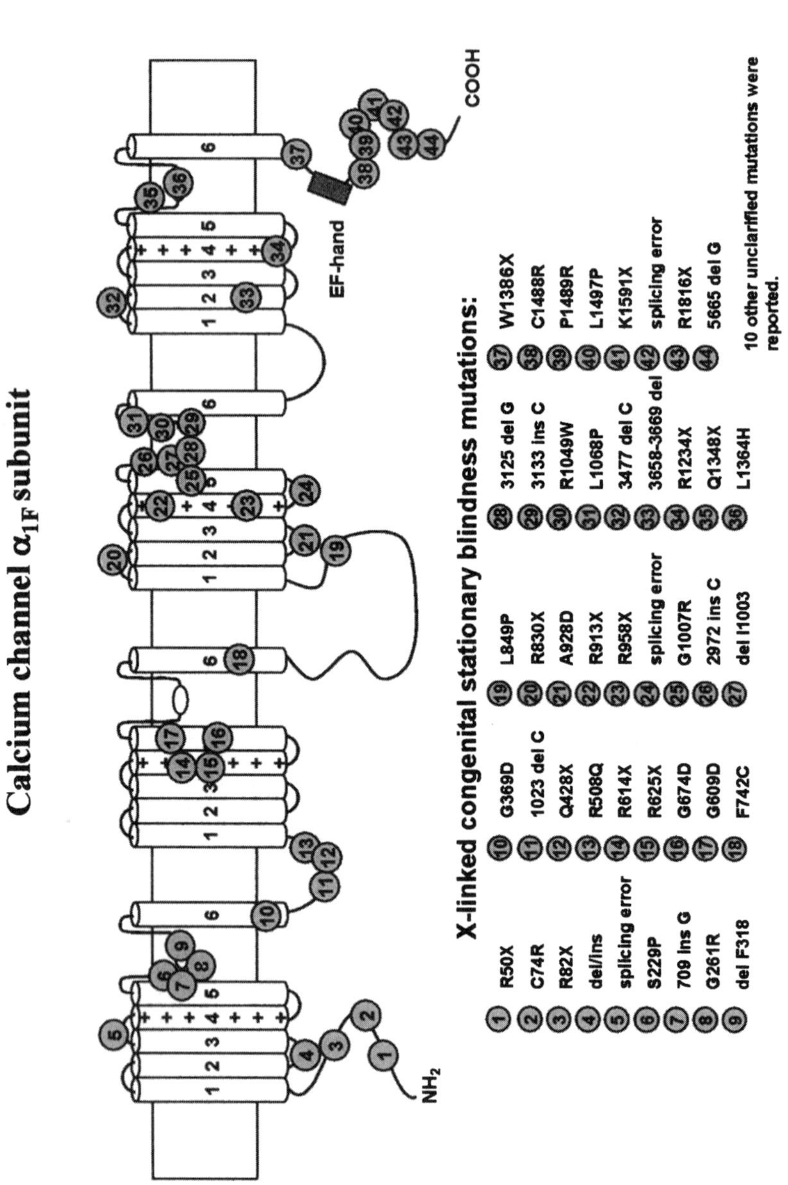

Figure 3. Schematic representation of CSNB mutations in the CACNA1F gene.

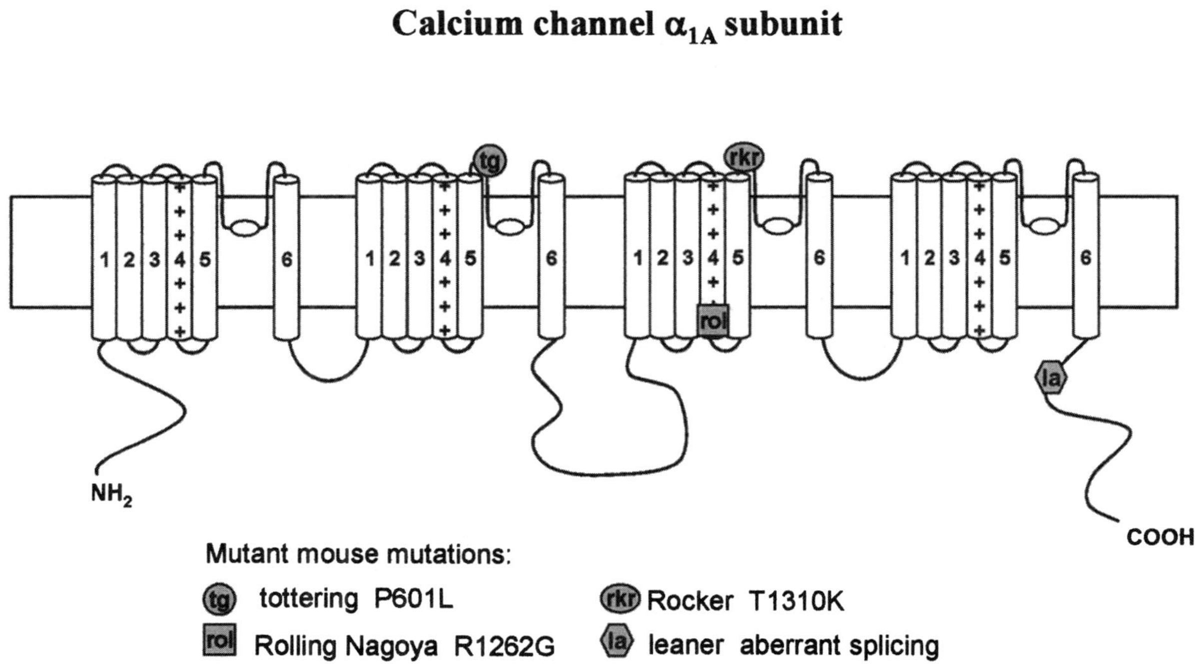

Figure 4. Schematic representation of mouse mutations in the CACNA1A gene.

Table 2. Effects of familial hemiplegic migraine (FHM) mutations in the neuronal α_{1A} channel

Mutation	Location	Biophysical properties	Predicted effect on Ca^{2+} influx	References
R192Q	I S4	$V_{1/2}$ activation ↓ or → $V_{1/2}$ inactivation → Inactivation time constant → Recovery from inactivation → Open probability ↑ Unitary conductance → Channel density ↑ or ↓	↑ or ↓ Single channel ↑	a, b, c, d
R583Q	II S4	$V_{1/2}$ activation ↓ $V_{1/2}$ inactivation ↓ Inactivation time constant ↓ Recovery from inactivation ↓	↓	e, f, g
T666M	II S5-S6	$V_{1/2}$ activation ↓ $V_{1/2}$ inactivation → Inactivation time constant ↓ Recovery from inactivation ↓ Open probability ↑ Unitary conductance ↓ (→) Channel density ↓	↓ Single channel ↑	a, b, c, d, h
V714A	II S6	$V_{1/2}$ activation ↓ $V_{1/2}$ inactivation → Inactivation time constant ↓ Recovery from inactivation ↑ Open probability ↑ Unitary conductance ↓ (→) Channel density ↓	↓ Single channel ↑	a, b, c, d
D715E	II S6	$V_{1/2}$ activation ↓ $V_{1/2}$ inactivation ↓ Inactivation time constant ↓ Recovery from inactivation →	↓	e, h
K1335E	III S3-S4	?	?	g
Y1384C	III S5	?	?	g, i
V1457L	III S5-S6	$V_{1/2}$ activation ↓ $V_{1/2}$ inactivation → Inactivation time constant ↑ Recovery from inactivation ↓ Open probability ↑ Unitary conductance ↓ Channel density ↓	↓ Single channel ↑	j, e, d
R1667W	IV S4	?	?	g
L1682P	IV S4-S5	?	?	k
W1683R	IV S4-S5	?	?	g
V1695I	IV S5	?	?	g
I1811L (I1819L)	IV S6	$V_{1/2}$ activation ↓ $V_{1/2}$ inactivation → Inactivation time constant → Recovery from inactivation ↑ Open probability ↑ Unitary conductance → (↓) Channel density ↓	↓ Single channel ↑	a, b, c, d

Table 2 ***(from preceding page)*** Symbols: ↑ = Increased, ↓ = Decreased, → = Unchanged, ? = Unknown, ( ) = observed in some patches. References: a, Ophoff et al., 1996; b, Kraus et al., 1998; c, Hans et al., 1999; d, Tottene et al., 2002; e, Kraus et al., 2000; f, Battistini et al., 1999; g, Ducros et al., 1999a; h, Ducros et al., 1999b; i, Vahedi et al., 1999; j, Carrera et al., 1999; k, Montagna, 2000.

Ophoff and colleagues first identified the α_{1A} (*Cacna1a*) gene as the gene responsible for FHM, EA-2, and possibly also for the more common types of migraine (Ophoff et al., 1996). So far, 13 different missense mutations have been reported for FHM (Ophoff et al., 2001; Pietrobon, 2002), all resulting in amino acid substitutions of conserved residues in functionally important regions of the P/Q-type α_{1A} Ca^{2+} channel (Figure 2; Table 2). Pietrobon's group and Striessnig's group have carried out extensive functional studies on recombinant P/Q-type channels with FHM mutations (Kraus et al., 1998; Hans et al., 1999; Kraus et al., 2000; Tottene et al., 2002). Recombinant P/Q-type Ca^{2+} channels carrying different mutations showed functional variability, reflecting variable effects of the mutations on P/Q-type channel function. The consequences of effects ranged from alteration of activation and inactivation gating to reduction of ion permeation. In regard to EA-2 (with and without progressive ataxia), at least 17 different mutations for EA-2 have been reported (Figure 2, Table 3) (Ophoff et al., 2001; Pietrobon, 2002). Many of the EA mutations lead to truncations of the α_{1A} protein, while the rest of them result in amino acid substitutions located in functionally important regions of the channel, similar to FHM mutations. Functional consequences of the EA mutations include complete loss-of-function without effects on channel expression (F1491S, R1279stop) (Guida et al., 2001; Wappl et al., 2002), and alteration of activation and inactivation gating (G293R, AY1593/1594D)(Wappl et al., 2002).

A link between human epilepsy and mutations in the P/Q-type Ca^{2+} channel α_{1A} subunit gene has been described. A point mutation that introduces a premature stop codon after R1820, resulting in a truncation of the C-terminus of the α_{1A} subunit, was identified in a child with early onset absence epilepsy and cerebellar ataxia (Jouvenceau et al., 2001). Expression studies indicated a robust dominant negative effect of the mutant on the wild type α_{1A} channel. This likely reflected a disturbance of surface expression of the channels. A recent report of a family with co-occurrence of idiopathic focal epilepsy, EA-2, and migraine (Holtmann et al., 2002) lends support to the idea that the human α_{1A} gene is linked to idiopathic generalized epilepsy.

CAG repeat expansions have been associated with a group of late onset, progressive neurodegenerative diseases. The CAG repeats are translated into polyglutamine (Q) tracts in the protein products, leading to a gain of function in the protein product accounting for the dominant inheritance. Expansion of a CAG repeat in the human α_{1A} gene 3'-region encoding the C-terminus for an α_{1A} splice isoform has also been identified in autosomal dominant spinocerebellar ataxia 6 (SCA6) (Figure 2) (Zhuchenko et al., 1997; Matsuyama et al., 1999). Surprisingly, the CAG repeat is exceptionally small in SCA6, ranging from 21 to 33, whereas a repeat of > 40 is what generally leads to neurodegeneration and neurological symptoms in other polyQ diseases. Our characterization of P/Q-type Ca^{2+} channels with an expanded stretch of 24, 30, or 40 glutamines revealed direct effects of polyglutamine expansion on channel properties (Matsuyama et al., 1999). The P/Q-type channels with 30 or 40 polyQ exhibited an 8 mV hyperpolarizing shift in the voltage dependence of inactivation. This is the first indication that expanded polyQ alters directly functions of the affected gene product, providing a hint for why relatively small Q repeats induce ataxia in SCA6. More recently, the above functional alteration has been shown to be specific for the SCA6 mutant of the Purkinje neuron-type α_{1A} splice variant, which lacks the asparagine-proline (NP) insertion in the S3-S4 linker of channel domain IV (Toru et al., 2000). In contrast to these reductions in channel availabilities, two reports using the β4 accessory subunit have displayed changes in α_{1A} channel gating or current densities consistent with enhanced Ca^{2+} entry (Restituito et al., 2000; Piedras-Renteria et al., 2001). These differences leave considerable uncertainty about the basis of SCA6 pathologies (Table 4). Since the functional studies are done exclusively in recombinant systems that necessarily involve a certain degree of non-physiological conditions, future studies using native neuronal preparations from mutant and transgenic mice should provide a clearer view of the pathophysiological bases of SCA6.

2.3. Congenital Stationary Night Blindness and Mutations in α_{1F} (Ca_v1.4: *Cacna1f*)

X-linked congenital stationary night blindness (CSNB) is a recessive, non-progressive retinal disorder characterized by night blindness, decreased visual activity, myopia, nystagmus, and strabismus. Complete and incomplete CSNB have been described. Patients with the incomplete form CSNB2 show moderate myopia to hyperopia and subnormal but measurable rod/ cone function. A retinal-specific Ca^{2+} channel gene, α_{1F} (Ca_v1.4) mapping to the CSNB2 region Xp11.23 was identified (Bech-Hansen et al., 1998; Strom et al., 1998). α_{1F} showed the closest sequence similarity to the α_{1S}, α_{1C}, and α_{1D} genes and so was classified as L-type. Mutation analysis in the families with incomplete CSNB2 revealed nonsense, frameshift, and splice site mutations predicted to cause premature protein truncation, which may be consistent with the recessive loss-of-function of the causative gene in incomplete CSNB2 (Figure 3).

3. MOUSE CALCIUM CHANNEL MUTATIONS

3.1. *Tottering* Mouse Strains and α_{1A} (Ca_v2.1) Mutations

To elucidate etiology and to develop methods for treatments of human genetic channelopathies such as SCA6, FHM, EA-2, and common forms of epilepsy, the spontaneous mouse mutants of the α_{1A} subunit gene are useful models. The *tottering* (*tg*) locus on mouse chromosome 8 is a well known epileptic mouse mutation that induces a delayed-onset, recessive disorder consisting of ataxia, paroxysmal dyskinesia, and absence seizures resembling petit mal epilepsy (Fletcher et al., 1996; Doyle et al., 1997). Fletcher and colleagues identified the mouse α_{1A} gene that is mutated in *tg* mutants (Figure 4) (Fletcher et al., 1996). *tg* is a missense mutation that causes substitution of L for P (P601L) at a position close to the conserved pore-lining region ("P" region) in the extracellular segment of the second of the four internal domain repeats of the channel. Mice with an allelic tottering mutation *leaner* (tg^{la}), which causes more severe symptoms, have a single nucleotide substitution at an exon/ intron junction, which results in skipping the exon or in failure to splice out the succeeding intron (Fletcher et al., 1996). In both cases, the tg^{la} mutation causes truncation of the normal open reading frame and expression of aberrant C-terminal sequences (Figure 4).

Table 3. Effects of episodic-ataxia type 2 (EA2) mutations in the α_{1A} channel

Mutation	Location	Biophysical properties	Predicted effect on Ca^{2+} influx	References
G293R	I S5-S6	Current density ↓ $V_{1/2}$ activation ↑ Inactivation time constant ↓ Unitary conductance → Mean open times ↓	↓	a, b
Splicing error	II S1	?	?	c
2259-60 delAG	II S6	?	?	c
3797delC	III S1	?	?	c
4073 delC	III S1	?	?	d
R1279X	III S1-S2	Complete loss of channel activity	↓	a, e, f
Splicing error	III S2	?	?	d
Splicing error	III S5	?	?	c
F1404C	III S5-S6	Complete loss of channel activity	↓	f
Y1443X	III S5-S6	?	?	c
F1491S	III S6	Complete loss of channel activity	↓	g
R1546X	IV S1	?	?	c
R1547X	IV S1	Complete loss of channel activity	↓	h, f
Deletion (Y1594) and A1593D	IV S1-S2	Complete loss of channel activity (tsA-201 cells) Current density ↓ (oocytes) $V_{1/2}$ activation ↑ (oocytes) Inactivation time constant ↓ (oocytes) Recovery from inactivation ↓ (oocytes)	↓	c, a
R1666H	IV S4	?	?	i
E1757R	IV S5-S6	?	?	j
R1820X	COOH terminal	Complete loss of channel activity (Dominant negative effect on wild-type allele)	↓	k
CAG23	COOH terminal	?	?	l

Symbols, ↑ = Increased, ↓ = Decreased, → = Unchanged, ? = Unknown. References: a, Wappl et al., 2002; b, Yue et al., 1997; c, Denier et al., 1999; d, Ophoff et al., 1996; e, Yue et al., 1998; f, Jen et al., 2001; g, Guida et al., 2001; h, Jen et al., 1999; i, Friend et al., 1999; j, Denier et al., 2001; k, Jouvenceau et al., 2001; l, Jodice et al., 1997.

Table 4. Effects of spinocerebellar ataxia type 6 (SCA6) CAG repeats in the α_{1A} channel

Number of CAG repeats	Biophysical properties	Predicted effect on Ca^{2+} influx	References
24 (-NP*)	$V_{1/2}$ inactivation ↓	↓	a
28 (-NP*)	$V_{1/2}$ inactivation ↓	↓	
28 (+NP*)	$V_{1/2}$ inactivation ↑	↑	
30	$V_{1/2}$ inactivation ↓	↓	b
40	$V_{1/2}$ inactivation ↓	↓	
30	$V_{1/2}$ activation ↓ Inactivation time constant ↑ (Only when expressed with β_4 subunit)	↑	c
23	Channel density ↑ $V_{1/2}$ activation ↓	↑	d
27	Channel density ↑	↑	
72	Channel density ↑ $V_{1/2}$ activation ↓	↑	

Symbols: asterisk (*) indicates the presence or absence (+NP or –NP) of an asparagine-proline (NP) insertion in the domain IV S3-S4 in the α_{1A} channels; see Bourinet et al., 1999. ↑ = Increased, ↓ = Decreased, → = Unchanged, ? = Unknown. References: a, Toru et al., 2000; b, Matsuyama et al., 1999; c, Restituito et al., 2000; d, Piedras-Rentería et al., 2001.

3.1.1. Characterization of Tottering *Mutants*

Comprehensive comparison of the mutant Ca^{2+} channel properties has been carried out for native cerebellar Purkinje neurons of *tg* and tg^{la} mice, in which many cell-specific factors can affect channel phenotype (Dove et al., 1998; Lorenzon et al., 1998; Wakamori et al., 1998), and for channels in recombinant expression systems, in which direct effects of the mutations can be evaluated (Wakamori et al., 1998). Causative relationships have been demonstrated among the *tg* mutations, the affected Ca^{2+} channel properties, and the neurological disorders. The most striking feature of Ca^{2+} channel currents of mutant Purkinje neurons was a marked reduction in current density, to ~ 60% and ~ 40% of control in *tg* and tg^{la} mice, respectively (Wakamori et al., 1998). The Ca^{2+} channel currents in the *tg* Purkinje cells showed an increase in the non-inactivating component of current during voltage-dependent inactivation. In addition to this change, neurons from tg^{la} mice showed a more distinct change in the voltage dependence of activation and inactivation, being shifted in the depolarizing direction by ~ 10 mV, with a broader voltage dependence of inactivation. In the recombinant expression system, the *tg* channel with a missense mutation (P601L) and the shorter form of the two possible tg^{la} products showed a significant reduction in current density, while the longer form of the tg^{la} channels had a normal current density. The shift in voltage dependence of activation and inactivation was observed only for the long tg^{la} channel. Thus, single *tg* mutations of the α_{1A} gene are directly responsible for the neuropathic phenotypes of P/Q-type channels in *tg* mice.

Mouse mutation *rolling Nagoya* (tg^{rol}) was identified as an allelic mutation of *tg* and tg^{la} (Oda, 1981). Homozygous tg^{rol} mice show poor motor coordination of hindlimbs, and sometimes stiffness of the hindlimbs and tail, but no seizures (Oda, 1973, 1981). We presented evidence that the tg^{rol} mutation causes a defect in the voltage-sensing mechanism of P/Q-type Ca^{2+} channels (Mori et al., 2000). The nucleotide sequence alteration in the tg^{rol} mouse strain leads to the charge-neutralizing R-to-G substitution at the amino acid position 1262 in the voltage sensing S4 domain of the third internal repeat of the α_{1A} subunit (Figure 4). Reduced voltage sensitivity and diminished P/Q-type channel activity are the direct functional consequences of the tg^{rol} mutation. Both the recombinant α_{1A} channel with the tg^{rol} mutation and the native P-type channels in Purkinje neurons acutely dissociated from tg^{rol} mouse cerebellum showed similar modified voltage dependence of activation: midpoint potential was shifted in the depolarizing direction, and the steepness of the activation curve was decreased. The valence of the apparent single-gate charge for activation (z_m), calculated from the slope factor (k_a) through limiting slope analysis with the equation $z_m = kT/k_a e_0 = 25.7/k_a$, was reduced from 4.2 to 3.2 by the mutation in the recombinant α_{1A} channel, and from 5.0 to 4.0 in the native P-type channel (Hille, 1992). The 1.0 decrease in z_m is exactly what one would predict if the positively charged guanidine group of R1262 constitutes part of the gating charge that senses membrane depolarization.

Our results also demonstrated an impairment of action potential generation in tg^{rol} Purkinje cells (Mori et al., 2000). Considering the essential role played by Ca^{2+} spikes in dendritic trees in the integration of synaptic inputs, it is natural that compromised Ca^{2+} spike generation in the tg^{rol} mice severely impairs function of the cerebellum. Reduced Ca^{2+} influx may further fail to activate the Ca^{2+}-activated K^+ channels, leading to depolarization block of Na^+ spikes in tg^{rol} Purkinje neurons. Thus, reduced P-type channel activity is a critical cerebellar defect that underlies the ataxic phenotype of tg^{rol} mice.

Detailed neurophysiology of the mutant mice has provided important clues in understanding mechanistically how mutations cause neurological symptoms and how symptom onset and severity are determined. Presynaptic Ca^{2+} currents and neurotransmitter release at hippocampal Schaffer collateral synapses have been investigated in *tg* mice (Qian and Noebels, 2000). The study revealed a profound decrease in presynaptic P/Q-type currents and increased dependence on N-type channels for synaptic transmission. This is a sharp contrast to synaptic transmission in *lethargic* (*lh*) mice, which have a defect in the β4 subunit predicted to form P/Q-type channels with α_{1A} (Section 3.2.2). In *tg* mice as well as *lh* mice, a significant decrease in glutamatergic synaptic transmission mediated by NMDA and non-NMDA receptors has been described in somatosensory thalamic neurons, which represent one of the populations critical for seizure activity in the strains (Caddick et al., 1999). GABAergic transmission, however, stayed intact in the mutant strains.

Fundamental properties of excitatory synaptic transmission in the cerebellum and roles of Ca^{2+} channel subtypes therein were characterized in *tg* and tg^{rol} mice (Matsushita et al., 2002). The amplitude of excitatory postsynaptic current (EPSC) of the parallel fiber-Purkinje cell (PF-PC) synapses was considerably reduced in tg^{rol}. By contrast, the amplitude of the parallel fiber-mediated EPSC was only mildly decreased in young non-ataxic *tg* mice but diminished drastically in adult ataxic *tg* mice (P28-P35 days). This shows a good correlation between the time of impairment of the PF-PC synaptic transmission and the time of manifestation of ataxia. In contrast, the EPSC amplitude of the climbing fiber-Purkinje cell (CF-PC) synapses was preserved in *tg* and was even increased in tg^{rol}, possibly due to enhanced AMPA sensitivity of postsynaptic glutamate receptors in Purkinje cells. Dependence of the climbing fiber-mediated EPSC on other

Ca^{2+} channel subtypes in mutant mice may be a compensatory mechanism that maintains the CF-PC synaptic transmission. Interestingly, increased expression has also been reported for the P/Q-type channel α_{1A} subunit itself in tg^{rol} (Sawada et al., 2001) but not in tg^{la} (Lau et al., 1998), while binding of L-type antagonists, which prevented motor episodes, was significantly enhanced in the *tg* mouse brain (Campbell and Hess, 1999). The results all indicate that different mutations of the P/Q-type Ca^{2+} channel not only cause primary effects of different severity on the calcium channel, but also elicit diverse additional secondary effects that disrupt the balance of neural networks.

With regard to intracellular Ca^{2+} homeostasis, in tg^{la} Purkinje neurons, rapid Ca^{2+} buffering ability was abolished by diminished endoplasmic reticulum (ER) Ca^{2+} uptake and altered expression of Ca^{2+} binding proteins such as calbindin and parvalbumin (Dove et al., 2000). An important secondary abnormality common to the above three *tg* mutants is ectopic expression of tyrosine hydroxylase (TH) (Hess and Wilson, 1991). In contrast to normal mice, in which TH expression in Purkinje neurons was transient and developmentally regulated, *tg* mutants showed persistent TH expression through adulthood (Hess and Wilson, 1991; Mori et al., 2000). The pattern of ectopic TH expression was coextensive with zebrin staining (Fletcher et al., 1996; Mori et al., 2000).

An ultrastructural study of granule and Purkinje neurons revealed multiple contacts of Purkinje cell dendritic spines with individual parallel fiber varicosities in tg^{la}, *tg*, and tg^{rol} mice (Rhyu et al., 1999a; Rhyu et al., 1999b). These mutant mice also showed ectopic spines that almost always made synaptic contact with a parallel fiber varicosity in secondary branches of Purkinje cell dendrites. tg^{la} cerebella are characterized by apoptotic loss of Purkinje neurons and granule cells (Herrup and Wilczynski, 1982; Fletcher et al., 1996). Apoptosis of granule cells was reported in tg^{rol} cerebella as well (Suh et al., 2002), although this is still controversial (Mori et al., 2000).

Recently, *rocker* (tg^{rkr}) was found in descendants of a chemically mutagenized male mouse (Zwingman et al., 2001). In addition to the tg^{rkr} locus mapped to mouse chromosome 8, within 2 centimorgans of the α_{1A} *tottering* locus, complementation tests with the tg^{la} mutant identified tg^{rkr} as a new allele of the α_{1A} gene. tg^{rkr} is a point mutation that causes an amino acid exchange: threonine-to-lysine substitution at residue 1310, between transmembrane regions 5 and 6 in the third repeat (Figure 4). tg^{rkr} is of increased value not for its similarity to the other *tg* mutants, but for the distinct differences it introduces. tg^{rkr} is the first allele to present absence seizures (spike/ wave type discharges) without the intermittent movement disorder (the paroxysmal dyskinesia phenotype). tg^{rkr} mice display an ataxic, unstable gait accompanied by an intention tremor, typical of cerebellar dysfunction. The mutant is the first reported *tg* allele to present structural abnormalities in the mature Purkinje cell dendrite. The overall cytoarchitecture of the young adult brain appears normal, including its gross cerebellar morphology, while Golgi-Cox staining revealed dendritic abnormalities in the mature cerebellar cortex characterized by a reduction of branching in the Purkinje cell dendritic arbor and a "weeping willow" appearance of the secondary branches. Important distinctions between tg^{rkr} and the previously characterized alleles of this locus also include the absence of aberrant TH expression in Purkinje cells and ectopic spines.

3.1.2. Comparative Studies of Altered P-type Channel Function and Neurological Phenotypes Among Tottering *Mutants*

Cerebellar ataxia has been identified as a common behavioral abnormality among the three α_{1A} mutant mice tg^{rol}, tg^{la}, and *tg*. However, severity of cerebellar ataxia differs significantly among the three forms of mice; the severity of ataxia of tg^{rol} falls somewhere in between that of tg^{la} and *tg*. tg^{la} and tg^{rol} suffer from loss/ degeneration of cerebellar neurons (Herrup and Wilczynski, 1982). The data on tg^{rol} mice (Mori et al., 2000) in combination with our work on tg^{la} and *tg* mice (Wakamori et al., 1998) suggest that severity of the cerebellar defect in the mutant strains is somewhat correlated with deviation of P-type channel properties in Purkinje cells. Reduction in P-type current amplitude was the severest in tg^{la} Purkinje cells (~ 60%) compared with *tg* and tg^{rol} cells (~ 40%). Voltage dependence of activation of tg^{la} and tg^{rol} P-type channels, but not *tg* P-type channels, showed a depolarizing shift. P-type currents in the tg^{rol} and tg^{la} mutant displayed a shifted voltage dependence of inactivation, while a decrease in fractional inactivation during a two second depolarization was observed for the tg^{la} and *tg* P-type channels. In the null mutant mice lacking expression of the α_{1A} subunit, a complete loss of P-type channel function induces an ataxia that progressively worsens up to the point of premature death (Jun et al., 1999). It is thus possible that severity of ataxia in tg^{rol} mice reflects deviated P-type channel function, which results in a corresponding impairment of integrative properties of Purkinje cells in motor function.

Additional neurological differences are seen among the three P/Q-type α_{1A} subunit mutants. tg^{la} and *tg* mice display absence epilepsy (Noebels, 1984), but tg^{rol} mice do not (Oda, 1981). Only *tg* mice suffer from paroxysmal dyskinesia (Green and Sidman, 1962). The differences may derive from different involvement of the P/Q-type channel in respective neuronal functions. Since apoptosis has been observed only in tg^{la} cerebella, which contain the most severely functionally impaired P/Q-type channels, P/Q-type channel activity is redundant in increasing the intracellular Ca^{2+} concentration required for cell survival in wild type mice (Yano et al., 1998). By contrast, P/Q-type channel activity should be barely sufficient for TH expression, which shows an ectopic pattern in the cerebella of the three mutants. It is, however, difficult to discuss the genesis of absence seizures in a similar context, since experimental studies and clinical observations indicate a central role of thalamocortical circuits, which comprise multiple neuronal populations. Furthermore, absence epilepsy can involve prolonged and inappropriate expression of developmentally immature complexation between the N-type α_{1B} subunit and β subunit isoforms (McEnery et al., 1998). In fact, tg^{rol} mice in which impairment of P/Q-type channel function is intermediate do not show absence seizures. Future work using slice preparations together with intracellular Ca^{2+} concentration measurements should allow us to more precisely correlate P/Q-type channel function with absence epilepsy.

Since tg^{rkr} shows a unique phenotype, as detailed above, study of functional properties should provide interesting information in correlating deviation of Ca^{2+} channel function with different phenotypes. In particular, the tg^{rkr} phenotypes of axon degeneration, reduction of branching in the Purkinje cell dendritic arbor, and the "weeping willow" appearance of the secondary branches are notable. tg^{rkr} should increase greatly our understanding of the contribution of P/Q-type channels to development and morphogenesis of dendritic trees in Purkinje cells (Zwingman et al.,

2001). Thus, distinct neurological and morphological phenotypes emphasize the value of allelic series in the study of α_{1A} gene function.

3.2. Auxiliary Subunit Mutations

3.2.1. ducky *Mutation and the $\alpha_2\delta$ Gene* (Cacna2d)

ducky (*du*) is an autosomal recessive mutation that shows epileptic seizures with features similar to those in human idiopathic generalized epilepsy, like the *tg* mutants. Homozygotes for the *du* allele were originally characterized by an ataxic, wide-based gait and paroxysmal dyskinesia (Snell, 1955). Neurological studies of *ducky* revealed region-selective dysgenesis of the central nervous system, including spinal cord and cerebellum (Meier, 1968). High-resolution genetic and physical mapping of the *du* locus was adopted (Barclay et al., 2001) to identify the $\alpha_2\delta$-2 gene (*Cacna2d*) (Gao et al., 2000) among the three $\alpha_2\delta$ genes (Klugbauer et al., 1999). Mutations in the original *du* and a newly identified strain, *du*^{2J}, were characterized. The *du* mutation is a large genomic rearrangement including duplication of some exons, resulting in aberrant RNA transcripts that encode the first 414 nucleotides plus 24 novel nucleotides and a stop codon. The *du*^{2J} mutation is a two base pair (TG) deletion within the ninth exon that causes premature truncation of the protein. In homozygous *du* mutants, the Ca^{2+} channel current was reduced in acutely dissociated Purkinje neurons, in which the $\alpha_2\delta$-2 protein is normally enriched, with no change in single channel conductance. Compared to the recombinant P/Q-type channels containing wild type $\alpha_2\delta$-2 ($\alpha_{1A}\beta4$ $\alpha_2\delta$-2), those P/Q-type channels which contained *du* $\alpha_2\delta$-2 produced reduced Ca^{2+} channel currents (Brodbeck et al., 2002). The *du* cerebellum showed multiple primary dendrites and a decrease in size of the dendritic trees of Purkinje cells, and the *du* cerebellum had an overall smaller size. Thus, not only the behavioral phenotypes, but also morphological abnormalities indicative of immature development of Purkinje neurons and reduced P/Q-type channel activity in *du* resemble those of the P/Q-type α_{1A} gene *tg* mutants. It is conceivable that reduced P/Q-type channel activity is the common basis for the neurological phenotypes shared by different epileptic mutant mice. However, the mechanism whereby the functional changes of P/Q-type channels containing the mutant $\alpha_2\delta$-2 are translated into morphological and neurophysiological abnormalities remains to be determined (Brodbeck et al., 2002).

3.2.2. lethargic *Mutation and the $\beta4$ Gene* (Cacnb4)

The mutation *lethargic* (*lh*) arose spontaneously in the mouse inbred strain BALB/cGn. Homozygous *lh* mice exhibit chronic ataxia, lethargic behavior (hypoactivity), and spontaneous focal motor seizures (Sidman et al., 1965). A recent paper examining motor syndrome of the *lh* mutants in detail has distinguished a baseline of ataxia with intermittent attacks that are reliably triggered by specific environmental or pharmacological influences and represent a form of paroxysmal dyskinesia (Khan and Jinnah, 2002). *lh* mice are also characterized by brief episodes of behavioral immobility accompanied by generalized cortical spike-wave discharges. This phenotype resembles absence seizures of human petit mal epilepsy (Noebels and Sidman, 1982). No pathological changes were recorded in the brain, spinal cord, or muscles in the mutant mice.

Among the four genes encoding the Ca^{2+} channel accessory β subunit isoforms, the gene encoding $\beta4$ was localized to proximal mouse chromosome 2 by interspecific back cross analysis (Chin et al., 1995). Burgess, Noebels, and colleagues (Burgess et al., 1997) have revealed that the *lh* mutation is a four nucleotide insertion into a splice donor site that results in exon skipping and translational frameshifts at the N-terminal side of the major α_1-binding site (Pragnell et al., 1994). Interestingly, and paradoxically, the P-type channel current in dissociated Purkinje neurons and the immunolocalization of α_{1A} and α_{1B} proteins in the *lh* brain were indistinguishable from wild type, whereas steady-state associations of α_{1A} and α_{1B} with the remaining β1-β3 were increased significantly, without significant changes in β subunit mRNA abundance (Burgess et al., 1999). This observation, which suggests a rescue of Ca^{2+} channel currents by the remaining β1-β3, was supported by the later observation that the ratio of Ca^{2+} channel subtypes, Ca^{2+} dependence of neurotransmitter release, and G protein-mediated Ca^{2+} channel inhibition were all unaltered at hippocampal Schaffer collateral synapses in the mutant (Qian and Noebels, 2000). Before these papers, it was originally speculated that the *lh* defect should affect Ca^{2+} currents dramatically, considering the many different effects of β subunits on Ca^{2+} channel function (Olcese et al., 1994; Walker et al., 1998; Bourinet et al., 1999; Brice and Dolphin, 1999). The β4 defect may exert its effects on untested aspects in the papers, such as coupling to the downstream cascades of Ca^{2+} influx. An alternative possibility is that the β4 subunit controls localization of α_1 proteins in neuronal microdomains, which may be only resolvable with high-resolution microscopic techniques.

Physiological impacts of the *lh* mutation were examined in somatosensory thalamic neurons, where a significant decrease in glutamatergic synaptic transmission, but not GABAergic transmission, was induced by the *lh* mutation. In *lh* thalamic neurons, increases in the amount of low voltage-activated Ca^{2+} currents was observed, without alteration in mRNA levels of the T-type genes. This same result was found for neurons of *tg* and *stargazer* mice (Zhang et al., 2002). Since the T-type α_1 subunits lack canonical β subunit binding sites in their I-II loop regions, the observed effect should be secondary to alterations in the high voltage-activated Ca^{2+} channels. As an additional consequence of the mutation, expression and assembly of N-type channel α_{1B} protein and β proteins was altered in the *lh* brain, with the population of N-type channels in *lh* resembling those present in immature rat brain (McEnery et al., 1998).

The role of the β4 subunit in producing the *lethargic* mouse phenotype may have direct relevance to human neurology. Screening of small pedigrees with familial epilepsy and ataxia has found two mutations in β4. The premature-termination β4 mutation R482X, resulting in the deletion of 38 C-terminal amino acids, was found in a patient with juvenile myoclonic epilepsy, and the missense β4 mutation C104F was found in a family with generalized epilepsy and praxis-induced seizures and in a family with episodic ataxia (Escayg et al., 2000). Since these mutations were undetectable in 225 unaffected individuals, and since activation kinetics of recombinant P/Q-type channels were accelerated by the R482X β4 mutant, these results suggest an important role for the β4 gene on chromosome 2a22-23 in human *lh* mutation-related disorders.

3.2.3. stargazer *Mutation and the γ2 Gene* (Cacng2)

The *stargazer* (*stg*) mutation was first described as a single recessive gene mutation mapping to the middle of mouse chromosome 15 (Noebels et al., 1990; Letts et al., 1997). The seizures in *stg* mice are present throughout life and are characterized by relatively long high-frequency neocortical spike-wave discharges (Qiao and Noebels, 1993). In fact, this model for petit mal epilepsy has been reported to show the most prevalent and prolonged seizures of all the mouse single-gene epilepsy models, including *tg* and *lh*.

After the initial study that constructed 1.1 physically contiguous Mb critical to the *stg* phenotype (Letts et al., 1997), Letts, Frankel, and colleagues identified a new gene disrupted by early transposon in *stg* and its allelic mutant *waggler* (stg^{wag}) (Letts et al., 1998). The gene encoded a membrane protein, stargazin, with a cytoplasmic N- and C-terminus and four transmembrane domains. Stargazin shared modest sequence similarity with the γ subunit of the skeletal muscle L-type channel. Together with the similarity in exon structure, this led to referring to stargazin as the second Ca^{2+} channel γ subunit, γ2. Stargazin mRNA was concentrated in the brain, and stargazin localized to synaptic membranes. Recombinant expression studies demonstrated that stargazin modified biophysical properties of the α_{1A} channel, by shifting the voltage dependence of inactivation to more hyperpolarized potentials. A predicted effect of the *stg* mutation on P/Q-type channels, therefore, would be a shift in the voltage dependence of activation to more depolarized potentials. This is exactly what is elicited by the *tg* mutations in α_{1A} P/Q-type channels (Wakamori et al., 1998; Mori et al., 2000).

So far, at least eight homologues of the γ subunit have been identified (Klugbauer et al., 2000; Chu et al., 2001; Moss et al., 2002). Recent reports suggest that these proteins are indeed Ca^{2+} channel subunits that exert various modulatory effects on different α_1 channels. The γ2 stargazin also caused a significant α_2/δ-dependent decrease in current amplitude of both α_{1A} and α_{1B} channels (Kang et al., 2001), while γ7 almost abolished α_{1B} currents via a large reduction of α_{1B} protein expression (Moss et al., 2002). The slowing effect of γ2, γ3, and γ4 on the inactivation speed of the α_{1A} channel was dependent on the accessory β subunit. γ4 shifted the voltage dependence of inactivation of the α_{1A} channel to hyperpolarizing potentials, as did γ2 (Klugbauer et al., 2000). In addition, γ2 slowed deactivation of the T-type α_{1I} channel (Green et al., 2001), while γ5 accelerated activation and inactivation of the T-type α_{1G} channel (Klugbauer et al., 2000).

Despite these important recombinant studies, the primary role of γ subunits is yet to be established in native Ca^{2+} channels, except for that of skeletal muscle (Held et al., 2002). Biochemical experiments have indicated that γ2 and γ3 do indeed co-sediment and co-immunoprecipitate with the neuronal Ca^{2+} channel subunits *in vivo* (Kang et al., 2001). Several interesting defects in *stg* mutant mice have already been disclosed. Stimulation of low voltage-activated currents has been observed in thalamic relay neurons from *stg* mice as well as those from *tg* and *lh* mice (Zhang et al., 2002). Selective failure of BDNF among neurotrophins (Qiao et al., 1996), and reduced levels of constitutive and neurotrophin 4/5-induced tyrosine phosphorylation in phospholipase Cγ1, erk1, and erk2, associated with impaired eyeblink conditioning, has also been demonstrated in *stg* mice (Qiao et al., 1998).

Thus stargazin may play a role more general than simply as an accessory subunit for voltage-dependent Ca^{2+} channels (Letts et al., 1998). Hashimoto and colleagues (Hashimoto et al., 1999) reported that AMPA-type glutamate receptors, but not NMDA-type glutamate receptors, were significantly impaired in *stg* cerebellar granule neurons. This was also demonstrated by Chen and colleagues (Chen et al., 1999), who further demonstrated that the interaction of stargazin with AMPA receptor subunits is essential for the delivery of functional AMPA receptors to the surface of the granule neurons

(Chen et al., 2000). Direct interactions between stargazin and PSD-95 (Schnell et al., 2002), regulated by PKA phosphorylation at the C-terminus (Choi et al., 2002), are suggested to be important for the synaptic targeting of AMPA receptors.

Table 5. Mutant mice

	Mouse	Mutation (Location)	Biophysical properties	Predicted effect on Ca^{2+} influx	References
α_{1A} subunit	tottering	R601L (II S5-S6)	Current density ↓ Non-inactivating component ↑ Unitary conductance →	↓	a, b
	rolling Nagoya	R1262G (III S4)	Current density ↓ $V_{1/2}$ activation ↑ Slope factor (activation) ↑ Activation time constant ↑	↓	c
	leaner	splicing error (C-terminal)	Current density ↓ $V_{1/2}$ activation ↑ or → $V_{1/2}$ inactivation ↑ or → Unitary conductance → Open probability ↓	↓	a, d, e, b
	rocker	T1310K (III S5-S6)	?	?	f
β_4 subunit	lethargic		P-type current density →	→(?)	g, h
γ_2 subunit	stargazer		Current density → $V_{1/2}$ activation → $V_{1/2}$ inactivation →	?	i, j
	waggler		?	?	i
α_2/δ_2 subunit	ducky		Current density ↓ Unitary conductance →	↓	k, l, m

Symbols, ↑ = Increased, ↓ = Decreased, → = Unchanged, ? = Unknown. References: a, Fletcher et al., 1996; b, Wakamori et al., 1998; c, Mori et al., 2000; d, Lorenzon et al., 1998; e, Dove et al., 1998; f, Zwingman et al., 2001; g, Burgess et al., 1997; h, Burgess et al., 1999; i, Letts et al., 1998; j, Chen et al., 2000; k, Gao et al., 2000; l, Barclay et al., 2001; m, Brodbeck et al., 2002.

3.3. General Mechanisms Underlying Neurological Phenotypes Determined From Mouse Mutants?

The reports on mouse α_{1A} mutants (Dove et al., 1998; Lorenzon et al., 1998; Wakamori et al., 1998; Matsushita et al., 2002), α_{1B} knock out mice (Ino et al., 2001), and mice with mutations in other subunits (Burgess et al., 1997; Letts et al., 1998; Burgess et al., 1999; Gao et al., 2000; Barclay et al., 2001) provide important information for discussing the general mechanisms by which Ca^{2+} channel defects underlie various neurological phenotypes (Table 5). Two important aspects can be pointed out. The direct effect of Ca^{2+} channel gene mutation is the loss of Ca^{2+} channel function: reduced P/Q-type currents in neurons from *tg*, tg^{la}, and tg^{rol} mice, and selective elimination of N-type currents in α_{1B}-deficient neurons. As a primary consequence of the loss of Ca^{2+} channel function, neuronal processes such as neurotransmitter release are impaired significantly. As a secondary and longer-term effect of the mutations, however, a paradoxical postsynaptic hyperactivity is induced, including increased sensitivity of postsynaptic excitatory transmitter receptors (Matsushita et al., 2002) or enhanced receptor-evoked smooth muscle contraction (Ino et al., 2001). It can be speculated that the primary suppression of GABAergic input and the secondary enhancement of postsynaptic glutamatergic receptors in the mutants synergistically elicit hyperexcitability in single central neurons of particular local neuronal circuits, eventually leading to seizures and neuronal death in affected mice. In general, neuronal activity and excitability that normally play a role only in early development may be evoked by calcium channel mutations in adult mice.

4. CALCIUM CHANNEL MUTATIONS IN NON-MAMMALS

4.1. *Drosophila*

The *Drosophila* Dmca1A voltage-dependent Ca^{2+} channel α_1 subunit is encoded by the courtship song defective mutation *cacophony* (*cac*) gene, and the visual defect mutation *nightblind-A* (*nbA*) and the mutations *lethal(1)L13* are allelic to *cac* (Smith et al., 1998). Association of *cac* mutations with various courtship and visual phenotypes may suggest that Dmca1A Ca^{2+} channels mediate multiple and separable biological functions. A temperature-sensitive (TS) paralytic allele of *cac* (cac^{TS2}) was identified as one of those modifiers of the TS paralytic mutant *comt*, which encodes N-ethylmaleimide-sensitive fusion protein (NSF), essential for neurotransmitter release (Dellinger et al., 2000). Electrophysiological analysis of cac^{TS2} mutants at neuromuscular synapses revealed a marked reduction in neurotransmitter release at elevated temperatures, indicating a primary role of *cac*-encoded Dmca1A Ca^{2+} channels in neurotransmitter release (Kawasaki et al., 2000; Kawasaki et al., 2002).

4.2. *C. elegans*

The homolog of voltage-dependent Ca^{2+} channel α_{1A}, α_{1B}, and α_{1E} subunits in *Caenorhabditis elegans* worm is encoded by the *unc-2* gene (Schafer and Kenyon, 1995). *unc-2* is required for adaptation to the neurotransmitters dopamine and serotonin. A loss of function *unc-2* mutant failed to adapt to dopamine-induced paralysis or to stimulation of egg laying by serotonin; this mutant instead exhibited constitutive egg-laying regulated by serotonergic neurons. It was suggested that the *unc-2* Ca^{2+} channel α_1 subunit participates in modulation of the response to serotonin by inhibiting the release of an egg-laying-potentiation neuropeptide. *unc-2* was also implicated in cell migration and pathfinding in the touch receptor neuron AMV (Tam et al., 2000). Another gene, *egl-19*, has been shown to encode the α_1 homolog for the L-type Ca^{2+} channel in *C. elegans* (Lee et al., 1997; Jospin et al., 2002). The null *egl-19* mutation is lethal, while a partial loss of function mutation enfeebled the contraction of body wall muscle cells and necessitated stronger depolarizations to raise muscle intracellular Ca^{2+}.

4.3. Yeast

Two genes, MID1 and CCH1, that likely represent a *Saccharomyces cerevisiae* homolog of the voltage-gated Ca^{2+} channel were discovered in the course of identifying recessive repressors of the *cdc1*(Ts) temperature-sensitive growth gene (Paidhungat and Garrett, 1997). MID1 and CCH1 controlled Ca^{2+} entry and accumulation, including that related to capacitative Ca^{2+} entry controlled by the Pmr1p pump of the secretory pathway (Fischer et al., 1997; Locke et al., 2000).

5. SUMMARY

To unravel the complex contribution of voltage-dependent Ca^{2+} channels to cellular function in the intact animal, subtype-selective blockers and toxins have been widely employed. However, such pharmacological compounds generally exert all-or-none blockade of currents through the target Ca^{2+} channel. Moreover, selective blockers are unavailable for many classes or splice forms of Ca^{2+} channel. An alternative and powerful approach that circumvents these problems is the molecular genetic approach. As discussed (Mori et al., 2000; Zwingman et al., 2001), the many mutants that disrupt Ca^{2+} channels not only illustrate the importance of these channels in a wide variety of biological processes, but also emphasize the value of an allelic series of mutations in a given gene. Respective mutations of the same allele can alter the activity of the target Ca^{2+} channel differentially, unlike pharmacological blockers. Continued functional identification of Ca^{2+} channels responsible for disease plus further physiological characterization of allelic series should elucidate the molecular and cellular pathways responsible for the different phenotypes. This will solidify the biological relevance of the great number of studies made on the structure-function relationships of voltage-gated Ca^{2+} channels (Mori et al., 1996; Walker and De Waard, 1998) and provide an important basis for the development of effective screening systems for drugs targeting particular neurological disorders.

ACKNOWLEDGMENTS

This work was supported by research grants from the Ministry of Education, Science, Sports, and Culture of Japan.

REFERENCES

Ahlijanian, M. K., Westenbroek, R. E. and Catterall, W. A., 1990, Subunit structure and localization of dihydropyridine-sensitive calcium channels in mammalian brain, spinal cord, and retina, *Neuron* **4**:819-832.

Artalejo, C. R., Adams, M. E. and Fox, A. P., 1994, Three types of Ca^{2+} channel trigger secretion with different efficacies in chromaffin cells, *Nature* **367**:72-76.

Barclay, J., Balaguero, N., Mione, M., Ackerman, S. L., Letts, V. A., Brodbeck, J., Canti, C., Meir, A., Page, K. M., Kusumi, K., Perez-Reyes, E., Lander, E. S., Frankel, W. N., Gardiner, R. M., Dolphin, A. C. and Rees, M., 2001, Ducky mouse phenotype of epilepsy and ataxia is associated with mutations in the Cacna2d2 gene and decreased calcium channel current in cerebellar Purkinje cells, *J Neurosci* **21**:6095-6104.

Battistini, S., Stenirri, S., Piatti, M., Gelfi, C., Righetti, P. G., Rocchi, R., Giannini, F., Battistini, N., Guazzi, G. C., Ferrari, M. and Carrera, P., 1999, A new CACNA1A gene mutation in acetazolamide-responsive familial hemiplegic migraine and ataxia, *Neurology* **53**:38-43.

Bean, B. P., 1989, Classes of calcium channels in vertebrate cells, *Annu Rev Physiol* **51**:367-384.

Bech-Hansen, N. T., Naylor, M. J., Maybaum, T. A., Pearce, W. G., Koop, B., Fishman, G. A., Mets, M., Musarella, M. A. and Boycott, K. M., 1998, Loss-of-function mutations in a calcium-channel alpha1-

subunit gene in Xp11.23 cause incomplete X-linked congenital stationary night blindness, *Nat Genet* **19**:264-267.

Bourinet, E., Soong, T. W., Sutton, K., Slaymaker, S., Mathews, E., Monteil, A., Zamponi, G. W., Nargeot, J. and Snutch, T. P., 1999, Splicing of α1A subunit gene generates phenotypic variants of P- and Q-type calcium channels, *Nat Neurosci* **2**:407-415.

Brice, N. L. and Dolphin, A. C., 1999, Differential plasma membrane targeting of voltage-dependent calcium channel subunits expressed in a polarized epithelial cell line, *J Physiol* **515 (Pt 3)**:685-694.

Brodbeck, J., Davies, A., Courtney, J. M., Meir, A., Balaguero, N., Canti, C., Moss, F. J., Page, K. M., Pratt, W. S., Hunt, S. P., Barclay, J., Rees, M. and Dolphin, A. C., 2002, The ducky mutation in Cacna2d2 results in altered Purkinje cell morphology and is associated with the expression of a truncated α2δ-2 protein with abnormal function, *J Biol Chem* **277**:7684-7693.

Burgess, D. L., Jones, J. M., Meisler, M. H. and Noebels, J. L., 1997, Mutation of the Ca^{2+} channel beta subunit gene Cchb4 is associated with ataxia and seizures in the *lethargic* (lh) mouse, *Cell* **88**:385-392.

Burgess, D. L., Biddlecome, G. H., McDonough, S. I., Diaz, M. E., Zilinski, C. A., Bean, B. P., Campbell, K. P. and Noebels, J. L., 1999, beta subunit reshuffling modifies N- and P/Q-type Ca2+ channel subunit compositions in lethargic mouse brain, *Mol Cell Neurosci* **13**:293-311.

Caddick, S. J., Wang, C., Fletcher, C. F., Jenkins, N. A., Copeland, N. G. and Hosford, D. A., 1999, Excitatory but not inhibitory synaptic transmission is reduced in lethargic ($Cacnb4^{lh}$) and tottering ($Cacna1a^{tg}$) mouse thalami, *J Neurophysiol* **81**:2066-2074.

Campbell, D. B. and Hess, E. J., 1999, L-type calcium channels contribute to the *tottering* mouse dystonic episodes, *Mol Pharmacol* **55**:23-31.

Campbell, K. P., Leung, A. T. and Sharp, A. H., 1988, The biochemistry and molecular biology of the dihydropyridine-sensitive calcium channel, *Trends Neurosci* **11**:425-430.

Carrera, P., Piatti, M., Stenirri, S., Grimaldi, L. M., Marchioni, E., Curcio, M., Righetti, P. G., Ferrari, M. and Gelfi, C., 1999, Genetic heterogeneity in Italian families with familial hemiplegic migraine, *Neurology* **53**:26-33.

Chen, L., Bao, S., Qiao, X. and Thompson, R. F., 1999, Impaired cerebellar synapse maturation in *waggler*, a mutant mouse with a disrupted neuronal calcium channel gamma subunit, *Proc Natl Acad Sci U S A* **96**:12132-12137.

Chen, L., Chetkovich, D. M., Petralia, R. S., Sweeney, N. T., Kawasaki, Y., Wenthold, R. J., Bredt, D. S. and Nicoll, R. A., 2000, Stargazin regulates synaptic targeting of AMPA receptors by two distinct mechanisms, *Nature* **408**:936-943.

Chin, H., Kwon, O. J., Jung, H. H., Kim, D. S. and Kozak, C. A., 1995, Genetic mapping of the mouse genes encoding the voltage-sensitive calcium channel subunits, *Genomics* **28**:592-595.

Choi, J., Ko, J., Park, E., Lee, J. R., Yoon, J., Lim, S. and Kim, E., 2002, Phosphorylation of stargazin by protein kinase A regulates its interaction with PSD-95, *J Biol Chem* **277**:12359-12363.

Chu, P. J., Robertson, H. M. and Best, P. M., 2001, Calcium channel gamma subunits provide insights into the evolution of this gene family, *Gene* **280**:37-48.

Clapham, D. E., 1995, Calcium signaling, *Cell* **80**:259-268.

Dellinger, B., Felling, R. and Ordway, R. W., 2000, Genetic modifiers of the Drosophila NSF mutant, *comatose*, include a temperature-sensitive paralytic allele of the calcium channel α1-subunit gene, *cacophony*, *Genetics* **155**:203-211.

Denier, C., Ducros, A., Vahedi, K., Joutel, A., Thierry, P., Ritz, A., Castelnovo, G., Deonna, T., Gerard, P., Devoize, J. L., Gayou, A., Perrouty, B., Soisson, T., Autret, A., Warter, J. M., Vighetto, A., Van Bogaert, P., Alamowitch, S., Roullet, E. and Tournier-Lasserve, E., 1999, High prevalence of CACNA1A truncations and broader clinical spectrum in episodic ataxia type 2, *Neurology* **52**:1816-1821.

Denier, C., Ducros, A., Durr, A., Eymard, B., Chassande, B. and Tournier-Lasserve, E., 2001, Missense CACNA1A mutation causing episodic ataxia type 2, *Arch Neurol* **58**:292-295.

Dove, L. S., Abbott, L. C. and Griffith, W. H., 1998, Whole-cell and single-channel analysis of P-type calcium currents in cerebellar Purkinje cells of *leaner* mutant mice, *J Neurosci* **18**:7687-7699.

Dove, L. S., Nahm, S. S., Murchison, D., Abbott, L. C. and Griffith, W. H., 2000, Altered calcium homeostasis in cerebellar Purkinje cells of *leaner* mutant mice, *J Neurophysiol* **84**:513-524.

Doyle, J., Ren, X., Lennon, G. and Stubbs, L., 1997, Mutations in the Cacnl1a4 calcium channel gene are associated with seizures, cerebellar degeneration, and ataxia in *tottering* and *leaner* mutant mice, *Mamm Genome* **8**:113-120.

Ducros, A., Joutel, A., Labauge, P., Denier, C., Joutel, A., Vahedi, K., Bousser, M. G. and Tournier-Lasserve, E., 1999a, Characterization of CACNA1A mutations in familial and sporadic hemiplegic migraine, *Neurology* **52**:A273.

Ducros, A., Denier, C., Joutel, A., Vahedi, K., Michel, A., Darcel, F., Madigand, M., Guerouaou, D., Tison, F., Julien, J., Hirsch, E., Chedru, F., Bisgard, C., Lucotte, G., Despres, P., Billard, C., Barthez, M. A., Ponsot, G., Bousser, M. G. and Tournier-Lasserve, E., 1999b, Recurrence of the T666M calcium channel CACNA1A gene mutation in familial hemiplegic migraine with progressive cerebellar ataxia, *Am J Hum Genet* **64**:89-98.

Ertel, E. A., Campbell, K. P., Harpold, M. M., Hofmann, F., Mori, Y., Perez-Reyes, E., Schwartz, A., Snutch, T. P., Tanabe, T., Birnbaumer, L., Tsien, R. W. and Catterall, W. A., 2000, Nomenclature of voltage-gated calcium channels, *Neuron* **25**:533-535.

Escayg, A., De Waard, M., Lee, D. D., Bichet, D., Wolf, P., Mayer, T., Johnston, J., Baloh, R., Sander, T. and Meisler, M. H., 2000, Coding and noncoding variation of the human calcium-channel beta4-subunit gene CACNB4 in patients with idiopathic generalized epilepsy and episodic ataxia, *Am J Hum Genet* **66**:1531-1539.

Fischer, M., Schnell, N., Chattaway, J., Davies, P., Dixon, G. and Sanders, D., 1997, The Saccharomyces cerevisiae CCH1 gene is involved in calcium influx and mating, *FEBS Lett* **419**:259-262.

Fletcher, C. F., Lutz, C. M., O'Sullivan, T. N., Shaughnessy, J. D., Jr., Hawkes, R., Frankel, W. N., Copeland, N. G. and Jenkins, N. A., 1996, Absence epilepsy in *tottering* mutant mice is associated with calcium channel defects, *Cell* **87**:607-617.

Fontaine, B., Vale-Santos, J., Jurkat-Rott, K., Reboul, J., Plassart, E., Rime, C. S., Elbaz, A., Heine, R., Guimaraes, J., Weissenbach, J. and et al., 1994, Mapping of the hypokalaemic periodic paralysis (HypoPP) locus to chromosome 1q31-32 in three European families, *Nat Genet* **6**:267-272.

Friend, K. L., Crimmins, D., Phan, T. G., Sue, C. M., Colley, A., Fung, V. S., Morris, J. G., Sutherland, G. R. and Richards, R. I., 1999, Detection of a novel missense mutation and second recurrent mutation in the CACNA1A gene in individuals with EA-2 and FHM, *Hum Genet* **105**:261-265.

Gao, B., Sekido, Y., Maximov, A., Saad, M., Forgacs, E., Latif, F., Wei, M. H., Lerman, M., Lee, J. H., Perez-Reyes, E., Bezprozvanny, I. and Minna, J. D., 2000, Functional properties of a new voltage-dependent calcium channel $\alpha_2\delta$ auxiliary subunit gene (CACNA2D2), *J Biol Chem* **275**:12237-12242.

Glossmann, H. and Striessnig, J., 1990, Molecular properties of calcium channels, *Rev Physiol Biochem Pharmacol* **114**:1-105.

Green, M. C. and Sidman, R. L., 1962, Tottering - a neuromuscular mutation in the mouse, *Journal of Heredity* **53**:79.

Green, P. J., Warre, R., Hayes, P. D., McNaughton, N. C., Medhurst, A. D., Pangalos, M., Duckworth, D. M. and Randall, A. D., 2001, Kinetic modification of the α_{1I} subunit-mediated T-type Ca^{2+} channel by a human neuronal Ca^{2+} channel gamma subunit, *J Physiol* **533**:467-478.

Guida, S., Trettel, F., Pagnutti, S., Mantuano, E., Tottene, A., Veneziano, L., Fellin, T., Spadaro, M., Stauderman, K., Williams, M., Volsen, S., Ophoff, R., Frants, R., Jodice, C., Frontali, M. and Pietrobon, D., 2001, Complete loss of P/Q calcium channel activity caused by a CACNA1A missense mutation carried by patients with episodic ataxia type 2, *Am J Hum Genet* **68**:759-764.

Hans, M., Luvisetto, S., Williams, M. E., Spagnolo, M., Urrutia, A., Tottene, A., Brust, P. F., Johnson, E. C., Harpold, M. M., Stauderman, K. A. and Pietrobon, D., 1999, Functional consequences of mutations in the human α1A calcium channel subunit linked to familial hemiplegic migraine, *J Neurosci* **19**:1610-1619.

Hashimoto, K., Fukaya, M., Qiao, X., Sakimura, K., Watanabe, M. and Kano, M., 1999, Impairment of AMPA receptor function in cerebellar granule cells of ataxic mutant mouse *stargazer*, *J Neurosci* **19**:6027-6036.

Held, B., Freise, D., Freichel, M., Hoth, M. and Flockerzi, V., 2002, Skeletal muscle L-type Ca^{2+} current modulation in γ1-deficient and wildtype murine myotubes by the γ1 subunit and cAMP, *J Physiol* **539**:459-468.

Herrup, K. and Wilczynski, S. L., 1982, Cerebellar cell degeneration in the *leaner* mutant mouse, *Neuroscience* 7:2185-2196.

Hess, E. J. and Wilson, M. C., 1991, *Tottering* and *leaner* mutations perturb transient developmental expression of tyrosine hydroxylase in embryologically distinct Purkinje cells, *Neuron* **6**:123-132.

Hille, B., 1992, *Ionic Channels of Excitable Membranes*, Sinauer Associates, Sunderland, MA.

Hirning, L. D., Fox, A. P., McCleskey, E. W., Olivera, B. M., Thayer, S. A., Miller, R. J. and Tsien, R. W., 1988, Dominant role of N-type Ca2+ channels in evoked release of norepinephrine from sympathetic neurons, *Science* **239**:57-61.

Holtmann, M., Opp, J., Tokarzewski, M. and Korn-Merker, E., 2002, Human epilepsy, episodic ataxia type 2, and migraine, *Lancet* **359**:170-171.

Ino, M., Yoshinaga, T., Wakamori, M., Miyamoto, N., Takahashi, E., Sonoda, J., Kagaya, T., Oki, T., Nagasu, T., Nishizawa, Y., Tanaka, I., Imoto, K., Aizawa, S., Koch, S., Schwartz, A., Niidome, T., Sawada, K. and Mori, Y., 2001, Functional disorders of the sympathetic nervous system in mice lacking the α 1B subunit (Cav 2.2) of N-type calcium channels, *Proc Natl Acad Sci U S A* **98**:5323-5328.

Jen, J., Yue, Q., Nelson, S. F., Yu, H., Litt, M., Nutt, J. and Baloh, R. W., 1999, A novel nonsense mutation in CACNA1A causes episodic ataxia and hemiplegia, *Neurology* **53**:34-37.

Jen, J., Wan, J., Graves, M., Yu, H., Mock, A. F., Coulin, C. J., Kim, G., Yue, Q., Papazian, D. M. and Baloh, R. W., 2001, Loss-of-function EA2 mutations are associated with impaired neuromuscular transmission, *Neurology* **57**:1843-1848.

Jodice, C., Mantuano, E., Veneziano, L., Trettel, F., Sabbadini, G., Calandriello, L., Francia, A., Spadaro, M., Pierelli, F., Salvi, F., Ophoff, R. A., Frants, R. R. and Frontali, M., 1997, Episodic ataxia type 2 (EA2) and spinocerebellar ataxia type 6 (SCA6) due to CAG repeat expansion in the CACNA1A gene on chromosome 19p, *Hum Mol Genet* **6**:1973-1978.

Jospin, M., Jacquemond, V., Mariol, M. C., Segalat, L. and Allard, B., 2002, The L-type voltage-dependent Ca^{2+} channel EGL-19 controls body wall muscle function in *Caenorhabditis elegans*, *J Cell Biol* **159**:337-348.

Jouvenceau, A., Eunson, L. H., Spauschus, A., Ramesh, V., Zuberi, S. M., Kullmann, D. M. and Hanna, M. G., 2001, Human epilepsy associated with dysfunction of the brain P/Q-type calcium channel, *Lancet* **358**:801-807.

Jun, K., Piedras-Renteria, E. S., Smith, S. M., Wheeler, D. B., Lee, S. B., Lee, T. G., Chin, H., Adams, M. E., Scheller, R. H., Tsien, R. W. and Shin, H. S., 1999, Ablation of P/Q-type Ca^{2+} channel currents, altered synaptic transmission, and progressive ataxia in mice lacking the α1A-subunit, *Proc Natl Acad Sci U S A* **96**:15245-15250.

Jurkat-Rott, K., Lehmann-Horn, F., Elbaz, A., Heine, R., Gregg, R. G., Hogan, K., Powers, P. A., Lapie, P., Vale-Santos, J. E., Weissenbach, J. and et al., 1994, A calcium channel mutation causing hypokalemic periodic paralysis, *Hum Mol Genet* **3**:1415-1419.

Jurkat-Rott, K., Uetz, U., Pika-Hartlaub, U., Powell, J., Fontaine, B., Melzer, W. and Lehmann-Horn, F., 1998, Calcium currents and transients of native and heterologously expressed mutant skeletal muscle DHP receptor α1 subunits (R528H), *FEBS Lett* **423**:198-204.

Kang, M. G., Chen, C. C., Felix, R., Letts, V. A., Frankel, W. N., Mori, Y. and Campbell, K. P., 2001, Biochemical and biophysical evidence for gamma 2 subunit association with neuronal voltage-activated Ca^{2+} channels, *J Biol Chem* **276**:32917-32924.

Kawasaki, F., Felling, R. and Ordway, R. W., 2000, A temperature-sensitive paralytic mutant defines a primary synaptic calcium channel in *Drosophila*, *J Neurosci* **20**:4885-4889.

Kawasaki, F., Collins, S. C. and Ordway, R. W., 2002, Synaptic calcium-channel function in *Drosophila*: analysis and transformation rescue of temperature-sensitive paralytic and lethal mutations of *cacophony*, *J Neurosci* **22**:5856-5864.

Khan, Z. and Jinnah, H. A., 2002, Paroxysmal dyskinesias in the *lethargic* mouse mutant, *J Neurosci* **22**:8193-8200.

Klugbauer, N., Lacinova, L., Marais, E., Hobom, M. and Hofmann, F., 1999, Molecular diversity of the calcium channel $\alpha2\delta$ subunit, *J Neurosci* **19**:684-691.

Klugbauer, N., Dai, S., Specht, V., Lacinova, L., Marais, E., Bohn, G. and Hofmann, F., 2000, A family of gamma-like calcium channel subunits, *FEBS Lett* **470**:189-197.

Kobayashi, T. and Mori, Y., 1998, Ca^{2+} channel antagonists and neuroprotection from cerebral ischemia, *Eur J Pharmacol* **363**:1-15.

Kraus, R. L., Sinnegger, M. J., Glossmann, H., Hering, S. and Striessnig, J., 1998, Familial hemiplegic migraine mutations change α1A Ca^{2+} channel kinetics, *J Biol Chem* **273**:5586-5590.

Kraus, R. L., Sinnegger, M. J., Koschak, A., Glossmann, H., Stenirri, S., Carrera, P. and Striessnig, J., 2000, Three new familial hemiplegic migraine mutants affect P/Q-type Ca^{2+} channel kinetics, *J Biol Chem* **275**:9239-9243.

Lapie, P., Goudet, C., Nargeot, J., Fontaine, B. and Lory, P., 1996, Electrophysiological properties of the hypokalaemic periodic paralysis mutation (R528H) of the skeletal muscle alpha 1s subunit as expressed in mouse L cells, *FEBS Lett* **382**:244-248.

Lau, F. C., Abbott, L. C., Rhyu, I. J., Kim, D. S. and Chin, H., 1998, Expression of calcium channel α1A mRNA and protein in the *leaner* mouse (tg^{la}/tg^{la}) cerebellum, *Brain Res Mol Brain Res* **59**:93-99.

Lee, J. H., Daud, A. N., Cribbs, L. L., Lacerda, A. E., Pereverzev, A., Klockner, U., Schneider, T. and Perez-Reyes, E., 1999, Cloning and expression of a novel member of the low voltage-activated T-type calcium channel family, *J Neurosci* **19**:1912-1921.

Lee, R. Y., Lobel, L., Hengartner, M., Horvitz, H. R. and Avery, L., 1997, Mutations in the α1 subunit of an L-type voltage-activated Ca^{2+} channel cause myotonia in *Caenorhabditis elegans*, *Embo J* **16**:6066-6076.

Lerche, H., Klugbauer, N., Lehmann-Horn, F., Hofmann, F. and Melzer, W., 1996, Expression and functional characterization of the cardiac L-type calcium channel carrying a skeletal muscle DHP-receptor mutation causing hypokalaemic periodic paralysis, *Pflugers Arch* **431**:461-463.

Letts, V. A., Valenzuela, A., Kirley, J. P., Sweet, H. O., Davisson, M. T. and Frankel, W. N., 1997, Genetic and physical maps of the *stargazer* locus on mouse chromosome 15, *Genomics* **43**:62-68.

Letts, V. A., Felix, R., Biddlecome, G. H., Arikkath, J., Mahaffey, C. L., Valenzuela, A., Bartlett, F. S., 2nd, Mori, Y., Campbell, K. P. and Frankel, W. N., 1998, The mouse *stargazer* gene encodes a neuronal Ca^{2+}-channel gamma subunit, *Nat Genet* **19**:340-347.

Llinas, R., Sugimori, M., Lin, J. W. and Cherksey, B., 1989, Blocking and isolation of a calcium channel from neurons in mammals and cephalopods utilizing a toxin fraction (FTX) from funnel-web spider poison, *Proc Natl Acad Sci U S A* **86**:1689-1693.

Llinas, R., Sugimori, M., Hillman, D. E. and Cherksey, B., 1992, Distribution and functional significance of the P-type, voltage-dependent Ca^{2+} channels in the mammalian central nervous system, *Trends Neurosci* **15**:351-355.

Locke, E. G., Bonilla, M., Liang, L., Takita, Y. and Cunningham, K. W., 2000, A homolog of voltage-gated Ca^{2+} channels stimulated by depletion of secretory Ca^{2+} in yeast, *Mol Cell Biol* **20**:6686-6694.

Lorenzon, N. M., Lutz, C. M., Frankel, W. N. and Beam, K. G., 1998, Altered calcium channel currents in Purkinje cells of the neurological mutant mouse *leaner*, *J Neurosci* **18**:4482-4489.

Matsushita, K., Wakamori, M., Rhyu, I. J., Arii, T., Oda, S., Mori, Y. and Imoto, K., 2002, Bidirectional alterations in cerebellar synaptic transmission of *tottering* and *rolling* Ca^{2+} channel mutant mice, *J Neurosci* **22**:4388-4398.

Matsuyama, Z., Wakamori, M., Mori, Y., Kawakami, H., Nakamura, S. and Imoto, K., 1999, Direct alteration of the P/Q-type Ca^{2+} channel property by polyglutamine expansion in spinocerebellar ataxia 6, *J Neurosci* **19**:RC14.

May, A., Ophoff, R. A., Terwindt, G. M., Urban, C., van Eijk, R., Haan, J., Diener, H. C., Lindhout, D., Frants, R. R., Sandkuijl, L. A. and et al., 1995, Familial hemiplegic migraine locus on 19p13 is involved in the common forms of migraine with and without aura, *Hum Genet* **96**:604-608.

McEnery, M. W., Copeland, T. D. and Vance, C. L., 1998, Altered expression and assembly of N-type calcium channel alpha1B and beta subunits in epileptic *lethargic* (lh/lh) mouse, *J Biol Chem* **273**:21435-21438.

Meier, H., 1968, The neuropathology of *ducky*, a neurological mutation of the mouse. A pathological and preliminary histochemical study, *Acta Neuropathol (Berl)* **11**:15-28.

Minaker, K. L., Meneilly, G. S., Flier, J. S. and Rowe, J. W., 1988, Insulin-mediated hypokalemia and paralysis in familial hypokalemic periodic paralysis, *Am J Med* **84**:1001-1006.

Mintz, I. M., Venema, V. J., Swiderek, K. M., Lee, T. D., Bean, B. P. and Adams, M. E., 1992, P-type calcium channels blocked by the spider toxin ω-Aga-IVA, *Nature* **355**:827-829.

Montagna, P., 2000, Molecular genetics of migraine headaches: a review, *Cephalalgia* **20**:3-14.

Mori, Y., Friedrich, T., Kim, M. S., Mikami, A., Nakai, J., Ruth, P., Bosse, E., Hofmann, F., Flockerzi, V., Furuichi, T. and et al., 1991, Primary structure and functional expression from complementary DNA of a brain calcium channel, *Nature* **350**:398-402.

Mori, Y., Mikala, G., Varadi, G., Kobayashi, T., Koch, S., Wakamori, M. and Schwartz, A., 1996, Molecular pharmacology of voltage-dependent calcium channels, *Jpn J Pharmacol* **72**:83-109.

Mori, Y., Wakamori, M., Oda, S., Fletcher, C. F., Sekiguchi, N., Mori, E., Copeland, N. G., Jenkins, N. A., Matsushita, K., Matsuyama, Z. and Imoto, K., 2000, Reduced voltage sensitivity of activation of P/Q-type Ca^{2+} channels is associated with the ataxic mouse mutation *rolling Nagoya* (tg^{rol}), *J Neurosci* **20**:5654-5662.

Morrill, J. A., Brown, R. H., Jr. and Cannon, S. C., 1998, Gating of the L-type Ca channel in human skeletal myotubes: an activation defect caused by the hypokalemic periodic paralysis mutation R528H, *J Neurosci* **18**:10320-10334.

Morrill, J. A. and Cannon, S. C., 1999, Effects of mutations causing hypokalaemic periodic paralysis on the skeletal muscle L-type Ca^{2+} channel expressed in Xenopus laevis oocytes, *J Physiol* **520 Pt 2**:321-336.

Moss, F. J., Viard, P., Davies, A., Bertaso, F., Page, K. M., Graham, A., Canti, C., Plumpton, M., Plumpton, C., Clare, J. J. and Dolphin, A. C., 2002, The novel product of a five-exon stargazin-related gene abolishes $Ca_V2.2$ calcium channel expression, *Embo J* **21**:1514-1523.

Noebels, J. L. and Sidman, R. L., 1982, Three mutations causing spike-wave seizures in the mouse, *Neuroscience* **Supplement** 7:159.

Noebels, J. L., 1984, A single gene error of noradrenergic axon growth synchronizes central neurones, *Nature* **310**:409-411.

Noebels, J. L., Qiao, X., Bronson, R. T., Spencer, C. and Davisson, M. T., 1990, *Stargazer*: a new neurological mutant on chromosome 15 in the mouse with prolonged cortical seizures, *Epilepsy Res* 7:129-135.

Oda, S., 1973, [The observation of rolling mouse Nagoya (rol), a new neurological mutant, and its maintenance (author's transl)], *Jikken Dobutsu* **22**:281-288.

Oda, S., 1981, A new allele of the tottering locus, rolling mouse Nagoya, on chromosome no. 8 in the mouse, *Japan Journal of Genetics* **56**:295.

Olcese, R., Qin, N., Schneider, T., Neely, A., Wei, X., Stefani, E. and Birnbaumer, L., 1994, The amino terminus of a calcium channel beta subunit sets rates of channel inactivation independently of the subunit's effect on activation, *Neuron* **13**:1433-1438.

Ophoff, R. A., Terwindt, G. M., Vergouwe, M. N., van Eijk, R., Oefner, P. J., Hoffman, S. M., Lamerdin, J. E., Mohrenweiser, H. W., Bulman, D. E., Ferrari, M., Haan, J., Lindhout, D., van Ommen, G. J., Hofker, M. H., Ferrari, M. D. and Frants, R. R., 1996, Familial hemiplegic migraine and episodic ataxia type-2 are caused by mutations in the Ca^{2+} channel gene CACNL1A4, *Cell* **87**:543-552.

Ophoff, R. A., van den Maagdenberg, A. M., Roon, K. I., Ferrari, M. D. and Frants, R. R., 2001, The impact of pharmacogenetics for migraine, *Eur J Pharmacol* **413**:1-10.

Paidhungat, M. and Garrett, S., 1997, A homolog of mammalian, voltage-gated calcium channels mediates yeast pheromone-stimulated Ca^{2+} uptake and exacerbates the cdc1(Ts) growth defect, *Mol Cell Biol* **17**:6339-6347.

Piedras-Renteria, E. S., Watase, K., Harata, N., Zhuchenko, O., Zoghbi, H. Y., Lee, C. C. and Tsien, R. W., 2001, Increased expression of alpha 1A Ca^{2+} channel currents arising from expanded trinucleotide repeats in spinocerebellar ataxia type 6, *J Neurosci* **21**:9185-9193.

Pietrobon, D., 2002, Calcium channels and channelopathies of the central nervous system, *Mol Neurobiol* **25**:31-50.

Pragnell, M., De Waard, M., Mori, Y., Tanabe, T., Snutch, T. P. and Campbell, K. P., 1994, Calcium channel beta-subunit binds to a conserved motif in the I-II cytoplasmic linker of the alpha 1-subunit, *Nature* **368**:67-70.

Ptacek, L. J., Tawil, R., Griggs, R. C., Engel, A. G., Layzer, R. B., Kwiecinski, H., McManis, P. G., Santiago, L., Moore, M., Fouad, G. and et al., 1994, Dihydropyridine receptor mutations cause hypokalemic periodic paralysis, *Cell* **77**:863-868.

Qian, J. and Noebels, J. L., 2000, Presynaptic Ca^{2+} influx at a mouse central synapse with Ca^{2+} channel subunit mutations, *J Neurosci* **20**:163-170.

Qiao, X. and Noebels, J. L., 1993, Developmental analysis of hippocampal mossy fiber outgrowth in a mutant mouse with inherited spike-wave seizures, *J Neurosci* **13**:4622-4635.

Qiao, X., Hefti, F., Knusel, B. and Noebels, J. L., 1996, Selective failure of brain-derived neurotrophic factor mRNA expression in the cerebellum of *stargazer*, a mutant mouse with ataxia, *J Neurosci* **16**:640-648.

Qiao, X., Chen, L., Gao, H., Bao, S., Hefti, F., Thompson, R. F. and Knusel, B., 1998, Cerebellar brain-derived neurotrophic factor-TrkB defect associated with impairment of eyeblink conditioning in *Stargazer* mutant mice, *J Neurosci* **18**:6990-6999.

Regan, L. J., Sah, D. W. and Bean, B. P., 1991, Ca^{2+} channels in rat central and peripheral neurons: high-threshold current resistant to dihydropyridine blockers and ω-conotoxin, *Neuron* **6**:269-280.

Regehr, W. G. and Mintz, I. M., 1994, Participation of multiple calcium channel types in transmission at single climbing fiber to Purkinje cell synapses, *Neuron* **12**:605-613.

Restituito, S., Thompson, R. M., Eliet, J., Raike, R. S., Riedl, M., Charnet, P. and Gomez, C. M., 2000, The polyglutamine expansion in spinocerebellar ataxia type 6 causes a beta subunit-specific enhanced activation of P/Q-type calcium channels in *Xenopus* oocytes, *J Neurosci* **20**:6394-6403.

Rhyu, I. J., Abbott, L. C., Walker, D. B. and Sotelo, C., 1999a, An ultrastructural study of granule cell/Purkinje cell synapses in *tottering* (tg/tg), *leaner* (tg^{la}/tg^{la}) and compound heterozygous *tottering/leaner* (tg/tg^{la}) mice, *Neuroscience* **90**:717-728.

Rhyu, I. J., Oda, S., Uhm, C. S., Kim, H., Suh, Y. S. and Abbott, L. C., 1999b, Morphologic investigation of rolling mouse Nagoya (tg^{rol}/tg^{rol}) cerebellar Purkinje cells: an ataxic mutant, revisited, *Neurosci Lett* **266**:49-52.

Rudel, R., Lehmann-Horn, F., Ricker, K. and Kuther, G., 1984, Hypokalemic periodic paralysis: in vitro investigation of muscle fiber membrane parameters, *Muscle Nerve* **7**:110-120.

Ruff, R. L., 1999, Insulin acts in hypokalemic periodic paralysis by reducing inward rectifier K^+ current, *Neurology* **53**:1556-1563.

Russell, M. B., Iselius, L. and Olesen, J., 1995, Inheritance of migraine investigated by complex segregation analysis, *Hum Genet* **96**:726-730.

Sather, W. A., Tanabe, T., Zhang, J. F., Mori, Y., Adams, M. E. and Tsien, R. W., 1993, Distinctive biophysical and pharmacological properties of class A (BI) calcium channel alpha 1 subunits, *Neuron* **11**:291-303.

Sawada, K., Sakata-Haga, H., Ando, M., Takeda, N. and Fukui, Y., 2001, An increased expression of Ca^{2+} channel α_{1A} subunit immunoreactivity in deep cerebellar neurons of *rolling mouse Nagoya*, *Neurosci Lett* **316**:87-90.

Schafer, W. R. and Kenyon, C. J., 1995, A calcium-channel homologue required for adaptation to dopamine and serotonin in *Caenorhabditis elegans*, *Nature* **375**:73-78.

Schnell, E., Sizemore, M., Karimzadegan, S., Chen, L., Bredt, D. S. and Nicoll, R. A., 2002, Direct interactions between PSD-95 and stargazin control synaptic AMPA receptor number, *Proc Natl Acad Sci U S A* **99**:13902-13907.

Sidman, R. L., Green, M. C. and Apple, S. H., 1965, Catalog of the neurological mutants of the mouse, Harvard University Press, Cambridge, Massachusetts.

Sipos, I., Jurkat-Rott, K., Harasztosi, C., Fontaine, B., Kovacs, L., Melzer, W. and Lehmann-Horn, F., 1995, Skeletal muscle DHP receptor mutations alter calcium currents in human hypokalaemic periodic paralysis myotubes, *J Physiol* **483 (Pt 2)**:299-306.

Smith, L. A., Peixoto, A. A., Kramer, E. M., Villella, A. and Hall, J. C., 1998, Courtship and visual defects of *cacophony* mutants reveal functional complexity of a calcium-channel $\alpha 1$ subunit in Drosophila, *Genetics* **149**:1407-1426.

Snell, G. D., 1955, Ducky, a new second chromosome mutation in the mouse, *Journal of Heredity* **46**:27.

Stea, A., Tomlinson, W. J., Soong, T. W., Bourinet, E., Dubel, S. J., Vincent, S. R. and Snutch, T. P., 1994, Localization and functional properties of a rat brain alpha 1A calcium channel reflect similarities to neuronal Q- and P-type channels, *Proc Natl Acad Sci U S A* **91**:10576-10580.

Strom, T. M., Nyakatura, G., Apfelstedt-Sylla, E., Hellebrand, H., Lorenz, B., Weber, B. H., Wutz, K., Gutwillinger, N., Ruther, K., Drescher, B., Sauer, C., Zrenner, E., Meitinger, T., Rosenthal, A. and Meindl, A., 1998, An L-type calcium-channel gene mutated in incomplete X-linked congenital stationary night blindness, *Nat Genet* **19**:260-263.

Suh, Y. S., Oda, S., Kang, Y. H., Kim, H. and Rhyu, I. J., 2002, Apoptotic cell death of cerebellar granule cells in rolling mouse Nagoya, *Neurosci Lett* **325**:1-4.

Takahashi, T. and Momiyama, A., 1993, Different types of calcium channels mediate central synaptic transmission, *Nature* **366**:156-158.

Tam, T., Mathews, E., Snutch, T. P. and Schafer, W. R., 2000, Voltage-gated calcium channels direct neuronal migration in *Caenorhabditis elegans*, *Dev Biol* **226**:104-117.

Toru, S., Murakoshi, T., Ishikawa, K., Saegusa, H., Fujigasaki, H., Uchihara, T., Nagayama, S., Osanai, M., Mizusawa, H. and Tanabe, T., 2000, Spinocerebellar ataxia type 6 mutation alters P-type calcium channel function, *J Biol Chem* **275**:10893-10898.

Tottene, A., Fellin, T., Pagnutti, S., Luvisetto, S., Striessnig, J., Fletcher, C. and Pietrobon, D., 2002, Familial hemiplegic migraine mutations increase Ca^{2+} influx through single human $Ca_V2.1$ channels and decrease maximal $Ca_V2.1$ current density in neurons, *Proc Natl Acad Sci U S A* **99**:13284-13289.

Tricarico, D., Servidei, S., Tonali, P., Jurkat-Rott, K. and Camerino, D. C., 1999, Impairment of skeletal muscle adenosine triphosphate-sensitive K^+ channels in patients with hypokalemic periodic paralysis, *J Clin Invest* **103**:675-682.

Tsien, R. W., Ellinor, P. T. and Horne, W. A., 1991, Molecular diversity of voltage-dependent Ca^{2+} channels, *Trends Pharmacol Sci* **12**:349-354.

Turner, T. J., Adams, M. E. and Dunlap, K., 1992, Calcium channels coupled to glutamate release identified by ω-Aga-IVA, *Science* **258**:310-313.

Vahedi, K., Denier, C. and Cros, A., 1999, Sporadic hemiplegic migraine with de novo CACNA1A missense mutation, *Neurology* **52**:A274.

Wakamori, M., Yamazaki, K., Matsunodaira, H., Teramoto, T., Tanaka, I., Niidome, T., Sawada, K., Nishizawa, Y., Sekiguchi, N., Mori, E., Mori, Y. and Imoto, K., 1998, Single *tottering* mutations responsible for the neuropathic phenotype of the P-type calcium channel, *J Biol Chem* **273**:34857-34867.

Walker, D., Bichet, D., Campbell, K. P. and De Waard, M., 1998, A beta 4 isoform-specific interaction site in the carboxyl-terminal region of the voltage-dependent Ca^{2+} channel alpha 1A subunit, *J Biol Chem* **273**:2361-2367.

Walker, D. and De Waard, M., 1998, Subunit interaction sites in voltage-dependent Ca^{2+} channels: role in channel function, *Trends Neurosci* **21**:148-154.

Wappl, E., Koschak, A., Poteser, M., Sinnegger, M. J., Walter, D., Eberhart, A., Groschner, K., Glossmann, H., Kraus, R. L., Grabner, M. and Striessnig, J., 2002, Functional consequences of P/Q-type Ca^{2+} channel $Ca_v2.1$ missense mutations associated with episodic ataxia type 2 and progressive ataxia, *J Biol Chem* **277**:6960-6966.

Witcher, D. R., De Waard, M., Sakamoto, J., Franzini-Armstrong, C., Pragnell, M., Kahl, S. D. and Campbell, K. P., 1993, Subunit identification and reconstitution of the N-type Ca^{2+} channel complex purified from brain, *Science* **261**:486-489.

Yano, S., Tokumitsu, H. and Soderling, T. R., 1998, Calcium promotes cell survival through CaM-K kinase activation of the protein-kinase-B pathway, *Nature* **396**:584-587.

Yue, Q., Jen, J. C., Nelson, S. F. and Baloh, R. W., 1997, Progressive ataxia due to a missense mutation in a calcium-channel gene, *Am J Hum Genet* **61**:1078-1087.

Yue, Q., Jen, J. C., Thwe, M. M., Nelson, S. F. and Baloh, R. W., 1998, De novo mutation in CACNA1A caused acetazolamide-responsive episodic ataxia, *Am J Med Genet* **77**:298-301.

Zhang, J. F., Randall, A. D., Ellinor, P. T., Horne, W. A., Sather, W. A., Tanabe, T., Schwarz, T. L. and Tsien, R. W., 1993, Distinctive pharmacology and kinetics of cloned neuronal Ca^{2+} channels and their possible counterparts in mammalian CNS neurons, *Neuropharmacology* **32**:1075-1088.

Zhang, Y., Mori, M., Burgess, D. L. and Noebels, J. L., 2002, Mutations in high-voltage-activated calcium channel genes stimulate low-voltage-activated currents in mouse thalamic relay neurons, *J Neurosci* **22**:6362-6371.

Zhuchenko, O., Bailey, J., Bonnen, P., Ashizawa, T., Stockton, D. W., Amos, C., Dobyns, W. B., Subramony, S. H., Zoghbi, H. Y. and Lee, C. C., 1997, Autosomal dominant cerebellar ataxia (SCA6) associated with small polyglutamine expansions in the alpha 1A-voltage-dependent calcium channel, *Nat Genet* **15**:62-69.

Zwingman, T. A., Neumann, P. E., Noebels, J. L. and Herrup, K., 2001, *Rocker* is a new variant of the voltage-dependent calcium channel gene Cacna1a, *J Neurosci* **21**:1169-1178.

MODULATION OF HIGH VOLTAGE-ACTIVATED CALCIUM CHANNELS BY G PROTEIN-COUPLED RECEPTORS

Aaron M. Beedle[1] and Gerald W. Zamponi[1]

1. INTRODUCTION

Many endogenous substances such as neurotransmitters and peptides bind G protein-coupled receptors (GPCR), thereby initiating signaling cascades that lead to calcium channel modulation. Agonist activation of the GPCR superfamily releases a G protein heterotrimer and promotes dissociation of a G protein α subunit from a G protein $\beta\gamma$ heterodimer. Each of these components (Gα and G$\beta\gamma$) regulates calcium channel activity. Gα mediates voltage-independent calcium channel phosphorylation or dephosphorylation to increase or decrease calcium current. G$\beta\gamma$ binds directly to the calcium channel and inhibits current in a voltage-dependent manner. In this review, we discuss the structural features of GPCR ligand binding and activation, and typical Gα signaling cascades are presented briefly. We examine the molecular determinants of G$\beta\gamma$ binding to presynaptic calcium channels and the modulation of G$\beta\gamma$ effects by other converging signaling pathways. Mechanisms for desensitization of GPCR activity and new evidence of receptor dimerization will be discussed.

Voltage-dependent calcium channels (VDCCs) are essential for cellular processes such as neurotransmitter release, hormone secretion, and muscle contraction (Marks, 1992; Dunlap et al., 1995). Thus, up- or downregulation of these channels results in significant physiological consequences. In fact, many current pharmaceutical agents, including analgesics, antihypertensives, and antipsychotics, exert their therapeutic effects by interacting with plasma membrane receptors that couple to VDCCs through GTP-binding proteins (G proteins). The development of selective agonists and antagonists (see Table 1) has enabled the study of calcium channel modulation by many GPCRs in native cells, including neurons and cardiac myocytes.

---

[1] Department of Physiology and Biophysics, University of Calgary, Calgary, Alberta, Canada T2N 4N1

Calcium Channel Pharmacology, Edited by McDonough
Kluwer Academic/Plenum Publishers, 2004

Calcium current modulation by neurotransmitters acting through GPCRs was first identified by Dunlap and Fischbach (1981) more than twenty years ago, with the finding that activation of adrenergic or serotonin receptors in chick neurons inhibited VDCCs. Membrane delimited regulation of calcium channels by these and other identified receptors, such as opioid, muscarinic, and somatostatin receptors, was found to be sensitive to applications of *Bordatella pertussis* toxin (PTX) and other G protein-interacting compounds (e.g., GTPγS, GDPβS) (Forscher et al., 1986; Holz et al., 1986; Bean, 1989b; Ikeda and Schofield, 1989; Lipscombe et al., 1989; Beech et al., 1992; Bernheim et al., 1992; Golard and Siegelbaum, 1993). Typical effects of these receptors included inhibition of peak calcium current and a slowing of current activation kinetics, which could be at least partially relieved by depolarizing test potentials (Bean, 1989a; Lipscombe et al., 1989). Consequently, regulatory effects on calcium currents could be classified into voltage-dependent and voltage-independent inhibition. But GPCRs do not just inhibit calcium currents; activation of receptors such as the β_2 adrenergic receptor may also facilitate or upregulate channel activity (Hartzell et al., 1991). Thus, physiological data indicated that GPCRs modulate calcium channels through several distinct pathways.

While early data provided descriptive accounts of receptor-mediated calcium channel modulation, the precise second messengers involved in calcium channel potentiation or inhibition were often not delineated. In this regard, advances in molecular and biochemical techniques have greatly enhanced our knowledge of receptor signaling by allowing the coexpression of known receptor and channel subtypes for analysis in heterologous systems. Following twenty years of intense discovery, we now have a better understanding of the molecular events that underlie GPCR modulation of VDCCs. Solving the structure of GPCRs, and determining mechanisms of GPCR receptor/channel colocalization, aspects of receptor dimerization in GPCR function, and requirements in promiscuous versus specific receptor-G protein coupling, however, remain hot topics on the frontiers of GPCR research.

2. OVERVIEW

2.1. G Protein-Coupled Receptors

G protein-coupled receptors form the largest superfamily of identified proteins, with greater than 1,000 members. GPCRs can be divided into three types: families A, B, and C (Gether, 2000). Family A (rhodopsin family) is the largest and best-characterized family and includes GPCRs for biogenic amines (e.g. acetylcholine, dopamine, and histamine), some peptides (e.g. neuropeptide Y, opioids, and substance P), and other molecules (e.g. nucleotides, cannabinoids). Family B GPCRs are activated by peptides (e.g., VIP), and family C receptors include metabotropic glutamate and GABA receptors (among others). GPCRs share several conserved features including: i) a general structure of seven alpha-helical transmembrane segments, with extracellular N- and intracellular C-termini, ii) extracellular agonist binding sites, iii) intracellular interactions with a G protein αβγ heterotrimer, and iv) guanine nucleotide exchange factor-activity in response to agonist binding. Structural analyses of family A GPCRs suggest that the seven transmembrane segments (TM 1 through 7) are arranged in a counterclockwise configuration (as viewed from the extracellular side) with all but TM 4 and TM 7 tilted relative to the plane of the membrane (Gether, 2000). The three-dimensional apposition of these tilted helices contributes to the formation of hydrophobic ligand binding pockets for some GPCRs. The arrangement of extracellular loops (ECL) is structurally constrained by the formation of a disulfide bridge between ECL 1 and 2 in many GPCRs, despite considerable sequence divergence. Intracellular loops (ICL) are involved in receptor activation and are relatively conserved among GPCR subfamilies (Thompson et al., 1993). Figure 1 depicts some structural characteristics of various family members. Features involved in GPCR signaling will be discussed below (Section 3).

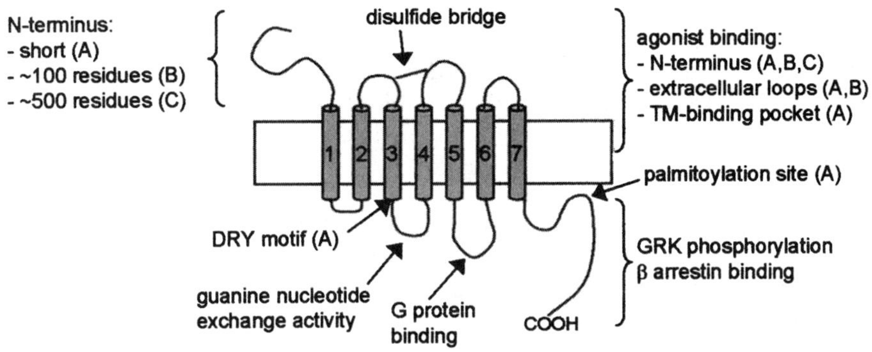

Figure 1. Putative GPCR secondary structure. Regions implicated in GPCR activity are indicated. A, B, and C, indicated GPCR families A, B, and C, respectively. Sizes of intracellular and extracellular domains are not to scale.

2.2. G Proteins

Historically, only G protein α subunits (Gα) were thought to initiate GPCR signaling. However, it is now apparent that both Gα and Gβγ can interact with effectors. To date, molecular cloning has identified sixteen Gα, five Gβ, and eleven Gγ subunits (Mumby, 2000). Gα subunits contain intrinsic GTPase activity and are classified into four families according to sequence homology and effector coupling: $G\alpha_s$, $G\alpha_i$, $G\alpha_q$, $G\alpha_{12}$. The $G\alpha_s$ family ($G\alpha_s$, $G\alpha_{olf}$), sensitive to ADP-ribosylation by cholera toxin (CTX), stimulates all nine isoforms of adenylyl cyclase leading to increased cAMP levels and activation of cAMP-dependent protein kinase (PKA). The $G\alpha_i$ family ($G\alpha_{i1}$, $G\alpha_{i2}$, $G\alpha_{i3}$, $G\alpha_{oA/B}$, $G\alpha_{t1}$, $G\alpha_{t2}$, $G\alpha_g$, $G\alpha_z$) is inhibited by pertussis toxin (PTX), which ADP-ribosylates C-terminal cysteine residues (except $G\alpha_z$; $G\alpha_t$ also CTX-sensitive). Members of the $G\alpha_i$ family typically inhibit adenylyl cyclase activity, although coupling to other effectors has also been reported (Mumby, 2000). $G\alpha_q$ ($G\alpha_q$, $G\alpha_{11}$, $G\alpha_{14}$, $G\alpha_{15/16}$) and $G\alpha_{12}$ ($G\alpha_{12}$, $G\alpha_{13}$) are both PTX- and CTX-insensitive. $G\alpha_q$ subunits stimulate phosphoinositide-specific phospholipase C-β (PLCβ) leading to the formation of inositol-1,4,5-triphosphate (IP_3) and diacylglycerol (DAG). IP_3 triggers the release of Ca^{2+} from intracellular stores that, with DAG, activates protein kinase C (PKC). $G\alpha_{12}$ are poorly understood, but apparently stimulate the Rho family of GTPases leading to activation of c-*Jun*-N-terminal kinase (Jun) (Glick et al., 2000). $G\alpha_{olf}$, $G\alpha_{t1}$, $G\alpha_{t2}$, and $G\alpha_g$ have a limited tissue distribution and are not examined here. All Gα subunits undergo posttranslational myristoylation or palmitoylation to facilitate protein-membrane interactions. Many of the phosphorylation events triggered by activation of different Gα subunits (such as PKC and PKA) ultimately regulate calcium channel activity. This will be discussed further in Section 4.1.

Gβ subunits (Gβ_{1-4}, highly homologous; and Gβ_5, less similar) are characterized by repeated tryptophan-aspartate motifs (WD repeats) of unknown function and an N-terminal amphipathic helix involved in Gγ-binding (Wickman and Clapham, 1995). Gγ subunits (G$\gamma_{1-5,7-12}$) undergo isoprenylation at the C-terminus, which promotes G$\beta\gamma$ localization to the plasma membrane (Wickman and Clapham, 1995; Mumby, 2000). Gβ and Gγ exist as a heterodimer that binds Gα-GDP but has low affinity for Gα-GTP. This property is exploited experimentally to stimulate or inhibit G protein activity. GTPγS (guanosine 5′-O-(3-thiotriphosphate)) is a GTP analog resistant to Gα hydrolysis and therefore stimulates G proteins by preventing their deactivation (reassociation of Gα and G$\beta\gamma$). Conversely, GDPβS (guanosine 5′-[β-thio]diphosphate), an analog that cannot be phosphorylated, is often used to inhibit G proteins.

2.3. Voltage-Dependent Calcium Channels

Voltage-dependent calcium currents are classified according to both biophysical and pharmacological criteria. High voltage-activated (HVA) channels are opened by strong depolarizations ($\geq$ -20 mV) and include L-type (dihydropyridine-sensitive), N-type (ω-conotoxin GVIA-sensitive), P/Q-type (ω-agatoxin IVA-sensitive), and R-type (insensitive to traditional blockers) channels. By contrast, low voltage-activated (LVA), or T-type, calcium channels require much smaller depolarizations from resting membrane potential ($\geq$ -60 mV) for opening. At the molecular level, HVA calcium channels are created by the association of an α_1 pore-forming subunit (Ca_v), and accessory α_2-δ, β, and γ subunits (Catterall, 2000). The α_1 subunit belongs to the superfamily of voltage-gated ion channels and consists of 4 repeat domains, each with 6 transmembrane-spanning segments and a short re-entrant P-loop. The intracellular N- and C-termini, and cytoplasmic linkers (I-II linker, II-III linker, and III-IV linker) are important in calcium channel modulation and protein-protein binding.

To date, ten α_1, four $\alpha_2\delta$ (Ellis et al., 1988; Klugbauer et al., 1999; Qin et al., 2002), four β (Castellano and Perez-Reyes, 1994), and eight γ (Chu et al., 2001) subunits have been identified. α_{1S}/Ca_v1.1, α_{1C}/Ca_v1.2, α_{1D}/Ca_v1.3, and α_{1F}/Ca_v1.4 correspond to native L-type channels (Mikami et al., 1989; Snutch et al., 1991; Williams et al., 1992b; Tomlinson et al., 1993; Bech-Hansen et al., 1998); α_{1A}/Ca_v2.1, α_{1B}/Ca_v2.2, and α_{1E}/Ca_v2.3 resemble endogenous P/Q-type, N-type, and R-type channels, respectively (Mori et al., 1991; Dubel et al., 1992; Williams et al., 1992a; Fujita et al., 1993; Sather et al., 1993; Soong et al., 1993; Zhang et al., 1993; Stea et al., 1994; Williams et al., 1994; Bourinet et al., 1999); and α_{1G}/Ca_v3.1, α_{1H}/Ca_v3.2, and α_{1I}/Ca_v3.3 form T-type channels (Cribbs et al., 1998; Perez-Reyes et al., 1998; Lee et al., 1999; Monteil et al., 2000a; Monteil et al., 2000b; McRory et al., 2001; Beedle et al., 2002a). Most calcium channels are expressed in the brain and demonstrate varied subcellular localization. In particular, P/Q- and N-type channels are located at presynaptic sites where they control neurotransmitter release (Westenbroek et al., 1992; McCleskey, 1994; Dunlap et al., 1995; Westenbroek et al., 1995; Westenbroek et al., 1998). This chapter deals with modulation of high voltage-activated calcium channels; for modulation of low voltage-activated T-type channels, see Ertel (this volume).

3. GPCR LIGAND BINDING AND RECEPTOR ACTIVATION

3.1. Ligand Binding

The structural determinants of ligand binding to GPCRs vary among the different families of receptors (Gether, 2000). For example, the large N-terminus of metabotropic glutamate receptors (GPCR family C) comprises virtually the entire glutamate-binding domain (O'Hara et al., 1993; Takahashi et al., 1993; Rosemond et al., 2002). For family B vasoactive intestinal polypeptide receptors (VIPR/VPAC), however, both the N-terminus and extracellular loops (particularly ECL1) are important in VIP interactions (Couvineau et al., 1995; Du et al., 1997). Ligand binding to family A GPCRs often involves the formation of a binding pocket within the membrane. Small ligands (e.g., epinephrine) bind solely within this pocket, while larger ligands (e.g., opioids) also interact with extracellular structures (Frielle et al., 1989; Dietrich et al., 1998). Thus, the specific ligand interaction sites differ between GPCRs.

As opioid receptors are targets for clinically active analgesics, the specificity of opioid ligands has been given considerable study and serves as an example of ligand binding to family A GPCRs. Data from modeling studies suggest that, in all opioid receptors (OR), a binding core is formed within the membrane by TM 3, 4, 5, 6 and 7 that is reinforced by intrahelical hydrogen bonds (Pogozheva et al., 1998; Law et al., 1999). Nonselective ligands appear to interact with this pocket but not extracellular structures (Law et al., 1999). The presence of a disulfide bond between ECL 1 and ECL 2 is important in the selecitivity of ligands for particular receptors as it may physically constrain access to the membranous binding pocket (Pogozheva et al., 1998). Considerable evidence demonstrates the importance of extracellular loop regions in the specificity of all opioid ligands. ECL 1 and ECL 3 are implicated in the selectivity of μ-OR ligand binding (Fukuda et al., 1995a; Fukuda et al., 1995b; Minami et al., 1995; Dietrich et al., 1998; Lu et al., 1998). For δ-ORs, ECL 3 is important in antagonist binding while ECL 3, TM 2, and TM 5 through TM 7 may be involved in agonist selectivity (Kong et al., 1993; Fukuda et al., 1995a; Pepin et al., 1997; Morikawa et al., 1998a). The endogenous κ-OR ligand, dynorphin, binds to regions of the ECL 2 via direct helix-helix interactions between the peptide and the receptor (Wang et al., 1994; Xue et al., 1994; Meng et al., 1995; Paterlini et al., 1997; Ferguson et al., 2000; Zhang et al., 2002). However, κ-OR antagonists do not appear to interact with ECL 2, but may bind to the N-terminus (naloxone) or ECL 3 (nor-BNA) (Kong et al., 1994; Xue et al., 1994). Thus, the determinants of ligand specificity for GPCRs vary depending on the receptor subtype, and agonist/antagonist activity of ligands. Ultimately, differences in the intramembranous binding core and extracellular loop regions of family A GPCRs determine the ability of different ligands to initiate calcium channel modulation through GPCR activation.

3.2. GPCR Activation of Heterotrimeric G Proteins

Agonist binding to GPCRs initiates conformational changes that stimulate bound G proteins and promote their dissociation from the receptor. It is increasingly believed that GPCRs, in the absence of ligand, are maintained in a structurally restricted "inactive" conformation. In this state, the intracellular domains of TM 3 and TM 6 are thought to be in close association, as mutations that disrupt this interaction promote a constitutively active state (Huang et al., 2001). Thus, "repulsive" movements of TM 3 and TM 6 result in GPCR activation. In particular, residues within an "XBBXXB" motif at the junction of ICL3 and TM6 are implicated in receptor activation of G proteins (Cotecchia et al., 1990; Huang et al., 2001). It is perhaps intuitive that TM 3 and TM 6 should be associated with conformational changes in receptor activation, as they are connected to ICL 2 and ICL 3, which possess guanine nucleotide exchange activity and G protein binding affinity, respectively. The membrane-associated C-terminus may also be involved in receptor activation; however, its precise role is not clear (Law et al., 1999). Specific molecular events transducing receptor activation are indicated by molecular dynamics simulations. In particular, D and R of the DRY motif in TM 3 (common to many group 1 GPCRs) may form a salt bridge that is disrupted during receptor activation. Also, bonding between TM 2 and 7 may be initiated with GPCR activation (Filizola et al., 1999). Precisely how such GPCR conformational changes induce exchange of GDP for GTP by the Gα subunit are unknown and may require solving of the receptor crystal structure to be elucidated.

Although the intracellular loop 3 is thought to mediate GPCR-Gα binding and thus tether the G protein heterotrimer to the GPCR (Luttrell et al., 1993; Liu et al., 1995; Liu and Wess, 1996), there is little indication as to how GPCRs couple to specific Gα subtypes. For example, opioid receptor activity is typically PTX-sensitive, indicating a coupling to $G\alpha_{i/o}$ proteins (Bourinet et al., 1996; Piros et al., 1996a; Toth et al., 1996; Connor and Christie, 1999), but no $G\alpha_{i/o}$ consensus motif for GPCR binding has been discovered. Furthermore, a number of GPCRs are now being implicated in "promiscuous" G protein coupling; that is, a specific GPCR will couple to one Gα in a certain cell type or environment, yet couple to a different class of Gα in a different cell type or environment (Laugwitz et al., 1996; Hamm, 1998; McCool et al., 1998; Wade et al., 1999). Within the $G\alpha_{i/o}$ family, varying reports suggest that i) GPCRs may couple indiscriminately to virtually all $G\alpha_{i/o}$ subtypes, ii) GPCRs preferentially interact with only one $G\alpha_{i/o}$ subtype, and iii) GPCR to $G\alpha_{i/o}$ coupling depends on the activation state of the receptor (Hescheler et al., 1987; Taussig et al., 1992; Laugwitz et al., 1993; Prather et al., 1995; Law and Reisine, 1997). In general, it is not known whether the receptors themselves solely determine the interaction with different G protein subtypes, or whether environmental factors are involved. Potential mechanisms to determine the specificity of Gα-GPCR interactions might include: i) structural recognition by a receptor for specific G proteins; ii) coexpression of a receptor with particular G proteins; iii) cotargeting of a receptor and G proteins; iv) G protein specificity for a particular phosphorylation or conformational state of the receptor; or iv) a presence of inhibitory factors to prevent binding of nontarget G proteins to a receptor. Identifying common mechanisms for selective G protein-GPCR coupling are critical to understand the cellular/environmental factors regulating GPCR signaling and subsequent calcium channel modulation.

Table 1. Modulation of high voltage activated calcium channels by G-protein coupled receptors

GPCR family	Receptor gene	Second messenger coupling	Endogenous ligands	Agonists	Nonpeptide antagonists	VDCC modulation (native cells)	VDCC modulation (expression systems)[a]	Selected References
Adrenergic α	α1[b]	$G\alpha_{q/11}$, $G\alpha_o$	NE[c], EPI	phenylephrine	prazosin,	-↑ L-type	-↑ $Ca_v1.2$	1,2
	α2	$G\alpha_o$ ($G\alpha_{o1}$), may also couple to $G\alpha_q$	NE, EPI	UK14304, clonidine	idazoxan, yohimbine (a2 spec)	-↓ N-type (VD & VI)[d] -↓ HVA	-↓ $Ca_v2.3$	3-8
β	β1, β2	$G\alpha_s$	NE, EPI	Iso, albuterol (β2)	propranolol	-↑ L-type -↑ P/Q-type	-↑ $Ca_v1.2$	9-13
Angiotensin (ANG)	AT_1	$G\alpha_{13}$, $G\alpha_{i/o}$ (less common)	ANG II	L-162,313 (partial agonist)	losartan (AT_{1A} > AT_{1B})	-↑ L-type -↓ L-type (PTX-sens and –insens) -↓ N-type (VI, PTX-insens) -↓ N-type (PTX-sens)		14-17
Bradykinin (BK)	B_1, B_2	$G\alpha_{q/11}$ B2: $G\alpha_{11}$, $G\alpha_{13}$, $G\alpha_{i/o}$?	BK	desArg[9]-BK (B1)		B1: -↓ L-type B2: -↓ N-type -↓ L-type		18-20

GPCR family	Receptor gene	Second messenger coupling	Endogenous ligands	Agonists	Nonpeptide antagonists	VDCC modulation (native cells)	VDCC modulation (expression systems)[a]	Selected References
Cannabinoids	CB1, CB2	$G\alpha_{i/o}$	anand-amide (partial agonist)	WIN55,212-2; CP55,940; THC	SR141716A (CB1), SR144528 (CB2)	-↓ N-type (VD & VI) -↓ L-type	CB1: -↓ N-type (VD & VI) -↓ Q-type (VD & VI) - L-type no effect	21-24
Cholinergic (muscarinic) mAChR's M1 class	m1, m3, m5	$G\alpha_q$	ACh	muscarine, carbachol, MCh	atropine, pirenzepine (m1), *p*-FHHSiD (m3)	-↓ N-type (VI) -↓ L-type	m1: -↓ $Ca_v2.2$ ($G\beta\gamma$-fast & VI-slow) -↑ $Ca_v2.3$ -↓ $Ca_v1.2$	25-30
M4 class	m2, m4	$G\alpha_i/G_o$ ($G\alpha_{o1}$)	ACh	muscarine, carbachol, Oxo-M	atropine, himbacine, meth-octramine (m2)	-↓ N-type (VD & small VI) -↓ Q-type -↓ R-type -↓ L-type (?)	m2: -↓ $Ca_v2.2$ (VD) -↓ $Ca_v2.1$ (VD) –↑ $Ca_v2.3$	28, 30-35
Dopaminergic D1 family	D1, D5	$G\alpha_s$	dopamine	SKF-38393, apo-morphine	SCH 23390	-↑ L-type -↓ N-type (VI, small VD) -↓ P-type		36, 37

GPCR family	Receptor gene	Second messenger coupling	Endogenous ligands	Agonists	Nonpeptide antagonists	VDCC modulation (native cells)	VDCC modulation (expression systems)[a]	Selected References
Dopaminergic cont. D2 family	D2, D3, D4	$G\alpha_{i/o}$, $G\alpha_z$	dopamine	quinpirole, NPA	sulpiride, spiperone, NB: many "D2" antag's block VDCC directly	-↓ L-type (D2) -↓ P/Q-type -↓ N-type -↓ R-type	D2: -↓ Cav2.2 (VD & VI, VI only) -↓ $Ca_v2.3_{long}$ (VD & VI) -no effect – $Ca_v1.3$ -↓ Q-type D3: -↓ P/Q-type -↓ N-type (VD & VI) -↓ P/Q-type -↓ R-type	38-44
$\mathbf{GABA_B}$	$GABA_{B1}$, $GABA_{B2}$ (hetero-dimer)	$G\alpha_{i/o}$	GABA	baclofen, APMPA	CGP35348, CGP55845, SCH50911	-↓ HVA -↓ N-type (VD) -↓ P/Q-type (VD) -↓ R-type -↑ L-type (VI)		3, 32, 45-48
Glutamate (metabotropic) Group I	mGluR1, mGluR5	$G\alpha_{q/11}$, $G\alpha_{i/o}$	glutamate	quisqualate > ACPD, DHPG	MCPG, L-AP3, MPEP (mGluR5), LY367385 (mGluR1), CPCCOEt (mGluR1), 4C3HPG	-↑ I_{Ba}, -↓ N-type (VD & VI) -↑ L-type	-↓ $Ca_v2.2$ (VD & VI) -↓ $Ca_v2.1$ (VD & VI) mGluR1α: -↑ $Ca_v2.3$	49-53

GPCR family	Receptor gene	Second messenger coupling	Endogenous ligands	Agonists	Nonpeptide antagonists	VDCC modulation (native cells)	VDCC modulation (expression systems)[a]	Selected References
Glutamate (metabotropic) cont. Group II	mGluR2, mGluR3	$G\alpha_{i/o}$	glutamate	ACPD, LY379268, DCG-IV, 4C3HPG	MCPG, MCCG, EGLU, LY341495	-↓ P/Q-type -↓ N-type (VD & VI) -↓ L-type	mGluR2: -↓ N-type (mostly VD) mGluR3: -↓ N-type (VD & VI)	53-57
Group III	mGluR4, mGluR6, mGluR7, mGluR8	$G\alpha_{i/o}$	glutamate	ACPD, L-AP4, L-SOP	MSOP, CPPG	-↓ P/Q-type (VD & VI) -↓ N-type -↓ R-type -↑ L-type	mGluR7: -↓ P/Q-type (VI)	47, 58-61
Histamine	H_1	$G\alpha_q$	HA	2-methyl-HA	mepyramine, ReN 1869	-↓ N-type (VD & VI) -↓ P/Q-type (VD & VI)		62
	H_2	$G\alpha_s$, $G\alpha_{q/11}$	HA	dimaprit	ranitidine, cimetidine	-↑ L-type		63
	H_3	$G\alpha_{i/o}$	HA	R(-)-α-methyl HA;	thioperamide	-↓ N-type -↓ P/Q-type		64
Neuropeptide Y	Y1, Y2, Y5	$G\alpha_{i/o}$	NPY, PYY (Y2: PYY>NPY)	LPNPY (Y1), NPY_{13-36} (Y2), C2-NPY (Y2), [D-Trp^{32}]-NPY (Y5)	BIBP3226 (Y1), BIIE0246 (Y2), CGP71683A (Y5)	-↓ N-type (VD & VI) -↓ P/Q-type (Y2) -↑ non-L-type? (Y1)	Y1, Y2: -↓ $Ca_v2.2$ (VD & VI) -↓ $Ca_v2.3$	65-70

GPCR family	Receptor gene	Second messenger coupling	Endogenous ligands	Agonists	Nonpeptide antagonists	VDCC modulation (native cells)	VDCC modulation (expression systems)[a]	Selected References
Neuropeptide Y cont.	Y4/PP1	$G\alpha_{i/o}$	PP >> NPY	LPNPY (Y1 > Y4)		-↓ N-type (VD & VI) -↓ P/Q-type	-↓ $Ca_v2.2$ (VD & VI) -↓ $Ca_v2.3$	69, 71, 72
Opioid/opioid receptor-like: δ	DOR1	$G\alpha_{i/o}$ ($G\alpha_{o1}$, $G\alpha_{i1}$, $G\alpha_{i2}$) $G\alpha_z$, $G\alpha_{16}$	deltorpin-2, β-endorphin, enkephalin	etorphine, DPDPE, DADLE, DSLET, (-)-TAN-67 (δ1)	naltrindole, BNTX (δ1)	-↓ N-type (VD & VI) -↓ P/Q-type -↓ L-type	-↓ L-type (VD)	18, 73-77
κ	KOR1	$G\alpha_{i/o}$, $G\alpha_z$, $G\alpha_{16}$	dynorpin A	etorphine, U69,593, U50,488, β-FNA,	norBNA, naloxone, naltrexone, bremazocine	-↓ N-type -↓ P/Q-type	-↓ $Ca_v2.2$ (VD & VI) -↓ $Ca_v2.1$ -no effect - $Ca_v2.3$	66, 78, 79
μ	MOR1	$G\alpha_{i/o}$ ($G\alpha_{o2}$?, $G\alpha_{i3}$), $G\alpha_z$, $G\alpha_{16}$	endo-morpins, β-endorphin, enkephalins	etorphine, morphine, DAMGO, fentanyl	β-FNA, naloxone, naltrexone	-↓ N-type (VD) -↓ P/Q-type	-↓ $Ca_v2.2$ (VD & VI) -↓ $Ca_v2.1$ (VD & VI) -↑ Cav1.3 - no effect Cav1.2, $Ca_v2.3$ -↓ L-type (VD)	80-84

GPCR family	Receptor gene	Second messenger coupling	Endogenous ligands	Agonists	Nonpeptide antagonists	VDCC modulation (native cells)	VDCC modulation (expression systems)[a]	Selected References
Opioid/opioid receptor-like: cont. ORL	ORL1	$G\alpha_{i/o}$, $G\alpha_z$, $G\alpha_{12}$, $G\alpha_{14}$, $G\alpha_{16}$	nociceptin/orphanin FQ	Ro 64-6198, [Arg^{14},Lys^{15}]nociceptin,	JTC-801?	-↓ N-type (VD & VI) -↓ P/Q-type -↓ R-type -↓ L-type	-↓ $Ca_v2.2$ (VD & VI) -↓ $Ca_v2.1$ -no effect on $Ca_v1.2$, $Ca_v1.3$, $Ca_v2.3$	85-90
Prostaglandin	EP1	$G\alpha_{q/11}$?	PGE_2, PGE_1	17PT-PGE2, ONO-DI-004, sulprostone	SC-51089, SC-19220;	-↑ L-type -↓ L-type -↑ N-type		91-93
	EP2	$G\alpha_s$	PGE_1, PGE_2	butaprost, misoprostol, 11-deoxyPGE_1, ONO-AE1-259		-↑ L-type		94
	EP3	$G\alpha_{i/o}$, (PTX-insens too?)	PGE_1, PGE_2	sulprostone, ONO-AE-248, misoprostol, 11-dPGE_1		-↓ N-type (VD & small VI) -↓ P/Q-type (VD) -↓ R-type, -↓ L-type (VI)		7, 92, 95-98

GPCR family	Receptor gene	Second messenger coupling	Endogenous ligands	Agonists	Nonpeptide antagonists	VDCC modulation (native cells)	VDCC modulation (expression systems)[a]	Selected References
Purine and pyrimidine receptors P1	A_1	$G\alpha_{i/o}$, $G\alpha_z$	adenosine	(R)-PIA, CCPA, CHA, 2-CA (slightly selective), APNEA	CPT, DPCPX, 8-PT	-↓ N-type (VD &VI) -↓ P/Q-type -↓ R-type -↓ L-type	-↓ $Ca_v2.3$	7, 67, 99-102
	A_2	$G\alpha_s$, $G\alpha_{q/11}$ (A_{2B})	adenosine	CGS21680 (A_{2A} selective), DPMA, 2-CA	DMPX, alloxazine, 8-PT, ZM241385 (A_{2A}), DPCPX (A_{2B})	-↑ L-type -↑ P-type -↑ N-type -↓ L-type -↓ N-type		103-106
	A_3	$G\alpha_{i/o}$, $G\alpha_s$?	adenosine	IB-MECA, APNEA		-↑ HVA		107
P2	$P2Y_1$, $P2Y_{11}$	$G\alpha_q$	$P2Y_1$: ADP, $P2Y_{11}$: ATP	$P2Y_1$: 2-MeSADP, $P2Y_{11}$: 2-MeSATP	$P2Y_1$: PPADS, suramin, ABP	-↓ I_{Ca} (VD, $P2Y_1$-like) -↓ L-type -↓ N-type -↓ non-L, non-N-type; *P2 receptor class is often unclear	$P2Y_1$: -↓ N-type	108-111
	$P2Y_2$, $P2Y_4$	$P2Y_4$: $G\alpha_{i/o}$, $G\alpha_{q/11}$	$P2Y_2$: UTP, ATP; $P2Y_4$: UTP>>ATP	ATPγS	suramin ($P2Y_2$), RB-2 ($P2Y_4$)	-↑ L-type -↓ L-type -↓ N-type	$P2Y_2$: -↓ N-type (VD & VI)	112-114

GPCR family	Receptor gene	Second messenger coupling	Endogenous ligands	Agonists	Nonpeptide antagonists	VDCC modulation (native cells)	VDCC modulation (expression systems)[a]	Selected References
Purine and pyrimidine receptors P2 cont.	P2Y_6	$G\alpha_{i/o}$, and PTX-insens	UDP, UTP	UDPβS	RB-2		-↓ N-type	115
Serotonin (5-HT)	5-HT_{1A}, 5-HT_{1C}, 5-HT_{1D}	$G\alpha_{i/o}$	5-HT	OH-DPAT (5-HT_{1A}), L-694,247 (5-HT_{1D}), 5-CT	NAN 190, (+)WAY1001 35, methyl-sergide, GR127935 (5-HT_{1B}), methiothepin	-↓ N-type (VD & VI) -↓ P/Q-type (VD & VI)	5-HT_{1A}: -↓ N-type -↓ non-L, non-P-type	116-120
	5-HT2	$G\alpha_{q/11}$	5-HT	α-m-5-HT, DOI	methyl-sergide, clozapine, ketanserin (may block VDCCs)	-↓ L-type (Ca_v1.2)		121
	5-HT4	$G\alpha_s$	5-HT	5-MeOT, partial agonists: zacopride, BIMU 8, renzapride	GR113808	-↑ L-type		122, 123
Somatostatin (SRIF)	sst_1, sst_2, sst_4	$G\alpha_{i/o}$, sst_2: $G\alpha_{o2}$	SRIF	SOM, CH-275 (sst_1), BIM23027 (sst_2), MK 678 (sst_2)		-↓ N-type (VD & VI) -↓ P/Q-type sst_1: -↓ I_{Ba} sst_2: -↓ L-type -↑ L-type (?)	sst_2: -↓ Ca_v2.2 (VD & VI) -↓ or no effect – Ca_v1.3, Ca_v2.3	7, 35, 78, 124-126

GPCR family	Receptor gene	Second messenger coupling	Endogenous ligands	Agonists	Nonpeptide antagonists	VDCC modulation (native cells)	VDCC modulation (expression systems)[a]	Selected References
Somatostatin (SRIF) cont.	5-HT2	$G\alpha_{q/11}$	5-HT	α-m-5-HT, DOI	methyl-sergide, clozapine, ketanserin (may block VDCCs directly)	-↓ L-type (Ca_v1.2)		121
Tachykinin receptors: Substance P	NK_1	$G\alpha_{q/11}$, $G\alpha_s$?	SP	[Sar9]-SP	GR 82334, CP 99994, L703606	-↓ N-type (VI) -↓ Q-type (VI) -↓ R-type -↑ L-type -↑ HVA		127-132
VIP	$VPAC_1$ (VIPR1), $VPAC_2$ (VIPR2)	$G\alpha_s$	VIP, PACAP	PACAP 38, PACAP 27, secretin ($VPAC_1$), RO 25-1553 ($VPAC_2$)		-↓ N-type (VD) -↓ non-N/non-L-type -↑ L-type		133, 134

[a] Either channel cDNAs, receptor cDNA, or both, expressed in a heterologous system.

[b] only receptor subtypes shown to couple to VDCC are included.

[c] Abbreviations: ABP, adenosine 3′,5′-bis-phosphate; AC, adenylyl cyclase; ACh, acetylcholine; ACPD, (1*S*,3*R*)-1-aminocyclopentane-1,3-dicarboxylic acid; L-AP4, l-2-amino-4-phosphonobutyrate; APMPA, 3-amino-propyl-(methyl)phosphonic acid dihydrochloride; APNEA, N^6-2-(4-aminephenyl) ethyladenosine; ATPγS, adenosine 5′-[γ-thio]triphosphate; BIBP3226, [(*R*)-*N*2-(diphenylacetyl)-*N*-(4-hydroxyphenylmethyl)-D-arginine amide]; BIIE0246, (S)-N2-[[1-[2-[4-[(R,S)-5,11-dihydro-6(6h)-oxodibenz[b,e]azepin-11-yl]-1-piperazinyl]-2-oxoethyl]cyclopentyl]acetyl]-N-[2-[1,2-dihydro-3,5-(4H)-dioxo-1,2-diphenyl-3H-1,2,4-triazol-4-yl]ethyl]-argininamide; BIM-23027, c[*N*-Me-Ala-Tyr-D-Trp-Lys-Abu-Phe]; BIM-23056, D-Phe-Phe-Tyr-D-Trp-Lys-Val-Phe-D-Nal-NH_2; BIMU 8, *N*-(endo-8-methyl-8-azabicyclo[3.3.1]-2,3-dihydro-3-isopropyl-2-oxo-1H-benzimidazol-1-carboxamide)hydrochloride; BNTX, 7-benzylidenaltrexone; 4C3HPG, *S*-4-carboxy-3-hydroxy-phenylglycine; 2-CA, 2-chloroadenosine; CCPA, chloro-*N*-cyclopentyladenosine; CGP71683A, trans-naphtalene-1-sulphonic acie[4-[(4-amino-quinazolin-2-ylamino)-methyl]-cyclohexylmethyl]-

amide hydrochloride; CGS21680, 2-[4-(2-carboxyethyl)phenylethylamino]-5′-*N*-ethylcarboxamidoadenosine; CH-275, desAA(1,2,5)-[DTrp[8],IAmp[9]]-somatostatine-14; CHA, *N*-6-cyclohezyladenosine; CK1, casein kinase 1; CP 55,940, [1α,2β-(*R*)-5α]-(-)-5-(1,1-dimethylheptyl)-2-[5-hydroxy-2-(3-hydroxypropyl)cyclohexyl]phenol; CP 99994, ((+)-2*S*,3*S*)-3-(2-methoxybenzylamino)-2-phenylpiperidine; CPCCOEt, 7-cyclopropan[b]chromen-1a-carboxylic acid ethyl ester; CPP-1, cyclo(7-aminoheptanoyl-phe-d-try-lys-O-benzyl-thr)acetate; CPPG, (*R*,*S*)-α-cyclopropyl-4-phosphonophenyl glycine; CPT, 8-cyclopentyltheophylline; 5-CT, 5-carboxamidotryptamin; CTX, cholera toxin; DADLE, [D-Ala[2],D-Leu[5]]-enkephalin; DAMGO, [D-Ala[2],*N*-Me-Phe[4],Gly[5]-ol]-enkephalin; DCG-IV, (2*S*,1′*R*,2′*R*,3′*R*)-2-(2,3-dicarboxycyclopropyl)glycine; DHPG, (*S*)-3,5-dihidroxyphenylglycine; DMPX, 3,7-dimethyl-1-propargylxanthine; DOI, (±)-2,5-dimethoxy-4-iodoamphetamine hydrochloride; DPCPX, 8-cyclopentyl-1,3-dipropylxanthine; DPDPE, [D-Pen[2],D-Pen[5]]-enkephalin; 11-dPGE$_1$, 11-deoxy-prostaglandin E$_1$, DPMA, N^6-[2-(3,5-dimethyloxyphenyl)-2-(2-methylphenyl) ethyl)adenosine; DSLET, [D-Ser[2],D-Leu[5]]-enkephalin-Thr; EGLU, (*S*)-α-ethylglutamate; EPI, epinephrine; *p*-FHHSiD, hexahydro-silo-difenidol hydrochloride, *p*-fluoro analog; β-FNA, β-funaltrexamine; GABA, γ-aminobutyric acid; GR113808, [1-[2-(methylsulfonyl)amino]ethyl]-4-piperidinyl]methyl-1-methyl-1*H*-indole-3-carboxylate; GR127935, *N*-[4-methoxy-3-(4-methyl-1-piperazinyl)phenyl]-2′-methyl-4′-(5-methyl-1,2,4-oxadiazol-3-yl)[1,1-biphenyl]-4-carboxamide; GR82334, [D-Pro[9][spiro-g-lactam]Leu[10]]-physalaemin-(1-11); HA, histamine; IB-MECA, 1-deoxy-1-[6-[[(3-iodophenyl)methyl]amino]-9H-purin-9-yl]-*N*-methyl-β-D-ribofuranuronamide; Iso, isoproterenol; L-162,313, 5,7-dimethyl-2-ethyl-3-[[4-[2(n-butyloxycarbonylsulfonamido)-5-isobutyl-3-thienyl]phenyl]methyl]imidazol[4,5,6]-pyridine; L-372,588, c[Phe-Trp-D-Trp-Lys-Thr-Tyr-AHA]; L-694,247, 2-[5-[3-(4-methylsulfonylamino)benzyl-1,4-oxadiazol-5-yl[-1-H-indole-3-yl]ethylamine; L703606, (cis)-2-(Diphenylmethyl)-N-[(2-iodophenyl)-methyl]-1-azabicyclo[2.2.2]octan-e-amine; L-AP3, L-2-amino-3-phosphonopropionate; LPNPY, Leu31-Pro34-neuropeptide Y; L-SOP, L-serine-O-phosphate; LY341495, (2*S*)-2-amino-2-(1*S*,2*S*-2-carboxycycloprop-1-yl)-3(xanthy-9-yl)propanoic acid; LY367385, (+)-2-methyl-4-carboxyphenylglycine; LY379268, (--)-2-oxa-4-aminobicyclo[3.1.0]hexane-4.6-dicarboxylate; α-m-5-HT, α-methyl-5-hydroxytryptamine maleate; MCh, acetyl-beta-methylcholine; MCCG, (2*S*,3*S*,4*S*)-2-methyl-2-(carboxycyclopropyl)glycine; MCPG, α-methyl-4-carboxyphenolglycine; 5-MeOT, 5-methoxytryptamine; 2-MeSADP, 2-Methylthio-ADP; MK-678, c[*N*-Me-Ala-Tyr-D-Trp-Lys-Thr-Phe]; MPEP, 2-methyl-6-(phenylethynyl)pyridine; MSOP, (*R*,*S*)-α-methylserine-O-phosphate; NAN 190, 1-(2-methoxyphenyl)-4-[4-(2-phthalimido)butyl]piperazine hydrobromide; NE, norepinephrine; norBNA, norbinaltorphimine; NPA, *R*(-)-propylnorapomorphine; OH-DPAT, R(+)-8-hydroxydipropylaminotetralin hydrobromide; ONO-AE-248, (16*S*)-9-deoxy-9beta-chloro-15-deoxy-16-hyfroxy-17,17-trimethylene-19,20-didehydro prostaglandin F2; ONO-AE1-259, 11,15-O-dimethyl prostaglandin E2; ONO-DI-004, (17S)-2,5-ethano-6-oxo-17,20-dimethyl prostaglandin E1; Oxo-M, oxotremorine methiodide; PACAP, pituitary adenylate cyclase activating peptide; (R)-PIA, N^6-(*R*)-phenyl-2-propyladenosine; PLC, phospholipase C; PPADS, pyridoxalphosphate-6-azophenyl-2′,4′-disulphonate; 17PT-PGE$_2$, 17-phenyl-trinor-prostaglandin E$_2$; PP, pancreatic polypeptide; 8-PT, 8-phenyltheuphylline; PTX, pertussis toxin; PYY, peptide YY; RB-2, reactive blue-2; ReN 1869, (*R*)-1-(3-(10,11-dihydro-5H-dibenzo[a,d]cyclohepten-5-ylidene)-1-propyl)-3-piperidine carboxylic acid; Ro 64-6198, [(1*S*,3a*S*)-8-(2,3,3a,4,5,6-hexahydro-1H-phenalen-1-yl)-1-phenyl-1,3,8-triaza-spiro[4.5]decan-4-one]; [Sar9]-SP, [Sar9]-substance P sulfone; SC-19220, 8-chlorodibenz[*b*,*f*][1,4]oxazepine-10(11*H*)-carboxylic acid, 2-acetylhydrazide; SC51089, 8-chlorodibenz[*b*,*f*][1,4]oxazepine-10(11*H*)-carboxylic acid, 2-[1-oxo-3-(4-pyridinyl)propyl]hydrazide, monohydrochloride; SCH 23390, *R*-(+)-7-chloro-8-hydroxy-3-methyl-1-phenyl-2,3,4,5-tetrahydro-1H-3-benzazepine hydrochloride; SCH50911, (-)-(R)-5,5-dimethylmorpholinyl-2-acetic acid ethyl ester HCl; SKF-38393, 2,3,4,5-tetrahydro-7,8-dihydroxy-1-phenyl-1H-3-benzazepine; SOM, [D-Trp[8]]somatostatin; SP, substance P; SR141716A, *N*-piperidino-5-(4-chlorophenyl)-1-(2,4-dichlorophenyl)-4-methyl-3-pyrazole-carboxamide; SR144528, [*N*-[1*S*]-endo-1,3,3-trimethylbicyclo[2.2.1]heptan-2-yl]-5-(4-chloro-3-methylphenyl)-1-(4-methylbenzyl)-pyrazole-3-carbocamide; (-)-TAN-67, (4aS*, 12aR*)-4a-(3-hydroxyphenyl)-2-methyl-1,2,3,4,4a,5,12,12a-octahydropyrido[3,4-b]acridine; THC, Δ^9-tetrahydrocannabinol; U50,488, (-)-(*trans*)-3,4-dichloro-*N*-methyl-*N*-[2-(1-pyrrolidiny)-cyclohexyl] benzene acetamide; U69,593, (+)-(5α,7α,8β)-*N*-methyl-*N*-(7-(1-pyrrolidinyl)-1-oxaspiro(4,5)dec-8-yl) benzene acetamide; UDPβS, uridine 5′-O-(2-thiodiphosphate; UK14304, 5-bromo-*N*-(4,5-dihydro-1H-imidazol-2-yl)-6-quinoxalinamine; VIP, vasoactive intestinal polypeptide; (+)WAY 100135, *N*-tert-butyl 3,4-(2-methoxyphenyl)piperazin-1-yl-2phenylpropanamide dihydrochloride; WIN 55,212-2, *R*(+)-[2,3-dihydro-5-methyl-3-[(morpholinyl)methyl]pyrolo[1,2,3-de]-1,4-benzoxazin-*N*-(1-napthalenyl)methanone mesylate; ZM241385, 4-(2-[7-amino-2-(2-furyl)[1,2,4]triazolo[2,3-a][1,3,5]triazin-5-ylamino]ethyl)phenol.

[d] Voltage-independent modulation of VDCCs is, in many cases, not directly indicated but rather inferred by a failure of a strongly depolarizing prepulse to completely restore current amplitude to control. It is possible in such instances that the apparent voltage-independent component is merely evidence that the prepulse was not of long enough duration or large enough amplitude or that $G\beta\gamma$ has already begun to reassociate with the channel as the prepulse and interpulse parameters are often different.

References: 1 Liu et al., 1994; 2 Yoshinaga et al., 1999; 3 Holz et al., 1986; 4 Lipscombe et al., 1989; 5 Ruiz-Velasco et al., 2000; 6 Ikeda, 1996; 7 Jeong and Wurster, 1997; 8 Caulfield et al., 1994; 9 Hartzell et al., 1991; 10 Davare et al., 2001; 11 Huang eta l., 1998; 12 Gao et al., 1997; 13 Yatani et al., 1995; 14 Shapiro et al., 1994; 15 Ikegami et al., 2001; 16 Marcez-Lepretre et al., 1996; 17 Bacal and Kunze, 1994; 18 Taussig et al., 1992; 19 Connor and Henderson, 1997; 20 Wilk-Blaszczak et al., 1994; 21 Mackie et al., 1995; 22 Pan et al., 1996; 23 Caulfield and Brown, 1992; 24 Mackie et al., 1993; 25 Beech et al., 1992; 26 Bernheim et al., 1992; 27 Haley et al., 2000; 28 Kammermeier et al., 2000; 29 Melliti et al., 2001; 30 Melliti et al., 2000; 31 Patil et al., 1996; 32 Brody and Yue, 2000b; 33 Brody et al., 1997; 34 Jeong and Ikeda, 2000; 35 Kleuss et al., 1991; 36 Surmeier et al., 1995; 37 Artalejo et al., 1990; 38 Bigornia et al., 1990; 39 Lledo et al., 1992; 40 Hernandez-Lopez et al., 2000; 41 Page et al., 1998; 42 Kuzhikandathil and Oxford, 1999; 43 Yan et al., 1997; 44 Marchetti et al., 1986; 45 Shen and Slaughter, 1999; 46 Mintz and Bean, 1993; 47 Wu et al., 1998; 48 Wheeler et al., 1994; 49 McCool et al., 1998; 50 Stea et al., 1995; 51 Salt et al., 1999; 52 Chavis et al., 1995; 53 Choi and Lovinger, 1996; 54 McCool et al., 1996; 55 Chavis et al., 1994; 56 Pisani et al., 2002; 57 Ikeda et al., 1995; 58 Takahashi et al., 1996; 59 Stefani et al., 1998; 60 Linn and Gafka et al., 1999; 61 Perroy et al., 2000; 62 Currie and Fox, 2000; 63 Levi and Alloatti, 1988; 64 Takashita et al., 1998; 65 Sun and Miller, 1999; 66 Su et al., 1998; 67 Qian et al., 1997; 68 Bleakmann et al., 1991; 69 King et al., 1999; 70 Wiley et al., 1993; 71 Foucart et al., 1993; 72 Sun et al., 1998; 73 Hescheler et al., 1987; 74 Acosta and Lopez, 1999; 75 Piros et al., 1996b; 76 Tsunoo et al., 1986; 77 Morikawa et al., 1999; 78 Toth et al., 1996; 79 Tallent et al., 1994; 80 Bourinet et al., 1996; 81 Laugwitz et al., 1993; 82 Schroeder et al., 1991; 83 Soldo et al., 1998; 84 Scholze et al., 2001; 85 Yung et al., 1999; 86 Beedle et al., 2002b; 87 Yamada et al., 2002; 88 Morikawa et al., 1998b; 89 Larsson et al., 2000; 90 Knoflach et al., 1996; 91 Asboth et al., 1996; 92 Tanaka et al., 1998; 93 White et al., 1996; 94 Csukas et al., 1998; 95 Ito et al., 2000; 96 Borgland et al., 2002; 97 Yamamoto et al., 1999; 98 Ikeda, 1992; 99 Kasai and Aosaki, 1989; 100 Yawo and Chuhma, 1993; 101 McCool and Farroni, 2001; 102 Brown and Dale, 2000; 103 Liang and Morley, 1996; 104 Mogul et al., 1993; 105 Stella et al., 2002; 106 Park et al., 1998; 275 Fleming and Mogul, 1997; 108 Filippov et al., 2000; 109 King et al., 1998; 110 Vartian and Boehm, 2001; 111 Gandia et al., 1993; 112 Filippov et al., 1998; 113 Bidet et al., 2000; 114 Filippov and Brown, 1996; 115 Filippov et al., 1999; 116 Bayliss et al., 1997; 117 Sun and Dale, 1999; 118 Penington et al., 1991; 119 Wu et al., 2002; 120 Foehring, 1996; 121 Day et al., 2002; 122 Ouadid et al., 1992; 123 Blondel et al., 1997; 124 Golard and Siegelbaum, 1993; 125 Ikeda and Schofield, 1989; 126 Roosterman et al., 1998; 127 Shapiro and Hille, 1993; 128 Margeta-Mitrovic et al., 1997; 129 Gilbert et al., 1998; 130 Holst et al., 2001; 131 Gallacher et al., 1990; 132 Ryu and Randic, 1990; 133 Zhu and Ikeda, 1994; 134 Tiaho and Nerbonne, 1996.

4. G PROTEINS MODULATE CALCIUM CHANNELS

GPCR activation stimulates the exchange of GDP for GTP at the Gα subunit of the G protein heterotrimer. GTP binding promotes the release of G proteins from the receptor, and Gα dissociation from the stable Gβγ dimer. Both active G proteins are involved in calcium channel modulation. Table 1 categorizes relevant signaling pathways, with representatives from each family (A, B, and C) of GPCRs, in calcium channel modulation. Receptor subunits, associated Gα subunits, pharmacological profiles, and down or upregulation of calcium current types are indicated. Early data from physiological systems (i.e., native cells) often did not indicate the particular GPCR subtype involved; these data are classified tentatively according to pharmacological or G protein signaling information. There are many other GPCRs linked to calcium channel regulation, but these are generally not well characterized and are not described here. Table 1 does, however, demonstrate the wide array of receptor types implicated in calcium channel modulation. Specifics of some of these pathways are detailed below.

4.1. Gα and Voltage-Independent Inhibition

G protein coupled receptors interact with Gα subunits to activate distinct signaling cascades. Standard $G\alpha_s$/PKA and $G\alpha_q$/PLCβ second messenger pathways upregulate a variety of HVA calcium currents by direct channel phosphorylation. However, uncommon cascades also exist, resulting in entirely opposite effects on calcium currents. Here, we present examples of standard and atypical $G\alpha_s$ and $G\alpha_q$-mediated second messenger systems as samples of the wide diversity among GPCR signaling mechanisms.

Activation of β-adrenergic receptors (βAR) in the heart is associated with positive inotropic effects through interactions with L-type calcium channels. The βAR couples to $G\alpha_s$ to activate adenylyl cyclase conversion of ATP to cAMP. Increased intracellular concentrations of cAMP activate the cAMP-dependent protein kinase (PKA). PKA then directly phosphorylates the L-type channel, leading to current facilitation (Hartzell et al., 1991). It is important to note that in transient expression systems and isolated myocytes, the coexpression of an A Kinase anchoring protein (AKAP) is required to permit functional βAR receptor coupling to L-type calcium channels (Johnson et al., 1997; Zhong et al., 1999).

In contrast, activation of D_1 dopaminergic receptors leads to inhibition of presynaptic calcium channels in a voltage-independent manner. D_1 also couples to the stimulatory $G\alpha_s$ and, like the βAR, activates adenylyl cyclase, increases cAMP formation, and activates PKA. However, instead of interacting with calcium channels, D_1-stimulated PKA phosphorylates protein phosphatase 1 (PP1) which, now in its activated state, dephosphorylates residues on N- and P/Q-type calcium channels, resulting finally in current inhibition (Surmeier et al., 1995). In D_1-expressing striatal neurons, basal kinase activity is high (Surmeier et al., 1995). This may serve as the molecular switch resulting in PKA inhibitory actions rather than PKA stimulatory actions at HVA calcium channels.

PLCβ modulatory pathways also show divergence. Activation of metabotropic glutamate receptor 1 (mGluR1) leads to a "common" PKC effect. Agonist binding to mGluR1 induces activation of the associated $G\alpha_q$ subunit. $G\alpha_q$ activates phospholipase Cβ (PLCβ) to promote the formation of IP_3 and diacylglycerol (DAG). IP_3 stimulates the release of calcium from intracellular stores which, combined with DAG, activates protein kinase C (PKC). PKC then phosphorylates ser/thr residues within the I-II linker of

$Ca_v2.1$ (N-type) and $Ca_v2.3$ (R-type) calcium channels to facilitate channel activity (Stea et al., 1995; Hamid et al., 1999). In contrast, serotonin receptor (5-HT_2) signaling through PLCβ can lead to L-type current inhibition. Agonist binding to 5-HT_2 stimulates $G\alpha_q$ activation of PLCβ and generation of IP_3. However, the IP_3-induced increase in intracellular calcium activates calcineurin, a protein phosphatase (PP2B), rather than PKC. Subsequent dephosphorylation of $Ca_v1.2$ by calcineurin inhibits L-type current (Day et al., 2002). Therefore, GPCR coupling, while often conforming to standard signaling pathways, can lead to alternate regulation of voltage-dependent calcium current. See Table 1 for Gα-coupling of other GPCRs.

Recent reports indicate that the specificity of different GPCRs for particular signaling cascades may represent a physical colocalization of the signaling components with calcium channels. For example, i) β_2 adrenergic receptors exist in a macromolecular signaling complex containing calcium channels ($Ca_v1.2$), $G\alpha_s$, adenylyl cyclase, PKA, and calcineurin (Davare et al., 2001), and ii) mGluR inhibition of calcium channels requires the presence of a scaffolding protein (Perroy et al., 2002). In addition, calcium channel inhibition by the galanin receptor 2 may involve translocation of the $Ca_v2.2$ C-terminus to the plasma membrane and subsequent direct interactions with $G\alpha_q$ and PKC (Simen et al., 2001). Therefore, we may find that only signaling components present as a complex with the GPCR and the calcium channel are important in channel modulation.

4.2. Gβγ Targets Presynaptic Calcium Channels

The Gβγ heterodimer was originally believed to be an inactive subunit in G protein signal transduction. However, it is now accepted that Gβγ itself is an active signaling molecule that interacts directly with both inwardly rectifying K^+ channels and VDCCs (Wickman and Clapham, 1995). Starting in 1996, a number of convincing reports provided evidence that Gβγ binds directly to presynaptic calcium channels (Herlitze et al., 1996; Zhang et al., 1996; De Waard et al., 1997; Zamponi et al., 1997). Zhang et al. (1996) identified domain 1 and the C-terminus as potential interaction sites. Subsequently, two distinct binding sites in the N- and P/Q-type calcium channel I-II linkers were implicated in Gβγ modulation (see Figure 2) (De Waard et al., 1997; Zamponi et al., 1997). One site, on the N-terminal side of the I-II linker, contains a QXXER motif that has previously been suggested to involve Gβγ binding (Chen et al., 1995; Weng et al., 1996; Herlitze et al., 1997). This Gβγ binding site overlaps with the calcium channel β subunit binding site (alpha interaction domain, or AID), explaining the dependence of Gβγ signaling on coexpression of calcium channel β subunit subtypes (see below). The second Gβγ site is approximately 20 residues downstream of the QXXER-containing region in the I-II linker (De Waard et al., 1997; Zamponi et al., 1997). A number of studies have subsequently confirmed a role for the I-II linker in Gβγ modulation of presynaptic calcium channels (Herlitze et al., 1997; Page et al., 1997; Simen and Miller, 1998, 2000) (Qin et al., 1997). Chimeras between $Ca_v2.2$ and $Ca_v2.3$ have also suggested that the I-II linker and regions of domain I may transduce the inhibitory effects of Gβγ onto the calcium channel, while N- and C-termini may be important in the voltage-dependence of Gβγ modulation (Page et al., 1997; Qin et al., 1997; Simen and Miller, 1998; Stephens et al., 1998; Canti et al., 1999; Hamid et al., 1999; Simen and Miller, 2000).

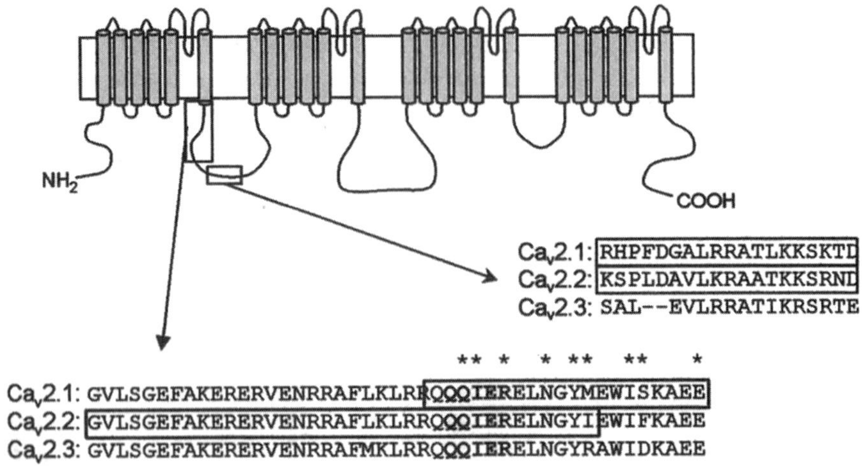

Figure 2. Two regions of the I-II linker of presynaptic calcium channels are implicated in Gβγ modulation. The putative secondary structure of the α_1 calcium channel subunit is shown with the approximate location of the Gβγ binding sites in the I-II linker boxed. Alignments of the P/Q-type $Ca_v2.1$, N-type $Ca_v2.2$, and R-type $Ca_v2.3$ protein sequence are shown (GenBank Accession Nos. NM_012918, M92905, and L15453). The first binding region includes the QXXER motif (in bold) and overlaps with the α interaction domain where the calcium channel β subunit binds (asterisks). Notice that there is considerable sequence divergence among calcium channel subunits in the second binding domain; this may explain why R-type channels are poor substrates for Gβγ regulation. Boxed sequences indicate regions involved in Gβγ binding identified by Zamponi et al. (1997). The N- and C-termini are also involved in G protein binding (sequences not shown).

Gβγ modulation of calcium channels is membrane delimited, meaning that no soluble cytoplasmic messenger is involved (Hille, 1994). This inhibition occurs in a voltage-dependent manner with an associated slowing of current activation kinetics (Bean, 1989a). As shown by Elmslie et al., a strongly depolarizing prepulse administered 5 ms prior to a test pulse can relieve both Gβγ inhibition and kinetic slowing (Elmslie et al., 1990). Notably, N-type channels typically exhibit greater inhibition of current by Gβγ, and greater relief by a depolarizing prepulse, than P/Q-type channels (Colecraft et al., 2000). The evidence of voltage-dependent Gβγ inhibition led to the proposal of a "willing" and "reluctant" model of calcium channel gating in the presence of Gβγ (Bean, 1989a). In this model, unbound channels can open readily, and are thus "willing"; whereas channels bound to Gβγ require stronger depolarizations to activate, making them "reluctant" to open. It is believed that Gβγ must physically dissociate to permit channel activation (Patil et al., 1996; Jones et al., 1997), although it has been suggested more recently that reluctant openings may also occur with $Ca_v2.2$ channels (Colecraft et al., 2000; Colecraft et al., 2001). Despite some earlier controversy, it is now accepted that Gβγ inhibition of N-type channels occurs with a 1:1 stoichiometry (Zamponi and Snutch, 1998).

It is important to note that, when it comes to calcium channel modulation, not all Gβ subunits are created equal. All five Gβ subunits plus $G\gamma_2$, $G\gamma_3$, or $G\gamma_4$ can inhibit N-type channels in superior cervical ganglion (SCG) cells. However, $G\beta_4$ exhibited the greatest channel inhibition and $G\beta_5$ the least (Herlitze et al., 1996; Ikeda, 1996; Garcia et al., 1998; Ruiz-Velasco and Ikeda, 2000). Heterologous expression of different $G\beta$ subunits and $G\gamma_2$ with $Ca_v2.2$ (N-type) calcium channels in tsA-201 cells revealed that $G\beta_1$ and $G\beta_3$ inhibit these channels most effectively, followed by $G\beta_4$ and $G\beta_2$ (Arnot et al., 2000). $G\beta_5\gamma_2$ mediated little to no calcium channel modulation. In contrast, a different pattern is observed with transiently expressed $Ca_v2.2$ (P/Q-type) calcium channels, which are most sensitive to $G_{\beta 4}$ (Arnot et al., 2000). Furthermore, $G\beta$ binding is influenced by the species of calcium channel β subunit coexpressed. In general, $G\beta\gamma$ interactions with N-type calcium channels are least effective in the presence of the calcium channel β_4 subunit, and β_{2a}-expressing channels are most susceptible to $G\beta\gamma$-induced kinetic slowing (Feng et al., 2001; Meir and Dolphin, 2002).

Increasing evidence now suggests that in addition to voltage-dependent interactions with the I-II linker of N- and P/Q-type calcium channels, $G\beta\gamma$ may also stimulate voltage-independent signaling cascades in calcium channel modulation. $G\beta\gamma$ can directly activate some species of adenylyl cyclases and phospholipases independent of the $G\alpha$ subunit (Mumby, 2000). For example, angiotensin (AT_1) receptor activation liberates $G\beta\gamma$ to stimulate phospholipase C, leading to PKC-mediated upregulation of L-type calcium current (Macrez et al., 1997). How a cell determines whether the $G\alpha$ or the $G\beta\gamma$ is the active signaling species is not entirely clear, but may involve phosphorylation of the GPCR (Daaka et al., 1997). Thus, both $G\beta\gamma$ and $G\alpha$ subunits must now be considered when identifying the mechanisms of voltage-independent calcium channel inhibition.

4.3. Convergence of Cell Signaling at Presynaptic Calcium Channels

Simultaneous activation of more than one signaling pathway at a time is likely a common physiological phenomenon, particularly at presynaptic sites where multiple transmitters can be co-released from the same vesicles and several inputs can converge onto the same synapse. In such cases, where parallel pathways are co-activated, how is activation of separate signaling mechanisms resolved at the level of the calcium channel? Crosstalk between $G\beta\gamma$ and PKC signaling can occur at the level of the presynaptic N-type calcium channel. PKC phosphorylates the I-II linker of $Ca_v2.2$ (N-type) calcium channels at Thr422 and Ser425 (Zamponi et al., 1997; Hamid et al., 1999). While phosphorylation of Thr422 or Ser425 upregulates N-type current, phosphorylation of only Thr422 interferes with $G\beta\gamma$-calcium channel binding and thus, antagonizes $G\beta\gamma$-induced current inhibition (Hamid et al., 1999). Recent evidence indicates that this integration of $G\beta\gamma$ and PKC signaling is specific for dimers created with the $G\beta_1$ subunit and is not observed with $G\beta_{2,3,\text{ or }4}$ -mediated inhibition (Cooper et al., 2000). Thus, GPCRs that preferentially couple to $G\beta_1$ are most susceptible to crosstalk with PKC pathways (Swartz, 1993; Swartz et al., 1993; Cooper et al., 2000). PKC-independent convergence of signaling pathways has also been demonstrated, although the molecular determinants are not well defined. In ciliary and dorsal root ganglion, Polo-Parada and Pilar (1999) discovered that either κ-OR or μ-OR activation can occlude responses to somatostatin, regardless of whether the opioids are applied before or during a somatostatin treatment, a response they termed "demodulation." Similarly, opioids can demodulate inhibitory responses to GABA and norepinephrine (Polo-Parada and Pilar, 1999). Essentially, the opioid receptor signaling (voltage-independent in this instance) becomes "dominant," while the somatostatin receptor modulation (voltage-dependent) becomes "submissive." Given that

GPCRs may exist in a macromolecular signaling complex containing calcium channels, and other effectors (see above), one may speculate that the ability of some GPCR signaling cascades to dominate over others in calcium channel modulation may reflect an underlying structural association.

Other molecules, including several proteins of the synaptic vesicle release complex, can modulate calcium currents (Stanley and Mirotznik, 1997; Magga et al., 2000; Jarvis and Zamponi, 2001). Syntaxin 1A and cysteine string protein (CSP) interact with the "synprint" domain located within the II-III linker of N- and P/Q-type calcium channels (Sheng et al., 1994; Rettig et al., 1996; Walker and De Waard, 1998; Lin and Scheller, 2000). Syntaxin 1A (but not syntaxin 1B; Jarvis and Zamponi, 2001; Lu et al., 2001) and CSP binding promote G protein inhibition of presynaptic calcium channels by acting as chaperones between Gβγ and the channel (Jarvis et al., 2000; Magga et al., 2000). Thus, a calcium channel's proximity to primed transmitter vesicles may influence its modulatory state. Of particular note, during this process, syntaxin 1A appears to displace the Gγ subunit from the Gβγ dimer (Jarvis et al., 2002), in a manner reminiscent of certain RGS proteins (Snow et al., 1998; Posner et al., 1999; Zhang and Simmonds, 2000) (see Section 5). This suggests the intriguing possibility that syntaxin 1A can serve as a membrane anchoring domain of Gβ. Consistent with a physical interaction between G proteins and vesicle release proteins, Blackmer et al. (2001) recently reported a calcium channel independent inhibition of synaptic transmission in reticulospinal/motoneuron synapses of the lamprey.

Changes in cell excitability can also feed back to alter modulation of calcium channels by Gβγ. As binding of Gβγ to the I-II linker is voltage-dependent, modeling and physiological studies suggest that depolarizing action potential trains may provide enough of a stimulus to disrupt Gβγ-channel interactions and evoke short term changes in synaptic function (Bertram and Behan, 1999; Brody and Yue, 2000a). Thus, the calcium channel may act as a coincidence detector coupling cell activity to intracellular signaling cascades.

Considerable evidence now suggests that GPCRs can activate tyrosine kinase (TK) receptor signaling cascades (e.g., epidermal growth factor receptor, Trk neurotrophin receptors and glucocorticoid receptors) in the absence of TK receptor agonists (Daub et al., 1997; Castagliuolo et al., 2000; Lee and Chao, 2001; Schmidt et al., 2001). Such pathways often involve GPCR-mediated activation of phosphoinositide 3-kinase (PI3-K) and mitogen-activated protein kinase (MAPK). The implications of GPCR transactivation of growth factor receptors with regard to calcium channel modulation are not presently clear. It is likely, however, that these signaling pathways will regulate calcium channel currents. Peptide-induced potentiation of L-type currents involving MAPK and upregulation of N-type currents through MAPK interactions with the calcium channel β subunit have already been described (Dziema and Obrietan, 2002; Fitzgerald, 2002).

5. RECEPTOR DESENSITIZATION AND DIMERIZATION

5.1. Desensitization

Termination of GPCR signaling cascades in the continued presence of agonist is termed "desensitization." Some GPCRs undergo rapid desensitization (e.g., β_2 adrenergic receptor, δ opioid receptor, and somatostatin receptors), while others may be relatively resistant (e.g., κ opioid receptor) (Beaumont et al., 1998; Jordan et al., 2001; Pfeiffer et al., 2002). Desensitization typically involves a combination of G protein-RGS (regulators of G protein signaling) interactions and phosphorylation steps that deactivate G proteins, uncouple GPCRs from G proteins, and induce receptor internalization.

RGS proteins are GTPase-activating proteins (GAP) that increase the rate of GTP hydrolysis by Gα (Watson et al., 1996). By stimulating the conversion of GTP to GDP by Gα, RGS proteins turn off Gα activity and promote the reassociation of Gα with Gβγ (e.g., RGS 2 and RGS 9) (Potenza et al., 1999; Rahman et al., 1999). Members of the RGS family bind to different Gα subtypes (Saugstad et al., 1998; Zhou et al., 2001), but may also interact with Gβγ or Gβ5 to alter voltage-dependent calcium channel inhibition (Mark et al., 2000; Zhou et al., 2000).

G protein coupled receptor kinases (GRKs) directly phosphorylate the C-terminus of opioid receptors and other GPCRs in the first step of receptor desensitization (Zhang et al., 1998; Kovoor et al., 1999). This phosphorylation promotes translocation of β arrestin proteins to the membrane, where they bind to the GPCR IL 3 and C-terminus to antagonize GPCR/G protein coupling (Cheng et al., 1998; Zhang et al., 1998; Schulz et al., 1999; Cen et al., 2001). β arrestins can then interact with dynamin molecules in clathrin coated pits to induce internalization of GPCRs (Whistler and von Zastrow, 1998; Li et al., 1999). The GRK/β arrestin pathway is active in opioid receptor desensitization, as dominant negative constructs of several proteins in the pathway (GRK 2, β arrestin 1, dynamin 1) inhibit receptor internalization (Raynor et al., 1994; Zhang et al., 1998; Li et al., 1999). Of note, GPCRs may also undergo receptor desensitization via direct internalization that does not involve GRK activity (Beaumont et al., 1998).

How does receptor desensitization impact calcium channel activity? It is obvious that once a particular GPCR pathway has desensitized, a new or continued presence of the agonist will be less efficacious in regulating calcium channel activity. For example, somatostatin receptor mediated inhibition of N-type calcium channels desensitizes within 2 minutes, and subsequent reapplication of agonist is less efficacious (Zamponi, unpublished observations). Desensitization in one pathway, however, may also impact other GPCRs. For example, in "heterologous desensitization," desensitization of receptor "A" by agonist "A" also causes desensitization of receptor "B," even though no agonist "B" is present (Bundey and Nahorski, 2001; Mandyam et al., 2002). Thus, agonist "B" modulation of calcium channels is also dampened. The fate of the deactivated G protein αβγ heterotrimer associated with the signal by receptor "A" is unclear. If Gαβγ reassociates with the desensitized receptor, calcium channel modulation is still reduced. But if Gαβγ can now promote the coupling of receptor "C" to channel activity, agonist "C" modulation of calcium channels may actually be enhanced. To our knowledge, the fate of the Gαβγ of desensitized receptors has not been studied.

5.2. Dimer- or Oligomerization

Increasing evidence supports the notion that GPCRs may exist as functional monomers, dimers, and oligomers. Perhaps the most arresting example of dimerization involves the metabotropic $GABA_B$ receptor. In this case, both subunits, $GABA_{B1}$ and $GABA_{B2}$, must be present to create a functional GPCR, with $GABA_{B1}$ binding receptor ligands and $GABA_{B2}$ interacting with G proteins to transduce receptor activation (Kaupmann et al., 1997; Jones et al., 1998; Robbins et al., 2001). GPCR homodimers have been suggested for β_2 adrenergic receptors (Hebert et al., 1996; Hebert et al., 1998), μ, γ, and κ opioid receptors (Cvejic and Devi, 1997; Jordan and Devi, 1999; George et al., 2000), and somatostatin receptors (Pfeiffer et al., 2001), among others. Functional heterodimers/heteromultimers between subfamilies of receptors (μ/δ opioid receptors), or between distantly related GPCRs (β_2 AR/δOR, β_2 AR/κOR, sst_{2A}/μOR, and AT_1/B_2) have also been demonstrated (AbdAlla et al., 2000; George et al., 2000; Pfeiffer et al., 2001; Pfeiffer et al., 2002). In most cases, these heterodimers retain normal pharmacological profiles and signaling cascades for both receptor types (Pfeiffer et al., 2001; Pfeiffer et al., 2002). However, one report suggests that μ and δ opioid receptor heterodimers display divergent ligand selectivity, indicating that heteromultimers may possess distinct binding pockets (George et al., 2000).

What is the function of GPCR oligomerization? The largest role of oligomerization seems to involve receptor trafficking. Dimerization may precede receptor internalization of homodimers (Cvejic and Devi, 1997), may induce heterodimer internalization in the presence of agonists for one or both receptor types (Pfeiffer et al., 2001; Pfeiffer et al., 2002), or may inhibit internalization of a typically desensitized receptor. For example, β_2AR/δOR dimers are internalized in the presence of isoproterenol or opioids alone, but neither agonist induces internalization of β_2AR/κOR dimers (Pfeiffer et al., 2001). As β_2AR signaling through MAP kinase requires receptor internalization, heterodimerization with κOR functionally uncouples this pathway. Therefore, hetero-oligomerization of GPCRs may serve as a means of determining the specific G protein-coupling of "promiscuous receptors" in a given cell type/environment. Heteromultimer formation may also alter the specificity of Gα coupling (George et al., 2000) or alter the mechanism of receptor internalization (AbdAlla et al., 2000). Thus, dimer/ oligomer formation may have a significant impact on calcium channel modulation. In some cases, dimerization is required for calcium channel modulation ($GABA_B$) while in other cases, dimerization (e.g., β_2AR/δOR) ultimately reduces the ability of GPCRs to regulate calcium currents.

6. CONCLUSIONS

Pathways governing the interaction between G protein-coupled receptors and calcium channels are becoming clearer. In particular, the structural correlates of G$\beta\gamma$ inhibition of calcium channels, perhaps the most universal of all GPCR signaling cascades, is well described. However, it is increasingly evident that calcium channel modulation is influenced by the convergence of many signaling pathways. Thus, calcium channel regulation by GPCRs in any given cell is an exceedingly complicated process.

ACKNOWLEDGMENTS

Research in G protein modulation of calcium channels is supported by the Canadian Institutes of Health Research (CIHR, operating grant to GWZ). GWZ is funded by the Alberta Heritage Foundation for Medical Research (AHFMR), CIHR, and the National Alliance for Research on Schizophrenia and Depression (NARSAD). AMB is the recipient of AHFMR and Natural Science and Engineering Research Council of Canada (NSERC) studentship awards.

REFERENCES

AbdAlla, S., Lother, H., and Quitterer, U., 2000, AT1-receptor heterodimers show enhanced G-protein activation and altered receptor sequestration, *Nature* **407**:94-98.

Acosta, C. G. and Lopez, H. S., 1999, δ opioid receptor modulation of several voltage-dependent Ca^{2+} currents in rat sensory neurons, *J Neurosci* **19**:8337-8348.

Arnot, M. I., Stotz, S. C., Jarvis, S. E., and Zamponi, G. W., 2000, Differential modulation of N-type 1B and P/Q-type 1A calcium channels by different G protein subunit isoforms, *J Physiol* **527 Pt 2**:203-212.

Artalejo, C. R., Ariano, M. A., Perlman, R. L., and Fox, A. P., 1990, Activation of facilitation calcium channels in chromaffin cells by D1 dopamine receptors through a cAMP/protein kinase A-dependent mechanism, *Nature* **348**:239-242.

Asboth, G., Phaneuf, S., Europe-Finner, G. N., Toth, M., and Bernal, A. L., 1996, Prostaglandin E2 activates phospholipase C and elevates intracellular calcium in cultured myometrial cells: involvement of EP1 and EP3 receptor subtypes, *Endocrinology* **137**:2572-2579.

Bacal, K. and Kunze, D. L., 1994, Dual effects of angiotensin II on calcium currents in neonatal rat nodose neurons, *J Neurosci* **14**:7159-7167.

Bayliss, D. A., Li, Y. W., and Talley, E. M., 1997, Effects of serotonin on caudal raphe neurons: inhibition of N- and P/Q-type calcium channels and the afterhyperpolarization, *J Neurophysiol* **77**:1362-1374.

Bean, B. P., 1989a, Neurotransmitter inhibition of neuronal calcium currents by changes in channel voltage dependence, *Nature* **340**:153-156.

Bean, B. P., 1989b, Classes of calcium channels in vertebrate cells, *Annu Rev Physiol* **51**:367-384.

Beaumont, V., Hepworth, M. B., Luty, J. S., Kelly, E., and Henderson, G., 1998, Somatostatin receptor desensitization in NG108-15 cells. A consequence of receptor sequestration, *J Biol Chem* **273**:33174-33183.

Bech-Hansen, N. T., Naylor, M. J., Maybaum, T. A., Pearce, W. G., Koop, B., Fishman, G. A., Mets, M., Musarella, M. A., and Boycott, K. M., 1998, Loss-of-function mutations in a calcium-channel α1-subunit gene in Xp11.23 cause incomplete X-linked congenital stationary night blindness, *Nat Genet* **19**:264-267.

Beech, D. J., Bernheim, L., and Hille, B., 1992, Pertussis toxin and voltage dependence distinguish multiple pathways modulating calcium channels of rat sympathetic neurons, *Neuron* **8**:97-106.

Beedle, A. M., Hamid, J., and Zamponi, G. W., 2002a, Inhibition of transiently expressed low- and high-voltage-activated calcium channels by trivalent metal cations, *J Membr Biol* **187**:225-238.

Beedle, A. M., Poirot, O., Barrere, C., Hamid, J., Nargeot, J., Zamponi, G. W., and Bourinet, E., 2002b, Inhibition of voltage-gated calcium channels by nociceptin/hORL1, *Soc Neurosci Abstr* 341.5.

Bernheim, L., Mathie, A., and Hille, B., 1992, Characterization of muscarinic receptor subtypes inhibiting Ca^{2+} current and M current in rat sympathetic neurons, *Proc Natl Acad Sci U S A* **89**:9544-9548.

Bertram, R. and Behan, M., 1999, Implications of G-protein-mediated Ca^{2+} channel inhibition for neurotransmitter release and facilitation, *J Comput Neurosci* **7**:197-211.

Bidet, M., De Renzis, G., Martial, S., Rubera, I., Tauc, M., and Poujeol, P., 2000, Extracellular ATP increases $[Ca^{2+}]_i$ in distal tubule cells. I. Evidence for a P2Y2 purinoceptor, *Am J Physiol Renal Physiol* **279**:F92-F101.

Bigornia, L., Allen, C. N., Jan, C. R., Lyon, R. A., Titeler, M., and Schneider, A. S., 1990, D2 dopamine receptors modulate calcium channel currents and catecholamine secretion in bovine adrenal chromaffin cells, *J Pharmacol Exp Ther* **252**:586-592.

Blackmer, T., Larsen, E. C., Takahashi, M., Martin, T. F., Alford, S., and Hamm, H. E., 2001, G protein βγ subunit-mediated presynaptic inhibition: regulation of exocytotic fusion downstream of Ca^{2+} entry, *Science* **292**:293-297.

Bleakman, D., Colmers, W. F., Fournier, A., and Miller, R. J., 1991, Neuropeptide Y inhibits Ca^{2+} influx into cultured dorsal root ganglion neurones of the rat via a Y2 receptor, *Br J Pharmacol* **103**:1781-1789.

Blondel, O., Vandecasteele, G., Gastineau, M., Leclerc, S., Dahmoune, Y., Langlois, M., and Fischmeister, R., 1997, Molecular and functional characterization of a 5-HT_4 receptor cloned from human atrium, *FEBS Lett* **412**:465-474.

Borgland, S. L., Connor, M., Ryan, R. M., Ball, H. J., and Christie, M. J., 2002, Prostaglandin E_2 inhibits calcium current in two sub-populations of acutely isolated mouse trigeminal sensory neurons, *J Physiol* **539**:433-444.

Bourinet, E., Soong, T. W., Stea, A., and Snutch, T. P., 1996, Determinants of the G protein-dependent opioid modulation of neuronal calcium channels, *Proc Natl Acad Sci U S A* **93**:1486-1491.

Bourinet, E., Soong, T. W., Sutton, K., Slaymaker, S., Mathews, E., Monteil, A., Zamponi, G. W., Nargeot, J., and Snutch, T. P., 1999, Splicing of α1A subunit gene generates phenotypic variants of P- and Q-type calcium channels, *Nat Neurosci* **2**:407-415.

Brody, D. L., Patil, P. G., Mulle, J. G., Snutch, T. P., and Yue, D. T., 1997, Bursts of action potential waveforms relieve G-protein inhibition of recombinant P/Q-type Ca^{2+} channels in HEK 293 cells, *J Physiol* **499 (Pt 3)**:637-644.

Brody, D. L. and Yue, D. T., 2000a, Relief of G-protein inhibition of calcium channels and short-term synaptic facilitation in cultured hippocampal neurons, *J Neurosci* **20**:889-898.

Brody, D. L. and Yue, D. T., 2000b, Release-independent short-term synaptic depression in cultured hippocampal neurons, *J Neurosci* **20**:2480-2494.

Brown, P. and Dale, N., 2000, Adenosine A1 receptors modulate high voltage-activated Ca^{2+} currents and motor pattern generation in the *xenopus* embryo, *J Physiol* **525 Pt 3**:655-667.

Bundey, R. A. and Nahorski, S. R., 2001, Homologous and heterologous uncoupling of muscarinic M_3 and α_{1B} adrenoceptors to $G\alpha_{q/11}$ in SH-SY5Y human neuroblastoma cells, *Br J Pharmacol* **134**:257-264.

Canti, C., Page, K. M., Stephens, G. J., and Dolphin, A. C., 1999, Identification of residues in the N terminus of α1B critical for inhibition of the voltage-dependent calcium channel by Gβγ, *J Neurosci* **19**:6855-6864.

Castagliuolo, I., Valenick, L., Liu, J., and Pothoulakis, C., 2000, Epidermal growth factor receptor transactivation mediates substance P-induced mitogenic responses in U-373 MG cells, *J Biol Chem* **275**:26545-26550.

Castellano, A. and Perez-Reyes, E., 1994, Molecular diversity of Ca^{2+} channel beta subunits, *Biochem Soc Trans* **22**:483-488.

Catterall, W. A., 2000, Structure and regulation of voltage-gated Ca^{2+} channels, *Annu Rev Cell Dev Biol* **16**:521-555.

Caulfield, M. P., and Brown, D. A., 1992, Cannabinoid receptor agonists inhibit Ca current in NG108-15 neuroblastoma cells via a pertussis toxin-sensitive mechanism, *Br J Pharmacol* **106**:231-232.

Caulfield, M. P., Jones, S., Vallis, Y., Buckley, N. J., Kim, G. D., Milligan, G., and Brown, D. A., 1994, Muscarinic M-current inhibition via Gαq/11 and α-adrenoceptor inhibition of Ca^{2+} current via Gαo in rat sympathetic neurones, *J Physiol* **477 (Pt 3)**:415-422.

Cen, B., Xiong, Y., Ma, L., and Pei, G., 2001, Direct and differential interaction of β-arrestins with the intracellular domains of different opioid receptors, *Mol Pharmacol* **59**:758-764.

Chavis, P., Shinozaki, H., Bockaert, J., and Fagni, L., 1994, The metabotropic glutamate receptor types 2/3 inhibit L-type calcium channels via a pertussis toxin-sensitive G-protein in cultured cerebellar granule cells, *J Neurosci* **14**:7067-7076.

Chavis, P., Nooney, J. M., Bockaert, J., Fagni, L., Feltz, A., and Bossu, J. L., 1995, Facilitatory coupling between a glutamate metabotropic receptor and dihydropyridine-sensitive calcium channels in cultured cerebellar granule cells, *J Neurosci* **15**:135-143.

Chen, J., DeVivo, M., Dingus, J., Harry, A., Li, J., Sui, J., Carty, D. J., Blank, J. L., Exton, J. H., Stoffel, R. H., and et al., 1995, A region of adenylyl cyclase 2 critical for regulation by G protein βγ subunits, *Science* **268**:1166-1169.

Cheng, Z. J., Yu, Q. M., Wu, Y. L., Ma, L., and Pei, G., 1998, Selective interference of β-arrestin 1 with κ and δ but not μ opioid receptor/G protein coupling, *J Biol Chem* **273**:24328-24333.

Choi, S. and Lovinger, D. M., 1996, Metabotropic glutamate receptor modulation of voltage-gated Ca^{2+} channels involves multiple receptor subtypes in cortical neurons, *J Neurosci* **16**:36-45.

Chu, P. J., Robertson, H. M., and Best, P. M., 2001, Calcium channel γ subunits provide insights into the evolution of this gene family, *Gene* **280**:37-48.

Colecraft, H. M., Patil, P. G., and Yue, D. T., 2000, Differential occurrence of reluctant openings in G-protein-inhibited N- and P/Q-type calcium channels, *J Gen Physiol* **115**:175-192.

Colecraft, H. M., Brody, D. L., and Yue, D. T., 2001, G-protein inhibition of N- and P/Q-type calcium channels: distinctive elementary mechanisms and their functional impact, *J Neurosci* **21**:1137-1147.

Connor, M. and Henderson, G., 1997, Bradykinin inhibition of N- and L-type calcium channel currents in NG108-15 cells, *Neuropharmacology* **36**:115-124.

Connor, M. and Christie, M. D., 1999, Opioid receptor signalling mechanisms, *Clin Exp Pharmacol Physiol* **26**:493-499.

Cooper, C. B., Arnot, M. I., Feng, Z. P., Jarvis, S. E., Hamid, J., and Zamponi, G. W., 2000, Cross-talk between G-protein and protein kinase C modulation of N-type calcium channels is dependent on the G-protein β subunit isoform, *J Biol Chem* **275**:40777-40781.

Cotecchia, S., Exum, S., Caron, M. G., and Lefkowitz, R. J., 1990, Regions of the α1-adrenergic receptor involved in coupling to phosphatidylinositol hydrolysis and enhanced sensitivity of biological function, *Proc Natl Acad Sci U S A* **87**:2896-2900.

Couvineau, A., Gaudin, P., Maoret, J. J., Rouyer-Fessard, C., Nicole, P., and Laburthe, M., 1995, Highly conserved aspartate 68, tryptophane 73 and glycine 109 in the N-terminal extracellular domain of the human VIP receptor are essential for its ability to bind VIP, *Biochem Biophys Res Commun* **206**:246-252.

Cribbs, L. L., Lee, J. H., Yang, J., Satin, J., Zhang, Y., Daud, A., Barclay, J., Williamson, M. P., Fox, M., Rees, M., and Perez-Reyes, E., 1998, Cloning and characterization of α1H from human heart, a member of the T-type Ca^{2+} channel gene family, *Circ Res* **83**:103-109.

Csukas, S., Hanke, C. J., Rewolinski, D., and Campbell, W. B., 1998, Prostaglandin E2-induced aldosterone release is mediated by an EP2 receptor, *Hypertension* **31**:575-581.

Currie, K. P. and Fox, A. P., 2000, Voltage-dependent, pertussis toxin insensitive inhibition of calcium currents by histamine in bovine adrenal chromaffin cells, *J Neurophysiol* **83**:1435-1442.

Cvejic, S. and Devi, L. A., 1997, Dimerization of the delta opioid receptor: implication for a role in receptor internalization, *J Biol Chem* **272**:26959-26964.

Daaka, Y., Luttrell, L. M., and Lefkowitz, R. J., 1997, Switching of the coupling of the β2-adrenergic receptor to different G proteins by protein kinase A, *Nature* **390**:88-91.

Daub, H., Wallasch, C., Lankenau, A., Herrlich, A., and Ullrich, A., 1997, Signal characteristics of G protein-transactivated EGF receptor, *Embo J* **16**:7032-7044.

Davare, M. A., Avdonin, V., Hall, D. D., Peden, E. M., Burette, A., Weinberg, R. J., Horne, M. C., Hoshi, T., and Hell, J. W., 2001, A β2 adrenergic receptor signaling complex assembled with the Ca^{2+} channel $Ca_v1.2$, *Science* **293**:98-101.

Day, M., Olson, P. A., Platzer, J., Striessnig, J., and Surmeier, D. J., 2002, Stimulation of 5-HT_2 receptors in prefrontal pyramidal neurons inhibits $Ca_v1.2$ L type Ca^{2+} currents via a PLCβ/IP_3/calcineurin signaling cascade, *J Neurophysiol* **87**:2490-2504.

De Waard, M., Liu, H., Walker, D., Scott, V. E., Gurnett, C. A., and Campbell, K. P., 1997, Direct binding of G-protein βγ complex to voltage-dependent calcium channels, *Nature* **385**:446-450.

Dietrich, G., Gaibelet, G., Capeyrou, R., Butour, J. L., Pontet, F., and Emorine, L. J., 1998, Implication of the first and third extracellular loops of the μ-opioid receptor in the formation of the ligand binding site: a study using chimeric μ-opioid/angiotensin receptors, *J Neurochem* **70**:2106-2111.

Du, K., Nicole, P., Couvineau, A., and Laburthe, M., 1997, Aspartate 196 in the first extracellular loop of the human VIP1 receptor is essential for VIP binding and VIP-stimulated cAMP production, *Biochem Biophys Res Commun* **230**:289-292.

Dubel, S. J., Starr, T. V., Hell, J., Ahlijanian, M. K., Enyeart, J. J., Catterall, W. A., and Snutch, T. P., 1992, Molecular cloning of the α-1 subunit of an ω-conotoxin-sensitive calcium channel, *Proc Natl Acad Sci U S A* **89**:5058-5062.

Dunlap, K. and Fischbach, G. D., 1981, Neurotransmitters decrease the calcium conductance activated by depolarization of embryonic chick sensory neurones, *J Physiol* **317**:519-535.

Dunlap, K., Luebke, J. I., and Turner, T. J., 1995, Exocytotic Ca^{2+} channels in mammalian central neurons, *Trends Neurosci* **18**:89-98.

Dziema, H. and Obrietan, K., 2002, PACAP potentiates L-type calcium channel conductance in suprachiasmatic nucleus neurons by activating the MAPK pathway, *J Neurophysiol* **88**:1374-1386.

Ellis, S. B., Williams, M. E., Ways, N. R., Brenner, R., Sharp, A. H., Leung, A. T., Campbell, K. P., McKenna, E., Koch, W. J., Hui, A., and et al., 1988, Sequence and expression of mRNAs encoding the α1 and α2 subunits of a DHP-sensitive calcium channel, *Science* **241**:1661-1664.

Elmslie, K. S., Zhou, W., and Jones, S. W., 1990, LHRH and GTP-γ-S modify calcium current activation in bullfrog sympathetic neurons, *Neuron* **5**:75-80.

Feng, Z. P., Arnot, M. I., Doering, C. J., and Zamponi, G. W., 2001, Calcium channel β subunits differentially regulate the inhibition of N-type channels by individual Gβ isoforms, *J Biol Chem* **276**:45051-45058.

Ferguson, D. M., Kramer, S., Metzger, T. G., Law, P. Y., and Portoghese, P. S., 2000, Isosteric replacement of acidic with neutral residues in extracellular loop-2 of the κ-opioid receptor does not affect dynorphin A_{1-13} affinity and function, *J Med Chem* **43**:1251-1252.

Filippov, A. K. and Brown, D. A., 1996, Activation of nucleotide receptors inhibits high-threshold calcium currents in NG108-15 neuronal hybrid cells, *Eur J Neurosci* **8**:1149-1155.

Filippov, A. K., Webb, T. E., Barnard, E. A., and Brown, D. A., 1998, $P2Y_2$ nucleotide receptors expressed heterologously in sympathetic neurons inhibit both N-type Ca^{2+} and M-type K^+ currents, *J Neurosci* **18**:5170-5179.

Filippov, A. K., Webb, T. E., Barnard, E. A., and Brown, D. A., 1999, Dual coupling of heterologously-expressed rat P2Y6 nucleotide receptors to N-type Ca^{2+} and M-type K^+ currents in rat sympathetic neurones, *Br J Pharmacol* **126**:1009-1017.

Filippov, A. K., Brown, D. A., and Barnard, E. A., 2000, The P2Y$_1$ receptor closes the N-type Ca^{2+} channel in neurones, with both adenosine triphosphates and diphosphates as potent agonists, *Br J Pharmacol* **129**:1063-1066.

Filizola, M., Laakkonen, L., and Loew, G. H., 1999, 3D modeling, ligand binding and activation studies of the cloned mouse δ, μ, and κ opioid receptors, *Protein Eng* **12**:927-942.

Fitzgerald, E. M., 2002, The presence of Ca^{2+} channel beta subunit is required for mitogen-activated protein kinase (MAPK)-dependent modulation of α1B Ca^{2+} channels in COS-7 cells, *J Physiol* **543**:425-437.

Fleming, K. M. and Mogul, D. J., 1997, Adenosine A3 receptors potentiate hippocampal calcium current by a PKA-dependent/PKC-independent pathway, *Neuropharmacology* **36**:353-362.

Foehring, R. C., 1996, Serotonin modulates N- and P-type calcium currents in neocortical pyramidal neurons via a membrane-delimited pathway, *J Neurophysiol* **75**:648-659.

Forscher, P., Oxford, G. S., and Schulz, D., 1986, Noradrenaline modulates calcium channels in avian dorsal root ganglion cells through tight receptor-channel coupling, *J Physiol* **379**:131-144.

Foucart, S., Bleakman, D., Bindokas, V. P., and Miller, R. J., 1993, Neuropeptide Y and pancreatic polypeptide reduce calcium currents in acutely dissociated neurons from adult rat superior cervical ganglia, *J Pharmacol Exp Ther* **265**:903-909.

Frielle, T., Caron, M. G., and Lefkowitz, R. J., 1989, Properties of the β1- and β2-adrenergic receptor subtypes revealed by molecular cloning, *Clin Chem* **35**:721-725.

Fujita, Y., Mynlieff, M., Dirksen, R. T., Kim, M. S., Niidome, T., Nakai, J., Friedrich, T., Iwabe, N., Miyata, T., Furuichi, T., and et al., 1993, Primary structure and functional expression of the ω-conotoxin-sensitive N-type calcium channel from rabbit brain, *Neuron* **10**:585-598.

Fukuda, K., Kato, S., and Mori, K., 1995a, Location of regions of the opioid receptor involved in selective agonist binding, *J Biol Chem* **270**:6702-6709.

Fukuda, K., Terasako, K., Kato, S., and Mori, K., 1995b, Identification of the amino acid residues involved in selective agonist binding in the first extracellular loop of the δ- and μ-opioid receptors, *FEBS Lett* **373**:177-181.

Gallacher, D. V., Hanley, M. R., Petersen, O. H., Roberts, M. L., Squire-Pollard, L. G., and Yule, D. I., 1990, Substance P and bombesin elevate cytosolic Ca^{2+} by different molecular mechanisms in a rat pancreatic acinar cell line, *J Physiol* **426**:193-207.

Gandia, L., Garcia, A. G., and Morad, M., 1993, ATP modulation of calcium channels in chromaffin cells, *J Physiol* **470**:55-72.

Gao, T., Yatani, A., Dell'Acqua, M. L., Sako, H., Green, S. A., Dascal, N., Scott, J. D., and Hosey, M. M., 1997, cAMP-dependent regulation of cardiac L-type Ca^{2+} channels requires membrane targeting of PKA and phosphorylation of channel subunits, *Neuron* **19**:185-196.

Garcia, D. E., Li, B., Garcia-Ferreiro, R. E., Hernandez-Ochoa, E. O., Yan, K., Gautam, N., Catterall, W. A., Mackie, K., and Hille, B., 1998, G-protein β-subunit specificity in the fast membrane-delimited inhibition of Ca^{2+} channels, *J Neurosci* **18**:9163-9170.

George, S. R., Fan, T., Xie, Z., Tse, R., Tam, V., Varghese, G., and O'Dowd, B. F., 2000, Oligomerization of μ- and δ-opioid receptors. Generation of novel functional properties, *J Biol Chem* **275**:26128-26135.

Gether, U., 2000, Uncovering molecular mechanisms involved in activation of G protein-coupled receptors, *Endocr Rev* **21**:90-113.

Gilbert, R., Ryan, J. S., Horackova, M., Smith, F. M., and Kelly, M. E., 1998, Actions of substance P on membrane potential and ionic currents in guinea pig stellate ganglion neurons, *Am J Physiol* **274**:C892-903.

Glick, J. L., Meigs, T. E., and Casey, P. J., 2000, in *GTPases*, A. Hall, ed., Oxford University Press, New York, p. 35-66.

Golard, A. and Siegelbaum, S. A., 1993, Kinetic basis for the voltage-dependent inhibition of N-type calcium current by somatostatin and norepinephrine in chick sympathetic neurons, *J Neurosci* **13**:3884-3894.

Haley, J. E., Delmas, P., Offermanns, S., Abogadie, F. C., Simon, M. I., Buckley, N. J., and Brown, D. A., 2000, Muscarinic inhibition of calcium current and M current in Gαq-deficient mice, *J Neurosci* **20**:3973-3979.

Hamid, J., Nelson, D., Spaetgens, R., Dubel, S. J., Snutch, T. P., and Zamponi, G. W., 1999, Identification of an integration center for cross-talk between protein kinase C and G protein modulation of N-type calcium channels, *J Biol Chem* **274**:6195-6202.

Hamm, H. E., 1998, The many faces of G protein signaling, *J Biol Chem* **273**:669-672.

Hartzell, H. C., Mery, P. F., Fischmeister, R., and Szabo, G., 1991, Sympathetic regulation of cardiac calcium current is due exclusively to cAMP-dependent phosphorylation, *Nature* **351**:573-576.

Hebert, T. E., Moffett, S., Morello, J. P., Loisel, T. P., Bichet, D. G., Barret, C., and Bouvier, M., 1996, A peptide derived from a β2-adrenergic receptor transmembrane domain inhibits both receptor dimerization and activation, *J Biol Chem* **271**:16384-16392.

Hebert, T. E., Loisel, T. P., Adam, L., Ethier, N., Onge, S. S., and Bouvier, M., 1998, Functional rescue of a constitutively desensitized β2AR through receptor dimerization, *Biochem J* **330 (Pt 1)**:287-293.

Herlitze, S., Garcia, D. E., Mackie, K., Hille, B., Scheuer, T., and Catterall, W. A., 1996, Modulation of Ca^{2+} channels by G-protein βγ subunits, *Nature* **380**:258-262.

Herlitze, S., Hockerman, G. H., Scheuer, T., and Catterall, W. A., 1997, Molecular determinants of inactivation and G protein modulation in the intracellular loop connecting domains I and II of the calcium channel α1A subunit, *Proc Natl Acad Sci U S A* **94**:1512-1516.

Hernandez-Lopez, S., Tkatch, T., Perez-Garci, E., Galarraga, E., Bargas, J., Hamm, H., and Surmeier, D. J., 2000, D2 dopamine receptors in striatal medium spiny neurons reduce L-type Ca^{2+} currents and excitability via a novel $PLC_{\beta}1$-IP3-calcineurin-signaling cascade, *J Neurosci* **20**:8987-8995.

Hescheler, J., Rosenthal, W., Trautwein, W., and Schultz, G., 1987, The GTP-binding protein, Go, regulates neuronal calcium channels, *Nature* **325**:445-447.

Hille, B., 1994, Modulation of ion-channel function by G-protein-coupled receptors, *Trends Neurosci* **17**:531-536.

Holst, B., Hastrup, H., Raffetseder, U., Martini, L., and Schwartz, T. W., 2001, Two active molecular phenotypes of the tachykinin NK1 receptor revealed by G-protein fusions and mutagenesis, *J Biol Chem* **276**:19793-19799.

Holz, G. G., Rane, S. G., and Dunlap, K., 1986, GTP-binding proteins mediate transmitter inhibition of voltage-dependent calcium channels, *Nature* **319**:670-672.

Huang, C. C., Wang, S. J., and Gean, P. W., 1998, Selective enhancement of P-type calcium currents by isoproterenol in the rat amygdala, *J Neurosci* **18**:2276-2282.

Huang, P., Li, J., Chen, C., Visiers, I., Weinstein, H., and Liu-Chen, L. Y., 2001, Functional role of a conserved motif in TM6 of the rat μ opioid receptor: constitutively active and inactive receptors result from substitutions of Thr6.34(279) with Lys and Asp, *Biochemistry* **40**:13501-13509.

Ikeda, S. R. and Schofield, G. G., 1989, Somatostatin blocks a calcium current in rat sympathetic ganglion neurones, *J Physiol* **409**:221-240.

Ikeda, S. R., 1992, Prostaglandin modulation of Ca^{2+} channels in rat sympathetic neurones is mediated by guanine nucleotide binding proteins, *J Physiol* **458**:339-359.

Ikeda, S. R., Lovinger, D. M., McCool, B. A., and Lewis, D. L., 1995, Heterologous expression of metabotropic glutamate receptors in adult rat sympathetic neurons: subtype-specific coupling to ion channels, *Neuron* **14**:1029-1038.

Ikeda, S. R., 1996, Voltage-dependent modulation of N-type calcium channels by G-protein βγ subunits, *Nature* **380**:255-258.

Ikegami, H., Endoh, T., and Suzuki, T., 2001, Angiotensin II-induced inhibition of calcium currents in hamster submandibular ganglion neurons, *Neurosci Res* **41**:227-232.

Ito, Y., Murai, Y., Ishibashi, H., Onoue, H., and Akaike, N., 2000, The prostaglandin E series modulates high-voltage-activated calcium channels probably through the EP3 receptor in rat paratracheal ganglia, *Neuropharmacology* **39**:181-190.

Jarvis, S. E., Magga, J. M., Beedle, A. M., Braun, J. E., and Zamponi, G. W., 2000, G protein modulation of N-type calcium channels is facilitated by physical interactions between syntaxin 1A and Gβγ, *J Biol Chem* **275**:6388-6394.

Jarvis, S. E. and Zamponi, G. W., 2001, Distinct molecular determinants govern syntaxin 1A-mediated inactivation and G-protein inhibition of N-type calcium channels, *J Neurosci* **21**:2939-2948.

Jarvis, S. E., Barr, W., Feng, Z. P., Hamid, J., and Zamponi, G. W., 2002, Molecular determinants of syntaxin 1 modulation of N-type calcium channels, *J Biol Chem* **277**:44399-44407.

Jeong, S. W. and Wurster, R. D., 1997, Muscarinic receptor activation modulates Ca^{2+} channels in rat intracardiac neurons via a PTX- and voltage-sensitive pathway, *J Neurophysiol* **78**:1476-1490.

Jeong, S. W. and Ikeda, S. R., 2000, Effect of G protein heterotrimer composition on coupling of neurotransmitter receptors to N-type Ca^{2+} channel modulation in sympathetic neurons, *Proc Natl Acad Sci U S A* **97**:907-912.

Johnson, B. D., Brousal, J. P., Peterson, B. Z., Gallombardo, P. A., Hockerman, G. H., Lai, Y., Scheuer, T., and Catterall, W. A., 1997, Modulation of the cloned skeletal muscle L-type Ca^{2+} channel by anchored cAMP-dependent protein kinase, *J Neurosci* **17**:1243-1255.

Jones, K. A., Borowsky, B., Tamm, J. A., Craig, D. A., Durkin, M. M., Dai, M., Yao, W. J., Johnson, M., Gunwaldsen, C., Huang, L. Y., Tang, C., Shen, Q., Salon, J. A., Morse, K., Laz, T., Smith, K. E.,

Nagarathnam, D., Noble, S. A., Branchek, T. A., and Gerald, C., 1998, $GABA_B$ receptors function as a heteromeric assembly of the subunits $GABA_BR1$ and $GABA_BR2$, *Nature* **396**:674-679.

Jones, L. P., Patil, P. G., Snutch, T. P., and Yue, D. T., 1997, G-protein modulation of N-type calcium channel gating current in human embryonic kidney cells (HEK 293), *J Physiol* **498 (Pt 3)**:601-610.

Jordan, B. A. and Devi, L. A., 1999, G-protein-coupled receptor heterodimerization modulates receptor function, *Nature* **399**:697-700.

Jordan, B. A., Trapaidze, N., Gomes, I., Nivarthi, R., and Devi, L. A., 2001, Oligomerization of opioid receptors with β2-adrenergic receptors: a role in trafficking and mitogen-activated protein kinase activation, *Proc Natl Acad Sci U S A* **98**:343-348.

Kammermeier, P. J., Ruiz-Velasco, V., and Ikeda, S. R., 2000, A voltage-independent calcium current inhibitory pathway activated by muscarinic agonists in rat sympathetic neurons requires both Gαq/11 and Gβγ, *J Neurosci* **20**:5623-5629.

Kasai, H. and Aosaki, T., 1989, Modulation of Ca-channel current by an adenosine analog mediated by a GTP-binding protein in chick sensory neurons, *Pflugers Arch* **414**:145-149.

Kaupmann, K., Huggel, K., Heid, J., Flor, P. J., Bischoff, S., Mickel, S. J., McMaster, G., Angst, C., Bittiger, H., Froestl, W., and Bettler, B., 1997, Expression cloning of $GABA_B$ receptors uncovers similarity to metabotropic glutamate receptors, *Nature* **386**:239-246.

King, B. F., Townsend-Nicholson, A., and Burnstock, G., 1998, Metabotropic receptors for ATP and UTP: exploring the correspondence between native and recombinant nucleotide receptors, *Trends Pharmacol Sci* **19**:506-514.

King, P. J., Widdowson, P. S., Doods, H. N., and Williams, G., 1999, Regulation of neuropeptide Y release by neuropeptide Y receptor ligands and calcium channel antagonists in hypothalamic slices, *J Neurochem* **73**:641-646.

Kleuss, C., Hescheler, J., Ewel, C., Rosenthal, W., Schultz, G., and Wittig, B., 1991, Assignment of G-protein subtypes to specific receptors inducing inhibition of calcium currents, *Nature* **353**:43-48.

Klugbauer, N., Lacinova, L., Marais, E., Hobom, M., and Hofmann, F., 1999, Molecular diversity of the calcium channel α2δ subunit, *J Neurosci* **19**:684-691.

Knoflach, F., Reinscheid, R. K., Civelli, O., and Kemp, J. A., 1996, Modulation of voltage-gated calcium channels by orphanin FQ in freshly dissociated hippocampal neurons, *J Neurosci* **16**:6657-6664.

Kong, H., Raynor, K., Yasuda, K., Moe, S. T., Portoghese, P. S., Bell, G. I., and Reisine, T., 1993, A single residue, aspartic acid 95, in the δ opioid receptor specifies selective high affinity agonist binding, *J Biol Chem* **268**:23055-23058.

Kong, H., Raynor, K., Yano, H., Takeda, J., Bell, G. I., and Reisine, T., 1994, Agonists and antagonists bind to different domains of the cloned κ opioid receptor, *Proc Natl Acad Sci U S A* **91**:8042-8046.

Kovoor, A., Celver, J., Abdryashitov, R. I., Chavkin, C., and Gurevich, V. V., 1999, Targeted construction of phosphorylation-independent β-arrestin mutants with constitutive activity in cells, *J Biol Chem* **274**:6831-6834.

Kuzhikandathil, E. V. and Oxford, G. S., 1999, Activation of human D3 dopamine receptor inhibits P/Q-type calcium channels and secretory activity in AtT-20 cells, *J Neurosci* **19**:1698-1707.

Larsson, K. P., Olsen, U. B., and Hansen, A. J., 2000, Nociceptin is a potent inhibitor of N-type Ca^{2+} channels in rat sympathetic ganglion neurons, *Neurosci Lett* **296**:121-124.

Laugwitz, K. L., Offermanns, S., Spicher, K., and Schultz, G., 1993, μ and δ opioid receptors differentially couple to G protein subtypes in membranes of human neuroblastoma SH-SY5Y cells, *Neuron* **10**:233-242.

Laugwitz, K. L., Allgeier, A., Offermanns, S., Spicher, K., Van Sande, J., Dumont, J. E., and Schultz, G., 1996, The human thyrotropin receptor: a heptahelical receptor capable of stimulating members of all four G protein families, *Proc Natl Acad Sci U S A* **93**:116-120.

Law, P. Y., Wong, Y. H., and Loh, H. H., 1999, Mutational analysis of the structure and function of opioid receptors, *Biopolymers* **51**:440-455.

Law, S. F. and Reisine, T., 1997, Changes in the association of G protein subunits with the cloned mouse δ opioid receptor on agonist stimulation, *J Pharmacol Exp Ther* **281**:1476-1486.

Lee, F. S. and Chao, M. V., 2001, Activation of Trk neurotrophin receptors in the absence of neurotrophins, *Proc Natl Acad Sci U S A* **98**:3555-3560.

Lee, J. H., Daud, A. N., Cribbs, L. L., Lacerda, A. E., Pereverzev, A., Klockner, U., Schneider, T., and Perez-Reyes, E., 1999, Cloning and expression of a novel member of the low voltage-activated T-type calcium channel family, *J Neurosci* **19**:1912-1921.

Levi, R. C. and Alloatti, G., 1988, Histamine modulates calcium current in guinea pig ventricular myocytes, *J Pharmacol Exp Ther* **246**:377-383.

Li, J. G., Luo, L. Y., Krupnick, J. G., Benovic, J. L., and Liu-Chen, L. Y., 1999, U50,488H-induced internalization of the human kappa opioid receptor involves a β-arrestin- and dynamin-dependent mechanism. Kappa receptor internalization is not required for mitogen-activated protein kinase activation, *J Biol Chem* **274**:12087-12094.

Liang, B. T. and Morley, J. F., 1996, A new cyclic AMP-independent, Gs-mediated stimulatory mechanism via the adenosine A2a receptor in the intact cardiac cell, *J Biol Chem* **271**:18678-18685.

Lin, R. C. and Scheller, R. H., 2000, Mechanisms of synaptic vesicle exocytosis, *Annu Rev Cell Dev Biol* **16**:19-49.

Linn, C. L. and Gafka, A. C., 1999, Activation of metabotropic glutamate receptors modulates the voltage-gated sustained calcium current in a teleost horizontal cell, *J Neurophysiol* **81**:425-434.

Lipscombe, D., Kongsamut, S., and Tsien, R. W., 1989, α-adrenergic inhibition of sympathetic neurotransmitter release mediated by modulation of N-type calcium-channel gating, *Nature* **340**:639-642.

Liu, J., Conklin, B. R., Blin, N., Yun, J., and Wess, J., 1995, Identification of a receptor/G-protein contact site critical for signaling specificity and G-protein activation, *Proc Natl Acad Sci U S A* **92**:11642-11646.

Liu, J. and Wess, J., 1996, Different single receptor domains determine the distinct G protein coupling profiles of members of the vasopressin receptor family, *J Biol Chem* **271**:8772-8778.

Liu, Q. Y., Karpinski, E., and Pang, P. K., 1994, L-channel modulation by α-1 adrenoceptor activation in neonatal rat ventricular cells: intracellular mechanisms, *J Pharmacol Exp Ther* **271**:944-951.

Lledo, P. M., Homburger, V., Bockaert, J., and Vincent, J. D., 1992, Differential G protein-mediated coupling of D2 dopamine receptors to K^+ and Ca^{2+} currents in rat anterior pituitary cells, *Neuron* **8**:455-463.

Lu, Q., AtKisson, M. S., Jarvis, S. E., Feng, Z. P., Zamponi, G. W., and Dunlap, K., 2001, Syntaxin 1A supports voltage-dependent inhibition of α1B Ca^{2+} channels by Gβγ in chick sensory neurons, *J Neurosci* **21**:2949-2957.

Lu, Y. F., Xu, H., Liu-Chen, L. Y., Chen, C., Partilla, J. S., Brine, G. A., Carroll, F. I., Rice, K. C., Lai, J., Porreca, F., Sadee, W., and Rothman, R. B., 1998, Opioid peptide receptor studies. 7. The methylfentanyl congener RTI-4614-4 and its four enantiomers bind to different domains of the rat μ opioid receptor, *Synapse* **28**:117-124.

Luttrell, L. M., Ostrowski, J., Cotecchia, S., Kendall, H., and Lefkowitz, R. J., 1993, Antagonism of catecholamine receptor signaling by expression of cytoplasmic domains of the receptors, *Science* **259**:1453-1457.

Mackie, K., Devane, W. A., and Hille, B., 1993, Anandamide, an endogenous cannabinoid, inhibits calcium currents as a partial agonist in N18 neuroblastoma cells, *Mol Pharmacol* **44**:498-503.

Mackie, K., Lai, Y., Westenbroek, R., and Mitchell, R., 1995, Cannabinoids activate an inwardly rectifying potassium conductance and inhibit Q-type calcium currents in AtT20 cells transfected with rat brain cannabinoid receptor, *J Neurosci* **15**:6552-6561.

Macrez, N., Morel, J. L., Kalkbrenner, F., Viard, P., Schultz, G., and Mironneau, J., 1997, A betagamma dimer derived from G13 transduces the angiotensin AT1 receptor signal to stimulation of Ca2+ channels in rat portal vein myocytes, *J Biol Chem* **272**:23180-23185.

Macrez-Lepretre, N., Morel, J. L., and Mironneau, J., 1996, Angiotensin II-mediated activation of L-type calcium channels involves phosphatidylinositol hydrolysis-independent activation of protein kinase C in rat portal vein myocytes, *J Pharmacol Exp Ther* **278**:468-475.

Magga, J. M., Jarvis, S. E., Arnot, M. I., Zamponi, G. W., and Braun, J. E., 2000, Cysteine string protein regulates G protein modulation of N-type calcium channels, *Neuron* **28**:195-204.

Mandyam, C. D., Thakker, D. R., Christensen, J. L., and Standifer, K. M., 2002, Orphanin FQ/nociceptin-mediated desensitization of opioid receptor-like 1 receptor and mu opioid receptors involves protein kinase C: a molecular mechanism for heterologous cross-talk, *J Pharmacol Exp Ther* **302**:502-509.

Marchetti, C., Carbone, E., and Lux, H. D., 1986, Effects of dopamine and noradrenaline on Ca channels of cultured sensory and sympathetic neurons of chick, *Pflugers Arch* **406**:104-111.

Margetà-Mitrovic, M., Grigg, J. J., Koyano, K., Nakajima, Y., and Nakajima, S., 1997, Neurotensin and substance P inhibit low- and high-voltage-activated Ca^{2+} channels in cultured newborn rat nucleus basalis neurons, *J Neurophysiol* **78**:1341-1352.

Mark, M. D., Wittemann, S., and Herlitze, S., 2000, G protein modulation of recombinant P/Q-type calcium channels by regulators of G protein signalling proteins, *J Physiol* **528 Pt 1**:65-77.

Marks, A. R., 1992, Calcium channels expressed in vascular smooth muscle, *Circulation* **86**:III61-67.

McCleskey, E. W., 1994, Calcium channels: cellular roles and molecular mechanisms, *Curr Opin Neurobiol* 4:304-312.

McCool, B. A., Pin, J. P., Brust, P. F., Harpold, M. M., and Lovinger, D. M., 1996, Functional coupling of rat group II metabotropic glutamate receptors to an ω-conotoxin GVIA-sensitive calcium channel in human embryonic kidney 293 cells, *Mol Pharmacol* **50**:912-922.

McCool, B. A., Pin, J. P., Harpold, M. M., Brust, P. F., Stauderman, K. A., and Lovinger, D. M., 1998, Rat group I metabotropic glutamate receptors inhibit neuronal Ca^{2+} channels via multiple signal transduction pathways in HEK 293 cells, *J Neurophysiol* **79**:379-391.

McCool, B. A. and Farroni, J. S., 2001, A1 adenosine receptors inhibit multiple voltage-gated Ca^{2+} channel subtypes in acutely isolated rat basolateral amygdala neurons, *Br J Pharmacol* **132**:879-888.

McRory, J. E., Santi, C. M., Hamming, K. S., Mezeyova, J., Sutton, K. G., Baillie, D. L., Stea, A., and Snutch, T. P., 2001, Molecular and functional characterization of a family of rat brain T-type calcium channels, *J Biol Chem* **276**:3999-4011.

Meir, A. and Dolphin, A. C., 2002, Kinetics and Gβγ modulation of $Ca_v2.2$ channels with different auxiliary β subunits, *Pflugers Arch* **444**:263-275.

Melliti, K., Meza, U., and Adams, B., 2000, Muscarinic stimulation of α_{1E} Ca channels is selectively blocked by the effector antagonist function of RGS2 and phospholipase C-β1, *J Neurosci* **20**:7167-7173.

Melliti, K., Meza, U., and Adams, B. A., 2001, RGS2 blocks slow muscarinic inhibition of N-type Ca^{2+} channels reconstituted in a human cell line, *J Physiol* **532**:337-347.

Meng, F., Hoversten, M. T., Thompson, R. C., Taylor, L., Watson, S. J., and Akil, H., 1995, A chimeric study of the molecular basis of affinity and selectivity of the κ and the δ opioid receptors. Potential role of extracellular domains, *J Biol Chem* **270**:12730-12736.

Mikami, A., Imoto, K., Tanabe, T., Niidome, T., Mori, Y., Takeshima, H., Narumiya, S., and Numa, S., 1989, Primary structure and functional expression of the cardiac dihydropyridine-sensitive calcium channel, *Nature* **340**:230-233.

Minami, M., Onogi, T., Nakagawa, T., Katao, Y., Aoki, Y., Katsumata, S., and Satoh, M., 1995, DAMGO, a μ-opioid receptor selective ligand, distinguishes between μ-and κ-opioid receptors at a different region from that for the distinction between μ- and δ-opioid receptors, *FEBS Lett* **364**:23-27.

Mintz, I. M. and Bean, B. P., 1993, $GABA_B$ receptor inhibition of P-type Ca^{2+} channels in central neurons, *Neuron* **10**:889-898.

Mogul, D. J., Adams, M. E., and Fox, A. P., 1993, Differential activation of adenosine receptors decreases N-type but potentiates P-type Ca^{2+} current in hippocampal CA3 neurons, *Neuron* **10**;327-334.

Monteil, A., Chemin, J., Bourinet, E., Mennessier, G., Lory, P., and Nargeot, J., 2000a, Molecular and functional properties of the human α_{1G} subunit that forms T-type calcium channels, *J Biol Chem* **275**:6090-6100.

Monteil, A., Chemin, J., Leuranguer, V., Altier, C., Mennessier, G., Bourinet, E., Lory, P., and Nargeot, J., 2000b, Specific properties of T-type calcium channels generated by the human α1I subunit, *J Biol Chem* **275**:16530-16535.

Mori, Y., Friedrich, T., Kim, M. S., Mikami, A., Nakai, J., Ruth, P., Bosse, E., Hofmann, F., Flockerzi, V., Furuichi, T., Mikoshiba, K., Imoto, K., Tanabe, T., and Numa, S., 1991, Primary structure and functional expression from complementary DNA of a brain calcium channel, *Nature* **350**:398-402.

Morikawa, H., Fukuda, K., Mima, H., Shoda, T., Kato, S., and Mori, K., 1998a, Desensitization and resensitization of δ-opioid receptor-mediated Ca^{2+} channel inhibition in NG108-15 cells, *Br J Pharmacol* **123**:1111-1118.

Morikawa, H., Fukuda, K., Mima, H., Shoda, T., Kato, S., and Mori, K., 1998b, Nociceptin receptor-mediated Ca^{2+} channel inhibition and its desensitization in NG108-15 cells, *Eur J Pharmacol* **351**:247-252.

Morikawa, H., Mima, H., Uga, H., Shoda, T., and Fukuda, K., 1999, Opioid potentiation of N-type Ca^{2+} channel currents via pertussis-toxin-sensitive G proteins in NG108-15 cells, *Pflugers Arch* **438**:423-426.

Mumby, S. M., 2000, in *GTPases*, A. Hall, ed., Oxford University Press, New York, p. 1-34.

O'Hara, P. J., Sheppard, P. O., Thogersen, H., Venezia, D., Haldeman, B. A., McGrane, V., Houamed, K. M., Thomsen, C., Gilbert, T. L., and Mulvihill, E. R., 1993, The ligand-binding domain in metabotropic glutamate receptors is related to bacterial periplasmic binding proteins, *Neuron* **11**:41-52.

Ouadid, H., Seguin, J., Dumuis, A., Bockaert, J., and Nargeot, J., 1992, Serotonin increases calcium current in human atrial myocytes via the newly described 5-hydroxytryptamine4 receptors, *Mol Pharmacol* **41**:346-351.

Page, K. M., Stephens, G. J., Berrow, N. S., and Dolphin, A. C., 1997, The intracellular loop between domains I and II of the B-type calcium channel confers aspects of G-protein sensitivity to the E-type calcium channel, *J Neurosci* **17**:1330-1338.

Page, K. M., Canti, C., Stephens, G. J., Berrow, N. S., and Dolphin, A. C., 1998, Identification of the amino terminus of neuronal Ca^{2+} channel α1 subunits α1B and α1E as an essential determinant of G-protein modulation, *J Neurosci* **18**:4815-4824.

Pan, X., Ikeda, S. R., and Lewis, D. L., 1996, Rat brain cannabinoid receptor modulates N-type Ca^{2+} channels in a neuronal expression system, *Mol Pharmacol* **49**:707-714.

Park, T. J., Chung, S., Han, M. K., Kim, U. H., and Kim, K. T., 1998, Inhibition of voltage-sensitive calcium channels by the A2A adenosine receptor in PC12 cells, *J Neurochem* **71**:1251-1260.

Paterlini, G., Portoghese, P. S., and Ferguson, D. M., 1997, Molecular simulation of dynorphin A-(1-10) binding to extracellular loop 2 of the κ-opioid receptor. A model for receptor activation, *J Med Chem* **40**:3254-3262.

Patil, P. G., de Leon, M., Reed, R. R., Dubel, S., Snutch, T. P., and Yue, D. T., 1996, Elementary events underlying voltage-dependent G-protein inhibition of N-type calcium channels, *Biophys J* **71**:2509-2521.

Penington, N. J., Kelly, J. S., and Fox, A. P., 1991, A study of the mechanism of Ca^{2+} current inhibition produced by serotonin in rat dorsal raphe neurons, *J Neurosci* **11**:3594-3609.

Pepin, M. C., Yue, S. Y., Roberts, E., Wahlestedt, C., and Walker, P., 1997, Novel "restoration of function" mutagenesis strategy to identify amino acids of the δ-opioid receptor involved in ligand binding, *J Biol Chem* **272**:9260-9267.

Perez-Reyes, E., Cribbs, L. L., Daud, A., Lacerda, A. E., Barclay, J., Williamson, M. P., Fox, M., Rees, M., and Lee, J. H., 1998, Molecular characterization of a neuronal low-voltage-activated T-type calcium channel, *Nature* **391**:896-900.

Perroy, J., Prezeau, L., De Waard, M., Shigemoto, R., Bockaert, J., and Fagni, L., 2000, Selective blockade of P/Q-type calcium channels by the metabotropic glutamate receptor type 7 involves a phospholipase C pathway in neurons, *J Neurosci* **20**:7896-7904.

Perroy, J., El Far, O., Bertaso, F., Pin, J. P., Betz, H., Bockaert, J., and Fagni, L., 2002, PICK1 is required for the control of synaptic transmission by the metabotropic glutamate receptor 7, *Embo J* **21**:2990-2999.

Pfeiffer, M., Koch, T., Schroder, H., Klutzny, M., Kirscht, S., Kreienkamp, H. J., Hollt, V., and Schulz, S., 2001, Homo- and heterodimerization of somatostatin receptor subtypes. Inactivation of sst_3 receptor function by heterodimerization with sst_{2A}, *J Biol Chem* **276**:14027-14036.

Pfeiffer, M., Koch, T., Schroder, H., Laugsch, M., Hollt, V., and Schulz, S., 2002, Heterodimerization of somatostatin and opioid receptors cross-modulates phosphorylation, internalization, and desensitization, *J Biol Chem* **277**:19762-19772.

Piros, E. T., Hales, T. G., and Evans, C. J., 1996a, Functional analysis of cloned opioid receptors in transfected cell lines, *Neurochem Res* **21**:1277-1285.

Piros, E. T., Prather, P. L., Law, P. Y., Evans, C. J., and Hales, T. G., 1996b, Voltage-dependent inhibition of Ca^{2+} channels in GH3 cells by cloned μ- and δ-opioid receptors, *Mol Pharmacol* **50**:947-956.

Pisani, A., Bonsi, P., Catania, M. V., Giuffrida, R., Morari, M., Marti, M., Centonze, D., Bernardi, G., Kingston, A. E., and Calabresi, P., 2002, Metabotropic glutamate 2 receptors modulate synaptic inputs and calcium signals in striatal cholinergic interneurons, *J Neurosci* **22**:6176-6185.

Pogozheva, I. D., Lomize, A. L., and Mosberg, H. I., 1998, Opioid receptor three-dimensional structures from distance geometry calculations with hydrogen bonding constraints, *Biophys J* **75**:612-634.

Polo-Parada, L. and Pilar, G., 1999, κ- and μ-opioids reverse the somatostatin inhibition of Ca^{2+} currents in ciliary and dorsal root ganglion neurons, *J Neurosci* **19**:5213-5227.

Posner, B. A., Gilman, A. G., and Harris, B. A., 1999, Regulators of G protein signaling 6 and 7. Purification of complexes with Gβ5 and assessment of their effects on G protein-mediated signaling pathways, *J Biol Chem* **274**:31087-31093.

Potenza, M. N., Gold, S. J., Roby-Shemkowitz, A., Lerner, M. R., and Nestler, E. J., 1999, Effects of regulators of G protein-signaling proteins on the functional response of the μ-opioid receptor in a melanophore-based assay, *J Pharmacol Exp Ther* **291**:482-491.

Prather, P. L., McGinn, T. M., Claude, P. A., Liu-Chen, L. Y., Loh, H. H., and Law, P. Y., 1995, Properties of a κ-opioid receptor expressed in CHO cells: interaction with multiple G-proteins is not specific for any individual G α subunit and is similar to that of other opioid receptors, *Brain Res Mol Brain Res* **29**:336-346.

Qian, J., Colmers, W. F., and Saggau, P., 1997, Inhibition of synaptic transmission by neuropeptide Y in rat hippocampal area CA1: modulation of presynaptic Ca^{2+} entry, *J Neurosci* **17**:8169-8177.

Qin, N., Platano, D., Olcese, R., Stefani, E., and Birnbaumer, L., 1997, Direct interaction of gβγ with a C-terminal gbetagamma-binding domain of the Ca^{2+} channel α1 subunit is responsible for channel inhibition by G protein-coupled receptors, *Proc Natl Acad Sci U S A* **94**:8866-8871.

Qin, N., Yagel, S., Momplaisir, M. L., Codd, E. E., and D'Andrea, M. R., 2002, Molecular cloning and characterization of the human voltage-gated calcium channel $\alpha_2\delta$-4 subunit, *Mol Pharmacol* **62**:485-496.

Rahman, Z., Gold, S. J., Potenza, M. N., Cowan, C. W., Ni, Y. G., He, W., Wensel, T. G., and Nestler, E. J., 1999, Cloning and characterization of RGS9-2: a striatal-enriched alternatively spliced product of the RGS9 gene, *J Neurosci* **19**:2016-2026.

Raynor, K., Kong, H., Hines, J., Kong, G., Benovic, J., Yasuda, K., Bell, G. I., and Reisine, T., 1994, Molecular mechanisms of agonist-induced desensitization of the cloned mouse κ opioid receptor, *J Pharmacol Exp Ther* **270**:1381-1386.

Rettig, J., Sheng, Z. H., Kim, D. K., Hodson, C. D., Snutch, T. P., and Catterall, W. A., 1996, Isoform-specific interaction of the α1A subunits of brain Ca^{2+} channels with the presynaptic proteins syntaxin and SNAP-25, *Proc Natl Acad Sci U S A* **93**:7363-7368.

Robbins, M. J., Calver, A. R., Filippov, A. K., Hirst, W. D., Russell, R. B., Wood, M. D., Nasir, S., Couve, A., Brown, D. A., Moss, S. J., and Pangalos, M. N., 2001, $GABA_{B2}$ is essential for G-protein coupling of the $GABA_B$ receptor heterodimer, *J Neurosci* **21**:8043-8052.

Roosterman, D., Glassmeier, G., Baumeister, H., Scherubl, H., and Meyerhof, W., 1998, A somatostatin receptor 1 selective ligand inhibits Ca^{2+} currents in rat insulinoma 1046-38 cells, *FEBS Lett* **425**:137-140.

Rosemond, E., Peltekova, V., Naples, M., Thogersen, H., and Hampson, D. R., 2002, Molecular determinants of high affinity binding to group III metabotropic glutamate receptors, *J Biol Chem* **277**:7333-7340.

Ruiz-Velasco, V. and Ikeda, S. R., 2000, Multiple G-protein βγ combinations produce voltage-dependent inhibition of N-type calcium channels in rat superior cervical ganglion neurons, *J Neurosci* **20**:2183-2191.

Ryu, P. D. and Randic, M., 1990, Low- and high-voltage-activated calcium currents in rat spinal dorsal horn neurons, *J Neurophysiol* **63**:273-285.

Salt, T. E., Binns, K. E., Turner, J. P., Gasparini, F., and Kuhn, R., 1999, Antagonism of the mGlu5 agonist 2-chloro-5-hydroxyphenylglycine by the novel selective mGlu5 antagonist 6-methyl-2-(phenylethynyl)-pyridine (MPEP) in the thalamus, *Br J Pharmacol* **127**:1057-1059.

Sather, W. A., Tanabe, T., Zhang, J. F., Mori, Y., Adams, M. E., and Tsien, R. W., 1993, Distinctive biophysical and pharmacological properties of class A (BI) calcium channel α1 subunits, *Neuron* **11**:291-303.

Saugstad, J. A., Marino, M. J., Folk, J. A., Hepler, J. R., and Conn, P. J., 1998, RGS4 inhibits signaling by group I metabotropic glutamate receptors, *J Neurosci* **18**:905-913.

Schmidt, P., Holsboer, F., and Spengler, D., 2001, $Beta_2$-adrenergic receptors potentiate glucocorticoid receptor transactivation via G protein beta gamma-subunits and the phosphoinositide 3-kinase pathway, *Mol Endocrinol* **15**:553-564.

Scholze, A., Plant, T. D., Dolphin, A. C., and Nurnberg, B., 2001, Functional expression and characterization of a voltage-gated $Ca_V1.3$ (α_{1D}) calcium channel subunit from an insulin-secreting cell line, *Mol Endocrinol* **15**:1211-1221.

Schroeder, J. E., Fischbach, P. S., Zheng, D., and McCleskey, E. W., 1991, Activation of μ opioid receptors inhibits transient high- and low-threshold Ca^{2+} currents, but spares a sustained current, *Neuron* **6**:13-20.

Schulz, R., Wehmeyer, A., Murphy, J., and Schulz, K., 1999, Phosducin, β-arrestin and opioid receptor migration, *Eur J Pharmacol* **375**:349-357.

Shapiro, M. S. and Hille, B., 1993, Substance P and somatostatin inhibit calcium channels in rat sympathetic neurons via different G protein pathways, *Neuron* **10**:11-20.

Shapiro, M. S., Wollmuth, L. P., and Hille, B., 1994, Angiotensin II inhibits calcium and M current channels in rat sympathetic neurons via G proteins, *Neuron* **12**:1319-1329.

Shen, W. and Slaughter, M. M., 1999, Metabotropic GABA receptors facilitate L-type and inhibit N-type calcium channels in single salamander retinal neurons, *J Physiol* **516 (Pt 3)**:711-718.

Sheng, Z. H., Rettig, J., Takahashi, M., and Catterall, W. A., 1994, Identification of a syntaxin-binding site on N-type calcium channels, *Neuron* **13**:1303-1313.

Simen, A. A. and Miller, R. J., 1998, Structural features determining differential receptor regulation of neuronal Ca channels, *J Neurosci* **18**:3689-3698.

Simen, A. A. and Miller, R. J., 2000, Involvement of regions in domain I in the opioid receptor sensitivity of α_{1B} Ca^{2+} channels, *Mol Pharmacol* **57**:1064-1074.

Simen, A. A., Lee, C. C., Simen, B. B., Bindokas, V. P., and Miller, R. J., 2001, The C terminus of the Ca channel α_{1B} subunit mediates selective inhibition by G-protein-coupled receptors, *J Neurosci* **21**:7587-7597.

Snow, B. E., Krumins, A. M., Brothers, G. M., Lee, S. F., Wall, M. A., Chung, S., Mangion, J., Arya, S., Gilman, A. G., and Siderovski, D. P., 1998, A G protein γ subunit-like domain shared between RGS11 and other RGS proteins specifies binding to Gβ5 subunits, *Proc Natl Acad Sci U S A* **95**:13307-13312.

Snutch, T. P., Tomlinson, W. J., Leonard, J. P., and Gilbert, M. M., 1991, Distinct calcium channels are generated by alternative splicing and are differentially expressed in the mammalian CNS, *Neuron* **7**:45-57.

Soldo, B. L. and Moises, H. C., 1998, μ-opioid receptor activation inhibits N- and P-type Ca^{2+} channel currents in magnocellular neurones of the rat supraoptic nucleus, *J Physiol* **513 (Pt 3)**:787-804.

Soong, T. W., Stea, A., Hodson, C. D., Dubel, S. J., Vincent, S. R., and Snutch, T. P., 1993, Structure and functional expression of a member of the low voltage-activated calcium channel family, *Science* **260**:1133-1136.

Stanley, E. F. and Mirotznik, R. R., 1997, Cleavage of syntaxin prevents G-protein regulation of presynaptic calcium channels, *Nature* **385**:340-343.

Stea, A., Tomlinson, W. J., Soong, T. W., Bourinet, E., Dubel, S. J., Vincent, S. R., and Snutch, T. P., 1994, Localization and functional properties of a rat brain α1A calcium channel reflect similarities to neuronal Q- and P-type channels, *Proc Natl Acad Sci U S A* **91**:10576-10580.

Stea, A., Soong, T. W., and Snutch, T. P., 1995, Determinants of PKC-dependent modulation of a family of neuronal calcium channels, *Neuron* **15**:929-940.

Stefani, A., Spadoni, F., and Bernardi, G., 1998, Group III metabotropic glutamate receptor agonists modulate high voltage-activated Ca^{2+} currents in pyramidal neurons of the adult rat, *Exp Brain Res* **119**:237-244.

Stella, S. L., Jr., Bryson, E. J., and Thoreson, W. B., 2002, A2 adenosine receptors inhibit calcium influx through L-type calcium channels in rod photoreceptors of the salamander retina, *J Neurophysiol* **87**:351-360.

Stephens, G. J., Canti, C., Page, K. M., and Dolphin, A. C., 1998, Role of domain I of neuronal Ca^{2+} channel α1 subunits in G protein modulation, *J Physiol* **509 (Pt 1)**:163-169.

Su, X., Wachtel, R. E., and Gebhart, G. F., 1998, Inhibition of calcium currents in rat colon sensory neurons by K- but not mu- or delta-opioids, *J Neurophysiol* **80**:3112-3119.

Sun, L., Philipson, L. H., and Miller, R. J., 1998, Regulation of K^+ and Ca^{++} channels by a family of neuropeptide Y receptors, *J Pharmacol Exp Ther* **284**:625-632.

Sun, L. and Miller, R. J., 1999, Multiple neuropeptide Y receptors regulate K^+ and Ca^{2+} channels in acutely isolated neurons from the rat arcuate nucleus, *J Neurophysiol* **81**:1391-1403.

Sun, Q. Q. and Dale, N., 1999, G-proteins are involved in 5-HT receptor-mediated modulation of N- and P/Q-but not T-type Ca^{2+} channels, *J Neurosci* **19**:890-899.

Surmeier, D. J., Bargas, J., Hemmings, H. C., Jr., Nairn, A. C., and Greengard, P., 1995, Modulation of calcium currents by a D1 dopaminergic protein kinase/phosphatase cascade in rat neostriatal neurons, *Neuron* **14**:385-397.

Swartz, K. J., 1993, Modulation of Ca^{2+} channels by protein kinase C in rat central and peripheral neurons: disruption of G protein-mediated inhibition, *Neuron* **11**:305-320.

Swartz, K. J., Merritt, A., Bean, B. P., and Lovinger, D. M., 1993, Protein kinase C modulates glutamate receptor inhibition of Ca^{2+} channels and synaptic transmission, *Nature* **361**:165-168.

Takahashi, K., Tsuchida, K., Tanabe, Y., Masu, M., and Nakanishi, S., 1993, Role of the large extracellular domain of metabotropic glutamate receptors in agonist selectivity determination, *J Biol Chem* **268**:19341-19345.

Takahashi, T., Forsythe, I. D., Tsujimoto, T., Barnes-Davies, M., and Onodera, K., 1996, Presynaptic calcium current modulation by a metabotropic glutamate receptor, *Science* **274**:594-597.

Takeshita, Y., Watanabe, T., Sakata, T., Munakata, M., Ishibashi, H., and Akaike, N., 1998, Histamine modulates high-voltage-activated calcium channels in neurons dissociated from the rat tuberomammillary nucleus, *Neuroscience* **87**:797-805.

Tallent, M., Dichter, M. A., Bell, G. I., and Reisine, T., 1994, The cloned kappa opioid receptor couples to an N-type calcium current in undifferentiated PC-12 cells, *Neuroscience* **63**:1033-1040.

Tanaka, K., Shibuya, I., Kabashima, N., Ueta, Y., and Yamashita, H., 1998, Inhibition of voltage-dependent calcium channels by prostaglandin E2 in rat melanotrophs, *Endocrinology* **139**:4801-4810.

Taussig, R., Sanchez, S., Rifo, M., Gilman, A. G., and Belardetti, F., 1992, Inhibition of the ω-conotoxin-sensitive calcium current by distinct G proteins, *Neuron* **8**:799-809.

Thompson, R. C., Mansour, A., Akil, H., and Watson, S. J., 1993, Cloning and pharmacological characterization of a rat μ opioid receptor, *Neuron* **11**:903-913.

Tiaho, F. and Nerbonne, J. M., 1996, VIP and secretin augment cardiac L-type calcium channel currents in isolated adult rat ventricular myocytes, *Pflugers Arch* **432**:821-830.

Tomlinson, W. J., Stea, A., Bourinet, E., Charnet, P., Nargeot, J., and Snutch, T. P., 1993, Functional properties of a neuronal class C L-type calcium channel, *Neuropharmacology* **32**:1117-1126.

Toth, P. T., Shekter, L. R., Ma, G. H., Philipson, L. H., and Miller, R. J., 1996, Selective G-protein regulation of neuronal calcium channels, *J Neurosci* **16**:4617-4624.

Tsunoo, A., Yoshii, M., and Narahashi, T., 1986, Block of calcium channels by enkephalin and somatostatin in neuroblastoma-glioma hybrid NG108-15 cells, *Proc Natl Acad Sci U S A* **83**:9832-9836.

Vartian, N. and Boehm, S., 2001, P2Y receptor-mediated inhibition of voltage-activated Ca^{2+} currents in PC12 cells, *Eur J Neurosci* **13**:899-908.

Wade, S. M., Lim, W. K., Lan, K. L., Chung, D. A., Nanamori, M., and Neubig, R. R., 1999, G_i activator region of α_{2A}-adrenergic receptors: distinct basic residues mediate G_i versus G_s activation, *Mol Pharmacol* **56**:1005-1013.

Walker, D. and De Waard, M., 1998, Subunit interaction sites in voltage-dependent Ca^{2+} channels: role in channel function, *Trends Neurosci* **21**:148-154.

Wang, J. B., Johnson, P. S., Wu, J. M., Wang, W. F., and Uhl, G. R., 1994, Human κ opiate receptor second extracellular loop elevates dynorphin's affinity for human μ/κ chimeras, *J Biol Chem* **269**:25966-25969.

Watson, N., Linder, M. E., Druey, K. M., Kehrl, J. H., and Blumer, K. J., 1996, RGS family members: GTPase-activating proteins for heterotrimeric G-protein α-subunits, *Nature* **383**:172-175.

Weng, G., Li, J., Dingus, J., Hildebrandt, J. D., Weinstein, H., and Iyengar, R., 1996, Gβ subunit interacts with a peptide encoding region 956-982 of adenylyl cyclase 2. Cross-linking of the peptide to free Gβγ but not the heterotrimer, *J Biol Chem* **271**:26445-26448.

Westenbroek, R. E., Hell, J. W., Warner, C., Dubel, S. J., Snutch, T. P., and Catterall, W. A., 1992, Biochemical properties and subcellular distribution of an N-type calcium channel α1 subunit, *Neuron* **9**:1099-1115.

Westenbroek, R. E., Sakurai, T., Elliott, E. M., Hell, J. W., Starr, T. V., Snutch, T. P., and Catterall, W. A., 1995, Immunochemical identification and subcellular distribution of the α1A subunits of brain calcium channels, *J Neurosci* **15**:6403-6418.

Westenbroek, R. E., Hoskins, L., and Catterall, W. A., 1998, Localization of Ca^{2+} channel subtypes on rat spinal motor neurons, interneurons, and nerve terminals, *J Neurosci* **18**:6319-6330.

Wheeler, D. B., Randall, A., and Tsien, R. W., 1994, Roles of N-type and Q-type Ca^{2+} channels in supporting hippocampal synaptic transmission, *Science* **264**:107-111.

Whistler, J. L. and von Zastrow, M., 1998, Morphine-activated opioid receptors elude desensitization by β-arrestin, *Proc Natl Acad Sci U S A* **95**:9914-9919.

White, D. M., 1996, Mechanism of prostaglandin E2-induced substance P release from cultured sensory neurons, *Neuroscience* **70**:561-565.

Wickman, K. and Clapham, D. E., 1995, Ion channel regulation by G proteins, *Physiol Rev* **75**:865-885.

Wiley, J. W., Gross, R. A., and MacDonald, R. L., 1993, Agonists for neuropeptide Y receptor subtypes NPY-1 and NPY-2 have opposite actions on rat nodose neuron calcium currents, *J Neurophysiol* **70**:324-330.

Wilk-Blaszczak, M. A., Singer, W. D., Gutowski, S., Sternweis, P. C., and Belardetti, F., 1994, The G protein G_{13} mediates inhibition of voltage-dependent calcium current by bradykinin, *Neuron* **13**:1215-1224.

Williams, M. E., Brust, P. F., Feldman, D. H., Patthi, S., Simerson, S., Maroufi, A., McCue, A. F., Velicelebi, G., Ellis, S. B., and Harpold, M. M., 1992a, Structure and functional expression of an ω-conotoxin-sensitive human N-type calcium channel, *Science* **257**:389-395.

Williams, M. E., Feldman, D. H., McCue, A. F., Brenner, R., Velicelebi, G., Ellis, S. B., and Harpold, M. M., 1992b, Structure and functional expression of α1, α2, and β subunits of a novel human neuronal calcium channel subtype, *Neuron* **8**:71-84.

Williams, M. E., Marubio, L. M., Deal, C. R., Hans, M., Brust, P. F., Philipson, L. H., Miller, R. J., Johnson, E. C., Harpold, M. M., and Ellis, S. B., 1994, Structure and functional characterization of neuronal α1E calcium channel subtypes, *J Biol Chem* **269**:22347-22357.

Wu, L. G., Borst, J. G., and Sakmann, B., 1998, R-type Ca^{2+} currents evoke transmitter release at a rat central synapse, *Proc Natl Acad Sci U S A* **95**:4720-4725.

Wu, X., Kushwaha, N., Albert, P. R., and Penington, N. J., 2002, A critical protein kinase C phosphorylation site on the 5-HT_{1A} receptor controlling coupling to N-type calcium channels, *J Physiol* **538**:41-51.

Xue, J. C., Chen, C., Zhu, J., Kunapuli, S., DeRiel, J. K., Yu, L., and Liu-Chen, L. Y., 1994, Differential binding domains of peptide and non-peptide ligands in the cloned rat κ opioid receptor, *J Biol Chem* **269**:30195-30199.

Yamada, H., Nakamoto, H., Suzuki, Y., Ito, T., and Aisaka, K., 2002, Pharmacological profiles of a novel opioid receptor-like1 (ORL_1) receptor antagonist, JTC-801, *Br J Pharmacol* **135**:323-332.

Yamamoto, T., Habuchi, Y., Tanaka, H., Suto, F., Morikawa, J., Kashima, K., and Yoshimura, M., 1999, EP receptor-mediated inhibition by prostaglandin E_1 of cardiac L-type Ca_{2+} current of rabbits, *Am J Physiol* **277**:H1369-1374.

Yan, Z., Song, W.-J., and Surmeier, D. J., 1997, D_2 dopamine receptors reduce N-type Ca^{2+} currents in rat neostriatal cholinergic interneurons through a membrane-delimited, protein-kinase-C-insensitive pathway, *Journal of Neurophysiology* **77**:1003-1015.

Yatani, A., Wakamori, M., Niidome, T., Yamamoto, S., Tanaka, I., Mori, Y., Katayama, K., and Green, S., 1995, Stable expression and coupling of cardiac L-type Ca^{2+} channels with β1-adrenoceptors, *Circ Res* **76**:335-342.

Yawo, H. and Chuhma, N., 1993, Preferential inhibition of ω-conotoxin-sensitive presynaptic Ca^{2+} channels by adenosine autoreceptors, *Nature* **365**:256-258.

Yoshinaga, T., Zhang, S., Niidome, T., Hiraoka, M., and Hirano, Y., 1999, Potentiation of recombinant L-type Ca channel currents by α1-adrenoceptors coexpressed in baby hamster kidney (BHK) cells, *Life Sci* **64**:1643-1651.

Yung, L. Y., Joshi, S. A., Chan, R. Y., Chan, J. S., Pei, G., and Wong, Y. H., 1999, GαL1 (Gα14) couples the opioid receptor-like1 receptor to stimulation of phospholipase C, *J Pharmacol Exp Ther* **288**:232-238.

Zamponi, G. W., Bourinet, E., Nelson, D., Nargeot, J., and Snutch, T. P., 1997, Crosstalk between G proteins and protein kinase C mediated by the calcium channel α1 subunit, *Nature* **385**:442-446.

Zamponi, G. W. and Snutch, T. P., 1998, Decay of prepulse facilitation of N type calcium channels during G protein inhibition is consistent with binding of a single Gβ subunit, *Proc Natl Acad Sci U S A* **95**:4035-4039.

Zhang, J., Ferguson, S. S., Barak, L. S., Bodduluri, S. R., Laporte, S. A., Law, P. Y., and Caron, M. G., 1998, Role for G protein-coupled receptor kinase in agonist-specific regulation of μ-opioid receptor responsiveness, *Proc Natl Acad Sci U S A* **95**:7157-7162.

Zhang, J. F., Randall, A. D., Ellinor, P. T., Horne, W. A., Sather, W. A., Tanabe, T., Schwarz, T. L., and Tsien, R. W., 1993, Distinctive pharmacology and kinetics of cloned neuronal Ca^{2+} channels and their possible counterparts in mammalian CNS neurons, *Neuropharmacology* **32**:1075-1088.

Zhang, J. F., Ellinor, P. T., Aldrich, R. W., and Tsien, R. W., 1996, Multiple structural elements in voltage-dependent Ca^{2+} channels support their inhibition by G proteins, *Neuron* **17**:991-1003.

Zhang, J. H. and Simmonds, W. T., 2000, Copurification of brain G-protein beta5 with RGS6 and RGS7, *J Neurosci* **20RC59**:1-5.

Zhang, L., DeHaven, R. N., and Goodman, M., 2002, NMR and modeling studies of a synthetic extracellular loop II of the κ opioid receptor in a DPC micelle, *Biochemistry* **41**:61-68.

Zhong, J., Hume, J. R., and Keef, K. D., 1999, Anchoring protein is required for cAMP-dependent stimulation of L-type Ca^{2+} channels in rabbit portal vein, *Am J Physiol* **277**:C840-844.

Zhou, J., Moroi, K., Nishiyama, M., Usui, H., Seki, N., Ishida, J., Fukamizu, A., and Kimura, S., 2001, Characterization of RGS5 in regulation of G protein-coupled receptor signaling, *Life Sci* **68**:1457-1469.

Zhou, J. Y., Siderovski, D. P., and Miller, R. J., 2000, Selective regulation of N-type Ca channels by different combinations of G-protein β/γ subunits and RGS proteins, *J Neurosci* **20**:7143-7148.

Zhu, Y. and Ikeda, S. R., 1994, VIP inhibits N-type Ca^{2+} channels of sympathetic neurons via a pertussis toxin-insensitive but cholera toxin-sensitive pathway, *Neuron* **13**:657-669.

ALTERNATIVE SPLICING IN VOLTAGE GATED CALCIUM CHANNELS

Diane Lipscombe[1] and Andrew J. Castiglioni[1]

1. INTRODUCTION

The mammalian nervous system utilizes extensive alternative splicing to support an astounding level of structural and functional diversity in the voltage-gated calcium channel family. All major calcium channel subunits, $Ca_v\alpha_1$, $Ca_v\beta$, and $Ca_v\alpha_2\delta$, are subject to alternative splicing. The structural changes generated by this form of RNA processing are often quite subtle but clearly critical for highly effective coupling to specific signaling pathways. The inclusion or exclusion of certain alternatively spliced exons can be differentially regulated according to tissue type, cell type, and stage of development. In certain instances, structural and functional differences generated by alternative splicing translate into unique pharmacological phenotypes. Pharmacologically distinct splice isoforms preferentially expressed in specific tissues or cell types offer exciting new targets for the design of highly specific drugs.

2. STRUCTURAL DIVERSITY AMONG VOLTAGE GATED CALCIUM CHANNELS

Mammals utilize several different mechanisms to optimize the behavior of voltage-gated calcium channels and to couple these channels efficiently to a variety of signaling pathways. This is particularly evident in the nervous system, where calcium channels link membrane depolarization to a wide spectrum of signaling pathways, thereby regulating diverse neuronal functions. The nervous system contains the widest array of calcium channel subtypes, although significant diversity exists among calcium channels expressed in other tissues, particularly the cardiovascular system.

---

[1] Department of Neuroscience, 192 Thayer Street, Brown University, Providence, RI 02912. Address correspondence to Dr. Lipscombe.

Calcium Channel Pharmacology, Edited by McDonough
Kluwer Academic/Plenum Publishers, 2004

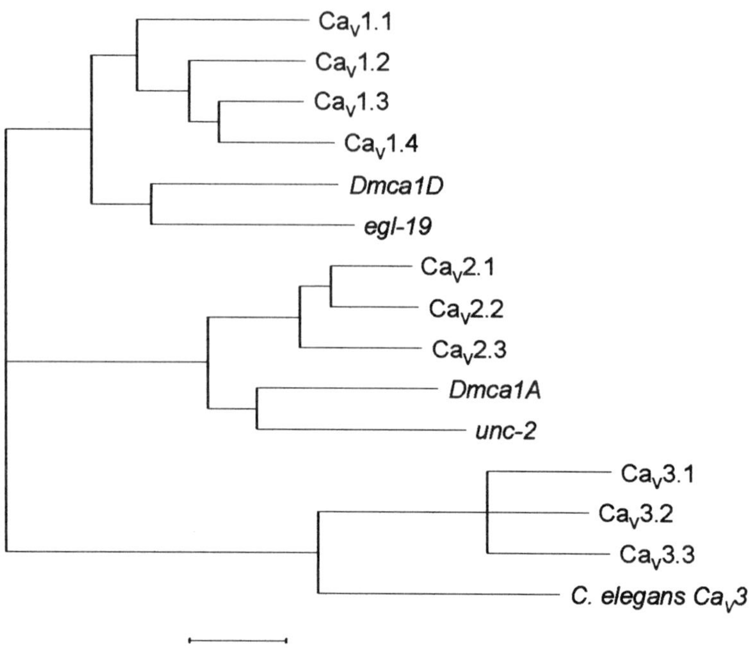

Figure 1. Calcium channel $Ca_v\alpha_1$ subunit gene tree. The sequences for the human genes, $Ca_v1.1$ (GenBank Accession #L33798), $Ca_v1.2$ (#AJ224873), $Ca_v1.3$ (#M76558), $Ca_v1.4$ (#AJ224874), $Ca_v2.1$ (#AB035727), $Ca_v2.2$ (#M94173), $Ca_v2.3$ (#L27745), $Ca_v3.1$ (AF190860), $Ca_v3.2$ (#AF05196), and $Ca_v3.3$ (#AF211189) were aligned along with gene sequences of *Drosophila DmcalA* (#U55776) and *DmcalD* (#U00690), and *C. elegans unc-2* (#AAB36868), *egl-19* (#AAC47755), and Ca_v3 (#AAA79201). Alignments were carried out using transmembrane domains and short linkers, but excluding the variable N- and C-termini and intracellular loops I-II, II-III, and III-IV. A branch-and-bound tree search using maximum parsimony was performed to find the optimal tree (Genetic Computer Group, paupsearch and paupdisplay programs). A highly similar tree was also obtained using minimum evolutionary distance as the optimality criterion. The original tree is unrooted. The representation has been rooted using the midpoint method. The scale bar corresponds to 1 substitution per 100 amino acids.

Human, rat, and mouse genomes contain ten known $Ca_v\alpha_1$ subunit genes, nine of which are expressed in the nervous system (Figure 1) (Ertel et al., 2000). Analyses to date suggest that each $Ca_v\alpha_1$ gene contains multiple sites of alternative splicing. Other voltage-gated calcium channel genes encode the major auxiliary subunits, including four $Ca_v\beta$, four $Ca_v\alpha_2\delta$, and several γ subunit genes (Birnbaumer et al., 1998; Walker and De Waard, 1998; Klugbauer et al., 1999; Burgess et al., 2001; Qin et al., 2002). The majority of $Ca_v\alpha_1$ and auxiliary Ca_v subunits undergo alternative pre-mRNA splicing. It is likely that distinct tissues and functionally specialized cell types will express a unique subset of splice isoforms. This process supports a staggering level of diversity among voltage-gated calcium channels (Koch et al., 1990; Perez-Reyes et al., 1990; Hui et al., 1991; Snutch et al., 1991; Kim et al., 1992; Soldatov, 1994; Williams et al., 1994; Murakami et al., 1996; Lin et al., 1997; Birnbaumer et al., 1998; Qin et al., 1998; Bourinet et al., 1999; Klugbauer et al., 1999; Mittman et al., 1999a, b; Chemin et al., 2001a; Helton and Horne, 2002; Helton et al., 2002; Lipscombe et al., 2002).

In Ca_v subunits, alternative splicing may modify the biophysical properties of the channel and/or affect the coupling of the channel to downstream proteins and intracellular signaling pathways. Regulated alternative splicing events must contribute some functional advantage to the organism and specific cells in which they occur. In some cases, variations in channel structure and function generated by alternative splicing of Ca_v pre-mRNAs influence channel pharmacology. This chapter will highlight such examples (Welling et al., 1997; Zuhlke et al., 1998; Beam, 1999; Bourinet et al., 1999; Hans et al., 1999; Fearon et al., 2000; Jimenez et al., 2000; Krovetz et al., 2000; Lacinova et al., 2000b; Yatani and Kamp, 2000; Helton et al., 2002; Kaneko et al., 2002).

2.1. Multiple Genes

Voltage-gated calcium channels couple to individual signaling pathways defined primarily by the $Ca_v\alpha_1$ subunit. The *L-type Ca_v1* gene family links excitation to contraction in muscle cells in a variety of species including mammals (Tanabe et al., 1987; Bean, 1989; Catterall, 1991; Mori et al., 1993). L-type Ca_v1 calcium channels also couple excitation to gene expression in neurons, excitation to secretion in endocrine cells and in cochlear hair cells, drive pacemaking in the SA node of the heart, and regulate intrinsic rhythmic firing in neurons of the central nervous system (Ghosh and Greenberg, 1995; Fuchs, 1996; Liljelund et al., 2000; Platzer et al., 2000; Pennartz et al., 2002; Zhang, 2002). L-type channels are also present in fibroblasts and testis (Soldatov, 1992; Goodwin et al., 1998). *Ca_v2 genes* comprise the core of P/Q-type, N-type, and R-type channels. These channels link excitation to neurotransmitter release in neurons (Dunlap et al., 1995), but they have also been implicated in synaptogensis and regulation of gene expression (Jones et al., 1997; Brosenitsch and Katz, 2001). Finally, *Ca_v3 genes* comprise the functional core of T-type calcium channels that contribute to rhythmic firing in pacemaker cells of the heart and in thalamic neurons of the brain (Huguenard, 1998; Triggle, 1998; Perez-Reyes, 2003). T-type calcium channels activate at relatively hyperpolarized voltages and possess slow deactivation kinetics (Perez-Reyes, 2003). The slow deactivation speed of Ca_v3 T-type channels is unique among $Ca_v\alpha_1$ subunits. T-type calcium channels have also been implicated in controlling hormone release from glomerulosa cells (Cohen et al., 1988).

Expression patterns of Ca_v subunits are modified during development. The $Ca_v3.3$ gene, for example, is expressed early in development throughout the rat brain, but is reported by some to be restricted to the striatum in the adult (McRory et al., 2001), although other groups report a more widespread distribution (Talley et al., 1999). Temporal changes in the expression of $Ca_v\beta$ subunits in brain have been reported that are thought to contribute to the maturation of Ca^{2+} channel currents (Vance et al., 1998; Walker and De Waard, 1998). Furthermore, evidence from knockout studies suggests that different mammalian Ca_v genes can substitute for each other functionally to varying degrees. $Ca_v2.1^{-/-}$ mice missing the P/Q-type channel, for example, are not viable to adulthood, but survive for 3-4 weeks after birth (Jun et al., 1999); in these mice, N-type and R-type currents are up-regulated at the CNS synapses where P/Q-type currents normally dominate (Jun et al., 1999).

2.2. An Overview of Alternative Splicing in $Ca_v\alpha_1$ Genes

Molecular cloning studies provided the first evidence for the presence of splice isoforms in the three $Ca_v\alpha_1$ gene families (Koch et al., 1990; Perez-Reyes et al., 1990; Hui et al.,

1991; Snutch et al., 1991; Diebold et al., 1992; Soldatov, 1992; Schneider et al., 1994; Williams et al., 1994; Marubio et al., 1996; Lin et al., 1997; Takimoto et al., 1997; Mittman et al., 1999a; Cribbs et al., 2000; Chemin et al., 2001a). Multiple, highly homologous clones were isolated for the various $Ca_v\alpha_1$ subunits. These clones contained discrete regions of sequence divergence highly indicative of alternative splicing. The present availability of public genome sequence databases greatly facilitates proper identification of a true splice isoform, as distinct from a cloning artifact. Studies of $Ca_v1.2$ splice isoforms benefited from the efforts of Soldatov, who published the structure of the $Ca_v1.2$ gene almost 10 years ago (Soldatov, 1994). An updated and revised structure of the human $Ca_v1.2$ gene is shown in Figure 3.

The $Ca_v\alpha_1$ subunit contains four repeats of the canonical six transmembrane motif common to all voltage-gated channels. This is illustrated together with the human $Ca_v1.2$ gene structure in Figure 3 (similar figures are shown for three other $Ca_v\alpha_1$ genes in Figures 5, 6, and 9). Certain sites of alternative splicing are conserved among different mammalian $Ca_v\alpha_1$ genes, although the precise amino acid sequences of the alternatively spliced exons differ (Lipscombe et al., 2002). In regions with more sequence variance among $Ca_v\alpha_1$ genes, sites of alternative splicing tend to be conserved only between close relatives.

2.2.1. *The Extracellular Domain IV S3-S4 Linker*

One common area of alternative splicing is the IVS3-IVS4 extracellular linker. Alternative splicing in this domain is present in several mammalian families and is present even in the *Drosophila* homolog of the mammalian $Ca_v2.1$ and $Ca_v2.2$ genes (Smith et al., 1996; Peixoto et al., 1997; Smith et al., 1998; Lipscombe et al., 2002). The conservation of alternative splicing in this region is consistent with the idea that the IVS3-IVS4 linker plays a critical role in modulating Ca^{2+} channel activity (exon 33 in $Ca_v1.2$, exon 32 in $Ca_v1.3$, exon 31a in $Ca_v2.1$, and exon 31a in $Ca_v2.2$; see Figures 3, 5, 6, and 9) (Perez-Reyes et al., 1990; Snutch et al., 1991; Starr et al., 1991; Diebold et al., 1992; Barry et al., 1995; Ihara et al., 1995; Lin et al., 1997; Ligon et al., 1998; Bourinet et al., 1999; Hans et al., 1999; Lin et al., 1999; Krovetz et al., 2000). The IVS3-IVS4 linkers of a number of voltage-gated and structurally related ion channels are important in determining the time course and voltage dependence of channel activation (Perez-Reyes et al., 1990; Nakai et al., 1994; Lin et al., 1997; Mathur et al., 1997; Tang and Papazian, 1997; Hans et al., 1999).

2.2.2. *The Intracellular Domain II-III Loop*

The composition and size of the intracellular loop connecting domains II and III (L_{II-III}) are highly variable among different $Ca_v\alpha_1$ subunits. These cytoplasmic L_{II-III} domains play a key role in coupling $Ca_v\alpha_1$ subunits to their respective downstream effector proteins. In skeletal muscle, the L_{II-III} region of $Ca_v1.1$ couples the voltage-gated Ca^{2+} channel to the ryanodine receptor, creating a physical bridge to the sarcoplasmic reticulum (Mori et al., 1993). The L_{II-III} sequences of $Ca_v2.1$ and $Ca_v2.2$ subunits bind the synaptic proteins syntaxin and SNAP-25. Consequently, L_{II-III} is thought to be critical for coupling P/Q-type and N-type Ca^{2+} channels to neurotransmitter release in presynaptic nerve terminals (Leveque et al., 1994; Sheng et al., 1994; Rettig et al., 1996; Tobi et al., 1998; Catterall, 1999). The L_{II-III} loop region of all three Ca_v2 genes contains sites of alternative splicing. The other interesting feature of this region is the presence of a long exon, exon 19 in $Ca_v2.1$ and $Ca_v2.2$ (Figures 6, 9).

2.2.3. The C-Terminal Domain

The C-terminus of $Ca_v\alpha_1$ constitutes about one fourth of the channel protein and has been implicated in regulating various aspects of Ca^{2+} channel function, including inactivation, modulation by G proteins, and protein-protein interactions that regulate activity and/or target the channel to specific cellular compartments (Wei et al., 1994; Maximov et al., 1999; Zuhlke et al., 1999; Gao et al., 2000; Hering et al., 2000; Ivanina et al., 2000; Kepplinger et al., 2000; Peterson et al., 2000; Simen et al., 2001; Staes et al., 2001; Maximov and Bezprozvanny, 2002). The sequence of the C-terminus is also the least conserved among different $Ca_v\alpha_1$ genes. Biochemical analyses have provided evidence for the presence of at least two size forms of most $Ca_v\alpha_1$ subunits in the mammalian brain (Hell et al., 1994; Schneider et al., 1994; Sakurai et al., 1996). Good evidence now exists that alternative splicing in the C-terminus is an important mechanism for generating some of the different size forms of $Ca_v\alpha_1$ subunits (Hui et al., 1991; Schneider et al., 1994; Williams et al., 1994; Klockner et al., 1997; Soldatov et al., 1997; Zhuchenko et al., 1997; Ligon et al., 1998; Pereverzev et al., 1998; Bourinet et al., 1999; Lu and Dunlap, 1999; Maximov et al., 1999; Vigues et al., 1999; Hering et al., 2000; Krovetz et al., 2000; McRory et al., 2001; Xu and Lipscombe, 2001). Mutually exclusive exons 37a and 37b are also conserved in $Ca_V2.1$, $Ca_V2.2$, and $Ca_V2.3$ genes, though the amino acid compositions of these exons differ between genes (Section 4.1.4.)

Although the functional consequences of alternative splicing in the C-terminus are not fully understood, effects on inactivation kinetics have been reported for $Ca_v1.2$ and $Ca_v2.1$ (Figure 4) (Soldatov et al., 1997; Bourinet et al., 1999; Krovetz et al., 2000; Restituito et al., 2000). Splicing in the C-terminal domain has also been reported to affect the ability of G proteins to inhibit the $Ca_v2.2$ N-type Ca^{2+} channel (Simen et al., 2001), and to affect sub-cellular targeting of $Ca_v2.1$ and $Ca_v2.2$ subunits (Maximov et al., 1999; Maximov and Bezprozvanny, 2002).

2.3. How Many Splice Variants Can Be Generated From a Single $Ca_v\alpha_1$ Gene?

Each $Ca_v\alpha_1$ gene is large, containing upwards of 50 exons and spanning between 50 kb and 800 kb of human genome sequence (Figures 3, 5, 6, 9, 11) (Soldatov, 1994; Yamada et al., 1995; Ophoff et al., 1996; Mittman et al., 1999a, b; Abernethy and Soldatov, 2002). While there are indications that certain alternatively spliced exons may be linked or invariably co-expressed in specific tissues (Monteil et al., 2000a; Restituito et al., 2000; Tsunemi et al., 2002), many other alternatively spliced exons are likely to be independently regulated (Lipscombe et al., 2002). Alternative splicing enables the possibility that hundreds, if not thousands, of distinct mRNAs may be generated from each $Ca_v\alpha_1$ gene. There is precedence for this level of diversity in neuronal genes. For example, as a result of alternative splicing in neurons, three neurexin genes, each containing nearly 20 constitutively expressed exons, can potentially generate more than 1000 isoforms (Missler and Sudhof, 1998; Graveley, 2001). Cell-specific expression of several hundred variants of the calcium-activated potassium channel *Slo* gene in hair cells of the inner ear has been documented (Navaratnam et al., 1997; Rosenblatt et al., 1997). Likewise, the *Drosophila* axon guidance gene, *DSCAM*, is predicted to generate more than 38,000 splice variants, a number that exceeds current estimates of the total gene number in *Drosophila* (Black, 2000).

In summary, each Ca_v gene may have the potential to generate thousands of functionally distinct splice isoforms. Confirming that each is expressed and of functional importance, however, poses a significant challenge. Expression patterns of certain splice isoforms are likely to be restricted to specific cell populations that may not be represented in pools of RNA isolated from whole tissue preparations.

2.4. Identifying and Confirming Splice Isoforms of Ca_v Genes

Investigators initially identified splice isoforms of the various $Ca_v\alpha_1$ genes by comparing cDNAs isolated from a variety of different tissues and different species. However, traditional methods of molecular cloning are labor intensive and only identify the most abundant splice isoforms of a given $Ca_v\alpha_1$ family. Additional sites of alternative splicing were identified through sequence analyses of human $Ca_v\alpha_1$ genes in conjunction with large scale RT-PCR and the use of exon-finding algorithms (Mittman et al., 1999a, b; Monteil et al., 2000a; Soong et al., 2002). Microarray technology offers a way to rapidly screen mRNA from a wide variety of tissues. This technology has the potential to automate and improve the process significantly. It allows for the identification of even very short exons whose expression is tissue-specific. Exon arrays have already been used successfully to map the tissue- and disease-specific expression patterns of genes on chromosome 22, which contains > 8,000 exons (Shoemaker et al., 2001). RT-PCR targeted at the level of single identified cell-types provides critical information on the cell-specific expression of splice isoforms of Ca_v genes (Tsunemi et al., 2002).

Each method of analysis has its advantages and disadvantages. The sensitivity of PCR permits amplification of rare mRNAs, but aberrant or incompletely spliced mRNAs may also be amplified and misidentified. Aberrant splicing appears to be particularly prevalent in tumor cells (Ge et al., 1999) (A. J. C. and D. L., unpublished observations) and in cells that express a low level of target mRNAs (W. Xu and D. L., unpublished observations). Analysis of genomic sequence in conjunction with studies outlined above is essential to discern aberrant splice products from true splice isoforms. The presence of previously unrecognized alternatively spliced exons can be predicted from analysis of genomic sequence (e.g., exon 37a of $Ca_v2.2$, Figure 9) if exons are large enough in size. However, particularly small exons that reside in long intronic stretches such as exon 31a of $Ca_v2.1$ and exon 31a of $Ca_v2.2$, which contain only 6 nucleotides, cannot be identified from analysis of genomic sequence alone (Figures 6, 9) (Lin et al., 1997; Lin et al., 1999). Even smaller variations arising from RNA editing of single nucleotides in $Ca_v2.1$ have been documented (Tsunemi et al., 2002).

Expressed Sequence Tag (EST) databases, derived from sequencing cDNAs from a variety of normal and diseased tissues, will be invaluable adjuncts in identifying alternatively spliced exons. Currently, however, expressed sequences derived *in silico* from the EST database are 3' biased, and large genes such as $Ca_v\alpha_1$ are not well represented. Bioinformatic-based analyses using ESTs suggest that 35% to 60% of human genes are subject to alternative splicing in at least one site (Black, 2000; Modrek and Lee, 2002). These estimates will only increase as the database expands to include an even wider diversity of tissues and cell types.

In summary, the identification of additional sites of alternative splicing in the Ca_v gene family, and the confirmation of putative sites already reported, requires a multi-

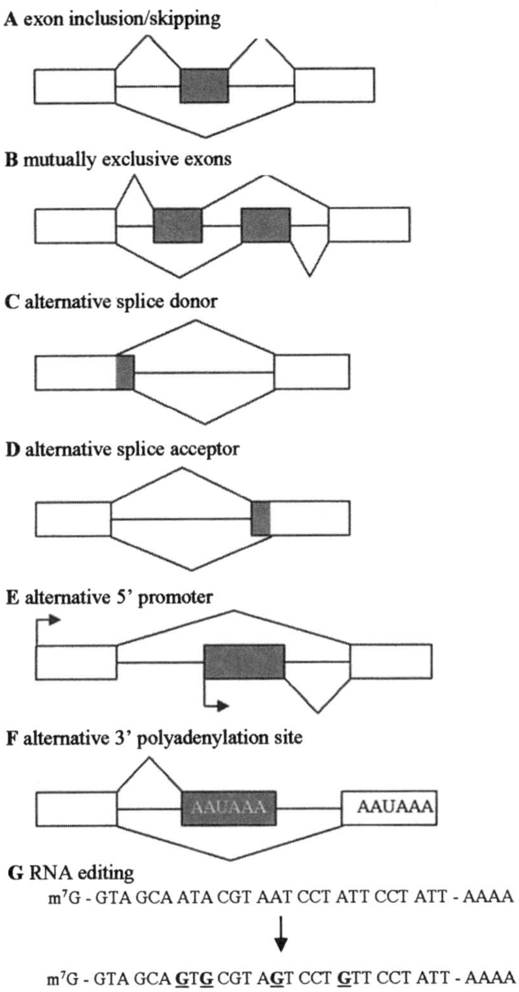

Figure 2. Different mechanisms, including alternative splicing, generate diversity among mRNAs. Various post-transcriptional mechanisms for generating diversity among mRNAs presented in schematic form. Diversity among mRNAs expressed in the nervous system arises from forms of RNA processing including: **(A)** exon inclusion/skipping; **(B)** mutually exclusive exon expression; the use of **(C)** alternate 5' splice donor sites or **(D)** alternate 3' splice acceptor sites; the use of **(E)** alternate 5' promoters or **(F)** alternate 3' polyadenylation sites (Schorge et al., 1999); and **(G)** RNA editing. There is evidence for RNA editing in *Drosophila* Ca channel *DmcalA* (Smith et al., 1998). Reprinted with permission from Lipscombe et al., 2002.

pronged approach. This endeavor combines mRNA analysis from a variety of tissues and/or cell types together with genomic and functional analyses. At least two criteria should be utilized to validate that a given sequence variation among Ca_v clones arises from bona fide alternative splicing and is not an artifact arising from cloning errors or aberrant splicing events:

1) The specific Ca_v sequence should be expressed at significant levels and present in RNA from a defined tissue or cell type that is not an immortalized cell line. This criterion may be difficult to establish when a particular splice isoform is restricted to a small subset of cells, but it is necessary and essential to establish functional significance.

2) Consensus splice junctions should be identified in the genomic sequence flanking the alternatively spliced exon. These will most commonly involve ag – EXON – gt, although other dinucleotide junctions are used in mammals (Sharp and Burge, 1997; Modrek and Lee, 2002). The numerous genome sequences available permit rapid and easy cross-species comparison of alternatively spliced exons.

2.5. Processing of Pre-mRNA

The major patterns of alternative splicing observed in mammals are shown in Figure 2A-2D. Most eukaryotic genes consist of several discontinuous coding sequences (exons) that average 50 to 300 nucleotides in length. Non-coding sequences (introns) lie between exons and are more variable in length, ranging from 200 to several thousand nucleotides. The pre-messenger RNA initially transcribed from the genomic DNA is processed to generate stable messenger RNAs for protein translation. This RNA processing removes the introns and splices the exons together. The spliceosome, a large complex that contains nuclear ribonucleoprotein particles, numerous proteins, and specific splicing factors, usually mediates this process (Sharp and Burge, 1997). Signature dinucleotide sequences, gt and ag, reside at the respective 5' and 3' boundaries of more than 99% of introns (Modrek and Lee, 2002). Other recognition sites are also used (Sharp and Burge, 1997), and examples of introns whose boundaries are not defined by gt-ag have been documented for $Ca_v\alpha_1$ genes (Kaneko et al., 2002).

Certain exons are invariant in mRNAs transcribed from a specific gene. These constitutively expressed exons are essential for basic protein function and structure. The exons that are only found in a subset of mRNAs from a specific gene are alternatively expressed. These exons generally modify protein function and fine-tune the protein for specialized tasks. The inclusion of alternatively expressed exons in the final mRNA may be influenced by a number of different factors such as tissue type, cell type, developmental stage, and cellular activity (Black, 1998; Grabowski, 1998; Xie and Black, 2001). There is a large family of known factors that direct constitutive RNA splicing, but only a handful of those that direct alternative splicing have been identified and characterized (Grabowski, 1998, 2000; Jensen et al., 2000).

Examples of each pattern of alternative splicing, as well as other forms of RNA processing illustrated in Figure 2, have been observed in Ca_v genes. Exons may be included or excluded (Figure 2A), or a series of homologous exons might be mutually exclusive, whereby only one will be selected and included in the final fully spliced mRNA (Figure 2B). Examples of alternative splicing also arise from the use of multiple

5' donor and/or 3' acceptor splice sites (Figures 2C, 2D). Even more subtle variations arise from the direct editing of specific RNA nucleotides (Figure 2G). In RNA editing, adenosine deaminase catalyses the conversion of a specific adenosine residue in the pre-mRNA transcript to ionosine, which is read as a guanosine by the translation machinery (Reenan, 2001). One of the most notable examples of mammalian RNA editing is of the AMPA receptor. In this case, RNA editing of a single adenosine in the GluR2 subunit changes an amino acid regulating permeation of calcium through the pore (Sommer et al., 1991). RNA editing has been documented in voltage-gated calcium channels of *Drosophila* (Smith et al., 1998) and in $Ca_v2.1$ P/Q-type channels expressed in mouse Purkinje cells (Tsunemi et al., 2002). The number of identified sites of RNA editing in mammalian Ca_v mRNAs will almost certainly increase with the advent of better techniques to enable screening for single base changes in RNA from identified cell populations. The use of genome sequence databases to compare cDNA sequence to genomic DNA sequence will also facilitate the identification of RNA editing sites.

Diversity also exists in 5' and 3' untranslated regions of mRNAs transcribed from the same gene. This arises from the use of alternate 5' promoters and alternate 3' polyadenylation/cleavage sites (Figures 2E, 2F). In $Ca_v1.2$ and $Ca_v\beta_4$ genes, the use of alternate 5' promoters is thought to generate isoforms that differ at the N-terminus (Figures 3, 11). In the 3' untranslated region, the N-type $Ca_v2.2$ gene contains three polyadenylation sites, two of which are used in the mammalian brain to generate two differently-sized forms of the $Ca_v2.2$ mRNA that have different stabilities (Schorge et al., 1999). Multiple sites of 3' polyadenylation have also been observed in Ca_v3 genes (Mittman et al., 1999a). It is estimated that at least 25% of human genes give rise to mRNAs that use alternate polyadenylation sites (Modrek and Lee, 2002). The length and composition of the 3' untranslated region can modify mRNA stability and its targeting to specific regions of the cell.

2.6. Nomenclature of Splice Variants

Voltage-gated calcium channel nomenclature has evolved several times over the past 10 years. This chapter adopts the most recent naming strategy for identifying members of the major Ca_v subunits (Figure 1) (Ertel et al., 2000). The nomenclature is logical, with the exception of labeling individual splice isoforms based solely on sequential letters. The number of splice sites and possible permutations is now too large. Therefore, where applicable, we identify each splice isoform based on the specific alternatively spliced exons expressed. Exon numbering is based on the structures of the human Ca_v genes presented here (Figures 3, 5, 6, 9, 11). Comparisons among studies would be greatly facilitated if exon numbering were adopted universally to identify isoforms of Ca_v subunits. This, however, will require standardizing the naming of exons in the various Ca_v genes. In this chapter we maintain the numbering schemes used in previously published Ca_v gene structures. We use the following notation "e[x]" after the Ca_v symbol to note the presence or absence of specific exons. For example the splice isoform of $Ca_v2.2$ that lacks exon 24a and contains exon 31a would be denoted $Ca_v2.2e[\Delta24a/31a]$. We also use superscript case letters to the left and right of exons to indicate the use of alternate 3' acceptor and 5' donor splice junctions, respectively. For example, three splice isoforms of $Ca_v2.1$ can be generated from the use of three alternate 3' acceptors at the junction of exon 10. We denote these splice isoforms as $Ca_v2.1e[10]$, $Ca_v2.1e[^a10]$,

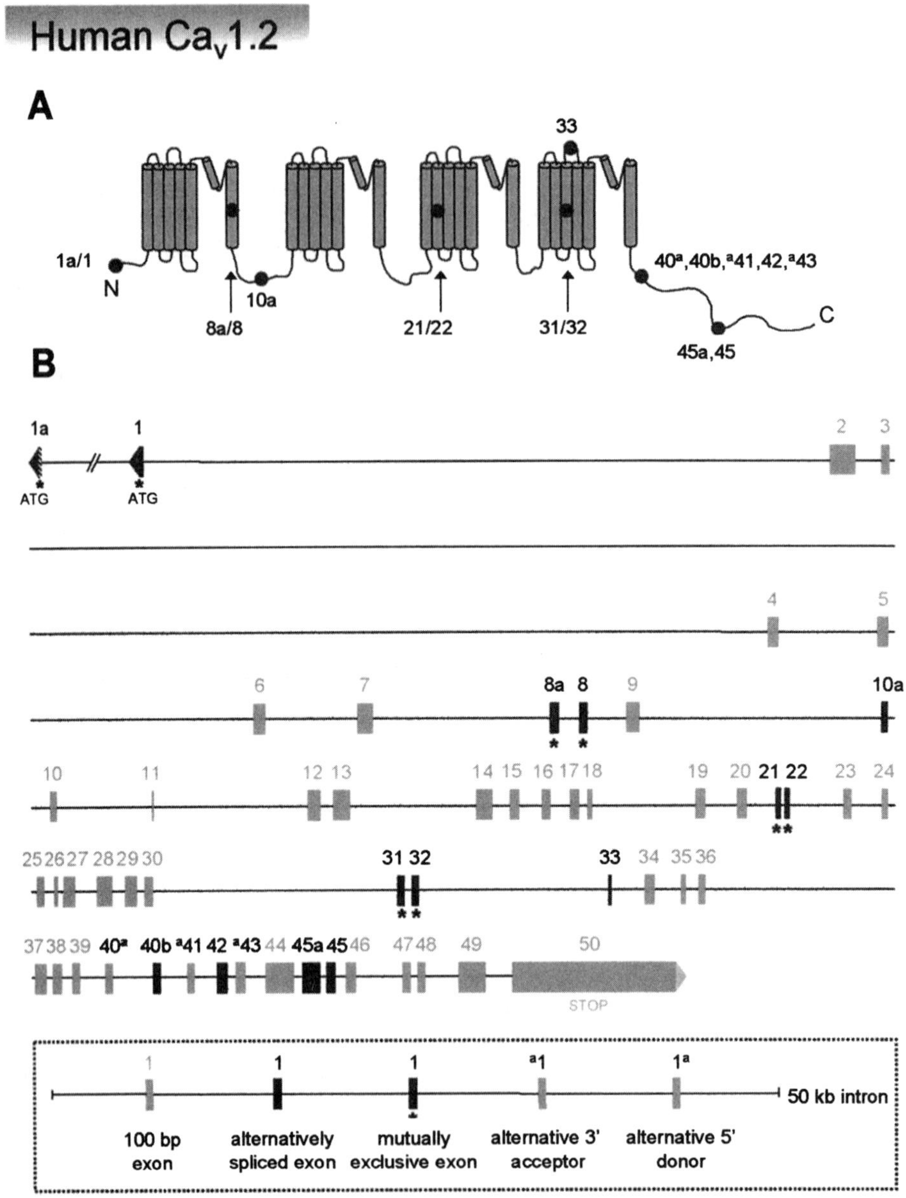

Figure 3. Structure of the human Ca$_V$1.2 gene. **A)** Putative topology of membrane spanning domains of the Ca$_V$1.2 subunit. Approximate locations of sites of alternative splicing are highlighted with black circles. **B)** The gene structure of human Ca$_V$1.2. Exons are denoted as solid boxes and introns as thin lines. Constitutive exons are shown in gray and alternatively spliced exons are shown in black. The gene structure was constructed from analysis of human genome sequence contigs NT_037806.2 and NT_009759.14. Exons are numbered based on the previously published human Ca$_V$1.2 gene structure (Soldatov, 1994). Hatched lines indicate that exon 1a could not be found in currently available human genome sequence though its presence is predicted from analyses of RNA isolated from rabbit and rat tissues.

and $Ca_v2.1e[^b10]$ (Figure 6). In some cases the pattern of alternative splicing can be quite complex.

3. ALTERNATIVE SPLICING IN L-TYPE Ca_v1 GENES

Four genes encode the $Ca_v\alpha_1$ subunits of L-type Ca^{2+} channels (Figure 1). The $Ca_v1.2$ and $Ca_v1.3$ genes together account for the majority of L-type calcium channels expressed in mammals. At least one of these genes is found in most excitable cells, including smooth muscle, cardiac muscle, endocrine cells, and neurons (Tsien et al., 1991; Ertel et al., 2000). The other two L-type calcium channel genes, $Ca_v1.1$ and $Ca_v1.4$, are localized to skeletal muscle and retina respectively (Catterall, 1991; Mori et al., 1993; Bech-Hansen et al., 1998). Recent research highlights significant functional and pharmacological differences between $Ca_v1.2$ and $Ca_v1.3$ L-type channels. Both are sensitive to dihydropyridine agonists and antagonists, the hallmark pharmacological feature of an L-type channel. However, $Ca_v1.3$ channels are significantly less sensitive to dihydropyridine antagonists compared to $Ca_v1.2$ (Koschak et al., 2001; Xu and Lipscombe, 2001; Lipscombe, 2002). Further, the two L-type channels are functionally different. Most notably, $Ca_v1.3$ L-type channels activate at voltages about 20 mV more hyperpolarized compared to $Ca_v1.2$ (Koschak et al., 2001; Scholze et al., 2001; Xu and Lipscombe, 2001). Additional functional diversity arises from alternative splicing, which has been observed in all $Ca_v\alpha_1$ subunits including $Ca_v1.1$ (Malouf et al., 1992) and $Ca_v1.4$ (C. Thaler and D. L., unpublished observations). A more detailed account of splice isoforms of $Ca_v1.2$ and $Ca_v1.3$ follows.

3.1. Splice Isoforms of $Ca_v1.2$

The human $Ca_v1.2$ gene is located at chromosome position 12p13, and it spans greater than 800 kb of human genomic sequence. It can be encoded by at least 55 exons, and at least 15 of these exons can be alternatively spliced (Soldatov, 1994; Abernethy and Soldatov, 2002). We constructed the $Ca_v1.2$ gene structure from the human genome database and highlighted exons proposed to undergo alternative splicing by at least one group of investigators (Figure 3). These include exons 1a/1 in the N-terminus, exons 8a/8 in IS6, exons 21/22 in IIIS2, exons 31/32 in IVS3, exon 33 in the IVS3-IVS4 linker, and exons 40, 40b, 41, 42, 43, 45a, and 45 in the C-terminus.

Splice isoforms of $Ca_v1.2$ have been reported in a variety of tissues including brain, smooth muscle, heart, testis, fibroblasts, and osteoblasts (Koch et al., 1990; Perez-Reyes et al., 1990; Snutch et al., 1991; Diebold et al., 1992; Kim et al., 1992; Soldatov, 1992; Feron et al., 1994; Klockner et al., 1997; Takimoto et al., 1997; Welling et al., 1997; Goodwin et al., 1998; Khan et al., 1998; Bielefeldt, 1999; Fearon et al., 2000; Liu et al., 2000; Abernethy and Soldatov, 2002). Functional changes associated with alternative splicing include modification of voltage-dependent and calcium-dependent inactivation (Soldatov et al., 1997; Soldatov et al., 1998; Hering et al., 2000; Abernethy and Soldatov, 2002). Furthermore, alternative splicing of $Ca_v1.2$ modifies the sensitivity of the L-type channel to dihydropyridine antagonists (Soldatov et al., 1995; Welling et al., 1997; Zuhlke et al., 1998; Lacinova et al., 2000b; Abernethy and Soldatov, 2002) and to oxygen (Fearon et al., 2000; Yatani and Kamp, 2000). The expression pattern of $Ca_v1.2$ splice isoforms is influenced by a number of factors, including tissue and cell type (Koch et al.,

1990; Snutch et al., 1991; Feron et al., 1994; Welling et al., 1997; Goodwin et al., 1998), developmental stage (Diebold et al., 1992; Feron et al., 1994), tissue stress (Gidh-Jain et al., 1995; Yang et al., 2000), and electrical activity (Pereon et al., 1997).

3.1.1. Alternative Splicing Modifies Inactivation

Voltage-gated calcium channels inactivate in response to depolarization. This inactivation can be mediated by voltage and, in the case of certain channels, by calcium itself passing through the channel pore and acting on cytoplasmic domains of $Ca_v\alpha_1$ subunits (Brehm and Eckert, 1978; de Leon et al., 1995; Hering et al., 2000). Splice isoforms of Ca_v subunits that exhibit different inactivation properties have aided in the identification of functional domains in the Ca_v complex that contribute to inactivation. In 1997, Soldatov and colleagues reported on $Ca_v1.2$ isoforms that arise from a complex pattern of alternative splicing involving exons 40-43 in the C-terminus. These splice forms, originally isolated from hippocampus, generated L-type channels with different inactivation profiles (Soldatov, 1994). Isoform $Ca_v1.2e[40^a/40b/\Delta41/\Delta42/^a43]$, referred to as $\alpha_{1C,86}$ in Figure 4 by Soldatov and colleagues, uses an alternate 5' donor in exon 40, includes exon 40b, lacks exons 41 and 42 completely, and uses an alternative 3' acceptor in exon 43. This splice isoform underlies L-type channels with unusually rapid voltage-dependent inactivation kinetics with barium as the permeant ion (Figure 4) (Soldatov et al., 1997). Normally, L-type Ca_v channels inactivate relatively slowly with barium as the charge carrier but much faster when calcium is the permeant ion because of calcium-dependent inactivation (Figure 4) (Brehm and Eckert, 1978). The specific splicing pattern in the C-terminus that defines isoform $Ca_v1.2e[40^a/40b/\Delta41/\Delta42/^a43]$ confers rapid voltage-dependent inactivation compared to other isoforms (Soldatov et al., 1997). At the same time, this particular splicing pattern eliminated or reduced the contribution of calcium-dependent inactivation because no significant change in inactivation kinetics of L-type currents was observed when calcium replaced barium as the charge carrier (Figure 4) (Soldatov et al., 1997). Beyond knowing it is expressed in hippocampus, the expression pattern of the rapidly inactivating $Ca_v1.2e[40^a/40b/\Delta41/\Delta42/^a43]$ splice isoform is not known.

3.1.2. Alternative Splicing Modifies Pharmacology of $Ca_v1.2$

Alternative splicing modifies the sensitivity of $Ca_v1.2$ channels to dihydropyridine antagonists (Welling et al., 1993; Soldatov et al., 1995; Welling et al., 1997; Zuhlke et al., 1998; Lacinova et al., 2000b; Abernethy and Soldatov, 2002). $Ca_v1.2$ isoforms isolated from smooth muscle and cardiac muscle differ in four discrete regions: the N-terminus, transmembrane domains IS6 and IVS3, and the C terminus. Most significantly, when studied in heterologous expression systems, the pharmacological profiles of the $Ca_v1.2$ isoforms parallel those of L-type channels in smooth muscle and cardiac muscle respectively (Welling et al., 1993). Chimeric analysis together with other evidence points to the importance of mutually exclusive exons 8a and 8 in contributing to the different dihydropyridine sensitivities of cardiac compared to smooth muscle L-type channels (Welling et al., 1997; Zuhlke et al., 1998). $Ca_v1.2e[8]$ isoforms generate L-type channels that are smooth muscle-like. They are highly sensitive to inhibition by nisoldipine at rest (resting-state block) and exhibit slow or incomplete recovery from inhibition at more depolarized voltages (inactive-state block). In contrast, the cardiac-like $Ca_v1.2e[8a]$ channels undergo significantly less inhibition from the resting state and recover relatively

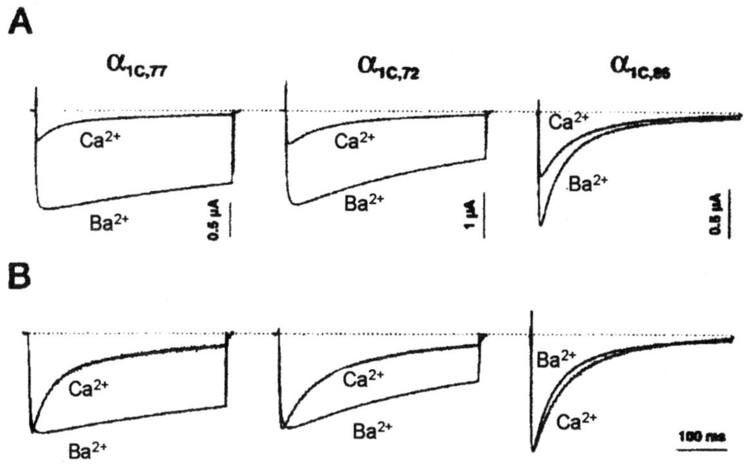

Figure 4. Splice forms of $Ca_v1.2$ that do and do not display Ca^{2+}-dependent inactivation. I_{Ba} and I_{Ca} through $Ca_v1.2$ splice forms $\alpha_{1C,77}$, $\alpha_{1C,72}$, and $\alpha_{1C,86}$, each coexpressed with β_1 and $\alpha_2\delta$, were recorded in *Xenopus* oocytes after injection of 50 nL of 40 mM BAPTA. Current traces were evoked by 400 ms depolarizing steps from V_h = -90 mV to +20 mV. The oocytes were superfused by a bath solution containing 40 mM Ba^{2+} or 40 mM Ca^{2+}. **A)** I_{Ba} and I_{Ca} recorded from the same oocyte expressing the indicated splice variant. **B)** traces from panel A normalized to peak current to show Ca^{2+}-dependent inactivation. Reprinted with permission from Soldatov et al., 1997.

rapidly from block of the inactive state (Welling et al., 1997). Analysis of isradipine action on $Ca_v1.2e[8a]$ and $Ca_v1.2e[8]$ splice isoforms from human fibroblasts yields consistent results (Zuhlke et al., 1998).

Dihydropyridines are state-dependent inhibitors that preferentially act on the inactive state of the L-type calcium channel (Bean, 1984; Sanguinetti and Kass, 1984; Triggle, 1999). Alternative splicing might modify dihydropridine binding directly, or indirectly by modifying fractional channel inactivation. Lacinova and colleagues found that smooth muscle and cardiac isoforms of $Ca_v1.2$ had similar channel gating properties and favor the hypothesis that alternative splicing of exons 8a and 8 in domain IS6 affects dihydropyridine binding directly (Lacinova et al., 2000b). The differential distribution of exon 8a provides support for the idea that this splice locus contributes to cardiac and smooth muscle phenotypes. The $Ca_v1.2e[8a]$ splice isoform that is less sensitive to dihydropyridine antagonists is expressed in heart but not in aortic tissue. In contrast, $Ca_v1.2e[8]$, characterized by a higher sensitivity to dihydropyridines, is predominantly found in vascular smooth muscle (Welling et al., 1997).

3.1.3. Domains IVS3 and IVS3-IVS4

There are significant differences in biophysical properties between cardiac and skeletal muscle L-type calcium channels. Notably, cardiac L-type channels open with faster kinetics compared to skeletal muscle channels. Sequence variations between $Ca_v1.1$ and $Ca_v1.2$ genes in the S3-S4 linkers of domain I appear to play a critical role in conferring the slow activation kinetics typical of skeletal muscle Ca_v currents and the relatively fast activation kinetics of cardiac Ca_v currents (Nakai et al., 1994). These results suggested early on that the extracellular loops connecting S3 and S4 transmembrane domains could govern channel opening kinetics.

Interestingly, $Ca_v1.2$ splice isoforms that either contain or lack exon 33 in the S3-S4 extracellular linker of domain IV have been found in mammalian tissues, including human heart (Perez-Reyes et al., 1990; Diebold et al., 1992). A second site of alternative splicing involving the mutually exclusive expression of exons 31 and 32 that encode the IVS3 transmembrane domain of $Ca_v1.2$ has also been described (Figure 3) (Perez-Reyes et al., 1990; Diebold et al., 1992). Alternative splicing at the exon 31/32 locus is differentially regulated in the heart during development. Both $Ca_v1.2$e[31] and $Ca_v1.2$e[32] splice isoforms are expressed in the heart of newborns, but this splicing pattern changes during development. In the adult heart the $Ca_v1.2$e[32] isoform predominates (Diebold et al., 1992). The functional significance of alternative splicing at the exon 31/32 and exon 33 loci remains to be determined.

3.1.4. Alternative Splicing Confers Oxygen Sensing to $Ca_v1.2$ Channels

Three additional isoforms of $Ca_v1.2$ have been isolated from human heart that generate L-type calcium channels with very similar biophysical properties, based on standard protocols for characterizing calcium channel behavior (Klockner et al., 1997). Significantly, isoforms that contain the 71 amino acids encoded by exon 45a, referred to as the hHT isoform, are uniquely sensitive to oxygen (Klockner et al., 1997). Specifically, $Ca_v1.2$e[45a] is inhibited under hypoxic conditions (Fearon et al., 2000; Yatani and Kamp, 2000). Research suggests that the oxygen-sensing form of $Ca_v1.2$ underlies the relaxation of certain smooth muscle cells that respond selectively to hypoxia (Yatani and Kamp, 2000). Determining the cell-specific exon distribution and correlating this with the ability to sense oxygen will be of great importance. The presence of $Ca_v1.2$e[45a] isoforms might in turn provide a particularly useful marker for identifying cells uniquely sensitive to hypoxic conditions. It is also notable that the specific oxygen sensing capabilities of $Ca_v1.2$e[45a] were not immediately obvious. This underscores the difficulty in attributing function to splice isoforms whose biophysical properties are at first inspection indistinguishable in heterologous expression systems.

3.1.5. Truncated Forms of $Ca_v1.2$ Channels

The question of functional significance remains unresolved for the set of splice isoforms that encode severely truncated $Ca_v1.2$ proteins (Soldatov, 1994; Wielowieyski et al., 2001). These isoforms arise because either whole exons are skipped, for example exon 19 in the II-III intracellular linker (Wielowieyski et al., 2001), or because alternative acceptors sites are used, for example in exon 15 (Soldatov, 1994). Both cause a frameshift and the introduction of a premature stop codon (Abernethy and Soldatov, 2002). No evidence yet exists that such hemi-channels (containing only domains I and II and variable C-termini) can reconstitute conducting ion channels. However, truncated versions of other $Ca_v\alpha_1$ genes have been described (Malouf et al., 1992; Scott et al., 1998; Okagaki et al., 2001; Raghib et al., 2001). In skeletal muscle a truncated isoform of $Ca_v1.1$ dominates at birth but is down regulated in adult skeletal muscle (Malouf et al., 1992). Evidence exists to support the hypothesis that these truncated isoforms may function as dominant-negative proteins to curtail calcium influx (Okagaki et al., 2001; Raghib et al., 2001).

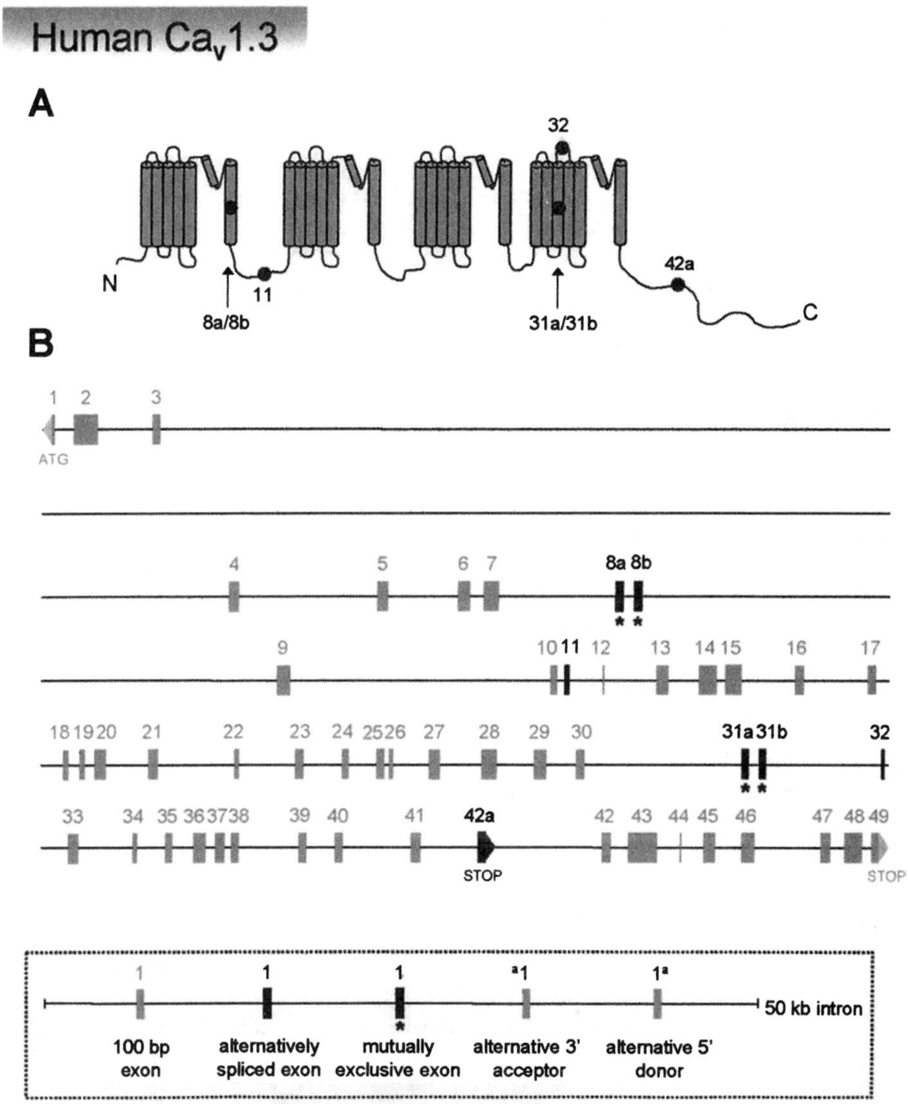

Figure 5. Structure of the human $Ca_v1.3$ gene. **A)** 2D topology of the $Ca_v1.3$ protein. Approximate locations of alternative splicing are highlighted with black circles. **B)** The gene structure of human $Ca_v1.3$. Exons are denoted as solid boxes and introns as thin lines. Constitutive exons are shown in gray and alternatively spliced exons are shown in black. The gene structure was constructed from analysis of human genome sequence contig NT_005986.14. Exons are numbered according to the previously published gene structure (Yamada et al., 1995).

3.1.6. N-Terminal Variants of $Ca_v1.2$ Channels

Other splice isoforms of $Ca_v1.2$ include those that differ in the length and sequence of the N-terminus determined by the inclusion of either exon 1a or 1 (Figure 3). Exon 1a has been observed in rat and rabbit mRNA, but sequence corresponding to exon 1a has not been identified in the human $Ca_v1.2$ gene. The very 5' end of the human $Ca_v1.2$ gene is not available in the database as of this writing, perhaps because it is located in a hard to sequence region at the tip of chromosome 12 next to the telomere.

3.1.7. Therapeutic Considerations

The higher sensitivity of smooth muscle L-type calcium channels to dihydropyridine antagonists compared to cardiac L-type channels is thought critical for the clinical effectiveness of antihypertensives (Triggle, this volume). The relatively lower effectiveness of antihypertensives on heart L-type calcium channels helps minimize unwanted side effects from L-type channel blockade of cardiac function (Triggle, 1999; Angus et al., 2000). As discussed above (Section 3.1.2), alternative splicing at the exon 8a/8 locus in $Ca_v1.2$ is likely to underlie at least part of the therapeutically advantageous pharmacological differences between smooth muscle and cardiac L-type channels. The other contributing factor is the state-dependent nature of dihydropyridine inhibition (Bean, 1984; Sanguinetti and Kass, 1984; Hering et al., 2000) that predicts more effective inhibition of Ca_v channels in smooth muscle cells, which rest at more depolarized membrane potentials compared to cardiac cells (Triggle, 1999). The development of L-type calcium channel blockers with an even higher selectivity for the $Ca_v1.2e[8]$ smooth muscle splice isoform might provide more effective antihypertensive agents with minimal effects on cardiac function (Angus et al., 2000).

3.2. Splice Isoforms of $Ca_v1.3$

The unique properties of $Ca_v1.3$ channels - activation at low voltages and incomplete inhibition by high concentrations of dihydropyridine antagonists - have only recently been highlighted (Platzer et al., 2000; Koschak et al., 2001; Xu and Lipscombe, 2001). The functional contributions of $Ca_v1.3$ L-type channels as a whole are not yet fully appreciated. Consequently, even less is known about the specific properties and importance of $Ca_v1.3$ splice isoforms.

3.2.1. Sites of Splicing

The human $Ca_v1.3$ gene is located at chromosome position 3p14 and spans approximately 400 kb of human genomic sequence. It can be encoded by at least 52 exons, and at least seven of these exons can be alternatively spliced (Figure 5). The structure of the human $Ca_v1.3$ gene highlights exons that have been shown to be subject to alternative splicing. These sites include mutually exclusive exons 8a/8b in transmembrane IS6, exon 11 in the intracellular loop I-II, exons 31a/31b of transmembrane IVS3, exon 32 in the IVS3-IVS4 extracellular linker, and exon 42a in the C-terminus. A limited number of splice isoforms of $Ca_v1.3$ have been described, and there is some evidence for cell and tissue-specific alternative splicing in hair cells, heart, neurons, and endocrine cells (Kollmar et al., 1997; Takimoto et al., 1997; Koschak et al., 2001; Safa et al., 2001; Lipscombe, 2002). Presently, only a limited number of studies address the functional significance of alternative splicing of $Ca_v1.3$, and significant differences in biophysical properties have not been reported (Safa et al., 2001; Xu and Lipscombe, 2001). Alternative splicing of exon 32, which encodes a 15 amino acid segment in domain IVS3-IVS4 of $Ca_v1.3$, does not appear to affect inhibition by dihydropyridine antagonists (Xu and Lipscombe, 2001).

3.2.2. Therapeutic Implications

There is good evidence that $Ca_v1.3$ channels are important in regulating transmitter release in hair cells and endocrine cells (Seino et al., 1992; Kollmar et al., 1997; Platzer et al., 2000) and contribute to pacemaking in the sinoatrial node (Zhang et al., 2002). $Ca_v1.3$ L-type calcium channels are also obvious candidates for underlying rhythmic firing in the suprachiasmatic nucleus and Purkinje neurons (Liljelund et al., 2000; Pennartz et al., 2002). The $Ca_v1.3$ L-type calcium channel is therefore a potentially interesting therapeutic target for regulating pacemaking in the heart (Platzer et al., 2000; Lipscombe, 2002; Zhang, 2002) and controlling epileptic activity in the central nervous system. Although the value of $Ca_v1.3$ as a therapeutic target has yet to be proven, the presence of tissue-specific splicing would suggest an approach to achieve specificity of drug action. Pharmacological tools, however, are now needed to facilitate the separation of different L-type currents.

4. ALTERNATIVE SPLICING IN THE Ca_v2 GENES

Ca_v2 genes encode the high voltage-activated P/Q-type, N-type, and R-type calcium channels that control the majority of transmitter release from neurons of the mammalian nervous system (Figure 1). As discussed in other chapters, these channels represent important drug targets (see Cohen and Kraus, this volume). The P/Q-type channel is linked to human disorders including familial hemiplegic migraine, ataxia and epilepsy (Fletcher et al., 1996; Ophoff et al., 1996; Zhuchenko et al., 1997; Fletcher et al., 1998; Miller, 2001). Splice isoforms of all three Ca_v2 genes have been identified and have been characterized to various extents (Schneider et al., 1994; Williams et al., 1994; Sakurai et al., 1995; Marubio et al., 1996; Lin et al., 1997; Welling et al., 1997; Zhuchenko et al., 1997; Kamphuis and Hendriksen, 1998; Ligon et al., 1998; Pereverzev et al., 1998; Vigues et al., 1998; Beam, 1999; Bourinet et al., 1999; Ghasemzadeh et al., 1999; Hans et al., 1999; Lin et al., 1999; Lu and Dunlap, 1999; Schramm et al., 1999; Vigues et al., 1999; Cahill et al., 2000; Krovetz et al., 2000; Pan and Lipscombe, 2000; Lu et al., 2001; Raghib et al., 2001; Vajna et al., 2001; Jagannathan et al., 2002; Kaneko et al., 2002; Lipscombe et al., 2002; Maximov and Bezprozvanny, 2002; Pereverzev et al., 2002; Soong et al., 2002; Tsunemi et al., 2002; Vigues et al., 2002). Several splice isoforms of Ca_v2 genes give rise to functionally and in some cases pharmacologically distinct channels (Bourinet et al., 1999; Hans et al., 1999; Jimenez et al., 2000; Kaneko et al., 2002).

4.1. Splice Isoforms of P/Q-type $Ca_v2.1$ Channels

The human $Ca_v2.1$ gene is located at chromosome position 19p13 and spans approximately 400 kb of human genomic sequence. It can be encoded by at least 49 exons, and at least nine of these can be alternatively spliced (Figure 6). Sites of alternative splicing include exon 10 in the I-II intracellular loop, exons 16 and 17 in the II-III intracellular loop, exon 31a in the IVS3-IVS4 extracellular linker, and exons

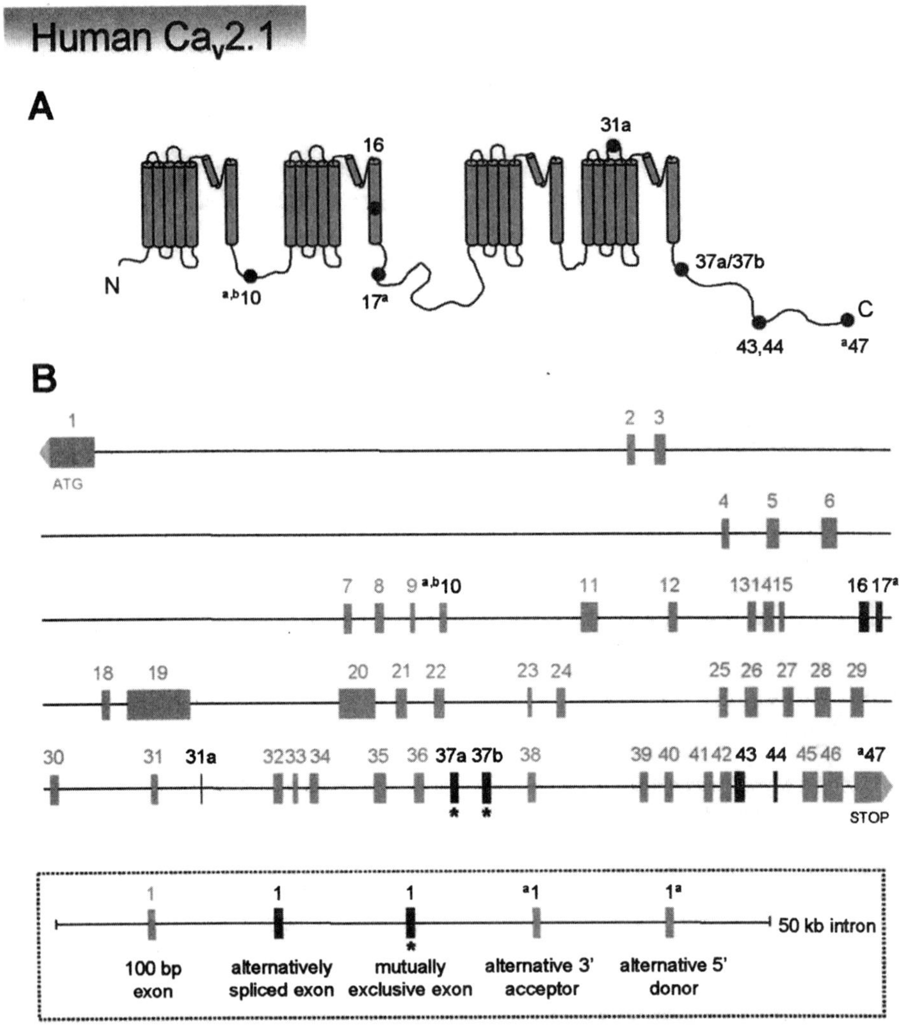

Figure 6. Structure of the human Ca$_V$2.1 gene. **A)** 2D topology of the Ca$_v$2.1 protein. Approximate locations of alternative splicing are highlighted with black circles. **B)** The gene structure of human Ca$_v$2.1. Exons are denoted as solid boxes and introns as thin lines. Constitutive exons are shown in gray and alternatively spliced exons are shown in black. The gene structure was constructed from analysis of human genome sequence contig NT_011295.9. Exons are numbered according to the previously published gene structure (Ophoff et al., 1996).

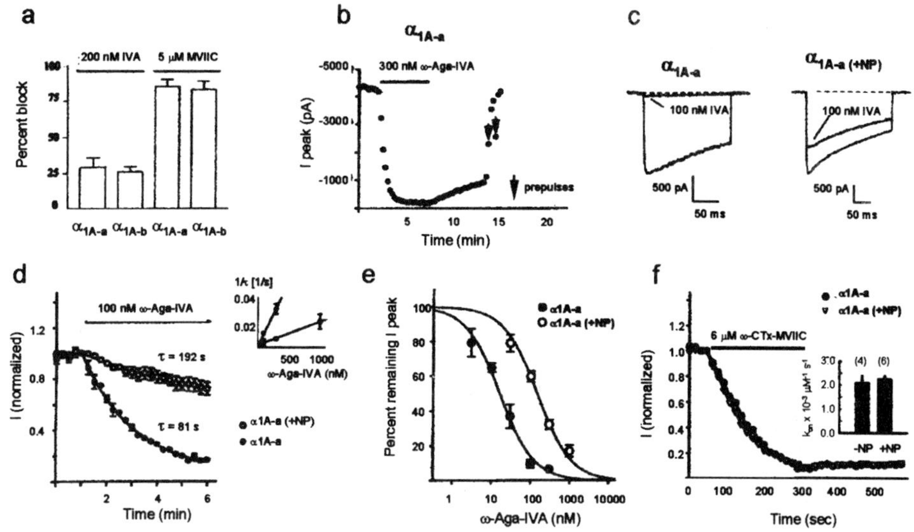

Figure 7. Sensitivities of splice forms of $Ca_v2.1$ to ω-Aga-IVA and ω-CTx-MVIIC. A) Whole-cell current records obtained from α_{1A-a} and α_{1A-b} (expressed transiently in *Xenopus* oocytes) after application of 200 nM ω-Aga-IVA or 6 μM ω-CTx-MVIIC. Mean ± s.e., n = 5-8. **B-F)** Channels expressed transiently in HEK cells. **B)** Time course of development of ω-Aga-IVA (300 nM) block of α_{1A-a} and reversal of toxin action with repetitive depolarizations (10 x 10 ms steps to +120 mV at 1 Hz). Test pulses to 0 mV were elicited from a holding potential of –100 mV every 15 seconds. **C)** Whole-cell current records obtained from α_{1A-a} and α_{1A-a} (+NP) with or without 100 nM ω-Aga-IVA. **D)** Average normalized time course comparing the rate of development of ω-Aga-IVA (100 nM) block of α_{1A-a} and α_{1A-a} (+NP). Inset, on-rate of block of α_{1A-a} and α_{1A-a} (+NP) varies linearly with ω-Aga-IVA concentration. Mean data (± s.e., n = 3-5) were fit according to the equation $1/\tau_{on} = k_{on} * [\omega\text{-Aga-IVA}] + k_{off}$. The K_d values were 14.8 nM for α_{1A-a} and 167 nM for α_{1A-a} (+NP). **E)** Dose dependence of the effects of ω-Aga-IVA for α_{1A-a} and α_{1A-a} (+NP). The solid lines reflect a fit with the Hill equation with the Hill coefficient arbitrarily held at 1.0. The IC_{50} values obtained were 16.3 nM for α_{1A-a} and 146 nM for α_{1A-a} (+NP). **F)** Time course of development of ω-CTx-MVIIC (6 μM) block of α_{1A-a} and α_{1A-a} (+NP). Inset, blocking rate constants obtained for α_{1A-a} and α_{1A-a} (+NP). Test pulses to +10 mV were elicited from a holding potential of –100 mV every 15 seconds. Reprinted with permission from Bourinet et al., 1999.

37a/37b, 43, 44, and 47 in the C-terminus (Ophoff et al., 1996; Bourinet et al., 1999; Hans et al., 1999; Krovetz et al., 2000; Soong et al., 2002).

4.1.1. The Intracellular I-II Loop (L_{I-II}) Alternative splicing at exon 10 utilizes 3 different 3' acceptor sites, giving rise to three sequence variations in $Ca_v2.1$ that either contain a two amino acid valine-glycine insert (VG), a single amino acid glycine insert (G), or no insert ($Ca_v2.1e[^a10]$, $Ca_v2.1e[^b10]$, $Ca_v2.1e[10]$; Figure 6) (Bourinet et al., 1999; Krovetz et al., 2000; Soong et al., 2002). The expression of valine (V_{421}) in $Ca_v2.1$ is implicated in slowing dramatically the time course of P-type Ca^{2+} channel inactivation (Bourinet et al., 1999) (Figure 7) and increasing the degree of G protein inhibition (Bourinet et al., 1999). The presence of V_{421} in the L_{I-II} of $Ca_v2.1$ also affects the pharmacological sensitivity of the P/Q-type channel to mibefradil (Figure 8) (Jimenez et al., 2000). Specifically, the rapidly inactivating $Ca_v2.1$ isoform that lacks V_{421} is more sensitive to inhibition by mibefradil compared to the slowly inactivating isoform that contains V_{421}.

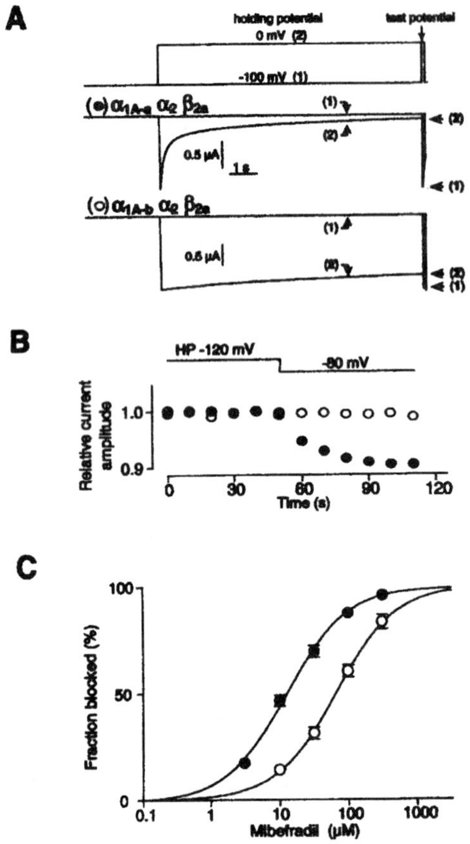

Figure 8. Greater sensitivity to mibefradil of a faster-inactivating splice form of $Ca_v2.1$. A) Differential inactivation properties of the two $Ca_v2.1$ (α_{1A}) splice variants. A 15 second pulse to 0 mV almost wholly inactivated $\alpha_{1A\text{-}a}$, but only produce a ~ 20% reduction in $\alpha_{1A\text{-}b}$ currents. B) Fractional channel inactivation at –80 mV under conditions close to steady-state. Note that for $\alpha_{1A\text{-}b}$, the inactivated fraction is essentially zero, while for $\alpha_{1A\text{-}a}$, around 10% of the current is inactivated. C) Dose-response of mibefradil blockade of the two splice forms. $\alpha_{1A\text{-}b}$ channels are less sensitive. Currents were evoked by 100 ms step depolarizations from –80 mV to 0 mV, delivered at 0.1 Hz. Reprinted from Jiménez et al., 2000, with permission from Elsevier.

The influence of this single amino acid insertion in the intracellular $L_{I\text{-}II}$ of $Ca_v2.1$ is best explained by preferential binding of mibefradil to the inactive state of the P/Q-type channel (Bezprozvanny and Tsien, 1995; Jimenez et al., 2000). The differential effects of mibefradil on splice isoforms of $Ca_v2.1$ serve to highlight that alternative splicing need not occur at sites involved directly in drug binding to have profound effects on channel pharmacology.

Little is known about the differential distribution of splice isoforms of $Ca_v2.1$ that arise from the use of alternate 3' acceptor splice sites at exon 10. Alternative splicing in exon 10 of $Ca_v2.1$, however, does not appear to underlie the characteristically slow inactivation kinetics typical of P-type currents of cerebellar Purkinje neurons. $Ca_v2.1$ clones isolated from mouse Purkinje cells by single cell RT-PCR lack V_{421} (Tsunemi et al., 2002). A similar pattern of alternative splicing is also observed in the homologous region of the N-type $Ca_v2.2$ gene. In this case, alternative use of a 3' acceptor at exon 10 of the N-type $Ca_v2.2$ gene determines expression of alanine 415 (A_{415}) (Lin et al., 1997). We have found no evidence that alternative splicing of exon 10 in $Ca_v2.2$ modifies channel function or that its expression is tissue-specific.

4.1.2. The IVS3-IVS4 Extracellular Linker

The alternatively spliced cassette exon in domain IVS3-IVS4 of $Ca_v2.1$ is only 6 nucleotides in length (Figure 6) and encodes the dipeptide sequence asparagine-proline (NP). A homologous 6 nucleotide exon is also present in the closely related $Ca_v2.2$ gene (Figure 9). The NP-encoding exon of the human $Ca_v2.1$ gene is located within approximately 9 kb of intron sequence, and it is differentially expressed in the mammalian nervous system (Bourinet et al., 1999; Lin et al., 1999; Krovetz et al., 2000; Soong et al., 2002). The functional consequences of exon 31a in the IVS3-IVS4 extracellular linker of $Ca_v2.1$ differ somewhat among the various studies. The presence of exon 31a has been reported to slow activation and inactivation kinetics of the P/Q-type current (Hans et al., 1999; Krovetz et al., 2000), whereas others report no effects on channel inactivation (Bourinet et al., 1999).

Interestingly, and of special significance pharmacologically, alternative splicing in the IVS3-IVS4 linker of the P/Q-type $Ca_v2.1$ gene modifies ω-agatoxin IVA sensitivity (Bourinet et al., 1999; Hans et al., 1999). The inclusion of NP ($Ca_v2.1$e[31a]) decreases toxin affinity 11-fold compared to splice isoforms of $Ca_v2.1$ that lack the NP encoding exon ($Ca_v2.1$e[Δ31a]) (Figure 7). This difference in ω-Aga-IVA toxin sensitivity between IVS3-IVS4 splice variants of $Ca_v2.1$ has led to the interesting suggestion that alternative splicing of exon 31a underlies at least some of the pharmacological differences between P-type and Q-type channels described in various neurons (Sather et al., 1993; Stea et al., 1994; Randall and Tsien, 1995; Bourinet et al., 1999). The implication is that P-type channels that are highly sensitive to ω-Aga-IVA contain the $Ca_v2.1$e[Δ31a] subunit, whereas Q-type channels that are less sensitive to ω-Aga-IVA contain $Ca_v2.1$e[31a] (Bourinet et al., 1999). Additional factors need to be considered and likely to contribute to the final phenotype of native currents. (i) The difference in ω-Aga-IVA sensitivity between native P-type and Q-type currents is ~ 100-fold, compared to the 10-fold difference between $Ca_v2.1$e[Δ31a] and $Ca_v2.1$e[31a] isoforms. (ii) The absolute levels of ω-Aga-IVA required to inhibit cloned $Ca_v2.1$e[Δ31a] channels are also higher than those required to block native P-type currents (Sather et al., 1993; Bourinet et al., 1999; Tsunemi et al., 2002). (iii) Cell-specific factors are known to influence the pharmacology and functional properties of the $Ca_v2.1$ subunit (Bourinet et al., 1999; Tsunemi et al., 2002). For example, Snutch and colleagues note that the pharmacological differences in ω-Aga-IVA-sensitivities between $Ca_v2.1$e[Δ31a] and $Ca_v2.1$e[31a] observed in mammalian kidney cells are not reproduced when the clones are expressed in *Xenopus* oocytes (Figure 7) (Bourinet et al., 1999).

The distribution of exon 31a of $Ca_v2.1$ has been analyzed by ribonuclease protection (Lin et al., 1999) and by RT-PCR (Bourinet et al., 1999). However, single cell RT-PCR analysis combined with functional analysis of the P/Q-type current in individual neurons is ultimately needed to link the presence of exon 31a to a specific channel phenotype. Analysis of RNA isolated from Purkinje cells suggests that exon 31a of $Ca_v2.1$ is absent from these cells, at least consistent with the hypothesis that $Ca_v2.1$ subunits lacking exon 31a correlate with expression of P-type currents (Toru et al., 2000; Tsunemi et al., 2002). The pharmacological difference generated by alternative splicing of the NP encoding exon in the IVS3-IVS4 domain, which is close to one of the channel voltage sensors, is also consistent with the proposal that ω-Aga-IVA inhibits P-type channels by modifying channel gating (McDonough et al., 1997). Voltage-gated channels other than calcium channels are also inhibited by toxins that target their respective S3-S4 linkers (Rogers et al., 1996; Swartz and MacKinnon, 1997; Cestele et al., 1998). In the case of the voltage-gated Na channel α subunit, a single amino acid (glycine) within the IIS3-IIS4 linker plays a major role in β-scorpion toxin binding (Cestele et al., 1998).

4.1.3. The Intracellular Domain II-III Loop

The presence of L_{II-III} isoforms of $Ca_v2.1$ was first suggested from studies with isoform-specific antibodies. Two L_{II-III} isoforms of $Ca_v2.1$ were found to be present in presynaptic nerve terminals based on immunohistochemical studies (Sakurai et al., 1996; Westenbroek et al., 1998; Wu et al., 1999), but their precise expression patterns differ (Sakurai et al., 1996). Presynaptic SNARE proteins bind to the L_{II-III} of Ca_v2 subunits, and interestingly, L_{II-III} isoforms were found to interact differently with syntaxin, SNAP-25, and synaptotagmin (Rettig et al., 1996; Kim et al., 1997; Zhong et al., 1999). It is thus possible that L_{II-III} isoforms of $Ca_v2.1$ might differ in their ability to couple excitation to secretion at synapses (Catterall, 1999). The mechanisms that give rise to these so-called rbA and B1 isoforms have not yet been determined. A putative 72 nucleotide open reading frame, flanked by consensus ag-gt splice junctions that might represent a 24 amino acid encoding exon, can be located in the L_{II-III} region of the human $Ca_v2.1$ gene. However, this exon was not detected in recent PCR screening of various regions of the human brain, and its location is outside of the variable region of rbA and BI isoforms (Rettig et al., 1996). Two other splice sites have been identified in the L_{II-III} region of the human $Ca_v2.1$ gene that involve exons 16 and 17, although splicing at this locus is rare and the functional significance is presently unclear (Soong et al., 2002).

4.1.4. The C-Terminus

The C-terminus of $Ca_v2.1$ can be encoded by 13 possible exons (exons 36-47) (Ophoff et al., 1996). There is strong evidence that exons 37a/37b, 43, 44, and 47 are subject to alternative splicing (Figure 6) (Zhuchenko et al., 1997; Bourinet et al., 1999; Krovetz et al., 2000; Soong et al., 2002).

Exons 37a and 37b are both 97 nucleotides in length and are spliced into fully processed $Ca_v2.1$ mRNAs in a mutually exclusive pattern. Nine amino acids differ between exons 37a and 37b and encode different forms of an EF-hand motif (Krovetz et al., 2000; Soong et al., 2002). Both isoforms are expressed in the human and rat brain and in human spinal cord. The functional significance of splicing at this locus in $Ca_v2.1$ is as yet undetermined (Bourinet et al., 1999; Krovetz et al., 2000; Soong et al., 2002), although splice junctions are conserved in the $Ca_v2.2$ gene, suggesting functional significance (Figure 9).

Alternative splicing of exon 44 in $Ca_v2.1$ has been reported to regulate inactivation kinetics of the human spinal cord P/Q-type channel (Krovetz et al., 2000). Others have shown that exon 43 is also subject to alternative splicing in human brain and, together with exon 44, modifies the efficiency of P/Q-type channel expression (Soong et al., 2002).

The most substantial changes in sequence generated by alternative splicing in the C-terminus, however, arise from the use of an alternate 3' acceptor site in the intron upstream of exon 47. Splicing at this site introduces a frameshift and early truncation of the C-terminus that lacks exon 47. Exon 47 encodes >230 amino acids in the long splice isoform of $Ca_v2.1$. Both of these $Ca_v2.1$ splice isoforms are expressed in human brain (Soong et al., 2002), spinal cord (Krovetz et al., 2000), and neuroblastoma cells (Hans et al., 1999). Interestingly, only certain splice isoforms of $Ca_v2.1$ that contain exon 47 express the SCA6 polyglutamine expansion that underlies spinocerebellar ataxia type 6 (Zhuchenko et al., 1997). Some speculate that an up-regulation of exon 47 inclusion during development might help explain the delayed onset spinocerebellar ataxia phenotype (Soong et al., 2002). The other well-documented feature of the C-termini of voltage-gated calcium channel $Ca_v\alpha_1$ subunits is sub-cellular targeting. Splice isoform-specific interactions between Mint1 and CASK and the C-termini of both $Ca_v2.1$ and $Ca_v2.2$ subunits have been reported (Maximov et al., 1999). Mint1 and CASK only interact with the long C-terminal splice isoform that contains exon 47, suggesting that alternative splicing at this site might play a role in regulating the density of Ca^{2+} channels at presynaptic nerve terminals (Maximov et al., 1999).

4.2. Splice Isoforms of N-type $Ca_v2.2$ Channels

The human $Ca_v2.2$ gene is located at chromosome position 9q34 and spans approximately 300 kb of genomic sequence. It can be encoded by at least 50 exons, at least ten of which can be alternatively spliced (Figure 9). Sites of alternative splicing include exon 10 in the I-II intracellular loop, exons 18a, 19, 20, and 21 in the II-III intracellular loop, exon 31a in the IVS3-IVS4 extracellular linker, and exons 37a/37b, and 46 in the C-terminus. Presently, there is no evidence for alternative splicing in the transmembrane domains of the $Ca_v2.2$ gene, although splicing in the transmembrane domains of other $Ca_v\alpha_1$ genes has been reported (Snutch et al., 1991; Diebold et al., 1992).

4.2.1. The IVS3-IVS4 Extracellular Linker

Exon 31a of $Ca_v2.2$ is present in human, mouse, and rat genes and encodes the dipeptide sequence glutamate-threonine (ET). Exon 31a of $Ca_v2.2$ is expressed in a tissue-specific pattern, and its presence alters Ca^{2+} channel function by slowing channel activation (Lin et al., 1997). In rat and human, we showed that exon 31a is preferentially expressed in neurons of the peripheral nervous system but suppressed within the central nervous system (Figure 10). Analysis of the $Ca_v2.2$ genomic sequence in this region reveals that exon 31a is buried within 14 kb of intron sequence and flanked by consensus ag-gt splice sequences (Figure 9) (Lin et al., 1997; Lin et al., 1999).

The mechanism by which modification of the IVS3-IVS4 linker in voltage-gated ion channels affects the kinetics and voltage-dependence of channel opening is not known. Curiously, the IVS3-IVS4 alternative exons of $Ca_v2.1$ and $Ca_v2.2$ encode different dipeptides (ET and NP, respectively), yet both slow Ca^{2+} channel activation kinetics to a similar degree (Lin et al., 1999). The close proximity of the IVS3-IVS4 extracellular linker to one of four putative voltage sensors (IVS4) is consistent with the fact that modifications in the composition of this region influence channel activation. It is possible that the length or the secondary structure of the IVS3-IVS4 linker places restraints on the movement of the adjoining IVS4 helix, thereby influencing the kinetics of channel activation. Consistent with this hypothesis, we observed differences in the time course of gating currents, thought to arise from the coordinated movement of the S4 helices, in cells expressing IVS3-IVS4 splice variants of $Ca_v2.2$ (S. I. McDonough and D. L., unpublished observations). Irrespective of the mechanism by which alterations in the IVS3-IVS4 linkers influence the gating of voltage-gated ion channels, it is clear that this region is an important site for regulating the efficacy of coupling voltage-sensing to channel opening (Nakai et al., 1994; Mathur et al., 1997)

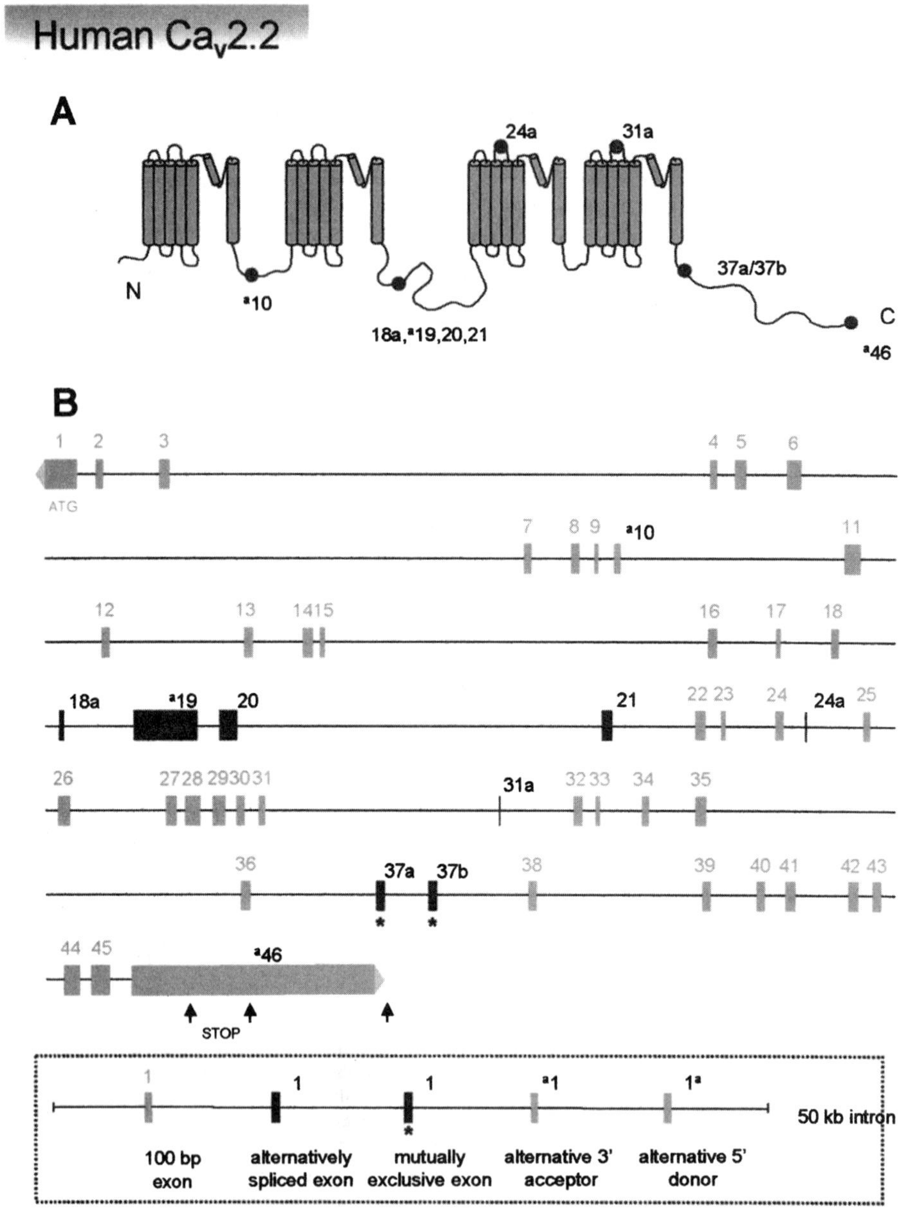

Figure 9. Structure of the human Ca$_v$2.2 gene. A) 2D topology of the Ca$_v$2.2 protein. Approximate locations of alternative splicing are highlighted with black circles. **B)** The gene structure of human Ca$_v$2.2. Exons are denoted as solid boxes and introns as thin lines. Constitutive exons are shown in gray and alternatively spliced exons are shown in black. The gene structure was constructed from analysis of human genome sequence contig NT_023929.14. Small arrows indicate the location of alternative polyadenylation sites. Exons are numbered according to the previously published gene structure (Lipscombe et al., 2002).

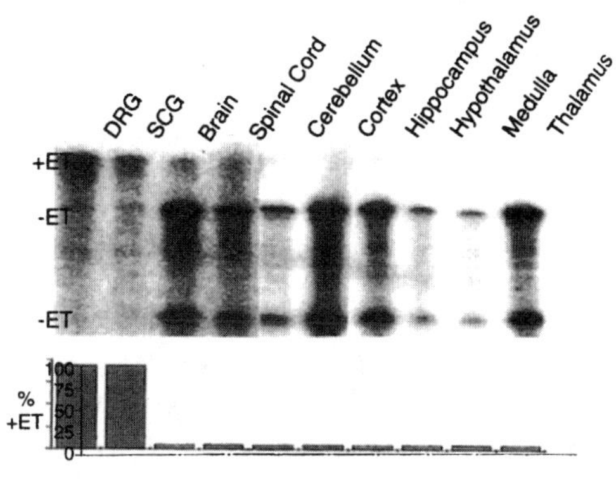

Figure 10. Dominant expression of $Ca_v2.2$ exon 31a (encoding ET) in peripheral neurons. RNase protection analysis of $Ca_v2.2$ RNA isolated from dorsal root ganglia (DRG), superior cervical ganglia (SCG), and various brain regions of adult rats, using a complementary probe that extends from nt 4379 to nt 4836 and contains the ET insert at nt 4675. **Top,** gel separation of RNase digested ^{32}P-labeled probe hybridized to the various RNA samples. +ET corresponds to the fully protected probe (463 bases), indicating high levels of ET-containing $Ca_v2.2$ RNA; the two shorter bands (-ET, 296 and 161 bases) most prominent in the CNS lanes correspond to cleaved probe, indicating high levels of ET-lacking $Ca_v2.2$ RNA. **Bottom,** average levels of +ET RNA; band intensities were calculated with a phosphorimager. Reprinted with permission from Lin et al., 1999; copyright 1999 by the Society for Neuroscience.

Currently, there is no drug or toxin that discriminates acceptably between $Ca_v2.2e[\Delta 31a]$ and $Ca_v2.2e[31a]$ splice isoforms. The widely used *Conus* toxin, ω-conotoxin GVIA, inhibits IVS3-IVS4 splice variants of $Ca_v2.2$ equally well (Z. Lin, Y. Lin, and D. L., unpublished observations), consistent with structure-function studies that place the main site of ω-conotoxin GVIA binding at sites distant from the IVS3-IVS4 extracellular linker of $Ca_v2.2$ (Ellinor et al., 1994). There are suggestions that certain *Conus* toxins can discriminate between the IVS3-IVS4 splice isoforms of Ca_v2 (Lewis et al., 2000); these pharmacological differences, however, are not observed when $Ca_v2.2$ isoforms are co-expressed with exogenous $Ca_v\beta$ subunits (Lewis et al., 2000).

4.2.2. The Long Intracellular Domain II-III Loop

The $L_{II\text{-}III}$ region of the $Ca_v2.2$ gene is made up of exons 18 through 21 and is dominated by exon 19, which encodes 266 amino acids (Figure 9). $L_{II\text{-}III}$ isoforms of the N-type $Ca_v2.2$ subunit, identified in mRNA from the mammalian nervous system (Coppola et al., 1994; Ghasemzadeh et al., 1999; Pan and Lipscombe, 2000; Kaneko et al., 2002), have been shown at the genomic level to arise from cell type-specific alternative splicing of exon 18a (Pan and Lipscombe, 2000) and from large deletions of several exons that encode regions of $L_{II\text{-}III}$ (Kaneko et al., 2002). The deletion isoforms have been found in human brain, and both lack critical elements in $L_{II\text{-}III}$ thought to mediate the binding of synaptic proteins. One $Ca_v2.2$ isoform is missing all of exons 19 through 21 (Kaneko et al., 2002).

The alternatively spliced exon 18a of $Ca_v2.2$, which encodes a 21 amino acid segment close to the N-terminus of $L_{II\text{-}III}$, is of interest to our laboratory because its expression appears to be cell type-specific and developmentally regulated. The majority of $Ca_v2.2$ mRNA in adult sympathetic ganglia contains exon 18a (>80%). However, in the adult central nervous system, exon 18a is lowest in more rostral structures such as neocortex (where < 20% of $Ca_v2.2$ mRNAs contain exon 18a) (Pan and Lipscombe, 2000). There is evidence that exon 18a is preferentially expressed in regions of the adult rat brain enriched in monoaminergic neurons, based on *in situ* hybridization and RT-PCR analysis (Ghasemzadeh et al., 1999). More studies are needed to map the cell-specific splicing patterns of the L_{II-III} region of $Ca_v2.2$, but the close proximity of exon 18a to the putative site of SNARE binding, together with the results of Ghasemzadeh and colleagues (Ghasemzadeh et al., 1999), raise the possibility that splicing in this region of $Ca_v2.2$ may be important for targeting specific splice variants to release sites that contain distinct neurotransmitters. Clearly, it will also be interesting to know whether the presence of exon 18a, which is enriched in serine and threonine residues, influences the binding of synaptic proteins. It should be noted that the significance of the alternative spliced exon 18a of $Ca_v2.2$ in the human brain was recently challenged (Kaneko et al., 2002). Based on their analyses, Kaneko and colleagues failed to amplify any $Ca_v2.2$ cDNAs that contained exon 18a from human brain. We have shown that exon 18a is indeed expressed in human brain and that its expression is up-regulated in the adult (A. C. Gray and D. L., unpublished observation). Analysis of the rat and human $Ca_v2.2$ gene structures reveals that exon 18a is a cassette exon located within approximately 9 kb of genomic sequence and flanked by consensus ag/gt dinucleotide splicing sequences (Figure 9). The 3' intron also contains an alternative dinucleotide splice-acceptor, ag, that encodes the amino acid arginine (R_{758}). Both splice junctions are utilized (Coppola et al., 1994; Cahill et al., 2000; Pan and Lipscombe, 2000; Kaneko et al., 2002). The functional significance of R_{758} in $Ca_v2.2$ is not known.

Although it is still unclear whether inclusion of exon 18a in $Ca_v2.2$ influences the efficacy of neurotransmitter release, its presence in L_{II-III} of $Ca_v2.2$ does have a functional effect on the biophysical properties of the N-type Ca^{2+} channel. We reported that exon 18a shifts the voltage dependence of steady-state inactivation of the N-type Ca^{2+} channel in the depolarizing direction, depending on which $Ca_v\beta$ subunit is expressed (Pan and Lipscombe, 2000). Other effects of exon 18a on N-type channel function, however, appear to dominate when $Ca_v2.2$ subunits are studied in mammalian cells. Specifically, exon 18a protects the N-type channel from cumulative inactivation induced by trains of action potential waveforms (C. Thaler and D. L., unpublished observations).

Recently identified $Ca_v2.2$ isoforms that are missing large segments of L_{II-III} have also been shown to have altered channel properties (Kaneko et al., 2002). The voltage-dependence of inactivation of two deletion isoforms was each shifted towards more depolarized potentials. In addition, the L_{II-III} deletion isoforms recovered more rapidly from inactivation compared to $Ca_v2.2$ channels containing a complete L_{II-III} (Kaneko et al., 2002). An unexpected but especially interesting observation from this recent study suggests that the L_{II-III} isoform lacking exons 19 though 21 is less sensitive to the *Conus* toxins ω-conotoxin MVIIA and GVIA (see McDonough, this volume, and Norton et al., this volume). This pharmacological difference may reflect the fact that both these toxins preferentially inhibit the inactive state of the N-type Ca^{2+} channel (Stocker et al., 1997), and that the deletion isoform of $Ca_v2.2$ requires stronger depolarizations for its inactivation. Genomic analysis has provided insight into the mechanism by which these $Ca_v2.2$ splice isoforms are generated (Kaneko et al., 2002). The shortest L_{II-III} splice isoform is generated by skipping exons 19, 20 and 21 and involves the use of consensus *gt-ag* dinucleotide splice junctions. The splicing pattern required to generate the second $Ca_v2.2$ L_{II-III} isoform reported in the Kaneko study is atypical (Kaneko et al., 2002).

4.2.3. The C-Terminus

In $Ca_v2.2$, the C-terminus can be encoded by at least 12 exons (exons 36 – 46). There are two putative sites of alternative splicing in the C-terminus, exons 37a/37b and exon 46. The proximal end of the C-terminus is encoded by either exon 37a or exon 37b. Both exons are present in the human, rat, and mouse genomes and appear to be expressed in a mutually exclusive pattern (Figure 9). As discussed above, the exon 37a/37b locus in $Ca_v2.2$ is homologous to that of the $Ca_v2.1$ gene. Exons 37a and 37b of $Ca_v2.2$ are approximately 60% identical at the nucleotide level and differ in 14 of their 32 amino acids. The presence of exon 37a was deduced first from analysis of the $Ca_v2.2$ genomic sequence (Mittman and Agnew, 2000). Until recently, all $Ca_v2.2$ clones isolated from the mammalian nervous system contained only exon 37b. We designed exon-specific primers because the identical size of exons 37a and 37b precluded distinction between splice isoforms of $Ca_v2.2$ containing these exons on the basis of size. We found that exon 37a was restricted to dorsal root ganglia and essentially absent from all other regions of the mammalian nervous system (T. J. Bell, C. Thaler, A. J. C., and D. L., manuscript in submission). It will be of great interest to determine whether exon 37a encodes a unique protein interaction site that has relevance to nociception.

Exon 46 in the C-terminus of $Ca_v2.2$ is expansive. It is greater than 3000 nucleotides in length and contains a putative alternative 3' acceptor site that is hypothesized to generate two different size forms of $Ca_v2.2$ (Williams et al., 1992; Lu and Dunlap, 1999; Maximov and Bezprozvanny, 2002). Exon 46 also contains three sites of polyadenylation, two of which are used to generate different size $Ca_v2.2$ mRNAs in the nervous system (Schorge et al., 1999). Although we have been unable to detect splice isoforms of $Ca_v2.2$ in the rat nervous system arising from alternative splicing in exon 46, others have done so (Maximov and Bezprozvanny, 2002). The relative levels of the short and long form of $Ca_v2.2$ remain to be established, but functional studies suggest that they are differentially localized in neurons. Specifically, the long form of $Ca_v2.2$ targets to synapses, via its interaction with Mint1 and CASK, whereas the shorter isoform of $Ca_v2.2$ predominantly localizes to soma and dendrites (Maximov and Bezprozvanny, 2002).

4.3. Splice Isoforms of $Ca_v2.3$ Channels

The third member of the Ca_v2 gene family is thought to encode at least part of what is generally referred to as the R-type current. This is defined operationally as the component of the non-T-type calcium current in neurons that is resistant to blockers of the N-type, P/Q-type and L-type channels (usually ω-CTx-GVIA, ω-Aga-IVA or ω-CTx-MVIIC, and nimodipine or nifedipine). The properties of the R-type current vary among different neurons, and some of this variability likely results from alternative splicing of $Ca_v2.3$ (Pereverzev et al., 2002). The $Ca_v2.3$ channel is less well characterized than N-type and P/Q-type channels because of the lack of selective toxins to inhibit its activity. However, there is no doubt that $Ca_v2.3$ has important roles in calcium signaling in excitable cells (see Saegusa and Tanabe, this volume, and Mori et al., this volume). $Ca_v2.3$ is expressed in brain as well as kidney and heart (Weiergraber et al., 2000) and couples to insulin secretion in an insulinoma cell line (Vajna et al., 2001).

It is important to note that a fraction of the resistant current generally termed R-type, which might be quite substantial in certain neurons, may not be carried by $Ca_v2.3$ channels, but rather by $Ca_v1.3$ L-type channels, since current arising from $Ca_v1.3$ is incompletely inhibited by dihydropyridines (Xu and Lipscombe, 2001). Support for this hypothesis comes from studies of neurons isolated from $Ca_v2.3$ knockout mice, which still express substantial current resistant to the standard inhibitors (Wilson et al., 2000; Miller, 2001).

The $Ca_v2.3$ subunit is subject to alternative splicing in the $L_{II\text{-}III}$ and C-terminus. Some studies demonstrate that this has functional consequences (Pereverzev et al., 2002). In the $L_{II\text{-}III}$ region of the $Ca_v2.3$ gene, there is a cassette exon (exon 18a) that encodes 19 amino acids. Multiple $Ca_v2.3$ mRNA sequences have been reported, some containing and others lacking this 57 nucleotide sequence. RT-PCR analysis also indicates that this exon is differentially expressed in neocortex and cerebellum (Pereverzev et al., 1998; Schramm et al., 1999). The presence of exon 18a appears to modify inactivation of the $Ca_v2.3$ channel when calcium is the charge carrier (Pereverzev et al., 2002). Approximately 100 amino acids downstream of the exon 18a locus, an additional 7 amino acid insertion has been identified in $L_{II\text{-}III}$ of $Ca_v2.3$, suggesting the presence of a second site of alternative splicing (Schneider et al., 1994; Pereverzev et al., 1998).

4.4. Therapeutic Considerations

Presynaptic calcium channels are potentially valuable therapeutic targets (Cohen and Kraus, this volume). Interest in blockers of $Ca_v2.1$ P/Q-type channels is somewhat limited. These channels mediate neuromuscular transmission (Bowersox et al., 1995) and play a dominant role in controlling neurotransmitter release in the central nervous system (Dunlap et al., 1995), functions that could complicate therapeutic treatment. $Ca_v2.2$ N-type and $Ca_v2.3$ R-type channels in general play more of a supporting role in regulating transmitter release in the central nervous system. Splice variants of these genes make potentially attractive drug targets (Cox, 2000; Vanegas and Schaible, 2000). N-type Ca^{2+} channel blockers in particular have been shown to be effective analgesics, though their clinical usefulness is complicated by unwanted side effects in part from inhibition of sympathetic tone (Miljanich and Ramachandran, 1995; Vanegas and Schaible, 2000). A drug with high selectivity for the N-type calcium channel in the pain pathway compared to the N-type calcium channel that drives sympathetic tone would be of tremendous therapeutic value. Cell-specific expression of $Ca_v2.2$ isoforms offers an approach to target N-type calcium channels of specific cell populations that control specific signaling pathways, including nociception.

5. ALTERNATIVE SPLICING IN THE Ca_v3 GENES

The Ca_v3 genes encode the T-type calcium channel family. T-type currents have several distinctive features. They activate at relatively hyperpolarized voltages and have been referred to as low voltage-activated (LVA); they close slowly as the membrane potential rapidly repolarizes in response to both square pulse depolarizations and also action potential waveforms; and they are more or less resistant to toxins and drugs that inhibit Ca_v1 and Ca_v2 channels (see Ertel, this volume). A variety of excitable cells including neurons, cardiac cells, and pituitary cells, in addition to liver and kidney, express T-type channels (Nilius et al., 1985; Huguenard, 1998; Triggle, 1998; Perez-Reyes, 2003). T-type channels underlie rhythmic firing in neurons and in the SA node of the heart and regulate hormone release.

There is a relatively high degree of variability among T-type currents studied in different tissues and cell-types (see Ertel, this volume). Much of this variability originates from the expression of three distinct Ca_v3 genes with functionally distinct properties including activation, deactivation, and inactivation kinetics; voltage-dependence of activation and inactivation; and calcium and barium permeability (Lacinova et al., 2000a; Monteil et al., 2000b; McRory et al., 2001; Chemin et al., 2002; Perez-Reyes, 2003). Perhaps most distinctive are T-type channels that contain the $Ca_v3.3$ subunit. These inactivate with significantly slower kinetics compared to $Ca_v3.1$ and $Ca_v3.2$ (Monteil et al., 2000b; McRory et al., 2001; Chemin et al., 2002; Perez-Reyes, 2003). The three Ca_v genes have distinct expression profiles (Talley et al., 1999; McRory et al., 2001; Perez-Reyes, 2003). All three genes are expressed in the mammalian brain. $Ca_v3.3$ is not found significantly in other tissues, whereas $Ca_v3.2$ is expressed in particular abundance in kidney. $Ca_v3.1$ and $Ca_v3.2$ are both expressed in heart.

Additional functional diversity among the T-type channel family arises from alternative splicing (Mittman et al., 1999a; Cribbs et al., 2000; Zhuang et al., 2000; Chemin et al., 2001a, b; Murbartian et al., 2002). Mittman and colleagues reported the structures of the human $Ca_v3.1$ and $Ca_v3.3$ genes (Mittman et al., 1999a). The $Ca_v3.1$ gene, located on chromosome 17q22, spans >66 kb and contains at least 38 exons. Alternative splicing occurs in at least 6 sites, including $L_{II\text{-}III}$, $L_{III\text{-}IV}$, and the C-terminus of $Ca_v3.1$ (Mittman et al., 1999a; Cribbs et al., 2000; Monteil et al., 2000a; Chemin et al., 2001a). There are also at least two polyadenylation signal sequences within the 3' untranslated region (Mittman et al., 1999a). Functional analyses of isoforms of $Ca_v3.1$ that arise from alternative splicing in intracellular loops $L_{II\text{-}III}$ and $L_{III\text{-}IV}$ involving exon 16 and exons 25/26 respectively demonstrate effects on the voltage-dependence and kinetics of T-type channel gating (Chemin et al., 2001a).

The $Ca_v3.3$ gene is located on chromosome 22q12-13, spans > 115 kb, and contains at least 36 exons. Based on analysis of the gene structure, 3 sites of alternative splicing were predicted. These involve exon 9 encoding 35 amino acids in $L_{I\text{-}II}$, an alternate splice acceptor within exon 33 that determines the expression of 13 amino acids in the C-terminus (Mittman et al., 1999b), and an additional site in exon 36 that involves expression of a single amino acid (Chemin et al., 2001b). Exon 9 of $Ca_v3.3$ has not been detected in rat tissue (Mittman et al., 1999b; Murbartian et al., 2002), although other splice isoforms of $Ca_v3.3$ have been isolated from rat brain (McRory et al., 2001; Murbartian et al., 2002). These involve 3 amino acids in domain IS6, 6 amino acids in $L_{II\text{-}III}$, the intracellular linker IVS4-IVS5, and several C-terminal splice isoforms of different size arising from the use of alternate 3' splice acceptor sites in exon 34 (McRory et al., 2001; Murbartian et al., 2002). Functional studies comparing different C-terminus splice isoforms of $Ca_v3.3$ suggest that alternative splicing in this region of the channel modifies activation, inactivation and deactivation properties (Chemin et al., 2001b). Alternative splicing appears to affect the transition rates between different states of the $Ca_v3.3$ channel rather than the voltage-dependence of these transitions (Murbartian et al., 2002).

At this writing, little is known about how alternative splicing might modify the pharmacological phenotypes of T-type calcium channels. However, as discussed previously (Section 3.1.2, Section 4.1.1.), modification of channel gating can be expected to have significant effects on the effectiveness of drugs whose mechanism of actions are state dependent. T-type channels are targets of antiepileptic agents such as ethosuximide, antihypertensives such as mibefradil, and are potential targets for antiarrhythmics (Triggle, 1998).

6. AUXILIARY SUBUNITS OF VOLTAGE-GATED CALCIUM CHANNELS

At least three other proteins associate with and modify the behavior of $Ca_v\alpha_1$ subunits in the macromolecular calcium channel complex: $Ca_v\beta$, $Ca_v\alpha_2\delta$, and γ. Multiple genes encode each subunit, and several genes are already known to be subject to alternative splicing (Birnbaumer et al., 1998; Walker and De Waard, 1998; Klugbauer et al., 1999; Burgess et al., 2001; Helton et al., 2002; Qin et al., 2002). In a few instances, the functional consequences of alternative splicing among these auxiliary subunits have been addressed.

6.1. Splice Isoforms of $Ca_v\beta_2$ and $Ca_v\beta_4$ subunits

$Ca_v\beta$ subunits are the best studied of the Ca_v auxiliary subunits. Their effects on channel gating can be substantial. Differential pairing among different $Ca_v\alpha_1$ and $Ca_v\beta$ subunits contributes greatly to channel diversity (Birnbaumer et al., 1998; Walker and De Waard, 1998). Alternative splicing, however, also plays a key role in fine-tuning channel gating and, at least in $Ca_v\beta_4$, channel pharmacology.

All $Ca_v\beta$ subunits are subject to alternative splicing, and two examples, involving the N-terminus of $Ca_v\beta_2$ and $Ca_v\beta_4$, are particularly notable. The $Ca_v\beta_2$ subunit is subject to alternative splicing in the N-terminal region that generates $Ca_v\beta_{2a}$ and $Ca_v\beta_{2b}$ isoforms. The $Ca_v\beta_{2a}$ isoform uniquely and dramatically slows the inactivation kinetics of associated $Ca_v\alpha_1$ proteins (Walker and De Waard, 1998). The $Ca_v\beta_{2a}$ splice isoform also contains two cysteines in its N-terminus that serve as targets for palmitoylation (Chien et al., 1996). The presence of these cysteines in $Ca_v\beta_{2a}$, but not in other $Ca_v\beta$ subunits, has been shown to be important in conferring some of the unique properties of $Ca_v\beta_{2a}$ including slowing of channel inactivation, decreasing the voltage dependence of inactivation, and suppressing pre-pulse facilitation (Birnbaumer et al., 1998; Qin et al., 1998). Evidence also indicates that palmitoylation of $Ca_v\beta_{2a}$ is dynamically regulated at the cellular level. Studies of native N-type and P/Q-type currents in chromaffin cells show that the degree of palmitoylation of $Ca_v\beta_{2a}$ correlates with the sensitivity to voltage-dependent inactivation (Hurley et al., 2000).

Alternative splicing of the $Ca_v\beta_4$ subunit also modifies the properties of certain $Ca_v\alpha_1$ subunits (Helton and Horne, 2002; Helton et al., 2002). The structure of the human $Ca_v\beta_4$ gene is shown in Figure 11. It is located at chromosome position 2q21 and spans greater than 300 kb of human genomic sequence. $Ca_v\beta_4$ can be encoded by at least 15 exons, three of which are known to be subject to alternative splicing. A splice isoform of $Ca_v\beta_4$ has been identified in human spinal cord, characterized by a short N-terminus and generated by the inclusion of exon 1c (Helton and Horne, 2002). When compared, $Ca_v\beta_4$e[1c] and $Ca_v\beta_4$e[1a/1b] (which contains a long N-terminus) were found to have different voltage dependences of channel activation and inactivation, and different rates of inactivation. These functional differences between splice isoforms of $Ca_v\beta_4$ were only observed upon co-expression with $Ca_v2.1$ and $Ca_v2.2$. In contrast, when co-expressed with $Ca_v1.2$ or $Ca_v2.3$ subunits, $Ca_v\beta_4$e[1c] and $Ca_v\beta_4$e[1a/1b] splice isoforms were not functionally distinct (Helton and Horne, 2002). Overall, the short splice isoform $Ca_v\beta_4$e[1c] reduced sensitivity of $Ca_v2.1$ and $Ca_v2.2$ channels to inactivation compared to

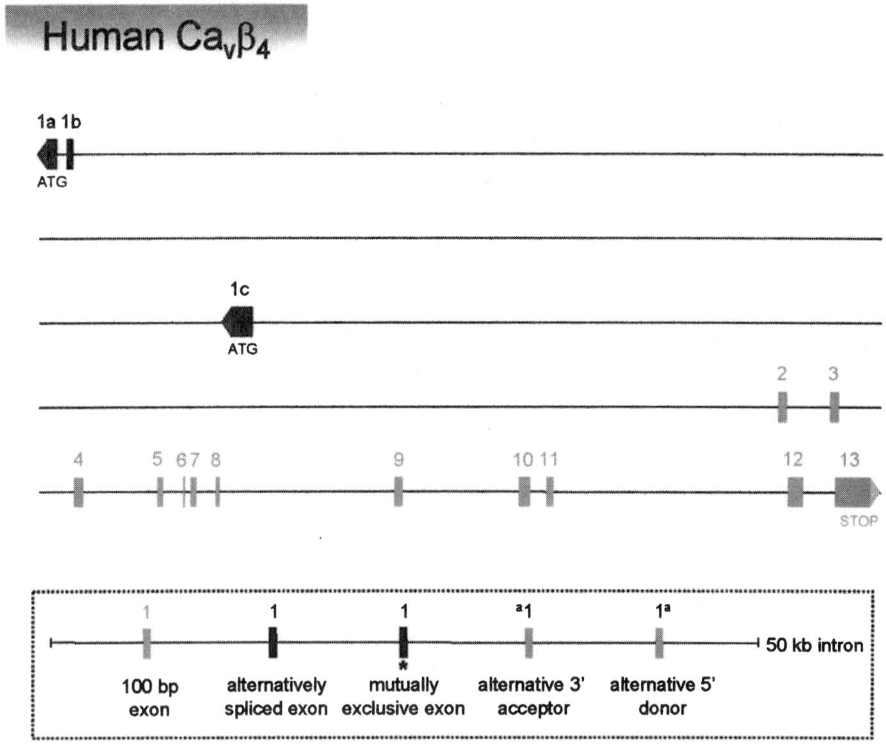

Figure 11. The gene structure of human $Ca_v\beta_4$. Exons are denoted as solid boxes and introns as thin lines. Constitutive exons are shown in gray and alternatively spliced exons are shown in black. The gene structure was constructed from analysis of human genome sequence contig NT_005273.9. Exons are numbered according to the previously published gene structure (Escayg et al., 1998).

those associated with $Ca_v\beta_4$e[1a/1b]. The presence of $Ca_v\beta_4$e[1c] slowed entry of the P/Q-type $Ca_v2.1$ channel into slow inactivation.

Most significantly, $Ca_v\beta_4$e[1c] dramatically reduced the rate of $Ca_v2.1$ channel inhibition by the *Conus* toxin ω-CTx-MVIIC (Helton et al., 2002). The major determinants of conotoxin binding to Ca_v2 channels reside in the extracellular domain close to the ion pore (Ellinor et al., 1994). However, the effects of ω-CTx-MVIIC on channels are known to be state dependent: ω-CTx-MVIIC binds preferentially to the inactive state of the channel. The $Ca_v\beta_4$ splice isoforms are differentially expressed in the human nervous system. $Ca_v\beta_4$e[1c] (named "β_{4a}" by Horne and colleagues; Figure 12) is found in many regions of the brain, whereas $Ca_v\beta_4$e[1a/1b] (named "β_{4b}" by Horne and colleagues; Figure 12) is restricted to cerebellum and is particularly prominent in the frontal cortex (Helton et al., 2002). These data predict that the P/Q-type channel of frontal cortex will have a higher sensitivity to inhibition by ω-CTx-MVIIC and will undergo stronger voltage-dependent inactivation. Helton and Horne's findings are reminiscent of Kaneko and colleagues, who reported that alternative splicing in the intracellular L_{II-III} of $Ca_v2.2$ modifies inactivation and the rate of inhibition of N-type channels by ω-conotoxins MVIIA and GVIA (Kaneko et al., 2002).

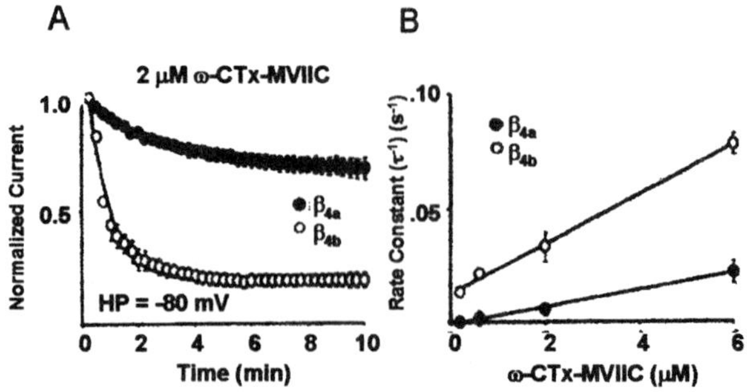

Figure 12. Splicing of β_4 alters inhibition of $Ca_v2.1$ channels by ω-CTx-MVIIC. Studies were performed on $Ca_v2.1$ channels expressed in *Xenopus* oocytes with $\alpha_2\delta$-1 and either β_{4a} or β_{4b}. **A)** Onset and degree of block by a ten minute exposure to 2 μM ω-CTx-MVIIC at a holding potential of –80 mV; each point is an average of 7 (β_{4a}) or 8 (β_{4b}) experiments. Fit is a single exponential decay function fit to the timecourse of inhibition. **B)** Rate constants for the time course of onset of toxin inhibition at four different toxin concentrations, determined from the toxin on-rate and from the steady-state degree of block. Reprinted with permission from Helton et al., 2002; copyright 2002 by the Society for Neuroscience.

7. CONSIDERATIONS FOR DRUG DEVELOPMENT

The potential number of alternative splice isoforms within the voltage gated calcium channel family is staggering. An unknown number of these isoforms likely generate pharmacologically distinct voltage-gated calcium channels. This poses a daunting short-term problem to those involved in drug screening: to select the one target clone for positive hits and an array of other clones for counter-screens. Nonetheless, the potential for achieving a greater degree of drug specificity once a target has been chosen and lead compounds identified is real. Though analyses of voltage-gated calcium channel splice isoforms are in their infancy, there are already examples of how alternative splicing modifies drug action. These include: 1) modification of dihydropyridine binding to splice isoforms of the $Ca_v1.2$ subunit; 2) ω-Aga-IVA and mibefradil inhibition of splice isoforms of $Ca_v2.1$; and 3) the effects of $Ca_v\beta_4$ subunit splice isoforms on toxin inhibition of $Ca_v2.1$. Emphasis should now focus on identifying specific splice isoforms expressed in specific target tissues and cell populations of interest: for example, $Ca_v2.2$ splice isoforms of nociceptive neurons of the dorsal root ganglia. Finally, as is often discussed, but frequently overlooked, alternative splicing may modify drug specificity via sites distant from the actual site of drug binding. Examples discussed in this chapter highlight this point. Studies of alternative splicing in voltage-gated calcium channels enhance our appreciation of genetic diversity, offer insight into domains of the channel important for modifying channel function, and provide strategies for the design of drugs with greater selectivity and efficacy.

ACKNOWLEDGMENTS

This work was supported by National Institutes of Health grants NS29967 and NS01927 (D.L.) and GM07601 (A.J.C.). The authors thank Annette C. Gray of the Lipscombe lab for the sequence alignment of Figure 1.

REFERENCES

Abernethy, D. R. and Soldatov, N. M., 2002, Structure-functional diversity of human L-type Ca^{2+} channel: perspectives for new pharmacological targets, *J Pharmacol Exp Ther* **300**:724-728.

Angus, J. A., Wright, C. E., and Xi, Q., 2000, Targetting voltage-gated calcium channels in cardiovascular therapy, *Lancet* **356**:1287-1289.

Barry, E. L., Gesek, F. A., Froehner, S. C., and Friedman, P. A., 1995, Multiple calcium channel transcripts in rat osteosarcoma cells: selective activation of α1D isoform by parathyroid hormone, *Proc Natl Acad Sci U S A* **92**:10914-10918.

Beam, K., 1999, Calcium channel splicing: mind your Ps and Qs, *Nat Neurosci* **2**:393-394.

Bean, B. P., 1984, Nitrendipine block of cardiac calcium channels: high-affinity binding to the inactivated state, *Proc Natl Acad Sci U S A* **81**:6388-6392.

Bean, B. P., 1989, Classes of calcium channels in vertebrate cells, *Annu Rev Physiol* **51**:367-384.

Bech-Hansen, N. T., Naylor, M. J., Maybaum, T. A., Pearce, W. G., Koop, B., Fishman, G. A., Mets, M., Musarella, M. A., and Boycott, K. M., 1998, Loss-of-function mutations in a calcium-channel α1-subunit gene in Xp11.23 cause incomplete X-linked congenital stationary night blindness, *Nat Genet* **19**:264-267.

Bezprozvanny, I. and Tsien, R. W., 1995, Voltage-dependent blockade of diverse types of voltage-gated Ca^{2+} channels expressed in Xenopus oocytes by the Ca^{2+} channel antagonist mibefradil (Ro 40-5967), *Mol Pharmacol* **48**:540-549.

Bielefeldt, K., 1999, Molecular diversity of voltage-sensitive calcium channels in smooth muscle cells, *J Lab Clin Med* **133**:469-477.

Birnbaumer, L., Qin, N., Olcese, R., Tareilus, E., Platano, D., Costantin, J., and Stefani, E., 1998, Structures and functions of calcium channel beta subunits, *J Bioenerg Biomembr* **30**:357-375.

Black, D. L., 1998, Splicing in the inner ear: a familiar tune, but what are the instruments?, *Neuron* **20**:165-168.

Black, D. L., 2000, Protein diversity from alternative splicing: a challenge for bioinformatics and post-genome biology, *Cell* **103**:367-370.

Bourinet, E., Soong, T. W., Sutton, K., Slaymaker, S., Mathews, E., Monteil, A., Zamponi, G. W., Nargeot, J., and Snutch, T. P., 1999, Splicing of α1A subunit gene generates phenotypic variants of P- and Q-type calcium channels, *Nat Neurosci* **2**:407-415.

Bowersox, S. S., Miljanich, G. P., Sugiura, Y., Li, C., Nadasdi, L., Hoffman, B. B., Ramachandran, J., and Ko, C. P., 1995, Differential blockade of voltage-sensitive calcium channels at the mouse neuromuscular junction by novel ω-conopeptides and ω-agatoxin-IVA, *J Pharmacol Exp Ther* **273**:248-256.

Brehm, P. and Eckert, R., 1978, Calcium entry leads to inactivation of calcium channel in Paramecium, *Science* **202**:1203-1206.

Brosenitsch, T. A. and Katz, D. M., 2001, Physiological patterns of electrical stimulation can induce neuronal gene expression by activating N-type calcium channels, *J Neurosci* **21**:2571-2579.

Burgess, D. L., Gefrides, L. A., Foreman, P. J., and Noebels, J. L., 2001, A cluster of three novel Ca^{2+} channel γ subunit genes on chromosome 19q13.4: evolution and expression profile of the γ subunit gene family, *Genomics* **71**:339-350.

Cahill, A. L., Hurley, J. H., and Fox, A. P., 2000, Coexpression of cloned α_{1B}, β_{2a}, and α_2/δ subunits produces non-inactivating calcium currents similar to those found in bovine chromaffin cells, *J Neurosci* **20**:1685-1693.

Catterall, W. A., 1991, Excitation-contraction coupling in vertebrate skeletal muscle: a tale of two calcium channels, *Cell* **64**:871-874.

Catterall, W. A., 1999, Interactions of presynaptic Ca^{2+} channels and snare proteins in neurotransmitter release, *Ann N Y Acad Sci* **868**:144-159.

Cestele, S., Qu, Y., Rogers, J. C., Rochat, H., Scheuer, T., and Catterall, W. A., 1998, Voltage sensor-trapping: enhanced activation of sodium channels by beta-scorpion toxin bound to the S3-S4 loop in domain II, *Neuron* **21**:919-931.

Chemin, J., Monteil, A., Bourinet, E., Nargeot, J., and Lory, P., 2001a, Alternatively spliced α_{1G} ($Ca_V3.1$) intracellular loops promote specific T-type Ca^{2+} channel gating properties, *Biophys J* **80**:1238-1250.

Chemin, J., Monteil, A., Dubel, S., Nargeot, J., and Lory, P., 2001b, The α_{1I} T-type calcium channel exhibits faster gating properties when overexpressed in neuroblastoma/glioma NG 108-15 cells, *Eur J Neurosci* **14**:1678-1686.

Chemin, J., Monteil, A., Perez-Reyes, E., Bourinet, E., Nargeot, J., and Lory, P., 2002, Specific contribution of human T-type calcium channel isotypes (α_{1G}, α_{1H} and α_{1I}) to neuronal excitability, *J Physiol* **540**:3-14.

Chien, A. J., Carr, K. M., Shirokov, R. E., Rios, E., and Hosey, M. M., 1996, Identification of palmitoylation sites within the L-type calcium channel β_{2a} subunit and effects on channel function, *J Biol Chem* **271**:26465-26468.

Cohen, C. J., McCarthy, R. T., Barrett, P. Q., and Rasmussen, H., 1988, Ca channels in adrenal glomerulosa cells: K^+ and angiotensin II increase T-type Ca channel current, *Proc Natl Acad Sci U S A* **85**:2412-2416.

Coppola, T., Waldmann, R., Borsotto, M., Heurteaux, C., Romey, G., Mattei, M. G., and Lazdunski, M., 1994, Molecular cloning of a murine N-type calcium channel $\alpha1$ subunit. Evidence for isoforms, brain distribution, and chromosomal localization, *FEBS Lett* **338**:1-5.

Cox, B., 2000, Calcium channel blockers and pain therapy, *Curr Rev Pain* **4**:488-498.

Cribbs, L. L., Gomora, J. C., Daud, A. N., Lee, J. H., and Perez-Reyes, E., 2000, Molecular cloning and functional expression of $Ca_v3.1c$, a T-type calcium channel from human brain, *FEBS Lett* **466**:54-58.

de Leon, M., Wang, Y., Jones, L., Perez-Reyes, E., Wei, X., Soong, T. W., Snutch, T. P., and Yue, D. T., 1995, Essential Ca^{2+}-binding motif for Ca^{2+}-sensitive inactivation of L-type Ca^{2+} channels, *Science* **270**:1502-1506.

Diebold, R. J., Koch, W. J., Ellinor, P. T., Wang, J. J., Muthuchamy, M., Wieczorek, D. F., and Schwartz, A., 1992, Mutually exclusive exon splicing of the cardiac calcium channel $\alpha1$ subunit gene generates developmentally regulated isoforms in the rat heart, *Proc Natl Acad Sci U S A* **89**:1497-1501.

Dunlap, K., Luebke, J. I., and Turner, T. J., 1995, Exocytotic Ca^{2+} channels in mammalian central neurons, *Trends Neurosci* **18**:89-98.

Ellinor, P. T., Zhang, J. F., Horne, W. A., and Tsien, R. W., 1994, Structural determinants of the blockade of N-type calcium channels by a peptide neurotoxin, *Nature* **372**:272-275.

Ertel, E. A., Campbell, K. P., Harpold, M. M., Hofmann, F., Mori, Y., Perez-Reyes, E., Schwartz, A., Snutch, T. P., Tanabe, T., Birnbaumer, L., Tsien, R. W., and Catterall, W. A., 2000, Nomenclature of voltage-gated calcium channels, *Neuron* **25**:533-535.

Escayg, A., Jones, J. M., Kearney, J. A., Hitchcock, P. F., and Meisler, M. H., 1998, Calcium channel $\beta4$ (CACNB4): human ortholog of the mouse epilepsy gene *lethargic*, *Genomics* **50**:14-22.

Fearon, I. M., Varadi, G., Koch, S., Isaacsohn, I., Ball, S. G., and Peers, C., 2000, Splice variants reveal the region involved in oxygen sensing by recombinant human L-type Ca^{2+} channels, *Circ Res* **87**:537-539.

Feron, O., Octave, J. N., Christen, M. O., and Godfraind, T., 1994, Quantification of two splicing events in the L-type calcium channel α-1 subunit of intestinal smooth muscle and other tissues, *Eur J Biochem* **222**:195-202.

Fletcher, C. F., Lutz, C. M., O'Sullivan, T. N., Shaughnessy, J. D., Jr., Hawkes, R., Frankel, W. N., Copeland, N. G., and Jenkins, N. A., 1996, Absence epilepsy in *tottering* mutant mice is associated with calcium channel defects, *Cell* **87**:607-617.

Fletcher, C. F., Copeland, N. G., and Jenkins, N. A., 1998, Genetic analysis of voltage-dependent calcium channels, *J Bioenerg Biomembr* **30**:387-398.

Fuchs, P. A., 1996, Synaptic transmission at vertebrate hair cells, *Curr Opin Neurobiol* **6**:514-519.

Gao, T., Bunemann, M., Gerhardstein, B. L., Ma, H., and Hosey, M. M., 2000, Role of the C terminus of the $\alpha1C$ ($Ca_V1.2$) subunit in membrane targeting of cardiac L-type calcium channels, *J Biol Chem* **275**:25436-25444.

Ge, K., DuHadaway, J., Du, W., Herlyn, M., Rodeck, U., and Prendergast, G. C., 1999, Mechanism for elimination of a tumor suppressor: aberrant splicing of a brain-specific exon causes loss of function of Bin1 in melanoma, *Proc Natl Acad Sci U S A* **96**:9689-9694.

Ghasemzadeh, M. B., Pierce, R. C., and Kalivas, P. W., 1999, The monoamine neurons of the rat brain preferentially express a splice variant of α_{1B} subunit of the N-type calcium channel, *J Neurochem* **73**:1718-1723.

Ghosh, A. and Greenberg, M. E., 1995, Calcium signaling in neurons: molecular mechanisms and cellular consequences, *Science* **268**:239-247.

Gidh-Jain, M., Huang, B., Jain, P., Battula, V., and el-Sherif, N., 1995, Reemergence of the fetal pattern of L-type calcium channel gene expression in non infarcted myocardium during left ventricular remodeling, *Biochem Biophys Res Commun* **216**:892-897.

Goodwin, L. O., Leeds, N. B., Hurley, I., Cooper, G. W., Pergolizzi, R. G., and Benoff, S., 1998, Alternative splicing of exons in the $\alpha 1$ subunit of the rat testis L-type voltage-dependent calcium channel generates germ line-specific dihydropyridine binding sites, *Mol Hum Reprod* **4**:215-226.

Grabowski, P. J., 1998, Splicing regulation in neurons: tinkering with cell-specific control, *Cell* **92**:709-712.

Grabowski, P. J., 2000, Genetic evidence for a Nova regulator of alternative splicing in the brain, *Neuron* **25**:254-256.

Graveley, B. R., 2001, Alternative splicing: increasing diversity in the proteomic world, *Trends Genet* **17**:100-107.

Hans, M., Urrutia, A., Deal, C., Brust, P. F., Stauderman, K., Ellis, S. B., Harpold, M. M., Johnson, E. C., and Williams, M. E., 1999, Structural elements in domain IV that influence biophysical and pharmacological properties of human alpha1A-containing high-voltage-activated calcium channels, *Biophys J* **76**:1384-1400.

Hell, J. W., Appleyard, S. M., Yokoyama, C. T., Warner, C., and Catterall, W. A., 1994, Differential phosphorylation of two size forms of the N-type calcium channel α_1 subunit which have different COOH termini, *J Biol Chem* **269**:7390-7396.

Helton, T. D. and Horne, W. A., 2002, Alternative splicing of the β_4 subunit has α_1 subunit subtype-specific effects on Ca^{2+} channel gating, *J Neurosci* **22**:1573-1582.

Helton, T. D., Kojetin, D. J., Cavanagh, J., and Horne, W. A., 2002, Alternative splicing of a $\beta 4$ subunit proline-rich motif regulates voltage-dependent gating and toxin block of $Ca_v2.1$ Ca^{2+} channels, *J Neurosci* **22**:9331-9339.

Hering, S., Berjukow, S., Sokolov, S., Marksteiner, R., Weiss, R. G., Kraus, R., and Timin, E. N., 2000, Molecular determinants of inactivation in voltage-gated Ca^{2+} channels, *J Physiol* **528 Pt 2**:237-249.

Huguenard, J. R., 1998, Low-voltage-activated (T-type) calcium-channel genes identified, *Trends Neurosci* **21**:451-452.

Hui, A., Ellinor, P. T., Krizanova, O., Wang, J. J., Diebold, R. J., and Schwartz, A., 1991, Molecular cloning of multiple subtypes of a novel rat brain isoform of the $\alpha 1$ subunit of the voltage-dependent calcium channel, *Neuron* **7**:35-44.

Hurley, J. H., Cahill, A. L., Currie, K. P., and Fox, A. P., 2000, The role of dynamic palmitoylation in Ca^{2+} channel inactivation, *Proc Natl Acad Sci U S A* **97**:9293-9298.

Ihara, Y., Yamada, Y., Fujii, Y., Gonoi, T., Yano, H., Yasuda, K., Inagaki, N., Seino, Y., and Seino, S., 1995, Molecular diversity and functional characterization of voltage-dependent calcium channels (CACN4) expressed in pancreatic β-cells, *Mol Endocrinol* **9**:121-130.

Ivanina, T., Blumenstein, Y., Shistik, E., Barzilai, R., and Dascal, N., 2000, Modulation of L-type Ca^{2+} channels by $G\beta\gamma$ and calmodulin via interactions with N and C termini of α1C, *J Biol Chem* **275**:39846-39854.

Jagannathan, S., Punt, E. L., Gu, Y., Arnoult, C., Sakkas, D., Barratt, C. L., and Publicover, S. J., 2002, Identification and localization of T-type voltage-operated calcium channel subunits in human male germ cells. Expression of multiple isoforms, *J Biol Chem* **277**:8449-8456.

Jensen, K. B., Dredge, B. K., Stefani, G., Zhong, R., Buckanovich, R. J., Okano, H. J., Yang, Y. Y., and Darnell, R. B., 2000, Nova-1 regulates neuron-specific alternative splicing and is essential for neuronal viability, *Neuron* **25**:359-371.

Jimenez, C., Bourinet, E., Leuranguer, V., Richard, S., Snutch, T. P., and Nargeot, J., 2000, Determinants of voltage-dependent inactivation affect Mibefradil block of calcium channels, *Neuropharmacology* **39**:1-10.

Jones, O. T., Bernstein, G. M., Jones, E. J., Jugloff, D. G., Law, M., Wong, W., and Mills, L. R., 1997, N-Type calcium channels in the developing rat hippocampus: subunit, complex, and regional expression, *J Neurosci* **17**:6152-6164.

Jun, K., Piedras-Renteria, E. S., Smith, S. M., Wheeler, D. B., Lee, S. B., Lee, T. G., Chin, H., Adams, M. E., Scheller, R. H., Tsien, R. W., and Shin, H. S., 1999, Ablation of P/Q-type Ca^{2+} channel currents, altered synaptic transmission, and progressive ataxia in mice lacking the α_{1A}-subunit, *Proc Natl Acad Sci U S A* **96**:15245-15250.

Kamphuis, W. and Hendriksen, H., 1998, Expression patterns of voltage-dependent calcium channel α_1 subunits (α_{1A}-α_{1E}) mRNA in rat retina, *Brain Res Mol Brain Res* **55**:209-220.

Kaneko, S., Cooper, C. B., Nishioka, N., Yamasaki, H., Suzuki, A., Jarvis, S. E., Akaike, A., Satoh, M., and Zamponi, G. W., 2002, Identification and characterization of novel human $Ca_v2.2$ (α_{1B}) calcium channel variants lacking the synaptic protein interaction site, *J Neurosci* **22**:82-92.

Kepplinger, K. J., Kahr, H., Forstner, G., Sonnleitner, M., Schindler, H., Schmidt, T., Groschner, K., Soldatov, N. M., and Romanin, C., 2000, A sequence in the carboxy-terminus of the α1C subunit important for targeting, conductance and open probability of L-type Ca^{2+} channels, *FEBS Lett* **477**:161-169.

Khan, I., Ahmad, S., and Thomas, N., 1998, Differential translation of the α-1 isoforms of L-type calcium channel in rat brain and other tissues, *Biochem Mol Biol Int* **45**:895-904.

Kim, H. L., Kim, H., Lee, P., King, R. G., and Chin, H., 1992, Rat brain expresses an alternatively spliced form of the dihydropyridine-sensitive L-type calcium channel $\alpha2$ subunit, *Proc Natl Acad Sci U S A* **89**:3251-3255.

Kim, M., Baro, D. J., Lanning, C. C., Doshi, M., Farnham, J., Moskowitz, H. S., Peck, J. H., Olivera, B. M., and Harris-Warrick, R. M., 1997, Alternative splicing in the pore-forming region of *shaker* potassium channels, *J Neurosci* **17**:8213-8224.

Klockner, U., Mikala, G., Eisfeld, J., Iles, D. E., Strobeck, M., Mershon, J. L., Schwartz, A., and Varadi, G., 1997, Properties of three COOH-terminal splice variants of a human cardiac L-type Ca^{2+}-channel $\alpha1$-subunit, *Am J Physiol* **272**:H1372-1381.

Klugbauer, N., Lacinova, L., Marais, E., Hobom, M., and Hofmann, F., 1999, Molecular diversity of the calcium channel $\alpha_2\delta$ subunit, *J Neurosci* **19**:684-691.

Koch, W. J., Ellinor, P. T., and Schwartz, A., 1990, cDNA cloning of a dihydropyridine-sensitive calcium channel from rat aorta. Evidence for the existence of alternatively spliced forms, *J Biol Chem* **265**:17786-17791.

Kollmar, R., Fak, J., Montgomery, L. G., and Hudspeth, A. J., 1997, Hair cell-specific splicing of mRNA for the α_{1D} subunit of voltage-gated Ca^{2+} channels in the chicken's cochlea, *Proc Natl Acad Sci U S A* **94**:14889-14893.

Koschak, A., Reimer, D., Huber, I., Grabner, M., Glossmann, H., Engel, J., and Striessnig, J., 2001, α_{1D} ($Ca_v1.3$) subunits can form L-type Ca^{2+} channels activating at negative voltages, *J Biol Chem* **276**:22100-22106.

Krovetz, H. S., Helton, T. D., Crews, A. L., and Horne, W. A., 2000, C-Terminal alternative splicing changes the gating properties of a human spinal cord calcium channel α_{1A} subunit, *J Neurosci* **20**:7564-7570.

Lacinova, L., Klugbauer, N., and Hofmann, F., 2000a, Low voltage activated calcium channels: from genes to function, *Gen Physiol Biophys* **19**:121-136.

Lacinova, L., Klugbauer, N., and Hofmann, F., 2000b, State- and isoform-dependent interaction of isradipine with the $\alpha1C$ L-type calcium channel, *Pflugers Arch* **440**:50-60.

Leveque, C., el Far, O., Martin-Moutot, N., Sato, K., Kato, R., Takahashi, M., and Seagar, M. J., 1994, Purification of the N-type calcium channel associated with syntaxin and synaptotagmin. A complex implicated in synaptic vesicle exocytosis, *J Biol Chem* **269**:6306-6312.

Lewis, R. J., Nielsen, K. J., Craik, D. J., Loughnan, M. L., Adams, D. A., Sharpe, I. A., Luchian, T., Adams, D. J., Bond, T., Thomas, L., Jones, A., Matheson, J. L., Drinkwater, R., Andrews, P. R., and Alewood, P. F., 2000, Novel ω-conotoxins from *Conus catus* discriminate among neuronal calcium channel subtypes, *J Biol Chem* **275**:35335-35344.

Ligon, B., Boyd, A. E., 3rd, and Dunlap, K., 1998, Class A calcium channel variants in pancreatic islets and their role in insulin secretion, *J Biol Chem* **273**:13905-13911.

Liljelund, P., Netzeband, J. G., and Gruol, D. L., 2000, L-Type calcium channels mediate calcium oscillations in early postnatal Purkinje neurons, *J Neurosci* **20**:7394-7403.

Lin, Z., Haus, S., Edgerton, J., and Lipscombe, D., 1997, Identification of functionally distinct isoforms of the N-type Ca^{2+} channel in rat sympathetic ganglia and brain, *Neuron* **18**:153-166.

Lin, Z., Lin, Y., Schorge, S., Pan, J. Q., Beierlein, M., and Lipscombe, D., 1999, Alternative splicing of a short cassette exon in α_{1B} generates functionally distinct N-type calcium channels in central and peripheral neurons, *J Neurosci* **19**:5322-5331.

Lipscombe, D., 2002, L-type calcium channels: highs and new lows, *Circ Res* **90**:933-935.

Lipscombe, D., Pan, J. Q., and Gray, A. C., 2002, Functional diversity in neuronal voltage-gated calcium channels by alternative splicing of $Ca_v\alpha_1$, *Mol Neurobiol* **26**:21-44.

Liu, R., Li, W., Karin, N. J., Bergh, J. J., Adler-Storthz, K., and Farach-Carson, M. C., 2000, Ribozyme ablation demonstrates that the cardiac subtype of the voltage-sensitive calcium channel is the molecular transducer of 1, 25-dihydroxyvitamin D(3)-stimulated calcium influx in osteoblastic cells, *J Biol Chem* **275**:8711-8718.

Lu, Q. and Dunlap, K., 1999, Cloning and functional expression of novel N-type Ca^{2+} channel variants, *J Biol Chem* **274**:34566-34575.

Lu, Q., AtKisson, M. S., Jarvis, S. E., Feng, Z. P., Zamponi, G. W., and Dunlap, K., 2001, Syntaxin 1A supports voltage-dependent inhibition of α_{1B} Ca^{2+} channels by $G\beta\gamma$ in chick sensory neurons, *J Neurosci* **21**:2949-2957.

Malouf, N. N., McMahon, D. K., Hainsworth, C. N., and Kay, B. K., 1992, A two-motif isoform of the major calcium channel subunit in skeletal muscle, *Neuron* **8**:899-906.

Marubio, L. M., Roenfeld, M., Dasgupta, S., Miller, R. J., and Philipson, L. H., 1996, Isoform expression of the voltage-dependent calcium channel $\alpha1E$, *Receptors Channels* **4**:243-251.

Mathur, R., Zheng, J., Yan, Y., and Sigworth, F. J., 1997, Role of the S3-S4 linker in Shaker potassium channel activation, *J Gen Physiol* **109**:191-199.

Maximov, A., Sudhof, T. C., and Bezprozvanny, I., 1999, Association of neuronal calcium channels with modular adaptor proteins, *J Biol Chem* **274**:24453-24456.

Maximov, A. and Bezprozvanny, I., 2002, Synaptic targeting of N-type calcium channels in hippocampal neurons, *J Neurosci* **22**:6939-6952.

McDonough, S. I., Mintz, I. M., and Bean, B. P., 1997, Alteration of P-type calcium channel gating by the spider toxin ω-Aga-IVA, *Biophys J* **72**:2117-2128.

McRory, J. E., Santi, C. M., Hamming, K. S., Mezeyova, J., Sutton, K. G., Baillie, D. L., Stea, A., and Snutch, T. P., 2001, Molecular and functional characterization of a family of rat brain T-type calcium channels, *J Biol Chem* **276**:3999-4011.

Miljanich, G. P. and Ramachandran, J., 1995, Antagonists of neuronal calcium channels: structure, function, and therapeutic implications, *Annu Rev Pharmacol Toxicol* **35**:707-734.

Miller, R. J., 2001, Rocking and rolling with Ca^{2+} channels, *Trends Neurosci* **24**:445-449.

Missler, M. and Sudhof, T. C., 1998, Neurexins: three genes and 1001 products, *Trends Genet* **14**:20-26.

Mittman, S., Guo, J., and Agnew, W. S., 1999a, Structure and alternative splicing of the gene encoding α_{1G}, a human brain T calcium channel α1 subunit, *Neurosci Lett* **274**:143-146.

Mittman, S., Guo, J., Emerick, M. C., and Agnew, W. S., 1999b, Structure and alternative splicing of the gene encoding α1I, a human brain T calcium channel α1 subunit, *Neurosci Lett* **269**:121-124.

Mittman, S. and Agnew, W. S., 2000, Three new alternatively spliced exons of the calcium channel α1 subunit gene CACNA1B, *Soc Neurosci Abstr* **40.4**:

Modrek, B. and Lee, C., 2002, A genomic view of alternative splicing, *Nat Genet* **30**:13-19.

Monteil, A., Chemin, J., Bourinet, E., Mennessier, G., Lory, P., and Nargeot, J., 2000a, Molecular and functional properties of the human α_{1G} subunit that forms T-type calcium channels, *J Biol Chem* **275**:6090-6100.

Monteil, A., Chemin, J., Leuranguer, V., Altier, C., Mennessier, G., Bourinet, E., Lory, P., and Nargeot, J., 2000b, Specific properties of T-type calcium channels generated by the human α1I subunit, *J Biol Chem* **275**:16530-16535.

Mori, Y., Niidome, T., Fujita, Y., Mynlieff, M., Dirksen, R. T., Beam, K. G., Iwabe, N., Miyata, T., Furutama, D., Furuichi, T., and et al., 1993, Molecular diversity of voltage-dependent calcium channel, *Ann N Y Acad Sci* **707**:87-108.

Murakami, M., Wissenbach, U., and Flockerzi, V., 1996, Gene structure of the murine calcium channel β3 subunit, cDNA and characterization of alternative splicing and transcription products, *Eur J Biochem* **236**:138-143.

Murbartian, J., Arias, J. M., Lee, J. H., Gomora, J. C., and Perez-Reyes, E., 2002, Alternative splicing of the rat $Ca_v3.3$ T-type calcium channel gene produces variants with distinct functional properties, *FEBS Lett* **528**:272-278.

Nakai, J., Adams, B. A., Imoto, K., and Beam, K. G., 1994, Critical roles of the S3 segment and S3-S4 linker of repeat I in activation of L-type calcium channels, *Proc Natl Acad Sci U S A* **91**:1014-1018.

Navaratnam, D. S., Bell, T. J., Tu, T. D., Cohen, E. L., and Oberholtzer, J. C., 1997, Differential distribution of Ca^{2+}-activated K^+ channel splice variants among hair cells along the tonotopic axis of the chick cochlea, *Neuron* **19**:1077-1085.

Nilius, B., Hess, P., Lansman, J. B., and Tsien, R. W., 1985, A novel type of cardiac calcium channel in ventricular cells, *Nature* **316**:443-446.

Okagaki, R., Izumi, H., Okada, T., Nagahora, H., Nakajo, K., and Okamura, Y., 2001, The maternal transcript for truncated voltage-dependent Ca^{2+} channels in the ascidian embryo: a potential suppressive role in Ca^{2+} channel expression, *Dev Biol* **230**:258-277.

Ophoff, R. A., Terwindt, G. M., Vergouwe, M. N., van Eijk, R., Oefner, P. J., Hoffman, S. M., Lamerdin, J. E., Mohrenweiser, H. W., Bulman, D. E., Ferrari, M., Haan, J., Lindhout, D., van Ommen, G. J., Hofker, M. H., Ferrari, M. D., and Frants, R. R., 1996, Familial hemiplegic migraine and episodic ataxia type-2 are caused by mutations in the Ca^{2+} channel gene CACNL1A4, *Cell* **87**:543-552.

Pan, J. Q. and Lipscombe, D., 2000, Alternative splicing in the cytoplasmic II-III loop of the N-type Ca channel α1B subunit: functional differences are beta subunit-specific, *J Neurosci* **20**:4769-4775.

Peixoto, A. A., Smith, L. A., and Hall, J. C., 1997, Genomic organization and evolution of alternative exons in a Drosophila calcium channel gene, *Genetics* **145**:1003-1013.

Pennartz, C. M., de Jeu, M. T., Bos, N. P., Schaap, J., and Geurtsen, A. M., 2002, Diurnal modulation of pacemaker potentials and calcium current in the mammalian circadian clock, *Nature* **416**:286-290.

Pereon, Y., Navarro, J., Hamilton, M., Booth, F. W., and Palade, P., 1997, Chronic stimulation differentially modulates expression of mRNA for dihydropyridine receptor isoforms in rat fast twitch skeletal muscle, *Biochem Biophys Res Commun* **235**:217-222.

Pereverzev, A., Klockner, U., Henry, M., Grabsch, H., Vajna, R., Olyschlager, S., Viatchenko-Karpinski, S., Schroder, R., Hescheler, J., and Schneider, T., 1998, Structural diversity of the voltage-dependent Ca^{2+} channel α1E-subunit, *Eur J Neurosci* **10**:916-925.

Pereverzev, A., Leroy, J., Krieger, A., Malecot, C. O., Hescheler, J., Pfitzer, G., Klockner, U., and Schneider, T., 2002, Alternate splicing in the cytosolic II-III loop and the carboxy terminus of human E-type voltage-gated Ca^{2+} channels: electrophysiological characterization of isoforms, *Mol Cell Neurosci* **21**:352-365.

Perez-Reyes, E., Wei, X. Y., Castellano, A., and Birnbaumer, L., 1990, Molecular diversity of L-type calcium channels. Evidence for alternative splicing of the transcripts of three non-allelic genes, *J Biol Chem* **265**:20430-20436.

Perez-Reyes, E., 2003, Molecular physiology of low-voltage-activated T-type calcium channels, *Physiol Rev* **83**:117-161.

Peterson, B. Z., Lee, J. S., Mulle, J. G., Wang, Y., de Leon, M., and Yue, D. T., 2000, Critical determinants of Ca^{2+}-dependent inactivation within an EF-hand motif of L-type Ca^{2+} channels, *Biophys J* **78**:1906-1920.

Platzer, J., Engel, J., Schrott-Fischer, A., Stephan, K., Bova, S., Chen, H., Zheng, H., and Striessnig, J., 2000, Congenital deafness and sinoatrial node dysfunction in mice lacking class D L-type Ca^{2+} channels, *Cell* **102**:89-97.

Qin, N., Platano, D., Olcese, R., Costantin, J. L., Stefani, E., and Birnbaumer, L., 1998, Unique regulatory properties of the type 2a Ca^{2+} channel β subunit caused by palmitoylation, *Proc Natl Acad Sci U S A* **95**:4690-4695.

Qin, N., Yagel, S., Momplaisir, M. L., Codd, E. E., and D'Andrea, M. R., 2002, Molecular cloning and characterization of the human voltage-gated calcium channel $\alpha_2\delta$-4 subunit, *Mol Pharmacol* **62**:485-496.

Raghib, A., Bertaso, F., Davies, A., Page, K. M., Meir, A., Bogdanov, Y., and Dolphin, A. C., 2001, Dominant-negative synthesis suppression of voltage-gated calcium channel $Ca_v2.2$ induced by truncated constructs, *J Neurosci* **21**:8495-8504.

Randall, A. and Tsien, R. W., 1995, Pharmacological dissection of multiple types of Ca^{2+} channel currents in rat cerebellar granule neurons, *J Neurosci* **15**:2995-3012.

Reenan, R. A., 2001, The RNA world meets behavior: A-->I pre-mRNA editing in animals, *Trends Genet* **17**:53-56.

Restituito, S., Thompson, R. M., Eliet, J., Raike, R. S., Riedl, M., Charnet, P., and Gomez, C. M., 2000, The polyglutamine expansion in spinocerebellar ataxia type 6 causes a beta subunit-specific enhanced activation of P/Q-type calcium channels in *Xenopus* oocytes, *J Neurosci* **20**:6394-6403.

Rettig, J., Sheng, Z. H., Kim, D. K., Hodson, C. D., Snutch, T. P., and Catterall, W. A., 1996, Isoform-specific interaction of the α1A subunits of brain Ca^{2+} channels with the presynaptic proteins syntaxin and SNAP-25, *Proc Natl Acad Sci U S A* **93**:7363-7368.

Rogers, J. C., Qu, Y., Tanada, T. N., Scheuer, T., and Catterall, W. A., 1996, Molecular determinants of high affinity binding of alpha-scorpion toxin and sea anemone toxin in the S3-S4 extracellular loop in domain IV of the Na^+ channel alpha subunit, *J Biol Chem* **271**:15950-15962.

Rosenblatt, K. P., Sun, Z. P., Heller, S., and Hudspeth, A. J., 1997, Distribution of Ca^{2+}-activated K^+ channel isoforms along the tonotopic gradient of the chicken's cochlea, *Neuron* **19**:1061-1075.

Safa, P., Boulter, J., and Hales, T. G., 2001, Functional properties of $Ca_v1.3$ α_{1D} L-type Ca^{2+} channel splice variants expressed by rat brain and neuroendocrine GH3 cells, *J Biol Chem* **276**:38727-38737.

Sakurai, T., Hell, J. W., Woppmann, A., Miljanich, G. P., and Catterall, W. A., 1995, Immunochemical identification and differential phosphorylation of alternatively spliced forms of the α1A subunit of brain calcium channels, *J Biol Chem* **270**:21234-21242.

Sakurai, T., Westenbroek, R. E., Rettig, J., Hell, J., and Catterall, W. A., 1996, Biochemical properties and subcellular distribution of the BI and rbA isoforms of α1A subunits of brain calcium channels, *J Cell Biol* **134**:511-528.

Sanguinetti, M. C. and Kass, R. S., 1984, Voltage-dependent block of calcium channel current in the calf cardiac Purkinje fiber by dihydropyridine calcium channel antagonists, *Circ Res* **55**:336-348.

Sather, W. A., Tanabe, T., Zhang, J. F., Mori, Y., Adams, M. E., and Tsien, R. W., 1993, Distinctive biophysical and pharmacological properties of class A (BI) calcium channel α1 subunits, *Neuron* **11**:291-303.

Schneider, T., Wei, X., Olcese, R., Costantin, J. L., Neely, A., Palade, P., Perez-Reyes, E., Qin, N., Zhou, J., Crawford, G. D., and et al., 1994, Molecular analysis and functional expression of the human type E neuronal Ca^{2+} channel α1 subunit, *Receptors Channels* **2**:255-270.

Scholze, A., Plant, T. D., Dolphin, A. C., and Nurnberg, B., 2001, Functional expression and characterization of a voltage-gated $Ca_V1.3$ α1D calcium channel subunit from an insulin-secreting cell line, *Mol Endocrinol* **15**:1211-1221.

Schorge, S., Gupta, S., Lin, Z., McEnery, M. W., and Lipscombe, D., 1999, Calcium channel activation stabilizes a neuronal calcium channel mRNA, *Nat Neurosci* **2**:785-790.

Schramm, M., Vajna, R., Pereverzev, A., Tottene, A., Klockner, U., Pietrobon, D., Hescheler, J., and Schneider, T., 1999, Isoforms of α1E voltage-gated calcium channels in rat cerebellar granule cells--detection of major calcium channel α1-transcripts by reverse transcription-polymerase chain reaction, *Neuroscience* **92**:565-575.

Scott, V. E., Felix, R., Arikkath, J., and Campbell, K. P., 1998, Evidence for a 95 kDa short form of the α1A subunit associated with the ω-conotoxin MVIIC receptor of the P/Q-type Ca^{2+} channels, *J Neurosci* **18**:641-647.

Seino, S., Chen, L., Seino, M., Blondel, O., Takeda, J., Johnson, J. H., and Bell, G. I., 1992, Cloning of the alpha 1 subunit of a voltage-dependent calcium channel expressed in pancreatic beta cells, *Proc Natl Acad Sci U S A* **89**:584-588.

Sharp, P. A. and Burge, C. B., 1997, Classification of introns: U2-type or U12-type, *Cell* **91**:875-879.

Sheng, Z. H., Rettig, J., Takahashi, M., and Catterall, W. A., 1994, Identification of a syntaxin-binding site on N-type calcium channels, *Neuron* **13**:1303-1313.

Shoemaker, D. D., Schadt, E. E., Armour, C. D., He, Y. D., Garrett-Engele, P., McDonagh, P. D., Loerch, P. M., Leonardson, A., Lum, P. Y., Cavet, G., Wu, L. F., Altschuler, S. J., Edwards, S., King, J., Tsang, J. S., Schimmack, G., Schelter, J. M., Koch, J., Ziman, M., Marton, M. J., Li, B., Cundiff, P., Ward, T., Castle, J., Krolewski, M., Meyer, M. R., Mao, M., Burchard, J., Kidd, M. J., Dai, H., Phillips, J. W., Linsley, P. S., Stoughton, R., Scherer, S., and Boguski, M. S., 2001, Experimental annotation of the human genome using microarray technology, *Nature* **409**:922-927.

Simen, A. A., Lee, C. C., Simen, B. B., Bindokas, V. P., and Miller, R. J., 2001, The C terminus of the Ca channel α1B subunit mediates selective inhibition by G-protein-coupled receptors, *J Neurosci* **21**:7587-7597.

Smith, L. A., Wang, X., Peixoto, A. A., Neumann, E. K., Hall, L. M., and Hall, J. C., 1996, A *Drosophila* calcium channel α1 subunit gene maps to a genetic locus associated with behavioral and visual defects, *J Neurosci* **16**:7868-7879.

Smith, L. A., Peixoto, A. A., and Hall, J. C., 1998, RNA editing in the *Drosophila* DMCA1A calcium-channel α1 subunit transcript, *J Neurogenet* **12**:227-240.

Snutch, T. P., Tomlinson, W. J., Leonard, J. P., and Gilbert, M. M., 1991, Distinct calcium channels are generated by alternative splicing and are differentially expressed in the mammalian CNS, *Neuron* **7**:45-57.

Soldatov, N. M., 1992, Molecular diversity of L-type Ca^{2+} channel transcripts in human fibroblasts, *Proc Natl Acad Sci U S A* **89**:4628-4632.

Soldatov, N. M., 1994, Genomic structure of human L-type Ca^{2+} channel, *Genomics* **22**:77-87.

Soldatov, N. M., Bouron, A., and Reuter, H., 1995, Different voltage-dependent inhibition by dihydropyridines of human Ca^{2+} channel splice variants, *J Biol Chem* **270**:10540-10543.

Soldatov, N. M., Zuhlke, R. D., Bouron, A., and Reuter, H., 1997, Molecular structures involved in L-type calcium channel inactivation. Role of the carboxyl-terminal region encoded by exons 40-42 in α1C subunit in the kinetics and Ca^{2+} dependence of inactivation, *J Biol Chem* **272**:3560-3566.

Soldatov, N. M., Oz, M., O'Brien, K. A., Abernethy, D. R., and Morad, M., 1998, Molecular determinants of L-type Ca^{2+} channel inactivation. Segment exchange analysis of the carboxyl-terminal cytoplasmic motif encoded by exons 40-42 of the human α1C subunit gene, *J Biol Chem* **273**:957-963.

Sommer, B., Kohler, M., Sprengel, R., and Seeburg, P. H., 1991, RNA editing in brain controls a determinant of ion flow in glutamate-gated channels, *Cell* **67**:11-19.

Soong, T. W., DeMaria, C. D., Alvania, R. S., Zweifel, L. S., Liang, M. C., Mittman, S., Agnew, W. S., and Yue, D. T., 2002, Systematic identification of splice variants in human P/Q-type channel α1(2.1) subunits: implications for current density and Ca^{2+}-dependent inactivation, *J Neurosci* **22**:10142-10152.

Staes, M., Talavera, K., Klugbauer, N., Prenen, J., Lacinova, L., Droogmans, G., Hofmann, F., and Nilius, B., 2001, The amino side of the C-terminus determines fast inactivation of the T-type calcium channel α1G, *J Physiol* **530**:35-45.

Starr, T. V., Prystay, W., and Snutch, T. P., 1991, Primary structure of a calcium channel that is highly expressed in the rat cerebellum, *Proc Natl Acad Sci U S A* **88**:5621-5625.

Stea, A., Tomlinson, W. J., Soong, T. W., Bourinet, E., Dubel, S. J., Vincent, S. R., and Snutch, T. P., 1994, Localization and functional properties of a rat brain α1A calcium channel reflect similarities to neuronal Q- and P-type channels, *Proc Natl Acad Sci U S A* **91**:10576-10580.

Stocker, J. W., Nadasdi, L., Aldrich, R. W., and Tsien, R. W., 1997, Preferential interaction of ω-conotoxins with inactivated N-type Ca^{2+} channels, *J Neurosci* **17**:3002-3013.

Swartz, K. J. and MacKinnon, R., 1997, Mapping the receptor site for hanatoxin, a gating modifier of voltage-dependent K^+ channels, *Neuron* **18**:675-682.

Takimoto, K., Li, D., Nerbonne, J. M., and Levitan, E. S., 1997, Distribution, splicing and glucocorticoid-induced expression of cardiac α1C and α1D voltage-gated Ca^{2+} channel mRNAs, *J Mol Cell Cardiol* **29**:3035-3042.

Talley, E. M., Cribbs, L. L., Lee, J. H., Daud, A., Perez-Reyes, E., and Bayliss, D. A., 1999, Differential distribution of three members of a gene family encoding low voltage-activated (T-type) calcium channels, *J Neurosci* **19**:1895-1911.

Tanabe, T., Takeshima, H., Mikami, A., Flockerzi, V., Takahashi, H., Kangawa, K., Kojima, M., Matsuo, H., Hirose, T., and Numa, S., 1987, Primary structure of the receptor for calcium channel blockers from skeletal muscle, *Nature* **328**:313-318.

Tang, C. Y. and Papazian, D. M., 1997, Transfer of voltage independence from a rat olfactory channel to the Drosophila ether-a-go-go K^+ channel, *J Gen Physiol* **109**:301-311.

Tobi, D., Wiser, O., Trus, M., and Atlas, D., 1998, N-type voltage-sensitive calcium channel interacts with syntaxin, synaptotagmin and SNAP-25 in a multiprotein complex, *Receptors Channels* **6**:89-98.

Toru, S., Murakoshi, T., Ishikawa, K., Saegusa, H., Fujigasaki, H., Uchihara, T., Nagayama, S., Osanai, M., Mizusawa, H., and Tanabe, T., 2000, Spinocerebellar ataxia type 6 mutation alters P-type calcium channel function, *J Biol Chem* **275**:10893-10898.

Triggle, D. J., 1998, The physiological and pharmacological significance of cardiovascular T-type, voltage-gated calcium channels, *Am J Hypertens* **11**:80S-87S.

Triggle, D. J., 1999, The pharmacology of ion channels: with particular reference to voltage-gated Ca^{2+} channels, *Eur J Pharmacol* **375**:311-325.

Tsien, R. W., Ellinor, P. T., and Horne, W. A., 1991, Molecular diversity of voltage-dependent Ca^{2+} channels, *Trends Pharmacol Sci* **12**:349-354.

Tsunemi, T., Saegusa, H., Ishikawa, K., Nagayama, S., Murakoshi, T., Mizusawa, H., and Tanabe, T., 2002, Novel $Ca_v2.1$ splice variants isolated from Purkinje cells do not generate P-type Ca^{2+} current, *J Biol Chem* **277**:7214-7221.

Vajna, R., Klockner, U., Pereverzev, A., Weiergraber, M., Chen, X., Miljanich, G., Klugbauer, N., Hescheler, J., Perez-Reyes, E., and Schneider, T., 2001, Functional coupling between 'R-type' Ca^{2+} channels and insulin secretion in the insulinoma cell line INS-1, *Eur J Biochem* **268**:1066-1075.

Vance, C. L., Begg, C. M., Lee, W. L., Haase, H., Copeland, T. D., and McEnery, M. W., 1998, Differential expression and association of calcium channel α1B and β subunits during rat brain ontogeny, *J Biol Chem* **273**:14495-14502.

Vanegas, H. and Schaible, H., 2000, Effects of antagonists to high-threshold calcium channels upon spinal mechanisms of pain, hyperalgesia and allodynia, *Pain* **85**:9-18.

Vigues, S., Chabret, C., Valentin, S., and Valmier, J., 1998, Rat embryonic hippocampal neurons express a new class A calcium channel variant, *Neurosci Lett* **258**:37-40.

Vigues, S., Gastaldi, M., Chabret, C., Massacrier, A., Cau, P., and Valmier, J., 1999, Regulation of calcium channel α_{1A} subunit splice variant mRNAs in kainate-induced temporal lobe epilepsy, *Neurobiol Dis* **6**:288-301.

Vigues, S., Gastaldi, M., Massacrier, A., Cau, P., and Valmier, J., 2002, The α_{1A} subunits of rat brain calcium channels are developmentally regulated by alternative RNA splicing, *Neuroscience* **113**:509-517.

Walker, D. and De Waard, M., 1998, Subunit interaction sites in voltage-dependent Ca^{2+} channels: role in channel function, *Trends Neurosci* **21**:148-154.

Wei, X., Neely, A., Lacerda, A. E., Olcese, R., Stefani, E., Perez-Reyes, E., and Birnbaumer, L., 1994, Modification of Ca^{2+} channel activity by deletions at the carboxyl terminus of the cardiac α1 subunit, *J Biol Chem* **269**:1635-1640.

Weiergraber, M., Pereverzev, A., Vajna, R., Henry, M., Schramm, M., Nastainczyk, W., Grabsch, H., and Schneider, T., 2000, Immunodetection of α_{1E} voltage-gated Ca^{2+} channel in chromogranin-positive muscle cells of rat heart, and in distal tubules of human kidney, *J Histochem Cytochem* **48**:807-819.

Welling, A., Kwan, Y. W., Bosse, E., Flockerzi, V., Hofmann, F., and Kass, R. S., 1993, Subunit-dependent modulation of recombinant L-type calcium channels. Molecular basis for dihydropyridine tissue selectivity, *Circ Res* **73**:974-980.

Welling, A., Ludwig, A., Zimmer, S., Klugbauer, N., Flockerzi, V., and Hofmann, F., 1997, Alternatively spliced IS6 segments of the α1C gene determine the tissue-specific dihydropyridine sensitivity of cardiac and vascular smooth muscle L-type Ca^{2+} channels, *Circ Res* **81**:526-532.

Westenbroek, R. E., Hoskins, L., and Catterall, W. A., 1998, Localization of Ca^{2+} channel subtypes on rat spinal motor neurons, interneurons, and nerve terminals, *J Neurosci* **18**:6319-6330.

Wielowieyski, P. A., Wigle, J. T., Salih, M., Hum, P., and Tuana, B. S., 2001, Alternative splicing in intracellular loop connecting domains II and III of the alpha 1 subunit of $Ca_v1.2$ Ca^{2+} channels predicts two-domain polypeptides with unique C-terminal tails, *J Biol Chem* **276**:1398-1406.

Williams, M. E., Brust, P. F., Feldman, D. H., Patthi, S., Simerson, S., Maroufi, A., McCue, A. F., Velicelebi, G., Ellis, S. B., and Harpold, M. M., 1992, Structure and functional expression of an ω-conotoxin-sensitive human N-type calcium channel, *Science* **257**:389-395.

Williams, M. E., Marubio, L. M., Deal, C. R., Hans, M., Brust, P. F., Philipson, L. H., Miller, R. J., Johnson, E. C., Harpold, M. M., and Ellis, S. B., 1994, Structure and functional characterization of neuronal α1E calcium channel subtypes, *J Biol Chem* **269**:22347-22357.

Wilson, S. M., Toth, P. T., Oh, S. B., Gillard, S. E., Volsen, S., Ren, D., Philipson, L. H., Lee, E. C., Fletcher, C. F., Tessarollo, L., Copeland, N. G., Jenkins, N. A., and Miller, R. J., 2000, The status of voltage-dependent calcium channels in α1E knock-out mice, *J Neurosci* **20**:8566-8571.

Wu, L. G., Westenbroek, R. E., Borst, J. G., Catterall, W. A., and Sakmann, B., 1999, Calcium channel types with distinct presynaptic localization couple differentially to transmitter release in single calyx-type synapses, *J Neurosci* **19**:726-736.

Xie, J. and Black, D. L., 2001, A CaMK IV responsive RNA element mediates depolarization-induced alternative splicing of ion channels, *Nature* **410**:936-939.

Xu, W. and Lipscombe, D., 2001, Neuronal $Ca_V1.3\alpha_1$ L-type channels activate at relatively hyperpolarized membrane potentials and are incompletely inhibited by dihydropyridines, *J Neurosci* **21**:5944-5951.

Yamada, Y., Masuda, K., Li, Q., Ihara, Y., Kubota, A., Miura, T., Nakamura, K., Fujii, Y., Seino, S., and Seino, Y., 1995, The structures of the human calcium channel α1 subunit (CACNL1A2) and β subunit (CACNLB3) genes, *Genomics* **27**:312-319.

Yang, Y., Chen, X., Margulies, K., Jeevanandam, V., Pollack, P., Bailey, B. A., and Houser, S. R., 2000, L-type Ca^{2+} channel α1C subunit isoform switching in failing human ventricular myocardium, *J Mol Cell Cardiol* **32**:973-984.

Yatani, A. and Kamp, T. J., 2000, Tails of the L-type Ca^{2+} channel: to sense oxygen or not, *Circ Res* **87**:535-536.

Zhang, Z., Xu, Y, Song, H, Rodriguez, J, Tuteja, D, Namkung, Y, Shin, H-W, Chiamvimonvat, N, 2002, Functional role of $Ca_v1.3$ (α_{1D}) calcium channel in sinoatrial nodes: Insight gained using gen-targeted null mutant mice, *Circ Res* **90**:981-987.

Zhong, H., Yokoyama, C. T., Scheuer, T., and Catterall, W. A., 1999, Reciprocal regulation of P/Q-type Ca^{2+} channels by SNAP-25, syntaxin and synaptotagmin, *Nat Neurosci* **2**:939-941.

Zhuang, H., Bhattacharjee, A., Hu, F., Zhang, M., Goswami, T., Wang, L., Wu, S., Berggren, P. O., and Li, M., 2000, Cloning of a T-type Ca^{2+} channel isoform in insulin-secreting cells, *Diabetes* **49**:59-64.

Zhuchenko, O., Bailey, J., Bonnen, P., Ashizawa, T., Stockton, D. W., Amos, C., Dobyns, W. B., Subramony, S. H., Zoghbi, H. Y., and Lee, C. C., 1997, Autosomal dominant cerebellar ataxia (SCA6) associated with small polyglutamine expansions in the α1A-voltage-dependent calcium channel, *Nat Genet* **15**:62-69.

Zuhlke, R. D., Bouron, A., Soldatov, N. M., and Reuter, H., 1998, Ca^{2+} channel sensitivity towards the blocker isradipine is affected by alternative splicing of the human α1C subunit gene, *FEBS Lett* **427**:220-224.

Zuhlke, R. D., Pitt, G. S., Deisseroth, K., Tsien, R. W., and Reuter, H., 1999, Calmodulin supports both inactivation and facilitation of L-type calcium channels, *Nature* **399**:159-162.

INDEX